Genetics

A Guide to Basic Concepts and Problem Solving

Richard P. Nickerson

Simmons College

HarperCollins*Publishers*

To My Parents

ISBN 0-673-39684-3

Transferred to digital print on demand, 2002
Printed and bound by Antony Rowe Ltd, Eastbourne

6—MKN—94 93 92

Preface

My goal in writing *Genetics: A Guide to Basic Concepts and Problem Solving* has been to develop a self-instruction guide which allows students to address, on their own, many of the difficulties they experience in mastering the basic concepts of genetics. Working problems is an important way of mastering the principles of genetics and, as such, is an important aspect of any genetics course. Although virtually all texts provide extensive problem sets at the ends of chapters, most devote little attention to the specifics of how to solve genetics problems. Many students of genetics experience major difficulty with problem solving. The source of the difficulty varies—it may be a lack of familiarity with the necessary factual groundwork, an inability to identify the factual material that is relevant to a problem, difficulty in applying that material to problem solving, or other reasons. If students are to achieve mastery of the material and be successful in the course, it is essential that these difficulties be overcome.

This book is suitable for use in all introductory genetics courses as a supplement to the standard genetics texts. Topics are presented in the sequence found in many of these texts, but the order in which the chapters are used may be changed to suit the needs of individual courses. Each chapter deals with a discrete topic and is written to be as self-contained as possible.

TO THE STUDENT

The *Guide* is designed to engage you as an active participant in the learning process—as the basic factual material is presented, as you are guided through sample problems, and as you solve problems on your own. Key features of this book are as follows.

Chapter Outline An outline providing an overview of the material is included at the beginning of each chapter to assist you in locating information within the chapter.

Review of Concepts Each chapter supplies the basic information necessary to comprehend one of the major topics of genetics and to solve problems related to that topic.

Workbook-style Questions Throughout each chapter, there is a series of workbook-style questions that guides you through important concepts. Answering these questions allows you to use and expand upon the information presented. Spaces are provided for you to record your answers, and all answer blanks within a chapter are numbered consecutively. Answers are keyed to the numbers and are footnoted at the bottoms of the pages so you can quickly check the accuracy of your responses.

Sample Problems and Solutions The factual information presented is applied within each chapter as you are guided, in stepwise fashion, through the solutions to sample problems. This approach helps you consolidate and build upon the knowledge you have acquired. During this process, you are alerted to potential pitfalls and techniques for avoiding them. In some instances, alternative approaches to solving the problem are explored as a reminder that problems often can be solved in different ways.

Chapter Summaries Each chapter ends with a brief summary which serves to review the key concepts presented.

Problem Sets Following each chapter is a set of problems to work through on your own. Some of the initial problems may be simpler than later ones and are designed to verify that you have a grasp of the basic factual material and are using a stepwise, logical approach to problem solving. All the problems within a set are designed to illustrate and clarify the principles of genetics, as well as to test for mastery of the topic being covered.

Solutions to Problems Complete solutions to all problems are included in Appendix B. These explanations, unlike those included in most texts, are designed to give insight to the rationale and methods for solving problems. The detailed solutions reinforce the self-guided learning process and, if you have gone astray, should allow you to pinpoint where errors were made, to discover why you made them, and to sense how they could have been avoided. If you need assistance from your instructor, you should be able to ask very specific questions.

Comprehensive Problem Set for Chapters 1–13 A set of comprehensive problems, designed to test your comprehension of many of the basic concepts of Mendelian genetics, is found in Appendix A. These problems, which consider the simultaneous inheritance of two or more traits that show different methods of inheritance, should be worked only after you have mastered the material in Chapters 1 through 13 and successfully worked the problem sets at the ends of those chapters. Complete solutions to this problem set are included in Appendix B.

Correlation Guide for Several Genetics Texts To assist you in determining where topics covered in this book may be found in your text and vice versa, a correlation guide is included in Appendix C. This guide indicates the chapters in this book that correlate with the chapters in several popular genetics texts.

ACKNOWLEDGMENTS

The many individuals who have helped in the preparation of this book have my gratitude and thanks. Chief among them are the reviewers who have provided valuable suggestions and criticisms at different stages in this project. They include Anna W. Berkovitz, Purdue University; James B. Boyd, University of California, Davis; Brian Bradley, University of Maryland; Sheldon Broedel, University of Maryland–Baltimore County; Diane Dodd, University of North Carolina–Wilmington; Gordin Edlin, University of California, Davis; Lee Ehrman, State University of New York, Purchase; Ted Emigh, North Carolina State University; David J. Hicks, Manchester College; Donald Hurst, Brooklyn College; Martin J. Michaelson, Texas A&M University; John Osterman, University of Nebraska; R. H. Richardson, University of Texas at Austin; Allen F. Sherald, George Mason University; R. W. Siegel, University of California, Los Angeles; Edward Simon, Purdue University; Judith Van Houten, University of Vermont; Robert Wiggers, Texas A&M University; and C. Richard Wrathall, American University.

I also wish to thank Sharon Soltzberg for her encouragement and many suggestions during the early stages of this project; Gregory Gazaway for his assistance and good cheer; Marsha Watson, Carl Wolf, Cathleen Haggerty, and Sarah Koolsbergen for help with the typing; Elizabeth Stoltz, Kimberly A. Stieglitz, and Kim A. Tetreault for checking problems and proofreading; Bruce A. Ledbetter and Mark Rhynsburger for proofreading; and the many students who have used the chapters, worked through the problem sets, and made many helpful suggestions for improvement. I am also pleased to acknowledge Professor W. D. Russell-Hunter (Syracuse University) who taught me much about biology and writing.

I am grateful for the support and encouragement of the staff members at Scott, Foresman/Little, Brown who helped produce this book. In particular, Bonnie Roesch, Life Science Editor; Adam P. Bryer, Technical Editor; Ann Buesing, Project Editor; and Ellen Pettengell, Designer. It has been a pleasure to work with each of these outstanding professionals.

Richard P. Nickerson
Simmons College
Boston, Massachusetts

Contents

CHAPTER 13

CHAPTER 14

CHAPTER 15

CHAPTER 16

CHAPTER 17

CHAPTER 18

CHAPTER 19

CHAPTER

1

Mitosis and Meiosis

INTRODUCTION

Prokaryotes, Eukaryotes, and Viruses Compared

Living organisms can be classified as either prokaryotic or eukaryotic. **Prokaryotes** are simpler organisms, bacteria and blue-green algae, whose cells lack a membrane-bound nucleus. Their genetic material is found in a single chromosome which consists of one double-stranded molecule of DNA (deoxyribonucleic acid) associated, or complexed, with a few protein molecules. **Eukaryotes** are more complex single- and multicellular forms whose cells possess a membrane-bound nucleus. This nucleus contains most of the cell's genetic material apportioned among a few to several hundred chromosomes. Each **eukaryotic chromosome** is made of **chromatin** (a macromolecular complex consisting of one double-helix DNA molecule which runs continuously from one end of the chromosome to the other), appreciable amounts of protein, and some RNA (ribonucleic acid).

Viruses are small particles that generally consist of a protein coat enclosing a single viral chromosome of either DNA or RNA. Viruses lack the metabolic machinery essential for energy transformations and protein synthesis, and consequently they are intracellular parasites that reproduce only in susceptible prokaryotic or eukaryotic host cells.

Chromosome Number in Eukaryotes

Each eukaryotic species has a characteristic number of chromosomes in the nuclei of its cells. For example, human cells have 46 chromosomes, cells of the fruit fly *Drosophila melanogaster* have 8 chromosomes, and cells of the corn plant *Zea mays* have 20 chromosomes. In the nuclei of many eukaryotic cells, chromosomes occur in pairs in what is designated as a **diploid,** or $2n$, chromosome complement. The chromosome numbers listed above for humans, fruit flies, and corn are diploid numbers, and the nuclei of **somatic,** or body, cells of these species contain 23, 4, and 10 *pairs* of chromosomes, respectively. In the nuclei of other eukaryotic cells, the chromosome complement consists of one chromosome from each pair and is referred to as **haploid,** or n. The haploid number for a species is always half of its diploid number. The genetic information carried by a haploid set of chromosomes is designated as the **genome** for a species.

Sexually reproducing eukaryotic species experience an alternation between haploid and diploid stages in their life cycles. In many plant species, organisms may spend a significant part of their life cycle in the haploid stage. The specialized reproductive cells known as **spores;** the gamete-producing stage, or **ga-**

metophyte; and the **gametes** or reproductive cells known as eggs (ova) and sperm are all haploid stages of the life cycles of these plants. In most animals, since only the gametes are *n*, the haploid stage is very brief. In the human life cycle, for example, only the gametes are haploid with their 23 individual chromosomes, while all other stages of the life cycle are diploid.

In sexual reproduction, the union of a haploid egg with a haploid sperm forms a diploid fertilized egg, or **zygote.** One member of each chromosome pair present in the zygote comes from the male parent, via the sperm, while the other comes from the female parent, via the egg. To produce a multicellular organism, the zygote passes through many rounds of cell division, each of which is preceded by a faithful duplication of the chromosomes. As a result, each somatic cell nucleus possesses the diploid number of chromosomes and each chromosome pair in each of these somatic cell nuclei is made up of one chromosome of paternal origin and one of maternal origin.

The chromosomes making up each pair are said to be **homologous** to each other: they usually are alike in size and shape and carry, in the same linear sequence, the hereditary determinants, or **genes,** for the same traits. The notable exception to this involves the pair of **sex chromosomes,** found in some animals and plants, which are involved with sex determination and may differ from each other in size, shape, and gene content. Note that any chromosome that is not a sex chromosome is designated as an **autosome.** Any **nonhomologous chromosomes** of a species may differ considerably in their appearance and gene content.

Mitosis and Meiosis: An Overview

Since genes are subunits of chromosomes, an understanding of the duplication and distribution of chromosomes during cell reproduction is necessary to understand how genes are transmitted from one generation to the next in eukaryotes. Both mitosis and meiosis are types of *nuclear* division that distribute the genetic material during the process of cell division. Mitosis and meiosis are always preceded by a duplication of chromosomes of the cell that is to divide. In **mitosis,** a single nucleus divides to produce two nuclei, each containing exactly the same genetic information as the original nucleus. **Meiosis** is a specialized type of nuclear division that produces certain reproductive cells; it begins with a diploid nucleus and produces nuclei with the haploid number of chromosomes. The following overview minimizes the detail and focuses on the major events of each process. Keep in mind that mitosis and meiosis occur only in eukaryotic cells.

MITOSIS AND ITS PLACE IN THE CELL CYCLE

Cell Cycle

The growth of a new cell to a mature size—and its subsequent division to form two daughter cells—makes up a continuous series of events known as the **cell cycle.** The cell cycle occurs in both haploid and diploid cells and can be divided into three major parts: (1) **interphase,** (2) **mitosis,** and (3) **cytokinesis,** as shown in Figure 1-1.

Interphase

Throughout interphase, which makes up the major portion of the cell cycle, the cell grows to its full size and differentiates. Interphase can be subdivided into the three stages.

1. **G_1,** or gap-1, stage, during which the molecules required for DNA replication and chromosome duplication are synthesized and arranged.
2. **S,** or synthesis, stage, during which the DNA is synthesized, doubling the amount of DNA

FIGURE 1-1

Overview of the cell cycle.

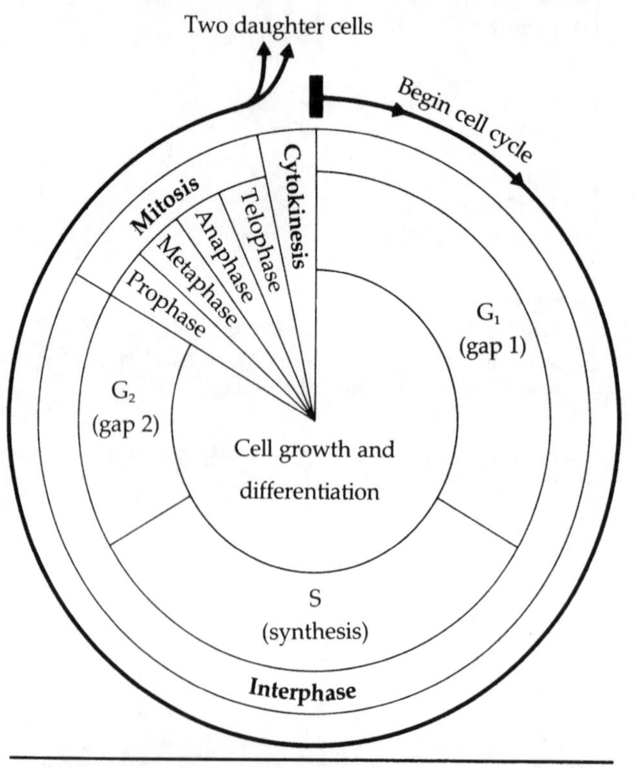

From *Concepts of Genetics*, Second Edition, by William S. Klug and Michael R. Cummings. Copyright © 1986, 1983 Scott, Foresman and Company.

in the nucleus, and chromosome duplication is completed.

3. **G_2,** or gap-2, stage, during which the final events leading to mitosis are completed.

Before its duplication in S, each chromosome exists in the form of a single, elongate strand of chromatin which has a constriction, or **centromere,** at some point along its length (Figure 1-2a). After duplication, each chromosome exists in the form of two genetically identical, elongate strands known as sister **chromatids,** each with its own centromere, which are joined at their centromere regions (Figure 1-2b). (Note: The duplication of the chromatin of the centromere lags behind that of the rest of the chromosome but nonetheless is completed by the end of the S stage. Some texts indicate that centromere duplication is delayed until mitosis or meiosis is underway, but this is incorrect.)

Stages of Mitosis

At the end of interphase, the duplicated chromosomes must be separated into two sets. This is accomplished by the process of mitosis, which is divided for ease of study into the series of stages discussed next. Each stage is illustrated in Figure 1-3.

Prophase The physical state of the chromosomes at the end of interphase poses an obstacle to their precise separation: the long, delicate strands of chromatin (Figure 1-2b) could easily be tangled and broken. During prophase, this potential problem is overcome by the coiling and supercoiling (coiling of coils) of the chromatids, condensing them into short, thick strands that remain joined at their centromeres (Figure 1-4). After this condensation in late prophase, the duplicated chromosomes are visible (by light microscopy) (Figure 1-3a).

The spindle apparatus, a collection of microtubules or fibers which plays a role in the separation of the sister chromatids of each duplicated chromosome, begins to assemble in the cytoplasm. The nuclear membrane and nucleolus break down, and in cells that possess them, cell organelles known as centrioles position themselves at the poles of the spindle apparatus (note that centrioles do not occur in higher plants). Two spindle fibers, one from each pole of the spindle apparatus, attach to each duplicated chromosome, such that the fiber from one pole is attached to the centromere of one chromatid and the fiber from the opposite pole is attached to the centromere of the sister chromatid. These fibers attach to the centromeres at sites known as **kinetochores.**

FIGURE 1-3

Stages of mitosis in an animal cell with a diploid number of 4: (a) late prophase, (b) metaphase, (c) anaphase, (d) telophase.

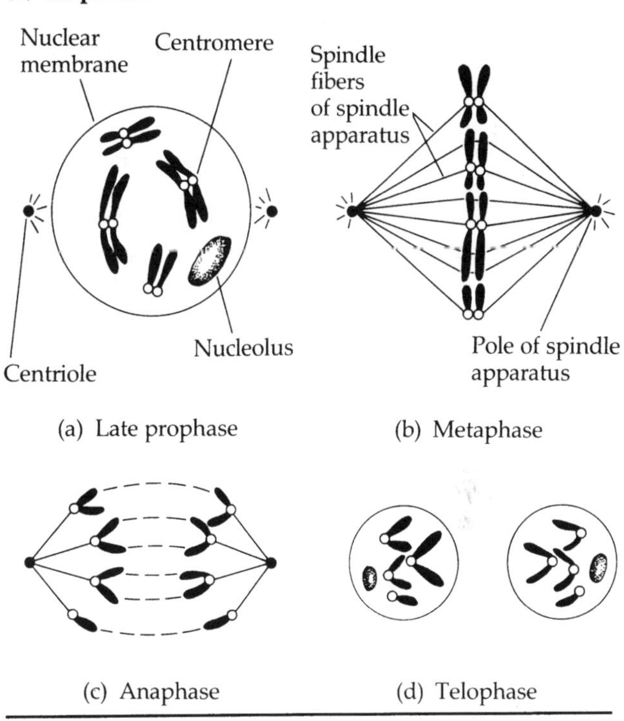

FIGURE 1-2

(a) Chromosome before duplication. (b) Chromosome after duplication (two-chromatid stage). Each chromatid contains one double-stranded helical DNA molecule.

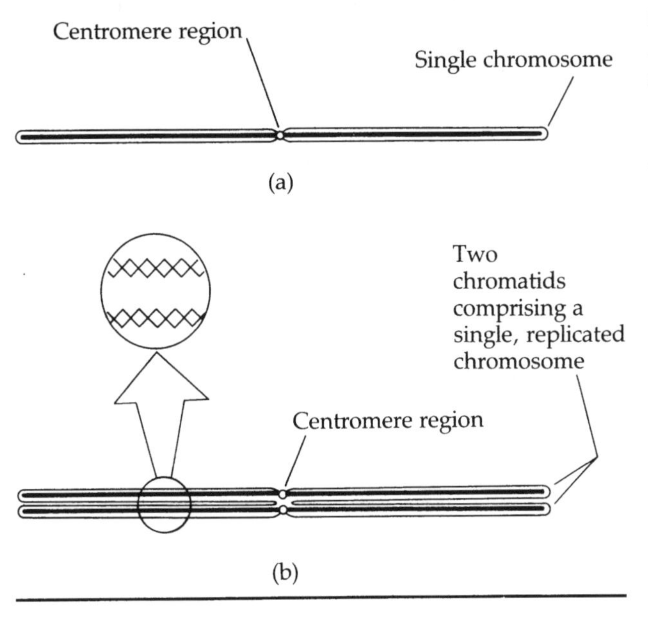

FIGURE 1-4

Chromosome after condensation of sister chromatids. Details of supercoiling not shown.

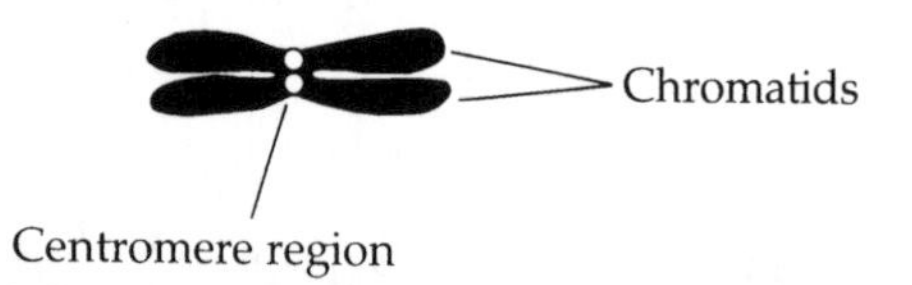

Metaphase During metaphase (Figure 1-3b), the duplicated chromosomes align on a plane, or equator, midway between the poles of the spindle apparatus.

Anaphase During anaphase (Figure 1-3c), the centromeres joining the two sister chromatids of each duplicated chromosome separate and the two sister chromatids separate. Once this separation, or **disjunction,** occurs, each chromatid is called a **daughter chromosome.** As the spindle fibers shorten, the daughter chromosomes of each pair move towards opposite poles of the spindle.

Telophase During telophase (Figure 1-3d), the spindle begins to break down as the separated chromosomes begin to uncoil and lose the shapes by which they could be identified in the earlier stages of mitosis. They return to the extended-chromatin state which characterizes the interphase period of the cell cycle, a nuclear membrane forms around each set of chromosomes, and a nucleolus appears within each nuclear membrane. At the end of telophase, two identical nuclei exist within the cell.

Cytokinesis

Mitosis is usually, but not always, followed by **cytokinesis,** the division of the cytoplasm of the parental cell to produce two daughter cells, each of which receives one of the new nuclei. After cytokinesis, the daughter cells begin interphase and, in the case of continuously dividing cells, the cycle repeats itself.

Mitosis Summarized

Mitosis is a process of nuclear division which occurs within the eukaryotic cell cycle. It begins with one nucleus and produces two nuclei containing chromosomes identical in number, type, and genetic content to those found in the original nucleus. This means that mitosis must be preceded by a precise replication of the DNA and a duplication of the chromosomes. One copy of each chromosome is distributed to each daughter nucleus. Note that mitosis *maintains* in the daughter cells the chromosome number of the parent cell: if the parent cell is haploid, the nuclei of the daughter cells are haploid, and if it is diploid, the nuclei of the daughter cells are diploid.

FIGURE 1-5

Four chromosomes (two homologous pairs). Maternal chromosomes—replicas of those contributed by the mother at fertilization—are solid; paternal chromosomes—replicas of those contributed by the father at fertilization—are dotted.

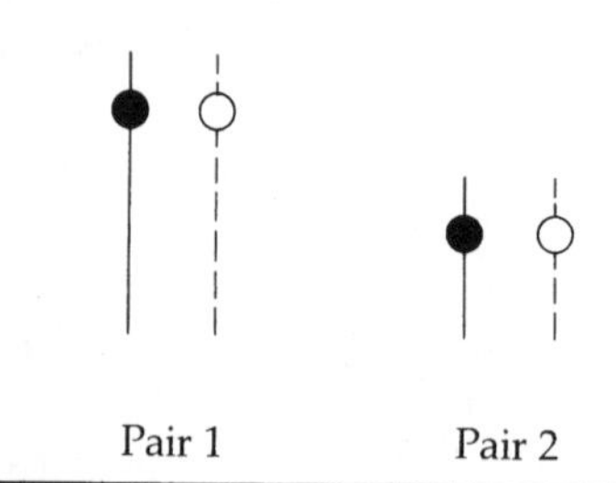

Diagraming Mitosis

To clarify the key stages of mitosis, we will use the example of a hypothetical organism whose cells contain two chromosome pairs (diploid, or $2n$, number of 4) (Figure 1-5).

The two members of each pair of chromosomes carry many gene pairs. Let us assume that the gene pair determining the height of the organism is located on homologous chromosome pair 1. The gene for this trait occurs in two forms, or **alleles.** One allele, designated by the symbol S, is for tallness and its counterpart allele, s, is for shortness. We will assume that the maternal homolog carries the s allele while its paternal chromosome carries the S allele (Figure 1-6a). The gene pair determining another trait of interest to us, say, fur color, is carried on homologous chromosome pair 2.

FIGURE 1-6

Four chromosomes (two homologous pairs). One chromosome pair carries a gene pair for height and the other, a gene pair for fur color. Within each chromosome pair, each homolog carries a different allele.

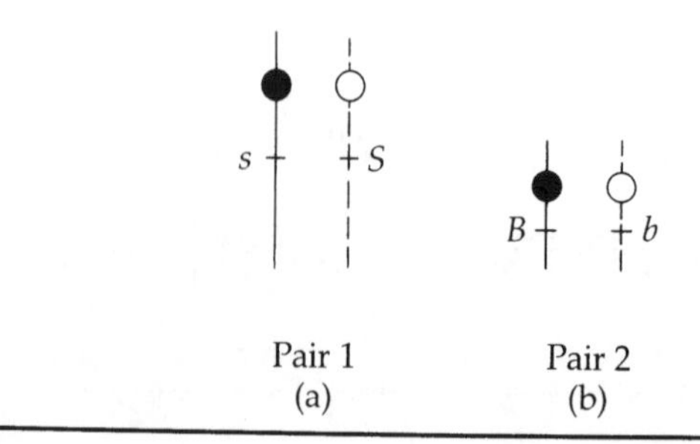

FIGURE 1-7

Mitotic nuclear division.

(a) This diagram shows the diploid number of chromosomes ($2n = 4$) in the nucleus. These chromosomes have not yet duplicated. (Note that the chromosomes cannot be seen at this stage but experimental evidence indicates that this is how they would appear.)

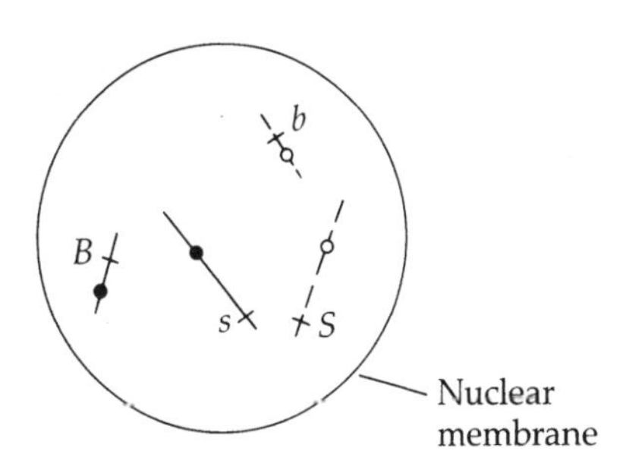

(b) Make a sketch showing the chromosomes of nucleus (a) after their duplication. Remember that after duplication, each chromosome consists of two chromatids joined at their centromeres. For example, chromosome

B

when duplicated would appear as:

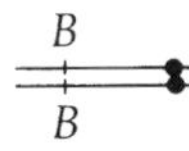

(c) Make a sketch showing the chromosomes of the two diploid nuclei produced once the two chromatids of each duplicated chromosome of nucleus (b) separate from each other.

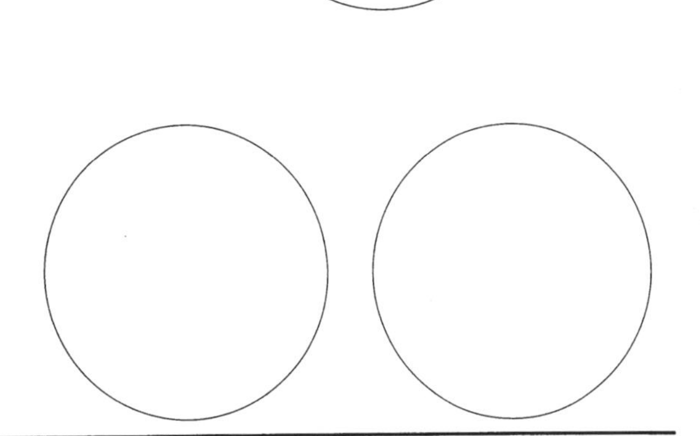

This gene occurs in two allelic forms: one, designated *B*, is for black fur and its counterpart allele, *b*, is for brown fur. Assume that the maternal chromosome carries *B* and its paternal homolog carries *b* (Figure 1-6b).

Using the two pairs of chromosomes from Figure 1-6, complete the diagrams in Figure 1-7 by showing the key stages in mitosis. Make sure your sketches of the two daughter nuclei contain the same number and kind of chromosomes as the parental nucleus.

MEIOSIS

Meiosis is a specialized type of nuclear division producing certain reproductive cells. It is preceded by DNA replication and chromosome duplication during a premeiotic S phase and consists of two successive but separate meiotic divisions, meiosis I and meiosis II. Meiosis begins with a specialized diploid ($2n$) nucleus and produces nuclei with the haploid (n) number of chromosomes. In other words, the process *reduces* the chromosome number of the original nucleus by one-half. In animals, meiosis occurs in the primary gametocytes, the diploid cells which give rise to eggs and sperm. In plants, meiosis takes place in spore-forming cells, which develop from the diploid spore-producing, or **sporophyte,** stage of the life cycle. Meiosis ensures that the chromosome number of a species is maintained from one generation to the next. If gametes were diploid, what would be the consequence for the chromosome number of the zygote and all the cells that subsequently arose mitotically from the zygote? ____________________ [1]

[1] They would be $4n$, or tetraploid.

If each generation's gametes were diploid, what would happen to the chromosome number in each successive generation? ____[2] Thus, for a diploid species to reproduce sexually and, at the same time, maintain a constant chromosome number generation after generation, its gametes must carry just one member of each pair of chromosomes present in its diploid cells. Put another way, the gametes must be haploid.

A cell undergoing meiosis I and meiosis II nuclear divisions has the potential to produce four haploid cells. Each division is subdivided into stages analogous to those described for mitosis: prophase, metaphase, anaphase, and telophase. The prophase of meiosis I is a lengthy stage which is divided into five substages: leptonema, zygonema, pachynema, diplonema, and diakinesis. Key events in most of these stages are shown in Figure 1-8, which begins with a cell nucleus of a male organism with a diploid number of four. Refer to this figure as you read the discussion of meiosis that follows.

Stages of Meiosis I

Prophase I (Figures 1-8a, b, c, d, and e) The major events occurring during this extended stage are considered below under the appropriate prophase I substage.

During **leptonema,** the first substage of prophase I (Figure 1-8a), the previously duplicated chromosomes begin to condense and the individual chromosomes become distinct as elongate strands; even though the chromosomes appear single, each consists of two chromatids joined at their centromere regions.

During **zygonema** (Figure 1-8b), the chromosomes continue to condense and the two members of each homologous pair align side by side. This lengthwise, point-by-point pairing of homologs is a distinctive feature of meiosis and is referred to as **synapsis,** or **homologous pairing.** The **synaptonemal complex** which forms between each homologous pair at this time apparently assists in bringing about and maintaining synapsis. The assemblage of two synapsed chromosomes is known as a **bivalent.** How many bivalents would occur in a nucleus that had three pairs of chromosomes ($2n = 6$)? ____________[3] How does the number of bivalents found in the nucleus compare with the haploid number for a species? ____[4] How does the number of bivalents compare with the diploid number for a species? ____[5]

In **pachynema** (Figure 1-8c), the two chromatids making up each chromosome become distinct, while remaining joined at their centromere regions. Consequently, each bivalent now appears as four parallel chromatids and may be called a **tetrad.** (Note: The term "tetrad" may also be used to refer to the four haploid cells which are formed at the end of a single meiosis.) During this substage, nonsister chromatids within a tetrad may make contact with each other at certain points, paving the way for **crossing over.** This important process involves the breakage of the two nonsister chromatids at their points of contact followed by the reciprocal exchange of corresponding segments between these chromatids. The points of contact will become apparent in the next substage as X-shaped configurations called **chiasmata** (singular: chiasma). If there are genetic differences between the exchanged segments, as is usually the case, the participating chromatids experience **genetic recombination.** Recombination is discussed in later chapters, but its importance to this discussion is that the process produces chromatids with combinations of maternal and paternal genes. (Keep in mind that before recombination, each chromatid consists solely of either paternal or maternal genes). Chromosomes formed during meiosis that possess these combinations are known as **recombinant chromosomes** and they make an important contribution to genetic variability.

In **diplonema** (Figure 1-8d), the homologous chromosomes making up each tetrad begin to repel each other, a process that begins at the centromere regions and progresses outward. The separation, however, is not complete since the nonsister chromatids (the homologous maternal and paternal components) of each tetrad remain in contact at the chiasmata. Near the end of diplonema, the chiasmata begin to move towards the tips of the chromatids.

In **diakinesis** (Figure 1-8e), the final stage of prophase I, the chromosomes continue their condensation, and the homologs within each tetrad are well separated except at the chiasmata. The chiasmata have completed their movement to the tips of the chromatids. The spindle apparatus now develops in the cytoplasm, the nucleolus and nuclear membrane break down, and the two centromere regions within each tetrad become attached to separate spindle fibers, one from each pole of the spindle apparatus.

Metaphase I (Figure 1-8f) The tetrads line up on the equator, equidistant from the two poles of the spindle.

[2] It would double with each generation. [3] Three. [4] It equals the haploid number. [5] It is one-half the diploid number.

FIGURE 1-8

Key events in meiosis. Events above the dotted line occur in meiosis I and events below it, in meiosis II.
(a) Prophase I, leptonema: Nucleus has two pairs of duplicated homologous chromosomes; those of maternal origin are solid and those of paternal origin are dotted. Each chromosome, although appearing single, consists of two sister chromatids joined at their centromere regions.
(b) Prophase I, zygonema: Homologous chromosomes are paired along their lengths, or synapsed, to show bivalent configurations.
(c) Prophase I, pachynema: Two chromatids making up each chromosome are distinct and each homologous chromosome pair shows the tetrad configuration. Crossing over occurs between nonsister chromatids within tetrads. Each chromatid participating in crossing over will show genetic recombination, a combination of maternal and paternal genetic material, as a consequence of this mutual exchange of homologous segments.
(d) Prophase I, diplonema: Two pairs of chromatids making up each tetrad spread apart from each other except at chiasmata (sites of crossing over).
(e) Prophase I, diakinesis: Chromosomes have condensed further and the chiasmata have moved to chromatid tips.
(f) Metaphase I: Homologous pairs line up on equator, midway between the poles of the spindle. The two duplicated homologs making up each tetrad separate in anaphase I (not shown) to become dyads and move to opposite poles.
(g) Telophase I: Each product nucleus contains one duplicated chromosome, or dyad, from each homologous pair present in the original nucleus.
(h) Metaphase II: Dyads line up on the equator with the centromere region of each attached to spindle fibers from each pole. During anaphase II (not shown), centromeres of the two sister chromatids of each dyad separate, allowing the sister chromatids of each dyad to separate and move to opposite poles. Each sister chromatid is now referred to as a chromosome.
(i) Telophase II: Four haploid product nuclei result from meiosis.

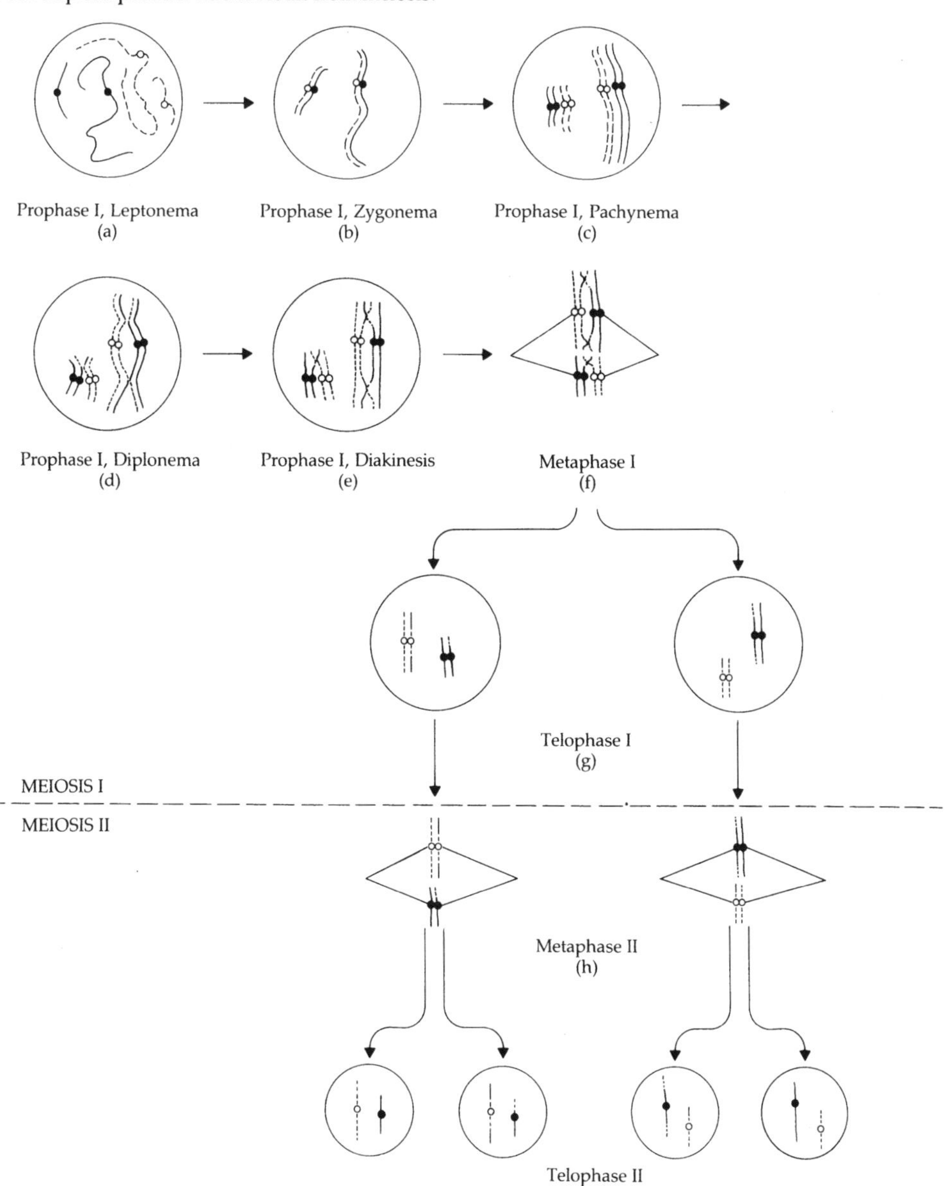

Contact between the nonsister chromatids within each tetrad is maintained by the chiasmata at the chromatid tips. Each tetrad orients itself randomly relative to the poles of the spindle apparatus and to the other tetrads. This random orientation is another important source of genetic variability, since it determines the spindle poles to which the maternal and paternal chromosomes of the tetrad will be pulled during anaphase I (described next) and, in turn, the combinations of maternal and paternal chromosomes that occur in the haploid product nuclei. This random orientation of the tetrads and subsequent separation of the maternal and paternal chromosomes of the tetrad provide the basis for Mendel's law of independent assortment discussed in Chapter 2. Note again that the chromosomes are lined up at the equator of the spindle apparatus as duplicated homologous pairs and that the centromere region of each homolog is attached to spindle fibers from either one or the other pole of the spindle apparatus. This is an important difference from mitosis where the duplicated chromosomes line up independently of their homologs and where the centromere region of each homolog is attached to spindle fibers from both poles.

Anaphase I The two duplicated homologs making up each tetrad separate from each other and move toward opposite poles of the spindle. (Note that the two centromeres in the centromere region of each duplicated chromosome do *not* separate.) After this disjunction, each duplicated chromosome still consists of two chromatids joined at their centromere region, a configuration referred to as a **dyad.** If there were no crossing over, both chromatids of each dyad would be entirely of either maternal or paternal genetic material. Because of their participation in crossing over, some chromatids within these dyads are combinations of maternal and paternal genes.

Telophase I (Figure 1-8g) The dyads are positioned at opposite poles of the spindle apparatus which now begins to break down. Often but not always, a nuclear membrane forms around the group of chromosomes at each spindle pole.

The events just reviewed make up the first meiotic division. If the nucleus undergoing meiosis I contains two homologous pairs, or four chromosomes, how many chromosomes will be grouped at each of the spindle poles at the end of meiosis I? ________[6] The first meiotic division is often referred to as a **reduction division** since it reduces the diploid number of chromosomes by one-half. Are the chromosomes present at the spindle poles at the end of meiosis I single-chromatid or double-chromatid chromosomes? ________ ________________________________[7] What further step do you think will be necessary to reduce each chromosome to a single-chromatid chromosome in order to form haploid nuclei? ________________ ________________________[8]

Stages of Meiosis II

The second meiotic division separates the two chromatids making up each dyad. During prophase II (not shown in Figure 1-8), spindle fibers from each pole attach to the centromere region of each dyad. In metaphase II, the dyads are positioned on the equator of the cell, equidistant from the two poles (Figure 1-8h). During anaphase II (not shown in Figure 1-8), the centromeres joining each pair of chromatids separate and the two chromatids of the dyad separate and are pulled to opposite poles. Note that once the two chromatids of each dyad separate, each chromatid is referred to as a chromosome. During telophase II, a nuclear membrane forms around the group of chromosomes at each pole and meiosis II is complete (Figure 1-8i). Note that the chromosomes at the beginning of meiosis II are double-chromatid, or dyad, chromosomes, while those in the haploid product nuclei are single-stranded. How many chromosomes were present in each of the nuclei at the beginning of meiosis II? ________________________________ ______[9] How many chromosomes are present in each of the nuclei formed at the end of meiosis II? ________ ________________________________[10] Meiosis II is often referred to as an **equational division** since it maintains the chromosome number at the haploid level.

Meiosis Summarized

Meiosis begins with a diploid nucleus in which the genetic material has already replicated to produce two copies of each chromosome. Two meiotic divisions result in the formation of haploid nuclei. During meiosis I, after homologous pairing, or synapsis, crossing over between nonsister chromatids generates combinations of maternal and paternal genes within individual chromatids. The two double-chromatid chromosomes making up each homologous pair then separate, producing nuclei with half the number of chromosomes found in the original nucleus. During meiosis II, the

[6] Two. [7] Double-chromatid chromosomes. [8] Separation of the chromatids making up each double-chromatid chromosome; this occurs during the second meiotic division. [9] Two double-chromatid chromosomes. [10] Two single-chromatid chromosomes.

FIGURE 1-9

Meiotic nuclear division.

(a) This nucleus shows the diploid number of chromosomes ($2n = 4$).

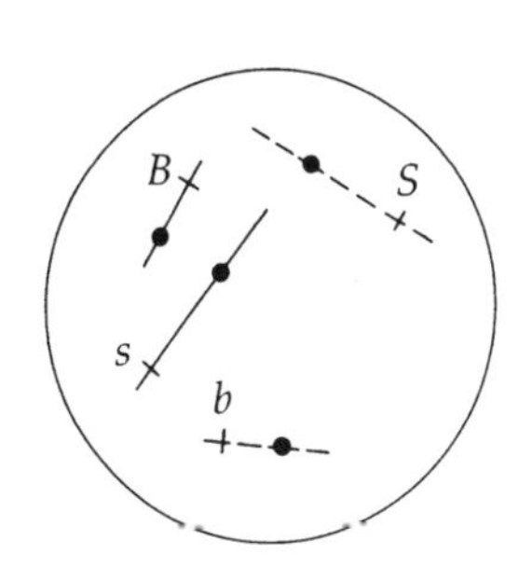

(b) Sketch each of the chromosomes in duplicated form (consisting of two chromatids joined by centromeres) synapsed with its homolog. Assume that no crossing over occurs. For example, the duplicated chromosome

B
B

paired side by side with its duplicated homolog would appear as

b
b
B
B

(c) Sketch the chromosomes after the first meiotic division, during which each duplicated chromosome separates from its homolog.

(d) Sketch the chromosomes after the second meiotic division, during which the two chromatids making up each duplicated chromosome separate from each other.

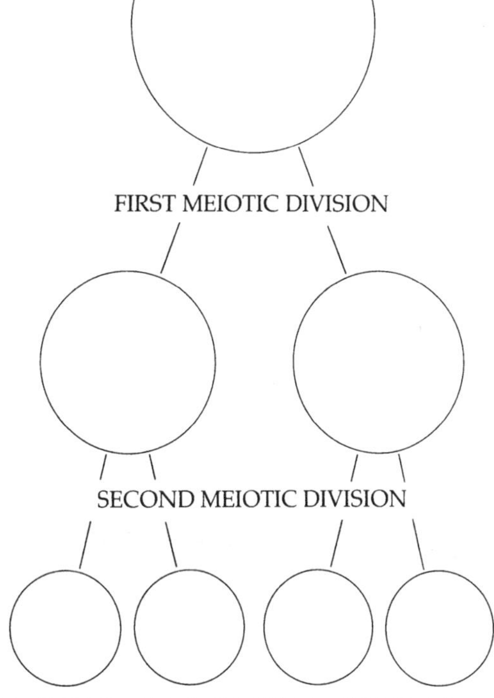

sister chromatids of each duplicated chromosome separate to opposite poles to make up the haploid nuclei of special reproductive cells—eggs or sperm in the case of animals, or spores in the case of plants.

Diagraming Meiosis

Using a hypothetical organism with two chromosome pairs ($2n = 4$), complete the series of diagrams in Figure 1-9 to show the key stages in meiosis. Assume that no crossing over occurs during this process. Make sure that your completed diagrams show that (1) meiosis reduces the chromosome number, with the diploid nucleus giving rise to haploid nuclei, and (2) each haploid nucleus has one member of each pair of homologous chromosomes present in the parental nucleus.

Chromosome and Gene Assortment

Assortment refers to the distribution of different combinations of maternal and paternal chromosomes to the haploid nuclei formed during meiosis. The term can also refer to the distribution of different combinations of the genes carried by maternal and paternal chromosomes. Examine the four haploid nuclei that result from your meiosis diagram (Figure 1-9d). Since the maternal and paternal chromosome of each homologous pair carry a different gene, the chromosomes in these nuclei can be identified by these genes. How many kinds of gene (or chromosome) combinations do you find in these nuclei? ________[11] What are these gene combinations? ______________________________[12] Is it possible to produce nuclei with chromosome and thus gene combinations which are different from those seen in your sketches above? ________[13] Examine Figure 1-9b and note that the particular combinations of chromosomes that end up in the haploid product nuclei depend on the way in which the homologous pairs position themselves during metaphase I. With two pairs of chromosomes, there are two possible ways that the tetrads could orient themselves relative to each other as shown in Figure 1-10. Each of these orientations produces different combinations of chromosomes and of genes in the haploid nuclei produced in meiosis.

To verify that chromosome assortment depends upon tetrad orientation, use Figure 1-11 to sketch the orientation that is *different* from the one you used in Figure 1-9b. Determine the chromosome and gene combinations that this orientation produces. How many kinds of gene combinations do you find in these four nuclei? ________[14] What are these gene combinations? ______________________________[15] How do they compare with the gene combinations found in the haploid product nuclei of your first drawings of meiosis in Figure 1-9d? ______________________________[16]

At this point you should realize that the particular chromosome combinations that assort into the haploid nuclei depend upon how the tetrads (the synapsed homologous chromosome pairs) position themselves relative to each other during metaphase of the first meiotic division. Chance determines the arrangement of tetrads within each cell, and once the chromosomes are positioned, the chromosome and gene combinations of the haploid nuclei arising from that cell are determined.

Any particular cell carrying two pairs of chromosomes can show one of two possible orientations during metaphase I and can make just two kinds of nuclei. Since most organisms produce their haploid reproductive cells (gametes or spores) in large numbers with many cells undergoing meiosis simultaneously, the laws of chance indicate that each alternative orientation of the two chromosome pairs during meiosis I would occur in half the cells. Following these cells through the second meiotic division would show that four combinations of chromosomes (and their genes) would occur with equal frequencies in the gametes.

For example, if we followed 100 diploid cells each with two chromosome pairs ($2n = 4$) that develop into sperm (spermatocytes), half of these cells would show one of the two possible orientations and the other half would show the other orientation. A total of 400 sperm cells would be produced (since each diploid cell produces four gametes during spermatogenesis). Four kinds of gametes (*SB*, *Sb*, *sB*, *sb*) would be produced, and we would expect to find them in equal frequencies (100 of each kind).

As we have seen, our hypothetical organism with two pairs of chromosomes can produce four different combinations of its maternal and paternal chromosomes. As the number of chromosome pairs possessed by a species increases, the number of possible, equally likely chromosome combinations increases tremendously and is given by the generalized expression 2^n, where n is the number of chromosome pairs. How many different combinations would be possible in the fruit fly *Drosophila melanogaster* which has a diploid number of 8? ________[17] In humans, where there are 23 pairs of chromosomes, the number of different chromosome combinations is 2^{23}, or over eight million. The chance that two gametes produced by the same individual will have the same chromosome combination is very slim indeed.

FIGURE 1-10

Possible chromosome orientations during metaphase I.

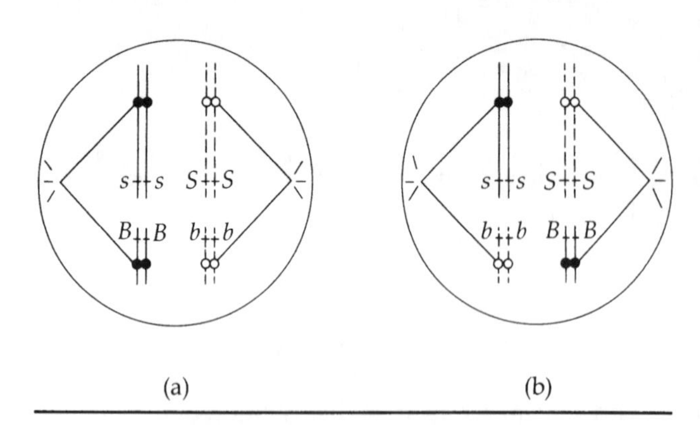

[11] Two. [12] Either *sB* and *Sb*, or *sb* and *SB*. [13] Yes. [14] Two. [15] Either *sB* and *Sb*, or *sb* and *SB*. [16] The two combinations are different.
[17] With a diploid number of 8, there are four pairs of chromosomes and $2^4 = 16$ different chromosome combinations.

FIGURE 1-11

Products of the alternative chromosome orientation.

(a) Sketch the two duplicated chromosomes of each homologous pair synapsed with each other. Be sure that their orientation differs from the one you used in your earlier meiosis sketch, Figure 1-9b.

(b) Sketch the nuclei produced in the first meiotic division.

(c) Sketch the nuclei produced in the second meiotic division.

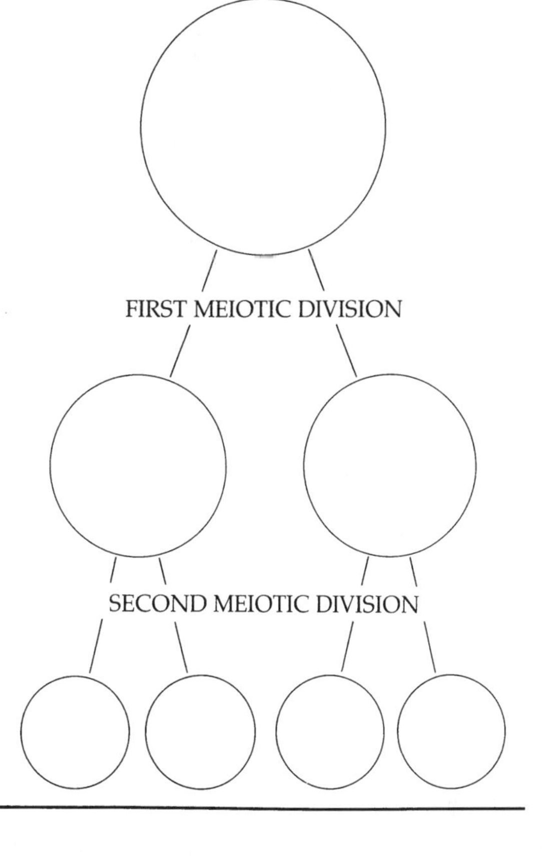

Meiosis and Fertilization in the Animal and Plant Life Cycle: Restoring the Diploid Number

Fertilization is the union of a haploid sperm nucleus with a haploid egg nucleus to produce a diploid, one-celled zygote. The zygote will undergo mitotic cell division and thus begin to form an embryo. Assume that the egg and sperm nuclei belonging to our hypothetical $2n$ species have the chromosome makeup shown in Figure 1-12. Sketch the chromosome makeup for the zygote nucleus as well as for the nuclei of the two cells formed mitotically from the zygote. Be sure that your sketches of fertilization and mitosis show that (1) fertilization involves the union of two haploid gametes, a male sperm with the female egg, and restores the diploid number, and (2) one member of each chromosome pair in the zygote and in each of the nuclei developing from the zygote is maternal in origin and the other is of paternal origin.

Note that in most organisms, the zygote undergoes an extended series of mitotic cell divisions. The first division produces two daughter cells with an identical chromosome makeup. Each of these then di-

FIGURE 1-12

Fertilization and first mitotic division.

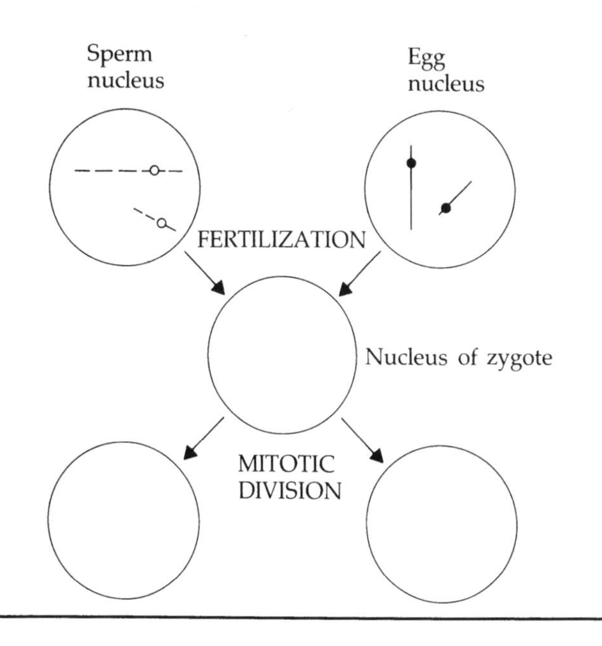

vides to produce four cells which give rise to eight cells, and so on. Within each diploid cell, one member of each chromosome pair is always of paternal origin and the other member is always of maternal origin. Each round of mitotic division faithfully produces daughter cells with the same chromosome makeup. Continuing mitotic divisions eventually produce an adult. Once the individual reaches maturity, some of its diploid cells will undergo meiosis, producing specialized reproductive cells—gametes in the case of animals, or spores in the case of plants. Each of these reproductive cells will contain a single chromosome from each chromosome pair present in the diploid cell.

In animals, the union of two gametes restores the diploid number and the life cycle repeats itself (Figure 1-13). In plants (Figure 1-14), individual spores develop mitotically into a multicellular, haploid gametophyte which mitotically produces gametes. As with animals, union of two haploid gametes restores the diploid number. The zygote that results grows mitotically into a multicellular, diploid sporophyte which will meiotically produce spores. This cycle of alternating haploid and diploid stages than repeats itself.

Significance of Meiosis

Meiosis is of key importance to sexually reproducing eukaryotic species in two ways. (1) In combination with fertilization, it maintains the chromosome number of a species. During the formation of specialized reproductive cells, meiosis reduces the chromosome number from diploid to haploid. When haploid gametes unite at fertilization, the diploid number of the species is restored. This ensures that the chromosome number of a species remains constant after each successive round of sexual reproduction. (2) Meiosis produces genetic variability by forming new combinations of genes in two ways. First, combinations of maternal and paternal genes on individual chromosomes arise through crossing over between maternal and paternal chromatids of each homologous pair during prophase I. Second, new combinations of maternal and paternal chromosomes arise through the random alignment and subsequent separation (assortment) of the tetrads during metaphase I.

FIGURE 1-13

Generalized animal life cycle.

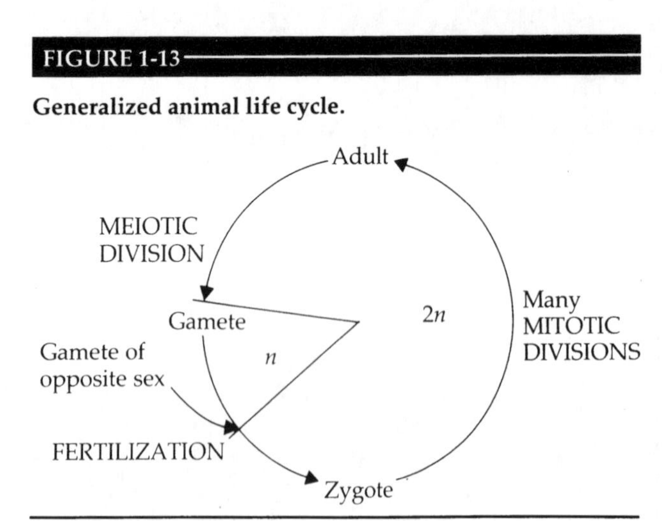

FIGURE 1-14

Generalized plant life cycle.

FERTILIZATION
Zygote
Many MITOTIC DIVISIONS
Gamete of opposite sex
$2n$
Gamete
Adult sporophyte
MITOTIC DIVISION
n
Adult gametophyte
MEIOTIC DIVISION
Spore
Many MITOTIC DIVISIONS

NO MITOSIS OR MEIOSIS IN PROKARYOTIC CELLS OR VIRUSES

Mitosis and meiosis occur only in eukaryotic cells. The simple cell division that occurs in prokaryotes such as bacteria is preceded by a replication of the single, circular chromosome, but no spindle apparatus is formed to separate the replicates. Each daughter chromosome is attached to a different section of the inner surface of the cell membrane and, during the subsequent cell division, the inward growth of the cell membrane and cell wall at the middle of the cell separates the two replicates and produces a constriction that subdivides the cell. The two daughter cells that result are of approximately equal size and are genetically identical. Nothing resembling meiotic cell division occurs in prokaryotes. Prokaryotes do exhibit mechanisms that produce genetic recombination and these will be discussed later.

Viruses are completely lacking in cellular organization and exhibit nothing that resembles cell division. Viral reproduction occurs only in a susceptible prokaryotic or eukaryotic host cell. Following infection, the viral chromosome directs the metabolic machinery of the host cell to produce a large number of progeny viruses. This involves making many replicates of the viral chromosome and synthesizing a new protein coat for each replicate. Following packaging of each new viral chromosome into a protein coat, the membrane of the host cell breaks down and progeny viruses are released. These virus particles can now go on to infect other host cells. Mechanisms that produce genetic recombination in viruses will be discussed later.

SUMMARY

Both mitosis and meiosis are types of nuclear division that distribute genetic material during the process of eukaryotic cell division, and both are preceded by DNA replication and chromosome duplication.

Mitosis begins with one nucleus and produces two nuclei each with exactly the same genetic information as the original nucleus, thus maintaining in the daughter nuclei the same chromosome number found in the original nucleus, regardless of whether mitosis occurs in haploid or diploid cells.

Meiosis begins with a diploid cell or nucleus and, after two divisions, produces four haploid reproductive cells or nuclei. Meiosis entails genetic variation in the reproductive cells in two ways: the pairing of homologous chromosomes in prophase I permits the occurrence of crossing over, generating new combinations of maternal and paternal genes within individual chromosomes; and the random alignment of tetrads at metaphase I followed by assortment generates new combinations of maternal and paternal chromosomes.

Meiosis, in combination with fertilization, guarantees that the chromosome number for a species remains constant after each round of sexual reproduction. Haploid gametes unite at fertilization to produce a diploid, one-celled zygote, which, in multicellular organisms, will undergo an extended series of mitotic cell divisions that will eventually produce an adult. In animals, this diploid adult goes on to meiotically produce gametes which unite to form a zygote, thus completing the life cycle. In plants, the diploid adult is a sporophyte which meiotically produces haploid spores. Each spore has the potential to grow through mitotic cell division into a multicellular haploid plant form known as the gametophyte which mitotically produces gametes; gamete union produces a zygote which completes the plant life cycle.

No mitosis or meiosis occurs in the reproduction of prokaryotic cells or viruses. The single, circular, prokaryotic chromosome replicates prior to cell division, and the replicates, attached to the cell membrane at different points, are separated as the cell elongates prior to cell division. Then the cell membrane and cell wall grow inward and subdivide the cytoplasm to form two genetically identical daughter cells.

Viruses lack cellular organization and reproduce only in a susceptible prokaryotic or eukaryotic host cell. Following infection, the viral chromosome directs the metabolic machinery of the host cell to produce a large number of genetically identical progeny viruses.

PROBLEM SET

1-1. **a.** Distinguish between chromatids and chromosomes. Indicate how they are similar and how they are different.
b. What is meant by the term homologous chromosomes? What is meant by the term nonhomologous chromosomes?

1-2. What is the significance of mitotic nuclear division for a multicellular organism?

1-3. Assume that a cell about to undergo mitosis has a diploid number of 8.
a. How many homologous pairs of chromosomes are present in the nucleus of this cell?
b. During what stage or stages of mitosis are the chromosomes most distinct?
c. Briefly describe how a chromosome appears when it becomes distinct.
d. How many chromatids are present during late prophase?
e. What is the greatest number of chromatids that could be present during mitosis?
f. Assume that both members of one homologous pair carry the allele *B*.
i. How many copies of *B* would exist in the nucleus during prophase?
ii. How many copies of *B* would exist in each of the daughter nuclei that result from mitosis?
g. Assume that one member of one homologous pair carries the allele *A* and the other member of the pair carries allele *a*.
i. How many copies of *A* would exist in the nucleus during prophase?
ii. How many copies of *A* would exist in each of the daughter nuclei that result from mitosis?

1-4. Assume that a cell about to undergo meiosis has a diploid number of 8.
a. How many homologous pairs of chromosomes are present in the nucleus of this cell?
b. Briefly compare the appearance of a chromosome when it first becomes distinctly visible in early prophase I with its appearance at the end of prophase I.
c. How many chromatids would be present during prophase I?
d. What is the greatest number of chromatids that would be present during meiosis?
e. Assume that both members of one homologous pair carry the allele *B*.
i. How many copies of *B* would exist in the nucleus during prophase I?
ii. How many copies of *B* would exist in each of the daughter nuclei that result from meiosis?
f. Assume that one member of one homologous pair carries the allele *A* and the other member of the pair carries allele *a*.
i. How many copies of *A* would exist in the nucleus during prophase I?
ii. How many copies of *A* would exist in each of the daughter nuclei produced by meiosis?

1-5. What major distinction would you expect to observe in comparing the chromosomes in the nuclei of two diploid cells from the same organism, one of which is in metaphase of mitosis and the other in metaphase of meiosis I?

1-6. A total of 12 bivalents are counted in a nucleus during prophase of meiosis I.
a. How many tetrads would occur in this nucleus?
b. What is the diploid number for this species?
c. What is the haploid number for this species?

PROBLEM SET

1-7. The separation of the centromeres allows the chromatids of a replicated chromosome to separate.
- **a.** When during mitosis do the centromeres separate?
- **b.** When during meiosis do the centromeres separate?

1-8. A human cell with a diploid number of 46 is undergoing meiosis I.
- **a.** How many chromosomes are present in each of the nuclei formed at the end of meiosis I?
- **b.** Is the number of chromosomes in each of these nuclei equal to the haploid or to the diploid number for the species?
- **c.** How many chromatids make up each of these chromosomes?
- **d.** What is the total number of chromatids found in each of these nuclei?
- **e.** If the same cell had undergone mitosis rather than meiosis I, would the chromosomes in the two resulting product nuclei be identical to those found in the nuclei at the end of meiosis I? Explain.

1-9. After examining microscope slides which show the various stages of mitosis and meiosis in the roundworm *Parascaris*, a student records that the fully condensed chromosomes at the start of both meiosis II and mitosis each consist of two chromatids.
- **a.** Is this an accurate observation?
- **b.** The student also notes that at the end of meiosis II and mitosis each chromosome is made up of a single strand. Is this observation accurate?
- **c.** Because of the similarity between the chromosomes at the start and at the end of meiosis II and of mitosis, the student concludes that meiosis II is a mitotic-like division. Is this correct? Explain why or why not.
- **d.** The student also notes that the chromatids making up each of the double chromosomes present at the start of meiosis II are not necessarily true sister chromatids, whereas those making up the replicated mitosis chromosomes are genuine sister chromatids. Is this statement correct? Explain why or why not.

1-10. The nuclei that are formed after a mitotic nuclear division have identical genetic content. This is not true of the nuclei that arise following meiosis. Identify two phenomena which give rise to the genetic variability associated with meiosis.

1-11. Crossing over, which involves the reciprocal exchange of corresponding segments between nonsister homologous chromatids, is a major source of genetic variability.
- **a.** Explain how this process produces new combinations of maternal and paternal genes.
- **b.** If the crossing over occurred between sister chromatids, would new combinations of genetic material arise? Explain.

1-12. In a general way, distinguish between the variability produced by the random assortment of chromosomes and that produced by crossing over.

1-13. Identify which of the following statements about meiosis are true and which are false. If a statement is false, explain why it is false.
- **a.** Following meiosis I, the group of chromosomes around each spindle pole contains half the number of chromosomes found in the nucleus of the original diploid cell undergoing meiosis.
- **b.** In meiosis, the number of chromatids in a late-prophase-I nucleus is equal to the number of chromatids making up the chromosomes grouped around one of the poles of the spindle near the end of meiosis I.

PROBLEM SET

- **c.** The number of centromeres in a late-prophase-I nucleus is equal to the number of centromeres in a late-prophase-II nucleus.
- **d.** The number of centromeres in a late-prophase-II nucleus is equal to the number of centromeres in one of the haploid nuclei produced by meiosis.
- **e.** The number of dyads carried by a single nucleus at the end of meiosis I is equal to the number of bivalents present in the "parent" nucleus before meiosis I.

1-14. The diploid cells of a hypothetical animal carry two pairs of chromosomes, one designated as 1*,1 and the other as 2*,2. Chromosomes marked with an asterisk (1* and 2*) are of maternal origin and those unmarked (1 and 2) are of paternal origin.

- **a.** Identify the different chromosome combinations that could occur in the gametes produced by this individual. How many are there?
- **b.** What is the probability of producing gametes that carry only chromosomes of paternal origin?
- **c.** What is the probability of producing gametes that carry only chromosomes of maternal origin?
- **d.** What is the probability of producing gametes with a combination of maternal and paternal chromosomes?
- **e.** If a male of this species produces a large number of gametes, say 5000 sperm, how many of each of the combinations identified in 1-14a would you expect to be produced?
- **f.** Assume that male gametes with all possible chromosome combinations unite at fertilization with female gametes carrying 1*,2 chromosomes only. Identify the different chromosome combinations that would occur in the zygotes.

1-15. In a diploid cell ($2n = 4$) that is undergoing meiotic cell division to produce four sperm cells, one crossover occurs between two nonsister chromatids in each tetrad during prophase of meiosis I. The products of these crossovers are shown in the following diagram in one possible alignment of these tetrads during metaphase of meiosis I; maternally derived chromosomes are drawn in solid lines and with solid centromeres while paternally derived chromosomes are drawn in dotted lines and with open centromeres.

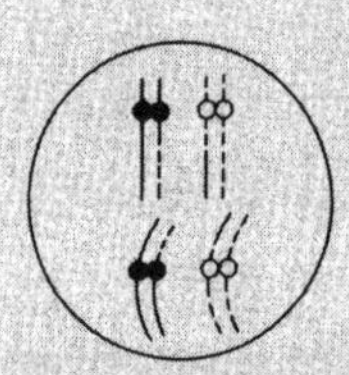

- **a.** With this particular metaphase-I alignment of the tetrads, how many different combinations of maternal and paternal chromosomes are possible in the sperm nuclei that result? Draw them.
- **b.** Could other combinations of maternal and paternal chromosomes occur in the sperm nuclei following these crossover events? If so, explain how they would arise and draw them.

1-16. Ferns have both a haploid and a diploid stage in their life cycle. Early in the life cycle, all the cells in the prothallus (the young fern plant) are haploid. The prothallus is the gamete-producing stage of the life cycle.

PROBLEM SET

a. What type of nuclear division must be involved in the production of gametes by the prothallus?
b. The union of two fern gametes produces a zygote, which grows into the more commonly recognized diploid fern plant which in turn produces haploid spores. What type of nuclear division must be involved in the production of spores by the diploid plant?

CHAPTER

2

Mendel's Laws

INTRODUCTION

The basic principles of genetics were announced by Gregor Mendel in a report published in 1866, well before anything was known about chromosome structure, mitosis, or meiosis. Mendel discovered that heritable characters are determined by discrete units of inheritance and that these units rather than the traits themselves are transmitted from parent to offspring. He predicted the behavior of these units during reproductive-cell formation and the manner of their transmission from one generation to another. In this chapter we look at Mendel's experiments and how they led him to these predictions.

MENDEL'S CROSSES

Mendel studied the inheritance of a number of characters in the garden pea (*Pisum sativum*). This plant is well suited for genetic studies because its life cycle is completed in a single growing season, large numbers can be easily grown, many progeny are produced, it differs in a variety of characters which can be studied, and it can readily be crossed. In nature, pea plants reproduce by **self-pollination** (also referred to as **self-fertilization** or **selfing**). Each flower contains both male (stamen) and female (pistil) parts; pollen produced by the stamen normally transfers to the pistil within the same flower to fertilize ovules that develop within the pistil. A plant breeder can achieve **cross-pollination,** or **cross-fertilization,** by removing the stamens from a flower before the pollen matures and transferring pollen obtained from another plant to the flower's pistil. Regardless of the method of pollination, the zygotes that result following fertilization develop into seeds which can be planted to grow into the next generation's plants.

Monohybrid Crosses and the Law of Segregation

Experimental Design and Results Mendel studied seven pea-plant characters (stem length, seed color, seed shape, flower placement, flower color, pod color, and pod shape), each of which exhibits two alternative, contrasting appearances, or **phenotypes.** For example, the character of stem length is tall in some plants and short in others. Most of Mendel's crosses focused on one or two of these characters at a time and the **parent,** or $\mathbf{P_1}$**,** plants he used were from **true-breeding,** or **pure-breeding,** strains. Such strains, when selfed, produce offspring with identical phenotypes generation after generation. Recognizing the contrasting expressions for these characters, considering these characters one at a time, and using pairs of parent plants which were true breeding for these contrasting expressions were the keys to much of Mendel's success.

In some experiments, designated as **monohybrid crosses,** Mendel crossed P_1 plants that exhibited contrasting phenotypes for a single character. For example, plants with tall stems were crossed with plants with short stems and, in other crosses, plants with yellow seeds were crossed with plants with green seeds. (Note that each of these examples deals with alternative expressions of a *single* character—stem length or seed color.) For each character studied, Mendel found that only one of its two contrasting expressions appeared in the progeny, designated as the **first filial,** or **F_1,** generation. For example, in crossing pure-breeding parents by pollinating tall plants with pollen derived from short plants, all the F_1 progeny were tall. The same type of progeny were formed when a **reciprocal cross** was carried out, that is, when short plants were pollinated with pollen derived from tall plants. Mendel introduced the terms **dominant** to designate the phenotype shown by the F_1 (tall) and **recessive** to designate its alternative (short) which was not exhibited by the F_1 progeny.

Mendel allowed F_1 plants to self-pollinate in order to produce the **second filial,** or **F_2,** generation. For each of the seven characters studied, both expressions of the character appeared in the F_2 in a ratio very close to 3 dominants to 1 recessive. In one cross, for example, the F_2 consisted of 678 tall and 218 short plants. This ratio of 678 to 218 can be simplified by dividing each value in the ratio by 218, the smallest value in the ratio: 678/218 = 3.11 and 218/218 = 1 to give a ratio of 3.11 to 1, or 3.11:1. The scheme followed by Mendel to produce this F_2 generation is diagramed in Figure 2-1.

For each character studied by Mendel, only one of the contrasting phenotypes exhibited by the pure-breeding parents appeared in the F_1, and this same phenotype occurred in the F_1 of reciprocal crosses. In addition, the parental phenotype not expressed in the F_1 reappeared in F_2, where the dominant and recessive phenotypes occurred with a ratio of approximately 3 to 1. Among the things Mendel sought to explain were the disappearance of the recessive parent's phenotype in the F_1, its reappearance in the F_2, and the 3:1 phenotypic ratio in the F_2.

FIGURE 2-1

P_1: Tall × Short (Cross-pollination)

↓

F_1: 100% tall

F_1 × F_1 (Self-pollination)

↓

F_2: 678 tall : 218 short (3.11-to-1 ratio)

Mendel's Postulates, Including the Law of Segregation Based on similar observations for all of the seven characters he studied, Mendel proposed the following postulates, or principles.

1. Alternative phenotypes of a character (for example, tall and short stem length) are due to inherited "factors" (which are now called **genes**). Factors occur in contrasting forms (now known as **alleles**), producing the alternative phenotypes. Thus the factor (gene) for stem length has two forms (alleles): one for long stems which could by symbolized by the letter *S* and one for short stems symbolized by *s*. (Note that the system of assigning symbols to genes and their alleles that is used here involves using an uppercase letter to designate the dominant allele and a lowercase letter to designate the recessive allele. The letter used as the symbol is the first letter of the word describing the phenotype resulting from expression of the recessive allele, in this case, *short*. Additional information on assigning gene symbols is presented in subsequent chapters.)

2. Each individual carries two factors (genes) which determine the expression of each character studied. Mendel came to this conclusion based on the F_1 plants: since they exhibited the dominant phenotype, they must have carried the dominant factor, and since they gave rise to F_2 plants, some of whom exhibited the recessive phenotype, the F_1 plants must also carry the recessive factor. (The genes carried by an individual make up its **genotype** for the character, in contrast to the character's expression or appearance which is part of the individual's phenotype). An individual could have one of three possible combinations of alleles or genotypes: two dominant alleles, two recessive alleles, or one dominant and one recessive allele. An individual carrying two identical alleles is **homozygous**; homozygotes can be homozygous dominant, for example, *SS*, or homozygous recessive, for example, *ss*. An individual carrying two different alleles, for example, *Ss*, is **heterozygous.** In heterozygous individuals, like those produced in his F_1 generations, Mendel proposed that the dominant allele masks over or prevents the expression of its recessive counterpart.

3. Factors (genes) retain their identity, that is, they are not altered, from one generation to the next. The integrity of the factors carried by an organism for a particular trait is not influenced by the way the organism expresses the trait. Mendel proposed this idea because the recessive allele, expressed in one of the parents but not in the F_1, reappeared unchanged in the F_2 generation. This outcome was inconsistent with the theory of blended inheritance prevalent during the nineteenth century which held that the hereditary characteristics of an organism are the result of the blending of the germinal influences of its parents.

4. Factors (genes) that make up a pair separate or **segregate** from each other when reproductive cells are formed. This concept, known as Mendel's **law of segregation,** means that each gamete carries one member of the gene pair for each character. (See the discussion of metaphase and anaphase of meiosis I in Chapter 1.) Thus, half the gametes produced by an individual will have one of that individual's alleles and the rest of the gametes will have the individual's other allele for any given character. For example, a pure-breeding tall parent has two *S* alleles and all of its gametes will carry *S*. Similarly, a pure-breeding short parent has two *s* alleles and all of its gametes will carry *s*. Half the gametes produced by an *Ss* plant will carry the *S* allele while the other half will carry *s*.

5. The union of gametes to form the zygotes of the next generation is in no way influenced by the alleles they carry; that is, fertilization is random. Each gamete produced by one parent has an equal chance of fertilizing any of the gametes produced by the other parent.

The Monohybrid Cross in Modern Symbols Figure 2-2 again shows Mendel's crossing scheme, in which pure-breeding tall and short P_1 plants were crossed. This time, however, allelic symbols are used to represent the genotypes for each generation and to show the types of gametes and the zygotes formed through random fertilization.

Another way of depicting the random combination of the F_1 gametes to form the F_2 generation is to set up a **Punnett square** (named after the geneticist R. C. Punnett), arranging the gametes of one parent (by convention, the male or pollen-supplying plant) across the top of a square and those of the other parent (the female or seed-producing plant) along the side. Combining the heading at the left end of each horizontal row with the heading at the top of each vertical column gives all possible allelic combinations for the F_2 progeny. For example, combining the *S* gamete heading the upper row with the *S* gamete heading the column on the left gives the *SS* genotype found in the upper-left box as shown in the following Punnett square.

Gametes from *Ss* parent (side) \ Gametes from *Ss* parent (top)	*S*	*s*
S	*SS*	*Ss*
s	*Ss*	*ss*

Note that the four boxes within the Punnet square represent all (100%) of the F_2 progeny and that each box within the Punnett square represents 1/4 (25%) of the progeny. Genotype *SS*, found in one box, makes up 25% of the progeny; genotype *Ss*, found in two boxes, makes up 50% of the progeny; and genotype *ss*, found in the remaining box, makes up 25% of the progeny. Put in ratio form, the genotypic outcome can be written as 25%:50%:25%, or 1:2:1. The dominant phenotype is expressed by both genotypes *SS* and *Ss* and occurs in 25% + 50% = 75% of the progeny, whereas the recessive phenotype is expressed by genotype *ss* and occurs in 25% of the progeny. Thus, the genotypic and phenotypic outcome derived using the Punnett square is identical to that which was diagramed in Figure 2-2.

Chromosomal Basis for the Law of Segregation As was pointed out in Chapter 1, each of the alleles mak-

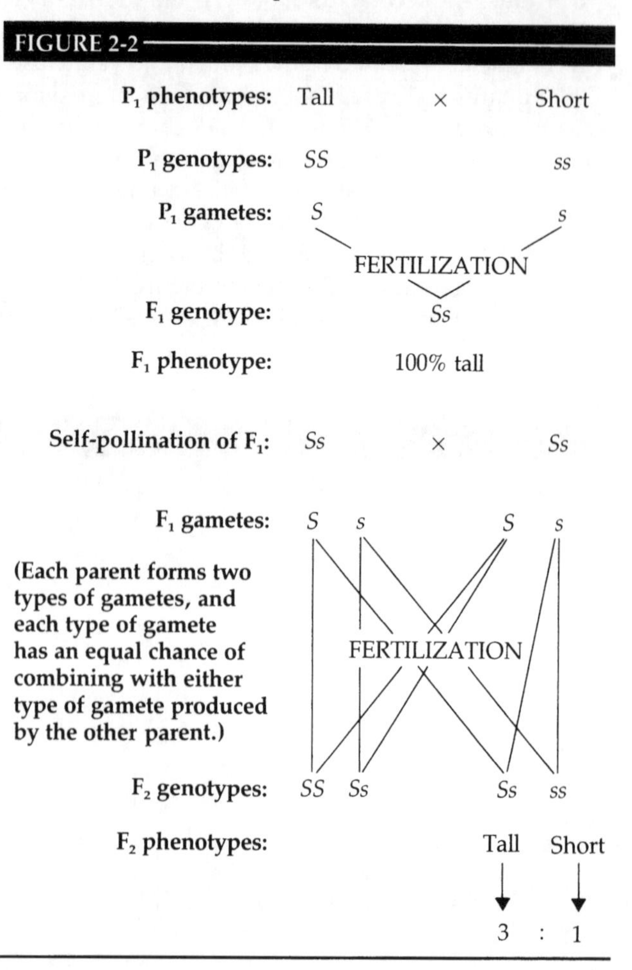

ing up a gene pair is carried on a different chromosome of a homologous pair. (The physical site on a chromosome where a particular gene is located is known as the gene's **locus.** Since homologous chromosomes separate, or segregate, during meiosis, each reproductive cell ends up with one member of each homologous pair and, therefore, one allele of each gene pair. Thus, gene segregation occurs because the homologous chromosomes segregate during meiosis.

Dihybrid Crosses and the Law of Independent Assortment

Experimental Design and Results In addition to the monohybrid crosses considered so far, Mendel also carried out **dihybrid crosses** which simultaneously studied the inheritance of two different characters. One cross, for example, involved the characters of seed shape and seed color. Pure-breeding plants with smooth and yellow seeds were crossed with pure-breeding plants with wrinkled and green seeds. Mendel found that all of the F_1 had smooth and yellow seeds, indicating that smooth is dominant to wrinkled and that yellow is dominant to green.

We can set up a scheme for this cross by using symbols to represent the two pairs of genes. For the seed-shape gene, let *W* represent the dominant allele for smooth and *w* the recessive allele for wrinkled. For the seed-color gene, let *G* represent the dominant allele for yellow and *g* represent the recessive allele for green. The genotype of the pure-breeding smooth and yellow plant can be symbolized as *WWGG* and that of the pure-breeding wrinkled and green plant as *wwgg*. The cross can be represented as *WWGG* × *wwgg*. The law of segregation tells us that the gametes made by each parent will carry one allele of each gene pair. Since the alleles making up each pair in each parent are identical, each parent makes a single type of gamete: *WWGG* forms *WG* gametes and *wwgg* forms *wg* gametes. The union of these gametes produces an F_1 generation with the genotype *WwGg* and a phenotype of smooth and yellow. This cross is summarized in Figure 2-3.

FIGURE 2-3

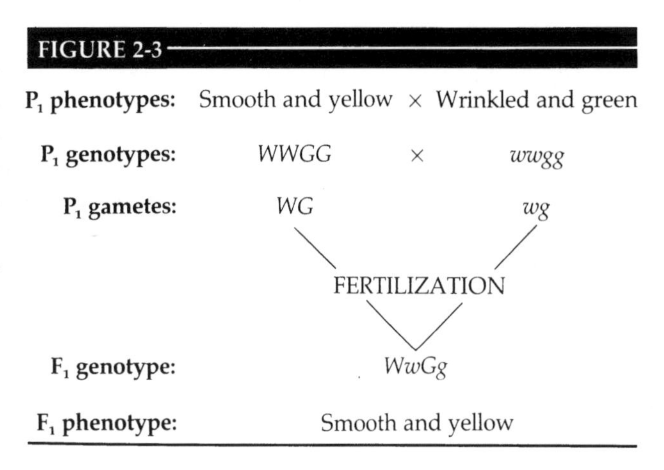

When Mendel allowed F_1 plants to self-pollinate, he found four phenotypic classes among the F_2 progeny in the following approximate frequencies: 9/16 smooth and yellow, 3/16 wrinkled and yellow, 3/16 smooth and green, and 1/16 wrinkled and green. These phenotypic frequencies can also be expressed in a ratio of 9:3:3:1.

Explaining the Dihybrid Outcome To explain these results, Mendel considered two alternative explanations, or hypotheses, with regard to the seed-shape and seed-color genes possessed by each of the original (P_1) parents. Either the particular color and shape genes possessed by each parent were inherited (1) *together,* so that only the combinations occurring in the parents were found in their gametes and in their F_1 and F_2 progeny, or (2) *independently,* so that new gene combinations could occur in their gametes and progeny. We will consider each of these hypotheses to see which is supported by Mendel's findings. We can use our knowledge of genetics, under the terms of each hypothesis, to predict the F_1 and F_2 progeny expected from Mendel's crosses and compare these predictions with Mendel's actual results. Agreement between the actual outcome and that predicted under a hypothesis will provide support for that hypothesis.

We will begin by evaluating the second hypothesis, which states that the genes for each character are inherited independently. If the genes are inherited independently, the inheritance of each character can be considered separately. With regard to seed color, crossing pure-breeding yellow and green P_1 plants gives an F_1 which, because of dominance, has the yellow phenotype. An F_1 plant carries both a dominant and recessive allele for this character and its genotype can be symbolized as *Gg*. The selfing of an F_1 plant to produce the F_2 can be represented as *Gg* × *Gg*. The alleles carried by an F_1 plant segregate at the time of reproductive-cell formation and half its gametes carry *G* and the other half carry *g*. The F_2 genotypes produced from the random union of these gametes are shown in the following Punnett square.

Gametes from *Gg* parent (rows) / Gametes from *Gg* parent (columns)	G	g
G	GG	Gg
g	Gg	gg

Since the four boxes in the square represent all of the progeny from this cross, each box represents 1/4 of the progeny. Three genotypes result: *GG*, found in one of four boxes, makes up 1/4 of the progeny; *Gg*, found in two boxes, makes up 2/4, or 1/2, of the progeny; and *gg*, found in the remaining box, makes up 1/4 of the progeny. The dominant yellow phenotype is expressed by genotypes *GG* and *Gg* and occurs in 1/4 + 2/4 = 3/4 of the progeny, and the recessive green phenotype is expressed by genotype *gg*, and occurs in 1/4 of the progeny. To summarize, if the genes responsible for seed color are inherited independently of those determining seed shape, then the F_2 progeny from the *WwGg* × *WwGg* cross would be expected to be yellow and green in a 3-to-1 ratio.

Now look back at the F_2 phenotypes produced in Mendel's dihybrid cross and determine the frequencies of progeny with yellow and green phenotypes. How many of the four phenotypic classes exhibit the yellow phenotype? ________[1] What fraction of all the progeny are yellow? ________[2] How many of the four phenotypic classes exhibit the green phenotype? ________[3] What fraction of all the progeny are green? ________[4] How do Mendel's results compare with those you have just predicted based on the hypothesis that seed color is inherited independently of seed shape? ________________________________[5]

Now consider seed shape. In crossing pure-breeding smooth and wrinkled P_1 plants, the F_1 progeny, because of dominance, would be smooth. An F_1 plant carries both a dominant and a recessive allele for this character and its genotype can be symbolized as *Ww*. The selfing of an F_1 plant to produce the F_2 can be represented as *Ww* × *Ww*. The alleles carried by an F_1 plant segregate at the time of reproductive-cell formation and half its gametes carry *W* and the other half carry *w*. The F_2 genotypes produced from the random union of these gametes are shown in the following Punnett square.

Gametes from *Ww* parent (rows) / Gametes from *Ww* parent (columns)	W	*w*
W	WW	*Ww*
w	*Ww*	*ww*

Three genotypes result: *WW*, found in one of four boxes, makes up 1/4 of the progeny; *Ww*, found in two boxes, makes up 2/4, or 1/2, of the progeny; and *ww*, found in the remaining box, makes up 1/4 of the progeny. The dominant smooth phenotype is expressed by genotypes *WW* and *Ww* and occurs in 1/4 + 2/4 = 3/4 of the progeny. The recessive wrinkled phenotype is expressed by genotype *ww* and occurs in 1/4 of the progeny. To summarize, if the genes responsible for seed shape are inherited independently of those determining seed color, then the F_2 progeny from the *WwGg* × *WwGg* cross would be expected to be smooth and wrinkled in a 3-to-1 ratio.

Now look back at the F_2 phenotypes from Mendel's dihybrid cross. How many of the four phenotypic classes exhibit the smooth phenotype? ________[6] What fraction of all the progeny are smooth? ________[7] How many of the four phenotypic classes exhibit the wrinkled phenotype? ________[8] What fraction of all the progeny are wrinkled? ________[9] How do Mendel's results compare with those you have just predicted based on the hypothesis that seed shape is inherited independently of seed color? ________________________________[10]

In summary, when we assume that the two characters are inherited independently and consider the inheritance of each separately, we find that the outcomes we predict for Mendel's crosses agree with the outcomes he observed. In other words, Mendel's results support the hypothesis that the two characters are inherited independently.

Although the data support the hypothesis of independent inheritance, we need to evaluate the alternative hypothesis which states that the genes involved in determining these two characters are inherited together. In the original parents, *W* occurs with *G*, and *w* with *g*. If the genes are inherited together, the genes would occur in the F_1 and F_2 progeny in the same combinations. The union of *WG* and *wg* gametes would produce F_1 progeny with genotype *WwGg* and the smooth and yellow phenotype. If *W* stayed with *G* and *w* stayed with *g*, only *WG* and *wg* gametes would be formed by the F_1 plants. The random union of these gametes to produce the F_2 is shown in the following Punnett square.

Gametes from *WwGg* plant (rows) / Gametes from *WwGg* plant (columns)	WG	*wg*
WG	WWGG	*WwGg*
wg	*WwGg*	*wwgg*

[1] Two: yellow and smooth, and yellow and wrinkled. [2] 9/16 + 3/16 = 12/16 = 3/4. [3] Two: green and smooth, and green and wrinkled. [4] 3/16 + 1/16 = 4/16 = 1/4. [5] Mendel's results agree with those predicted. [6] Two: yellow and smooth, and green and smooth. [7] 9/16 + 3/16 = 12/16 = 3/4. [8] Two: yellow and wrinkled, and green and wrinkled. [9] 3/16 + 1/16 = 4/16 = 1/4. [10] Mendel's results agree with those predicted.

Three genotypes result: *WWGG*, found in one of four boxes, makes up 1/4 of the progeny; *WwGg*, found in two boxes, makes up 2/4, or 1/2, of the progeny; and *wwgg*, found in the remaining box, makes up 1/4 of the progeny. Genotypes *WWGG* and *WwGg*, with a combined frequency of 1/4 + 1/2 = 3/4, produce the smooth and yellow phenotype, and genotype *wwgg*, with a frequency of 1/4, produces wrinkled and green seeds. Thus, if the genes for these two characters are inherited together, *two* F_2 phenotypes would be expected, in a ratio of 3 smooth and yellow seeds to 1 wrinkled and green seed. Since the F_2 generation observed by Mendel consisted of *four* phenotypic classes, his data do not support this hypothesis. In summary, Mendel's results support the hypothesis that the genes determining each character are inherited independently.

Explaining the 9:3:3:1 Phenotypic Ratio; The Law of Independent Assortment If the genes responsible for each character are inherited independently, it is possible not only to explain the occurrence of four phenotypic classes in the F_2, but also to explain why those classes occur in a 9:3:3:1 ratio. To do this, we need to be familiar with the **product law** of probability. This law deals with events that are **independent** of each other, where the occurrence (or nonoccurrence) of one has no bearing on the occurrence (or nonoccurrence) of another. The law states that the chance that two (or more) independent events will occur together is determined by multiplying their separate probabilities.

The product law can be applied to the dihybrid cross we are considering if each gene is inherited independently. By considering each character separately, we have already determined that the probability of producing yellow and green seeds in the F_2 is 3/4 and 1/4, respectively; and for the character of seed shape, the probability of producing smooth seeds and wrinkled seeds in the F_2 is 3/4 and 1/4, respectively. To determine the probability of yellow and smooth occurring together in the F_2, we multiply the probability of yellow (3/4) times the probability of smooth (3/4) to get $3/4 \times 3/4 = 9/16$. Determine the probability of each of the other F_2 dihybrid phenotypic classes.

Yellow and smooth: $3/4 \times 3/4 = 9/16$

Yellow and wrinkled: ____________[11]

Green and smooth: ____________[12]

Green and wrinkled: ____________[13]

To summarize, if the inheritance of each character is an independent event, the four possible F_2 phenotypic combinations would be expected in a ratio of 9:3:3:1. This ratio coincides with the F_2 ratio observed by Mendel in all of the dihybrid crosses he carried out and led him to the law of independent assortment.

The **law of independent assortment,** an additional postulate, states that the members of a segregating pair of factors (genes) assort, or distribute, themselves into the gametes independently of every other segregating pair.

The Dihybrid Cross in Modern Symbols At this point we can work through the dihybrid cross *WwGg* × *WwGg*, considering both characters simultaneously. The law of segregation tells us that each gamete will carry one of the parental alleles for each character. Thus, each gamete produced by a dihybrid *WwGg* plant will carry either *W* or *w* and will also carry either *G* or *g*. The law of independent assortment tells us that all possible combinations of these alleles will occur in the gametes produced by this plant; a gamete that carries *W* will have an equal chance of carrying either *G* or *g*, as will a gamete that carries *w*. Thus, four different combinations, *WG*, *Wg*, *wG*, and *wg*, will be produced in equal frequency by F_1 plants.

The random union of these four types of gametes is shown in the following 16-box Punnett square. Combining the type of gamete listed at the top of each vertical column with the type listed at the left end of each horizontal row gives the genotype found in each box. For example, combining the *WG* at the top of the column on the far left with the *WG* at the left end of the top row gives the *WWGG* genotype found in the upper-left box. The phenotype produced by the genotype is also given in each box. (Note that the boxes are numbered for reference in Table 2-1.)

Gametes from *WwGg* parent (rows) / Gametes from *WwGg* parent (columns)	*WG*	*Wg*	*wG*	*wg*
WG	*WWGG* smooth, yellow (1)	*WWGg* smooth, yellow (2)	*WwGG* smooth, yellow (3)	*WwGg* smooth, yellow (4)
Wg	*WWGg* smooth, yellow (5)	*WWgg* smooth, green (6)	*WwGg* smooth, yellow (7)	*Wwgg* smooth, green (8)
wG	*WwGG* smooth, yellow (9)	*WwGg* smooth, yellow (10)	*wwGG* wrinkled, yellow (11)	*wwGg* wrinkled, yellow (12)
wg	*WwGg* smooth, yellow (13)	*Wwgg* smooth, green (14)	*wwGg* wrinkled, yellow (15)	*wwgg* wrinkled, green (16)

[11] $3/4 \times 1/4 = 3/16$. [12] $1/4 \times 3/4 = 3/16$. [13] $1/4 \times 1/4 = 1/16$.

TABLE 2-1

F_2 Genotype	Found in boxes	Number of boxes	Total	Phenotype
WWGG	1	1	9	Smooth and yellow
WwGG	3, 9	2		
WWGg	2, 5	2		
WwGg	4, 7, 10, 13	4		
WWgg	6	1	3	Smooth and green
Wwgg	8, 14	2		
wwGG	11	1	3	Wrinkled and yellow
wwGg	12, 15	2		
wwgg	16	1	1	Wrinkled and green

The 16 boxes within the square represent all, or 16/16, of the progeny, and each box represents 1/16 of the progeny. Examining the 16 boxes in the square shows that nine different genotypes are produced. Because of dominance, these genotypes produce four different phenotypes which occur in a 9:3:3:1 ratio. These results are summarized in Table 2-1.

Chromosomal Basis for the Law of Independent Assortment The random orientation of the tetrads with respect to each other and to the poles of the spindle during metaphase I of meiosis provides the basis for Mendel's law of independent assortment. As was noted in Chapter 1, this orientation determines the pole of the spindle apparatus to which the duplicated maternal and paternal chromosomes making up each homologous pair separate during anaphase I. There is an equal chance that the maternal chromosome (or paternal chromosome) will end up at one or the other pole of the spindle apparatus. Furthermore, the pole to which the maternal chromosome from one homologous pair moves may receive either the maternal or paternal chromosome from each of the other homologous pairs present in the nucleus. In short, the manner in which each homologous pair positions itself is independent of the orientation of every other homologous pair.

The random assortment of homologous chromosomes (and the gene pairs they carry) during reproductive-cell formation means that a dihybrid parent produces gametes with all possible chromosome (or gene) combinations with regard to the two characters being studied, and forms them in equal numbers. Independent assortment generates genetic variability through the production of new combinations of maternal and paternal chromosomes in the reproductive cells. In organisms with a large number of chromosomes, an enormous number of these combinations can occur.

Well after Mendel's time, it was shown that the law of independent assortment applies only to genes carried on separate pairs of chromosomes, or carried very far apart on the chromosomes comprising a homologous pair. This is discussed in detail in Chapters 12 and 13 which deal with linkage and mapping.

ALTERNATIVES TO DOMINANCE: INCOMPLETE DOMINANCE AND CODOMINANCE

It is now recognized that there are other types of allelic interaction in heterozygotes besides the **complete** dominance of one allele over another that Mendel observed in the characters he studied. For example, in crossing snapdragons with white flowers and snapdragons with red flowers, all the F_1 progeny have pink flowers. This situation, where the phenotype of the heterozygous individual is intermediate to the phenotypes of the two types of homozygous individuals, is referred to as **incomplete,** or **partial, dominance.** In other instances, the phenotypes expressed in both parents are exhibited in the F_1 progeny. For example, humans homozygous for type A blood can mate with individuals homozygous for type B blood and the progeny inherit a gene for each blood type. These progeny produce the protein molecules (antigens) characterizing both blood type A and type B, giving them blood type AB. This situation, where both alleles are fully expressed in the heterozygote, is known as **codominance.**

GENETIC TERMINOLOGY

Avoiding Confusion: Allele vs. Gene vs. Locus

The terms *allele, gene,* and *locus* can sometimes lead to confusion, primarily because they can sometimes be used interchangeably. Alleles are alternative forms of a gene. The allele for tallness and the allele for shortness are alternative forms of the gene determining the character of height in pea plants. The "gene for tallness" and the "allele for tallness" mean the same thing. Locus refers to the actual physical site on a chromosome where a particular gene is located. The tallness allele or gene is located at the locus for height.

A Summary List of Essential Terms and Symbols

Learning the language of genetics is fundamental to its study and it will be necessary for you to learn the meaning and usage of many terms as you progress. Numerous terms and symbols with which you must be familiar have been introduced and defined in this and the preceding chapter. Be sure you can give the meaning of each of the following terms and symbols before going further in this book: allele, autosome, codominance, dihybrid cross, diploid, dominant, F_1, F_2, first filial, second filial, gene, genotype, haploid, heterozygous, homologous, homozygous, law of independent assorment, law of segregation, locus, monohybrid cross, n, $2n$, P_1, phenotype, recessive, and sex chromosome.

SUMMARY

Based on studies with the garden pea, Mendel proposed the law of segregation which states that the genes of a pair segregate at the time of reproductive-cell formation so that each gamete carries one allele for a particular character. The separation, or segregation , of the chromosomes making up each homologous pair during meiosis provides the basis for this law. By studying the simultaneous inheritance of two pairs of genes, where each pair determines a different character, Mendel proposed the law of independent assortment which states that different gene pairs are inherited independently of one another. The independent separation of the members of each pair of homologous chromosomes during meiosis provides the basis for this law. Independent assortment generates genetic variability through the production of new combinations of maternal and paternal chromosomes in the reproductive cells.

In addition to complete dominance, where one allele completely masks its counterpart in heterozygous individuals, two other types of allelic interaction are known. With incomplete, or partial, dominance, the heterozygote's phenotype is intermediate to the two corresponding homozygous phenotypes. With codominance, both alleles carried by the heterozygote exhibit their phenotypic effect.

Many basic terms are introduced in this chapter. An understanding of the meaning and usage of each is essential before going on to other chapters of this book. Many of the topics included in this chapter, including monohybrid and dihybrid crosses, will be discussed in greater detail in the chapters that follow.

PROBLEM SET

2-1. Distinguish between the terms
a. gene and allele.
b. genotype and phenotype.

2-2. Distinguish between the terms
a. homozygous and heterozygous.
b. dominant and recessive.

2-3. Assume that the gene for a particular character has two contrasting alleles, one dominant and the other recessive.
a. Can an organism's genotype for this character always be determined from its phenotype? Explain.
b. Under what circumstances can an organism's genotype for this character be directly determined from its phenotype? Explain.

2-4. Assume that the gene for a particular character has two contrasting alleles, one dominant and the other recessive. Under what circumstances would an individual homozygous for this character have a phenotype which is
a. the same as that of an individual heterozygous for this character?
b. different from that of an individual heterozygous for this character?

2-5. Why was it important for Mendel to use plants from pure-breeding strains as the parents in his crosses?

2-6. One of the pea-plant crosses carried out by Mendel is shown next. The character under study is flower color, and each parent is derived from a pure-breeding line.

P_1:	Purple	×	White
F_1:		100% purple	
$F_1 \times F_1$:	Purple	×	Purple
F_2:	Purple		White
	705 plants		224 plants

Identify each phenotype shown in the P_1, F_1, and F_2 generations as either homozygous or heterozygous.

2-7. Mendel proposed the existence of basic units of heredity, or factors, which determine the expression of each of the characters he studied in garden pea plants.
a. In the crossing of two pure-breeding plants with contrasting expressions of the same character, why is only one factor expressed in the F_1 generation?
b. What led Mendel to the conclusion that F_1 plants carry hereditary factors from each parent?
c. What evidence did Mendel have for his conclusion that the factors were unchanged during their transmission from generation to generation?

2-8. Mendel concluded that, during gamete production, the pair of factors (alleles) carried by an individual for a particular character separate randomly, with each gamete receiving one of the two parental factors.
a. If both factors (alleles) carried by a pea plant are for yellow seed color, what percentage of its gametes will have this factor?
b. Assume that a plant carries both the yellow and green factors (alleles). Identify the type or types of gametes produced and the frequencies.

PROBLEM SET

2-9. For each genotype, give the type or types of gametes produced and the frequencies.

a. *GG*

b. *Gg*

c. *aa*

d. *Aa*

2-10. How did Mendel explain the 3:1 phenotypic ratio, which occurred in the F_2 generation of his crosses that examined the inheritance of single characters.

2-11. A pea-plant cross carried out by Mendel is shown next. The character under study is seed color, and each parent is derived from a pure-breeding line.

P_1:	Yellow	×	Green
F_1:		100% yellow	
$F_1 \times F_1$:	Yellow	×	Yellow
F_2:	Yellow		Green
	6022 seeds		2001 seeds

Identify the type or types of gametes produced by each type of individual in the P_1, F_1, and F_2 generations.

2-12. A view prevalent in Mendel's time was that the genetic material carried by the two parents merged together so that the characters expressed in the offspring represented a blending of the characters of the parents.

a. If this view were true, what outcome would be predicted from the crossing of pure-breeding tall and short pea plants?

b. Cite two types of evidence that Mendel obtained indicating that the genetic material was not blended during the F_1 generation.

2-13. What experimental results led Mendel to his law of independent assortment?

2-14. For each genotype, give the type or types of gametes produced and the frequencies.

a. *aaGG*

b. *AaGG*

c. *AaGg*

d. *Aagg*

CHAPTER

3

Crosses Involving Single-Gene Inheritance; Basic Probability

INTRODUCTION

This chapter begins by describing two systems commonly used to assign symbols to genes and their alleles and provides a general plan of attack for the more common types of genetics problems. Several types of crosses are examined, including the testcross, involving autosomal genes with two alleles. Crosses of this type involve one character and fall under the heading *monohybrid crosses*, although strictly speaking, this label is reserved for crosses involving two parents who are heterozygous for the gene under consideration. Next we discuss how basic laws of probability may be used to predict the outcome of genetic crosses. Finally, to understand variation in gene expression, we consider penetrance and expressivity.

ASSIGNING GENE SYMBOLS

The first and simplest system for assigning symbols to genes and their alleles, introduced in Chapter 2, uses an uppercase letter to designate a dominant allele and a lowercase letter to designate a recessive allele. The letter used is often the first letter of the word describing the phenotype resulting from the expression of the recessive allele. For example, the expression of the recessive allele of a gene determining pigment production in corn plants produces an albino, or pigment-free, phenotype. Thus, the letter *A* is used to designate the allele for the normal green color, while *a* designates the recessive allele for the albino condition. This allelic pair could produce three diploid genotypes: *AA*, homozygous dominant, with the green phenotype; *Aa*, heterozygous, with the green phenotype; and *aa*, homozygous recessive, with the albino phenotype.

The second system which is more adaptable and therefore more widely used lends itself well to designating genes that have more than two alleles, as most genes do. This system designates the most common allele of a gene as the **wild-type** or normal form and the less common allele or alleles as **mutant** forms. The basic rules governing this system are as follows.

1. The symbol often consists of one, two, or three lowercase italic letters derived from the name of the mutant condition.
2. If the mutant allele is recessive, the first letter of the symbol is kept in lowercase, and if dominant, the first letter is capitalized.
3. The wild-type allele is designated by a superscript plus sign ($^+$) affixed to the symbol for the mutant allele.

For example, a mutant wing form known as *vestigial* occurs in the fruit fly *Drosophila melanogaster.* Two letters of vestigial, *vg*, are used as the allelic symbol, and since the mutant allele causing this condition is recessive, its symbol is *vg*. The symbol for wild-type counterpart allele which produces normal wing form is vg^+. The genotype for a homozygous dominant fly is vg^+vg^+, for a heterozygous fly is vg^+vg, and for a homozygous recessive fly is *vgvg*. As an additional example where the mutant allele is recessive to the wild-type allele, consider another mutant wing form found in *Drosophila* known as *blistery.* The symbol for the mutant allele for blistery wings is *bl* and the dominant counterpart allele for wild-type wings is bl^+. For each of the allelic combinations listed below, write out the genotype and phenotype for this trait.

	Genotype	Phenotype
Homozygous dominant	______[1]	______[2]
Heterozygous	______[3]	______[4]
Homozygous recessive	______[5]	______[6]

Note that each example that we have just considered uses a different symbol, vg^+ and bl^+, to designate the wild-type condition.

With the mutant wing form known as *wrinkled* which occurs in *Drosophila,* the mutant allele is dominant to the wild-type allele. Because this mutant allele is dominant, its symbol, *W*, is written in uppercase, while its recessive counterpart for wild-type wing form is written in uppercase with a superscript plus sign affixed: W^+. For each of the following allelic combinations, write out the genotype and phenotype for this trait.

	Genotype	Phenotype
Homozygous dominant	______[7]	______[8]
Heterozygous	______[9]	______[10]
Homozygous recessive	______[11]	______[12]

A variant of this system for assigning symbols uses only a plus sign to designate the wild-type allele, and a letter symbol to denote the mutant allele (with its first letter in uppercase if dominant and lowercase if recessive). For example, the wild-type counterpart for the *vg* allele considered earlier would be designated as +. However, using the plus sign unattached to a symbol may lead to confusion, especially where more than one gene is being considered. Consequently, the preferred designation for the wild-type allele is a superscript plus sign affixed to the symbol for the mutant allele.

When dominance is incomplete or partial, a capital letter is generally used to designate each allele, with a prime sign (′) or a superscript letter serving to differentiate between them. For example, in snapdragons, the two alleles for red and for white flower color show codominance. The letter *R* can be used as the basic symbol, with *R* and R' designating the alleles for red and white, respectively. Alternatively, superscripts combined with the symbol *C* (for color) could be used: C^R = red and C^W = white. Write out the genotypes and phenotypes for the following allelic combinations, using C^R and C^W as the allelic symbols.

	Genotype	Phenotype
Homozygous for the red allele	______[13]	______[14]
Heterozygous	______[15]	______[16]
Homozygous for the white allele	______[17]	______[18]

Other gene symbol systems are also in use; it is therefore absolutely essential that all symbols be clearly defined and listed in a key. A system for assigning symbols to alleles when there is more than one mutant form is covered in Chapter 8. Assigning symbols for sex-linked alleles is described in Chapter 10. The notational system used for designating alleles in bacteria and eukaryotic microorganisms is also covered later.

Two additional points about writing out genotypes should be mentioned before going on. (1) If a pair of alleles consists of a dominant and a recessive allele, always write the dominant allele first. If two or more pairs of genes are involved, always write the two symbols designating a particular gene pair side by side, for example, *AABb* not *ABab*, and vg^+vgSS not vg^+SvgS. (2) A pair of diagonal lines may be used between the two alleles of a gene as a reminder that they are carried on the same pair of homologous chromosomes, for example, $vg^+//vg^+$. Often the pair of slashes is reduced to a single slash, for example, vg^+/vg^+. In the discussion of linked loci in Chapter 12, this notation becomes particularly useful when simultaneously considering more than one gene pair carried on the same pair of homologous chromosomes.

[1] bl^+bl^+. [2] Wild-type wings. [3] bl^+bl. [4] Wild-type wings. [5] *blbl*. [6] Blistery wings. [7] *WW*. [8] Wrinkled wings. [9] WW^+. [10] Wrinkled wings. [11] W^+W^+. [12] Wild-type wings. [13] C^RC^R. [14] Red flowers. [15] C^RC^W. [16] Pink flowers. [17] C^WC^W. [18] White flowers.

SOLVING PROBLEMS: A PLAN OF ATTACK

Genetics problems differ considerably in their wording, in the traits and organisms involved, and in other factors. Nonetheless, most problems fall into one of a very limited number of basic categories and can be solved using the same fundamental approach. Some key steps in solving problems are listed next.

1. Begin by carefully reading the problem to determine the information presented and to get a general sense of what you are asked to do.

2. Determine the number of traits being considered. Usually just one or two will be involved. From this you can often tell how many sets of alleles are involved. Be careful to distinguish between a trait and its expression. For example, seed color in peas is *one* trait having *two* alternative expressions, yellow and green, and each expression is due to a different allele.

3. Identify, if possible, the type of interaction shown by the alleles that determine the trait. Complete dominance, incomplete dominance, or codominance is usually involved. It is also important to know whether the genes involved are carried on autosomes or sex chromosomes. This information may not be explicitly stated and may have to be deduced from the data presented.

4. Determine precisely what it is that you are asked to do. Problems often ask you to identify either the possible phenotypes of the progeny produced by a set of parents, or the genotypes of the parents who produce certain progeny.

5. Devise a general strategy for answering the question. If you are asked to identify the possible phenotypes of the progeny produced by a set of parents, it will be necessary to know the genotypes of both parents. Alternatively, if asked to determine the genotypes of the parents who produced certain progeny, it will be necessary to identify the genotypes of the progeny. Once you have a general strategy for answering the question, be alert to information included in the problem which will make it possible to carry out that strategy and answer the question.

6. Write out a key listing the symbols that will be used to designate alleles and indicate what each represents. If symbols for the alleles are not given in the problem, you will need to designate suitable symbols.

7. Use the symbols to describe each individual's genotype as best as you can from the information given in the problem. In many instances the genotypes of the parents or progeny will be given or at least implied. As you are reading problems, be alert to essential genetic information that may be presented in indirect form.

As an illustration, assume that the allele for straight hair (c^+) in guinea pigs shows complete dominance to the allele for curly hair (c) and that a problem includes the phrase "a guinea pig has straight hair and her male parent has curly hair." This phrase gives us phenotypic information about two individuals and allows us to make the following inferences about their genotypes. (1) The straight-haired guinea pig has at least one dominant c^+ allele, so her genotype is either c^+c^+ or c^+c. (2) Her curly-haired male parent must have two recessive alleles, so his genotype is cc. Note that we can go beyond these inferences. Since all of the male parent's gametes must carry the c allele, all of his offspring received a c allele from him, including the straight-haired guinea pig we are considering. Thus, her genotype must be c^+c. From this information we can also conclude that her female parent must have carried the c^+ allele and must therefore have had straight hair.

8. Identify all the kinds of gametes that will be produced by each parent and determine the relative proportion of each. This step is essential. If you have difficulty, review meiosis in Chapter 1. It is important to remember that each reproductive cell ends up with one member of every pair of homologous chromosomes, and therefore, one allele of every pair of alleles. (An exception to this occurs in some cases with sex-linked loci, as discussed in Chapter 10.)

Before going further with our list of steps in solving problems, it is useful to pause and look at an example in which the steps we have already considered are applied.

PROBLEM: In guinea pigs, the wild-type allele for straight hair is dominant to the allele for curly hair. Two heterozygous straight-haired individuals are mated. Identify the expected genotypes and phenotypes of their progeny.

SOLUTION:

Basic information given: The problem involves a single trait with two alternative expressions, straight and curly hair; and the alleles involved show complete dominance, with the wild-type allele for straight hair dominant to that for curly hair. The problem tells us that each parent exhibits the straight-hair phenotype and is heterozygous for this trait. No allelic symbols are given.

Question to be answered: What are the genotypes and phenotypes of the progeny?
General strategy: To answer this question, the genotypes of the parents must be identified.

At this point we have already taken care of the first five steps for solving a problem. The next step is to assign symbols to the alleles and list them in a key. We could use the letter c as the symbol for the curly-hair allele and c^+ as the symbol for the straight-hair allele.
Key:

c^+ = the allele for straight hair
c = the allele for curly hair

Now the symbols are used to write out the genotypes of the parents.

	Female	Male
Genotypes	c^+c	c^+c

The cross can be represented as $c^+c \times c^+c$. (Note that if the parents can be identified as to sex, the female genotype is always written first.)

Next, determine the types and relative frequencies of the gametes produced by each parent. One way to do this is to diagram the alleles possessed by each parent on homozygous chromosomes.

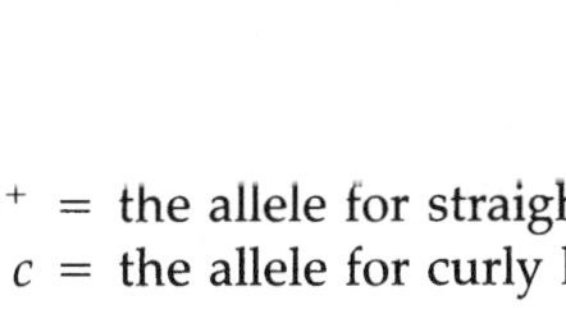

Female	Male
c^+ ‖ c	c^+ ‖ c

The homologs making up each pair separate from each other at the time of gamete formation. Half the eggs will carry c^+ and half will carry c; similarly, half the sperm will carry c^+ and half will carry c.

Now we are ready to resume our listing of key steps by considering the additional steps necessary to complete the solution of this type of problem.

9. To work out a cross, it is necessary to consider all the possible combinations of haploid gametes that unite to form diploid genotypes. One convenient way to examine these combinations is to set up a Punnett square. The square allows us to show all possible ways in which the gametes unite to form all possible genotypes. It is customary to arrange the female gametes along the side and the male gametes along the top. When completed, the empty boxes within the square will be filled in with all possible genotypes that result. Fill in the Punnett square by writing the female gamete (found on the left of each horizontal row) in each of the empty boxes to the right of that gamete; then write the male gamete (found at the top of each vertical column) in each of the boxes below it.

In the Punnett square for the guinea-pig problem, the box in the upper-left corner will contain c^+c^+. What will the remaining boxes contain?[19]

♀ \ ♂	c^+	c
c^+		
c		

10. State the results. Identify the progeny by genotype and phenotype, and determine the relative frequency for each.

For the guinea-pig mating, the genotypes and phenotypes of the progeny are as follows.
Genotypes: There are three genotypes: c^+c^+, c^+c, and cc. The four boxes within the Punnett square represent 4/4, or 100%, of the progeny, so each box represents 1/4, or 25%, of all the possible progeny. The box containing c^+c^+ represents 25% of the progeny that are homozygous for the normal allele, the two boxes containing c^+c represent 50% of the progeny that are heterozygous, and the remaining box containing cc represents 25% of the progeny that are homozygous for curly hair. The genotypic frequencies may be expressed in terms of percentages (25% c^+c^+, 50% c^+c, 25% cc), fractions (1/4 c^+c^+, 1/2 c^+c, 1/4 cc), or a ratio (1 c^+c^+ : 2 c^+c : 1 c^+c). Note that by convention, the genotypes are always written in order of decreasing number of dominant alleles. Progeny with two dominant alleles (in this case, c^+c^+) are written first, those with a single dominant allele (c^+c) are written next, and progeny with no dominant alleles (cc) are written last.
Phenotypes: Phenotypes are determined from the genotypes. In this example, because c^+ is dominant over c, genotypes c^+c^+ and c^+c both express the dominant trait, while genotype cc expresses the recessive trait. Phenotypic frequencies may also be expressed in terms of percentages (75% straight, 25% curly), fractions (3/4 straight, 1/4 curly), or a ratio (3 straight : 1 curly). ■

Note that these results give us the **expected statistical distribution** of offspring. They do not tell us that the first three offspring will always have straight hair

[19] Upper-right, c^+c; lower-left, c^+c; and lower-right, cc.

and that every fourth offspring will always have curly hair nor do they tell us that one out of every four offspring will always have curly hair. The results tell us that each offspring from this particular cross has a 0.75 (75%, 3 out of 4) chance of having straight hair and a 0.25 (25%, 1 out of 4) chance of having curly hair.

11. Once you have worked the problem, it is a good idea to reread it. Make sure you are providing what the problem asks for. Genotypes or phenotypes or both may be requested. If nothing specific is mentioned, provide both. Also give the expected frequencies of individuals in the form specified, either as fractions, percentages, or as a ratio. If the desired form is not indicated, you may select a form in which to present this information.

The steps that have just been applied will guide you through many genetics problems which ask you to determine the progeny produced by a given set of parents. However, not all problems are of this type. For example, you may be given information about the progeny and asked to determine the genotype of one or both parents, as in the following problem.

PROBLEM: Albinism in humans is characterized by the inability to synthesize melanin, the pigment which normally occurs in the skin and hair. This disorder is inherited as an autosomally based recessive trait. A normally pigmented woman and an albino man produce a family of three normal and two albino children. Identify the genotypes of the parents.

SOLUTION:

Basic information given: The problem involves a single trait with two alternative expressions, normal pigmentation and albino. The alleles involved are carried on a pair of autosomes and show complete dominance, with the allele for albinism recessive to that for normal pigmentation. The normally pigmented and albino children in this family occur in a 3:2 ratio. No allelic symbols are given.

Question to be answered: What are the genotypes of the parents?

Strategy: Some inferences can be made about the parents' genotypes from information which tells us that the mother's phenotype is normal and the father's phenotype is albino. Complete identification of the parents' genotypes will require identification of the progeny genotypes.

The next step is to designate symbols for the alleles and list them in a key. We could use the letter *a* as the symbol for the albino allele and *A* as the symbol for normal pigmentation.

Key:

A = the allele for normal pigmentation
a = the allele for albinism

These symbols are used to write out the genotypes of the parents as fully as possible, using the phenotypic information given in the problem. With complete dominance, the recessive allele is expressed only in the homozygous condition, and thus the albino father must have the genotype *aa*. Since the dominant allele is expressed in both the homozygous (*AA*) and heterozygous states (*Aa*), the normally pigmented mother could have either genotype, and this uncertainty is shown by representing her genotype as *A*__.

	Mother	**Father**
Genotypes	*A*__	*aa*

In order to determine whether the second allele carried by the mother is *A* or *a*, we need to determine the genotypes of the progeny. The albino children are genotype *aa*, and those with normal pigmentation may be either *AA* or *Aa* and are represented as *A*__. From the children's genotypes, we can reason backward to identify the gametes produced by the mother. Since some of the children have genotype *aa* and since each parent must have contributed an *a* allele to these children, one of the mother's alleles must be *a*. In light of this, the mother's genotype must be *Aa*.

Answer to the question: The father's genotype is *aa* and the mother's genotype is *Aa*. ■

To summarize the approach used for this type of problem, after determining the number of traits and the type of allelic interaction, we identified the question to be answered and a strategy for getting the answer. After designating symbols, we assigned genotypes to the parents and to the progeny to the extent possible. Then we worked backward from the progeny genotypes to determine the types of gametes made by each parent. Knowledge of the types of gametes made it possible to complete the identification of the genotype of each parent.

Note that the steps followed in solving the two previous problems will not solve all problems of these types, but will solve many of them. In any case, it is important to recognize the value of an organized, systematic approach to solving problems.

EXAMPLES OF SINGLE-GENE CROSSES

The simplest genetics crosses often involve one trait determined by a single gene with two alleles. As illustrations, we will consider a cross for each of the following frequently encountered types.

1. Both parents are heterozygous for the alleles determining the trait, and progeny genotypes and/or phenotypes are to be identified.
2. One parent is homozygous and the other is heterozygous, and the progeny genotypes and/or phenotypes are to be identified.
3. The genotypes and/or phenotypes of the progeny are given, and one or both of the parental genotypes are to be determined.

Both Parents Heterozygous

PROBLEM: In corn, *Zea mays*, the allele for tall height (s^+) is dominant to the allele for short height (s). Two heterozygous tall plants are crossed. What percentage of the F_1 progeny would be expected to be short?

SOLUTION:

Basic information given: One trait, plant height, is involved with two alternative expressions, tall and short. The allele for tall height shows complete dominance over the allele for short height. Allelic symbols are designated.

Question to be answered: What percentage of the F_1 plants are expected to be short?

Strategy: To solve this problem, we need to cross two heterozygous plants to determine the phenotypes of the F_1 progeny and identify the percentage that are short.

Symbols and key: Let s^+ be the tall-height allele and s be the short-height allele.

Parental genotypes: Since each is heterozygous, we infer that each has the genotype s^+s. The cross can be represented as $s^+s \times s^+s$.

Gametes: Each gamete will have one allele of the pair carried by the parent producing the gamete (because each gamete carries just one chromosome of each homologous pair of the diploid parent). Each parent will produce two types of gametes: half carry s^+ and the other half carry s.

Union of gametes: All possible random combinations of these eggs and sperm can be shown by completing the following Punnett square.

♀ \ ♂	s^+	s
s^+		
s		

Genotypes are produced as follows: upper-left box: s^+s^+; upper-right: s^+s; lower-left: s^+s; and lower-right: ss.

Resulting genotypes: Since the four boxes in the Punnett square represent 100% of the F_1, the box containing s^+s^+ represents 25% of the progeny that are homozygous for the tall-height allele, the two boxes containing s^+s represent 50% of the progeny that are heterozygous, and the remaining box containing ss represents 25% of the progeny that are homozygous for the short-height allele.

Phenotypes: Since s^+ is dominant, genotypes s^+s^+ and s^+s will be tall and ss will be short. Thus 75% of the F_1 is tall and 25% is short.

Answer: The proportion of short plants in the F_1 is 25%. ■

One Parent Heterozygous, the Other Homozygous

PROBLEM: In one variety of mouse, the wild-type allele for black fur is dominant to the allele for white fur. A pet-store owner crosses a black female mouse which is known to be heterozygous for these alleles with a white male mouse. Determine the progeny phenotypes and the expected percentage for each.

SOLUTION:

Basic information given: The problem indicates that a single trait, fur color, has two alternative expressions, black and white. The alleles involved show complete dominance, with the wild-type black-fur allele dominant to the white-fur allele. The female parent is heterozygous black and the male is white.

Question to be answered: Identify the phenotypes of the progeny and the expected percentage for each.

Strategy: To answer this question, we need to identify the genotypes of the parents and the types of gametes each forms, and we need to determine all possible combinations of those gametes to get the progeny genotypes. From these genotypes, we can identify the progeny phenotypes and their percentages.

Symbols and key: Use the letter w as the symbol for the allele that leads to white fur and w^+ as the symbol for the allele that leads to black fur.

Parental genotypes: The genotype of the black female parent is w^+w and that of the white male parent is ww. The cross is $w^+w \times ww$.

Gametes: Each gamete carries one member of the allelic pair present in the parent. The w^+w female parent will produce two kinds of gametes: half will carry w^+ and the other half will carry w. The ww male parent will produce one type of gamete carrying w.

Union of gametes: Complete the Punnett square to show the progeny genotypes.

♀ \ ♂	w	w
w^+		
w		

Genotypes are produced as follows: upper-left box: w^+w; upper-right: w^+w; lower-left: ww; and lower-right: ww.

Note that since the male produces one type of gamete, this Punnett square could be simplified so that it contains two boxes. When a parent produces one type of gamete, that gamete needs to be included in the Punnett square heading only once to get all possible combinations.

♀ \ ♂	w
w^+	
w	

With two boxes, each represents 50% of the progeny. Completion of this Punnett "square" gives the same results as the previous square.

Results: Fifty percent of the progeny have genotype w^+w and are black, and 50% will have genotype ww and are white.

Answer: Progeny phenotypes: 50% black and 50% white. ▬

One or Both Parental Genotypes Determined from Progeny

PROBLEM: A type of human bone disorder is due to a recessive allele carried on an autosome. A normal man and woman have five children, four who are normal and one who is afflicted. Identify the parents' genotypes.

SOLUTION:

Basic information given: We are dealing with a single, autosomally based trait, with the allele for the bone disorder recessive to the allele for the normal bones.

Question to be answered: What are the parents' genotypes?

Strategy: Identify the parents' genotypes from information given about their phenotypes and the phenotypes of their progeny. We can begin by assigning possible genotypes based on the parents' phenotypes and then attempt to narrow down those possibilities based on what can be inferred from the progeny phenotypes.

Symbols and key: Allelic symbols must be designated. Let d^+ be the allele for normal bones and let d be the allele for the bone disorder.

Parental genotypes: Based on their phenotype, we can infer that each parent is either homozygous dominant, d^+d^+, or heterozygous, d^+d.

Progeny genotypes: Each of their normal children could have either the d^+d^+ or d^+d genotype. The afflicted child must be homozygous for the recessive allele and have genotype dd. The dd child provides the key to definitively identifying the parental genotypes. We know that each parent contributes an allele to each child. To produce the dd child, each parent must have contributed a d allele. This means that each parent must possess the d allele and that each parent must be heterozygous for this trait.

Answer: Each parent is d^+d. ▬

USING THE TESTCROSS

Testcross Definition

In a testcross, an organism of unknown genotype that expresses the dominant phenotype for a trait is crossed with an organism homozygous recessive for the same trait. From the phenotypes of the progeny, the organism of unknown genotype can be identified as either homozygous dominant or heterozygous.

An Example

Assume that you are given instructions to find the genotype of a female rabbit with black fur. You are told that the allele for black fur (w^+) is dominant to that for white fur (w), but you know nothing about the parents of this particular black rabbit. Identify the two possible genotypes for the black rabbit. ____[20] Would it be possible to distinguish phenotypically between these two genotypes? ____________[21] Explain. ____[22] Testcrossing this black rabbit with a homozygous recessive white rabbit will allow us to identify her genotype. As you will see, different types of progeny will be produced depending on whether the organism with the dominant phenotype is homozygous for the dominant allele or heterozygous. In other words, the genotype of the genotypically unknown parent can be deduced from the progeny of the testcross mating.

PROBLEM: Identify the genotype of a black rabbit.

[20] Homozygous dominant, w^+w^+; or heterozygous, w^+w. [21] No. [22] Since w^+ is dominant to w, both genotypes would have the same phenotype.

SOLUTION:

Basic information given: We are dealing with a single trait, fur color, with two alternative expressions, black and white. The allele for black fur, w^+, is dominant to that for white fur, w. The female has black fur.

Question to be answered: What is the genotype of the black rabbit?

Strategy: To identify the genotype of the black rabbit, a testcross should be carried out by crossing it with a homozygous recessive white rabbit.

Symbols and key: Let w^+ be the black-fur allele and w be the white-fur allele.

Parental genotypes: From her phenotype, we can infer that the black rabbit has a genotype of either w^+w^+ or w^+w and this uncertainty can be expressed by representing the genotype as $w^+\_\_$. We can infer that the white rabbit has genotype ww.

Gametes: If the black rabbit is w^+w^+, all her gametes will carry w^+. If the black rabbit is w^+w, she will produce two types of gametes: half with w^+ and half with w. The white parent will form a single kind of gamete which carries w.

Union of gametes: Let us consider the two possibilities separately. (1) If the black female is w^+w^+, union of her w^+ gametes with w gametes from the male will produce heterozygous (w^+w) progeny which have black fur. (2) If the black female is w^+w, union of her w^+ and w gametes with w gametes from the male produces progeny with the two genotypes shown by filling in the Punnett square.

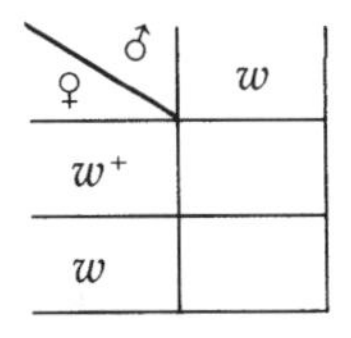

The cross produces progeny with *two* genotypes: w^+w which have black fur and ww which have white fur. These two phenotypes would be expected in equal frequencies. ■

Inferring from Testcross Outcomes

1. If all the testcross progeny are phenotypically alike, with all expressing the dominant allele, then the parent of undetermined genotype is homozygous dominant for the trait under study.
2. If approximately half of the testcross progeny express the dominant allele and half express the recessive allele, then the parent of unknown genotype is heterozygous for the trait under study.

Put another way, a homozygous dominant individual produces progeny showing the dominant phenotype while a heterozygous individual produces progeny showing the dominant and recessive phenotypes in a 1:1 ratio. Note that in order for a testcross to be useful, the number of progeny needs to be of sufficient size to provide a reliable indication of its true outcome. A large sample of progeny is more likely to reflect the true outcome than is a small one.

USING PROBABILITY TO PREDICT CROSS OUTCOMES

Probability is the chance that a particular event will occur. It is the expected number of times the particular event we are looking for occurs, relative to the total number of possible events (or the number of trials). This can be expressed in terms of the following formula.

$$\text{Probability} = \frac{\text{The number of looked-for events}}{\text{The number of possible events}}$$

For example, in flipping a coin, there are two possible outcomes or events: a head or a tail. The probability of getting a particular event, say a head, is one out of two, or 1/2. Similarly, the probability of getting the alternative event, a tail, is also one out of two, or 1/2.

A probability value may be expressed as a fraction, ratio, decimal, or percent, and may range from 0 (0%) to 1 (100%). If an event is certain to occur—for example, getting a head with the flip of a two-headed coin—its probability is 1. If an event is certain not to occur—for example, getting a tail with the flip of a two-headed coin—its probability is 0. If there is uncertainty as to whether or not an event will occur, its probability falls somewhere between 0 and 1.

The Punnett square provides a way of working genetic crosses by showing all possible ways in which gametes can unite to form all possible genotypes. As we will see, it is also a method of combining probabilities to give the expected frequencies for various kinds of offspring.

Independent Events and the Product Law

Events are considered **independent** if the occurrence (or nonoccurrence) of one has no bearing on the occurrence of another. If the probabilities of independent events are known, the chance that two or more such events will occur together (simultaneously or in sequence) equals the product of their separate probabilities. This is known as the **product law** of probability (or sometimes as the "law of independent events").

Let us look at how the product law is used in connection with predicting the outcome of a genetic cross. Consider a cross between two heterozygous tall pea plants where the allele for tall height, d^+, is dominant to the allele for dwarf height, d. The genotype of both of these plants is d^+d, and each will produce two kinds of gametes: half with d^+ and the other half with d. Since the two kinds of gametes are produced in equal numbers by each parent, the probability that any particular gamete will carry d^+ is 1/2 (= 50% = 0.5) and the probability that it will carry d is also 1/2.

Furthermore, the process which results in a particular allele ending up in a particular gamete of one parent is, of course, independent of the process which distributes the alleles to the gametes of the other parent. The probability of a sperm carrying d^+ combining with an egg carrying d to produce a d^+d genotype, for example, is equal to $1/2 \times 1/2 = 1/4$. In the following Punnett square, the probability for each type of gamete is indicated within the parentheses. Complete the square by multiplying the gametic probabilities to give the probability for each genotype.

	d^+ (1/2)	d (1/2)
d^+ (1/2)	d^+d^+ ( )	d^+d ( )
d (1/2)	d^+d ( )	dd ( )

The three categories of progeny occur with the following probabilities: d^+d^+ (1/4), d^+d (1/4 + 1/4 = 2/4 = 1/2), and dd (1/4). Since these three categories comprise all of the alternatives for the outcome of this cross, the sum of their separate probabilities equals 1. More will be said about this shortly.

Mutually Exclusive Events and the Law of the Sum

In arriving at the probability for the heterozygous individuals produced in this cross, you have, in fact, made use of another law of probability, the law of the sum. This law applies to events that are mutually exclusive. As an illustration of mutually exclusive events, consider the flip of a coin. The flip that comes up with a head obviously cannot, on the same flip, produce a tail. Similarly, the throw of a die (singular for "dice") has six possible outcomes and any one excludes the other five. Separate events are **mutually exclusive** if the occurrence of one of them prevents the occurrence of the other (or others).

The **law of the sum** of probabilities states that if two or more events are mutually exclusive, the probability that either one or another of them will occur equals the sum of their individual probabilities. (The law of the sum is sometimes referred to as the "either/or rule.") For example, the probability of getting either a head or a tail on the flip of a coin is the sum of their separate probabilities: 1/2 + 1/2 = 1. In throwing a die, the chance of coming up with a 3 is 1/6 (since a 3 is found on one of six sides of the die). The chance of coming up with a 4 is also 1/6. The probability of coming up with either a 3 or a 4 is 1/6 + 1/6 = 1/3. The probability of coming up with a 1, 2, 3, 4, 5, or 6 on the toss of a die is 1/6 + 1/6 + 1/6 + 1/6 + 1/6 + 1/6 = 1.

Note that the sum of the separate probabilities for *all* the mutually exclusive alternative events always equals 1. With only two mutually exclusive events, where "p" equals the probability of one event and "q" the probability of the other, this can be expressed as $p + q = 1$. With three mutually exclusive events whose respective probabilities are p, q, and r, the expression becomes $p + q + r = 1$. This pattern continues for any number of events. (With six mutually exclusive events, as occurs with the toss of a die, the applicable expression is $p + q + r + s + t + u = 1$.)

Let us look at how this law of the sum was used in connection with the heterozygous progeny in the $d^+d \times d^+d$ cross. There are two ways that a d^+d individual can be produced: the union of a d^+ sperm with a d egg or the union of a d sperm with a d^+ egg. For any particular zygote, these two types of fertilization are mutually exclusive. If a d^+d individual is produced through the first of these alternatives, it could not have been produced by the second, and vice versa. In determining the probability for the occurrence of heterozygous individuals from the cross $d^+d \times d^+d$, we have added together the probabilities of two mutually exclusive events: the probability of getting a d^+d individual, where d^+ is contributed by the sperm (1/4), plus the probability of getting a dd^+ individual, where d^+ is contributed by the egg (1/4), gives a probability of 1/4 + 1/4 = 1/2. In summarizing the phenotypic outcome of this cross, we make further use of the law of the sum. To find the probability that any one of the progeny from the $d^+d \times d^+d$ cross expressed the dominant trait, the probabilities of the three mutually exclusive outcomes, d^+d^+, d^+d, and dd^+, are added together: 1/4 (d^+d^+) + 1/4 (d^+d) + 1/4 (dd^+) = 3/4.

Probability Summary

Keep in mind the following key points relating to probability.

1. Probabilities may be expressed as fractions ranging from 0 (expressing an impossibility) to 1 (expressing a certainty). Decimals, percentages, and ratios may also be used to express probabilities.
2. Two or more events are independent if the occurrence (or nonoccurrence) of one has no effect on the occurrence of any of the other events.

3. If two or more events are independent, the probability of the occurrence of these events equals the product of their individual probabilities. This is the product law.
4. Two or more events are mutually exclusive if the occurrence of one excludes the occurrence of the other.
5. If two or more events are mutually exclusive, the probability that one or another will occur equals the sum of their individual probabilities. This is the law of the sum.
6. The sum of the separate probabilities for all the mutually exclusive outcomes equals 1.

A Cautionary Note Regarding Predicted Outcomes

It is important to remember that the solutions to genetics problems which give expected results based on the laws of probability do not tell us what will actually happen in a given situation. In the mating of two heterozygous organisms, for example, the laws of probability would predict that three offspring showing the dominant trait would be produced for each offspring showing the recessive trait. However, this predicted outcome reflects an infinitely large sample size. If only a few progeny are produced, the expected 3-to-1 ratio might not be found. For example, it is possible that all the progeny in a group of four could show the dominant trait. (The probability of this occurring is $3/4 \times 3/4 \times 3/4 \times 3/4 = 81/256$). It is also possible that all the progeny in a group of four could show the recessive trait. (The probability of this occurring is $1/4 \times 1/4 \times 1/4 \times 1/4 = 1/256$). As the number of progeny increases, the more likely it is that the phenotypic frequencies will approximate the predicted or expected 3-to-1 ratio.

A SUMMARY OF THE SIX TYPES OF CROSSES INVOLVING AN ALLELIC PAIR SHOWING COMPLETE DOMINANCE

Assume that a character is governed by a gene with two alleles, where *A* shows complete dominance to *a*. These two types of alleles can give rise to three different genotypes: *AA*, *Aa*, and *aa*. Organisms with these three genotypes can cross with each other in the six possible combinations listed in Table 3-1. Summarize these crosses by completing this table. For each cross, list the types of gametes produced by each parent, the frequencies in which these gametes arise, and the expected genotypes and phenotypes of the F_1 progeny. For example, each parent involved in the first cross will make a single type of gamete that carries allele *A*, and all progeny will have the genotype *AA* and express the dominant phenotype.

Although the loci under consideration will change from problem to problem involving a single allelic pair, keep in mind that as long as complete dominance operates between the two alleles, each of these six types of crosses will always yield the same types of results.

TABLE 3-1

Outcomes of the six types of crosses involving an allelic pair showing complete dominance.

	Gametes of		Proportions for F_1	
Cross Combinations	First parent	Second parent	Genotype(s)	Phenotype(s)
AA × *AA*	All *A*	All *A*	All *AA*	All dominant
AA × *Aa*	______[23]	______[24]	______[25]	______[26]
AA × *aa*	______[27]	______[28]	______[29]	______[30]
Aa × *Aa*	______[31]	______[32]	______[33]	______[34]
Aa × *aa*	______[35]	______[36]	______[37]	______[38]
aa × *aa*	______[39]	______[40]	______[41]	______[42]

PENETRANCE AND EXPRESSIVITY: VARIATION IN GENE EXPRESSION

Up to this point we have assumed that a dominant allele in the homozygous or heterozygous condition, or a recessive allele in the homozygous condition, will be (1) phenotypically expressed and (2) expressed in essentially the same way in all the organisms carrying it. These assumptions are valid for some genes, but others show considerable variation in their expression. This variability in expression may arise when organisms with the same allele are exposed to different environmental conditions or have a different genetic makeup (keep in mind that the expression of a gene can be influenced by genes carried at other loci possessed by the organism).

An allele may be expressed in some but not all the individuals that carry it. For example, there is a domi-

[23] All *A*. [24] 1/2 *A*, 1/2 *a*. [25] 1/2 *AA*, 1/2 *Aa*. [26] All dominant. [27] All *A*. [28] All *a*. [29] All *Aa*. [30] All dominant. [31] 1/2 *A*, 1/2 *a*. [32] 1/2 *A*, 1/2 *a*. [33] 1/4 *AA*, 1/2 *Aa*, 1/4 *aa*. [34] 3/4 dominant, 1/4 recessive. [35] 1/2 *A*, 1/2 *a*. [36] All *a*. [37] 1/2 *Aa*, 1/2 *aa*. [38] 1/2 dominant, 1/2 recessive. [39] All *a*. [40] All *a*. [41] All *aa*. [42] All recessive.

nant allele carried by some humans which causes the little finger to be bent and stiff. Some individuals carrying the allele express it while others carrying the allele are completely normal. The percentage of individuals that express the allele they are known to possess is designated as the allele's **penetrance.** An allele that is expressed in some individuals carrying it and not in others that carry it is said to be **incompletely penetrant.**

Penetrance, from the standpoint of an individual, is all or nothing: either the allele is expressed or it is not. However, within a population of individuals carrying the same allele, penetrance produces phenotypic variability. It is expected that a dominant gene with a penetrance of 65%, for example, would be expressed in 65 out of every 100 individuals carrying it. Since environmental conditions can influence the expression of an allele, the penetrance of an allele can be effectively assessed only when its expression is evaluated under a defined set of environmental conditions.

Within a population of individuals for whom an allele is penetrant, further phenotypic variation can arise if the allele is expressed to different degrees in different individuals. For example, a human disorder known as polydactyly causes individuals to develop extra toes or fingers. The condition is due to a dominant allele that shows considerable variation in its expression. Some individuals carrying the allele have rudimentary extra digits, while in others these digits are fully developed. Some carriers show extra digits only on the hands, others have them only on the feet, and some have extra digits on both hands and feet. A gene showing this type of effect is said to exhibit **variable expressivity.** Phenotypic differences arising from differences in penetrance or expressivity (or in both) can make analysis of inheritance patterns difficult. Geneticists often avoid studying situations where penetrance is less than 100% and expressivity is variable. As a consequence, they fail to study a large fraction of the genome.

LETHAL GENES

Certain genes may produce the death of organisms carrying them. These **lethal genes** are believed to operate by disrupting vital biochemical pathways; many exert their effect during embryological development while others produce death at later stages of life. The allele producing yellow coat color, *Y*, in mice is an example of a lethal allele. Wild-type mice have dark-colored coats produced by expression of the normal allele, *y*. In the heterozygous state, *Yy*, *Y* is dominant to *y*, producing yellow coat color. In the homozygous condition, *YY*, death occurs during embryological development. Thus in the mating of two heterozygous mice, *Yy* × *Yy*, genotypes *YY*, *Yy*, and *yy* would be formed with an expected ratio of 1:2:1, as shown in the following Punnett square. All of the *YY* individuals, however would die early in development, so that the litter would consist only of *Yy* (yellow coated) and *yy* (dark coated) mice in a 2:1 ratio.

	Y	*y*
Y	*YY* (die as embryos)	*Yy* (yellow)
y	*Yy* (yellow)	*yy* (dark)

Thus, rather than producing the expected Mendelian phenotypic ratio of 3:1 in the progeny, the mating of these two heterozygotes has resulted in a phenotypic ratio among the living progeny of 2:1. This 2:1 phenotypic ratio is characteristic of crosses between two individuals heterozygous for a lethal allele. Lethal genes have been detected in virtually all organisms that have been studied.

SUMMARY

Two systems are commonly used to assign symbols to genes and their alleles. One system uses upper- and lowercase versions of the same letter to designate a gene's dominant and recessive alleles, respectively. Often the letter used is the first letter of the word describing the phenotype resulting from the expression of the recessive allele. A more widely used system uses a symbol of one or more letters derived from the name of the mutant condition. If the mutant allele is recessive, the first letter of the symbol is kept in lowercase, and if dominant, the first letter is capitalized. The wild-type allele is designated by a superscript plus sign affixed to the symbol for the mutant allele.

Regardless of the system used, it is essential that the symbols used in connection with each problem be clearly defined. A general plan is suggested for approaching the more common types of genetics problems and is applied to crosses that involve an autosomal gene with a single pair of alleles.

A testcross makes it possible to determine whether an organism expressing dominance for a trait is homozygous or heterozygous for the dominant allele. This is done by crossing the organism with a homozygous recessive. If all the testcross progeny express the dominant allele, the organism with the dominant phenotype is homozygous. If approximately half the progeny express the dominant allele and half express the recessive allele, then the organism is heterozygous.

Probability is the chance that a particular event will occur. It is the expected number of times the particular event we are looking for occurs, relative to the total number of possible events. A Punnett square provides a method of combining probabilities to give the expected frequencies for the various genotypes produced in a cross. Any two (or more) events are considered independent if the occurrence (or nonoccurrence) of one has no bearing on the occurrence of others. The probability that two (or more) independent events will occur together equals the product of their separate probabilities (product law). Two or more events are mutually exclusive if the occurrence of one excludes the occurrence of the other. The probability that one or another mutually exclusive event will occur equals the sum of the individual probabilities (law of the sum).

Sometimes an allele with the potential for expression will not always be expressed in the same way in all the organisms carrying it. An allele that is expressed in some individuals and not in other carriers is said to be incompletely penetrant. Within a population of individuals carrying and expressing an allele (in other words, individuals for whom an allele is penetrant), further phenotypic variation can arise if the allele shows variable expressivity, that is, if the allele is expressed to different degrees in different individuals. Both penetrance and expressivity contribute to phenotypic variability within a population.

Lethal genes disrupt vital biochemical pathways and may produce the death of organisms carrying them. Lethal genes may cause departures from expected Mendelian ratios and, like penetrance and expressivity, can make analysis of inheritance patterns difficult.

PROBLEM SET

3-1. The allele for yellow fruit color, o^+, is dominant to the allele for orange, o, in certain species of squash. A heterozygous plant with yellow fruit is crossed with a plant with orange fruit.

a. Identify the genotypes of both parent plants.
b. Identify the kinds of gametes produced by each parent.
c. Give the genotypes and phenotypes of the progeny and the expected percentage of each type.

3-2. The allele for red feather color in pigeons, b^+, is dominant to the allele for brown feathers, b. A red pigeon who had a red parent and a brown parent is mated with a brown pigeon.

a. Give the genotypes of the two pigeons being mated.
b. Identify the gametes produced by each of the pigeons being mated.
c. What proportion of the F_1 progeny would be expected to have brown feathers?
d. Assume that these pigeons produce five young pigeons. What is the probability that all five will be red?

3-3. Polydactyly is a genetically based human trait which results in extra fingers and toes. With one form of this disorder, the polydactyly allele, P, is dominant to the allele for the normal number of digits, p. A man exhibits this type of polydactyly; his father had polydactyly and his mother did not. He marries a normal woman whose family has no history of this genetic abnormality.

a. What are the genotypes of the woman and her husband?
b. What is the chance of polydactyly in their first child?

3-4. Cystic fibrosis is a serious human disease caused by an autosomal allele, c, which is recessive to the allele for the normal condition, C. With appropriate medical care, affected individuals may live to reach early adulthood and beyond. Two phenotypically normal people have four children; three are normal and one is affected with cystic fibrosis.

a. Identify, to the extent possible, the genotypes of the parents.
b. Identify, to the extent possible, the genotypes of the children.

3-5. In corn, a food-storing part of the seed known as the endosperm may be starchy or waxy. The allele for starchy endosperm, Wx, is dominant to the allele for waxy endosperm, wx. For each of the following crosses, indicate the genotypes for the parents and F_1 progeny.

Parents	F_1 progeny
a. Starchy × starchy	682 starchy
b. Starchy × starchy	506 starchy : 162 waxy
c. Starchy × waxy	784 starchy
d. Starchy × waxy	425 starchy : 409 waxy

3-6. Human albinism is a condition arising from an inability to manufacture the pigment melanin which is responsible for normal eye, hair, and skin coloration. This disorder is due to the expression of an autosomal allele, a, which is recessive to the allele for normal pigmentation, A. Identify the possible genotypes of both parents and progeny in the following situations.

a. Albino parents who have only albino children.
b. An albino woman and a normal man who have two albino and two normal children.
c. Phenotypically normal parents who have both normal and albino children.

PROBLEM SET

3-7. Studies indicate that about 3 out of 100 phenotypically normal humans carry, in the heterozygous condition, the recessive allele for albinism. What is the probability that
a. both members of a phenotypically normal couple are heterozygous for this allele?
b. the wife of a male who is heterozygous for the allele will also be heterozygous?

3-8. In the absence of a particular human enzyme, the amino acid phenylalanine, which is a component of many types of protein found in the diet, cannot be utilized in its normal metabolic pathway. Instead, the phenylalanine is converted into phenylpyruvic acid which accumulates in the body and has an adverse effect on the development of the nervous system in infant children. If untreated, this disorder, known as phenylketonuria, or PKU, can produce serious mental retardation. PKU is due to the expression of an autosomal allele, p, which is recessive to the allele, p^+, for normal production of the enzyme. Identify the progeny genotypes and their expected frequencies if
a. both parents are heterozygous for the condition.
b. one parent is homozygous for the dominant allele and the other is heterozygous.

3-9. Refer to information presented in problem 3-8.
a. If both parents are heterozygous at the phenylketonuria locus, what is the probability of their producing three children all of whom are affected with PKU?
b. If one parent is homozygous for the dominant allele and the other carries the allele for phenylketonuria in the heterozygous condition, what is the probability of their producing three children all of whom are affected with PKU?

3-10. Flower color in snapdragons is a trait showing incomplete dominance. It is governed by two alleles, C^R and C^W, for red- and white-pigment production, respectively. Identify the F_1 progeny expected from crossing
a. two pink-flowered plants.
b. two white-flowered plants.
c. a pink-flowered plant and a white-flowered plant.

3-11. A mother and father have three girls, and relatives have told the couple that, in light of this, the chance that their fourth child will be a boy is very high. Are these relatives correct? What is the probability that a fourth child born to this couple will be a male?

3-12. A type of human bone disorder is due to an autosomal allele, d, which is recessive to the allele, d^+, for normal bones. A man affected with this disorder marries a normal woman whose father was affected. What is the chance that a child produced by this couple will escape the disease?

3-13. On a particular day, five babies are scheduled to be born in the delivery area of a hospital. What is the probability that
a. all five will be male?
b. all five will be female?
c. five will be either all male or all female?

3-14. Huntington's chorea is a human disorder characterized by a slow but progressive deterioration of the nervous system. The disease, which generally does not manifest itself until late in the reproductive period (the 30s or 40s), is due to the presence of an autosomal allele (H) which is dominant to the

PROBLEM SET

allele for the normal condition (H^+). A man whose mother was affected with Huntington's chorea marries a woman with no history of this disorder in her family.

a. Assume that the man's mother was heterozygous for the disorder. What is the probability that the man will be affected?

b. Assume that the husband does indeed carry the dominant allele for this disorder. What is the probability that

  i. the first child born to this couple will possess the allele for Huntington's chorea?

  ii. the first three children will possess this harmful allele?

  iii. the first three children will not possess this harmful allele?

3-15. The color of tomatoes is under genetic control, with the allele for red (Y) dominant to the allele for yellow (y).

a. A plant with red tomatoes is crossed with a plant with yellow tomatoes, and the F_1 consists of 42 plants with red tomatoes and 37 with yellow tomatoes. Based on this outcome, what can be concluded about the genotype of the red-fruited parent plant?

b. Identify the plants expected from crossing the yellow and red F_1 plants.

3-16. A recessive allele, i, in the fruit fly *Drosophila* alters the normal development of one of the wing veins so that the vein contains a break or interruption within it. The dominant counterpart of this allele, i^+, produces the wild-type phenotype with normal vein development.

a. If two flies with interrupted veins are mated, what prediction would you make regarding the vein pattern shown by their progeny?

b. Examination of several hundred F_1 progeny indicates that about 90% of the flies have an interrupted vein while the rest show a normal vein. How would you explain the occurrence of the normal phenotype among some of the F_1 flies?

c. If you were to cross F_1 flies showing the normal vein pattern, what vein pattern or patterns would you expect to see in their progeny?

3-17. A genetic disorder in humans is known to be due to a dominant gene at a single locus. There is considerable variation in the degree of severity of the disorder that is seen in affected individuals. This severity can be clinically graded on a scale from 1 (representing a very mild case) to 10 (denoting a very severe case). The distribution of severity found within a large sample of affected adults is shown in the following graph.

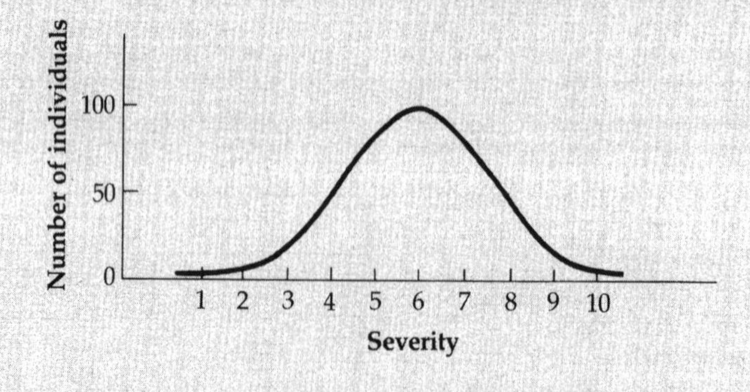

PROBLEM SET

a. Where, in the distribution, do most of the individuals in this sample fall?
b. Can you tell from the information provided here whether the gene that is responsible for this disorder has variable expressivity? Explain.
c. Can you tell from the information provided here whether the gene that is responsible for this disorder has variable penetrance? Explain.
d. Generally speaking, how might the wide range in the expression of this gene, that is, its variable expressivity, be explained?
e. If the allele responsible for the disorder exhibited variable penetrance, where, within the distribution, would the nonpenetrant individuals fall?
f. Where, within the distribution, would individuals lacking the dominant gene fall?

3-18. The allele for black fur color, *B*, in guinea pigs is dominant to the allele for brown fur, *b*.
a. How often would you expect to get a litter of three black and one brown from the mating of two heterozygous black guinea pigs?
b. How often would you expect to get a litter of one black and three brown from the mating of two heterozygous black guinea pigs?

3-19. Some dogs of the Mexican hairless breed are hairless while others have normal hair. Matings between hairless and normal dogs of this breed always produce hairless and normal progeny in a 1:1 ratio. In matings between two hairless dogs, hairless and normal dogs are always produced in a 2:1 ratio. How can these results be explained?

3-20. In chickens, the allele *C* is dominant to the wild-type allele, *c*, and alters embryological development to produce crooked, shortened legs and wings, a condition designated as "creeper." In the homozygous condition, allele *C* disrupts embryological development and results in death.
a. When two creepers are mated, what genotypes and phenotypes will be found among their progeny?
b. What mating could be carried out to verify that the creeper progeny are heterozygous at the *c* locus?
c. As a lethal allele, *C* is said to act as a recessive allele. Why do you think it is described in this way?

3-21. Wild-type mice have dark-colored coats produced by expression of the normal allele, *y*. The allele producing yellow coat color, *Y*, is dominant to *y*, and produces yellow coat color in heterozygotes. In the homozygous condition, *Y* is lethal. What phenotypic ratio would be expected among the progeny from the mating of a yellow-coated mouse and a dark-coated mouse?

CHAPTER

4

Crosses Involving Two Independently Assorting Traits (Dihybrid Crosses)

INTRODUCTION

This chapter considers crosses that deal simultaneously with the inheritance of two traits, where one trait is determined by alleles on one pair of chromosomes and the second trait by alleles on a different pair of chromosomes. Crosses of this type are often referred to as **dihybrid crosses,** although strictly speaking, a true dihybrid cross involves parents who are both heterozygous for each of the two genes under consideration. Since chance alone determines the orientation of the tetrads during meiosis and the subsequent distribution of the homologous chromosomes to each reproductive cell, genes on different pairs of chromosomes assort independently of each other. In other words, the segregation of one pair of alleles has nothing whatsoever to do with the segregation of the second pair of alleles. As was discussed in Chapters 1 and 2, this provides the basis for Mendel's law of independent assortment. We will consider basic approaches for solving problems of this type, review alternatives to the Punnett square as a way of combining gametes to identify progeny genotypes, and examine the two-locus testcross.

A BASIC APPROACH TO SOLVING TWO-TRAIT PROBLEMS

Two-trait problems generally fall into two categories. Either information is given about (1) the parents, and you are asked to identify the progeny, or (2) the progeny, and you are asked to identify the parents. Regardless of the type, the initial steps in solving the problems are the same as those discussed in Chapter 3 for crosses involving single autosomal genes. First carefully read the problem, identify what you are asked to do, and decide upon a general strategy for answering the question. Then identify the traits under consideration and determine the type of allelic interaction shown by each of the genes involved. Designate the symbols to identify the alleles of each gene and list them in a key. Then write down everything you can about the genotypes of the progeny and the parents. How you proceed from this point depends on the type of problem. We will look at examples of the two common types of problems and discuss approaches to each.

Identifying Progeny from Information about the Parents

The procedure for solving problems that give information about the parents and ask you to provide genotypes and/or phenotypes for the progeny is straightforward. It usually involves the following steps after the initial reading and analysis.

1. Identify all possible types of gametes produced by each parent.
2. Determine all possible genotypes that result from the random union of these gametes.
3. Tabulate the progeny genotypes, phenotypes, and expected frequencies and state the results.

An Example: Dihybrid × Dihybrid and the 9:3:3:1 Ratio
This example consists of a cross between two double-heterozygous individuals that produces progeny in a 9:3:3:1 phenotypic ratio.

PROBLEM: In guinea pigs, the genes determining the traits of coat texture and coat color are carried on separate pairs of autosomes. The allele for rough coat (*S*) is dominant to the allele for smooth coat (*s*) and the allele for black fur (*W*) is dominant to the allele for white fur (*w*). Two guinea pigs, each heterozygous at both of these loci, are mated. Identify the genotypes and phenotypes expected in the F_1 progeny.

SOLUTION:
Basic information given: The genes determining the two traits are on separate pairs of autosomes. Complete dominance operates at each locus, with the alleles for smooth coat and white fur recessive to their wild-type counterparts. Each parent is heterozygous at each locus. Gene symbols are given in the problem.
Question to be answered: What are the progeny genotypes and phenotypes?
Strategy: To determine the progeny genotypes and phenotypes, we must identify the parental genotypes.
Symbol key:

Coat-texture trait:	*S* =	allele for rough fur
	s =	allele for smooth coat
Fur-color trait:	*W* =	allele for black fur
	w =	allele for white fur

Parental genotypes: Each parent is heterozygous at both loci and the mating can be represented as *SsWw* × *SsWw*.

Next, we need to identify the types of gametes produced by each parent. The chromosome pairs carrying alleles for these two traits will assort independently of each other at the time of gamete formation. Because of this, each *SsWw* parent will produce four kinds of gametes, with each gamete carrying one allele for each trait. Four possible allelic combinations result: *SW*, *Sw*, *sW*, and *sw*. Notice that if we consider each gene separately, half the gametes carry the *S* allele and half carry *s* and, similarly, half carry *W* and the other half carry *w*. Chance determines the manner in which these independently assorting genes combine in the gametes. Consequently half of the gametes with *S* also carry *W* and the other half of them carry *w*. Similarly, half the gametes with *s* carry *W* and the other half of them carry *w*.

The random union of the four different kinds of gametes produced by each parent can be shown in a Punnett square with 16 boxes. The types of gametes are customarily listed across the top and down the side in order of decreasing numbers of dominant alleles; that is, the type of gamete with two dominant alleles comes first, then the two types with one dominant allele each, and finally the type with no dominant alleles. Determine the genotypes by filling in each box in the square. This is done by writing the gamete found on the left of each horizontal row in each of the empty boxes in the row, and writing the gamete found at the top of each vertical column in each of the boxes in the column. As you fill in each genotype, write the two alleles for each trait next to each other. For example, in the box in the upper-left corner, write *SSWW* rather than *SWSW*. (Note that the boxes are numbered for reference in the following discussion. Normally there would be no reason for you to number them.)

	SW	*Sw*	*sW*	*sw*
SW	(1)	(2)	(3)	(4)
Sw	(5)	(6)	(7)	(8)
sW	(9)	(10)	(11)	(12)
sw	(13)	(14)	(15)	(16)

Summarize the outcome beginning with the genotypes. When a mating involves multihybrid parents, the generalized expression 3^n, where n is the number of heterozygous gene pairs carried by the parents, gives the number of progeny genotypes. Since this cross involves dihybrids, $n = 2$, and the number of genotypes expected among the progeny is $3^2 = 9$. A list of these nine different genotypic combinations follows. To tabulate the results, record the number of each box of the Punnett square opposite the genotype which the box contains. For example, box 1 contains the genotype *SSWW* and a "1" is recorded opposite the *SSWW* genotype in the list. When you are finished, tally up the number of Punnett-square boxes showing each genotype to determine the genotypic ratio.

SSWW: ______________________________[1]

SSWw: ______________________________[2]

SSww: ______________________________[3]

SsWW: ______________________________[4]

SsWw: ______________________________[5]

Ssww: ______________________________[6]

ssWW: ______________________________[7]

ssWw: ______________________________[8]

ssww: ______________________________[9]

Genotypic ratio: ___ : ___ : ___ : ___ : ___ : ___ : ___ : ___ : ___[10]

A similar analysis will give us the phenotypic ratio. When a mating involves multihybrid parents and complete dominance operates at each locus, the generalized expression 2^n, where n is the number of heterozygous pairs carried by the parents, gives the number of progeny phenotypes. Since this cross involves dihybrids, $n = 2$, and the number of phenotypes expected among the progeny is $2^2 = 4$. A list of these four phenotypes follows. Determine the phenotype produced by the genotype in each box of the Punnett square and record the number of that box following the appropriate phenotype. For example, the *SSWW* genotype in box 1 has a phenotype of rough, black. Write the phenotype in that box and record the number "1" opposite that phenotype in the list. When you are finished, tally up the number of boxes showing each phenotype to determine the phenotypic ratio for the progeny.

Rough, black: ______________________________[11]

Rough, white: ______________________________[12]

Smooth, black: ______________________________[13]

Smooth, white: ______________________________[14]

Phenotypic ratio: ___ : ___ : ___ : ___[15]

The 9:3:3:1 phenotypic ratio observed in this example always occurs whenever (1) both parents are heterozygous for each of two genes carried on independently assorting autosomes; (2) the two genes have unrelated phenotypic effects; and (3) complete dominance operates at both loci, and each locus has a dominant allele present.

An Alternative Approach: Treating the Two-Trait Cross as Two Separate Single-Trait Crosses Another approach to solving two-trait (and more elaborate) problems involves treating the inheritance of each trait separately, that is, to approach the *SsWw* × *SsWw* cross as if it were two separate monohybrid crosses. This approach can be used as long as the traits involved are inherited independently of each other. To illustrate this, we will use the guinea-pig mating we have been considering. The coat-texture part of this mating involves crossing *Ss* × *Ss*. What fraction of the expected progeny of this mating would be *SS*? __________[16] What fraction would be *Ss*? __________[17] What fraction would be *ss*? __________[18] The coat-color part of the dihybrid mating involves crossing *Ww* × *Ww* to give progeny that are 1/4 *WW*, 1/2 *Ww*, and 1/4 *ww*.

Genotypic outcome: Up to this point, we have used Punnett squares to show how *gametes* from different parents combine to form all possible genotypes and to combine probabilities using the product law to give genotypic frequencies. A square can also be used to show the ways in which *genotypes* for different traits join to form all possible genotypic combinations, whose probabilities also can be combined to give the expected genotypic frequencies. This is done in this case by listing the *s*-locus genotypes and their probabilities along one side of the square and the *w*-locus genotypes and their probabilities along the other side.

Complete the following square by filling in the boxes using the standard method. For example, combining *SS* and *WW* in the upper-left box gives genotype *SSWW*. The probability of any two of these independently inherited genotypes occurring together is given by multiplying their separate probabilities. For example, the probability of the *SSWW* combination in the upper-left box is 1/4 × 1/4 = 1/16.

	1/4 *SS*	1/2 *Ss*	1/4 *ss*
1/4 *WW*	1/16 *SSWW*		
1/2 *Ww*			
1/4 *ww*			

[1] One box: 1. [2] Two boxes: 2, 5. [3] One box: 6. [4] Two boxes: 3, 9. [5] Four boxes: 4, 7, 10, 13. [6] Two boxes: 8, 14. [7] One box: 11. [8] Two boxes: 12, 15. [9] One box: 16. [10] 1:2:1:2:4:2:1:2:1. [11] Nine boxes: 1, 2, 3, 4, 5, 7, 9, 10, 13. [12] Three boxes: 6, 8, 14. [13] Three boxes: 11, 12, 15. [14] One box: 16. [15] 9:3:3:1. [16] 1/4 *SS*. [17] 1/2 *Ss*. [18] 1/4 *ss*.

Compare this genotypic outcome with that produced in the 16-box Punnett square considered earlier in this chapter to verify that they are exactly the same.

Phenotypic outcome: The phenotypic outcome can be determined in the same manner as the genotypic outcome. What fraction of the progeny of the mating *Ss* × *Ss* would be expected to have a rough coat? ________[19] What fraction would have a smooth coat? ________[20] The progeny from mating *Ww* × *Ww* are 3/4 black fur and 1/4 white fur. These individual phenotypes and their probabilities can be placed along two sides of a Punnett square and combined as follows.

	3/4 rough	1/4 smooth
3/4 black	9/16 rough, black	3/16 smooth, black
1/4 white	3/16 rough, white	1/16 smooth, white

Compare the phenotypic outcome with that from the 16-box Punnett square considered earlier in this chapter to verify that they are exactly the same.

Determining Probabilities for Specific Genotypic or Phenotypic Combinations The expected probability for a specific progeny genotype or phenotype can be determined without generating all possible genotypic or phenotypic combinations. This is done by simply multiplying the probabilities for the specific combination in which you are interested. For example, you might be asked the frequency with which genotype *Ssww* would be expected among the progeny of the dihybrid cross *SsWw* × *SsWw*. As you determined above, the probability of *Ss* from the mating *Ss* × *Ss* is 1/2 and probability of *ww* from the mating *Ww* × *Ww* is 1/4. The probability of genotype *Ssww* is given by using the product law of independent events and multiplying the separate probabilities: 1/2 × 1/4 = 1/8. This can be written symbolically, where p stands for probability, as follows:

$$p(Ssww) = p(Ss) \times p(ww) = 1/2 \times 1/4 = 1/8.$$

The same approach, of course, can be used to identify the probability with which specific phenotypes occur. For example, if asked to determine the expected frequency with which the rough, black phenotype occurs among the progeny of this cross, you would multiply the probability of rough, 3/4, by the probability of black, 3/4, to get 3/4 × 3/4 = 9/16. This can be written out symbolically as follows:

$$\text{p(rough, black)} = \text{p(rough)} \times \text{p(black)} = 3/4 \times 3/4 = 9/16.$$

[19] 3/4 rough. [20] 1/4 smooth.

An Alternative to the Punnett Square: Branch Diagrams Another way to solve genetics crosses is to use **branch diagrams.** Like Punnett squares, these can be set up to show all possible genotypic and phenotypic combinations. A direct approach to identifying the progeny is to consider the inheritance of each trait as a separate cross and then combine either the genotypes or phenotypes from each cross. If the probability for the formation of each genotype or phenotype is included in the branch diagram, the product law can be used to determine the expected frequency of each combination. Branch diagrams tend to be easier to use and less susceptible to error than the large Punnett squares required by two-trait (and more elaborate) crosses.

As an illustration, this method will be used to determine the genotypic outcome of the guinea-pig mating *SsWw* × *SsWw*. The expected genotypic outcome of the *Ss* × *Ss* cross is 1/4 *SS*, 1/2 *Ss*, and 1/4 *ss*, and that of the *Ww* × *Ww* cross is 1/4 *WW*, 1/2 *Ww*, and 1/4 *ww*. To set up the branch diagram, the genotypes produced from one cross and their probabilities are listed on the left. Then all the genotypes produced from the other cross and their probabilities are listed to the right of *each* of the genotypes produced by the first cross. (Note that if one cross produces fewer genotypes than the other, its genotypes are generally listed on the left; otherwise, the genotypes from either cross could be listed there). Combining the genotypes gives the genotypic combinations, and multiplying the probabilities gives the expected genotypic frequencies. Complete the following branch diagram to give the genotypic outcome of the *SsWw* × *SsWw* mating.

Genotypes from one cross	Genotypes from other cross		Progeny genotypes	Expected frequency
	WW (1/4)	=	*SSWW*	1/16
SS (1/4)	*Ww* (1/2)	=	________	________
	ww (1/4)	=	________	________
	WW (1/4)	=	________	________
Ss (1/2)	*Ww* (1/2)	=	________	________
	ww (1/4)	=	________	________
	WW (1/4)	=	________	________
ss (1/4)	*Ww* (1/2)	=	________	________
	ww (1/4)	=	________	________

Your results should give a 1:2:1:2:4:2:1:2:1 ratio and can be checked by referring back to the summary of genotypes from the 16-box Punnett square completed earlier.

Often problems involving two-trait crosses will ask only for the phenotypic outcome. The phenotypic outcome could, of course, be determined by first identifying all of the genotypes (in a branch diagram, for example), and then determining the phenotype for each genotype. A more direct way involves identifying the phenotypes expected when the inheritance of each trait is considered separately, and combining these phenotypes in a branch diagram. As you know, the phenotypic outcome of the $Ss \times Ss$ mating is 3/4 rough and 1/4 smooth, and that of the $Ww \times Ww$ mating is 3/4 black and 1/4 white. A branch diagram set up as follows will allow you to identify all possible combinations of these phenotypes and the expected frequency for each. Complete this branch diagram.

Phenotypes from one cross	Phenotypes from the second cross	Progeny phenotypes	Expected frequency
Rough (3/4)	Black (3/4) =	________	________
	White (1/4) =	________	________
Smooth (3/4)	Black (3/4) =	________	________
	White (1/4) =	________	________

Your results should give the standard 9:3:3:1 phenotypic ratio that was seen earlier with the 16-box Punnett square.

A Second Example: The 1:1:1:1 Ratio Let us look at another example of a cross involving two independently assorting traits.

PROBLEM: In the fruit fly *Drosophila melanogaster,* vestigial wings are considerably reduced in size and produced by an allele, vg, which is recessive to the allele for wild-type wing size, vg^+. Ebony body color is produced by an allele, e, which is recessive to that for wild-type body color, e^+. Genes for these two traits are carried on separate pairs of autosomes. A male fruit fly known to be heterozygous for both traits is mated with a female fly with vestigial wings and ebony body. Characterize the expected progeny with regard to genotypes, phenotypes, and frequencies.

SOLUTION:

Basic information given: Two traits are involved which are determined by independently assorting genes. Complete dominance operates at each locus, with vestigial and ebony alleles recessive to their wild-type counterparts. The male is heterozygous for both traits and the female has vestigial wings and ebony body. Gene symbols are provided.

Question to be answered: What are the progeny genotypes and phenotypes and their frequencies?

Strategy: To identify the progeny genotypes and phenotypes, we must determine the parental genotypes.

Symbol key:

Wing trait:	vg^+	= dominant allele for normal wings
	vg	= recessive allele for vestigial wings
Body-color trait:	e^+	= dominant allele for normal body color
	e	= recessive allele for ebony body color

Parental genotypes: We can infer that the female parent with vestigial wings and ebony body must be homozygous recessive for each trait and has genotype $vg/vg e/e$. (Note that, as was described in Chapter 3, a slash may be used when writing genotypes to represent a pair of homologous chromosomes, with the alleles carried by that pair written on either side of the slash. This notation is often used when writing *Drosophila* genotypes, and since each of the loci we are considering is on a different pair of chromosomes, the symbolism for the genotypes includes two slashes.) The genotype of the male parent who is heterozygous for each gene is $vg^+/vg e^+/e$ and the mating can be represented as $vg/vg e/e \times vg^+/vg e^+/e$.

Gamete production: The chromosome pairs carrying alleles for these two traits assort independently. Because of this, the $vg^+/vg e^+/e$ male parent will produce four types of gametes in equal numbers: vg^+e^+, vg^+e, $vg e^+$, and $vg e$. The $vg/vg e/e$ parent will produce a single type of gamete, $vg e$.

Union of gametes: The random union of these gametes can be shown in the following Punnett square. Complete the square to show the genotypes that are formed.

	vg^+e^+	vg^+e	$vg e^+$	$vg e$
$vg e$				

Progeny: Equal numbers of four different genotypes would be expected: (1) $vg^+/vg e^+/e$, (2) $vg^+/vg e/e$, (3) $vg/vg e^+/e$, and (4) $vg/vg e/e$, showing respective phenotypes of (1) normal wings, normal body, (2) normal wings, ebony body, (3) vestigial wings, normal body, and (4) vestigial wings, ebony body. This gives a phenotypic and genotypic ratio of 1:1:1:1. ■

Note that this 1:1:1:1 phenotypic ratio always occurs whenever (1) one parent is heterozygous for each of two genes carried on independently assorting chromosomes, and the other parent is homozygous recessive at these genes; (2) the two genes have unrelated phenotypic effects; and (3) complete dominance operates at both loci, and each locus has a dominant allele present in the dihybrid parent.

Identifying Parents from Information about the Progeny

Another type of two-trait problem asks you to identify genotypes and/or phenotypes for one or both of the parents based on information supplied about the progeny, such as progeny phenotypes and their frequencies. Some problems of this type also may supply you with partial genotypic or phenotypic information for one or both parents.

Tailoring a Strategy for Solving This Type of Problem Because problems of this type vary widely in the nature of the information supplied, it is impossible to design a standard strategy that can be routinely used when solving them. How you proceed depends on the amount and nature of the information available. As you tailor a strategy, keep the following points in mind.

1. You can work backward from the progeny genotypes to determine the types of gametes produced by the parents. If you can identify some or all of the types of gametes produced by a parent, part or all of its genotype can be inferred. Remember that when progeny exhibit the homozygous recessive phenotype for a trait, each parent must carry at least one recessive allele for that trait.

2. The phenotypic ratio, if it is provided or can be generated from the numerical outcome of a cross, may provide an important clue to the parents' genotypes. For example, as you demonstrated earlier in this chapter, a phenotypic ratio of 9:3:3:1 indicates that both parents are dihybrid, and a phenotypic ratio of 1:1:1:1 indicates that one parent is dihybrid while the other is homozygous recessive for both loci. If the ratio fails to tell what the parent's genotypes are, it still may be useful since it allows you to eliminate certain parental genotypes from further consideration. For example, data that differ significantly from a 9:3:3:1 ratio indicate that the parents are not a pair of dihybrids or, as will be discussed later, that the loci are linked or that epistasis is operating. (The statistical significance of deviations from expected ratios can be evaluated using the chi-square test described in Chapter 6.)

3. Often it is helpful to consider each trait separately. The phenotypic ratio for a single trait can be determined by identifying the number of progeny out of the total that show each phenotypic expression of that trait and calculating the ratio. For example, assume that you are considering the outcome of the cross between two corn plants where the 403 progeny consist of 102 tall, yellow; 115 short, yellow; 97 tall, green; and 89 short, green. Considering the trait of plant height, there are 102 + 97 = 199 tall plants and 115 + 89 = 204 short plants. The ratio of 199:204 is very close to a 1:1 ratio. As you demonstrated in Chapter 3, a phenotypic ratio of 3:1 indicates that both parents are heterozygous at the locus, and a 1:1 ratio indicates that one parent is heterozygous and the other is homozygous recessive.

4. The parental genotypes you arrive at must be consistent with both the progeny phenotypes and their phenotypic ratio. To illustrate this, consider a single trait of height in pea plants, where the allele for tall height (*S*) is dominant to the allele for short height (*s*). You are told that a cross produces a mixture of tall and short plants in a 3:1 ratio. Based on phenotypes alone and ignoring the phenotypic ratio, the mixture of tall and short could arise from either *Ss* × *Ss* or *Ss* × *ss* crosses. Considering the phenotypic ratio makes it possible to identify the correct set of parents. Since tall and short plants occur in a 3:1 ratio, we know the cross producing these progeny must be *Ss* × *Ss*. (The *Ss* × *ss* cross would produce progeny in a 1:1 ratio.)

5. If your strategy does not work, devise another. Working with one approach, even if it leads to a dead end, may suggest alternative approaches to solving the problem. As the next example will show, most problems of this type can be solved in more than one way.

6. Once you have identified the parental genotypes, your results can be checked by working out the cross between these parents to identify the expected progeny genotypes, phenotypes, and their expected frequencies. The outcome of this cross must be in general agreement with the information about the progeny given to you in the problem. If it is not, you have identified the parental genotypes incorrectly and need to backtrack to find your error.

An Example: Designing and Applying a Strategy; The 3:3:1:1 Ratio This example considers a cross where we are given the genotype of one parent and the phenotypic outcome of the cross and asked to identify the genotype of the other parent.

PROBLEM: In corn, *Zea mays*, the endosperm is a part of the seed which contains stored food, and the allele for waxy endosperm, *wx*, is recessive to the allele for starchy endosperm, *Wx*. The allele for yellow seedlings, *y*, is recessive to the allele for green seedlings, *Y*. The genes for these two traits assort independently. A plant heterozygous for both genes is crossed with another corn plant and the progeny consist of 212 starchy, green; 198 starchy, yellow; 65 waxy, green; and 70 waxy, yellow. Identify the genotype of the unknown parent plant.

SOLUTION:

Basic information given: The two loci assort independently, and complete dominance operates at each locus, with the alleles for waxy endosperm and for yellow seedlings recessive to their wild-type counterparts. One parent is heterozygous for both genes. The other parent's genotype is unknown. Four phenotypes occur among the progeny in a ratio of 212:198:65:70.
Question to be answered: What is the genotype of the unknown parent?
Strategy: To determine the genotype of the unknown parent, we use the three sources of information available to us: (1) progeny phenotypes, (2) progeny phenotypic frequencies, and (3) the dihybrid genotype of the known parent. From the progeny phenotypes, we can make some inferences about the progeny genotypes. Since we know that one parent is a dihybrid, we can identify the types of gametes it contributed to the progeny. Alleles carried by the progeny and not contributed by the dihybrid parent must come from the unknown parent. The phenotypic ratio may tell us something about the genotypes of the parents. Then we should separately examine the inheritance of each trait.
Symbol key:

Endosperm trait:	*Wx* =	dominant starchy-endosperm allele
	wx =	recessive waxy-endosperm allele
Seedling-color trait:	*Y* =	dominant green-seedling allele
	y =	recessive yellow-seedling allele

Identification of genotypes: Next, we should write out everything that we can about the genotypes of the progeny and parents. Inferences about progeny genotypes can be made from their phenotypes, as follows: (1) starchy, green must have at least one dominant allele at each locus: *Wx*_*Y*_; (2) starchy, yellow must have at least one dominant allele at the endosperm locus and two recessive alleles at the color locus: *Wx*_*yy*; (3) waxy, green must have two recessive alleles at the endosperm locus and at least one dominant allele at the color locus: *wxwxY*_; and (4) waxy, yellow must have two recessive alleles at each locus: *wxwxyy*. The dihybrid parent has genotype *WxwxYy*. From the *wxwxyy* genotype for the waxy, yellow progeny, we know that *both* parents must carry the *wx* and the *y* allele. The unknown parent's genotype can be represented as _*wx*_*y*.
Solving the problem: At this point, three genotypic combinations are possible for the unknown parent: (1) homozygous recessive at each locus, (2) heterozygous at each locus, or (3) homozygous recessive at one locus and heterozygous at the other. The phenotypic ratio of 212:198:65:70 can be simplified by dividing each frequency by the smallest frequency in the series (65): 212/65 = 3.26, 198/65 = 3.05, 65/65 = 1, and 70/65 = 1.08. This gives a ratio of 3.26:3.05:1:1.08 which is very close to a 3:3:1:1 ratio. On the basis of this ratio, we can eliminate the first two possibilities from further consideration since neither would give this ratio. (If the unknown parent were homozygous recessive at each locus, the cross would be *WxwxYy* × *wxwxyy* and the four progeny phenotypes would be expected in a 1:1:1:1 ratio; if it were heterozygous at each locus, the cross would be *WxwxYy* × *WxwxYy* and the four progeny phenotypes would be expected in a 9:3:3:1 ratio.)

We need to look more closely at the third possibility, specifically that the unknown parent is either *wxwxYy* or *Wxwxyy*. At this point it is wise to consider the two loci separately and take a look at the phenotypic ratio for each trait. First consider the endosperm locus. Of the total number of progeny, how many exhibit the starchy phenotype? ________[21] How many exhibit the waxy phenotype? ________[22] This ratio of 410:135 can be simplified by dividing each frequency by the smallest value (135): 410/135 = 3.04 and 135/135 = 1. This gives us a ratio of 3.04:1 which is very close to a 3:1 ratio.

Now let us predict the phenotypic ratio for the endosperm trait for each of the two parental genotypes under consideration. If the unknown parent is homozygous at the *wx* locus (with genotype *wxwxYy*), the cross with the known heterozygous parent would be *wxwx* × *Wxwx*. What ratio of starchy and waxy progeny would be expected from this mating? ________[23] Alternatively, if the unknown parent is heterozygous

[21] 212 + 198 = 410. [22] 65 + 70 = 135. [23] 1:1.

at the *wx* locus (with genotype *Wxwxyy*), the cross with the known heterozygous parent would be *Wxwx* × *Wxwx*. What ratio of starchy and waxy progeny would be expected from this mating? ________[24] Since the observed ratio is very close to 3:1, what can be inferred about the unknown parent? ____________

________________________________[25]

At this point we know that the unknown parent is homozygous at one locus and heterozygous at the other and we have just established that the heterozygosity occurs at the *wx* locus. What does this allow us to conclude about the unknown parent's genotype at the *y* locus? ____________________

__________[26] In light of this, the parental mating for this locus would be *yy* × *Yy*. What is the expected ratio of green and yellow progeny? ________[27] How many of the actual progeny are green? ________[28] How many are yellow? ________[29] This observed ratio of 277:268 is very close to the expected 1:1 ratio. Thus, the observed ratio confirms our conclusion that the unknown parent is homozygous recessive at the color locus.

Answer: The unknown parent is heterozygous at the endosperm locus and homozygous recessive at the color locus and has genotype *Wxwxyy*. ■

A more direct way of solving this problem involves considering the two traits separately right from the beginning. The phenotypic ratio for each trait is calculated to see what it reveals about the parents' genotypes. First consider the endosperm trait. Of the total number of progeny, how many are starchy? ____[30] How many are waxy? ________[31] Simplify this ratio of 410:135. ________[32] What type of cross would yield a 3:1 ratio? ____________________

__________[33] The problem tells us that the known parent is heterozygous at this locus. What can be inferred about the unknown parent's genotype at this locus? ____________________

______[34]

Next we calculate the phenotypic ratio for the seedling-color trait. Of the total number of progeny, how many are green? ________[35] How many are yellow? ________[36] This ratio of 277:268 is very close to a 1:1 ratio. What type of cross would give a 1:1 phenotypic ratio? ____________________

__________[37] The problem tells us that the known parent is heterozygous at this locus. What can we infer about the unknown parent's genotype at this locus?

____

____[38]

Answer: The unknown parent is heterozygous at the endosperm locus and homozygous recessive at the seedling-color locus and has genotype *Wxwxyy*.

Note that regardless of how we get this answer, it can be checked by working out the cross using the parental genotype we have identified to get the expected progeny phenotypes and phenotypic ratio. These can then be compared with information about the progeny given in the problem. Working this cross results in four progeny phenotypes in a 3:3:1:1 ratio, which is in accord with information presented in the problem. Our answer checks out.

Note that this 3:3:1:1 phenotypic ratio always occurs whenever (1) one parent is heterozygous for each of two genes carried on independently assorting chromosomes, and the other parent is heterozygous for one gene and homozygous recessive for the other; (2) the two genes have unrelated phenotypic effects; and (3) complete dominance operates at both loci, and each heterozygous locus has a dominant allele present.

TWO-LOCUS OR TWO-POINT TESTCROSS

Occasionally, in breeding studies involving two independently assorting traits, organisms are encountered where only the phenotype is known. This, of course, poses no problem if the individual expresses recessive characteristics for both traits, since the genotype can readily be inferred. Drawing inferences is more difficult, however, if the individual exhibits dominant characters. What genotypes would be possible for an individual expressing dominant characters for both traits? ____________________

______[39] The genotype of the double-dominant organism can be identified from the progeny it produces in a testcross. This involves crossing the individual of uncertain genotype with an organism homozygous recessive for the two loci under study.

[24] 3:1. [25] It is heterozygous at the *wx* locus. [26] It must be homozygous recessive. [27] 1:1. [28] 212 + 65 = 277. [29] 198 + 70 = 268. [30] 212 + 198 = 410. [31] 65 + 70 = 135. [32] Divide the smallest value, 135, into each frequency: 410/135 = 3.04 and 135/135 = 1, giving a simplified ratio of 3.04:1, which is very close to 3:1. [33] Heterozygote × heterozygote. [34] It is heterozygous. [35] 212 + 65 = 277. [36] 198 + 70 = 268. [37] Heterozygote × homozygous recessive. [38] It is homozygous recessive. [39] Four genotypes are possible: homozygous dominant at each locus, heterozygous at one locus and homozygous dominant at the other or vice versa, or heterozygous at both loci.

Relationship among the Genotype of a Double-Dominant Parent, Gamete Production, and Testcross Progeny

We will identify the testcross progeny for each of the four possible genotypes for a double-dominant parent, showing the relationship between the genotype of the double-dominant parent, the number of kinds of gametes produced by that parent, and the testcross progeny. We will cross a double-dominant corn plant with the starchy, green genotype and a waxy, yellow plant with genotype *wxwxyy*. How many kinds of gametes would be made by the double–homozygous recessive plant? ________[40] What alleles are carried by these gametes? ________[41]

The four possible genotypes for the double-dominant plant are as follows.

1. Heterozygous at each locus: The testcross is *WxwxYy* × *wxwxyy*. How many different types of gametes would be formed by the *WxwxYy* parent? ________[42] Identify the allelic makeup of these gametes. ________________________[43] How many different kinds of progeny would result from this testcross? ________[44] Identify the progeny genotype or genotypes. ________________________[45] Identify the progeny phenotype or phenotypes. ________________________[46]

2. Homozygous dominant at each locus: The testcross is *WxWxYY* × *wxwxyy*. How many different types of gametes would be formed by the *WxWxYY* parent? ________[47] Identify the gamete or gametes. ________[48] How many different kinds of progeny would result from this mating? ________[49] Identify the progeny genotype or genotypes. ________________________[50] Identify the progeny phenotype or phenotypes. ________________________[51]

3. Homozygous dominant at the *wx* locus and heterozygous at the *y* locus: The testcross is *WxWxYy* × *wxwxyy*. How many kinds of gametes would the *WxWxYy* parent produce? ________[52] Identify the gamete or gametes. ________[53] How many kinds of progeny would be expected from this testcross? ________[54] Identify the progeny genotype or genotypes. ________________________[55] Identify the progeny phenotype or phenotypes. ________________________[56]

4. Heterozygous at *wx* locus and homozygous dominant at the *y* locus: The testcross is *WxwxYY* × *wxwxyy*. How many kinds of gametes would the *WxwxYY* parent produce? ________[57] Identify the gamete or gametes. ________[58] How many kinds of progeny would be expected from this testcross? ________[59] Identify the progeny genotype or genotypes. ________________________[60] Identify the phenotype or phenotypes. ________________________[61]

Now that you have considered the outcomes of testcrosses with all possible genotypes for the double-dominant parent, compare the number of phenotypes shown by the testcross progeny to the number of different kinds of gametes produced by the double-dominant parent. ________________________[62] From the number of gametes produced by the double-dominant parent, we can, in turn, infer its genotype. All this is summarized in the following table.

Number of phenotypes in testcross progeny	Number of types of gametes formed by double-dominant parent	Genotype of double-dominant parent
1	1	Homozygous at both loci
2	2	Homozygous at one locus, heterozygous at other
4	4	Heterozygous at both loci

[40] One. [41] *wxy*. [42] Four. [43] *WxY*, *Wxy*, *wxY*, and *wxy*. [44] Four. [45] *WxwxYy*, *Wxwxyy*, *wxwxYy*, and *wxwxyy*.
[46] Starchy, green; starchy, yellow; waxy, green; and waxy, yellow, respectively. [47] One. [48] *WxY*. [49] One. [50] *WxwxYy*.
[51] Starchy, green. [52] Two. [53] *WxY* and *Wxy*. [54] Two. [55] *WxwxYy* and *Wxwxyy*. [56] Starchy, green and starchy, yellow.
[57] Two. [58] *WxY* and *wxY*. [59] Two. [60] *WxwxYy* and *wxwxYy*. [61] Starchy, green and waxy, green. [62] They are equal.

Inferring the Genotype of a Double-Dominant Parent from Testcross Progeny

If the testcross progeny exhibit one or four phenotypes, the genotype of the double-dominant parent can be readily inferred. However, when progeny exhibit two phenotypes, the genotype of the double-dominant parent is still unknown because we do not know which locus is heterozygous and which is homozygous dominant. This ambiguity can be resolved, however, if we consider the two loci separately. Begin with the locus which is homozygous dominant in the unknown parent. In general terms, what genotype or genotypes will the testcross progeny have at this locus? ________________[63] What phenotype or phenotypes will the testcross progeny exhibit for this trait? ________________[64] Now consider the locus which is heterozygous in the unknown parent. What genotype or genotypes will the testcross progeny have at this locus? ________________[65] What phenotype will the testcross progeny exhibit for this trait? ________________[66] Put another way, the locus that exhibits a single phenotype in the testcross progeny is homozygous dominant in the parent, and the locus that exhibits two phenotypes in the progeny is heterozygous in the parent.

SUMMARY

This chapter considers crosses that deal simultaneously with the inheritance of two traits, where one trait is determined by alleles on one pair of chromosomes and the second trait by alleles on a different pair of chromosomes. Two-trait problems generally fall into two categories where information is given either (1) about the parents and you are asked to identify the progeny or (2) about the progeny and you are asked to identify the parents. Each requires the identification of the types of gametes produced by each parent. If called upon to identify progeny, either a Punnett square or branch diagram may be used to show all possible random combinations of the gametes to form the genotypes and to combine probabilities using the product law to get the expected frequency for each genotype. Problems which ask you to identify the parents generally involve working backward from progeny phenotypes and phenotypic ratios to identify the gametes produced by the parents, from which the parental genotypes can be inferred. Key points to consider in devising a strategy for solving both types of problems are discussed.

In a two-point or two-locus testcross, the genotype of a double-dominant individual is determined by crossing the double-dominant individual of uncertain genotype with another that is homozygous recessive for the loci in question. The number of kinds of progeny arising from a testcross indicates the number of kinds of gametes produced by the double-dominant parent, and from that the genetic makeup of this parent can be inferred.

[63] Heterozygous. [64] Dominant phenotype. [65] Heterozygous and homozygous recessive in a 1:1 ratio.
[66] Dominant and recessive in a 1:1 ratio.

PROBLEM SET

4-1. Two independently assorting genes, *G* and *H*, each possess two alleles, one of which exhibits complete dominance over its recessive counterpart. For the following genotypes, identify the types of gametes made by each and the frequencies in which each type would be expected.

a. *GgHh*
b. *ggHH*
c. *ggHh*
d. *GGHh*
e. *gghh*

4-2. In the garden pea, the allele for green seed pods (*Y*) is dominant to the allele for yellow seed pods (*y*) and the allele for full pod shape (*C*) is dominant to the recessive allele (*c*) for constricted pod shape. The genes for these two traits are carried on different pairs of chromosomes. A plant homozygous for both green seed pods and full pod shape is crossed with a plant that has yellow seed pods and constricted pod shape.

a. Give the genotype for each parent plant.
b. Indicate the kind or kinds of gametes produced by each parent.
c. Give the genotypes and phenotypes of the F_1 progeny.

4-3. A cross between two of the F_1 plants arising from the cross described in problem 4-2 is carried out.

a. Indicate the kinds of gametes produced by each F_1 parent plant.
b. Determine the phenotypes of the F_2 and the ratio in which they would be expected.
c. If the F_2 consisted of 800 plants, how many plants would be expected to show each phenotype?

4-4. An F_1 plant produced from the cross described in problem 4-2 was pollinated by a plant having yellow pod color and constricted pod shape.

a. How many different genotypes would be found in the progeny?
b. If there are 1000 progeny plants, how many of each phenotype would be expected?

4-5. In rabbits, the allele for spotted coat, *S*, is dominant to the allele for solid, *s*. The allele for black coat color, *B*, is dominant to the allele for brown, *b*. The genes for these two traits are carried on different pairs of autosomes. All of the progeny produced by mating a solid, black rabbit with a spotted, brown rabbit were found to be spotted, black.

a. What is the genotype of the progeny?
b. Describe the expected phenotype or phenotypes of the F_2 progeny if two of the spotted, black F_1 rabbits are mated.

4-6. In the jimsonweed plant, the allele for smooth seed pods, *s*, is recessive to the allele for spiny pods, *S*. At another independently assorting locus, the allele for white flowers, *w*, is recessive to the allele for purple flowers, *W*.

a. Two plants of unknown genotype are crossed and the following F_1 progeny result: smooth, white: 27; spiny, white: 85; spiny, purple: 256; and smooth, purple: 93. Identify possible genotypes for the parent plants.
b. In another cross, two plants of unknown genotype are crossed and produce the following F_1 progeny: smooth, white: 75; spiny, white: 82; spiny, purple: 69; and smooth, purple: 77. Identify possible genotypes for the parent plants.

4-7. Refer to the information given in problem 4-6. A plant with genotype *SsWw* is crossed with a plant with genotype *ssWw*. Determine the probability with

PROBLEM SET

which each of the following progeny would be expected.

a. Genotype *ssww*.
b. Genotype *SsWw*.
c. Phenotype smooth pod, white flowers.
d. Phenotype smooth pod, purple flowers.

4-8. In tomato plants, hairy stems are produced by a dominant allele, *H*, and nonhairy stems arise from the expression of its recessive counterpart allele, *h*. For the trait of plant height, a dominant allele, *D*, produces tall plants while expression of the recessive allele, *d*, produces dwarf plants. Both loci are carried on separate pairs of chromosomes. You are given a tomato plant which has a hairy stem and is tall. Nothing is known about the specific genotype of this plant.

a. Identify all possible genotypes for this plant.
b. What procedure could be used to identify the genotype of this particular plant?
c. Assume that the plant in question is crossed with a plant with nonhairy stems and dwarfed height. Two phenotypes are found in roughly equal numbers among the 316 progeny plants. Based on this information, what can be concluded about the genetic makeup of the hairy, tall parental plant?
d. Of the 316 progeny plants, roughly half were hairy, tall and the rest were hairy, dwarf. With this additional information, what can be said about the genotype of the hairy, tall parental plant?

4-9. About half of the progeny from the testcross described in problem 4-8 were hairy, tall, and the other half were hairy, dwarf, and you should have identified the genotype of the double-dominant parent as *HHDd*. If the double-dominant parent involved in the testcross was *HhDD*, what phenotypes would occur in the testcross progeny? In what frequencies would these phenotypes be expected?

4-10. Refer to problem 4-8. Another hairy, tall tomato plant is crossed with a nonhairy, dwarf plant. The progeny consist of about 300 plants which fall in roughly equal numbers into four phenotypic categories.

a. How many kinds of gametes were produced by the double-dominant parent?
b. What are the four phenotypes found among the progeny?
c. What does the outcome of this cross tell us about the genotype of the hairy, tall parent?

4-11. In humans, nearsightedness (myopia) and the enzyme disorder PKU (phenylketonuria) are both inherited as independently assorting autosomal recessive traits. Use *N* to represent the dominant allele for normal vision and *n* for the recessive allele for nearsightedness. Use *P* to represent the dominant allele for the normal enzyme and *p* for the recessive allele for phenylketonuria. A phenotypically normal man who carries recessive alleles for both traits marries a woman who has the same genotype with regard to these traits.

a. What are the expected phenotypes and frequencies for their progeny?
b. What is the chance that their first child will have PKU?
c. What is the chance that any child they have will be nearsighted?
d. What is the chance that their first child will be nearsighted and have PKU?

4-12. A daughter of the couple described in problem 4-11 is nearsighted and

PROBLEM SET

heterozygous for the normal enzyme. Her husband has the normal phenotype for both traits, although his father was nearsighted and his mother had phenylketonuria.

a. What is the genotype of the husband?

b. What is the probability that a child of this couple is nearsighted and phenylketonuric?

c. What is the probability that a child of this couple is normal for both traits?

d. What is the probability that a child of this couple has normal vision and PKU?

e. What is the probability that a child of this couple is a girl who is nearsighted and phenylketonuric?

4-13. In pea plants the allele for tall plants, *S*, is dominant to the allele for short plants, *s*. The gene for smooth seeds, *W*, is dominant to the allele for wrinkled seeds, *w*. A cross between two pea plants produces a large number of progeny which fall into two categories, tall, smooth, and tall, wrinkled, in roughly equal frequencies. The students in a genetics class are asked to identify the genotypes of the parental plants and they suggest the following possibilities: (1) *SSww* × *SSww*, (2) *SsWw* × *SSww*, (3) *SsWw* × *ssww*, and (4) *Ssww* × *SSWw*.

a. Which of the possibilities could definitely *not* be the parents? Explain.

b. Of the remaining three possibilities, which is the least likely? Explain.

c. Identify the genotypes of the parental plants.

4-14. In the fruit fly *Drosophila*, the allele for vestigial wings, *vg*, is recessive to the allele for long wings, vg^+. An eye-color gene, carried on another pair of autosomes, has an allele for sepia eye color, *se*, which is recessive to the allele for red eye color, se^+. The progeny arising from a particular mating are as follows: 3/8 red, long; 3/8 red, vestigial; 1/8 sepia, long; and 1/8 sepia, vestigial. Members of a genetics class are asked to identify the genotypes of the parents of these progeny and suggest the following possibilities.

(1) $se^+/se^+vg^+/vg \times se/sevg/vg$
(2) $se^+/sevg^+/vg \times se^+/se^+vg/vg$
(3) $se^+/sevg^+/vg^+ \times se^+/se^+vg/vg$
(4) $se^+/sevg^+/vg \times se^+/sevg/vg$
(5) $se^+/sevg/vg \times se/sevg^+/vg$

Which of these sets gives the genotypes of the parents? Explain how you arrived at your answer.

4-15. A type of blindness known as aniridia occurs in humans and is due to an allele, *A*, which is dominant to the allele for normal vision, *a*. Migraine headaches are due to a dominant allele, *M*, which is dominant to the allele for the absence of migraine headaches, *m*. Each gene is found on a different pair of autosomes. A man with normal vision and without migraines marries a woman with normal vision and migraines. The man's father had aniridia. The woman's father had normal headaches, but never a migraine.

a. Was the man's father homozygous or heterozygous for aniridia?

b. Did the woman's mother suffer from migraines?

c. What is the probability of this couple producing a child who is normal for both traits?

PROBLEM SET

d. What is the probability that a child of this couple is afflicted with both traits?

e. What is the probability that a child of this couple has normal vision and migraines?

4-16. A locus governing feather color in chickens involves codominant alleles: F^B = allele for black, F^W = allele for splashed white. Genotypes for this trait give the following phenotypes: F^BF^B: black; F^BF^W: slate blue or Blue Andalusian; and F^WF^W: white. A second locus, carried on a different pair of autosomes, controls the presence or absence of feathers on the legs. The dominant allele, *F*, produces feathered legs while its recessive counterpart, *f*, produces featherless legs. A black-feathered chicken with feathered legs and derived from pure-breeding stock mates with a white-feathered chicken with featherless legs.

a. What are the genotypes and phenotypes of the F_1 progeny of this mating?

b. Several pairs of F_1 chickens are allowed to mate. Give the phenotypes of their offspring and the frequencies with which each phenotypic class would occur.

4-17. Refer to the basic information presented in problem 4-16. Assume that an F_2 blue chicken with featherless legs of genotype F^BF^Wff is backcrossed with its mother of genotype F^BF^WFf.

a. What is the probability that an offspring is blue with featherless legs?

b. What is the probability that an offspring is white with feathered legs?

4-18. Shape in radishes is controlled by a locus with two codominant alleles: L = long and L' = round. Combinations of these alleles produce the following phenotypes: LL = long, LL' = oval, and $L'L'$ = round. Color in radishes is controlled by another locus on a different pair of chromosomes which has two codominant alleles: R = red and R' = white. Combinations of these alleles produce the following phenotypes: RR = red, RR' = purple, and $R'R'$ = white.

a. In a mating between a long, purple plant and a round, red plant, what is the probability that an offspring will be oval, purple?

b. In a mating between a long, white plant and an oval, purple plant, what is the probability that an offspring will be oval, purple?

CHAPTER

5

Crosses Involving Three or More Independently Assorting Traits

INTRODUCTION

This chapter looks at crosses that simultaneously consider the inheritance of three or more traits, where each trait is determined by genes carried on independently assorting chromosomes. Approaches used in solving such crosses are extensions of the methods used in solving dihybrid crosses, described in Chapter 4. Presented are several generalized mathematical expressions used with crosses involving multihybrid parents to predict the number of kinds of gametes produced by parents, the number of gamete combinations arising from the union of the gametes, the number of genotypes and phenotypes among the progeny, and the proportion of homozygous recessive individuals among the progeny.

TRIHYBRID CROSSES

To illustrate a trihybrid cross, consider the three, independently assorting, autosomal loci controlling the traits of height, seed shape, and pod color in the garden pea, *Pisum sativum*. The gene for each trait has two alleles, with the allele for tall plants, *D*, dominant to the allele for dwarf, *d*; the allele for round seeds, *W*, dominant to the allele for wrinkled seeds, *w*; and the allele for green pods, *Y*, dominant to the allele for yellow color, *y*. Crossing two plants, each heterozygous for these three loci, could be represented as: *DdWwYy* × *DdWwYy*.

Identifying Gametes

Because of the segregation of alleles, each gamete produced by each of these trihybrids will carry one of the alleles found at each of the three loci. In other words, each gamete will carry either *D* or *d*, and *W* or *w*, and *Y* or *y*. All possible combinations of these genes in the gametes are shown in the branch diagram in Table 5-1; because of independent assortment, the eight different combinations generated by each parent will be produced with equal frequencies.

Identifying Progeny Phenotypes

You may have found that setting up a 16-box Punnett square for a dihybrid cross is a little cumbersome. Drawing a Punnett square for a trihybrid cross is even more unwieldy and time-consuming. Since each parent produces eight kinds of gametes, how many boxes are required to show all possible combinations of gametes? ________[1] A quicker and more direct ap-

[1] 8 × 8 = 64 boxes.

TABLE 5-1

Branch diagram showing types of gametes produced by *DdWwYy* parent.

Alleles at			
***d* locus**	***w* locus**	***y* locus**	**Gamete**
D	*W*	*Y*	= *DWY*
		y	= *DWy*
	w	*Y*	= *DwY*
		y	= *Dwy*
d	*W*	*Y*	= *dWY*
		y	= *dWy*
	w	*Y*	= *dwY*
		y	= *dwy*

proach to identifying the progeny phenotypes is to consider the trihybrid cross as if it were three separate monohybrid crosses. As described in Chapter 4, once the phenotypic outcome for each monohybrid cross is identified and the expected probability for each phenotype determined, probabilities for various phenotypic combinations found in the progeny can be calculated using the product law.

We will use this method to identify the phenotypes expected from the *DdWwYy* × *DdWwYy* cross. Considered by itself, the portion of this cross dealing with the plant-height trait, *Dd* × *Dd*, would be expected to yield tall and dwarf progeny with probabilities of 3/4 and 1/4, respectively; the portion dealing with seed-shape, *Ww* × *Ww*, would be expected to yield round and wrinkled progeny with probabilities of 3/4 and 1/4, respectively; and the portion dealing with pod-color, *Yy* × *Yy*, would be expected to yield green and yellow progeny with probabilities of 3/4 and 1/4, respectively.

All possible phenotypic combinations of height, seed shape, and pod color are shown in the branch diagram in Table 5-2. Complete the diagram by using the product law to determine the expected probability for each phenotypic combination.[2] Once you have done this, check your results by making sure that the probabilities of all eight phenotypic classes add up to 1. What is the phenotypic ratio? ____________________

____________________[3]

Identifying Progeny Genotypes

As described in Chapter 4, genotypic combinations and their expected probabilities can also be identified by considering the inheritance at each locus separately. The genotypic ratio expected among the progeny of the *Dd* × *Dd* cross is 1/4 *DD* : 1/2 *Dd* : 1/4 *dd*, and the crosses involving each of the other loci also generate 1/4:1/2:1/4 genotypic ratios. Although not shown here, a branch diagram could be used to show all possible genotypic combinations for these three loci. It would contain a total of 27 branches, giving a total of 27 different genotypic combinations. The product law could then be used to combine probabilities to give the expected probability for each combination.

MULTIHYBRID CROSSES

Some Generalized Mathematical Expressions

Through our consideration of monohybrid, dihybrid, and trihybrid crosses involving independently assorting loci, certain patterns emerge which lead us to some generalizations useful in dealing with the more elaborate multihybrid crosses. Once the number of gene pairs segregating two alleles has been identified for a cross between multihybrid parents, these generalized expressions make it easy to predict (1) the number of kinds of gametes produced by each parent, (2) the number of different combinations of male and female gametes arising through random fertilization, (3) the number of different genotypes among the progeny, (4) the number of different phenotypes among the progeny, and (5) the proportion of homozygous recessive individuals among the progeny. Keep in mind that these generalized mathematical expressions can be used only if each of the loci involved in the multihybrid cross is independently assorting and segregates two different alleles with those alleles separating into different gametes.

The values of these parameters for monohybrid, dihybrid, and trihybrid crosses are recorded in Table 5-3 (verify these, if necessary, by turning back to earlier chapters where you worked through crosses of these types). The column on the far right of the table gives the generalized mathematical expression for each pa-

[2] Probability values, listed from top to bottom, are 27/64, 9/64, 9/64, 3/64, 9/64, 3/64, 3/64, and 1/64. [3] 27:9:9:9:3:3:3:1.

TABLE 5-2

Branch diagram showing phenotypic outcome of *DdWwYy* × *DdWwYy* cross.

Traits of			
Plant height	**Seed shape**	**Pod color**	**Combined phenotype and its probability**
Tall (3/4)	Smooth (3/4)	Green (3/4)	= p(tall, smooth, green) = ___ × ___ × ___ = ___
		Yellow (1/4)	= p(tall, smooth, yellow) = ___ × ___ × ___ = ___
	Wrinkled (1/4)	Green (3/4)	= p(tall, wrinkled, green) = ___ × ___ × ___ = ___
		Yellow (1/4)	= p(tall, wrinkled, yellow) = ___ × ___ × ___ = ___
Dwarf (1/4)	Smooth (3/4)	Green (3/4)	= p(dwarf, smooth, green) = ___ × ___ × ___ = ___
		Yellow (1/4)	= p(dwarf, smooth, yellow) = ___ × ___ × ___ = ___
	Wrinkled (1/4)	Green (3/4)	= p(dwarf, wrinkled, green) = ___ × ___ × ___ = ___
		Yellow (1/4)	= p(dwarf, wrinkled, yellow) ___ × ___ × ___ = ___
			Total of probability values: ___

rameter, expressed in terms of n which represents the number of loci at which both parents considered in a cross are heterozygous. Use the generalized expression to determine the value for each parameter for a cross involving two tetrahybrid parents (for example, *AaBbCcDd* × *AaBbCcDd*) and record them in the appropriate column in Table 5-3.[4]

Using the Product Law to Determine the Probability of Specific Genotypes and Phenotypes

With crosses involving more than three independently assorting traits, Punnett squares are unthinkable and branch diagrams become elaborately complex. Fortunately, however, the probability with which a specific genotypic or phenotypic combination is expected to occur among the progeny of a multihybrid cross can be determined by using the product law. The procedure for doing this is similar to that used above with trihybrid crosses and is illustrated in the solutions to the three problems that follow.

PROBLEM: What proportion of the progeny from the cross *AaBBCcdd* × *AabbccDd* will have genotype *aaBbCcDd*? Assume that the four allelic pairs assort independently.

SOLUTION: Since the four loci assort independently, the inheritance of each trait can be considered separately and the product law can then be used to combine probabilities. A useful approach is to consider this problem as a series of separate questions.

1. In cross *Aa* × *Aa*, what is the probability (p) that an individual will have genotype *aa*? __________[5]
2. In cross *BB* × *bb*, what is the probability that an individual will have genotype *Bb*? __________[6]

[4] With a tetrahybrid cross, n = 4. Kinds of gametes = $2^n = 2^4 = 16$; gamete combinations = $4^n = 4^4 = 256$; progeny genotypes = $3^n = 3^4 = 81$; progeny phenotypes = $2^n = 2^4 = 16$; proportion of homozygous recessive progeny = $1/(4^n) = 1/(4^4) = 1/256$.
[5] p(*aa*) = 1/4. [6] p(*Bb*) = 1 (since all progeny would be *Bb*).

TABLE 5-3

Some generalized mathematical expressions useful with multihybrid crosses.

	Number of Loci at Which Both Parents Are Heterozygous				
	1 (monohybrid)	**2 (dihybrid)**	**3 (trihybrid)**	**4 (tetrahybrid)**	**n (multihybrid)**
Kinds of gametes per parent	2	4	8	______	2^n
Combinations produced by union of male and female gametes	4	16	64	______	4^n
Different genotypes among progeny	3	9	27	______	3^n
Different phenotypes among progeny (assuming complete dominance)	2	4	8	______	2^n
Proportion of homozygous recessives among progeny	1/4	1/16	1/64	______	$1/(4^n)$

3. In cross *Cc* × *cc*, what is the probability that an individual will have genotype *Cc*? ________[7]
4. In cross *dd* × *Dd*, what is the probability that an individual will have genotype *Dd*? ________[8]

The probability of getting genotype *aaBbCcDd* in one individual is given by multiplying the separate probabilities for each genotype:

p(*aaBbCcDd*) = 1/4 × 1 × 1/2 × 1/2 = 1/16. ■

PROBLEM: In the cross described in the preceding problem, *AaBBCcdd* × *AabbccDd*, what is the probability that an individual will express the dominant allele for all four traits (assume that complete dominance operates at each locus)?

SOLUTION: Since these loci assort independently, the inheritance of each trait can be considered by itself. The problem can be answered by breaking it down into a series of questions.

1. In the cross *Aa* × *Aa*, what is the probability that an individual will show the dominant phenotype, that is, have either genotype *AA* or *Aa*? ________[9]
2. In the cross *BB* × *bb*, what is the probability that an individual will show the dominant trait, that is, have genotype *Bb*? ________[10]
3. In the cross *Cc* × *cc*, what is the probability that an individual will show the dominant trait, that is, have genotype *Cc*? ________[11]
4. In the cross *dd* × *Dd*, what is the probability that an individual will show the dominant trait, that is, have genotype *Dd*? ________[12]

Now use the product law to determine the probability of these independent events occurring together:

p(dominance at *A*, *B*, *C*, and *D*) = ________________________________[13] ■

PROBLEM: In the cross *AaBBCcDdEeFF* × *AabbccddEeff*, what is the probability that an individual will have genotype *AaBbccDdeeFf*? Assume that the six gene pairs involved assort independently.

SOLUTION: Since we are dealing with independently inherited loci, we can determine the probability of producing the specified genotype for each of the six pairs and then use the product law to combine these probabilities. Consider each gene pair in a separate cross to get the probability of the particular genotypic outcome specified in the problem.

Aa × *Aa*, p(*Aa*) = 1/2
BB × *bb*, p(*Bb*) = 1
Cc × *cc*, p(*cc*) = 1/2
Dd × *dd*, p(*Dd*) = 1/2
Ee × *Ee*, p(*ee*) = 1/4
FF × *ff*, p(*Ff*) = 1

[7] p(*Cc*) = 1/2. [8] p(*Dd*) = 1/2. [9] p(dominant *A*) = 3/4. [10] p(dominant *B*) = 1 (since all the progeny would show the dominant trait). [11] p(dominant C) = 1/2. [12] p(dominant *D*) = 1/2. [13] 3/4 × 1 × 1/2 × 1/2 = 3/16.

Applying the product law,

$$p(AaBbccDdeeFf) = 1/2 \times 1 \times 1/2 \times 1/2 \times 1/4 \times 1 = 1/32.$$

SUMMARY

Approaches used in solving crosses involving three or more independently assorting traits are extensions of the methods used in solving crosses involving two independently assorting traits. Branch diagrams may be used to identify all of the various types of gametes produced by an individual as well as all the genotypic and phenotypic outcomes of a cross. These diagrams, however, become unwieldy and are increasingly prone to error as the number of loci involved increases. If the problem calls for the expected frequency of one or more specific progeny phenotypic or genotypic combinations, each locus can be considered independently, and the probability of producing the specified genotype or phenotype at that locus can be determined. Then the product law can be used to combine the independent probabilities. When dealing with crosses involving multihybrid parents, generalized mathematical expressions make it easy to predict the expected number of kinds of gametes produced by parents, the number of different combinations of male and female gametes, the number of different genotypes and phenotypes among the progeny, and the proportion of homozygous individuals among the progeny.

PROBLEM SET

5-1. Garden pea plants, *Pisum sativum*, are studied with regard to three independently assorting loci that control pod color (the allele for green, *Y*, is dominant to the allele for yellow, *y*), seed shape (the allele for round, *W*, is dominant to the allele for wrinkled, *w*, and pod shape (the allele for full, *C*, is dominant to the allele for constricted, *c*). Two plants, each heterozygous at all three loci, are crossed.

a. Identify the number of types of gametes produced by each parent and the expected proportion of each.

b. What proportion of the progeny would be expected to be phenotypically dominant for all three traits?

5-2. Refer back to the pea-plant cross described in problem 5-1. What proportion of the progeny would be expected

a. to be homozygous recessive for all three traits?

b. to be heterozygous at all three loci?

c. to have genotype *YYWwcc?*

d. to have genotype *yyWWcc?*

5-3. Add a fourth trait, plant height, to the three pea-plant traits considered in problem 5-1. Height in pea plants is inherited independently of the three seed traits, with the alleles for tall, *D*, dominant to the recessive allele for dwarf, *d*. Two tetrahybrid plants are crossed: *YyWwCcDd* × *YyWwCcDd*.

a. Identify the number of types of gametes produced by each parent and the expected proportion of each.

b. What proportion of progeny would be expected to be homozygous dominant at all four loci?

5-4. Refer to the cross described in problem 5-3. What proportion of the progeny would be expected to have

a. the dominant phenotype at all four loci?

b. genotype *YYwwccDd?*

c. the recessive genotype at all four loci?

5-5. A pea plant homozygous for the dominant allele at each of seven independently assorting loci is crossed with a plant homozygous for the recessive allele at each of these loci.

a. Identify the number of different types of gametes produced by each parent.

b. Identify the genotypes and the phenotypes of the F_1 progeny.

5-6. Assume that two F_1 plants from the cross described in problem 5-5 are crossed to produce an F_2 generation.

a. How many different genotypes would be expected among the F_2 generation?

b. How many different phenotypes would be expected among the F_2 generation?

5-7. Identify the types of gametes produced by an individual with the following genotypes. Assume that the three loci assort independently.

a. *AaBBCC*

b. *AabbCc*

c. *AaBBCcDd*

5-8. Assume that each of the three independently assorting, autosomally based human traits in the following list is controlled by a pair of alleles at a different locus. (1) Tongue rolling, the ability to roll the sides of the tongue together into a U-shape, is produced by the expression of the allele *T* which is dominant to the allele for nonrolling, *t*. (2) Alkaptonuria, a mild disorder

PROBLEM SET

caused by a liver-enzyme deficiency, is produced by the expression of the allele *a*, which is recessive to the allele for normal enzyme production, *A*. (3) Rh blood type, defined by the presence or absence of the Rh antigen on the surface of red blood cells, is controlled by the allele for antigen production, *Rh*, which is dominant to the allele for nonproduction, *rh*. (Note that individuals with the antigen are designated as Rh-positive and those lacking it as Rh-negative.) A man who cannot roll his tongue and who is heterozygous at both the alkaptonuria and Rh loci marries a woman who is heterozygous for the tongue-rolling trait, homozygous for normal production of the liver enzyme, and has Rh-negative blood.

a. What is the probability that a child of theirs will be a tongue roller with normal production of the liver enzyme and Rh-negative blood?

b. What is the probability that a child of theirs will be a nonroller with normal enzyme production and Rh-negative blood?

5-9. The six pairs of alleles involved in the mating *AaBBCcddeeFf* × *aabbccddEeFf* assort independently of each other. What is the probability of producing an offspring

a. with genotype *AaBbccddEeFf*?

b. expressing dominance at the first three loci and recessiveness at the last three loci?

c. expressing dominance at the second and sixth locus and recessiveness at the other four loci?

5-10. a. A partial study of the genetic makeup of an animal indicates that it is heterozygous at 11 different loci. How many different kinds of gametes would this animal be able to produce with regard to these 11 loci?

b. A survey of four additional loci indicates that all are homozygous. When these loci are considered along with the 11 loci that show heterozygosity, how many different kinds of gametes can be made?

5-11. a. A cross between two plants produces progeny which fall into 32 different phenotypic classes. Assuming that complete dominance operates at all of the loci involved and that each locus has two alleles, at how many different loci are the parents heterozygous?

b. If the total number of progeny was 3072, how many of them would be expected to be homozygous recessive at these loci?

5-12. a. A cross between two plants produces progeny which, through subsequent breeding experiments, are shown to fall into 27 genotypic classes. Assuming that complete dominance operates at all of the loci involved and that each locus has two alleles, at how many different loci are the parents heterozygous?

b. How many different phenotypes would these progeny be expected to show?

CHAPTER

6

The Chi-Square Test

INTRODUCTION

As we have shown, the laws of probability can be used to predict the expected outcome of a cross. For example, in crossing fruit flies homozygous for normal wings with flies homozygous for vestigial wings, where the allele for normal wings is dominant to the allele for vestigial, the entire F_1 generation would be expected to have normal wings. The predicted F_2 would have normal- and vestigial-winged flies in a phenotypic ratio of 3 to 1. With a total of, say, 200 F_2 flies, 150 would be expected with normal wings and 50 with vestigial wings. Rarely, however, does the outcome of a genetic cross perfectly fit the expected ratio; an array of factors, collectively referred to as *chance,* can produce a discrepancy between the expected and observed results. Assume that the 200 F_2 progeny consisted of normal- and vestigial-winged flies with frequencies of 160 and 40, respectively. These results are close to the predicted 3:1 ratio and most everyone would be willing to attribute the difference to chance. But what if the figures had been 170 and 30, or 180 and 20? Can these larger discrepancies also be attributed to chance? At what point is it appropriate to invoke something other than chance—observer error or a faulty hypothesis, perhaps—to explain a discrepancy?

CALCULATION OF THE CHI-SQUARE VALUE

The **chi-square (χ^2) test** (*chi,* written as χ, is the twenty-second letter of the Greek alphabet) is a statistical procedure which quantifies the difference between observed and expected results and determines the *likelihood* that the discrepancy is due to chance. As the details of the test are considered, keep in mind what we are trying to do: make a judgment about whether a set of data is reasonably close to the predicted outcome. The odds are high that a small discrepancy can be reasonably explained by chance. As the discrepancy gets bigger, the likelihood that it is due to chance decreases, and the odds increase that something other than chance is responsible.

Assume we wish to evaluate the discrepancy from expected values seen with our F_2 outcome of 160 wild and 40 vestigial flies. The approach to using the χ^2 test generally involves setting up a table like that shown in Table 6-1. Complete this table as we go through the steps for carrying out the test.

1. List the various categories, or classes, in which the progeny occur. Here we have two phenotypic classes, normal wings and vestigial wings, and these are listed in the table. (Note that in other

TABLE 6-1

Chi-square test.

Class	Number observed	Number expected	Deviation (Obs. − Exp.)	Deviation²	Deviation²/Exp.
Normal wings	______	______	______	______	______
Vestigial wings	______	______	______	______	______
				Chi-square = χ^2 =	______

cases, the categories we are concerned with might be genotypic classes.)

2. To the right of the classes, list the number of organisms observed in each class. There are 160 normal-winged flies and 40 vestigial-winged flies.
3. Next to each observed number, list the number of flies expected in each class. The F_2 flies were expected in a ratio of 3 normal to 1 vestigial. The expected numbers are the actual number of flies out of 200 that would be found in each class if the progeny had occurred in a perfect 3:1 ratio. These expected numbers can be figured out as follows. The 3:1 ratio involves 3 + 1 = 4 parts, and dividing 4 into the total number of F_2 flies, 200, indicates that 50 flies would be expected in each part. The expected numbers are: 50 × 1 = 50 vestigial- and 50 × 3 = 150 normal-winged flies.
4. Determine the deviation for each class by subtracting the expected number from the observed number. The deviations for the normal- and vestigial-winged flies are +10 and −10, respectively.
5. Square each deviation. This serves to eliminate the minus sign associated with negative deviations. (Remember that multiplication of a negative number by a negative number gives a positive number). The deviation squared for each of our classes is +100.
6. Divide each squared deviation by the expected number for its class. For the normal class, this is 100/150 = 0.667 and for the vestigial class, 100/50 = 2.0.
7. Sum the squared-deviation-divided-by-number-expected for each class to give the chi-square value. For our example, $\chi^2 = 0.667 + 2.0 = 2.667$.

Check your completed version of Table 6-1 for agreement with Table 6-2. The procedure just carried out is summarized by the formula below, where the symbol Σ means "take the sum of."

$$\chi^2 = \sum \frac{(\text{Obs.} - \text{Exp.})^2}{\text{Exp.}}$$

SENSITIVITY OF THE CHI-SQUARE VALUE TO DEVIATION SIZE

The size of the χ^2 value reflects the magnitude of the discrepancy between observed and expected results. To demonstrate this, we will consider three other possible outcomes for our F_2 generation that show a progressive increase in the number of normal flies and a decrease in the number of vestigial flies while holding the total number of flies constant at 200. The first outcome consists of 165 normal and 35 vestigial, the second, 170 normal and 30 vestigial, and the third, 175 normal and 25 vestigial. These outcomes and the χ^2 calculation for each are given in Table 6-3.

TABLE 6-2

Chi-square test.

Class	Number observed	Number expected	Deviation (Obs. − Exp.)	Deviation²	Deviation²/Exp.
Normal wings	160	150	+10	+100	100/150 = 0.667
Vestigial wings	40	50	−10	+100	100/50 = 2.0
				Chi-square value = χ^2 =	2.667

TABLE 6-3

A comparison of chi-square tests with different deviation sizes.

Outcome	Class	Obs.	Exp.	*Dev.*	Dev.^{2}	Dev.^{2}/Exp.
1	Normal	165	150	**+15**	225	225/150 = 1.50
	Vestigial	35	50	**−15**	225	225/50 = 4.50
						χ^2 = **6.00**
2	Normal	170	150	**+20**	400	400/150 = 2.667
	Vestigial	30	50	**−20**	400	400/50 = 8.000
						χ^2 = **10.667**
3	Normal	175	150	**+25**	625	625/150 = 4.167
	Vestigial	25	50	**−25**	625	625/50 = 12.500
						χ^2 = **16.667**

Examine Table 6-3 and compare the deviations for each of the three outcomes. What trend do you notice? ______[1] Compare the χ^2 values for each of these outcomes. What trend do you notice? ______[2] What conclusion can be drawn about the relationship between the deviation and the size of the χ^2 value? ______[3]

SENSITIVITY OF THE CHI-SQUARE VALUE TO SAMPLE SIZE

The χ^2 value is also influenced by the total number of progeny, or sample size, produced in a genetics cross. To demonstrate this, we consider three different outcomes for our F_2 generation that involve progeny totals of 160, 120, and 80, respectively. These outcomes and the χ^2 calculations for each are given in Table 6-4.

Compare the deviations for each of the three outcomes. What do you notice? ______[4] Compare the total number of progeny for each of the three outcomes. What trend is apparent? ______[5] Compare the χ^2 values for each of these outcomes. What trend do you notice? ______[6] What conclusion can be drawn about the relationship between the sample size and the size of the χ^2 value? ______[7]

The fact that the χ^2 value gets larger as the sample size decreases reflects the fact that, as populations get smaller relative to the infinitely large ideal population, they become more likely to show chance deviations from expected numbers. To summarize, the magnitude of the χ^2 value is influenced by both the size of the deviation and the size of the population being examined.

DEGREES OF FREEDOM

Before the χ^2 value can be interpreted, the degrees of freedom for the set of data must be determined. **Degrees of freedom** are defined as the number of classes that can vary independently, and in most cases this is 1 less than the number of classes in which progeny are expected to occur. (An exception to this definition is considered in Chapter 31 for chi-square tests that involve gene frequencies.) For the example under consideration, progeny are expected to fall into two classes, normal and vestigial wings. What are the degrees of freedom in this case? ______[8]

A brief explanation for why the degrees of freedom are 1 less than the number of classes in which progeny are expected is as follows. Each class obviously contributes to the total number of progeny aris-

TABLE 6-4

A comparison of chi-square tests with different sample sizes.

Outcome	*Total progeny*	Class	Obs.	Exp.	Dev.	Dev.^{2}	Dev.^{2}/Exp.
1	**160**	Normal	130	120	+10	100	100/120 = 0.833
		Vestigial	30	40	−10	100	100/40 = 2.500
							χ^2 = **3.333**
2	**120**	Normal	100	90	+10	100	100/90 = 1.111
		Vestigial	20	30	−10	100	100/30 = 3.333
							χ^2 = **4.444**
3	**80**	Normal	70	60	+10	100	100/60 = 1.667
		Vestigial	10	20	−10	100	100/20 = 5.000
							χ^2 = **6.667**

[1] It increases. [2] It increases. [3] The χ^2 value is directly related to the size of the deviation. [4] They are the same.
[5] Each example shows a successive reduction of 40. [6] It increases. [7] The χ^2 value is inversely related to the size of the sample.
[8] 1 less than two classes gives one degree of freedom.

ing from a cross. From a statistical standpoint, the number of individuals in either of the two classes, normal or vestigial, can vary. However, once the number of individuals in one of the two classes is identified, the number in the other class is immediately and automatically set since all the remaining members of the total must belong to that class. The class which has its frequency automatically set lacks the freedom to vary. Since only one of the two classes in our example has the freedom to vary, there is a single degree of freedom. Other genetics crosses may involve three, four, or more classes in which progeny would be expected to occur. If the progeny are expected to fall into four classes, for example, three would have the freedom to vary before the size of the fourth class would be set, giving three degrees of freedom.

THE CHI-SQUARE TABLE

Once the χ^2 value and the degrees of freedom have been determined, the deviation reflected in the χ^2 value can be interpreted using a table of chi-square values and their probabilities (Table 6-5) which indicates what the odds are that various deviations are due to chance. Note that the column on the far left is labeled "Degrees of freedom." Making up the body of this table to the right of each of the degrees of freedom is a series of χ^2 values. What happens to these χ^2 values in moving across the table? ____[9] Across the top of the table are a series of probability values, or p-values, which are expressed in the form of decimals. These decimal values can be readily converted to percents, fractions, or expressions of odds. For example, a p-value of 0.10 is the equivalent of 10%, or 10/100, or 1 out of 10. What happens to the probability values as you move across the table to the right? ____[10]

The following procedure describes how to use Table 6-5. Assume we wish to determine the probability value for a χ^2 value of 1.64 with one degree of freedom.

1. Begin by finding the degrees of freedom in the column on the far left.
2. Read across the table opposite the degrees of freedom until you find the χ^2 value. If the χ^2 value is

TABLE 6-5

Probabilities for different chi-square values.

Degrees of freedom	Probabilities									
	0.95	**0.90**	**0.70**	**0.50**	**0.30**	**0.20**	**0.10**	**0.05**	**0.01**	**0.001**
1	0.004	0.016	0.15	0.46	1.07	1.64	2.71	3.84	6.64	10.83
2	0.10	0.21	0.71	1.39	2.41	3.22	4.61	5.99	9.21	13.82
3	0.35	0.58	1.42	2.37	3.67	4.64	6.25	7.82	11.35	16.27
4	0.71	1.06	2.20	3.36	4.88	5.99	7.78	9.49	13.28	18.47
5	1.15	1.61	3.00	4.35	6.06	7.29	9.24	11.07	15.09	20.52
6	1.64	2.20	3.83	5.35	7.23	8.56	10.65	12.59	16.81	22.46
7	2.17	2.83	4.67	6.35	8.38	9.80	12.02	14.07	18.48	24.32
8	2.73	3.49	5.53	7.34	9.52	11.03	13.36	15.51	20.09	26.13
9	3.33	4.17	6.39	8.34	10.66	12.24	14.68	16.92	21.67	27.88
10	3.94	4.87	7.27	9.34	11.78	13.44	15.99	18.31	23.21	29.59
11	4.58	5.58	8.15	10.34	12.90	14.63	17.28	19.68	24.73	31.26
12	5.23	6.30	9.03	11.34	14.01	15.81	18.55	21.03	26.22	32.91
13	5.89	7.04	9.93	12.34	15.12	16.99	19.81	22.36	27.69	34.53
14	6.57	7.79	10.82	13.34	16.22	18.15	21.06	23.69	29.14	36.12
15	7.26	8.55	11.72	14.34	17.32	19.31	22.31	25.00	30.58	37.70
20	10.85	12.44	16.27	19.34	22.78	25.04	28.41	31.41	37.57	45.32
25	14.61	16.47	20.87	24.34	28.17	30.68	34.38	37.65	44.31	52.62
30	18.49	20.60	25.51	29.34	33.53	36.25	40.26	43.77	50.89	59.70
50	34.76	37.69	44.31	49.34	54.72	58.16	63.17	67.51	76.15	86.66

← Accept | Reject →
at 0.05 level

From Table IV of Fisher and Yates, *Statistical Tables for Biological, Agricultural and Medical Research,* published by Longman Group UK Ltd., London. Reprinted by permission.

[9] They get larger. [10] They decrease.

not listed, as often happens, locate the two values between which it falls.

3. Read up from the χ^2 value or the range of values to find the probability value or range of probability values. This p-value is the probability that chance is operating to produce the discrepancy.

The p-value for the χ^2 of 1.64 is 0.20, which could also be expressed as 20%, or 20/100, or 5 out of 100.

Note carefully what the probability value indicates. A probability value of 20% tells us that chance alone would be expected to produce 20% of the time a deviation at least as large as the one we observed. In other words, we could expect in 20 out of 100 cases a χ^2 value as large or larger than the one we have. In the other 80 out of 100 cases, we could expect the χ^2 value to be smaller than the one we observed.

As another example, determine the probability value for a χ^2 value of 2.17 with one degree of freedom. Can the p-value for this χ^2 value be directly determined from the table? ________[11] Within what p-value range does the probability lie? ____________ ________________________________[12] What is the estimated p-value? ________[13]

PASSING JUDGMENT ON THE DEVIATION: SIGNIFICANCE LEVEL

So far, we have learned how to calculate the χ^2 value, to figure out the degrees of freedom for a set of data, and to look up the χ^2 value in Table 6-5 to get a p-value. The final step involves the evaluation of the p-value. The p-value makes it possible to pass judgment on the deviation. Specifically, it allows us to decide whether the deviation is too large to reasonably be attributed to chance or small enough to be considered due to chance. In effect, the p-value allows us to decide whether the observed data support or do not support the underlying hypothesis. A small deviation will give a relatively high p-value. This indicates relatively high odds that the deviation is due to chance and that the data support the basic hypothesis under consideration. A large deviation, in contrast, will give a relatively small p-value. This indicates relatively low odds that the deviation is due to chance and low odds that the data support the hypothesis.

The terms "high" and "low" when applied to probability levels, however, are ambiguous, and a line must be drawn somewhere to distinguish between acceptable and unacceptable deviations. Biologists generally draw the line at the p = 0.05 level. With this cutoff, any χ^2 value with a probability value greater than 0.05, or 5%, (for example, 0.1, 0.4) is considered to reflect a deviation small enough to be reasonably attributed to chance. Put another way, any χ^2 value smaller than (to the left of) that found in the 0.05 p-value column indicates a deviation small enough to be attributed to chance. Such a deviation is labeled **nonsignificant** and the data are considered to support the hypothesis.

Any χ^2 value with a probability value of 0.05 (5%) or less (for example, 0.03 or 0.001) is considered to reflect a deviation large enough to regard with suspicion and to cause consideration of some factor other than chance to explain the discrepancy. Put another way, any χ^2 value equal to or larger than (in or to the right of) the p = 0.05 column indicates a deviation large enough to be attributed to some factor other than chance. These deviations are termed **significant,** and when they do occur, the data are considered not to support the hypothesis. Occasionally a biologist may elect to use a significance level of 0.01 (1%) rather than 0.05; the nature of the hypothesis being tested as well as the consequences arising from a mistaken decision are factors considered in selecting the level of significance.

The first example in this chapter considered an F_2 of 200 fruit flies, 160 with normal wings and 40 with vestigial wings, with a χ^2 value of 2.666 and one degree of freedom. Is it reasonable to attribute this deviation to chance? ________[14] Explain. ________________ ________________________________[15]

Table 6-6 lists a series of χ^2 values and degrees of freedom. Look up each chi-square value opposite the appropriate degrees of freedom in Table 6-5 and determine (by estimation, if necessary) the corresponding p-value. Then interpret each p-value by indicating whether it reflects a deviation that is significant or nonsignificant and whether the data from which the χ^2 value is derived support the underlying hypothesis. In each case, assume a significance level of 0.05.

USING THE CHI-SQUARE TEST WITH MORE THAN TWO CLASSES OF PROGENY

The χ^2 test can be used to evaluate the outcome of crosses involving more than two classes of progeny.

[11] This particular χ^2 value is not listed, although it obviously lies between 1.64 and 2.71. [12] The p-value falls between 0.20 and 0.10. [13] Since 2.17 is roughly at the midpoint of the 1.64–2.71 range, the p-value falls approximately halfway between 0.20 and 0.10, or at about 0.15. [14] Yes. [15] Since its p-value of 0.10 is greater than p = 0.05, we know that the deviation reflected in the χ^2 value of 2.666 is small enough to be reasonably attributable to chance.

TABLE 6-6

χ^2	Degrees of freedom	p-value	Deviation (significant or nonsignificant)	Hypothesis (support or nonsupport)
6.64	1			[16]
6.64	4			[17]
3.84	1			[18]
3.33	2			[19]
0.05	1			[20]
12.67	3			[21]

For example, consider a two-trait cross in pea plants where a pure-breeding stock with yellow and smooth seeds is crossed with a pure-breeding stock with green and wrinkled seeds. It is hypothesized that the allele for yellow color is dominant to that for green, that the allele for smooth seed coats is dominant to that for wrinkled, and that the alleles governing these two traits assort independently. Two members of the F_1 generation are crossed to produce an F_2 generation which exhibits four phenotypes. On the basis of the above hypothesis, these phenotypes would be expected to occur in a 9:3:3:1 ratio. The actual number of progeny showing each phenotype is tabulated as follows.

Phenotype	Number observed	Expected proportion
Yellow, smooth	330	9/16
Yellow, wrinkled	95	3/16
Green, smooth	108	3/16
Green, wrinkled	27	1/16
Total:	560	

Now we are ready to carry out a χ^2 test on this outcome. First, in the space provided, determine the number of progeny out of 560 that would be expected in each of the four phenotypic classes.[22]

Calculate the χ^2 value for these data by completing the following table.

Class	Obs.	Exp.	Dev.	Dev.^{2}	Dev.^{2}/Exp.
Yellow, smooth	330	315	______	______	______[23]
Yellow, wrinkled	95	105	______	______	______[24]
Green, smooth	108	105	______	______	______[25]
Green, wrinkled	27	35	______	______	______[26]
				χ^2 =	______[27]

How many degrees of freedom are there for these data? ______[28] What is the probability value for the χ^2 value of 3.581? ______[29] Can the deviation reflected in this χ^2 value be reasonably attributed to chance? ______[30] Is this deviation considered sig-

[16] 0.01, significant, nonsupport. [17] Between 0.20 and 0.10, nonsignificant, support. [18] Borderline at 0.05, significant, nonsupport. [19] Just under 0.20, nonsignificant, support. [20] Between 0.90 and 0.70, nonsignificant, support. [21] Less than 0.01, significant, nonsupport. [22] Since there are 9 + 3 + 3 + 1 = 16 parts to the 9:3:3:1 ratio and 560 progeny, each part involves 560/16 = 35 plants. The expected number of plants in each class is as follows: yellow, smooth, 9 × 35 = 315; yellow, wrinkled: 3 × 35 = 105; green, smooth, 3 × 35 = 105; green, wrinkled: 1 × 35 = 35. [23] +15, 225, 225/315 = 0.714. [24] −10, 100, 100/105 = 0.952. [25] +3, 9, 9/105 = 0.086. [26] −8, 64, 64/35 = 1.829. [27] 3.581. [28] 4 classes − 1 = 3 degrees of freedom. [29] Opposite three degrees of freedom in Table 6-5, 3.518 lies between χ^2 statistics with probability values of 0.50 and 0.30. [30] Since this probability value is greater than 0.05, the odds are high enough to reasonably attribute the deviation to chance.

nificant or nonsignificant? ________[31] Do these data support the hypothesis which predicts a phenotypic ratio of 9:3:3:1 in the F_2? ________[32]

PRECAUTIONS FOR USING THE CHI-SQUARE TEST

The χ^2 test is a very useful statistical tool but only if it is used and interpreted correctly. Keep the following precautions in mind as you use this test.

Actual Data Counts Must Be Used

The χ^2 test can only be used with actual data counts. Percents or decimal expressions may not be used.

Theoretical Ratios Must Be Used

The ratios that are tested in the χ^2 test are those obtained from theoretical expectations based upon the hypothesis and not those derived from the raw data.

Up to a Point, Corrections Can be Made for Small Samples

Although the χ^2 test is sensitive to sample size, there is a lower limit on the class size that can be tested. The reliability of any conclusion, of course, depends upon the amount of information used in drawing the conclusion. The greater the number of organisms in each category, the better the chance of an accurate conclusion. As a rule, the χ^2 test should not be used with any set of results where the expected number of individuals in *any* category is less than 5. In situations where the expected number of individuals in *any* class falls between 5 and 10, the **Yates correction factor** is usually used to adjust for the small sample size. This correction is made by subtracting 1/2 from the absolute value of the difference between the observed and expected numbers for *each* class. This is expressed symbolically as $|\text{Obs.} - \text{Exp.}| - 1/2$, where the vertical lines designate the absolute or positive value of the quantity they enclose. For example, if the observed number for a class is 6 and the expected number is 9, then Obs. − Exp. = 6 − 9 = −3. The Yates correction factor would adjust this difference to $|-3| - 1/2 = 2\frac{1}{2}$. In situations where this correction is appropriate, the expression for χ^2 becomes

$$\chi^2_{\text{corrected}} = \sum \frac{(|\text{Obs.} - \text{Exp.}| - 1/2)^2}{\text{Exp.}}.$$

[31] Nonsignificant. [32] Yes.

To improve accuracy, the Yates correction factor may also be used whenever the degrees of freedom equal 1, as occurs, for example, with two classes. The use of the Yates correction factor as well as the avoidance of the χ^2 test when a class in which progeny are expected has fewer than five individuals will help prevent inaccurate interpretation of results.

The remaining two points have been covered before, but since they can head off confusion, they warrant repeating.

The Test Cannot Pass Absolute Judgment on a Deviation

The χ^2 test cannot tell us with absolute certainty that a deviation is definitely due or not due to chance. The test only gives the probability that a deviation is due to chance. If the p-value is greater than 0.05, biologists usually feel the odds are good that chance is causing the deviation; that is, the deviation is small enough to be reasonably attributed it to chance. However, sometimes deviations of this size can be due to factors other than chance, but the likelihood of this happening is such that biologists are willing to take the risk. In contrast, if the p-value is at the 0.05 level or below, the odds that chance is responsible for the deviation are low; that is, the deviation is too large to be reasonably attributed to chance. Rarely, but once in a while, chance will produce a deviation that is this large, but the likelihood of this happening is so low, 1 out of 20 or less, that biologists are not willing to take the risk. Most of the time, very large deviations are due to factors other than chance. The key point is this: after using the χ^2 test, all that can be said is that the odds are high enough to reasonably attribute the deviation to chance or the odds are not high enough to reasonably attribute the deviation to chance.

The Test Cannot Pass Absolute Judgment on a Hypothesis

The χ^2 test cannot tell us whether a hypothesis is right or wrong. It can only tell us whether the data that we have collected support or do not support the hypothesis. If the hypothesis is not supported by the data, it does not necessarily mean that the hypothesis is faulty. It does mean that some factor other than chance should be considered to explain the discrepancy. As examples, the number of progeny may have been unusually small; maybe the experimenter had difficulty sorting out the phenotypes; perhaps the data were incorrectly recorded; or possibly environmental conditions fluctuated widely during the experiment and may have been more harmful to one phenotypic class than to another. Often it is better not to reject the hypothesis at once, but to repeat the experiment, paying careful attention to the elimination of factors other than chance that might distort the outcome.

SUMMARY

The key steps involved in applying the χ^2 test are as follows.

1. Formulate or identify the hypothesis being tested.
2. Inspect the results and figure out how many individuals fall into each class or category. These represent the observed numbers.
3. Determine the theoretical ratio in which progeny would be expected to occur, based on the hypothesis.
4. Based on the theoretical ratio, figure out how many of the total number of individuals produced would be expected to occur in each class. These represent the expected numbers.
5. Set up a chart to calculate the χ^2 value.
   - **a.** List each class into which progeny fall.
   - **b.** List the observed (Obs.) number for each class.
   - **c.** List the expected (Exp.) number for each class.
   - **d.** Calculate the deviation (Dev.) for each class by subtracting the expected number from the observed number (Dev. = Obs. − Exp.).
   - **e.** Square each deviation (Dev.^{2}).
   - **f.** Divide each squared deviation by its expected number (Dev.^{2}/Exp.).
   - **g.** Add up the Dev.^{2}/Exp. for each class. This total is the χ^2 value.
6. Figure out the degrees of freedom, which are 1 less than the number of classes or categories into which progeny fall.
7. Use Table 6-5 to determine the probability value, or p-value. Opposite the appropriate degrees of freedom, find the χ^2 value in the table, or if it is not there, find the range of values between which it falls. Above the χ^2 value or range of values you will find the probability value or range of values.
8. Interpret the probability value. If it is larger than 0.05, biologists consider the odds pretty good that the deviation is due to chance: classify the deviation as nonsignificant. If the p-value is equal to or smaller than 0.05, then the odds are slim that the deviation is due to chance, and it is considered significant.
9. Relate the probability to the hypothesis. If the p-value is greater than 0.05, the deviation is considered nonsignificant and the results support the hypothesis. If the p-value is equal to or less than 0.05, the deviation is considered significant and the results do not support the hypothesis.

PROBLEM SET

6-1. The allele for the albino condition in corn plants is recessive to the allele for normal pigmentation. A cross between two plants heterozygous at this locus produces 126 normal and 66 albino plants.
a. How many plants are expected in each phenotypic class?
b. Determine the χ^2 value for these data.

6-2. Refer to the information given in problem 6-1. In general, what would be the consequence for the χ^2 value if
a. the discrepancy between the observed and expected numbers had been smaller—for example, with 136 normal and 56 albino plants?
b. the sample size had been smaller—for example, with a total of 156 plants, but with the deviations remaining unchanged at $+18$ and -18?

6-3. Data from a series of genetics crosses give the following χ^2 values. For each, determine the degrees of freedom and, by consulting Table 6-5, the probability value for the χ^2 value. Indicate whether the deviation involved is significant or nonsignificant at the 0.05 level of significance and whether there is support for the hypothesis assumed for each cross.
a. Two classes of progeny, $\chi^2 = 3.020$.
b. Four classes of progeny, $\chi^2 = 10.360$.
c. Three classes of progeny, $\chi^2 = 1.555$.

6-4. Refer back to the information given in problem 6-3. Assume the significance level is changed from 0.05 to 0.01. How does this influence the interpretation of the deviations involved in each set of data?

6-5. Snapdragons show incomplete dominance with regard to the trait of flower color. Both the red and white alleles found in a heterozygous individual are expressed to produce a pink-flowered phenotype that is intermediate between the red and white phenotypes. A cross between red-flowered and white-flowered snapdragons produced an all pink-flowered F_1. Crossing members of the F_1 gave an F_2 of 42 red-, 110 pink-, and 48 white-flowered plants.
a. What is the theoretical ratio in which progeny would be expected to occur in this F_2?
b. How many of the 200 F_2 individuals would be expected in each of the three F_2 classes?
c. Calculate the χ^2 value for the F_2 data.
d. How many degrees of freedom are there for this set of data?
e. Using Table 6-5, determine the probability level for the χ^2 value.
f. What does the probability value indicate about the deviation between observed and expected numbers? Are the odds such that the deviation can most likely be attributed to chance?
g. Are the F_2 data in support of the basic hypothesis assumed to underlie this cross?

6-6. It is hypothesized that in the human species, the number of male and female births are equal. During a four-day period, 40 babies were born in a hospital. Of these, 13 were boys and 27 were girls.
a. What is the expected number of males and females in a sample of 40 births?
b. What is the χ^2 value for the hospital data?
c. Can the difference between observed and expected numbers reasonably be attributed to chance? Explain why or why not.
d. Do these results support the theoretical ratio of one male birth to one female birth? Explain.

PROBLEM SET

e. Is your belief in the existence of a 1:1 sex ratio in human births altered by these results? Explain.

6-7. In carrying out *Drosophila* crosses in the laboratory, a student hypothesized that the allele for the wild-type red eyes is dominant to the allele for sepia eye color, that the allele for wild-type long wings is dominant to that for short wings, and that the alleles governing these two traits assort independently of each other. The mating of a sepia-eyed, long-winged fly with a red-eyed, short-winged fly produced an F_1 with red eyes and long wings. The mating of F_1 flies produced an F_2 generation of 640 progeny: 344 red, long; 134 red, short; 128 sepia, long; and 34 sepia, short.

a. What is the expected ratio and number of flies for each of the four F_2 phenotypic classes?
b. Determine the χ^2 value for the F_2 outcome.
c. What is the probability value for this χ^2 value?
d. Can the difference between observed and expected numbers reasonably be attributed to chance? Why or why not?
e. Do these data support the theoretical ratio?

6-8. A researcher is attempting to determine the inheritance pattern exhibited by the alleles determining shell color in a species of freshwater snail. The trait is known to be controlled by alleles at a single autosomal locus, and the researcher hypothesizes that there are two alleles, one for brown which is completely dominant to the allele for yellow. A mating between a brown-shelled snail and a yellow-shelled snail, both from pure-breeding lines, produces progeny that are brown shelled. The mating of two F_1 snails produces an F_2 consisting of 54% with yellow shells and 46% with brown shells. The researcher wishes to carry out a χ^2 test on the F_2 outcome and decides to compare his observed values of 54% and 46% with a ratio of 1:1 which is close to the observed results. The resultant χ^2 value is small and, because of that, the researcher records in his notebook that "the deviation is definitely due to chance" and that "the hypothesis must be correct."

a. The researcher has made a number of errors in using the χ^2 test. Identify these errors.
b. What additional information would be required to carry out an accurate χ^2 test on these data?

6-9. Refer to the information in problem 6-8. Assume that the F_2 consisted of 27 yellow and 23 brown snails.

a. Based on the 1:1 F_2 ratio predicted by the researcher, calculate and interpret the χ^2 value for these data.
b. Based on the correct 3:1 F_2 ratio, calculate and interpret the χ^2 value for these data.
c. What would you suggest that the researcher do next?

6-10. Refer to the information given in problems 6-8 and 6-9. Assume the researcher repeated the mating with several pairs of parents to produce a larger F_2 generation. Due to adverse laboratory conditions, most of the eggs died and of the 44 F_2 progeny, 35 had brown shells and nine had yellow shells.

a. Based on the correct 3:1 F_2 ratio, calculate the χ^2 value for this data.
b. Do the data support the hypothesis of 3:1 ratio for the F_2 progeny of a Mendelian cross with complete dominance?
c. In light of the outcome of this χ^2 test, what would you suggest that the researcher do next?

CHAPTER

7

More on Probability: Unordered Events and Binomial and Multinomial Distributions

INTRODUCTION

Familiarity with the basic information on probability that is presented in Chapter 3 is essential to mastering the material presented in this chapter. A brief summary of some of the key points covered there is given in the two paragraphs that follow.

Independent Events and the Product Law

Events which do *not* influence each other are said to be *independent:* the occurrence (or nonoccurrence) of one in no way influences the occurrence of the other event or events. For example, the sex of a child at one birth is independent of the sex of a child at another birth. The probability of two or more independent events occurring together is given by the *product* of their separate probabilities. (This statement is known as the *product law*). For example, with the probability of a boy equal to 1/2 and the probability of a girl equal to 1/2, the probability of having a boy and a girl in that order is $1/2 \times 1/2 = 1/4$ and, similarly, the probability of having a girl and a boy in that order is $1/2 \times 1/2 = 1/4$.

Mutually Exclusive Events and the Law of the Sum

Events which *do* influence each other are said to be *mutually exclusive:* the occurrence of one *prevents* the occurrence of the other event or events. For example, if the first child born to a couple is a boy, it cannot be a girl, and vice versa. For a group of mutually exclusive events, the probability of either one or another of the group occurring at a particular time is given by the *sum* of their separate probabilities. (This statement is known as the *law of the sum*). For example, if a couple has two children, a boy and a girl, there are two mutually exclusive ways in which the family could arise: the boy could be born before the girl or the girl could be born before the boy. The probability of the occurrence of each sequence is 1/4, and the probability of producing either combination of a boy and a girl, without regard to order, is given by the sum of the probability values of the two sequences: $1/4 + 1/4 = 1/2$. Refer back to the appropriate parts of Chapter 3 if you need more information on independent or mutually exclusive events or the ways in which the probabilities of events of each type are combined.

This chapter focuses on mutually exclusive events and dwells on a rather obvious point: a specific combination of progeny arising from a particular mating can generally be produced in a number of different, mutually exclusive sequences. First, we will consider combinations involving two mutually exclusive events and then those involving three or more events.

COMBINATIONS INVOLVING TWO MUTUALLY EXCLUSIVE EVENTS

The two mutually exclusive events we will use to illustrate our discussion are the two sexes which are possible for a child.

An Example: Different Combinations of Three Children

Consider a family of three children made up of two girls and one boy. If a particular birth order is specified, such as a girl followed by a boy and then a girl, the probability of such a family is given by the *product* of the probabilities of each separate and *independent* birth. Since the probability of producing either a male or female child is the same (1/2), the probability of producing two girls and one boy in this particular sequence is $1/2 \times 1/2 \times 1/2 = 1/8$. (In reality, the odds slightly favor the male with about 106 boys being born to every 100 girls—but for our purposes here, we will consider them to be equal.) This particular sequence of girl, boy, girl is just one of the three possible sequences that could result in the *combination* of two girls and one boy. The other two are the boy followed by two girls, and two girls followed by the boy. If we are interested in the probability of the combination of two girls and one boy occurring, without regard to order in which the children are born, we must take into account all possible sequences which give us this combination. With each of the three possible mutually exclusive sequences having a probability of 1/8, the probability of producing this combination, without regard to order, is given by adding the probabilities of the three specific sequences: $1/8 + 1/8 + 1/8 = 3/8$.

The combination of two girls and one boy is obviously not the only combination of sexes that could arise in a family of three children. Another combination would be two boys and one girl. This combination, like the one just considered, could arise in three different sequences: boy, boy, girl; boy, girl, boy; and girl, boy, boy. Since the probability of each sequence is 1/8, the probability of the two-boy, one-girl combination is $1/8 + 1/8 + 1/8 = 3/8$. Two other combinations of sexes are possible in a family of three: three boys or three girls. Since each of these combinations arises through a single sequence, each would have a probability of 1/8.

These four combinations and the sequences that make up each are summarized in Table 7-1 using B and G to represent a boy and girl, respectively. Note that the probability of each specific sequence and of each combination is given in numeric terms as well as in generalized mathematical terms, where p and q are used to symbolize the probability of producing a boy or a girl, respectively. Note that the probabilities of all the possible combinations, expressed either in numerical value or in generalized expressions, add up to 1:

$$1/8 + 3/8 + 3/8 + 1/8 = 1$$
$$p^3 + 3p^2q + 3pq^2 + q^3 = 1.$$

Another Example: All Possible Combinations of Four Children

Let us look at the various sex combinations that could arise in a family of four children and the probabilities with which they would be expected to occur. The combinations of four boys or four girls would each occur

TABLE 7-1

All possible combinations of birth orders in a family of three children.

Combination	Sequence of births	Numeric probability of sequence	Numeric probability of combination	Generalized expression for probability of sequence	Generalized expression for probability of combination
3 boys	B, B, B	$1/2 \times 1/2 \times 1/2 = 1/8$	1/8	p^3	p^3
2 boys, 1 girl	B, B, G	$1/2 \times 1/2 \times 1/2 = 1/8$	3/8	p^2q	$3p^2q$
	B, G, B	$1/2 \times 1/2 \times 1/2 = 1/8$		p^2q	
	G, B, B	$1/2 \times 1/2 \times 1/2 = 1/8$		p^2q	
2 girls, 1 boy	G, G, B	$1/2 \times 1/2 \times 1/2 = 1/8$	3/8	pq^2	$3pq^2$
	G, B, G	$1/2 \times 1/2 \times 1/2 = 1/8$		pq^2	
	B, G, G	$1/2 \times 1/2 \times 1/2 = 1/8$		pq^2	
3 girls	G, G, G	$1/2 \times 1/2 \times 1/2 = 1/8$	1/8	q^3	q^3

with the same probability. What is that probability? _________ The combinations of three boys and one girl or three girls and one boy could each arise in four different sequences. Identify the four sequences through which the three-boy, one-girl combination could arise. _________________________________ _____________ What is the probability with which each of these sequences would be expected to occur? _____ What is the probability of the occurrence of the three-boy, one-girl combination? _________ Identify the four sequences through which the three-girl, one-boy combination could arise. ______________________ ______________________ What is the probability with which each of these sequences would be expected to occur? __________ What is the probability of the occurrence of the three-girl, one-boy combination? ______ Identify the six different sequences through which the combination of two boys and two girls could arise. ____ What is the probability with which each of these sequences would be expected to occur? __________ What is the probability of the occurrence of the two-boy, two-girl combination? _________

You can check your answers by consulting Table 7-2 which gives the five possible combinations for a family of four children and the various sequences which produce each combination. Note that the probability of each specific sequence and of each combination is given in numeric terms as well as in generalized terms, where p and q represent the probability of producing a male and female child, respectively. Note that the probabilities of all possible combinations, expressed either in numeric values or in generalized expressions, add up to 1:

$$1/16 + 4/16 + 6/16 + 4/16 + 1/16 = 1$$
$$p^4 + 4p^3q + 6p^2q^2 + 4pq^3 + q^4 = 1.$$

Binomial Expansion and Binomial Distributions

As you may realize, in both of the examples we have considered, the probabilities of all of the possible combinations, when expressed in terms of p and q, make up a **binomial distribution** produced by expanding the binomial, $(p + q)^n$, where n is the total number of children in the combination. For example, the expression for families with three children, $p^3 + 3p^2q + 3pq^2 + q^3$, is an expansion of the binomial $(p + q)$ raised to the third power, or $(p + q)^3$. The expression for families of four children, $p^4 + 4p^3q + 6p^2q^2 + 4pq^3 + q^4$, is an expansion of the same binomial raised to the fourth power, or $(p + q)^4$. If we had considered the various combinations involved in a two-child family (two boys; one boy, one girl; two girls), we would find that the expression summarizing the probabilities of each of the three combinations would be $p^2 + 2pq + q^2$, or the binomial $(p + q)$ raised to the second power: $(p + q)^2$.

TABLE 7-2

All possible combinations of birth orders in a family of four children.

Combination	Sequence of births	Numeric probability of sequence	Numeric probability of combination	Expression for probability of sequence	Expression for probability of combination
4 boys	B, B, B, B	1/16	1/16	p^4	p^4
3 boys, 1 girl	B, B, B, G	1/16	4/16	p^3q	$4p^3q$
	B, B, G, B	1/16		p^3q	
	B, G, B, B	1/16		p^3q	
	G, B, B, B	1/16		p^3q	
2 boys, 2 girls	B, B, G, G	1/16	6/16	p^2q^2	$6p^2q^2$
	B, G, B, G	1/16		p^2q^2	
	B, G, G, B	1/16		p^2q^2	
	G, B, B, G	1/16		p^2q^2	
	G, B, G, B	1/16		p^2q^2	
	G, G, B, B	1/16		p^2q^2	
1 boy, 3 girls	G, G, G, B	1/16	4/16	pq^3	$4pq^3$
	G, G, B, G	1/16		pq^3	
	G, B, G, G	1/16		pq^3	
	B, G, G, G	1/16		pq^3	
4 girls	G, G, G, G	1/16	1/16	q^4	q^4

The pattern illustrated by these three examples can be stated in general terms: expanding the binomial (p + q) to the *n*th power, where p and q represent the probability of mutually exclusive events and n represents the number of events occurring in a sequence, gives a binomial distribution that includes the probability of the occurrence of each of the several (or many) different combinations that occur when two mutually exclusive events occur n at a time.

Determining the Probability of a Specific Combination

Expanding a binomial, particularly to a higher power, can be tedious. Often, such an expansion, in providing us with an expression for the probability of *every* possible combination, gives more information than needed when problems ask for the probability of producing just *one* specific combination. For example, a problem might ask for the probability of producing a family of five children with the combination of four boys and one girl. To answer this question, we need the probability represented by just one of the terms in the binomial distribution arising from the expansion of the binomial $(p + q)^5$.

Fortunately, a shortcut method allows us to get the probability of only the specific combination we need. This probability is given by the expression

$$p(\text{specific combination}) = \frac{n!}{s!t!} p^s q^t$$

where

n = number of events in the combination (= 5 in our example)
p = probability of one mutually exclusive event (= 1/2, the probability of a boy)
q = probability of the other mutually exclusive event (= 1/2, the probability of a girl)
s = number of times the mutually exclusive event with probability p occurs in the combination (= 4, for the four boys)
t = number of times the mutually exclusive event with probability q occurs in the combination (= 1, for the one girl).

The exclamation point (!) following the symbols n, s, and t is read as **factorial.** It means that each term is to be multiplied by all the integers between itself and 1. (For example, $5! = 5 \times 4 \times 3 \times 2 \times 1 = 120$.) Note that 0! by definition equals 1.

Substituting the values for this specific combination into the equation gives

$$p(4\text{ boys}, 1\text{ girl}) = \frac{5!}{4!1!}(1/2)^4(1/2)^1 = \frac{5 \times 4 \times 3 \times 2 \times 1}{(4 \times 3 \times 2 \times 1)(1)}(1/2 \times 1/2 \times 1/2 \times 1/2)(1/2).$$

Note that in the fraction containing the factorials, the numbers appearing in the numerator cancel the same numbers appearing in the denominator. Thus, the 4, 3, 2, and 1 in the numerator and in the denominator may be canceled, leaving 5 in the numerator and 1 in the denominator.

$$= \frac{5}{1}(1/16)(1/2) = 5/32 = 0.16.$$

Thus, the probability of the expected occurrence of the combination of four boys and one girl is 0.16.

Now we are going to take a closer look at the expression to the right of the equal sign in the equation we have just used:

$$p(\text{specific combination}) = \frac{n!}{s!t!} p^s q^t.$$

This expression can be broken down into two parts. The $p^s q^t$ component gives the probability of the occurrence of just *one* of the several sequences of the specific combination in which we are interested. For example, the probability of producing the four-boy, one-girl combination through the sequence boy, boy, boy, boy, girl can be given by multiplying individual probabilities: $p \times p \times p \times p \times q = p^4q$. Alternatively, we can use the expression $p^s q^t$, substituting the values for s and t into the expression, and get the same thing: $(p)^4(q)^1 = p^4q^1 = p^4q$.

The other component of the expression to the right of the equal sign, n!/s!t!, gives the *total* number of sequences through which the specific combination could arise, and is designated as the "coefficient." Substituting the values for n, s, and t from our example gives us

$$\frac{5!}{4!1!} = \frac{5 \times 4 \times 3 \times 2 \times 1}{(4 \times 3 \times 2 \times 1)(1)} = \frac{5}{1} = 5.$$

That there are five sequences that could produce the four-boy, one-girl combination can readily be verified;

in addition to the boy, boy, boy, boy, girl sequence we have already considered, there are four others:

boy, boy, boy, girl, boy
boy, boy, girl, boy, boy
boy, girl, boy, boy, boy
girl, boy, boy, boy, boy.

By multiplying together the two components of the expression to the right of the equal sign (that is, the probability of the occurrence of *one* sequence by the *total* number of sequences), we get the probability of producing the specific combination.

Additional Examples of Determining the Probability of a Specific Combination

Albinism in humans is a heritable condition resulting in the inability to produce the pigment melanin which is responsible for hair, eye, and skin coloration. The allele for the albino condition, a, is recessive to the allele for melanin production, a^+.

PROBLEM: What is the probability of two heterozygous individuals producing a family of two normal children and one albino child?

SOLUTION: First, we must determine the probability with which the mating $a^+a \times a^+a$ would be expected to produce normal and albino children. With complete dominance operating, crossing these two heterozygotes produces normal and albino progeny with expected probabilities of 3/4 and 1/4, respectively. Since no particular birth order is specified, and since we are asked to determine the probability of a single combination of two normal and one albino, we can use the expression:

$$\text{p(specific combination)} = \frac{n!}{s!t!} p^s q^t$$

where

n = number of events in the combination (= 3)
p = probability of normal pigmentation (= 3/4)
q = probability of the albino condition (= 1/4)
s = number of normally pigmented individuals in the combination (= 2)
t = number of albino individuals in the combination (= 1).

Substituting values into the equation gives

$$\begin{aligned}\text{p(2 normal, 1 albino)} &= \frac{3!}{2!1!}(3/4)^2(1/4)^1 \\ &= \frac{3 \times 2 \times 1}{(2 \times 1)(1)}(9/16)(1/4) \\ &= \frac{3}{1}(9/16)(1/4) = 3(9/64) = 27/64 = 0.42.\end{aligned}$$

PROBLEM: What is the probability of the two heterozygous parents producing a family of three albinos?

SOLUTION: Since there is just one sequence (albino, albino, albino) that gives rise to this combination, the product law gives this probability just as easily as the equation used in the preceding problem. Using the product law, p(3 albinos) = 1/4 × 1/4 × 1/4 = 1/64. We can verify this by inserting the appropriate values into the expression:

$$\text{p(3 albinos)} = \frac{n!}{s!t!} p^s q^t$$

which gives

$$\text{p(3 albinos)} = \frac{3!}{0!3!}(3/4)^0(1/4)^3.$$

(Before solving this, remember that anything raised to the zeroth power equals 1 and that zero factorial equals 1.)

$$\begin{aligned}\text{p(3 albinos)} &= \frac{3 \times 2 \times 1}{(1)(3 \times 2 \times 1)}(1)(1/64) \\ &= (1)(1/64) = 1/64 = 0.016.\end{aligned}$$

COMBINATIONS INVOLVING THREE OR MORE MUTUALLY EXCLUSIVE EVENTS

Multinomial Expansion and Multinomial Distributions

So far, we have considered combinations involving *two* mutually exclusive events. With more than two events, the probabilities of all the possible combinations make up a multinomial distribution produced by expansion of a multinomial, $(p + q + r + \cdots)^n$, where the number of terms is defined by the number of mutually exclusive events; p, q, r, and subsequent terms represent the respective probabilities of these mutually exclusive events, and n represents the total number of events.

Determining the Probability of a Specific Combination

The probability of the occurrence of any specific term produced in expanding the multinomial (which provides the probability of a specific combination) is

$$\frac{n!}{s!t!u!\cdots} p^s q^t r^u \cdots$$

where $s + t + u + \cdots = n$, and $p + q + r + \cdots = 1$.

Let us take a look at a case involving four mutually exclusive events. Assume we are concerned with both the sex (male or female) and the pigmentation (normal or albino) of the progeny of two parents who are both heterozygous for the albino gene. Four mutually exclusive types of progeny could be produced: normal male, albino male, normal female, and albino female. What is the probability that such a mating will produce a normally pigmented child? ________[1] an albino child? ________[2] a male child? ________[3] a female child? ________[4]

Since sex and pigmentation are independent events, how, in general, would you determine the probability of the occurrence of each of the four mutually exclusive outcomes? ________________________[5] What is the probability of each mutually exclusive type: normal male? ________[6] albino male? ________[7] normal female? ________[8] albino female? ________[9] Now that we have the probability values of the occurrence of each of the mutually exclusive outcomes, we will consider some specific problems.

PROBLEM: What is the probability of these parents having a family of four children that consists of two normal males, one albino male, and one normal female?

SOLUTION: Since this specific combination involves just three of the four possible mutually exclusive outcomes, the generalized expression we need requires three terms, p, q, and r:

$$\frac{n!}{s!t!u!} p^s q^t r^u.$$

where

n = number of individuals in the combination = 4
p = probability of a normal male = 3/8
q = probability of an albino male = 1/8
r = probability of a normal female = 3/8
s = number of normal males in the combination = 2
t = number of albino males in the combination = 1
u = number of normal females in the combination = 1.

Substituting these values in the equation gives us

$$\begin{aligned} &p(\text{2 normal males, 1 albino male, 1 normal female}) \\ &= \frac{4!}{2!1!1!}(3/8)^2(1/8)^1(3/8)^1 \\ &= \frac{4 \times 3 \times 2 \times 1}{(2 \times 1)(1)(1)}(9/64)(1/8)(3/8) \\ &= (4 \times 3)(27/4096) = 12(27/4096) \\ &= 324/4096 = 0.079. \end{aligned}$$

PROBLEM: What is the probability of this couple having a family of six children that consists of two normal males, one albino male, two normal females, and one albino female?

SOLUTION: Substituting the appropriate values into the general expression gives

$$\begin{aligned} &p(\text{2 normal males, 1 albino male, 2 normal females, 1 albino female}) \\ &= \frac{6!}{2!1!2!1!}(3/8)^2(1/8)^1(3/8)^2(1/8)^1 \\ &= \frac{6 \times 5 \times 4 \times 3 \times 2 \times 1}{(2 \times 1)(1)(2 \times 1)(1)}(9/64)(1/8)(9/64)(1/8) \\ &= \frac{360}{2}(9/64)(1/8)(9/64)(1/8) \\ &= 180(81/262144) = 14580/262144 = 0.056. \end{aligned}$$

SUMMARY

A specific combination of progeny arising from a particular mating can generally be produced in a number of different sequences. Once all the different ways of forming a specific combination have been identified, their individual probabilities can be combined using the law of the sum to give the probability of producing that combination. The probabilities of all the possible combinations involving two mutually exclusive events

[1] 3/4. [2] 1/4. [3] 1/2. [4] 1/2.
[5] By multiplying their separate probabilities. [6] 3/4 × 1/2 = 3/8. [7] 1/4 × 1/2 = 1/8. [8] 3/4 × 1/2 = 3/8. [9] 1/4 × 1/2 = 1/8.

form a binomial distribution which is produced by expanding the binomial, $p + q$, to a power equal to the number of events in the combination. The probability of just one specific combination, represented by one term in the binomial distribution, is given by the expression

$$\frac{n!}{s!t!}p^s q^t$$

where

n = number of events in the combination
p = probability of one mutually exclusive event
q = probability of the other mutually exclusive event
s = number of times the mutually exclusive event with probability p occurs in the combination
t = number of times the mutually exclusive event with probability q occurs in the combination.

In cases where more than two mutually exclusive alternatives exist, the probabilities of all possible combinations involving these alternatives form a multinomial distribution given by expanding the multinomial $(p + q + r + \cdots)$ to a power equal to the number of events in the combination. The probability of a specific combination, represented by one term in the multinomial distribution, is given by the expression

$$\frac{n!}{s!t!u!\cdots}p^s q^t r^u \cdots.$$

PROBLEM SET

7-1. Explain the difference between mutually exclusive and independent events.

7-2. Explain the difference between the calculations involved in combining probabilities of mutually exclusive events and in combining probabilities of independent events.

7-3. Explain the difference between a sequence and a combination.

7-4. Explain the difference between the calculations involved in determining the probability of a sequence and determining the probability of a combination.

7-5. In guinea pigs, the allele for black fur (*B*) is dominant to the allele for brown fur (*b*). Assume that two heterozygous guinea pigs are mated.

- **a.** What is the probability of these parents producing a litter of three brown guinea pigs?
- **b.** What is the probability of these parents producing a litter of one brown and two black guinea pigs, born in the order of brown, black, black.
- **c.** Identify other sequences, in addition to brown, black, black, which would produce a litter of one brown and two black guinea pigs and indicate the probability with which each would be expected to occur.
- **d.** What is the probability of producing a litter of one brown and two black guinea pigs?
- **e.** What is the probability of producing a single litter with a black, black, brown sequence, or a black, brown, black sequence, or a brown, black, black sequence?

7-6. Refer back to the basic information given in problem 7-5.

- **a.** What is the probability of producing three successive litters consisting of a black, black, brown sequence followed by a black, brown, black sequence followed by a brown, black, black sequence?
- **b.** What is the probability of producing three successive litters, each consisting of one brown and two black guinea pigs?

7-7. Refer back to the basic information given in problem 7-5. Assume that p is the probability of producing a black guinea pig and q is the probability of producing a brown one.

- **a.** Write a generalized expression for the probability of a sequence of black, brown, black.
- **b.** Write a generalized expression for the probability of the combination of two black guinea pigs and a brown one.
- **c.** What do the expressions q^3 and $3pq^2$ represent in regard to possible outcomes of the guinea-pig cross?

7-8. Assume that a litter of five guinea pigs is produced from the mating described in problem 7-5.

- **a.** What is the probability that the litter will consist of four black guinea pigs and a brown one?
- **b.** What is the probability that the litter will consist of two black, one brown, and two black guinea pigs, born in that order.

7-9.
- **a.** What is the probability of a family of five children consisting of three boys and two girls? (Assume that there is an equal chance of producing a child of either sex.)
- **b.** What is the probability of this same family of five consisting of at least four boys?

7-10. The ability to taste the chemical substance phenylthiocarbamide (PTC) in humans is genetically based. The allele for tasting PTC, *T*, is dominant to the allele for nontasting, *t*. A couple consists of a taster woman and a taster man, both of whom are known to be heterozygous for this trait.

PROBLEM SET

a. What is the probability of the occurrence of one of each of the following among the offspring of this couple: male taster? male nontaster? female taster? female nontaster?
b. What is the probability of this couple having a family of five children consisting of one male taster, two male nontasters, and two female tasters?
c. What is the probability of this couple having a family of four children consisting of one male taster, one male nontaster, one female taster, and one female nontaster.

7-11. In the absence of a particular human enzyme, the sugar galactose is not metabolized and instead accumulates in the body, adversely affecting the development of the nervous system in infant children. This disorder, known as galactosemia, is due to the expression of an allele, g, which is recessive to the allele for normal enzyme production, g^+. Through an analysis of their families' genetic history, both phenotypically normal members of a couple are identified as heterozygous for the alleles governing this trait. What is the probability that
a. their first child will be galactosemic?
b. their third child will be normal for galactose utilization?
c. their third child will be a girl who is normal for galactose utilization?
d. in a family of five, there would be four normal males and one galactosemic female?

7-12. Assuming a 1:1 birth ratio, the laws of probability predict that half the children in a family will be boys and half will be girls.
a. What is the probability that this expected outcome would occur in a family of four children?
b. What is the probability that something other than the expected outcome would occur in a family of four children?

7-13. A man heterozygous at both the locus for PTC tasting (see problem 7-10) and at the locus for albinism marries an albino woman heterozygous at the PTC tasting locus. (Information regarding the inheritance of human albinism is given in this chapter.) These two loci assort independently.
a. What is the probability of producing three tasters with normal pigmentation and three nontaster albinos in a family of six children?
b. What is the probability of producing a family of six with three tasters with normal pigmentation, two nontaster albinos, and one taster albino?
c. What is the probability of producing a family of six with three tasters with normal pigmentation, two nontaster albinos, and one taster albino, in that order.

7-14. If the parents described in problem 7-13 had genotypes $Ttaa$ and tta^+a, what is the probability of producing a family of five consisting of three nontaster albino sons and two nontaster daughters with normal pigmentation?

CHAPTER

8

Multiple-Allelic Series

INTRODUCTION

For each of the traits discussed in the preceding chapters, we have considered only two different allelic forms. This situation represents the exception rather than the rule since many genes have more than two allelic forms. The alleles of such genes make up what is called a **multiple-allelic series.** Since any particular diploid individual carries only two alleles (which may be alike or different) for a trait, the existence of a multiple-allelic series can be detected only by surveying a group of individuals. The number of different combinations of alleles of a gene (that is, the number of genotypes) found among the individuals of a population will, of course, increase as the size of the multiple-allelic series increases. Multiple-allelic series occur in diploid and haploid organisms and are well known in prokaryotes as well as eukaryotes. This chapter considers those in diploid eukaryotes.

ASSIGNING SYMBOLS TO A MULTIPLE-ALLELIC SERIES

Assigning symbols to the alleles in a multiple-allelic series is relatively straightforward, provided we know whether individual alleles are dominant, codominant, or incompletely dominant with respect to each other. Sometimes there is one allele that shows complete dominance to all the others in the series, whereas another allele may be recessive to all others. The remaining alleles in the series may show intermediate levels of dominance, being dominant to some of the alleles and recessive to others. The same letter or combination of letters is used to designate all the alleles in a series. The allele that is recessive to all the others is designated by using the lowercase letter while the allele showing complete dominance is designated by the lowercase letter with a superscript plus sign ($^+$). Alternatively, the dominant allele may be represented by a capital letter. Alleles with intermediate levels of dominance are symbolized by using the letter with an appropriate superscript.

As an illustration, consider the white-eye locus in the fruit fly *Drosophila melanogaster.* The gene at this locus, which controls the degree of redness in the eye, has numerous alleles. The allele producing white eyes is recessive to all others in the series and is symbolized by w. The allele for red, the wild-type eye color, which is dominant to all others in the series, is designated by either w^+ or W. The other alleles in the series produce various amounts of red pigment yielding intensities of eye color ranging between white and the wild-type red. These alleles are designated by using w with an appropriate superscript. For example, alleles for eosin, cherry, honey, and pearl eye colors (given in order of decreasing intensity of the red hue) are assigned the symbols w^e, w^{ch}, w^h, and w^p, respectively. Although the wild-type allele (w^+) shows complete dominance to all the alleles in the series, all the other alleles exhibit incomplete dominance in relation to each other. For example, an individual heterozygous for the eosin and pearl alleles ($w^e w^p$) would have a phenotype intermediate to that shown by an individual homozygous for eosin ($w^e w^e$) and another individual homozygous for pearl ($w^p w^p$). This locus is carried on a sex, or X, chromosome and is therefore sex-linked. As explained in Chapter 10, inheritance patterns of such genes differ from those of genes carried on autosomes.

AN EXAMPLE WITH THREE MAJOR ALLELES: THE ABO BLOOD SYSTEM IN HUMANS

Assigning Symbols

Another example of a trait involving a multiple-allelic series is the ABO blood-type trait in humans. This trait is well known because of its clinical importance in determining whether blood transfusions can safely be made between individuals. The three major alleles involved in this multiple-allelic series determine four major phenotypes or blood types: O, A, B, and AB. The presence or absence of two antigen molecules (designated as antigen A and antigen B) on the surface of red blood cells serves to identify the blood type of an individual. Blood lacking both antigens A and B is designated as type O blood. Individuals with only antigen A are said to have type A blood; those with only antigen B have blood type B. If both antigens are present, the person has type AB blood.

The letter i is widely used as the basic symbol for the alleles of the ABO gene. The allele resulting in no antigen production is recessive to the others in the series and is designated as either i or I^O. The alleles that determine antigen A and antigen B are assigned the symbols I^A and I^B, respectively. Two types of allelic interaction are found among the three major alleles in this series. Alleles I^A and I^B are codominant to each other: the presence of both alleles means that both antigens A and B will be present on the red blood cells, and each is dominant to i. Before proceeding, complete the following table by listing the antigens present and the genotypes for each of the four phenotypes, keeping in mind that the same phenotype may be produced by more than one genotype.

Blood type (phenotype)	Antigens present	Genotypes
O	________[1]	________[2]
A	________[3]	________[4]
B	________[5]	________[6]
AB	________[7]	________[8]

[1] None. [2] ii. [3] A. [4] I^AI^A or I^Ai. [5] B. [6] I^BI^B or I^Bi. [7] A and B. [8] I^AI^B.

Solving Problems

To solve problems involving the identification of the progeny of a cross, it is important, whenever possible, to determine the genotypes of the individuals involved. Remember that blood types A and B may each have two possible genotypes. Because of this, it is sometimes necessary to work a problem in a number of ways unless some of the possibilities can be eliminated by logic.

PROBLEM: A man with type A blood marries a woman with type B blood. Their first child has type O blood. What does this information tell us about the genotype of each parent?

SOLUTION: The only possible genotype for the child is ii. Since each parent contributes one of these alleles, each must be heterozygous for the trait: the type A father is I^Ai and the type B mother is I^Bi. ■

PROBLEM: A mother with type AB blood has a child with type AB blood. The mother believes that an individual with type O blood is the father of the child. Can this man be the father of the child?

SOLUTION:
Given: Mother is type AB. Child is type AB. The man thought to be the father is type O.
Parental genotypes: The only possible genotype for the mother is I^AI^B. The only possible genotype for the father is ii.
Gametes: All gametes produced by the father will be of one type, carrying the i allele. The mother will produce equal quantities of two kinds of gametes: half will carry I^A and half will carry I^B, as shown in the following diagram.

Union of Gametes			
Male gamete	Female gametes		Progeny genotypes
i (1)	I^A (1/2)	=	I^Ai (1/2)
	I^B (1/2)	=	I^Bi (1/2)

Results: Progeny phenotypes will be type A and type B; thus, no type AB progeny could be produced by the mating of a type O father with a type AB mother. Therefore the man thought to be the father could not have fathered this child.

A second way of solving this problem involves a more abstract approach. The AB baby has an I^AI^B genotype. One of the baby's alleles came from the mother and the other from the father. Since the alleged father has neither the I^A nor the I^B allele, he could not have fathered the child. ▬

Additional Alleles in the ABO System

In addition to the three major alleles in the ABO multiple-allelic series, additional variants of the I^A and I^B alleles are known. If we designate the major I^A allele as I^{A_1}, we can assign the symbol I^{A_2} to the second variant. When the gene I^{A_2} is expressed, it produces an additional blood type: A_2. Allele I^{A_2} is recessive to I^{A_1}, but when paired with I^B it shows codominance. I^{A_1}, too, shows codominance when paired with I^B. If we consider the three major alleles, I^{A_1}, I^B, and i, along with I^{A_2}, how many different phenotypes could result from the various combinations of these alleles? ______[9]

AN EXAMPLE WITH FOUR MAJOR ALLELES: FUR COLOR IN RABBITS

Assigning Symbols

Another example of a gene with several alleles is the one responsible for fur color in rabbits. Colors range from wild-type dark-gray to white (albino). Between these extremes, in order of decreasing intensity of coloration, are silver-gray (chinchilla); light-gray; and white with black feet, ears, and nose (Himalayan). This multiple-allelic series involves at least four major alleles assigned the symbols: c^+, c^{ch}, c^h, and c. The wild-type allele, designated as c^+, is completely dominant to the other three alleles and produces the wild-type dark-gray phenotype. Homozygous c alleles produce the albino phenotype. Homozygous c^{ch} alleles give rise to the silver-gray phenotype. In heterozygous combination with either c^h or c, the c^{ch} allele shows incomplete dominance, resulting in light-gray fur color. Homozygous c^h alleles produce the Himalayan phenotype. This phenotype also occurs when c^h is combined with c because of the dominance of the c^h allele. Thus, the interactions among the four major alleles include both complete and incomplete dominance. List the genotypes that produce the five phenotypes in the table that follows.

Phenotype	Genotypes
Wild-type dark-gray	______[10]
Silver-gray (chinchilla)	______[11]
Light-gray	______[12]
White with black (Himalayan)	______[13]
White (albino)	______[14]

Solving Problems

PROBLEM: A silver-gray rabbit and a Himalayan rabbit with the genotype of c^hc are crossed. What are the possible genotypes and phenotypes of the offspring?

SOLUTION: First we must identify the genotype of the silver-gray rabbit. This phenotype can be produced only by the $c^{ch}c^{ch}$ genotype. Thus, the mating is $c^{ch}c^{ch} \times c^hc$. Complete the following branch diagram.

c^{ch} (1) → c^h (1/2) = ________
c^{ch} (1) → c (1/2) = ________

Outcome: Two genotypes, $c^{ch}c^h$ and $c^{ch}c$, are expected in a 1:1 ratio. Phenotypically, all the progeny would be light-gray. ▬

PROBLEM: A litter of rabbits includes albinos and silver-grays. What can be concluded about the genotypes and phenotypes of the parents? Would you expect to find other types of rabbits in this litter? Explain.

SOLUTION: First the genotypes of the progeny must be identified. The albino rabbits must be cc, while the silver-gray rabbits are $c^{ch}c^{ch}$. Since one allele of the cc pair comes from each parent, we can conclude that each parent carries the c allele. Similarly, an allele of the $c^{ch}c^{ch}$ pair must be contributed by each parent. Thus, each of the parents must have the $c^{ch}c$ genotype and the light-gray phenotype. This mating would also be expected to produce rabbits with the $c^{ch}c$ genotype and thus the light-gray phenotype. ▬

[9] Six: type B (genotypes I^BI^B and I^Bi); type O (genotype ii); Type A_1 (from genotypes $I^{A_1}I^{A_1}$, $I^{A_1}I^{A_2}$, and $I^{A_1}i$); type A_2 (from genotypes $I^{A_2}I^{A_2}$ and $I^{A_2}i$); type A_1B (genotype $I^{A_1}I^B$); type A_2B (genotype $I^{A_2}I^B$). [10] c^+c^+, c^+c^{ch}, c^+c^h, c^+c. [11] $c^{ch}c^{ch}$. [12] $c^{ch}c^h$, $c^{ch}c$. [13] c^hc^h, c^hc. [14] cc.

GENERALIZED MATHEMATICAL EXPRESSIONS GIVING THE NUMBER OF GENOTYPES IN A MULTIPLE-ALLELIC SERIES

As the number of alleles in a multiple-allelic series increases, the number of possible genotypes increases. In the ABO system, we found that three alleles taken in combinations of two at a time produce six different genotypes, three homozygous and three heterozygous. In considering fur color in rabbits, we found that the four major alleles taken in combinations of two at a time produce a total of ten different genotypes (verify this by counting the number of different genotypes you listed earlier). Of those ten genotypes, four were homozygous (c^+c^+, $c^{ch}c^{ch}$, c^hc^h, and cc) and the remaining six were heterozygous.

If the number of alleles in any particular multiple-allelic series is known, generalized mathematical expressions can be used to find the number of different genotypes arising from those alleles and the number of those genotypes which are homozygous and heterozygous. If the number of alleles in a multiple-allelic series is represented by n, then the total number of genotypes arising from those alleles is given by the expression $n(n + 1)/2$. The number of these genotypes that are homozygous equals n, while the number of heterozygous genotypes is given by the expression $n(n - 1)/2$. These relationships are summarized in Table 8-1.

TABLE 8-1

Relationships between the number of alleles in a multiple-allelic series and the number of genotypes.

Number of alleles in series	Total number of genotypes	Number of homozygous genotypes	Number of heterozygous genotypes
1	1	1	0
2	3	2	1
3	6	3	3
4	10	4	6
⋮	⋮	⋮	⋮
n	$\frac{n(n+1)}{2}$	n	$\frac{n(n-1)}{2}$

SUMMARY

In working problems involving multiple alleles, it is important to remember that regardless of the number of alleles in the series, an individual carries only two alleles. Before working a problem, you need to know the role of each allele in determining the phenotype and whether the interactions between these alleles involve complete dominance, codominance, or incomplete dominance. After assigning a symbol to each allele, try, whenever possible, to determine and designate the genotype for the individuals involved. If there are uncertainties about parental genotypes, it may be necessary to work a problem in a number of ways.

This chapter considers ABO blood types in humans and fur color in rabbits as examples of multiple-allelic traits. If the number of alleles in a multiple-allelic series is known, generalized mathematical expressions can be used to find the total number of genotypes as well as the number of genotypes that are homozygous and heterozygous.

PROBLEM SET

8-1. a. A silver-gray rabbit is mated with a rabbit from a pure-breeding line of wild-type dark-gray rabbits. Referring to the information presented earlier in this chapter, identify the genotype for each of these rabbits and the expected genotypes and phenotypes of the F_1.

b. In another cross, an albino rabbit is mated with a rabbit from a pure-breeding line showing the Himalayan pattern (white with black extremities.) Identify the genotype for each of these rabbits and the expected genotypes and phenotypes of the F_1.

c. A rabbit from the F_1 of the mating described in problem 8-1a is mated with a rabbit from the F_1 of the mating described in 8-1b. Determine the expected genotypes and phenotypes of their progeny.

8-2. a. A rabbit with light-gray fur is mated with a Himalayan rabbit. What genotypes would allow these parents to produce albino progeny?

b. If albinos were produced from the mating described in problem 8-2a, what other phenotype or phenotypes would be expected among the progeny?

8-3. A woman whose father had type O blood has type B blood. Her husband has type O blood.

a. What are the genotypes of the woman and her husband?

b. What types of gametes would be produced by the woman and her husband?

c. What blood types would you expect in their progeny and in what frequencies?

8-4. A woman has type A blood and is married to a man with type B blood.

a. What genotypes are possible for these individuals?

b. Their first child has type O blood. What can be concluded about the genotype of each parent?

c. What other phenotypes are possible among additional children this couple may have?

8-5. A paternity suit is filed against a man with type AB blood. The child has type A blood and the mother has type B. You are summoned to court as an expert in blood-group inheritance to answer the question "Could this man have fathered the child?" Explain your answer.

8-6. A woman with blood type AB marries a man with blood type B. The man's father is known to have type O blood. What proportion of their children, if any, would one expect to have blood type B?

8-7. During sexual reproduction in plants, pollen is transferred to the female part of the plant known as the stigma. Once transferred, a pollen grain may develop a pollen tube, through which a sperm nucleus from the pollen moves to unite with the egg nucleus. A number of plant species including *Nicotiana*, the tobacco plant, possess self-sterility loci ensuring that the pollen produced by a particular plant cannot fertilize egg cells produced by the same plant. The gene found at such a locus generally has many variant forms which make up a multiple-allelic series and may be designated as S_1, S_2, S_3, S_4, S_5, etc. The basis for the incompatibility is as follows: if the allele carried by a pollen grain is also possessed by the plant upon whose stigma the pollen grain lands, the pollen tube cannot develop and no fertilization occurs. For example, a plant with a genotype S_1S_2 will produce two kinds of pollen grain: half will carry S_1 while the other half will carry S_2. If an S_2 pollen grain from this or another plant, lands on a stigma of this S_1S_2 plant

PROBLEM SET

or on the stigma of any other plant carrying the S_2 allele, the pollen grain will not develop. Similarly, an S_1 pollen grain that lands on the stigma of this plant or on the stigma of any other plant carrying S_1 will not develop.

a. What outcome would you expect if pollen from an S_5S_6 plant was transferred to stigmas on an S_2S_3 plant?

b. What percentage of the pollen from an S_2S_3 plant could develop pollen tubes and achieve fertilization if transferred to stigmas of a plant with the S_1S_3 genotype?

c. Why would no plants be found which were homozygous for the same sterility allele?

d. Identify the genotype or genotypes produced by a cross between pollen grains from an S_3S_4 plant and egg cells from an S_1S_2 plant.

e. Identify the genotype or genotypes produced by a cross between pollen grains from an S_3S_4 plant and egg cells from an S_2S_3 plant.

8-8. Humans can be typed for the presence or absence of a number of blood antigens other than those of the ABO system. For example, antigens designated as M and N may be present on the surface of red blood cells. Antigen M is produced whenever allele *M* is present and antigen N is produced whenever allele *N* is present. An individual with genotype *MM* will produce antigen M and will have blood type M. An individual with genotype *NN* will produce antigen N and will have blood type N. In individuals with genotype *MN*, codominance results in both antigens being produced and blood type MN results. The alleles for the MN system assort independently of the alleles responsible for the ABO system. A child has blood type B, MN. Her mother is type AB, M. Determine whether each of the following blood types could belong to the father of this child. Explain each answer.

(1) Type B, MN
(2) Type A, N
(3) Type O, MN
(4) Type AB, N

8-9. The Rh antigen is another genetically determined antigen that may occur on the surface of human red blood cells. Individuals with this antigen are said to be Rh-positive and those without it are Rh-negative. The allele for the production of this antigen, *Rh*, is dominant to the *rh* allele for nonproduction of the antigen. The locus for the Rh antigen assorts independently of the locus determining ABO blood types. Determine whether parents with the following sets of genotypes could have a child with A, Rh-negative blood.

(1) $I^AiRhrh \times iirhrh$
(2) $I^AI^ARhRh \times I^Airhrh$
(3) $I^Airhrh \times I^BI^Brhrh$
(4) $I^Birhrh \times I^BiRhrh$

8-10. Refer to problems 8-8 and 8-9 for information on the MN and Rh blood types, respectively. A child has blood type B, MN, Rh-positive and her father is type O, N, Rh-negative. The genes involved in determining each trait are carried on different pairs of chromosomes. Indicate whether each of the following phenotypes is a possible phenotype for the mother of this child.

(1) A, N, Rh-negative
(2) AB, M, Rh-negative

PROBLEM SET

(3) AB, MN, Rh-positive
(4) A, N, Rh-positive
(5) O, M, Rh-positive

8-11. Eye color in the fruit fly *Drosophila* is controlled by a locus with a multiple-allelic series. A fly population which exists in the wild is known to possess six different alleles for the eye-color gene.

a. What is the minimum number of allelic types for this trait that could be carried by a single individual?
b. What is the maximum number of allelic types for this trait that could be carried by a single individual?
c. What is the number of different genotypes that could arise from these six alleles?
d. How many of these different genotypes would be homozygous?
e. How many of these different genotypes would be heterozygous?

8-12. Coat color in cattle involves a multiple-allelic series with at least four alleles: S, s^h, s^c, and s. Allele S causes a band of white to develop around the animal and is dominant to each of the other alleles in the series. Allele s^h causes the development of regular spots, such as those seen in Hereford cattle, and is dominant to alleles s^c and s. Allele s^c causes the development of solid color and is dominant to allele s. Allele s, which produces large irregular spots like those seen in Holstein cattle, is recessive to all the other alleles in the series. Predict the genotypes and phenotypes of the progeny of a mating between

a. an s^cs cow and an Ss bull.
b. an Ss^c cow and an s^hs^c bull.

8-13. Refer to the basic information given in problem 8-12. A cattle breeder wishes to buy from a cattle dealer a bull that is homozygous for regular spots and suspects that the particular bull she has in mind may not be homozygous for the s^h allele.

a. What genotypes besides s^hs^h are possible for this bull?
b. Is there a type of mating that could be carried out to provide some evidence one way or the other for the true genotype of the bull? Explain.

8-14. If the bull with regular spots referred to in problem 8-13 is mated with five different pure-breeding, solid-colored cows and each produces a calf that shows regular spots, what is the probability that the bull is homozygous for the s^h allele?

CHAPTER

9

Modified Dihybrid Ratios: Interaction of Products of Nonallelic Genes

INTRODUCTION

For most of the traits considered in earlier chapters of this book, the implication has been that the expression of each is determined by the interaction of a pair of alleles of a single gene. This is a simplistic view, since the expression of most traits arises from the interaction of the protein products of allelic pairs of at least several genes. This chapter examines several types of gene interactions including epistasis, complementation, and duplication. When a trait is controlled by the interaction of two independently assorting loci, each with two alleles, the crossing of two individuals heterozygous at both loci results in the modification of the classic 9:3:3:1 Mendelian dihybrid ratio.

AN EXAMPLE OF GENE INTERACTION: FUR COLOR IN MICE

As one example of gene interaction, consider fur color in mice. At least five loci make major contributions to fur color. Much of the variation that occurs in the expression of this trait is due to the fact that melanin, the pigment that gives fur its color, can be deposited in different amounts and in different locations on the individual hairs of the fur. The alleles at the locus designated as *C* determine whether or not melanin production occurs. The dominant allele, *C*, allows for melanin production, while the expression of its recessive counterpart, *c*, prevents melanin production. If no melanin is produced, the animal has a white coat and is designated as an albino. If at least one dominant allele is carried at the *C* locus, then the other loci involved in this trait can exert an effect on the final phenotype. One of these is the *B* locus, which influences the amount of melanin deposited in each strand of fur. The dominant allele at this locus, *B*, produces black strands while the expression of the recessive allele, *b*, results in less melanin deposition and produces brown strands. At another locus, designated as *A*, the expression of the dominant allele, *A*, produces a zone of yellow just below the tip of each strand of hair giving a "frosted" appearance to the coat, a condition referred to as *agouti*. Expression of the recessive allele, *a*, causes the yellow zone to be absent, resulting in a solid-color coat.

Let us consider some of the phenotypes that can result from the interaction of these three fur-color loci. A good approach is to consider each locus separately, and to illustrate this, we will look at the phenotype produced by the genotype *CcBbaa*. The presence of the dominant allele at the *C* locus causes melanin production, and the fur would be pigmented. At the *B* locus, the dominant allele would result in black fur. And at the *A* locus, the expression of the recessive allele would result in the black strands being solidly colored. Thus, the genotype *CcBbaa* would result in solid-black fur. To get some further practice, determine the fur color for each of the following genotypes.

Genotype	Phenotype
CCBBAA	________ 1
Ccbbaa	________ 2
CcbbAA	________ 3
ccBBAa	________ 4

EPISTATIC GENE INTERACTION

For the mouse fur trait, at least one dominant allele is necessary at the *C* locus in order for the alleles at the *B* and *A* loci to be expressed. Put another way, if the genotype at the *C* locus is *cc*, then the alleles at the other two loci, regardless of what they are, will *not* be expressed. This situation, where a particular allelic combination (for example, *cc*) at one locus blocks the expression of genes at one or more other loci, is known as **epistasis.** The locus that is doing the blocking is said to be **epistatic** to the locus or loci whose expression is being blocked, and the blocked loci are said to be **hypostatic** to the blocking locus. In this case, the *C* locus is epistatic to the *A* and *B* loci, while loci *A* and *B* are hypostatic to the *C* locus.

A Comparison of Crosses with and without Epistasis

Epistatic interactions among the loci determining a particular trait can have a significant effect on the phenotypic outcome of a cross. To illustrate this, we will compare the progeny expected from two fur-color crosses, one involving no epistasis and the other with epistasis. To keep our examples simple, genetic variability will occur at just two of the three fur-color loci we have considered up to this point.

Cross without Epistasis: The 9:3:3:1 Ratio Two individuals with the black agouti phenotype, each homozygous for the dominant allele at the *C* locus and heterozygous at both the *B* and the *A* loci, are mated. Since only the dominant allele occurs at the *C* locus, no epistasis will be involved in determining the phenotypes of the progeny of this *CCBbAa* × *CCBbAa* cross. Because the parents show no genetic variability at the *C* locus, we can ignore it in our cross, as long as we remember that each offspring will be *CC*; thus, the cross can be simplified to *BbAa* × *BbAa*. Since these two loci are inherited independently, their inheritance can be considered separately.

In the mating *Bb* × *Bb*, the expected probability of getting at least one dominant allele, and thus the black phenotype, is 3/4, and the expected probability of getting two recessive alleles, and thus the brown phenotype, is 1/4. Similarly, in the mating *Aa* × *Aa*, the expected probability of progeny getting at least one dominant allele, and thus the agouti phenotype, is 3/4, and the expected probability of getting two recessive alleles, and thus the solid phenotype, is 1/4. Determine the phenotypic outcome of this cross and the expected phenotypic probabilities for the progeny by completing the following branch diagram. Indicate the expected phenotypic ratio.

B-locus phenotype	*C*-locus phenotype		Progeny phenotype	Expected probability
Black (3/4)	Agouti (3/4)	=	________ 5	________ 6
	Solid (1/4)	=	________ 7	________ 8
Brown (1/4)	Agouti (3/4)	=	________ 9	________ 10
	Solid (1/4)	=	________ 11	________ 12

Expected phenotypic ratio: ________ . 13

Thus, in the absence of epistasis, a 9:3:3:1 phenotypic ratio occurs. When we have seen this Mendelian ratio before, it has reflected the relative frequencies of the four phenotypic combinations arising from simultaneous consideration of *two* different traits. How does that differ from what we are seeing here? ________________ 14

[1] Black agouti. [2] Brown. [3] Brown agouti. [4] Albino. [5] Black agouti. [6] 3/4 × 3/4 = 9/16. [7] Solid black. [8] 3/4 × 1/4 = 3/16. [9] Brown agouti. [10] 3/4 × 1/4 = 3/16. [11] Solid brown. [12] 1/16. [13] 9:3:3:1.
[14] Here it gives the relative frequencies of four phenotypic expressions of the *same* trait.

Cross with Epistasis: The 9:3:4 Ratio Now let us compare this outcome with that of a cross where epistasis is involved in determining the progeny phenotypes. Two individuals with the black phenotype, each heterozygous at both the *C* and the *B* loci and homozygous for the recessive allele at the *A* locus, are mated. Identify the phenotypes of the progeny expected from this mating. Epistasis will be involved in determining the phenotypes of the progeny of this *CcBbaa* × *CcBbaa* cross who end up homozygous for the *c* allele. Because the parents show no genetic variability at the *A* locus, we can ignore it in our cross, as long as we remember that each offspring will be *aa*; thus, the cross can be simplified to *CcBb* × *CcBb*. Since the *C* and *B* loci are inherited independently, their inheritance can be considered separately.

In the mating *Cc* × *Cc*, the expected probability of getting at least one dominant allele, and thus a pigmented phenotype, is 3/4, and the expected probability of getting two recessive alleles, and thus the albino phenotype, is 1/4. Similarly, in the mating *Bb* × *Bb*, the expected probability of progeny getting at least one dominant allele, and thus the black phenotype, is 3/4, and the expected probability of getting two recessive alleles, and thus the brown phenotype, is 1/4. Determine the phenotypic outcome and the expected phenotypic probabilities for this cross by completing the branch diagram that follows. Indicate the expected phenotypic ratio.

C-locus phenotype	*B*-locus phenotype		Progeny phenotype	Expected probability
Pigmented (3/4)	Black (3/4)	=	________[15]	________[16]
	Brown (1/4)	=	________[17]	________[18]
Albino (1/4)	Black (3/4)	=	________[19]	________[20]
	Albino (1/4)	=	________[21]	________[22]

Expected phenotypic ratio: ________.[23]

The phenotypic ratio of 9:3:4, a modification of the 9:3:3:1 ratio, is produced by the expression of recessive alleles at the epistatic *C* locus which masks alleles at the *B* locus in four out of 16 progeny. The epistasis seen here, arising from the double recessive at the epistatic locus, is referred to as **recessive epistasis.** In other cases, epistasis can arise from the expression of a dominant allele at an epistatic locus.

Additional Fur-Color Loci

In addition to the three loci discussed, two other loci are known to have a major influence on fur color in mice. Expression of the dominant allele at the *S* locus produces a nonspotted coat, while the expression of the recessive allele at this locus results in a spotted coat. The expression of a dominant allele at the *D* locus slightly reduces the color intensity to produce a milky appearance, while its recessive counterpart allows for the normal expression of fur color. As with the *A* and *B* loci, both the *S* and *D* loci are hypostatic to the *C* locus. Thus, at least five major loci interact to produce the fur phenotype in mice. Additional loci, known as **modifiers,** whose phenotypic effect are not as pronounced as the major loci we have been considering, may also be involved in influencing the fur phenotype in mice.

COMPLEMENTARY GENE INTERACTION

Another type of gene interaction involves genes at two or more loci which **complement** each other to produce a particular phenotype for a trait. The word "complement" refers to something that completes; at least one dominant allele at each locus is necessary to produce the particular phenotype.

Biochemical Basis

When studied, complementation, like other types of interaction, is often found to have a biochemical basis. For example, the production of the purple pigment which can color the kernels of the ears on a corn plant is known to involve a two-step chemical reaction. A precursor substance, acted upon by a particular enzyme, call it enzyme C, is converted into an intermediate substance which in turn is acted upon by another enzyme, call it enzyme P, to produce the final product, the purple pigment.

$$\text{Precursor} \xrightarrow{\text{Enzyme C}} \text{Intermediate} \xrightarrow{\text{Enzyme P}} \text{Purple pigment}$$

[15] Black fur. [16] 9/16. [17] Brown fur. [18] 3/16. [19] Albino. [20] 3/16. [21] Albino. [22] 1/16. [23] 9:3:4.

Both enzymes are essential for purple-pigment production, since one enzyme without the other blocks completion of the biochemical pathway. The production of each enzyme is controlled by a different locus: production of enzyme C is controlled by a dominant allele at locus C and production of enzyme P is controlled by a dominant allele at locus P. Recessive alleles at either of these two loci cause nonfunctional versions of their respective enzymes to be produced. Thus, in order for purple pigment to be made, at least one dominant allele must be present at *both* loci C and P. In the absence of purple-pigment production, kernels are white.

An Example: Kernel Color in Corn and the 9:7 Ratio

As an example involving this type of interaction, assume two corn plants, each with purple kernels and a genotype of *CcPp* are crossed: *CcPp* × *CcPp*. Since the two loci are inherited independently, their inheritance can be considered separately. In the cross *Cc* × *Cc*, the probability of getting at least one dominant allele is 3/4 and the probability of getting two recessive alleles is 1/4. Similarly in the cross *Pp* × *Pp*, the probability of getting at least one dominant allele is 3/4 and the probability of getting two recessive alleles is 1/4. In the space provided, use the product law to combine probabilities and determine the phenotypic outcome of the *CcPp* × *CcPp* cross.

Phenotypes and expected probabilities: ____.[24] Expected phenotypic ratio: __________.[25]

The 9:7 phenotypic ratio produced here represents a modification of the 9:3:3:1 dihybrid Mendelian ratio, and it arises whenever individuals heterozygous for two different loci are crossed and the dominant alleles at each locus complement each other to produce a specific phenotype.

DUPLICATE GENE INTERACTION AND THE 15:1 RATIO

Another type of gene interaction involves genes at two or more independently inherited loci which have a **duplicate** effect on the phenotype for a particular trait; such genes have the same effect on the phenotype.

Duplicate interaction occurs in the production of fruit shape in a plant called shepherd's purse: two loci, designated as A_1 and A_2, interact to produce either a round or narrow fruit. Round fruit results whenever there is at least one dominant allele at either the A_1 or A_2 locus; in other words, either one or two dominant alleles at the A_1 locus and/or one or two dominant alleles at the A_2 locus produce the round fruit shape. Narrow fruit results when there is no dominant allele at either locus. The symbols A_1 and A_2 can be used to represent the dominant alleles at the A_1 and A_2 loci, respectively, and a_1 and a_2 to represent their respective recessive counterparts. To illustrate this type of inheritance, we will cross two individuals, each heterozygous at both of these loci, that is, $A_1a_1A_2a_2 \times A_1a_1A_2a_2$. Since the two loci are inherited independently, the inheritance of each can be considered separately. In the space provided, use the product law to determine the expected phenotypes and phenotypic frequencies for the progeny.

Expected phenotypes and probabilities: ____.[26] Expected phenotypic ratio: __________.[27]

The 15:1 phenotypic ratio produced here is a modification of the 9:3:3:1 Mendelian dihybrid ratio and arises whenever individuals heterozygous for two different loci are crossed and the dominant alleles at each locus duplicate each other in producing a specific phenotype.

[24] 9/16 have at least one dominant allele at each locus, will make both enzymes, and synthesize the purple pigment; 7/16 lack a dominant allele at one or both loci and will lack one or both enzymes and, consequently, will be incapable of pigment production and will have white kernels. [25] 9:7. [26] 15/16 have at least one dominant allele at either the A_1 or A_2 locus and, consequently, have round fruit; 1/16 lack a dominant allele at both the A_1 locus and the A_2 locus and have the genotype $a_1a_1a_2a_2$, and thus have narrow fruit. [27] 15:1.

IDENTIFYING SOME COMMON PHENOTYPIC RATIOS PRODUCED THROUGH GENE INTERACTION

The following problems will allow you to determine the phenotypic ratios that arise from crosses between individuals heterozygous for two different loci, designated as *A* and *B*, when various types of gene interactions are operating. Assume that each locus has a gene pair, with one allele completely dominant to its recessive counterpart. Each problem involves a different type of interaction and gives a table listing the following four genotypic categories of progeny produced from the cross *AaBb* × *AaBb* and the expected frequency for each category.

1. At least one dominant allele at each locus: *A_B_*, 9/16.
2. At least one dominant allele at the *A* locus and two recessive alleles at the *B* locus: *A_bb*, 3/16.
3. At least one dominant allele at the *B* locus and two recessive alleles at the *A* locus: *aaB_*, 3/16.
4. A dominant allele at neither locus: *aabb*, 1/16.

In writing out the symbols for each genotypic category, a blank line is used to indicate the counterpart allele for a dominant allele. For example, *A_* designates either genotype *AA* or *Aa*. We can use this shorthand designation because we are concerned with *phenotypic* expression rather than the genotypes per se. Since complete dominance operates between the two types of alleles at each locus, the specific counterpart for a dominant allele is irrelevant. (When the alleles at a locus show codominance or incomplete dominance, this type of notation cannot be used.)

In each problem, you are given the total number of progeny phenotypes that result from the gene interaction and are asked to determine the ratio in which these phenotypes occur. In each case assume that (1) two loci interact to determine the expression of a single trait, (2) each locus has a pair of alleles which show complete dominance, (3) the two loci assort independently of each other, and (4) a total of 16 offspring are produced (corresponding to the 16 boxes in a 4 × 4 Punnett square). All the interactions involved in these exercises produce a modification of the classic 9:3:3:1 dihybrid phenotypic ratio by grouping the four genotypic categories in different ways to produce the different phenotypic ratios. To illustrate the procedure to be used, let us work through the first problem.

PROBLEM: The presence of a dominant genotype at one of the two loci, say locus *A*, masks the alleles at the other locus (the *B* locus) and thus works by itself in determining the phenotype. In the absence of a dominant genotype at the *A* locus, the alleles at the *B* locus determine the phenotype. Three phenotypes result; identify the ratio in which they are expected.

SOLUTION: We are told that three phenotypes result and we need to determine the ratio in which these phenotypes are expected with this type of interaction. *Genotypic categories 1 and 2:* Both have dominant alleles at the *A* locus which masks the alleles of the other locus; thus, both categories 1 and 2 show the same phenotype, say, phenotype 1. Since there are nine (of 16) individuals expected in genotypic category 1 and three in genotypic category 2, a total of 12 individuals are expected to show this phenotype.
Genotypic category 3: Dominance at the *B* locus produces the second phenotype which is expected in three of 16 progeny.
Genotypic category 4: The absence of a dominant allele from both loci produces the third phenotype which is expected in one of 16 progeny.

This expected outcome is summarized in the following table. The number of individuals (out of 16) expected in each genotypic category is listed under the appropriate phenotype. For example, the nine and three individuals in genotypic categories 1 and 2, respectively, exhibit the same phenotype and are listed under phenotype 1. The three individuals in genotypic category 3 exhibit a second phenotype and are listed in the table under phenotype 2. The single individual in genotypic category 4 exhibits the third phenotype and is listed under phenotype 3. Adding the number of individuals in each column gives the expected phenotypic ratio of 12:3:1.

Genotypic Categories			Phenotypes		
A locus	*B* locus	Expected frequencies	1	2	3
1. *A_*	*B_*	9/16	9		
2. *A_*	*bb*	3/16	3		
3. *aa*	*B_*	3/16		3	
4. *aa*	*bb*	1/16			1

Expected phenotypic ratio: 12 : 3 : 1 ■

Determine the phenotypic ratio expected in each of the following six problems.

PROBLEM: A recessive genotype at a particular locus (assume it is *A*) masks the alleles at the other locus and always results in the same phenotype. In the absence of a recessive genotype at the *A* locus, the alleles at both loci express themselves and influence the phenotype. Three phenotypes result; identify the ratio in which they are expected.

SOLUTION:

	Genotypic Categories			Phenotypes		
	A locus	*B* locus	Expected frequencies	1	2	3
1.	*A*__	*B*__	9/16			
2.	*A*__	*bb*	3/16			
3.	*aa*	*B*__	3/16			
4.	*aa*	*bb*	1/16			

Expected phenotypic ratio: ___ : ___ : ___ [28]

PROBLEM: The presence of a dominant genotype at either locus matched with a recessive genotype at the other locus results in the same phenotype. If both loci have dominant alleles or if both loci have recessive alleles, the alleles at each locus express themselves in the phenotype. Three phenotypes result; in what ratio are they expected?

SOLUTION:

	Genotypic Categories			Phenotypes		
	A locus	*B* locus	Expected frequencies	1	2	3
1.	*A*__	*B*__	9/16			
2.	*A*__	*bb*	3/16			
3.	*aa*	*B*__	3/16			
4.	*aa*	*bb*	1/16			

Expected phenotypic ratio: ___ : ___ : ___ [29]

PROBLEM: The presence of a dominant genotype at either locus or at both loci results in the same phenotype. Only when both loci have recessive genotypes is a different phenotype produced. Two phenotypes result; in what ratio are they expected?

SOLUTION:

	Genotypic Categories			Phenotypes	
	A locus	*B* locus	Expected frequencies	1	2
1.	*A*__	*B*__	9/16		
2.	*A*__	*bb*	3/16		
3.	*aa*	*B*__	3/16		
4.	*aa*	*bb*	1/16		

Expected phenotypic ratio: ___ : ___ [30]

PROBLEM: The presence of a recessive genotype at either locus or at both loci results in the same phenotype. Only when both loci have dominant genotypes is a different phenotype produced. Two phenotypes result; in what ratio are they expected?

SOLUTION:

	Genotypic Categories			Phenotypes	
	A locus	*B* locus	Expected frequencies	1	2
1.	*A*__	*B*__	9/16		
2.	*A*__	*bb*	3/16		
3.	*aa*	*B*__	3/16		
4.	*aa*	*bb*	1/16		

Expected phenotypic ratio: ___ : ___ [31]

PROBLEM: The presence of a dominant genotype at one particular locus, say the *A* locus, or a recessive genotype at the other locus results in the same phenotype. Only when the *A* locus lacks a dominant genotype while, in the same individual, the *B* locus lacks a recessive genotype can the alleles at each locus express themselves in the phenotype. Two phenotypes result; in what ratio are they expected?

SOLUTION:

	Genotypic Categories			Phenotypes	
	A locus	*B* locus	Expected frequencies	1	2
1.	*A*__	*B*__	9/16		
2.	*A*__	*bb*	3/16		
3.	*aa*	*B*__	3/16		
4.	*aa*	*bb*	1/16		

Expected phenotypic ratio: ___ : ___ [32]

Note that in each of these problems, the *AaBb* × *AaBb* cross produces the same genotypic categories in the same ratio. The phenotypic ratios, however, vary depending on the nature of the gene interaction between the two gene pairs.

SUMMARY

Gene interaction can produce a modification of Mendelian phenotypic ratios. Whenever the gene interac-

[28] 9 (category 1) : 3 (category 2) : 4 (categories 3 and 4). [29] 9 (category 1) : 6 (categories 2 and 3) : 1 (category 4).
[30] 15 (categories 1, 2, and 3) : 1 (category 4). [31] 9 (category 1) : 7 (categories 2, 3, and 4). [32] 13 (categories 1, 2, and 4) : 3 (category 3).

tion involves two loci each with two alleles, and two organisms heterozygous for these loci are crossed, the 9:3:3:1 phenotypic ratio is modified. For example, a 9:3:4 phenotypic ratio reflects recessive epistasis, while the 9:7 and the 15:1 ratios result from complementary and duplicate genes, respectively. Be alert for these modified phenotypic ratios. They provide a clue not only to the occurrence of gene interaction, but often to the nature of that interaction. (Note also that the 9:3:3:1 phenotypic ratio can also be modified by linkage—the presence of two loci on the same chromosome. Linkage in eukaryotes will be discussed in detail in Chapters 12 and 13.)

PROBLEM SET

9-1. Two independently assorting loci interact to exert a major influence on fur color in some varieties of dogs. Alleles at the *B* locus determine the amount of the pigment melanin deposited in the strands of fur. The dominant allele, *B*, results in more pigment deposition to produce black fur, while expression of the recessive allele, *b*, results in less pigment deposition to produce brown fur. At the *I* locus the dominant allele, *I*, inhibits the depositing of melanin and results in white fur, while expression of its recessive counterpart, *i*, allows melanin to be deposited normally. Determine the phenotypic ratio of the progeny produced by mating two phenotypically white dogs, each heterozygous at both loci.

9-2. Two pure-breeding and unrelated lines of pea plants with white flowers are crossed and all of the progeny have purple flowers. Crossing two members of the F_1 produces an F_2 of purple- and white-flowered plants in a ratio close to 9:7. Based on this ratio, what can be inferred about the number of independently assorting gene pairs involved in producing this trait and the manner in which these gene pairs interact? Assume that complete dominance operates at each of the loci involved in determining this trait.

9-3. The legs of chickens may or may not be feathered, a trait determined by two independently assorting loci. A cross was carried out between a pure-breeding rooster with featherless legs and a pure-breeding female with feathered legs. Mating members of the F_1, all of whom had feathered legs, produced an F_2 generation of 219 chickens with feathered legs and 15 chickens with featherless legs. Assume that the two loci involved in determining this trait are independently assorting and that complete dominance operates at each locus. From this outcome, what can be inferred about the manner in which these loci interact?

9-4. Assume that coat color in swine, which may be red, sandy, or white, is determined by two independently assorting loci. A mating between a red-coated male and a white-coated female, each of which comes from a pure-breeding line, produces a red-coated F_1. Mating members of the F_1 produces an F_2 which consists of some red, some sandy, and some white individuals in a ratio of 9:6:1. Explain the type of gene interaction which produces this phenotypic ratio. Assume that complete dominance operates at each of the loci involved in determining this trait.

9-5. The form of the comb, or fleshy crest, on the head of a chicken can vary considerably and is controlled by genes of at least two independently assorting loci designated as *P* and *R*. If one or two dominant alleles are found only at the *P* locus, a pea-type comb is produced. If one or two dominant alleles are found only at the *R* locus, a rose-type comb results. The presence of at least one dominant allele at each locus produces a walnut-type comb, and if recessive alleles are simultaneously expressed at both loci (that is, if the genotype is *pprr*), a single-type comb is produced. Determine the phenotypic outcome of the following matings.

a. *PPrr* × *ppRR*
b. *PpRr* × *PpRr*

9-6. Feather color in chickens involves the interaction of two independently assorting loci, each with a pair of genes that show complete dominance with regard to each other. At one of these, the color locus, the allele for colored feathers is dominant to the allele for noncolored, or white, feathers. At the other, the inhibitory locus, the allele for color inhibition is dominant to the allele for color expression. A dominant genotype at the inhibitory

PROBLEM SET

locus prevents the expression of the dominant genotype at the color locus. A recessive genotype at the color locus is always expressed, regardless of the genotype at the inhibitory locus.

a. Identify the genotypic combinations that produce white feathers.

b. Identify the genotypic combinations that produce colored feathers.

c. In a cross between a hen and a rooster, each heterozygous for both loci, determine the progeny phenotypes and the ratio in which they would occur.

9-7. The shape of yellow summer-squash fruit is determined by two independently assorting loci. Fruit occurs in three shapes: long, spherical, and oval. A plant with oval fruit is crossed with a plant with long fruit. Both of these plants come from different pure-breeding lines and produce an F_1 consisting entirely of plants with oval fruit. Crossing two F_1 plants produces seeds which grow into 295 F_2 plants, of which 19 have long fruit, 106 have spherical fruit, and the remainder have oval fruit. Based on the outcome of these two crosses, what hypothesis explains the interaction between these two loci? Assume that complete dominance operates at each of the loci involved in determining this trait.

9-8. In guinea pigs, the positioning of the hair follicles and, hence, the direction in which the coat hairs grow, is determined by the interaction of genes at two independently assorting loci. Guinea pigs show three different phenotypes with regard to this trait: follicles oriented in the same direction produce a smooth coat, irregularly positioned follicles produce a rough coat, and a mixture of both types of follicles produces a partly rough coat. Complete dominance operates at the *A* locus, with the allele for rough coat, *A*, being dominant to the allele for smooth coat, *a*. At the *B* locus incomplete or partial dominance operates according to the following rules.

(1) When the homozygous dominant genotype *BB* is coupled with a dominant genotype at the *A* locus, the expression of the *A*_ genotype is modified to produce a smooth coat.

(2) When the heterozygous dominant genotype *Bb* is coupled with a dominant genotype at the *A* locus, the expression of the *A*_ genotype is modified to produce a partly rough coat.

(3) When a homozygous recessive genotype *bb* is coupled with a dominant genotype at the *A* locus, the *A*_ genotype is not modified in any way and is expressed as a rough coat.

(4) When *any* genotype at the *B* locus is coupled with an *aa* genotype, there is no modification, and the *aa* genotype is expressed, producing a smooth coat.

a. In the mating of two smooth-coated guinea pigs of genotypes *AABB* and *aabb*, what will be the phenotype or phenotypes of the F_1 progeny?

b. Use the product law to determine the frequency with which the following phenotypes would be expected in the F_2.

- **i.** Smooth coat
- **ii.** Rough coat
- **iii.** Partly rough coat

c. Identify all the phenotypes that would be expected in the F_2 and the ratio in which they would occur.

9-9. The fur in a particular mammalian species may be spotted with either tan or black spots, or it may lack spots altogether. This trait is controlled by the interaction of two independently assorting loci. At the *S* locus, the dominant

PROBLEM SET

allele, *S*, results in the production of spots while its recessive counterpart, *s*, codes for the absence of spots. At the *T* locus, the dominant allele, *T*, results in the production of tan spots while its recessive counterpart, *t*, produces black spots. A male from a line of pure-breeding tan-spotted animals is mated with a female from a line of pure-breeding unspotted animals. All of their progeny are tan-spotted. When members of the F^1 are allowed to interbreed, their offspring consist of 144 tan-spotted, 63 unspotted, and 47 black-spotted animals. Based on the outcome of these two crosses, what hypothesis would you propose to explain the interaction between these two loci?

9-10. The seed color in a species of flowering plant can be either green or white, with the trait determined by two independently assorting loci. At one locus, the *G* allele produces green color and is dominant to the recessive *g* allele for white. At the other locus, the dominant *C* allele codes for color production and its recessive counterpart, *c*, codes for colorless or white seeds. A recessive genotype at either locus blocks the expression of the genes at the other locus. A pure-breeding green-seeded plant is crossed with a white-seeded plant homozygous for both loci, and all of the F_1 plants have green seeds.

a. Identify all possible genotypes for the white-seeded parent and identify the F_1 genotype or genotypes that each would produce.

b. Assume that the genotype of the original white-seeded parent is *ggcc*. F_1 plants from the cross described in problem 9-10a are crossed to produce the F_2. Identify the F_2 phenotypes and their expected ratio.

9-11. Based on the information given in problem 9-10, what would be the expected probability of producing a white-seeded plant in a cross between two white-seeded F_2 plants with the genotypes of *Ggcc* and *ggCc*?

9-12. Based on the information given in problem 9-10, what would be the expected probability of producing a white-seeded plant in a cross between two white-seeded plants with the genotypes of *Ggcc* and *ggcc*?

9-13. Body color in a species of beetle is determined by the interaction of two independently assorting loci. Four body colors are possible and they are, listed in order of increasing darkness, red, sooty, dark-sooty, and black. The two alleles at the *A* locus show complete dominance with the allele for red, *A*, dominant to the allele for sooty color, *a*. The two alleles at the *B* locus show partial or incomplete dominance: genotypes *BB*, *Bb*, and *bb* produce phenotypes of red, sooty, and black, respectively. The two loci interact according to the following rules.

(1) A dominant genotype at the *A* locus with any genotype at the *B* locus results in the *B* locus masking the *A* genotype and expressing the *B* genotype.

(2) The *aa* genotype with a dominant genotype at the *B* locus causes the color specified by the *BB* or *Bb* genotype to be shifted to the next darker body color. For example, *aa* combined with *Bb* causes the sooty specified by the *Bb* to darken to dark-sooty.

(3) *aa* combined with *bb* produces a black phenotype. Determine the phenotypes for each of the following genotypes.

a. *aaBb*

b. *aabb*

c. *Aabb*

d. *AaBB*

PROBLEM SET

e. *AaBb*
f. *aaBB*

9-14. A cross is carried out between a pair of organisms heterozygous for the two loci described in problem 9-13. Use the product law to determine the frequencies with which the following phenotypes would be expected among the progeny:

a. black.
b. dark-sooty.
c. red.
d. Determine the expected phenotypic ratio among the progeny produced in this mating.

9-15. Refer to the information given in problem 9-13. What are the probabilities of producing a black organism and a red organism, respectively, from the following matings?

a. *AaBb* × *aabb*
b. *aaBB* × *AAbb*

9-16. Wing color in an insect species shows four possible phenotypes, listed in order of increasing darkness: light-gray, medium-gray, dark-gray, and black. Two independently assorting loci interact to determine this color. The alleles at the *G* locus show complete dominance, with a dominant genotype producing light-gray wings and the recessive genotype producing medium-gray wings. The alleles at the *H* locus show incomplete dominance, with genotypes *HH*, *Hh*, and *hh* producing light-gray, dark-gray, and black wings, respectively. The mating of two dark-gray-winged individuals that are heterozygous at both loci produces 16 offspring whose genotypes are listed as follows, grouped according to their phenotypes.

Light-gray: *GGHH*(1), *GgHH*(2)
Medium-gray: *ggHH*(1), *ggHh*(2)
Dark-gray: *GGHh*(2), *GgHh*(4)
Black: *GGhh*(1), *Gghh*(2), *gghh*(1)

Specify the nature of the interaction between these two loci.

9-17. Refer to the information presented in this chapter that describes the *C*, *B*, and *A* loci whose products interact to determine fur color in mice.

a. Identify the expected genotypes and phenotypes of the F_2 progeny produced from the crossing of F_1 individuals derived from an albino parent with genotype *ccBBAa* and a brown agouti parent with genotype *CcbbAA*.
b. One of the albino mice from the F_2 is successively crossed with the same black agouti mouse to produce several litters. These litters include albino, black agouti, black, brown agouti, and brown mice. What can be concluded about the genotypes of the albino and black agouti parents?
c. With what type of mouse could the albino parent referred to in problem 9-17b be crossed to determine its genotype?

9-18. As described in this chapter, two loci in addition to those mentioned in problem 9-17 also play major roles in determining fur color in mice. At the *S* locus, expression of the dominant allele *S* produces a nonspotted coat, and its recessive counterpart *s* produces a spotted coat. At the *D* locus, the dominant allele *D* slightly reduces the color intensity to produce a milky fur color, and its recessive counterpart *d* produces a color of normal intensity. Assume that all five loci assort independently.

PROBLEM SET

a. Identify the fur-color phenotypes for mice with the following genotypes: *CcbbaaSsdd*, *ccBbAassDd*, and *CCBbAAssdd*.
b. Identify the expected phenotypes and their frequencies for the progeny from the cross *ccBbAassDd* × *ccBbAAssdd*.
c. Identify the expected phenotypes and their frequencies for the progeny from the cross *ccbbAaSsDD* × *CCBbAassdd*.

CHAPTER

10

Sex Chromosomes, Sex-Chromosome Systems, and Sex Linkage

INTRODUCTION

In most diploid organisms with distinct sexes, there are two categories of chromosomes: the **autosomes** and the **sex chromosomes.** In the human genome, for example, there are 23 pairs of chromosomes which consist of 22 pairs of autosomes and a pair of sex chromosomes. In the much studied fruit fly *Drosophila melanogaster,* there are three pairs of autosomes and one pair of sex chromosomes. The sex chromosomes usually play a key role in sex determination, and the loci carried on them show distinctive inheritance patterns. Whereas autosomal chromosome pairs are identical in size and shape, the sex chromosomes in species that possess two types, as we will see, generally differ from each other in shape, size, and in the loci they carry.

SEX-CHROMOSOME SYSTEMS

Three sex-chromosome systems, designated as XY, XO, and ZW, are generally encountered in different animal species.

The XY System

The XY system, the most common of the three, is found in humans, other mammals, and the fruit fly. This system involves two different types of sex chromosomes, one designated as X and the other as Y. The X and Y chromosomes generally differ from each other in shape and size. For example, in humans the Y chromosome is considerably smaller than the X chromosome. In fruit flies, the Y chromosome is slightly larger than the X and has a different shape. Each individual normally possesses a pair of sex chromosomes: a female has two X chromosomes (XX) and a male has one X and one Y chromosome (XY). The males are designated as **heterogametic** because they produce two types of gametes with regard to the sex chromosomes: half the sperm carry the X chromosome and half carry the Y. In contrast, the females are **homogametic** since they produce one type of egg, which carries an X chromosome. The sex of the offspring is determined by genes carried in the sperm. The fertilization of an egg by an X-carrying sperm produces an XX zygote which develops into a female, while its union with a Y-carrying sperm forms an XY zygote which develops into a male.

The XO System

The XO system, found in numerous insect species, involves a single sex chromosome, designated as X.

Females carry two of these chromosomes and are designated as XX, while males have a single X chromosome and are designated as XO, with the O serving to indicate that the X has no counterpart. Since each sex carries a different number of sex chromosomes, there is a difference in the total number of chromosomes carried by the females and males. For example, in the firefly genus *Pyrrhochoris,* males have a total of 23 chromosomes in the nuclei of their diploid cells while females have 24. With the absence of a counterpart for the X chromosome, males are heterogametic, forming two types of sperm cells: half carry the X chromosome and half lack it. In *Pyrrhochoris,* half the sperm have 12 chromosomes and half have 11, while the females are homogametic, with all eggs having 12 chromosomes. As with the XY system, the sex of the offspring is determined by the sperm. When a 12-chromosome sperm combines with an egg, how many X chromosomes will be in the zygote? ________[1] What sex will develop from this type of zygote? ________[2] When an 11-chromosome sperm combines with an egg, how many X chromosomes will be in the zygote? ________[3] What sex will develop from this type of zygote? ____[4]

The ZW System

The third system, known as ZW, operates in many birds, butterflies, moths, reptiles, and a few amphibians and fish. The ZW system shows some similarities to the XY system in that there are two kinds of sex chromosomes, but in this case one is designated as Z and the other as W. Using letters different from X and Y to label the chromosomes emphasizes a key difference associated with this system: it is the male rather than the female that carries a pair of sex chromosomes of the same type. Under the ZW system, males are designated as ZZ, and the females, with two different types of sex chromosomes, as ZW. Under the ZW system, then, males are homogametic and females are heterogametic, showing the reverse of the situation found under the XY system. Here it is the chromosomal makeup of the egg that determines the sex of the offspring. The union of a Z-carrying egg with a sperm produces a female while a W-carrying egg produces a male.

SEX-LINKAGE INHERITANCE PATTERNS WITH THE XY SYSTEM

In most species with the XY system, most of the loci carried on the X chromosome have no counterpart on the Y chromosome since much of the Y chromosome is genetically inert. Genes carried on the X chromosome were first identified by T. H. Morgan and are said to be **sex-linked** or **X-linked.** Such genes show dominance-recessiveness interactions only in females (XX) where two alleles are present. In males, dominance-recessive interactions cannot occur because there is only a single X chromosome and thus only a single allele for each sex-linked trait. Since one allele determines the male phenotype, males are said to be **hemizygous** with regard to sex-linked traits. Females, in contrast, are either homozygous or heterozygous for these traits. Since the Y chromosome carries no counterpart for the allele on the X chromosome, X-linked genes show a number of distinctive inheritance patterns, three of which are described next.

Absence of Father-to-Son Transmission for X-Linked Genes

The inheritance pattern of genes on the X chromosome is determined by the inheritance pattern of the X chromosomes. Since a father passes his X chromosome only to his daughters (with his Y chromosome going only to his sons), father-to-son transmission of X-linked loci does not occur. We can demonstrate this by considering the inheritance at the X-linked eye-color locus in the fruit fly and crossing a true-breeding red-eyed fly with a white-eyed male. The symbol for the recessive white-eye allele is w, and a w^+ (or sometimes just +) is used to designate its dominant counterpart, the allele for the wild-type red eye color. Often, the symbol for the gene is written as a superscript to the letter X which represents the X chromosome. In writing out the male genotype, the Y chromosome may be represented by the letter Y, by a bent slash, ⁄, or by a hooked dash, ⇀. Thus, the genotype for a white-eyed male could be written as X^wY, as X^w⁄, or as X^w⇀, and that for a red-eyed female as X^+X^+. (Some texts opt for a simpler system and delete the symbol for the X chromosome. Under that system, the white-eyed male parent in this cross could be written as wY, w⁄, or w⇀ and the red-eyed female as w^+w^+ or + +.) Identify the phenotypes of progeny that would be expected from the mating of X^+X^+ and X^wY by completing the following branch diagram. In specifying the outcome of the cross, list the females and males separately.

Female gametes	Male gametes		Progeny genotypes
	X^w	=	________
X^+			
	Y	=	________

[1] Two X chromosomes. [2] Female. [3] One X chromosome. [4] Male.

Expected outcome: ________________________________ ________________.[5] Note that all the female progeny are heterozygous for the trait and as such are called **carriers,** since they carry but do not express the recessive allele.

Now mate two members of the F_1. The expected progeny of this $X^+X^w \times X^+Y$ mating can be identified by completing the branch diagram that follows. In specifying the outcome of the cross, list the females and males separately.

Female gametes	Male gametes	Progeny genotypes
X^+	X^+	= ________
	Y	= ________
X^w	X^+	= ________
	Y	= ________

Expected phenotypic outcome: ________________ ________________________.[6]

Now that we have followed the transmission of this eye-color trait through three generations, look back at what has happened to the expression of the white-eye phenotype in going from parents to F_1 to F_2. What pattern do you observe? ________________ ________________________[7] This inheritance pattern, where the male parent expresses a recessive mutant phenotype that appears to skip the F_1 generation and then reappears in the F_2, is characteristic of loci that are X-linked. This skipping of a generation relates to the absence of father-to-son transmission of the X chromosome (and, of course, to the fact that the Y chromosome that is received by the male offspring carries no locus for this particular trait).

Differential Outcomes of Reciprocal Matings

The **reciprocal** of the original fruit-fly mating just described involves crossing a white-eyed female with a red-eyed male. Although reciprocal crosses involving autosomal traits give identical results, such crosses involving sex-linked traits do not. To demonstrate this, determine the expected phenotypic outcome for the reciprocal cross $X^wX^w \times X^+Y$, which is set up in the following branch diagram.

Female gametes	Male gametes	Progeny genotypes
X^w	X^+	= ________
	Y	= ________

Expected phenotypic outcome: ________________ ________________________.[8] Now compare these F_1 progeny with the F_1 progeny produced in the original cross ($X^+X^+ \times X^wY$) you worked through earlier. What do you notice? ________________ ________________[9] Why does this difference arise? ________________________ ________[10]

Crisscross Inheritance

Before leaving the $X^wX^w \times X^+Y$ cross you just worked, identify the sex of the parent with red eyes and compare it with the sex of the progeny with red eyes. What do you notice? ________________ ________________[11] Now compare the sex of the parent with white eyes and the sex of the progeny with white eyes. What do you notice? ________________ ________________________[12] This so-called **crisscross inheritance** pattern is demonstrated in the progeny in crosses involving X-linked loci whenever the female parent exhibits the recessive phenotype and the male parent shows the normal phenotype.

If you are working a cross and have no information as to whether a trait is X-linked or carried on an autosome, the three patterns that you have just demonstrated—that is, the absence of father-to-son transmission, crisscross pattern, and differential outcomes of reciprocal crosses—can be useful in identifying a locus as X-linked.

[5] All females are X^+X^w and red eyed; all males are X^+Y and red eyed. [6] Half the females are homozygous for the wild-type allele (X^+X^+) and half are heterozygous carriers (X^+X^w); all are red eyed. The males, half X^+Y and half X^wY, would be red eyed and white eyed, respectively. [7] The white-eye trait, expressed by the original male parent, is not shown by any F_1 flies, but reappears among half the male flies in the F_2 generation. [8] All females are red-eyed heterozygous carriers (X^+X^w) and all males are white eyed (X^wY). [9] The outcomes are different. [10] The locus is found on the X chromosome rather than on an autosome. [11] The red-eye phenotype of the male parent appears in the F_1 females. [12] The white-eye phenotype of the female parent appears in the F_1 males.

Examples of Crosses Involving Human X-Linked Disorders

Problems involving sex-linked traits generally ask you to identify progeny genotypes or phenotypes from information about the parents or to identify parental genotypes or phenotypes from information about the progeny. Among the numerous human X-linked traits, two are relatively common: hemophilia (a disorder resulting from a deficiency of proteins essential for blood clotting) and certain types of color blindness.

Example Involving Identification of Progeny Some problems ask you to draw a conclusion about the progeny based on information provided about the parents.

PROBLEM: A woman is heterozygous for the X-linked hemophilia allele which is recessive to the allele for normal blood clotting. Her husband's blood exhibits normal clotting. They plan to have a large family. Should they expect to have any children with hemophilia? If so, give the sex of those affected and their expected frequency.

SOLUTION:

Basic information given: The allele for hemophilia is X-linked and recessive to the allele for normal clotting. The woman is heterozygous for this allele. The husband's clotting is normal.

Question to be answered: Can this couple expect to have hemophiliac children? If so, identify the sex and expected frequency of the hemophiliac children.

Strategy: To answer this question, we need to identify the genotypes of the parents and the types of gametes each forms and determine all possible combinations of those gametes to get the progeny genotypes. From these genotypes we can identify the progeny phenotypes and their frequencies.

Key: Let H be the allele for normal clotting and h be the allele for hemophilia.

Parental genotypes: The mother is X^HX^h and the father is X^HY.

Gametes: The mother forms two kinds of gametes: X^H (1/2) and X^h (1/2). The father makes two kinds of gametes: X^H (1/2) and Y (1/2).

Union of gametes: Complete the following branch diagram.

Male gametes	Female gametes	Progeny genotypes
X^H (1/2)	X^H (1/2) =	________
	X^h (1/2) =	________
Y (1/2)	X^H (1/2) =	________
	X^h (1/2) =	________

Results: Expected progeny: 1/4 homozygous dominant (X^HX^H) normal female, 1/4 heterozygous (carrier) (X^HX^h) normal female, 1/4 normal (X^HY) male, and 1/4 hemophiliac (X^hY) male.

Answer: Yes, hemophiliac children could be produced: one out of two male children would be expected to be affected. ■

Example Involving Identification of Parents' Genotypes Some problems involving X-linked traits will ask you to infer parental genotypes or phenotypes based on information provided about the progeny as in the next example.

PROBLEM: A woman has given birth to two sons. Both are affected with the most common type of partial color blindness occurring in humans. This disorder, known as green weakness (deutan color blindness, or deuteranomaly), is caused by less than normal amounts of the green-sensitive vision pigment in certain cells of the retina of the eye and is responsible for altering the perception of both red and green colors. The trait is sex-linked and caused by an allele recessive to that for normal color vision. The woman and her husband have normal vision and you are asked to determine their genotypes.

SOLUTION:

Basic information given: Green-weakness color blindness is an X-linked trait and is caused by an allele that is recessive to that for normal color vision. The mother and father are phenotypically normal and have two sons, both of whom are affected with green-weakness color blindness.

Question to be answered: What are the parents' genotypes?

Strategy: To answer this question, we must make inferences from two sources of information: (1) the parents' phenotypes and (2) the sons' phenotypes.

Key: Let *G* be the allele for normal color vision and *g* be the allele for green weakness.

The following reasoning is involved in solving this problem.

1. Since the father is hemizygous and has normal vision, his X chromosome must carry the normal gene and his genotype must be X^GY.
2. The two sons must also be hemizygous for the trait. Since each is color blind, each must have the X^gY genotype.
3. Since the Y chromosome carried by each son could have come only from the father, each son must have received his X chromosome from his mother. This tells us that the mother has an X chromosome carrying the recessive allele for color blindness, that is, X^g.
4. Since the mother shows a normal phenotype, her other X chromosome must carry the normal allele (X^G). Thus, the mother's genotype is X^GX^g.

Answer: Mother's genotype: X^GX^g; father's genotype: X^GY. ■

Problems of this sort obviously need to be thought through on a step-by-step basis. Remember that under the XY system, the genotype of a male can readily be deduced from his phenotype since males are always hemizygous, and any X-linked gene present will be phenotypically expressed. The X-linked allele expressed by a male is always present on one of the mother's X chromosomes. If there are several male progeny, it is often possible to deduce from them the allele carried on each of the mother's X chromosomes. If you are only given phenotypes and proportions for the progeny and the parent genotypes are to be idenfied, begin with the male progeny to deduce the genotypes of the mother's X chromosomes. Once this information is available, it is often possible to deduce the genotype of the X chromosomes contributed by the father by looking at the female progeny.

Before concluding our discussion of the inheritance of traits under the XY system, two other types of traits should be mentioned.

1. Y-linked or **holandric** traits. Genes for these traits are restricted to the Y-chromosome and, since they lack counterparts on the X chromosome, normally only affect males. An affected father will transmit the trait to each of his sons but never to his daughters. The human trait of hairy ears, where tufts of hair of unusual length and coarseness grow from the rims of the ears, may be caused by a Y-linked allele.

2. X- and Y-linked traits. Loci that are carried by both X and Y chromosomes are known as **incompletely sex-linked.** This means that the locus is represented twice in both males and females, with a female carrying it on each X chromosome and a male carrying it on the X chromosome and on the Y chromosome. Either sex can be homozygous or heterozygous for such traits. In the fruit fly, *Drosophila,* a locus designated as bobbed, which influences bristle length, is carried on both the X and Y chromosomes.

SEX LINKAGE INHERITANCE PATTERNS WITH THE XO SYSTEM

The inheritance patterns for X-linked loci under the XO system are the same as those seen in the XY system. In a male (XO) there is a single X chromosome and, consequently, only a single allele to determine the phenotype for each X-linked trait. As with the XY system, a male (XO) is hemizygous for such traits while a female (XX) can be either homozygous or heterozygous. The absence of father-to-son transmission, the crisscross pattern, and the differential outcomes of reciprocal crosses which characterize the inheritance of X-linked loci under the XY system also characterize sex-linked loci under the XO system.

SEX-LINKAGE INHERITANCE PATTERNS WITH THE ZW SYSTEM

Under the ZW system there is a reversal of the situation seen with the XY and XO systems. Recall that under the ZW system, the males (ZZ) carry sex chromosomes of the same type and are homogametic and the females (ZW) have two different kinds of sex chromosomes and are heterogametic. Because much of the W chromosome is genetically inert, females have just one copy of sex-linked alleles and are hemizygous, whereas the males with two alleles may be homozygous or heterozygous. The inheritance patterns discussed under the XY and XO systems for X-linked loci also apply to Z-linked loci, but with the patterns reversed. For example, under the ZW system, there is an absence of mother-to-daughter transmission of Z-linked loci.

Example Involving Identification of Progeny

The inheritance pattern for a Z-linked locus is illustrated by the locus responsible for feather pattern in chickens. The dominant allele at this locus, *B*, produces bands of light and dark coloration, a phenotype referred to as barred. With the expression of the recessive allele, *b*, the feathers are uniformly colored (or nonbarred). Assume that a nonbarred female, Z^bW, is crossed with a barred male, Z^BZ^B. Determine the expected F_1 progeny by completing the following branch diagram.

Male gametes	Female gametes	Progeny genotypes
Z^B	Z^b	= ________
	W	= ________

Expected outcome: ____ .[13] Now cross two members of the F_1 in the space provided to produce the expected F_2 generation.

Expected outcome: ____ .[14]

POINTERS FOR SOLVING SEX-LINKED INHERITANCE PROBLEMS

Begin by identifying the sex-chromosome mechanism for the species involved. Basically this means identifying whether the heterogametic sex is male (XY and XO mechanisms) or female (ZW mechanism). (If you are uncertain whether the locus involved in the problem is sex-linked rather than autosomal, the following may provide clues: crisscross inheritance patterns, the lack of similar outcomes in reciprocal crosses, the absence of father-to-son transmission—which would occur with the XY or XO systems, and the absence of mother-to-daughter transmission—with the ZW system.) Then determine whether the mutant allele of the sex-linked gene is dominant or recessive to the normal, wild-type allele (most are recessive).

Remember the following points regarding X-linked genes under the XY or XO systems. The phenotype arising from the expression of an X-linked *recessive* allele will occur (1) more frequently in male than in female progeny, (2) in female progeny only when the responsible allele is expressed by the male parent *and* is carried or expressed in the female parent, and (3) in male progeny only if the female parent carries or expresses the allele.

The phenotype arising from the expression of an X-linked *dominant* allele will occur (1) in each generation, (2) in female progeny if either parent is affected, and (3) in male progeny if the female parent is affected.

Take a moment to note how these points for X-linked recessive and dominant alleles would be modified if the alleles involved were Z-linked. As a final point, be sure that in specifying the outcomes of crosses, you list the males and females separately.

SUMMARY

In diploid organisms with distinct sexes, there are two categories of chromosomes: the autosomes and the sex chromosomes. The sex chromosomes usually play a key role in sex determination and the loci carried on them show distinctive inheritance patterns. Three sex chromosome systems, XY, XO, and ZW, are encountered in different animal species. Under the XY and XO systems, males are hemizygous for X-linked traits since they have just one X chromosome and thus carry just one allele. Females, in contrast, are heterozygous or homozygous for these traits since they have two X chromosomes and thus carry two alleles. The reverse applies under the ZW system, where females are hemizygous and the males are homozygous or heterozygous. Sex-liked genes show a number of distinctive inheritance patterns: the absence of father-to-son transmission (the absence of mother-to-daughter transmission with the ZW system), different outcomes of reciprocal matings, and crisscross transmission. Pointers for solving inheritance problems involving sex linkage are summarized in the chapter.

[13] Males are Z^BZ^b and the females are Z^BW and all are barred. [14] Half the males are Z^BZ^B and half are Z^BZ^b and all are barred; half the females are Z^BW and barred and half are Z^bW and nonbarred.

PROBLEM SET

10-1. A type of partial color blindness in humans known as red weakness is less common than the green weakness discussed in this chapter and results in slightly different alterations in color perception. The locus for red weakness is carried on the X chromosome, and the allele producing the condition, *r*, is recessive to its wild-type counterpart, *R*, which produces normal color perception. A woman affected with red weakness marries a man with normal vision. Identify the expected genotypes and phenotypes of their children.

10-2. Miniature wing form in the fruit fly, *Drosophila*, is determined by the allele, *m*, which is carried on the X chromosome and which is recessive to the allele for wild-type normal wings, m^+. A miniature-winged female is mated with a normal-winged male.

a. What genotypes and phenotypes would be expected among their progeny?

b. What progeny phenotypes would be expected from mating an F_1 female with the father?

c. What progeny phenotypes would be expected from mating two F_1 flies?

10-3. Coat color in some varieties of cats is influenced in large measure by a pair of codominant alleles that are sex-linked. A black coat is produced by the allele *B*, and allele B' produces a yellow coat. When both alleles are present in the heterozygous condition, a tortoise-shell color is produced.

a. What progeny genotypes and phenotypes would be expected from the mating of a yellow male and tortoise-shell female?

b. What progeny genotypes and phenotypes would be expected from the mating of a black male and a homozygous yellow female?

c. Under normal circumstances, would you expect to find tortoise-shell males? Explain.

10-4. Refer to the basic information given in problem 10-3. A tortoise-shell cat has a litter of eight kittens—half male and half female. Among the males are two yellow and two black; among the females, two tortoise-shell and two black. What was the color of the tomcat who fathered this litter?

10-5. Hemophilia is a blood-clotting disorder caused by an allele, *h*, which is recessive to the allele for normal blood clotting, *H*. A woman whose father was a hemophiliac and whose mother was normal with no family history of hemophilia marries a man whose blood exhibits normal clotting.

a. What is the probability that a daughter will have hemophilia?

b. What is the probability that a son will have hemophilia?

10-6. A woman with normal blood clotting and with no family record of hemophilia marries a man whose mother was a hemophiliac. How many of their four children (two girls, two boys) would you expect to be affected with hemophilia?

10-7. Wing coloration in the magpie moth, a species exhibiting the ZW system of sex chromosomes, is determined by a locus carried on the Z chromosome. The dominant allele, *L*, at this locus produces dark wings while the expression of its recessive counterpart, *l*, results in wings which are light in color.

a. Cross a homozygous light-winged male with a dark-winged female and determine the expected F_1, indicating the sex and phenotype of the progeny.

PROBLEM SET

b. Cross a male and a female from the F_1 to determine the expected outcome of the F_2 generation.

c. For both of the crosses you have just worked through, compare the parental phenotypes with those of the male and female offspring. Which cross, if either, exhibits the crisscross inheritance pattern?

10-8. The bobbed locus in the fruit fly, *Drosophila*, is unusual in that it is carried on both the X and Y chromosomes. The recessive allele, designated by the two-letter symbol *bb*, produces bobbed or shortened bristles. Its dominant counterpart, bb^+, produces normal wild-type bristles.

a. Cross a male homozygous for wild-type bristles with a bobbed-bristled female and determine the genotypes and phenotypes expected among their progeny. Would the outcome of this cross be different if the bobbed locus were carried on a pair of autosomes?

b. Carry out the reciprocal of the cross described in problem 10-8a and compare its outcome with the outcome from 10-8a. Would the same outcome be expected if this locus were X-linked? Explain.

c. Cross two members of the F_1 produced from the cross described in 10-8a and determine the genotypes and phenotypes expected among the progeny. Would the outcome of this cross be different if the bobbed locus were carried on a pair of autosomes?

10-9. Green-weakness partial color blindness, or deuteranomaly, is a sex-linked recessive trait. The green-weakness allele, *g*, is recessive to the allele for normal color vision, *G*. Another form of abnormal vision is total color blindness which is due to an autosomal gene, *t*, which is recessive to its counterpart allele, *T*, for normal vision. A man with a normal phenotype marries a woman with green-weakness partial color blindness. Both individuals are heterozygous for autosomal color blindness. What phenotypes and phenotypic frequencies would be expected among their daughters and among their sons?

10-10. In *Drosophila*, the allele for white eyes, *w*, is inherited as a sex-linked trait and is recessive to the allele for red eyes, w^+. The allele for yellow body color, *y*, is also inherited as a sex-linked trait and is recessive to the allele for normal body color, y^+. The offspring of a cross between two flies gives the following results.

Females: 1/2 red eyes, yellow body; 1/2 white eyes, normal body. *Males*: 1/2 red eyes, yellow body; 1/2 white eyes, normal body. What were the genotypes of the parents? (Assume that no crossing over has taken place.) Note: In writing out the genotypes, be careful not to confuse *y* and y^+, which designate the alleles for body color, with Y, which symbolizes the sex chromosome in the male.

10-11. A rare sex-linked recessive allele, *i*, results in a cleft in the iris of the eye. Its dominant counterpart, *I*, is essential for the development of a normal iris. The dominant allele, *M*, at a certain autosomal locus is responsible for a particular type of migraine headache, and the expression of its recessive counterpart, *m*, results in the absence of this type of migraine. A mother, who is migraine-free, has a daughter who suffers from migraine headaches. In addition, this daughter has a cleft iris. Based on the information given here, what can be concluded about the genotype of

a. this daughter?

b. the father of this child?

c. the mother?

PROBLEM SET

10-12. Plumage color in chickens is controlled by a sex-linked gene. The dominant allele at this locus, *G*, produces silver plumage, while the expression of its recessive counterpart, *g*, results in gold plumage. (Remember that in chickens, the female is heterogametic.)

a. Identify the phenotypes and genotypes of the progeny which result from the mating of a gold-plumed rooster and a silver-plumed hen.

b. Identify the phenotypes and genotypes of the F_2 progeny which result when two members of the F_1 are mated.

10-13. Refer to the basic information given in problem 10-12.

a. Identify the phenotypes and genotypes of the progeny which result from the mating of a silver-plumed rooster, derived from a pure-breeding line, with a gold-plumed hen.

b. Identify the phenotypes and genotypes of the F_2 progeny which result when two members of the F_1 are mated.

10-14. Duchenne muscular dystrophy in humans is sex-linked, and the allele causing it, *d*, is recessive to the allele for the normal condition, *D*. This severe affliction usually begins to express itself between the ages of 3 and 5, and is characterized by muscle deterioration. The disease generally culminates in the death of affected individuals during the teenage years. A man and a woman have three young children, two boys, age 2 and 5, and one girl, age 1. The older son has just been diagnosed as having Duchenne muscular dystrophy. The parents have been told that this is an inherited condition and they are concerned about their other children developing this disorder. What is the likelihood that their other children will be affected?

10-15. In humans, the production of defective brown tooth enamel is due to a sex-linked dominant allele (*B*) that is dominant to its recessive counterpart (*b*). A man with brown tooth enamel marries a woman with normal teeth (and no history of brown tooth enamel in her family). Their daughter marries a man with normal tooth enamel.

a. What is the probability that a son from this latter marriage will have brown tooth enamel?

b. What is the probability that a daughter from this latter marriage will have normal tooth enamel?

10-16. The locus controlling production of the H-Y antigen is found on the Y chromosome in many mammals and has no counterpart on the X chromosome. In a cross between an H-Y antigen-producing male mouse and a female, what proportion of the male progeny would produce the antigen? What proportion of the female progeny would produce the antigen?

10-17. a. Basic information about the bobbed locus in the fruit fly, *Drosophila*, is given in problem 10-8. Expression of the allele bb^l, which is recessive to the wild-type allele bb^+, is lethal. Identify the genotypes and phenotypes of the progeny of a mating between a female heterozygous for this lethal allele and an $X^{bb^l}Y^{bb^+}$ male. What sex ratio would be expected among their progeny?

b. Would the outcome of the mating described in 10-17a differ if the lethal allele carried by the male were on his X chromosome rather than on his Y chromosome? Explain.

CHAPTER

11

Human Pedigree Analysis

INTRODUCTION

Much of what is known about single-gene inheritance patterns has been learned from the results of controlled experimental matings. Another approach to gathering this information is through analysis of family histories or **pedigrees** which trace the inheritance of a particular trait through successive generations. Pedigrees are often used with organisms that have long time intervals between generations and relatively small numbers of progeny. They have been particularly useful for identifying the inheritance patterns of human traits. Pedigree analysis may make it possible to hypothesize whether the allele responsible for a trait is inherited as an autosomal dominant, autosomal recessive, sex-linked recessive, or sex-linked dominant allele. Pedigrees play an essential role in genetic counseling, where a geneticist may be able to predict the chance that a trait will be transmitted from one generation to the next.

In this chapter we will look at the symbols used in pedigrees, construct a simple pedigree, and learn how the genotypes of individuals included in a pedigree may be deduced. We will review pedigree inheritance patterns shown by recessive and dominant autosomal and sex-linked traits and look at an approach for analyzing pedigrees when the nature of the allele responsible for the trait is unknown. At the end of this chapter we will consider how the risk of transmitting an allele may be determined from a pedigree.

PEDIGREE SYMBOLS

Pedigrees are constructed using a set of standard symbols: a male is represented by a square and a female by a circle. An open square or circle (□ or ○) indicates an individual with the normal or wild-type phenotype, and a shaded square or circle (■ or ●) denotes a mutant phenotype. A small solid dot (•) indicates an individual who died in infancy and a diamond (◇) designates an individual of unspecified sex. Two mating individuals are joined by a horizontal line. A vertical line from a mating line connects to a horizontal sibling line to which the offspring of these parents are connected.

A simple pedigree showing the inheritance history for a gene is shown in Figure 11-1. Note that each generation is shown on a different line of the pedigree that is numbered on the left with a roman numeral. Each individual within a particular generation is given an arabic number, assigned, if possible, according to the order of birth (if the birth order is unknown, the numbers are assigned arbitrarily). How many genera-

FIGURE 11-1

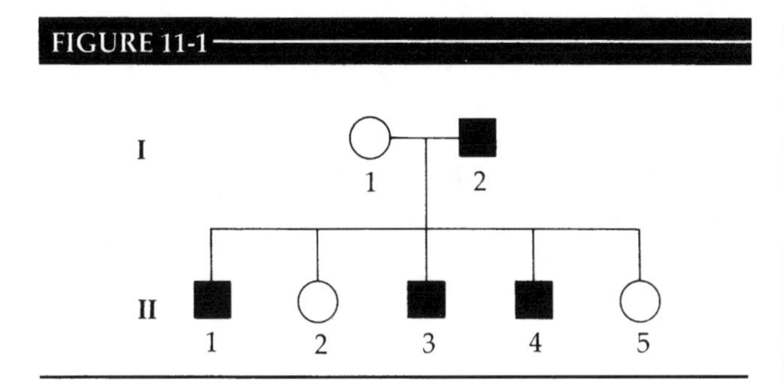

tions are shown in this pedigree? _________[1] How many matings are included in the pedigree? ________[2] How many offspring did the original parents produce? _________[3] How many generation-II males and females are affected with the mutant trait? _________[4]

CONSTRUCTING A SIMPLE PEDIGREE

With this basic information in hand, you are now ready to construct a pedigree. Albinism in humans is due to the absence of the pigment melanin—the skin is very lightly colored, the hair is white, and the irises of the eyes are pink (due to the exposure of blood vessels that are usually masked by the pigment). The condition is inherited as an autosomally based gene with two alleles: the normal pigment allele, c^+, is dominant to the albino allele, c. In the space provided, sketch a pedigree that summarizes the following family history. A normally pigmented man married an albino woman and they produced a family consisting of two sons, three daughters, and two sons, in that order. Their first two sons and last two daughters were normal, and the remaining children were albinos. The firstborn normal son married an albino woman and they had a normal son, an albino son, and two normal daughters, in that order. Be sure to number each generation and each individual within the generations.

Check your pedigree against Figure 11-2.

IDENTIFYING GENOTYPES FROM A PEDIGREE

It is often possible to deduce the genotypes for at least some of the individuals included in a pedigree. For example, the male parent in generation I of the pedigree in Figure 11-2 has normal pigmentation. Given this, what genotype or genotypes are possible for him? _________[5] What genotype or genotypes are possible for his wife? _________[6] Before considering their actual progeny, we will predict the progeny that would be produced with each of the father's possible geno-

FIGURE 11-2

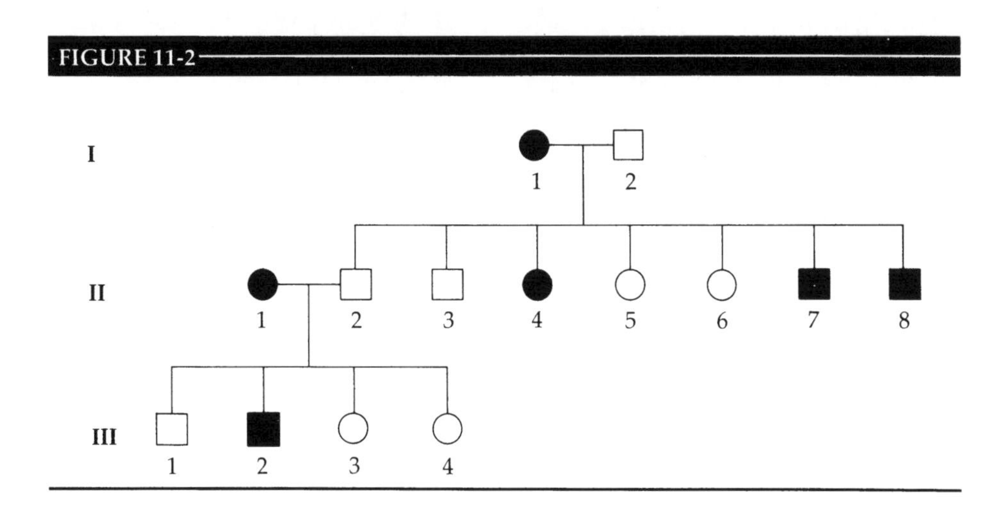

[1] Two. [2] One mating. [3] Five. [4] Three affected individuals, three males and no females. [5] c^+c^+ or c^+c. [6] cc.

types. If the father were c^+c^+, what progeny genotypes and phenotypes would be expected? ______________________________[7] If the father were c^+c, what progeny genotypes and phenotypes would be expected? ______________________________[8] Do the actual progeny of this mating make it possible to draw a definite conclusion about the genetic makeup of the father? ________[9] Explain. ______________________________[10]

Identify the genotypes of the individuals in generation II shown in Figure 11-2. ______________________________[11] Identify the genotypes of the generation-III individuals. ______________________________[12] In this pedigree it was possible to assign genotypes to all individuals. Note that in most pedigrees this may be impossible.

IDENTIFYING COMMON INHERITANCE PATTERNS IN PEDIGREES

Pedigree analysis may make it possible to identify, if it is not already known, the type of allele responsible for the trait under study. We will limit our consideration to the four most common types of alleles: autosomal dominant, autosomal recessive, sex-linked recessive, and sex-linked dominant. This analysis requires familiarity with the inheritance patterns shown by each of these types of alleles and we will examine a series of pedigrees in order to identify these patterns. However, before we look at these pedigrees, it is important to be aware of the following points.

1. In general, the greater the number of individuals and generations included in a pedigree, the more reliable the conclusions that can be drawn. However, some pedigrees, even large ones, may not yield enough information to permit even a tentative conclusion, while some small pedigrees may contain enough information to allow identification of the type of allele involved.
2. When the number of progeny from a mating is small, chance may cause their genotypes and phenotypes to differ significantly from the expected ratios. For example, with complete dominance the mating of an individual heterozygous for a trait with a homozygous recessive, say $Aa \times aa$, would be expected to produce 50% dominant (Aa) and 50% recessive (aa) progeny. A small number of progeny might, for example, consist of 75% or 100% dominant individuals or even 75% or 100% recessive individuals.
3. Many human disorders are caused by alleles that are very rare, that is, they occur infrequently in the human population at large. In light of this, the two following assumptions are routinely made when examining pedigrees involving rare alleles *unless there is evidence to the contrary.* (1) If the allele responsible for the trait is believed to be dominant, individuals expressing the trait are assumed to be heterozygous rather than homozygous dominant. (2) If the allele responsible for the trait is believed to be recessive, normal individuals marrying into a family that shows the disorder are assumed to be homozygous for the dominant allele rather than heterozygous. This assumption does not apply when parents are blood relatives such as first cousins; then there is a significant chance that normal individuals are heterozygous if they have relatives in earlier generations who expressed the trait.
4. It is generally assumed, unless there is evidence to the contrary, that the allele responsible for the trait is fully penetrant and expressivity is high.

EXAMPLES OF PEDIGREES SHOWING STANDARD TYPES OF INHERITANCE

Now we are ready to examine four human pedigrees which show traits caused by autosomal dominant, autosomal recessive, sex-linked recessive, and sex-linked dominant alleles, respectively. In each case we will assume that the allele causing the trait is very rare. Because the allele's inheritance type is known, we will be able to identify the genotypes of at least some of the individuals in each pedigree and predict the outcome of some matings. This, in turn, will allow us to identify the inheritance patterns that characterize the inheritance of each type of allele. A series of questions following each pedigree will focus your attention on the important inheritance patterns exhibited by that type of allele.

[7] All c^+c with normal pigmentation. [8] Half c^+c and half cc with normal and albino phenotypes, respectively. [9] Yes.
[10] Since both normal and albino progeny occur, the father must be c^+c. [11] Expected progeny from the mating producing generation II, $c^+c \times cc$, will be c^+c (normal) and cc (albino). Thus, all the normal progeny in generation II have genotype c^+c and all the albinos have genotype cc. [12] The mating producing generation III involves a c^+c (normal) male and a cc (albino) female; two types of progeny are expected: c^+c (normal) and cc (albino). Thus, all the normal progeny in generation III have genotype c^+c and all the albinos have genotype cc.

FIGURE 11-3

Pedigree showing the inheritance of an autosomal recessive trait.

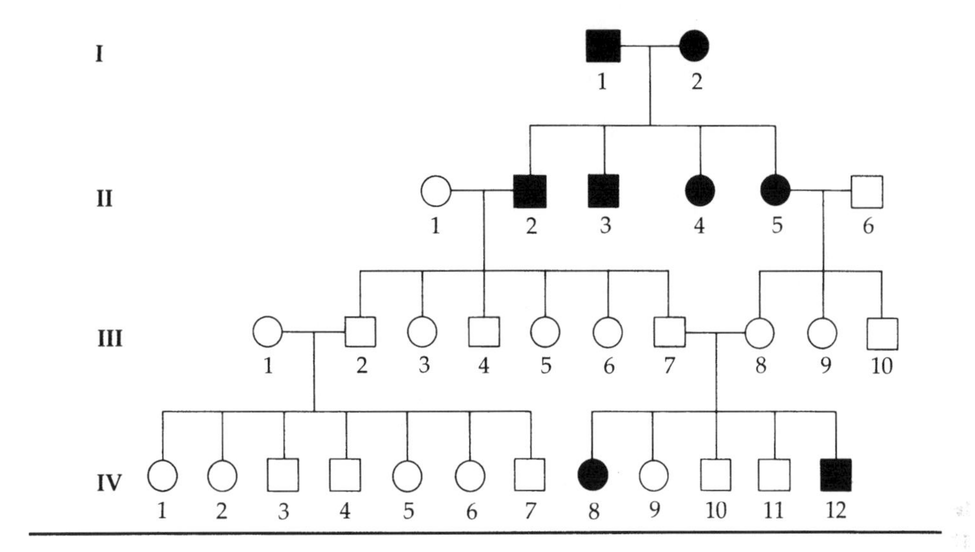

Once you are familiar with the inheritance patterns associated with these four types of alleles, you can go on to examine pedigrees where the nature of the allele responsible for the trait is unknown. Detecting inheritance patterns that are associated with a particular type of allele may allow you to identify how the allele causing the trait is inherited. For example, if you detect patterns associated with the inheritance of an autosomal recessive allele and fail to find patterns associated with any other type of allele, then you may conclude that the allele causing the trait is autosomal recessive. An approach for analyzing a pedigree where the nature of the allele causing the trait is unknown will be discussed later in this chapter.

Pedigrees Showing Autosomal Inheritance

Autosomal Recessive Trait Figure 11-3 presents a pedigree showing the inheritance of an autosomally based trait caused by the expression of a mutant allele which is recessive to its wild-type counterpart. An example of this type of trait is the human disorder galactosemia. This affliction is caused by a mutation that codes for a faulty enzyme that in turn prevents the breakdown of the sugar galactose and ultimately results in damage to the brain and other organs. Examine this pedigree for affected individuals and their distribution. Refer to this pedigree as you answer the questions that follow.

Is the mutant trait expressed in each generation? ________[13] Explain why this would be expected. ________________________________[14] Count the number of affected males and the number of affected females in the pedigree. What do you notice about the relative numbers of each sex that are affected? ________[15] If both parents in a mating express the mutant trait, what would be the expected phenotype or phenotypes of their progeny? ________________________[16] Identify a mating in the pedigree which is of this type. ________[17] Are the phenotypes of the actual progeny of this mating in general agreement with the predicted outcome? ________[18]

Since the recessive mutant allele is very rare, what is the most likely genotype for an individual with a normal phenotype from the population at large who marries into this family? ________[19] If such a normal individual were mated with an affected individual, what phenotype or phenotypes would be expected in the progeny? ________________________[20] Identify two matings in the pedigree which are of this type. ________________________[21] Are the phenotypes of the progeny of these matings in general agreement with the predicted outcome? ________[22] If an individual

[13] No. [14] The allele is recessive. [15] They are equal. [16] Both parents must be homozygous recessive and all their progeny would be homozygous recessive and affected. [17] I-1 × I-2. [18] Yes. [19] Homozygous dominant. [20] All progeny would be heterozygous and have the normal phenotype. [21] II-1 × II-2 and II-5 × II-6. [22] Yes.

FIGURE 11-4

Pedigree showing the inheritance of an autosomal dominant trait.

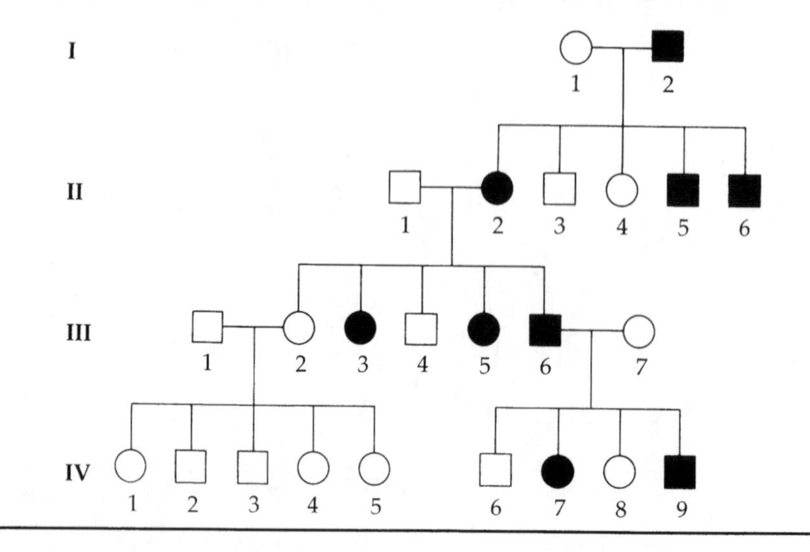

with a normal phenotype mated with an affected individual and affected offspring were produced, what would you conclude about the genotype of the normal parent? ________[23]

Autosomal Dominant Trait The pedigree in Figure 11-4 shows the inheritance of a mutant trait that is dominant to the wild-type allele and carried on an autosome. Huntington's chorea, a human disease characterized by a deterioration of the central nervous system and culminating in death, is an example of a trait of this type. Examine the pedigree for affected individuals and their distribution. Refer to this pedigree as you answer the questions that follow.

Is the mutant trait expressed in each generation? ________[24] Explain why this would be expected. ________[25] Count the number of affected males and the number of affected females in the pedigree. What do you notice about the relative numbers of affected individuals of each sex? ________[26] Since the allele for this mutant trait is very rare, what would be the most likely genotype for an individual such as I-2 who expresses the mutant trait? ________[27] If that individual mates with a normal individual, such as I-1, what phenotype or phenotypes would be predicted for their progeny and in what ratio? ________[28] Are the progeny of the I-1 × I-2 mating in general agreement with your prediction? ________[29] What progeny phenotype or phenotypes would you predict from a mating when neither parent expresses the trait? ________[30] Identify a mating from the pedigree where neither parent expresses the allele. ________[31] Are the actual progeny of this mating in general agreement with your predicted outcome for this type of mating? ________[32]

Comparing Inheritance Patterns for Recessive and Dominant Autosomal Alleles The questions you have answered in connection with the two previous pedigrees highlight some of the important inheritance patterns shown by dominant and recessive autosomal alleles. Use the following table to summarize key differences that allow you to distinguish between these two types of inheritance.

[23] Heterozygous. [24] Yes. [25] The mutant allele is dominant. [26] About equal. [27] Heterozygous.
[28] Heterozygous × homozygous recessive produces wild-type and mutant phenotypes in a 1:1 ratio. [29] Yes.
[30] Each parent is homozygous recessive; all progeny will be normal. [31] III-1 × III-2. [32] Yes.

	Autosomal recessive	Autosomal dominant
Must the trait occur in each generation?	________[33]	________[34]
Can a normal × normal mating produce affected progeny?	________[35]	________[36]
Expected progeny from normal × affected mating:	________[37]	________[38]
Minimum number of affected parents required to produce at least some affected progeny:	________[39]	________[40]

Summary of Pedigree Inheritance Patterns for Autosomal Alleles The following lists summarize key patterns that characterize the inheritance of rare recessive and dominant autosomal alleles.

Autosomal recessive inheritance patterns:

1. The trait may skip generations.
2. Most matings between normal and affected parents give rise to phenotypically normal progeny. (Note that when the allele for the trait is very rare, the phenotypically normal parent is assumed to be homozygous for the dominant allele unless there is information to the contrary.)
3. If both parents are affected, all their progeny are affected.
4. Normal parents can produce affected children (if both parents are heterozygous for the allele).
5. Affected progeny often occur when the parents are blood relatives.

Autosomal dominant inheritance patterns:

1. The trait appears in each generation (provided that penetrance is high and that families are large enough to reflect Mendelian ratios).
2. Matings between a normal and an affected individual are expected to produce normal and affected progeny in a 1:1 ratio. (Note that when the allele for the trait is very rare, an affected parent is assumed to be heterozygous unless there is information to the contrary.)
3. At least one parent must be affected in order for affected progeny to be produced.
4. Normal parents will always produce normal progeny.

Before going on, note that with both types of autosomal inheritance, there is an equal chance that the trait will affect *both* males and females. If you refer to each pedigree, you will find equal or approximately equal numbers of affected males and females.

Pedigrees Showing Sex-Linked Inheritance

Now we will look at pedigrees that show the inheritance of sex-linked recessive and dominant traits.

Sex-Linked Recessive Trait The pedigree in Figure 11-5 shows the inheritance history of a trait caused by an allele which is sex-linked and recessive to the allele for the normal condition. The human blood clotting disorder, hemophilia, is an example of this type of trait. Examine the pedigree for affected individuals and their distribution and answer the following questions.

Does the trait occur in each generation? ____[41] Explain why this would be expected? ________________________________[42] What is the sex of the affected individuals? ________[43] Why are no members of the opposite sex affected with the trait? ________________________________[44]

What type of mating would produce affected males? ________________________________[45] Identify matings which are of this type from the pedigree. ________________________________[46] What types of matings would produce affected females? ________________________________[47] Note that these two types of mat-

[33] May skip a generation. [34] Occurs in each generation. [35] Yes. [36] No. [37] (Because of allele rarity, the normal parent is assumed homozygous.) All normal. [38] (Because of allele rarity, the affected parent is assumed heterozygous.) Affected and normal in a 1:1 ratio. [39] None, provided that each parent is heterozygous. [40] One. [41] No (it skips generation II). [42] This is due to the recessive nature of the allele. [43] Male. [44] In order for a female to be affected, she must have received an allele for the trait from each parent; no female in the pedigree meets this requirement. [45] Carrier female × normal male would give rise to males, half of whom would be affected. (Note that affected female × normal male matings would also produce affected males; however, given the rarity of the allele, females of this type would seldom be encountered.) [46] II-5 × II-6, II-7 × II-8, and II-9 × II-10. [47] Carrier female × affected male would produce females, half of whom would be affected. Affected female × affected male would also produce affected females.

FIGURE 11-5

Pedigree showing the inheritance of a sex-linked recessive trait.

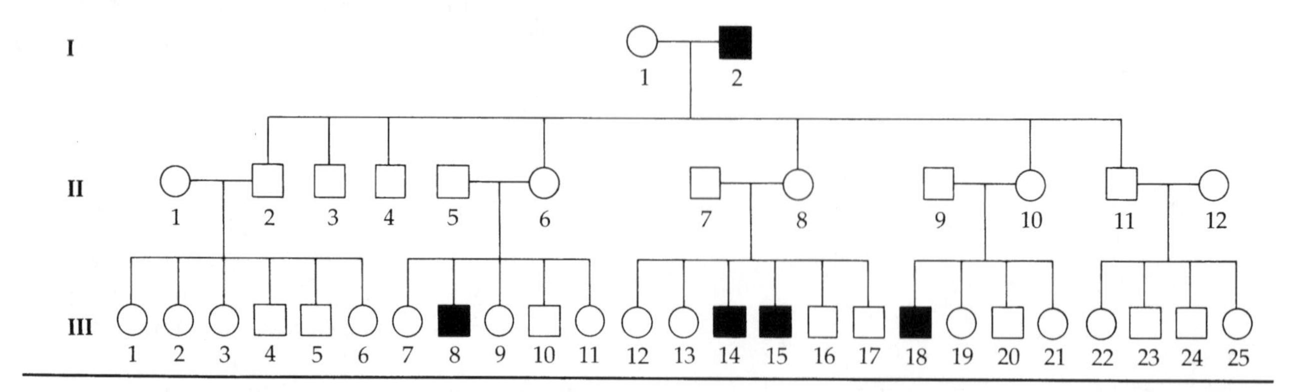

ings are not found in the pedigree and are rarely encountered. Explain why is this so? ______________ ______________[48]

Although carrier females cannot be phenotypically distinguished from those homozygous for the dominant allele, it is often possible, through pedigree studies, to identify carrier females. What clues would show up in a pedigree to help identify the carrier females? ______________

______________[49] What proportion of the male children produced by a carrier female and a normal male would you predict would be affected? ______________[50] Is the number of male progeny from carrier mother × normal father matings in the pedigree in general agreement with this prediction? ______________[51]

Sex-Linked Dominant Trait The pedigree in Figure 11-6 shows the inheritance history of a trait caused by a sex-linked allele which is dominant to the allele for the normal condition. Traits of this type are extremely rare, although one known example in humans is brown tooth enamel. Examine the pedigree for affected individuals and their distribution and answer the following questions.

What is the sex of the affected individuals? ______[52] In the pedigree involving a sex-linked recessive allele, no affected females were found. Why are affected females found in the pedigree we are currently considering? ______________

______________[53] Explain why the trait occurs in every generation. ______________

______________[54] What type of mating would produce affected females? ______________

______________[55] Identify matings of this type in the pedigree. ______________

______________[56] What type of mating would produce affected males? ______________

______________[57] Identify matings of this type in the pedigree. ______________

______________[58] What type of progeny would be expected from a mating between an affected father and a normal mother such as I-1 × I-2? ______________

______________[59] Is the outcome of the I-1 × I-2 mating in general agreement with your prediction? ______________[60] What type of progeny would be expected from a mating of an affected mother and a normal father? ______________

______________[61] Are the outcomes of matings II-1 × II-2 and II-9 × II-10 in general agreement with your prediction? ______________[62]

[48] The chance that two unrelated individuals will possess the same rare allele is very small. [49] Carrier females have affected fathers or carrier mothers, could have affected brothers, and could produce affected sons. [50] One half. [51] Yes. Half the males from matings II-5 × II-6, II-7 × II-8, and II-9 × II-10 are affected. [52] Male and female. [53] Since the allele is dominant, only a single copy is needed for any male or female to be affected. (Recall that with a sex-linked recessive allele, a female must possess two copies of the allele in order to be affected.) [54] The allele is dominant. [55] Since either the father or mother could transmit the X-linked allele to a daughter, any mating involving an affected parent of *either* sex would produce affected females. [56] In I-1 × I-2, the father is affected; in II-1 × II-2 and in II-9 × II-10, the mothers are affected. [57] Any matings involving an affected mother. [58] II-1 × II-2 and II-9 × II-10.
[59] All daughters would be affected because of father-to-daughter transmission of the allele, and all sons would be normal.
[60] Yes, since all daughters are affected. [61] Half of the offspring of each sex would be affected and half would be normal.
[62] Yes. For each mating, approximately half the progeny of each sex are normal and half are affected.

FIGURE 11-6

Pedigree showing the inheritance of a sex-linked dominant trait.

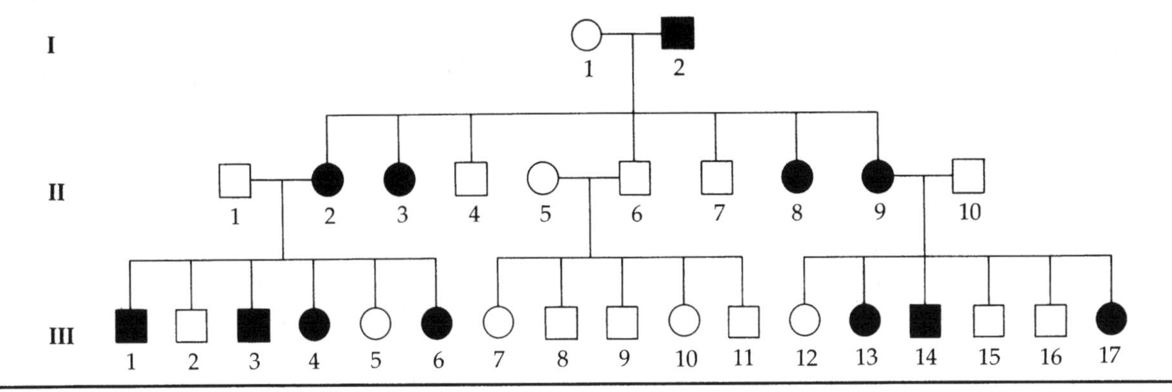

Comparing Inheritance Patterns for Recessive and Dominant Sex-Linked Alleles The questions you have answered in connection with the two preceding pedigrees highlight some of the important inheritance patterns shown by dominant and recessive sex-linked alleles. Use the following table to summarize key differences that allow you to distinguish between these two types of inheritance.

	Sex-linked recessive	Sex-linked dominant
Will the trait occur in each generation?	____[63]	____[64]
Sex of affected individuals:	____[65]	____[66]
Source of affected males:	____[67]	____[68]
Source of affected females:	____[69]	____[70]
Progeny of affected fathers (assume mothers are normal):	____[71]	____[72]
Progeny of affected mothers (assume fathers are normal):	____[73]	____[74]

Summary of Pedigree Inheritance Patterns for Sex-Linked Alleles The following lists summarize key patterns that characterize the inheritance of rare recessive and dominant sex-linked alleles.

Sex-linked recessive inheritance patterns:

1. Most affected individuals are males.
2. Affected males come from normal mothers who are carriers (that is, mothers whose fathers were affected) and not from affected fathers. In other words, there is no father-to-son transmission.
3. Affected fathers, provided they mate with normal, noncarrier females, produce normal sons and carrier daughters.
4. The mating of a normal father and a carrier mother produces normal and affected sons in an expected ratio of 1:1.
5. Affected females come from carrier-mother × affected-father matings and from affected-mother × affected-father matings.

Sex-linked dominant inheritance patterns:

1. Affected individuals appear in each generation.
2. Males and females are affected (since females have two X chromosomes, they have a greater chance of being affected than do males).
3. Affected males come from affected mothers.
4. Affected females come from either affected mothers or fathers.
5. Affected mothers and normal fathers produce sons and daughters, half of whom are affected. (Note that an affected mother, because of the rarity of the allele, is assumed to be heterozygous rather than homozygous for the dominant allele.)

[63] May skip a generation. [64] Occurs in each generation. [65] Mostly male (females very rarely). [66] Males and females.
[67] Any matings involving carrier mothers (50% chance of son being affected), or very rarely encountered affected mothers.
[68] From affected mothers (50% chance of son being affected). [69] Very rare matings of carrier mothers with affected fathers.
[70] Mating involving *either* affected mothers *or* fathers. [71] Carrier daughters and normal sons. [72] All daughters are affected and sons are normal. [73] Affected mothers are very rare but if they did occur, all their daughters would be carriers and all sons would be affected.
[74] Half of sons and half of daughters are affected.

6. Affected father × normal mother matings produce affected daughters and normal sons.

AN APPROACH TO PEDIGREE ANALYSIS

Occasionally it is necessary to analyze a pedigree when no information is available about the nature of the gene responsible for the trait. The key steps in carrying out such an analysis are as follows.

1. Determine whether the allele causing the trait is autosomally based or sex-linked by looking for a correlation between the expression of the trait and the sex of the individual. An approximately equal frequency of affected males and females supports an autosomal basis for the trait. If the trait appears mostly in males, then the trait may be carried on the X chromosome. (Other less frequently encountered possibilities which cause a trait to be expressed predominantly in one sex or the other include the allele being sex-limited, where expression occurs only in a single sex, sex-influenced, where expression is influenced by the sex of the individual, or carried on the Y chromosome. See your textbook for additional information on alleles of these types.)

2. Once the allele is tentatively identified as autosomal or sex-linked, determine whether it is dominant or recessive by examining the pedigree for patterns associated with the inheritance of a dominant or a recessive allele carried by that type of chromosome. If the patterns shown in the pedigree correspond to those associated with a particular type of allele, then the allele can be tentatively classified as that type.

3. Based on the first two steps, formulate a hypothesis about the type of allele responsible for the trait. You might, for example, arrive at the hypothesis that the allele for the trait is autosomal dominant or, alternatively, you might propose that it is sex-linked recessive.

4. Based on your hypothesis, you should be able to identify the genotypes for at least some of the parents and from that predict the phenotypes of the progeny produced from at least some of the matings shown in the pedigree.

5. Compare your predicted progeny phenotypes with the actual progeny phenotypes shown in the pedigree to see if they are in general agreement (remember that some families may not be large enough to approximate the expected ratio). If the pedigree fails to support your initial hypothesis, you need to develop a new one and start the process over again. Even if the pedigree provides support for your hypothesis, your work is not finished because you must then reexamine the pedigree for anything that is *not* consistent with your hypothesis. Only after carrying out these steps can you conclude that the pedigree supports your hypothesis.

Practice at interpreting pedigrees is provided in the problem set at the end of this chapter.

PREDICTING RISKS FROM PEDIGREES: GENETIC COUNSELING

Once the allele responsible for a trait has been classified as dominant or recessive and autosomal or sex-linked, it may be possible to predict the risk or chance that the allele will be transmitted from one generation to the next. Such risk prediction is usually carried out by genetic counselors, experts specially trained in human genetics. For example, a genetic counselor may be called upon when parents wish to know the risk of producing a child affected with a single-gene trait that is present in the family of one or both parents. The general strategy used by the counselor for answering this type of question is to prepare and examine the family pedigree to try to determine the genotypes of the prospective parents. If this can be done, the expected genotypes and phenotypes for their progeny can be predicted and the risk of producing an affected child can be determined.

SOME PRACTICE AT PREDICTING RISKS

Refer to the pedigree in Figure 11-4 that shows the inheritance pattern for an autosomal dominant trait. Assume that you are a genetic counselor who is called upon to advise members of the family whose inheritance history is shown in this pedigree. As you work the following problems, use *A* to symbolize the dominant allele for the trait and *a* to symbolize the recessive allele for the normal condition.

PROBLEM: Couple III-6 and III-7 has four children as shown in the pedigree and wants to know the chance that a fifth child will be affected.

SOLUTION: To answer this question you need to determine (1) the genotypes of the parents, (2) the expected genotypes and phenotypes of their children, and (3) the chance of producing an affected child. Solve the problem in the space provided.

Answer: Female III-7 has the normal phenotype and must therefore have genotype *aa*. Male III-6 is affected and thus must have either genotype *Aa* or *AA*. The man's father, II-1, was normal and must have had genotype *aa*, and therefore the man must have received an *a* allele from his father. Consequently, the man's genotype must be *Aa* and the mating is *aa* × *Aa*. The expected progeny are of two genotypes, half *Aa* and half *aa*, with affected and normal phenotypes, respectively. Thus, the chance that the next child of the couple will be affected is 1/2.

Problem: Affected female III-5 plans to marry a man affected with the same trait and wants to know the chance of their having a normal child.

Solution: To answer this question you need to determine (1) the genotypes of the prospective parents, (2) the expected genotypes and phenotypes of their children, and (3) the chance of producing an affected child. Solve the problem in the space provided.

Answer: Female III-5 is affected and therefore must have genotype *AA* or *Aa*. Since her father, II-1, was normal and must have had genotype *aa*, she must have received allele *a* from him and must have genotype *Aa*. Since the allele for the trait is rare, and since we have no information indicating anything to the contrary, we assume the affected man she wishes to marry has genotype *Aa*. Thus, the mating is *Aa* × *Aa* and genotypes *AA*, *Aa*, and *aa* would be produced in a 1:2:1 ratio. Only progeny with genotype *aa* have the normal phenotype and the chance of producing such a child would be 1/4.

Problem: What is the chance that woman IV-4 will pass on the trait if she marries?

Solution: This problem can be solved by determining the woman's genotype and identifying the chance that she will pass the trait on. Solve the problem in the space provided.

Answer: The allele causing the trait is dominant and the woman has the normal phenotype. Thus, her genotype must be *aa*. Since she does not carry the allele, her chance of transmitting it is 0.

Refer to the pedigree in Figure 11-3 that shows the inheritance pattern for an autosomal recessive trait. Work the following problems, using *E* to symbolize the dominant allele for the normal condition and *e* to symbolize the recessive allele for the abnormal condition.

Problem: Affected female IV-8 marries a normal male whose family shows no history of this trait. The couple wishes to know the chance of having an affected child.

Solution: Identify the genotypes of the parents, write out the mating, and determine the phenotypes of the expected progeny. Identify the probability of having an affected child. Solve this problem in the space provided.

Answer: Female IV-8 is affected, and since the allele causing the condition is recessive, she must have the homozygous recessive genotype, *ee*. Her normal husband must be either *EE* or *Ee*. Since the *e* allele is rare and since there is no history of individuals expressing this allele in his family, we assume that he is *EE*. The mating is $EE \times ee$. All progeny would have the genotype of *Ee* and the normal phenotype. The probability of producing an affected child is 0. ▬

PROBLEM: What is the probability that female IV-9 carries the allele for the trait?

SOLUTION:

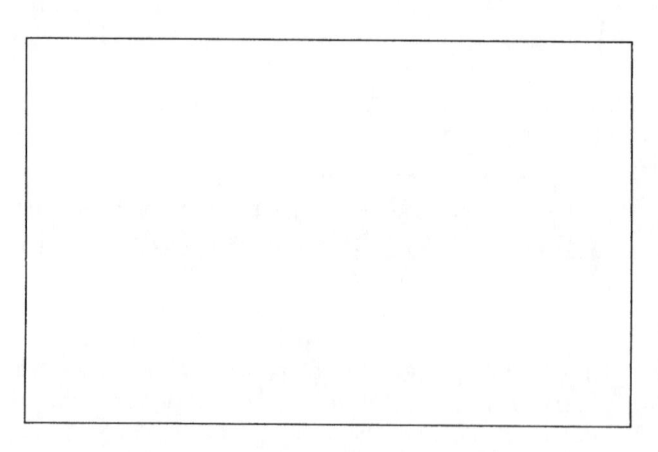

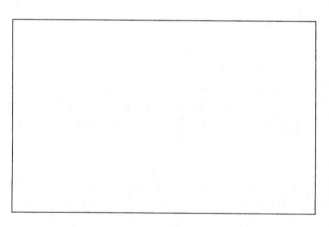

Answer: Female IV-9 has two affected siblings, IV-8 and IV-12, who must have the *ee* genotype. This tells us that each of her normal parents, III-7 and III-8, must carry the *e* allele and therefore have genotype *Ee*. The mating of her parents is $Ee \times Ee$ and their expected progeny would have genotypes *EE*, *Ee*, and *ee* in a 1:2:1 ratio. Thus, three of four expected progeny, that is, those with genotypes *EE* and *Ee*, would have the normal phenotype. Of these *normal* children expected from this mating, two out of every three will have genotype *Ee* and thus two out of every three normal individuals will be carriers of the *e* allele. Since she is normal, the chance that she is a carrier is 2/3. ▬

PROBLEM: If female IV-9 marries a male affected with this trait, what is the chance that they will have an affected child?

SOLUTION:

Answer: The affected male would have genotype *ee*. If IV-9 carries the *e* allele, the mating is $Ee \times ee$. The expected progeny would have genotypes *Ee* and *ee* in a 1:1 ratio. Thus, the chance of having an affected child is 1/2. The probability of IV-9 being a carrier *and* having an affected child from this marriage is given by the product of the separate probabilities (refer to the solution of the previous problem): $2/3 \times 1/2 = 2/6 = 1/3$. ▬

SUMMARY

Information about the manner in which traits due to single genes are inherited may be derived from pedigrees, or family histories, which use a set of standard symbols to trace the inheritance of a particular trait through successive generations. Pedigrees have been particularly helpful for identifying the inheritance patterns of human traits. Provided that a pedigree contains sufficient information, an analysis of the inheritance patterns it shows may make it possible to determine whether the allele responsible for a trait is inherited as an autosomal dominant, autosomal recessive, sex-linked recessive, or sex-linked dominant allele. Pedigrees showing a trait due to the inheritance of each of these types of alleles are examined in this chapter, and the inheritance patterns for each of these types of alleles are described. These patterns are summarized in the chapter. If the nature of the allele is known, the genotypes of at least some individuals within the pedigree can be identified and expected genotypes and phenotypes for their progeny can be determined. This process plays a key role in genetic counseling, where it may be possible to predict the chance that a trait will be transmitted from one generation to the next.

PROBLEM SET

11-1. A man and a woman have five children, as shown in the following pedigree.

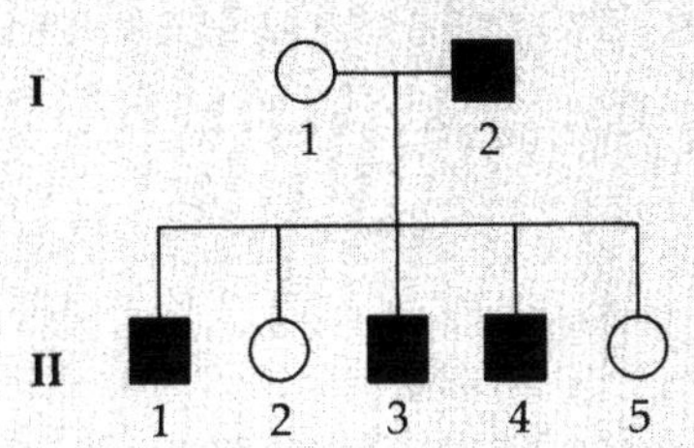

The allele responsible for the trait is very rare in the population at large and is carried on an autosome. Assume that the allele causing the trait is dominant and symbolized by *D* while its recessive counterpart is symbolized by *d*.

a. What is the mother's genotype?
b. What is the father's genotype?
c. Give the genotypes and their frequencies for the progeny expected from this mating.
d. Is this pedigree consistent with the hypothesis that the allele causing the trait is dominant?

11-2. Refer to the pedigree and the basic information given in problem 11-1. Assume that the allele responsible for the trait is recessive and symbolized by *d* while its dominant counterpart is symbolized by *D*.

a. What is the mother's genotype?
b. What is the father's genotype?
c. What is the genotype of the affected sons?
d. Give the genotypes and their frequencies for the progeny expected from this mating.
e. Is this pedigree consistent with the hypothesis that the allele causing the trait is recessive?

11-3. Refer to problems 11-1 and 11-2. Based on the available information, is the trait more likely to be due to a dominant or a recessive allele?

11-4. Assume that the parents in the pedigree in problem 11-1 were first cousins. Which hypothesis, dominance or recessiveness for the allele causing the trait, would you now be more likely to support?

11-5. Refer to the pedigree shown in problem 11-1. A geneticist, upon examining the pedigree, speculates that the allele responsible for the trait may be X-linked rather than autosomal. Is this pedigree consistent with the X-linkage of the allele? Explain.

11-6. Refer to the pedigree in Figure 11-3 which shows the inheritance pattern for an autosomal recessive trait. Use *e* to symbolize the recessive allele for the trait and *E* to symbolize the dominant allele for the normal condition. What is the probability that a sixth child born to couple III-7 and III-8 will be affected with the trait?

11-7. Refer to the information given in problem 11-6. Male III-2 and his brother, III-7, notice that although they are both normal and have normal wives, III-7 has two affected children while all of III-2's children are normal. Explain this difference.

11-8. Refer to the information given in problem 11-6. What is the probability that any given child of normal male III-10 will be normal if III-10 marries a woman who has the trait exhibited by his mother?

PROBLEM SET

11-9. Examine the following three-generation pedigree.

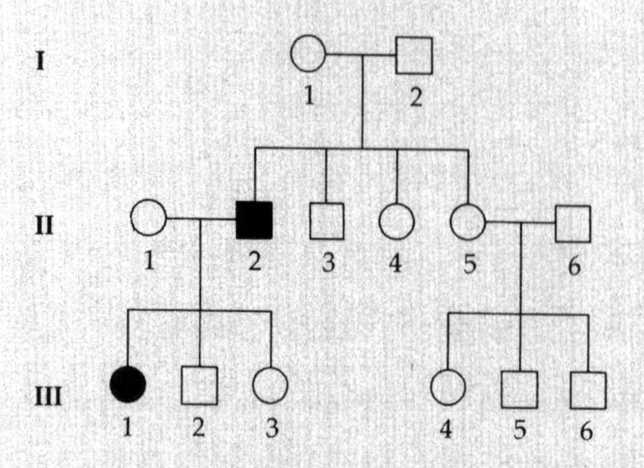

a. Draw a tentative conclusion about whether the allele for the trait is dominant or recessive to the normal allele. Explain.

b. What can be concluded about the genotype of female II-1?

c. Assume the allele for the trait is autosomal. What can be concluded about the genotypes of the original parents (generation I)?

d. Assume the allele for the trait is sex-linked. What can be concluded about the genotypes of the original parents (generation I)?

11-10. The phenotypes of the original parents in the following nonhuman pedigree are unknown.

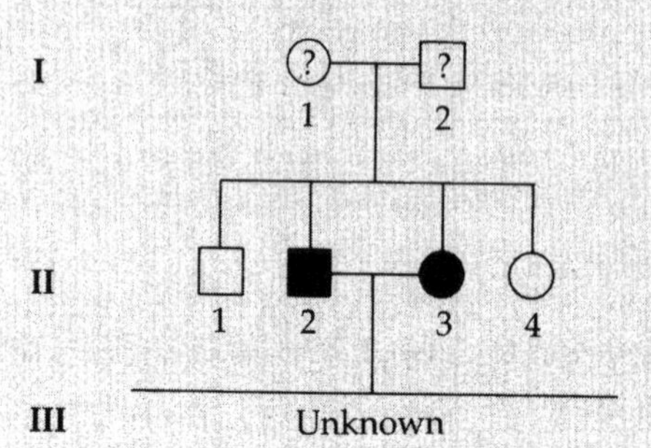

a. If the allele responsible for the trait is recessive, what, if anything, could be concluded about the genotype of one or both of the parents, I-1 and I-2?

b. If the allele responsible for the trait is dominant, what, if anything, could be concluded about the genotype of one or both of the parents, I-1 and I-2?

c. If the allele responsible for the trait is recessive, can we predict with certainty the phenotype or phenotypes of the offspring of the mating between II-2 and II-3? Explain.

d. If the allele responsible for the trait is dominant, what phenotype or phenotypes could we expect for the offspring of the mating between II-2 and II-3? Explain completely.

11-11. The following pedigree shows the inheritance history of a trait that is known to be due to a sex-linked allele. Decide whether the allele responsible for the trait is recessive or dominant, and give three reasons to justify your conclusion.

PROBLEM SET

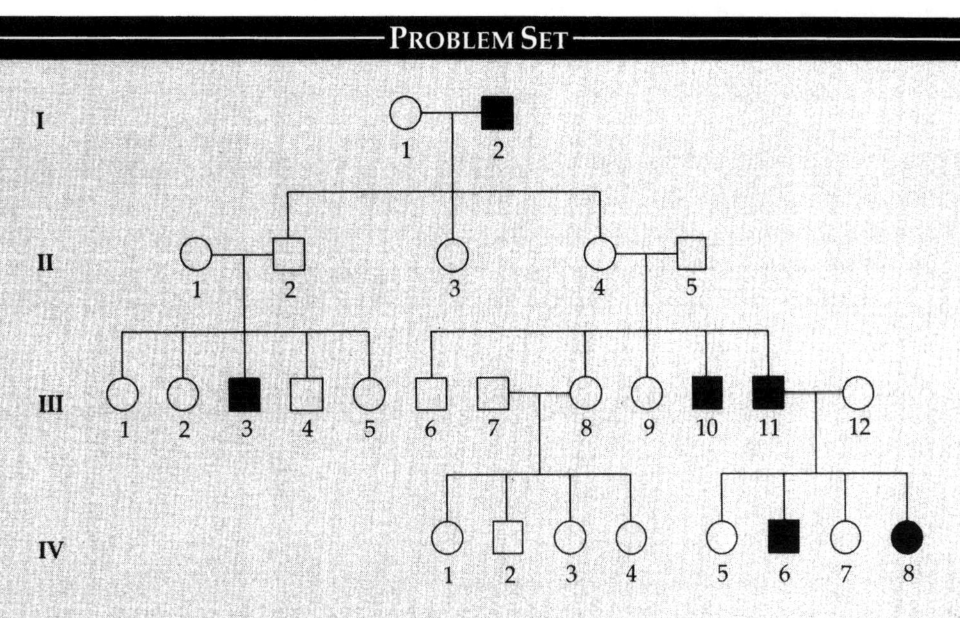

11-12. Refer to the pedigree and information included in problem 11-11 and verify that your answer to 11-11 is correct before proceeding.

a. Was female II-1 a carrier of the allele? Explain.

b. What was the genotype of female III-12?

c. What is the chance that a fifth child born to III-11 and III-12 will be affected?

d. What is the chance that a sixth child born to III-11 and III-12 will be a female carrier?

11-13. Refer to the pedigree and the information included in problem 11-11.

a. If IV-8 were to marry a normal male, what is the chance they would have an affected male child?

b. Can you say with certainty that the female III-8 was not a carrier of the allele responsible for the trait?

11-14. Refer to the pedigree presented in Figure 11-6 which shows the inheritance of a sex-linked dominant trait. Could the inheritance of this trait also be explained by an autosomal recessive allele? Explain why or why not.

11-15. Refer to the pedigree presented in Figure 11-6 which shows the inheritance of a sex-linked dominant trait. Could the inheritance of this trait also be explained by an autosomal dominant allele? Explain why or why not.

11-16. The following pedigree shows the inheritance history of a trait known to be due to a rare autosomal allele. Is the allele responsible for the trait recessive or dominant? Give three reasons to justify your answer.

PROBLEM SET

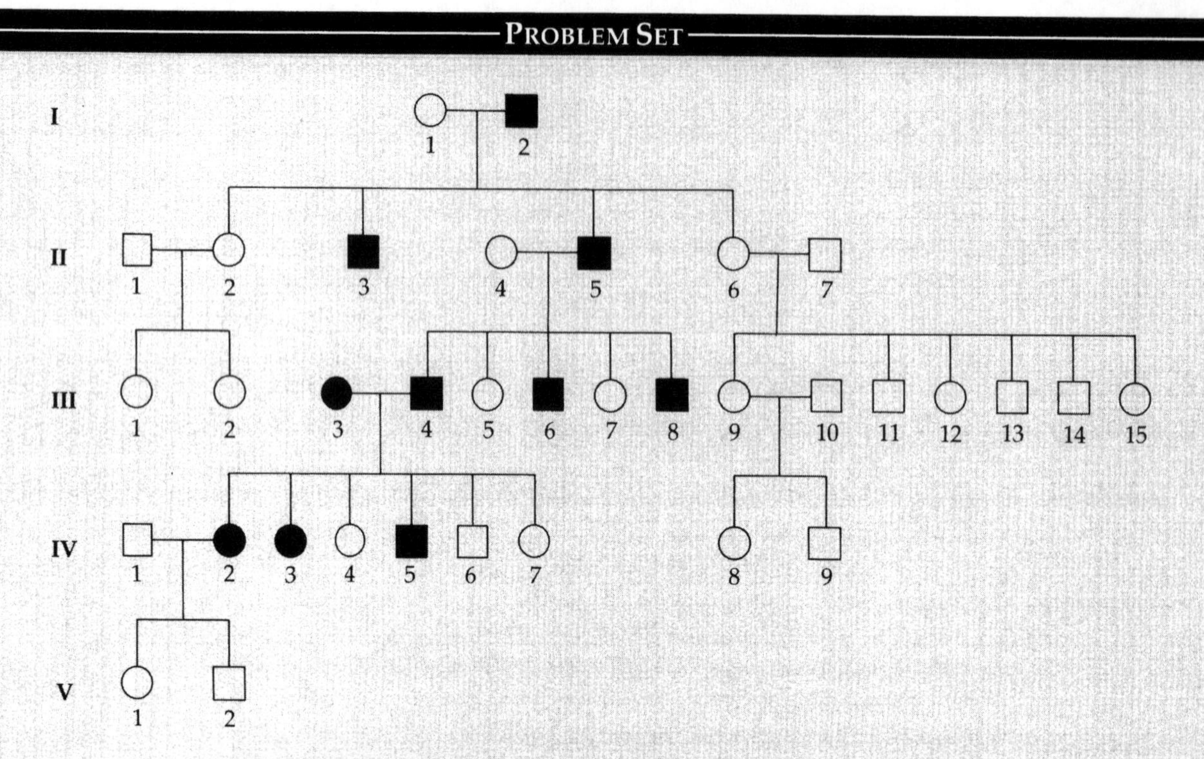

11-17. Refer to the pedigree in problem 11-16 and verify that your answer to 11-16 is correct before proceeding. Use *G* to represent the dominant allele and *g* to represent the recessive allele.

a. Identify the genotype of female IV-2.

b. What is the chance that a third child born to couple IV-1 and IV-2 would be an affected male?

c. What is the probability that female IV-7, who has affected brothers and sisters, will have an affected child if she marries a normal man?

11-18. Refer to the pedigree in problem 11-16. Use *G* to represent the dominant allele and *g* to represent the recessive allele.

a. What is the chance that affected male IV-5 is homozygous for the dominant allele?

b. If IV-5 marries a normal woman, what is the chance that they will have a normal child?

11-19. Refer to the pedigree presented in Figure 11-5 which shows the inheritance of a sex-linked recessive trait. Could the inheritance shown in this pedigree also be explained by an autosomal recessive allele? Explain why or why not.

11-20. Refer to the pedigree in Figure 11-5 which shows the inheritance of a sex-linked recessive trait.

a. What is the chance that a fifth child born to couple II-9 and II-10 will be normal?

b. If the fifth child is female, what is the chance she will be normal?

c. If the fifth child is male, what is the chance he will be normal?

CHAPTER

12

Linkage, Crossing Over, and the Two-Point Testcross

INTRODUCTION

Every organism possesses far more traits than it does chromosomes and, consequently, each chromosome carries genes for anywhere from a few to thousands of traits. Genes carried on the same chromosome are said to be **linked** and all the genes found on a chromosome comprise a **linkage group.** In Chapter 9 we saw how the classic Mendelian phenotypic ratio arising from a dihybrid mating involving independently assorting loci could be modified by gene interaction. In this chapter we will see how Mendelian ratios become modified when two loci involved in a cross are linked. In addition, we will consider how two-point testcrosses can be used to determine whether two loci are linked, how distances between two linked loci can be calculated, and how linkage maps are prepared.

DETECTION OF LINKAGE: THE TESTCROSS

A standard approach to detecting linkage of two genes is to carry out a testcross: an individual heterozygous for both of the loci under consideration is crossed with an organism homozygous recessive for these loci. Since the double–homozygous recessive parent produces a single type of gamete that carries recessive alleles, the genetic makeup of the gametes contributed by the dihybrid parent and the frequency with which each is formed can readily be identified by examining the phenotypes of the testcross progeny. As we will see, the types and frequencies of the gametes produced by the dihybrid and detected through the testcross progeny differ depending on whether the two loci are linked or not.

A COMPARATIVE LOOK AT THREE TWO-POINT TESTCROSSES

To see how the linkage of loci influences the outcome of a testcross, we will examine and compare three different testcrosses: one with unlinked loci, a second with completely linked loci, and a third with incompletely linked loci. Assume that each gene involved in these testcrosses has two alleles, with one showing complete dominance to its recessive counterpart.

Testcross with Unlinked Loci

We begin by reviewing a testcross involving two unlinked loci, designated as *d* and *e*, with the cross represented as *DdEe* × *ddee* (this type of testcross was discussed in detail in Chapter 4). With unlinked loci, independent assortment occurs. How many kinds of gametes will be produced by the dihybrid parent? ________[1] Identify the gene combinations carried by each of these kinds of gametes. ________[2] In what relative frequencies will these gametes be expected? ________[3] How many kinds of gametes will the double–homozygous recessive parent make? ________[4] Identify the gene combinations carried by these gametes. ________[5] Random union of these gametes is shown in the following branch diagram; identify the progeny genotypes by completing the diagram.

Gamete from *ddee* parent	Gametes from *DdEe* parent		Progeny genotypes
de	*DE*	=	________[6]
	De	=	________[7]
	dE	=	________[8]
	de	=	________[9]

How many phenotypes are found among the progeny? ________[10] In what ratio would these phenotypes be expected? ________[11]

Testcross with Completely Linked Loci

Next we consider a testcross involving two loci, designated as *f* and *g*, that are found on the same chromosome. The system commonly used to symbolize genotypes involving linked loci uses one or two diagonal lines to separate the genes carried by the two members of a homologous chromosome pair. The genes carried by one chromosome are written on one side of the diagonals and the genes of the homolog are written on the other side, with the genes on both sides always written in the same order. Using this method of notation, an individual heterozygous for both the *f* and *g* loci that carries the two dominant genes on one chromosome and the two recessive genes on the homolog can be represented as *FG//fg*, or more simply as *FG/fg*. A frequently encountered variant of this system uses one or two horizontal lines to represent the homologous pair with the genes carried on one chromosome written above the lines and the genes carried on the homolog written below the lines, as follows: $\frac{F\ \ G}{\overline{f\ \ g}}$, or more simply $\frac{F\ \ G}{f\ \ g}$. Combinations of linked genes found in gametes or other haploid cells may be written ahead of a diagonal line (for example, *FG/*) or above a horizontal line (for example $\underline{F\ \ G}$).

Either method of notation for diploid genotypes tells us at a glance (if we did not already know it) that the *f* and *g* loci are linked and the manner in which the alleles of the two genes are arranged on the chromosomes. The arrangement we have been considering, where both recessive alleles are carried on the same chromosome, is termed **coupling.** The alternative arrangement, with a recessive allele on each homolog is referred to as **repulsion** and can be represented as *Fg/fG* or as $\frac{F\ \ g}{f\ \ G}$.

Assume that the dihybrid involved in the testcross we are considering carries its recessive alleles in coupling. Furthermore, assume that there is no disruption of linkage of the *f* and *g* loci, that is, that there is no crossing over. How many kinds of gametes will be made by the *FG/fg* parent? ________[12] Identify the gene combinations carried by each of these kinds of gametes. ________[13] In what proportion will these gametes be expected? ________[14] The *fg/fg* testcross parent would, of course, make a single type of gamete carrying $\underline{f\ \ g}$. Random union of these gametes is shown in the following branch diagram; identify the progeny genotypes by completing the diagram.

Gamete from *fg/fg* parent	Gametes from *FG/fg* parent		Progeny genotypes
$\underline{f\ \ g}$	$\underline{F\ \ G}$	=	________[15]
	$\underline{f\ \ g}$	=	________[16]

How many different phenotypes would the progeny exhibit? ________[17] In what ratio would these phenotypes be expected? ________[18]

Testcross with Incompletely Linked Loci

In the testcross just considered, we assumed that the linkage was **complete,** that is, that nothing occurred to disrupt it. In general, this complete linkage exists with loci that are (1) positioned extremely close to each other on the chromosome, (2) located on opposite arms of the chromosomes and very close to their common centromere, or (3) separated by or part of an inversion of a section of the chromosome (see Chapter 16).

[1] Four. [2] *DE, De, dE,* and *de.* [3] Equal frequencies. [4] One type. [5] All will be *de.* [6] *DdEe* [7] *Ddee.* [8] *ddEe.* [9] *ddee.* [10] Four. [11] 1:1:1:1. [12] Two. [13] $\underline{F\ \ G}$ and $\underline{f\ \ g}$. [14] Equal frequencies. [15] *FG/fg.* [16] *fg/fg.* [17] Two. [18] 1:1.

Crossing Over Most loci on a chromosome are far enough apart so that their linkage can be disrupted by the process of crossing over which occurs during prophase of meiosis I (see Chapter 1). As you will recall, during crossing over, equal and corresponding segments are exchanged between two nonsister chromatids, that is, between maternal and paternal chromatids, in a tetrad (four-chromatid) configuration. Chromatids which have participated in crossing over show new combinations of linked genes which are called **recombinants.** Genes which participate in crossing over are said to be **incompletely,** or **partly, linked.**

The following sketch shows a pair of duplicated homologous chromosomes making up a tetrad. Two linked loci, *a* and *b*, are shown, with this tetrad belonging to the dihybrid *AB*/*ab*. The recessive alleles here are carried in coupling.

A B
A B

a b
a b

The following sketch shows the two inner chromatids of this tetrad configuration crossed over at a point between the *A* and *B* loci.

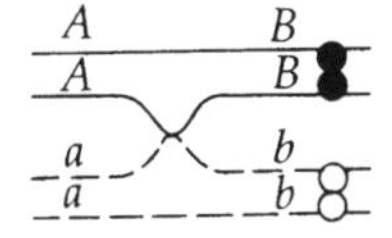

Make a sketch in the space provided showing how the four chromatids would appear following the reciprocal exchange of corresponding segments and after contact between the two inner strands of the tetrad has been resolved to produce two recombinant chromatids and two with the original linkage arrangements.[19]

What fraction of the four product chromatids is recombinants? ________[20] What fraction of the four product chromatids shows the original linkage arrangement? ________[21] Note that chromatids showing the original linkage arrangements are sometimes called the **parental,** or **nonrecombinant chromatids** to distinguish them from the recombinant chromatids. Be alert to the fact that when a single crossover event occurs, only *two* of the four chromatids in the tetrad are involved and consequently only *half* of the product chromatids are recombinants; the other two chromatids are noncrossovers and exhibit the original parental combinations.

Dihybrid Gamete Production Involving Crossing Over Now let us look at how crossing over influences the kinds and frequencies of gametes produced by our *AB*/*ab* dihybrid parent. This can be demonstrated by considering, say, 100 diploid cells known as primary spermatocytes that are about to undergo spermatogenesis in the gonads of a male to give rise to haploid sperm cells. Each primary spermatocyte, following meiosis, would be expected to give rise to four sperm cells to produce a total of 400 sperm. Assume that a crossover between the *a* and *b* loci occurs in 24 of these primary spermatocytes as they undergo meiosis (and that no crossing over occurs between these loci in the remaining 76 cells). The tetrads in each of these 24 spermatocytes can be represented by the diagram of a tetrad you drew previously, and each would give rise to the four product chromatids, *A B*, *A b*, *a B*, and *a b*. Following meiosis, each of these chromatids will be found in a different gamete. How many kinds of gametes will be produced from these four kinds of product chromatids? ________[22] How many of these kinds of gametes show the same gene combinations found in the original, parental chromosomes? ____[23] How many kinds show recombination? ________[24] Since there are 24 primary spermatocytes that experience this type of crossing over and each gives rise to four gametes for a total of 24 × 4 = 96 sperm cells, how many of each kind of gamete would you expect to be produced? ________[25]

Dihybrid Gamete Production without Crossing Over At this point we turn our attention to the 76 primary spermatocytes that do *not* experience crossing over. The tetrads in these spermatocytes would appear as follows.

[19] *A B*, *A b*, *a B*, and *a b*. [20] 1/2. [21] 1/2. [22] Four, with gene combinations *AB*, *Ab*, *aB*, and *ab*. [23] Two, with gene combinations *AB* and *ab*. [24] Two, with gene combinations *Ab* and *aB*. [25] 96/4 = 24.

A B
A B

a b
a b

How many kinds of gametes will be formed from a spermatocyte of this type? ________[26] How many of these kinds of gametes show the same gene combinations found in the original, parental chromosomes? ________[27] How many of these kinds of gametes carry recombinants? ________[28] Since there are 76 primary spermatocytes which experience no crossing over and each gives rise to four gametes for a total of 76 × 4 = 304 sperm cells, how many of each kind of gamete would you expect them to produce? ________[29]

Total Dihybrid Gamete Production Now consider all 400 of the gametes produced by combining the gametes formed from the 24 primary spermatocytes that experience crossing over with those from the 76 primary spermatocytes that do not. What is the number of gametes out of the total of 400 that are recombinant types? ________[30] What percentage of the 400 gametes are recombinants? ________[31] What is the number of gametes that are parental, or nonrecombinant, types? ________[32] What percentage of the 400 gametes are parental types? ________[33]

Testcross Outcome Now we are ready to determine the progeny produced from the *AB*/*ab* × *ab*/*ab* testcross. As you have just shown, the dihybrid parent will produce four kinds of gametes, divided as follows: 88% of the gametes will have parental-type chromosomes, equally divided between the *A B* and *a b* types, and 12% of the gametes will carry recombinants, equally divided between the *A b* and *a B* types. The homozygous recessive parent will, of course, produce gametes carrying just one type of chromosome: *a b*. Why are no recombinants formed when crossing over occurs in the *ab*/*ab* parent? ________
________[34]

Random union of these gametes is shown in the following branch diagram; identify the progeny genotypes by completing the diagram. Indicate the expected percentage of total progeny made up by each type of offspring.

Gametes from *ab*/*ab* Parent	Gametes from *AB*/*ab* Parent		Progeny genotypes	Genotypic frequencies
	A B (44%)	=	________	________
	a b (44%)	=	________	________
a b				
	A b (6%)	=	________	________
	a B (6%)	=	________	________

How many different phenotypes would the progeny exhibit? ________[35] Would these phenotypes be expected in equal frequencies? ________[36] Explain. ________[37]

Summary of the Three Types of Dihybrid Testcrosses

Now let us summarize what has been shown so far. We have examined three different testcrosses involving dihybrid organisms. In the first, the loci were not linked and thus assorted independently. In the second, the two loci were completely linked, with no crossing over to disrupt the linkage. In the third, the linkage was incomplete, with crossing over occurring between the two loci. Complete Table 12-1 to compare these three types of testcrosses.

To summarize, genes under study in a two-locus testcross are either linked or independently assorting. Any significant departure from the 1:1:1:1 phenotypic ratio (expected if the loci were independently assorting) that gives an excess of parental types and a shortage of recombinant types implies that the loci are linked and that crossing over has taken place between them.

Using the Chi-Square Test to Evaluate Departure from a 1:1:1:1 Ratio

Whether the departure from the 1:1:1:1 ratio is statistically significant may be determined by using the chi-square test. This procedure is described in detail in Chapter 6 and should be reviewed if necessary. A deviation from the ratio that results in a chi-square value with a probability value of 0.05 or less is generally considered statistically significant and suggests that the deviation is not due just to chance and that an

[26] Two, with gene combinations *AB* and *ab*. [27] Two. [28] None. [29] 304/2 = 152. [30] 48. [31] 48/400 = 0.12, or 0.12 × 100 = 12%. [32] 352. [33] 352/400 = 0.88, or 0.88 × 100 = 88%. [34] Any crossing over that occurs generates exactly the same gene combinations that existed before crossing over. [35] Four. [36] No. [37] Progeny with the parental-type chromosomes would occur with higher frequencies (44% for each type) than the progeny carrying the recombinant chromosomes (6% for each type).

TABLE 12-1

Comparison of three types of testcrosses.

	Gametes of Dihybrid Parents		Testcross Progeny	
Two-locus testcross involving:	Number of types	Relative frequency expected of each type	Number of phenotypes	Relative frequency expected of each type
Unlinked loci	______[38]	______[39]	______[40]	______[41]
Linked loci with no crossing over	______[42]	______[43]	______[44]	______[45]
Linked loci with crossing over	______[46]	______[47]	______[48]	______[49]

alternative to independent assortment is operating. Usually that alternative is linkage.

Absence of Crossing Over in *Drosophila* Males

An exception to the rule of crossing over and the production of recombinations of linked genes occurs in *Drosophila* males. This absence of crossing over means that in males the linkage for genes comprising each linkage group is complete. The detection of linkage in fruit flies has been facilitated because of this phenomenon. There is no suppression of crossing over in *Drosophila* females.

DETERMINING DISTANCES BETWEEN LOCI

Recombinants in a Two-Point Testcross Provide Map Distances

As we have just discussed, the outcome of a two-point testcross in conjunction with a chi-square test can often tell whether two loci are linked. Next, we will see how testcross results can be used to determine the distance between two linked loci. Assuming that two genes are linked, the occurrence of recombinants among the progeny of a testcross is evidence of what process? ________[50] What determines the frequency with which recombinants occur among testcross progeny? ____________________

_____[51] The frequency of crossing over is, in turn, directly related to the distance separating the linked loci. For loci that are extremely close together, crossing over between them is rare, and few, if any, recombinants are detected in a testcross. With a limited number of testcross progeny, recombination could go undetected. Such loci might fit the situation described earlier in this chapter for completely linked genes, producing a phenotypic ratio of 1:1. In contrast, two linked loci may be sufficiently far apart so that crossing over between them occurs in virtually all cells giving rise to gametes. Under such circumstances, half the gametes would carry recombinants, and testcross progeny would approach the 1:1:1:1 phenotypic ratio that results with unlinked, independently assorting loci. For loci separated by distances that fall between these extremes, the frequency of crossing over increases in proportion to the distance between the loci.

The frequency of crossing over can readily be determined from the frequency with which recombinants occur among testcross progeny and may, in turn, be used to set up a linkage map of a portion of the chromosome carrying the two loci. In setting up these maps, geneticists have set 1% of recombinants as equal to one unit of map distance. In the testcross example considered earlier that involved incomplete linkage, a total of 48 of the 400 progeny, or 12%, were recombinant for the parental genes. That tells us that crossing

[38] Four. [39] 1:1:1:1. [40] Four. [41] 1:1:1:1. [42] Two. [43] 1:1. [44] Two. [45] 1:1. [46] Four.
[47] Parental types generally formed in relatively high frequencies and recombinant types in relatively low frequencies. [48] Four.
[49] Two parental types generally formed in relatively high frequencies and two recombinant types in relatively low frequencies.
[50] Crossing over. [51] The frequency of crossing over.

FIGURE 12-1

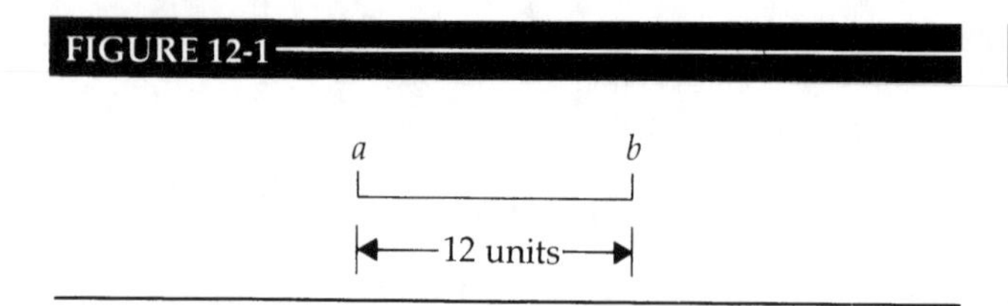

over occurred between the two loci under consideration with a frequency of 12% and that the two loci are separated on the chromosome by 12 map units, as shown in Figure 12.1.

Assume that in another two-point testcross involving the same species, linked loci *a* and *c* are studied and 6% of the progeny are found to be recombinants. A crossover frequency of 6% tells us that there are six units of map distance between the *a* and *c* loci. Combining the results of these two testcrosses gives us the distance between *a* and *b* and between *a* and *c*, but tells us nothing about the sequence, or linear order, of these three loci on the chromosome and nothing about the distance between *b* and *c*. Since distances are additive, the order *a-b-c* is excluded at once, but either of the sequences shown in Figure 12-2 is possible.

This ambiguity can be resolved only when the percentage of crossing over between *b* and *c* is determined by carrying out a testcross involving these loci. If, for example, this testcross tells us that the frequency of crossing over between *b* and *c* is about 18%, then the sequence shown in Figure 12-2a is correct. Alternatively, if the testcross indicates a crossover frequency of 6%, then the sequence shown in Figure 12-2b is correct. Assume that the testcross indicates that 50 of 287 progeny are recombinants. Which sequence is correct? __________[52]

Mapping Three Loci from Two-Point Testcross Results

In most instances where you are asked to set up a map of a region of a chromosome, you will be given either the frequencies with which at least three loci in that region crossover with each other or the testcross recombinant frequencies from which the crossover frequencies can readily be determined. For example, loci *s*, *t*, and *u* are known to be linked, and you are provided with testcross results indicating crossover frequencies of 16%, 35%, and 19% between *s* and *u*, *s* and *t*, and *t* and *u*, respectively. You are asked to prepare a map showing the relative positions of, and distance between, these three loci. A good approach is to start with the greatest distance, in this case, the 35 map units between *s* and *t*. Use a line to represent the chromosome and indicate the relative positions of these two loci as shown in Figure 12-3. Then select either of the remaining pairs of loci, say *s* and *u*. A crossover frequency of 16% indicates they are separated by 16 map units. Two alternatives exist regarding the position of *u*; it could be either to the left or to the right of the *s* locus, as shown in the two alternative maps in Figure 12-4. If *u* is to the left of *s*, as shown in Figure 12-4a, what would be the *u*-to-*t* map distance? __________[53] If *u* is to the right of *s*, as shown in Figure 12-4b, what would be the *u*-to-*t* distance? __________[54] Testcross data tell us that a distance of 19 units separates *u* and *t*. Which map is correct? __________[55]

FIGURE 12-2

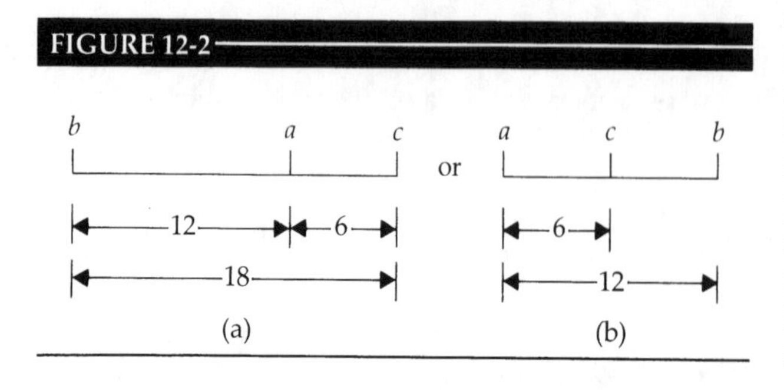

FIGURE 12-3

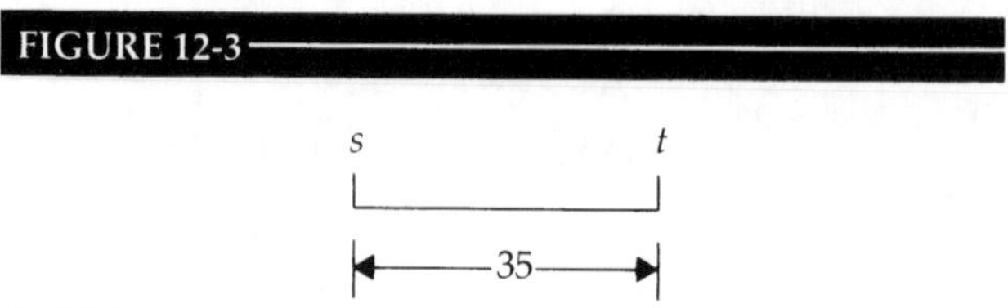

SUMMARY

The outcome of a testcross between an organism heterozygous for two loci and another homozygous recessive for the same two loci can generally tell us whether two loci are carried on the same or on separate chromosomes. Since the homozygous recessive testcross parent always produces the same type of gamete, variation in testcross outcome reflects variation in the kinds and frequencies of gametes produced by the double-heterozygous parent. If the loci assort independently, four kinds of gametes, expected in equal frequencies, give rise to four progeny phenotypes in a 1:1:1:1 ratio. If the loci are linked and positioned so that no crossing over occurs between them, two kinds of gametes are produced in equal frequencies, giving

[52] The recombinants make up 50/287 = 17.4% of the progeny, indicating that the distance separating *b* and *c* is about 17 map units. The sequence shown in Figure 12-2a is the correct one. [53] 16 + 35 = 51 units. [54] 35 − 16 = 19 units.
[55] The map in Figure 12-4b with the *u* locus between *s* and *t* is correct.

FIGURE 12-4

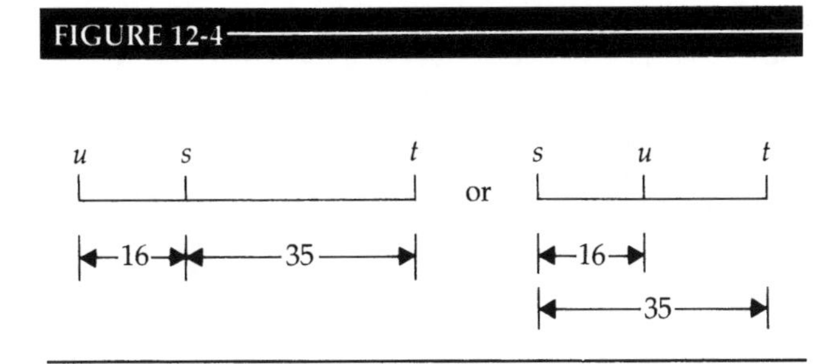

rise to two progeny phenotypes in a 1:1 ratio. If the loci are linked and crossing over occurs between them, then four kinds of gametes are expected in unequal frequencies, giving rise to four progeny phenotypes with more than half the progeny showing the phenotypes of one or the other of the testcross parents and less than half showing one or the other of the recombinant phenotypes. A testcross outcome that deviates from the 1:1:1:1 phenotypic ratio expected under the hypothesis of independent assortment and that is determined (using the chi-square test) to be statistically significant implies that the loci are linked.

The frequency of crossing over, which can readily be determined from the frequency with which recombinants occur among testcross progeny, increases in proportion to the distance between two linked loci and, consequently, serves as an index of the distance separating them. By equating 1% recombination frequency to one unit of map distance, the frequency of crossing over can be used to set up a linkage map of a portion of the chromosome carrying the two loci. Relating the outcomes of a series of two-point testcrosses that have a locus in common makes it possible to determine the sequence and distances separating a number of linked loci.

PROBLEM SET

12-1. The following sketch shows a pair of duplicated homologous chromosomes in a tetrad configuration with a crossover between loci *c* and *d*.

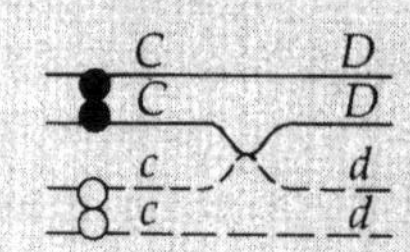

a. Sketch this chromosome pair at the end of the first meiotic division and at the end of the second meiotic division. Label each chromosome produced following the second meiotic division as either a recombinant or parental type.

b. Identify the linkage arrangement shown by the two loci on each of the chromosomes produced following the second meiotic division as either coupling or repulsion.

12-2. If the crossover site shown in problem 12-1 was at some point other than between the *c* and *d* loci, could the recombination be detected from the progeny of a *CcDd* × *ccdd* testcross? Explain.

12-3. In the tetrad shown in problem 12-1, the genes at the two loci show the coupling linkage arrangement. If these loci had been in repulsion, explain how the parental and recombinant chromosomes arising from meiosis would be different.

12-4. Refer to the sketch in problem 12-1. Assume that 18% of the tetrads that undergo meiosis experience crossing over in the region between the *c* and *d* loci.

a. What are the percentages of recombinant and nonrecombinant chromosomes produced following meiosis?

b. State, in general terms, the relationship between the percentage of tetrads experiencing crossing over in the region between two loci under study and the percentage of gametes that arise following the second meiotic division that carry chromosomes recombinant for those loci.

c. Explain why the frequency of recombinants is equal to half the frequency of crossover events.

12-5. a. State, in general terms, the relationship between the distance separating two loci and the frequency with which crossovers occur in the region between the two loci.

b. If two loci are far enough apart so that the probability of crossing over between them is 1, that is, a certainty, what will be the frequencies with which recombinant chromosomes and nonrecombinant chromosomes are produced by a dihybrid organism following meiosis? In what ratio would recombinant and nonrecombinant gametes be formed by the dihybrid parent?

12-6. Refer to the sketch shown in problem 12-1. After the crossover that is shown has occurred, assume that the same tetrad experiences a second crossover in the *c*-to-*d* region involving the same two chromatids.

a. What proportion of the chromosomes (or gametes) that result are crossover types? noncrossover types?

b. Considering just the *c* and *d* loci, could the two crossovers that occurred in the *c*-to-*d* region be detected from the final product chromosomes? Explain.

PROBLEM SET

c. If the probability of crossing over in the *c*-to-*d* region is 15% and if each crossover in this region is an independent event, what is the probability of two crossovers occurring in this region?

12-7. Refer to the information given in problem 12-6. Assume that the second crossover in the *c*-to-*d* region is followed by a third crossover in the same region that involves the same two chromatids.

a. What proportion of the chromosomes (or gametes) that result are crossover types? noncrossover types?

b. Considering just the *c* and *d* marker loci, could the number of crossover events that occurred in the *c*-to-*d* region be determined from the final product chromosomes?

c. If the probability of crossing over in the *c*-to-*d* region is 15%, and if each crossover in the region was an independent event, what is the probability of three crossovers occurring in this region?

12-8. In the fruit fly, *Drosophila melanogaster,* the black-body allele, b, is recessive to the wild-type gray allele, b^+, and the curved-wing allele, cu, is recessive to the wild-type straight-wing allele, cu^+. The loci for these two traits are linked. A gray-bodied, straight-winged fly, homozygous for each trait, was mated with a black-bodied, curved-winged fly. An F_1 female was then mated in a testcross with a male having a black body and curved wings.

a. Identify the phenotypes expected among the testcross progeny if there is no crossing over.

b. Identify the phenotypes expected among the testcross progeny if there is some crossing over between these two loci.

c. Would you expect the phenotypes listed in your answer to 12-8b to occur in equal frequencies? Explain why or why not.

12-9. The allele for black body color in fruit flies is recessive to the allele for wild-type normal color, and the allele for vestigial wings is recessive to the allele for normal wing form. A female fly heterozygous for both traits was crossed with a black-bodied male with vestigial wings. The progeny phenotypes and their frequencies were normal color, normal wings: 252; black color, normal wings: 52; normal color, vestigial wings: 48; and black color, vestigial wings: 248.

a. Do the results of this testcross support the hypothesis that the genes for these two traits are linked? Explain.

b. If these loci are linked, what is the percentage of recombination?

12-10. An F_1 corn plant is heterozygous for the traits of leaf color and seed shape. It exhibits red leaves and normal seeds and is crossed with another plant exhibiting the recessive phenotype for these two traits, green leaves and rounded seeds. The progeny were as follows: red, normal: 120; red, rounded: 138; green, normal: 116; and green, rounded: 126.

a. Do these data support the hypothesis that these two loci are linked? Explain your answer.

b. If these loci are linked, what is the percentage of recombination?

12-11. Two linked loci are studied in a testcross. Four phenotypes are found in the progeny. Two are parental phenotypes and each occurs with a frequency of 46%. The other two are recombinant phenotypes; each occurs with a frequency of 4%. How many units of map distance separate these loci?

12-12. In the fruit fly, *Drosophila melanogaster,* numerous loci are found on the second chromosome. Among them are the loci controlling cinnabar eye

PROBLEM SET

color and vestigial wings which are found 57.0 and 67.0 units of map distance, respectively, from one end of the chromosome. The allele for cinnabar, *cn*, is recessive to the wild-type red-eye allele, cn^+, and the allele for vestigial wings, *vg*, is recessive to that for normal wing form, vg^+. A female, homozygous for both red eyes and vestigial wings, is crossed with a male homozygous for both cinnabar eye color and normal wings.

a. What is the genotype of an F_1 female fly?

b. Identify the different kinds of gametes produced by an F_1 female and the expected proportion of each.

c. If an F_1 female is mated with a male with cinnabar eyes and vestigial wings, what types of progeny would be expected and in what proportions?

12-13. Loci *d*, *e*, and *f* are linked. A series of testcrosses yields the following information: genes *D* and *E* recombine with *d* and *e* in 6% of the offspring; *D* and *F* recombine with *d* and *f* in 22% of the offspring; and *E* and *F* recombine with *e* and *f* in 16% of the offspring. Draw a map showing the relative locations of these three loci and the distances between them.

12-14. In the fruit fly, *Drosophila melanogaster*, the allele for black body, *b*, is recessive to that for normal body color, b^+, and the allele for purple eyes, *p*, is recessive to that for normal, red eye color, p^+. The loci governing these two traits are found on the same chromosome, separated by a distance of six map units. A cross is set up between a fly heterozygous for each locus, b^+p/bp^+, and a homozygous recessive fly, *bp*/*bp*. What percentage of the progeny would be expected to exhibit both black bodies and purple eyes?

12-15. Determine the order of three linked loci, *s*, *t*, and *u*, and the map distances between them if the recombination frequencies are as follows.

a. *s*-to-*t* is 2%, *t*-to-*u* is 3%, and *s*-to-*u* is 1%.

b. *u*-to-*s* is 11%, *s*-to-*t* is 4%, and *u*-to-*t* is 7%.

c. *u*-to-*s* is 2%, *t*-to-*u* is 13%, and *s*-to-*t* is 15%.

12-16. In the fruit fly, the recessive allele at the *lt* locus produces light eyes and the recessive allele at the *vg* locus produces vestigial wings. The dominant alleles at these two loci, lt^+ and vg^+, produce wild-type eyes and wild-type wings, respectively. A female, heterozygous for both of these traits, is mated with a male who is homozygous for recessive alleles at both of these loci. The phenotypes found among their 260 offspring are as follows: wild-type: 114; wild-type eyes, vestigial wings: 9; light eyes, wild-type wings: 11; and light eyes, vestigial wings: 126.

a. Does the dihybrid parent carry alleles at these loci in the coupling or repulsion linkage arrangement? Explain.

b. Determine the recombination frequency between these two loci.

c. What percentage of the tetrads giving rise to the gametes produced by the dihybrid parent would have shown crossing over between the light-eye and the vestigial-wing loci?

12-17. The frequency of crossing over between the *m* and *s* loci in an organism is known to be 10.8%. In a testcross, a homozygous recessive individual, *mmss*, is mated with an individual heterozygous for both loci. The alleles carried by the dihybrid parent are in the repulsion linkage arrangement. Assume that the testcross mating produces a total of 298 progeny. Predict the phenotypes of the progeny and the frequency with which each would be expected to occur.

PROBLEM SET

12-18. Refer to the basic information given in problem 12-14. A mating is set up between a male fly who is heterozygous at the *b* and *p* loci and carries the alleles for these two traits in repulsion, b^+p/bp^+, and a female who is heterozygous at each locus and carries the alleles in coupling, b^+p^+/bp. Note that no crossing over occurs in male fruit flies but it does occur in females.

- **a.** Identify the types of male gametes and the expected frequency of each.
- **b.** Identify the types of female gametes and the expected frequency of each.
- **c.** Determine the phenotypes found among the progeny and the expected frequency of each.

12-19. The loci for striped body and cardinal eyes in *Drosophila* are linked and found at approximately 62 and 75 map units, respectively, from the same end of the third chromosome. The recessive alleles at these loci, *sr* and *cd*, respectively, are recessive to their counterpart alleles for wild-type body and eyes. Remember that crossing over is suppressed in *Drosophila* males.

- **a.** A female who is heterozygous at both of these loci and carries the genes in repulsion is crossed with a male homozygous recessive for both loci. From this mating, identify the genotypes and phenotypes of the progeny and their expected frequencies.
- **b.** Carry out the reciprocal of the cross described in 12-19a and identify the progeny and their expected frequencies.

12-20. Refer to the information given in problem 12-19. A male and a female, both heterozygous for each trait and both carrying the genes in repulsion, are crossed.

- **a.** Identify the progeny genotypes and phenotypes along with their expected frequencies.
- **b.** Is the outcome of this cross dependent upon the degree of linkage between the two loci under consideration? Explain.

12-21. In *Drosophila melanogaster*, the mutant genes for curly wings (*Cy*) and for plum eye color (*Pm*) are dominant to the wild-type alleles for normal wing shape and eye color. These two genes are linked, and homozygosity of the mutant allele at either locus is lethal. The inversion of a section of the chromosome carrying these loci prevents crossing over between them. A cross is carried out between flies that are heterozygous at both of these loci and that carry the genes in repulsion.

- **a.** Identify the progeny genotypes and phenotypes along with their expected frequencies.
- **b.** Would the curly, plum stock breed true if the heterozygous parents carried the genes for these two traits in coupling?

CHAPTER

13

The Three-Point Testcross and Chromosomal Mapping

INTRODUCTION

The material covered in this chapter builds upon that presented in Chapter 12 where three key points were established. (1) The occurrence of crossing over during meiosis is detected through the presence of recombinants among testcross progeny; the higher the percentage of recombinants, the greater the frequency of crossing over. (2) The frequency of crossing over between two loci is directly related to the distance separating the two loci; the greater the frequency of crossing over, the greater the map distance between the two loci. (3) There is a positive correlation between map distance and physical distance on the chromosome. An understanding of these points is essential before proceeding with this chapter.

LIMITATIONS OF TWO-POINT TESTCROSSES

Chromosome map distances derived from two-point testcross recombination frequencies may be underestimates if the two chromatids that participate in one crossover event also participate in a second event.

Double Crossovers: The Second Cancels the Effect of the First

The effect of a double crossover involving the same two chromatids can be shown by completing Figure 13-1. The first sketch, Figure 13-1a, represents a tetrad of the dihybrid *AaCc*, where the *a* and *c* loci are linked. In Figure 13-1b, two of the four chromatids are shown crossed over. In the space provided for Figure 13-1c, draw the four chromatids as they would appear after crossing over is complete and the two inner chromatids are no longer in contact. Verify the accuracy of your drawing by comparing it with Figure 13-1d. The next sketch, Figure 13-1e, shows the same two chromatids experiencing a second crossover in the same region. In the space provided for Figure 13-1f, draw the four chromatids as they would appear after this second crossover event is complete and the two inner chromatids are no longer in contact.

After you complete Figure 13-1, compare the final product chromatids in Figure 13-1f with the original precrossover chromatids in Figure 13-1a. With regard to the two loci we are considering, what do you notice?

______________________________[1]

What is the net consequence of the second crossover event? ______________________________

________[2] Would chromosomes recombinant for these two loci arise from the series of events shown in Figure 13-1? ________[3] (Note that our consideration of the double crossover event as two successive crossovers is

FIGURE 13-1

Effect of a double crossover involving the same two chromatids.

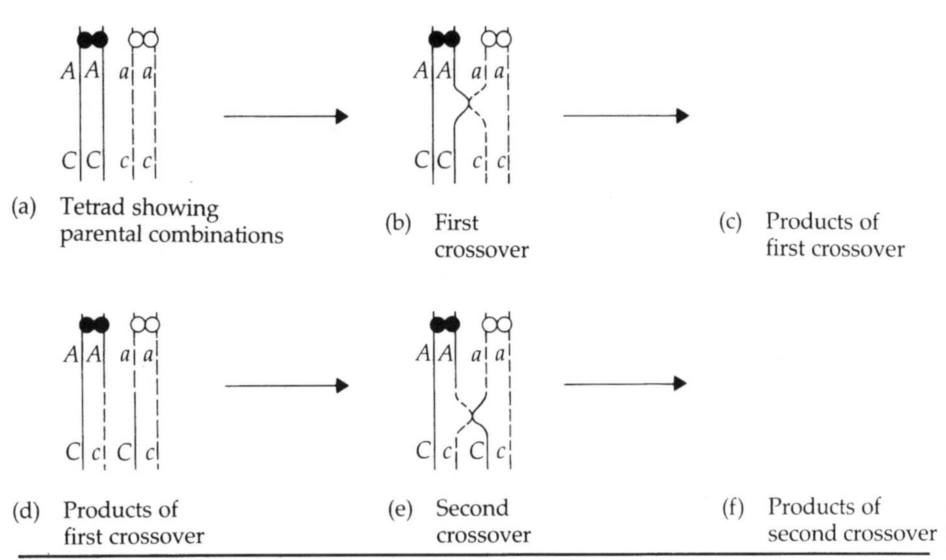

done for ease of understanding; in actuality, two such crossovers occur simultaneously. Regardless, the final outcome is the same.)

Underestimating Crossover Frequencies

As you just demonstrated, double-crossover events that have the second crossover occurring in the same chromosome region as the first and that involve the same two chromatids generate no recombinant chromatids. As a consequence, such double (and other even-numbered) crossovers do not generate recombinants among two-point testcross progeny and the crossovers go undetected. (Odd-numbered crossover events, in contrast, do yield recombinants and will be detected.) The absence of recombinants with even-numbered crossovers leads to an underestimate of the frequency of crossing over and, when maps are prepared, to an underestimate of the distance between loci. The magnitude of this underestimate increases with the distance separating the two loci on the chromosome: since the probability of a single crossover event increases as the distance between two loci increases, so, too, does the probability of a double-crossover event. (If the probability of a single crossover within a particular region is known, the likelihood of a second crossover in that region, assuming it is independent of the first, is given by the product law.) For this reason, greater accuracy is achieved in two-point mapping by using closely spaced loci. Distances between widely separated loci are best determined by adding the shorter distances separating intervening loci.

Another reason for using closely spaced loci is to avoid situations where distant, linked loci appear to be independently assorting. If the two loci under study are far enough apart so that the likelihood of crossing over between them is 100%, the recombination frequency will be 50%. As shown in Chapter 12, tetrads experiencing a single crossover during meiosis produce parental combinations in half of the product chromosomes and recombinant combinations in the other half.

It is also true that all the tetrads experiencing double crossovers, when taken collectively, produce parental- and recombinant-type gametes in a 1:1 ratio, as is summarized in Figure 13-2. As you just demonstrated, double crossovers involving the same two strands (two-stranded doubles) produce parental combinations in all the product chromosomes (Figure 13-2a). Other types of double crossovers are, of course, possible. With three-stranded double crossovers, shown in Figures 13-2b and c, one crossover occurs between two chromatids and the second crossover occurs between one of the chromatids participating in the first crossover and a third chromatid. Depending on the specific chromatids that participate, there are two ways in which three-stranded double-crossovers

[1] The product chromatids are identical to the original chromatids. [2] It cancels the effect of the first.
[3] No, since all the product chromosomes exhibit the parental combinations.

FIGURE 13-2

Collectively, the four types of double crossovers generate parental- and recombinant-type chromosomes in a 1:1 ratio.

Tetrad showing crossing over	Chromosomes produced	Number of parental chromosomes produced per tetrad	Number of recombinant chromosomes produced per tetrad
A A a a / C C c c →	A A a a / C C c c	4	0
(a) Two-stranded double crossover			
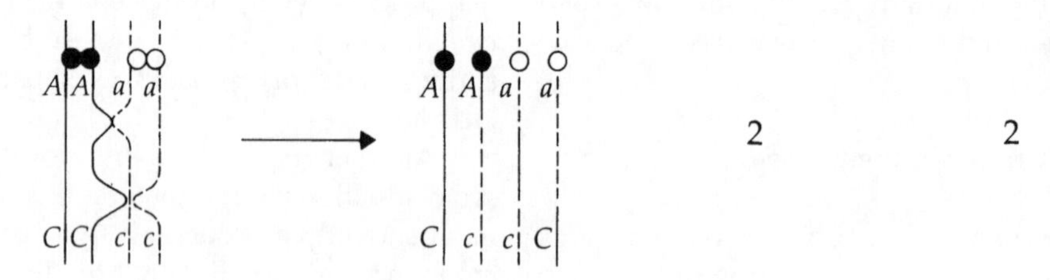		2	2
(b) Three-stranded double crossover			
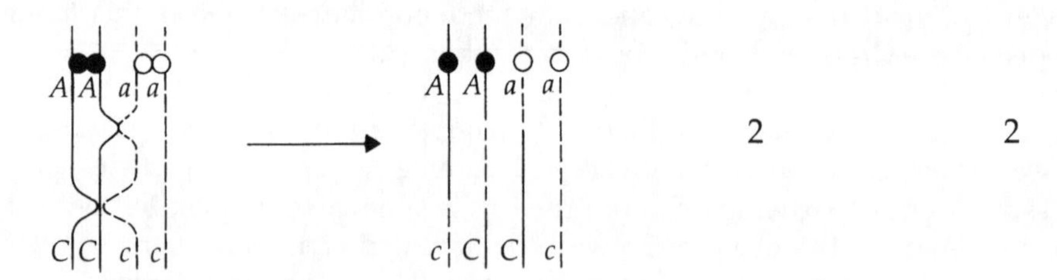		2	2
(c) Three-stranded double crossover			
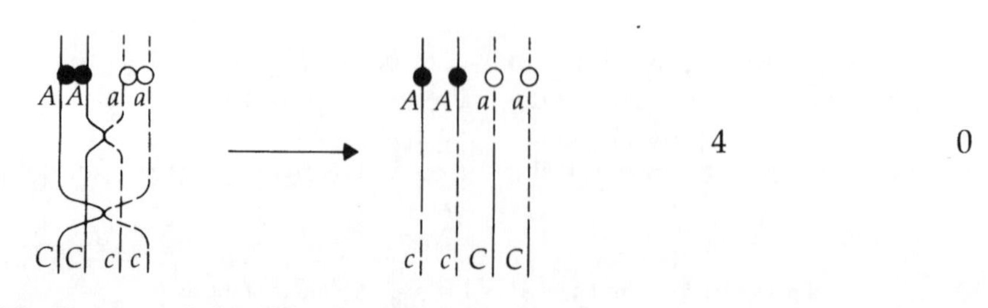		4	0
(d) Four-stranded double crossover			

can occur. In either case, two product chromosomes are recombinant and two are parental types. With four-stranded double crossovers (Figure 13-2d), two chromatids participate in one crossover and the other two chromatids participate in the second; all four of the product chromosomes are recombinant.

The occurrence of each of the four types of double crossovers shown in Figure 13-2 is equally probable and, with a large number of nuclei carrying tetrads that experience double crossovers, two-, three-, and four-stranded doubles would be expected to occur in a ratio of 1/4:1/2:1/4. With each two-, three-, and four-stranded double crossover producing 0, 2, and 4 recombinant chromosomes, respectively, the proportion of recombinant-carrying gametes derived from all types of double-crossover events is $(1/4)(0) + (1/2)(2) + (1/4)(4) = 2$ out of 4. The remaining two out of four gametes carry the parental combinations. Thus, a two-point testcross involving widely spaced loci could produce progeny in a phenotypic ratio of 1:1:1:1 which would lead to the conclusion that the loci are assorting independently. This misleading outcome can be avoided by using more closely spaced loci to determine map distances.

DETECTION OF DOUBLE CROSSOVERS: THE THREE-POINT TESTCROSS

Advantage of Considering a Third Locus

Fortunately, many of the double crossovers which go undetected in a two-point testcross can be detected if an additional locus, located between the two under study, is considered. The linkage relationships among the three loci are determined from a three-point testcross in which an organism heterozygous for the three loci is crossed with an individual homozygous recessive at these loci. In addition to providing measures of crossover or recombination frequencies and therefore map distances between the three loci, the three-point testcross also makes it possible to determine the relative position or sequence of these loci on the chromosome. In the discussions of three-point testcrosses that follow, assume that each locus controls a different trait and that two alleles occur at each locus with one allele exhibiting complete dominance to its recessive counterpart.

Consequence of a Double Crossover on the Middle Locus

Before looking at a specific three-point testcross, we need to consider the general procedure used to identify the middle locus and thereby establish the relative order of the three loci. The following is a sketch of two nonsister chromatids of a tetrad, each showing three loci with the parental combinations of alleles, *ABC* and *abc*. Note that the sister chromatid for each of the chromatids shown is omitted.

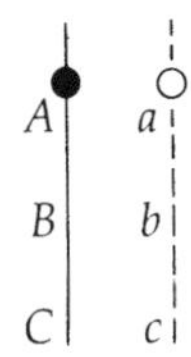

In the space provided, sketch these chromatids after a single crossover has occurred in the *a*-to-*b* region (region I).

Next, assume that these same two chromatids experience a crossover in the *b*-to-*c* region (region II). Sketch the final product chromatids in the space provided.

Now compare the original parental chromatids with the chromatids produced following the double crossover. At which locus do the alleles appear to have exchanged places relative to the other two loci? ____[4] What does the double crossover appear to do to the middle locus? ______[5] To summarize, a comparison of the parental and double-recombinant strands will reveal which of the three loci is in the middle, for it is always the middle locus which appears to be transposed.

Gamete Production in the Triple Heterozygote: Identifying the Parental and Double-Recombinant Chromosomes

Next we need to learn how to identify the parental and double-recombinant product chromosomes. The key

[4] The *b* locus. [5] To transpose it.

to doing this lies with the relative frequency with which gametes carrying the parental and double-recombinant chromosomes are produced by the triple-heterozygous parent. During the production of gametes in, say, a triple-heterozygous male of genotype *ABC/abc*, assume that any one of four possible situations could occur in a particular diploid cell (primary spermatocyte) that gives rise to haploid sperm cells: (1) no crossover in either of the regions separating these loci, (2) a single crossover in region I (*a*-to-*b*), (3) a single crossover in region II (*b*-to-*c*), or (4) a double crossover with one crossover in region I and a second in region II. Since this male has many primary spermatocytes undergoing meiosis at any given time, we need to consider each of these situations and its outcome. Table 13-1 identifies the product chromatids that arise in each case. As you examine this table, remember that each product chromatid becomes a chromo-

TABLE 13-1

Meiotic events producing the chromatids found in gametes of the triple heterozygote *ABC/abc*.

Meiotic event	Tetrad configuration	Chromatid Types: Parental (nonrecombinants)	Chromatid Types: Single-crossover recombinants	Chromatid Types: Double-crossover recombinants
No crossing over	A A a a / B B b b / C C c c	A B C; a b c	None	None
Single crossover in region I (between *a* and *b*)	A A a a / B B b b / C C c c	A B C; a b c	A b c; a B C	None
Single crossover in region II (between *b* and *c*)	A A a a / B B b b / C C c c	A B C; a b c	A B c; a b C	None
Double crossover in regions I and II (one between *a* and *b*, second between *b* and *c*)	A A a a / B B b b / C C c c	A B C; a b c	None	A b C; a B c

some and gets incorporated into a sperm cell produced by this triple-heterozygous parent.

Types and Relative Frequencies of Gametes

How many of the different product chromatid types shown in Table 13-1 are parental, or nonrecombinant, types? ________[6] How many are single-crossover recombinant types? ________[7] How many are double-crossover recombinant types? ________[8] What is the total number of the different types of chromatids produced here? ________[9] The exact number of gametes carrying each type of chromatid will depend on the amount of crossing over in each region. However, even without knowing the exact extent of crossing over, we can make some generalizations about the relative frequencies with which the different types of gametes are produced. What percentage of the gametes arising from each of the three crossover situations given in Table 13-1—that is, from the two types of single crossovers and from the double crossover—carry parental-type chromatids? ________[10] What percentage of the gametes arising from the noncrossover situation carry parental-type chromatids? ________[11] What does this lead you to conclude about the relative frequency of gametes bearing the parental-type combinations? ________________[12] Assuming that individual crossovers are independent events, state, in general terms, how the probability of a double crossover is determined? ________________[13] Since double-crossover events are considerably less frequent than either of the single-crossover events, what does this lead you to conclude about the relative frequency of gametes bearing the double-crossover recombinant strands? ________________[14]

To summarize, of the eight types of gametes produced by the triple heterozygote, those carrying the parental combinations will be the two most common types and those carrying the two double-crossover recombinants will be the two least common types. Although it is usually impossible to directly examine gametes of a triple heterozygote to determine the particular gene arrangement it carries, this can be inferred from the outcome of a three-point testcross. Since the homozygous recessive parent produces a single type of gamete carrying recessive alleles for the loci under study, the genetic makeup of the gametes contributed by the triple-heterozygous parent and the frequencies in which they are formed can readily be identified by examining the phenotypes of the testcross progeny. To illustrate this, we will analyze two three-point testcrosses.

Analysis of Three-Point Testcross Results: Example with Loci in Known Sequence

This testcross involves three loci, *a*, *b*, and *c*, that are carried on the same autosome. The sequence listed for the loci, with the *b* locus positioned between *a* and *c*, is the actual sequence. This testcross can be represented as *ABC*/*abc* × *abc*/*abc*. We have already identified the types of gametes produced by these two parents and they are shown in the Punnett square in Figure 13-3. Identify the progeny genotypes by completing this square.

Assume that there are a total of 1000 progeny produced, with the number in each of the eight classes given in Table 13-2. Notice that the eight classes are grouped in four **reciprocal pairs,** with the two members of a reciprocal pair produced through the same meiotic recombinant event. For example, classes 1 and 2 arise when no crossovers occur, classes 3 and 4 arise when a single crossover occurs in the *a*-to-*b* region, and so on. The members of a reciprocal pair generally occur in approximately equal frequencies. From the relative number of individuals in each class, we can work backward to infer the relative number of gametes of each type produced by the triple-heterozygous parent.

TABLE 13-2

The outcome of the testcross *ABC*/*abc* × *abc*/*abc*.

Class	Genotype	Number	Percentage of 1000		Type
1.	*ABC*/*abc*	422	42.2	84.9	Parentals
2.	*abc*/*abc*	427	42.7		
3.	*Abc*/*abc*	26	2.6	5.6	Single recombinants from crossovers in the *a*-to-*b* region
4.	*aBC*/*abc*	30	3.0		
5.	*ABc*/*abc*	44	4.4	9.0	Single recombinants from crossovers in the *b*-to-*c* region
6.	*abC*/*abc*	46	4.6		
7.	*AbC*/*abc*	2	0.2	0.5	Double recombinants from one crossover in the *a*-to-*b* region and a second crossover in the *b*-to-*c* region
8.	*aBc*/*abc*	3	0.3		

[6] Two. [7] Four. [8] Two. [9] Eight. [10] 50%. [11] 100%. [12] Gametes carrying the parental-type combinations are always the two most common types of gametes. [13] By multiplying the probabilities of the two types of single crossovers. [14] These gametes are always the two least common types of gametes.

FIGURE 13-3

Genotypes produced from the cross *ABC/abc* × *abc/abc*.

		Gametes from *ABC/abc*							
		Parental types		Single-crossover recombinants				Double-crossover recombinants	
				a-to-*b* crossovers		*b*-to-*c* crossovers		*a*-to-*b* and *b*-to-*c* crossovers	
		ABC	*abc*	*Abc*	*aBC*	*ABc*	*abC*	*AbC*	*aBc*
Gametes from *abc/abc*	*abc*	[15]	[16]	[17]	[18]	[19]	[20]	[21]	[22]

Verifying the Middle Locus Once the frequencies for the classes of progeny are known, a comparison of genotypes of the two most frequent classes, which are always the parentals, with the two least frequent classes, which are always the double recombinants, will identify the middle locus. Remember that a double crossover appears to switch the alleles of the middle locus. In this example, this step is unnecessary since we were told earlier that the *b* locus is in the middle; normally, however, this information is not provided and is something you would have to determine.

Estimating Crossover Frequencies and Map Distances Once the middle locus has been identified, the map distances can be readily determined from the total amount of crossing over occurring in each of the two chromosome regions. Recombinations arising from single crossovers in the *a*-to-*b* region are carried by the progeny in classes 3 and 4 (see Table 13-2) that make up a total of 5.6% of the progeny. This tells us that 5.6% of the gametes produced by the triple-heterozygous parent experienced a crossover in the *a*-to-*b* region. Are the progeny in classes 3 and 4 the only ones to have arisen from gametes with a crossover in the *a*-to-*b* region? ________[23] Explain. ____[24] The double-crossover progeny make up 0.5% of all the progeny. The total percentage of crossovers occurring in the *a*-to-*b* region is given by adding the percentages for both groups of progeny: 5.6% + 0.5% = 6.1%. Since, by definition, a 1% crossover frequency is equivalent to one unit of map distance, the *a* and *b* loci are separated by 6.1 map units, as shown in Figure 13-4.

The *b*-to-*c* distance is calculated in a similar way. Recombinants arising from a single crossover in the *b*-to-*c* region are found in classes 5 and 6 and make up 9.0% of the progeny. All of the double recombinants in classes 7 and 8, making up 0.5% of the progeny, also experienced a crossover in this region. What is the total crossover frequency in the *b*-to-*c* region? _____[25] What is the map distance between the *b* and *c* loci? __________[26] Our completed map appears in Figure 13-5.

Note that the double-crossover progeny have been counted twice in determining crossover frequencies. This is appropriate, however, since each individual is a product of *two* crossovers, one in each of the two chromosome regions being mapped.

Analysis of Three-Point Testcross Results: Example with Loci in Unknown Sequence

To make sure that you have mastered the basics of the three-point testcross, we will now look at an example

FIGURE 13-4

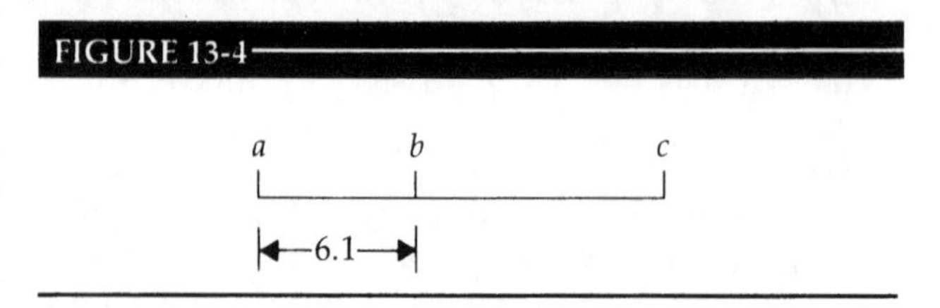

[15] *ABC/abc.* [16] *abc/abc.* [17] *Abc/abc.* [18] *aBC/abc.* [19] *ABc/abc.* [20] *abC/abc.* [21] *AbC/abc.* [22] *aBc/abc*. Each genotype in the square gives rise to a different phenotype. [23] No. [24] Each of the five double-crossover progeny in classes 7 and 8 have also come from a gamete with a crossover in this region. [25] 9.0% + 0.5% = 9.5%. [26] 9.5 map units.

FIGURE 13-5

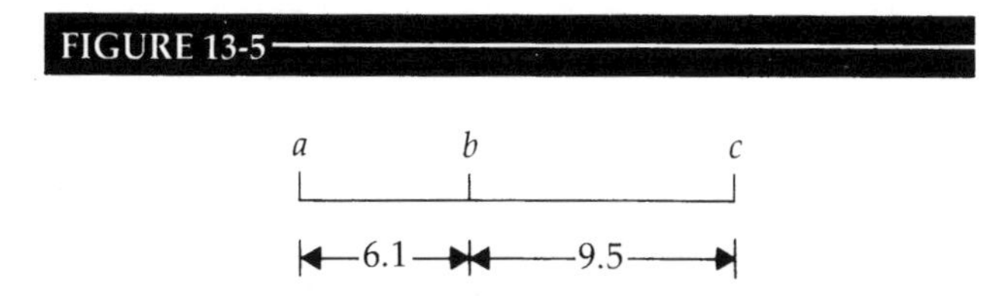

where both the sequence of loci and map distances need to be determined. The testcross involves three loci in the fruit fly *Drosophila melanogaster,* which are known to be carried on the same autosome and which influence the traits of (1) eye color, where the allele for the cinnabar color, *cn*, is recessive to its wild-type allele, cn^+; (2) wing form, where the allele for the vestigial, *vg*, is recessive to its wild-type counterpart, vg^+; and (3) body color, where the allele for black color, *b*, is recessive to the wild-type allele, b^+.

Since we know nothing about the relative order of the three loci, we can arbitrarily designate any order to begin our analysis. Let us use *cn-vg-b* with the understanding that this may or may not represent the correct sequence. The testcross may be represented as $cn^+vg^+b^+/cnvgb$ (female) × $cnvgb/cnvgb$ (male). The mating is carried out and a total of, say, 3200 progeny are collected. Inspection shows that eight different phenotypic classes are produced in the frequencies given in Table 13-3. Complete this table by deducing the genotype from each phenotype and then identifying the type of gamete contributed to it by the triple-heterozygous parent; the first one has been done for you. Remember that the frequency of each phenotypic class reflects the frequency of the gamete contributed to it by the triple-heterozygous parent.

Identifying the Parental- and Double-Crossover-Type Gametes The first step is to identify the two classes of progeny who received the parental-, or noncrossover-, type gametes from the triple-heterozygous parent. Since these two kinds of gametes will be the most common types, they will have produced the two largest classes of progeny. Refer to Table 13-3 and indicate, in the following table, the genotypes of these two classes and the type of triple-heterozygous gamete giving rise to each.

Parental Classes

Genotype	Triple-heterozygote gamete
______[41]	______[42]
______[43]	______[44]

We can now confirm the manner in which the alleles of the triple-heterozygous parent are associated. Since one gamete type is $cn^+vg^+b^+$ and the other is *cnvgb*, we know that all the recessive alleles were carried

TABLE 13-3

Deducing progeny genotypes and the types of gametes produced by the triple-heterozygous parent from the outcome of a three-point testcross.

Phenotype			Phenotypic frequency	Deduced genotype	Gamete from triple heterozygote
Eye	Wing	Body			
Normal	Normal	Black	140	$cn^+vg^+b/cnvgb$	cn^+vg^+b
Cinnabar	Normal	Normal	9	______[27]	______[28]
Normal	Normal	Normal	1330	______[29]	______[30]
Normal	Vestigial	Normal	147	______[31]	______[32]
Cinnabar	Vestigial	Normal	132	______[33]	______[34]
Normal	Vestigial	Black	6	______[35]	______[36]
Cinnabar	Vestigial	Black	1294	______[37]	______[38]
Cinnabar	Normal	Black	142	______[39]	______[40]
		Total:	3200		

[27] $cnvg^+b^+/cnvgb$. [28] $cnvg^+b^+$. [29] $cn^+vg^+b^+/cnvgb$. [30] $cn^+vg^+b^+$. [31] $cn^+vgb^+/cnvgb$. [32] cn^+vgb^+. [33] $cnvgb^+/cnvgb$. [34] $cnvgb^+$. [35] $cn^+vgb/cnvgb$. [36] cn^+vgb. [37] $cnvgb/cnvgb$. [38] *cnvgb*. [39] $cnvg^+b/cnvgb$. [40] $cnvg^+b$. [41] $cn^+vg^+b^+/cnvgb$. [42] $cn^+vg^+b^+$. [43] $cnvgb/cnvgb$. [44] *cnvgb*. Note that the answers to 41 and 42 are interchangeable with those to 43 and 44.

together on one chromosome and that all the dominant alleles were carried together on the other in the coupling arrangement.

Next we need to identify the double-crossover-, or double-recombinant-, type gametes. Since these two kinds of gametes will be the least common types, they will have produced the two smallest phenotypic classes. Indicate, in the following table, the genotypes of these two classes and the type of triple-heterozygous gamete giving rise to each.

Double-Crossover Classes

Genotype	Triple-heterozygous gamete
________[45]	________[46]
________[47]	________[48]

Identifying the Middle Locus Now compare the gametes contributed by the triple-heterozygous parent to the parental-type progeny and to the double-crossover-type progeny. Which one of the three loci in the double recombinants can be interchanged to produce the combinations seen in the parental strands? (Remember that a double crossover appears to switch the alleles at the middle locus.) ________[49] Which locus is in the middle? ________[50]

Once you have identified the middle locus, the placement of the other two loci is of no consequence. Either of these arrangements is correct: *vg-cn-b* or *b-cn-vg*. If you are puzzled by this, consider the following. A map of the northeastern United States will show the three major cities of New York, Philadelphia, and Washington, as shown in Figure 13-6. If this map is turned upside down (and the printing is righted), the

FIGURE 13-6

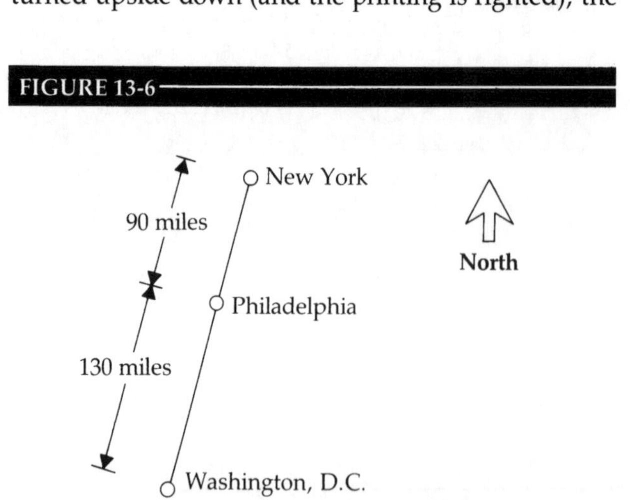

FIGURE 13-7

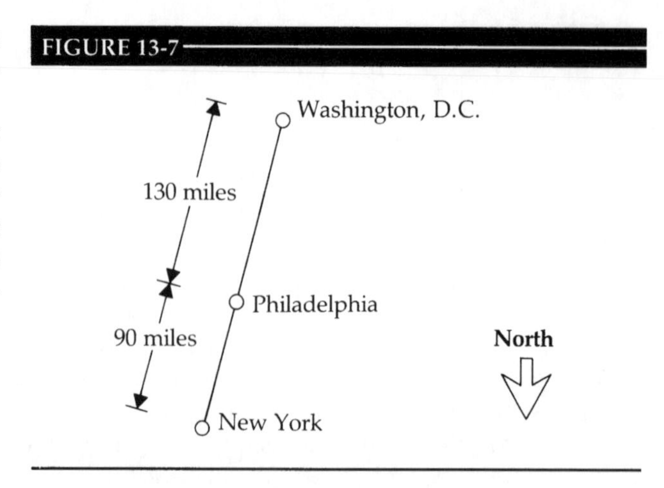

cities appear as in Figure 13-7. The first map has Washington, D.C., farthest to the left and at the bottom while the second map has New York in that position. Yet, in both cases, the distances and the relative sequence are the same; each represents the same map but with two different orientations. Similarly, *vg-cn-b* and *b-cn-vg* are the same arrangement with two different orientations.

Rewriting the Gene Sequence to Reflect the Actual Order Since the actual gene sequence differs from the one we arbitrarily selected when we began working with this testcross, it is useful at this point to rewrite the gene sequences to reflect their correct order. For example, after this rewriting, the triple-heterozygous parent could be represented as $vg^+cn^+b^+/vgcnb$ or, alternatively, as $b^+cn^+vg^+/bcnvg$. Use the space provided in Table 13-4 to rewrite the gene sequences to reflect the actual sequence; for consistency, list the *vg* locus first in the sequences.

Estimating Crossover Frequencies and Map Distances Since the *cn* locus is in the middle, the two regions within which crossovers occur can be defined: *vg*-to-*cn* (region I) and *cn*-to-*b* (region II). We need to determine the crossover frequency and map distance for each of these two regions.

In the space provided, sketch a crossover between two parental-type strands in the *vg*-to-*cn* region and show the recombinant strands that are produced.[58]

$$\frac{vg^+ \quad cn^+ \quad b^+}{vg \quad cn \quad b} \longrightarrow \qquad \longrightarrow$$

Parental chromatids — Crossed-over chromatids — Recombinant chromosomes

[45] $cnvg^+b^+/cnvgb$. [46] $cnvg^+b^+$. [47] $cn^+vgb/cnvgb$. [48] cn^+vgb. Note that the answers to 45 and 46 are interchangeable with those to 47 and 48. [49] *cn*. [50] *cn*.

TABLE 13-4

Rewriting the gene sequences in the triple-heterozygote gametes to reflect the actual sequence.

Phenotype				Gamete from Triple Heterozygote	
Eye	Wing	Body	Phenotypic frequency	Arbitrary sequence	Actual sequence
Normal	Normal	Black	140	cn^+vg^+b	vg^+cn^+b
Cinnabar	Normal	Normal	9	$cnvg^+b^+$	______[51]
Normal	Normal	Normal	1330	$cn^+vg^+b^+$	______[52]
Normal	Vestigial	Normal	147	cn^+vgb^+	______[53]
Cinnabar	Vestigial	Normal	132	$cnvgb^+$	______[54]
Normal	Vestigial	Black	6	cn^+vgb	______[55]
Cinnabar	Vestigial	Black	1294	$cnvgb$	______[56]
Cinnabar	Normal	Black	142	$cnvg^+b$	______[57]
		Total:	3200		

How many gametes bearing these two recombinant chromosomes are produced in the testcross? ______[59] Calculate the percentage of progeny out of a total of 3200 that carry each of these recombinant strands. ______________________[60]

What percentage of progeny carry one or the other of these recombinant strands? ________[61] The combined percentage of these two single-crossover products reflects most but not all the crossover events in the *vg*-to-*cn* region since each double-crossover product has also experienced a crossover in this region. You have already identified the double-recombinant gametes as vg^+cnb^+ and $vgcn^+b$ with respective frequencies of 9/3200 (0.28%) and 6/3200 (0.19%). Their combined percentage (0.47%) added to the percentage of single crossovers in the *vg*-to-*cn* region (9.0%) gives a total of 9.5%, indicating these two loci are separated by a map distance of 9.5 units, as shown in Figure 13-8.

We now need to repeat this procedure for the *cn*-to-*b* region of the chromosome. In the space provided, sketch a crossover between the two parental strands in the *cn*-to-*b* region and then show the recombinant strands that are produced.[62]

vg^+ cn^+ b^+ ⟶ ⟶
vg cn b

Parental chromatids — Crossed-over chromatids — Recombinant chromosomes

With what frequencies are the gametes bearing these two recombinant strands formed in the testcross? ______________________[63]

Calculate the percentage of progeny, out of a total of 3200, carrying one or the other of these recombinant strands? ______________________[64] Since the double-crossover recombinants, with a combined frequency of 0.47%, each experience a crossover in this region, the total crossover frequency in the *cn*-to-*b* region is 8.5% + 0.47% = 9.0%. Our completed map is shown in Figure 13-9.

FIGURE 13-8

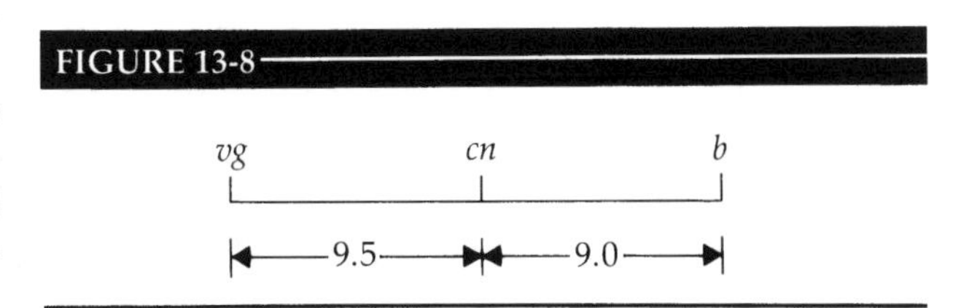

[51] vg^+cnb^+. [52] $vg^+cn^+b^+$. [53] $vgcn^+b^+$. [54] $vgcnb^+$. [55] $vgcn^+b$. [56] $vgcnb$. [57] vg^+cnb.
[58] Recombinant strands carry genes $vgcn^+b^+$ and vg^+cnb. [59] The numbers of $vgcn^+b^+$ and vg^+cnb gametes are 147 and 142, respectively. [60] 147/3200 = 0.0459, 0.0459 × 100 = 4.6% ($vgcn^+b^+$) and 142/3200 = 0.044, 0.044 × 100 = 4.4% (vg^+cnb). [61] 9.0%.
[62] Recombinant strands carry genes vg^+cn^+b and $vgcnb^+$. [63] vg^+cn^+b and $vgcnb^+$ have frequencies of 140 and 132, respectively.
[64] 140/3200 = 0.044, 0.044 × 100 = 4.4% (vg^+cn^+b) and 132/3200 = 0.041, 0.041 × 100 = 4.1% ($vgcnb^+$) for a combined value of 8.5%.

FIGURE 13-9

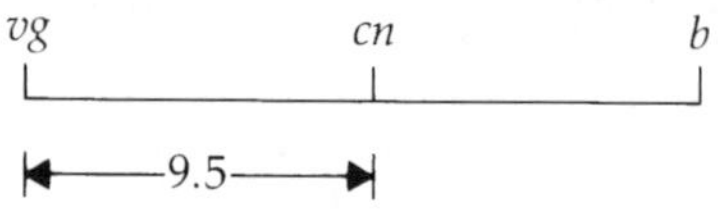

vg *cn* *b*

9.5

Summary of Key Steps in Analyzing a Three-Point Testcross

The following are key steps in analyzing a three-point testcross to determine the relative order of the three loci and the map distances between them.

1. If the sequence of the loci on the chromosome is unknown, arbitrarily select any sequence to begin with.
2. Identify the two most common classes of testcross progeny. This allows you to identify the parental-, or nonrecombinant-, type gametes produced by the triple heterozygote and tell whether this parent's genes are carried in coupling or in repulsion. (In the examples we have considered, the triple heterozygotes have carried the three wild-type alleles on one chromosome and the three mutant alleles on the homolog—that is, in the coupling arrangement. There is nothing sacred about this arrangement, and these parents could just as well have carried their recessive alleles spread between the two chromosomes in the repulsion arrangement.)
3. Identify the two least common phenotypic classes. This allows you to determine the double-crossover-, or double-recombinant-, type gametes produced by the triple heterozygote.
4. Compare the noncrossover- and the double-crossover-type gametes. The locus which appears to have its alleles interchanged is positioned between the other two loci.
5. If your arbitrarily selected gene sequence is the actual sequence, you are in luck. If it differs from the actual sequence, it will be very helpful to rewrite the sequence of loci in the gametes of the triple-heterozygous parent so they reflect the actual sequence.

FIGURE 13-10

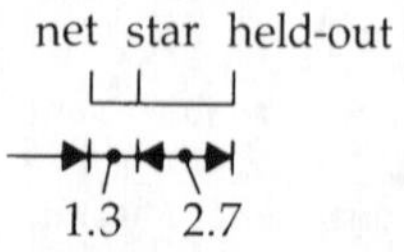

net star held-out

1.3 2.7

FIGURE 13-11

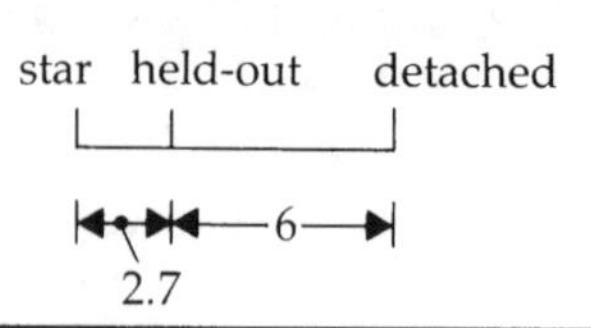

6. Determine the frequency of crossing over in each of the two chromosomal regions flanking the middle locus. This is done for each region by adding the percentage of progeny showing a single crossover in that region (derived from the combined frequencies of the two reciprocal classes of progeny which experience a single crossover in that region) and the frequency of double crossovers (derived from the combined frequencies of the two reciprocal classes which experience double crossovers). Since 1% of crossing over equals one unit of map distance, crossover frequencies can be readily converted to map distances.
7. Sketch a map showing the sequence of the three loci and the distances separating them.

COMPILING RECOMBINATION MAPS FROM THREE-POINT TESTCROSSES WITH COMMON LOCI

Recombination frequencies for an array of linked loci can be compiled from a series of three-point testcrosses and used to prepare a recombination map of a chromosome. Testcrosses with at least one common locus are used to relate various sections of the chromosome to each other in overlapping fashion. As an illustration of this, consider a testcross in the fruit fly *Drosophila melanogaster* involving loci for net-wing venation, star eyes, and held-out wings. Testcross results indicate that the star-eye locus is in the middle and separated from the other two loci by the distances indicated on the map in Figure 13-10.

A second testcross, involving the loci for star eyes, held-out wings, and detached veins, shows that held-out wings is the middle locus and gives the distances shown in Figure 13-11.

A third testcross involving loci for held-out wings, detached veins, and dachs legs places the detached-vein locus in the middle position with the distances shown in Figure 13-12.

Since these three testcrosses involve loci in overlapping portions of a chromosome, the three maps can be aligned to give the sequence of the five linked loci as shown in Figure 13-13. If the site of the net-wing locus is set at 0.0, distances between the loci can be designated.

FIGURE 13-12

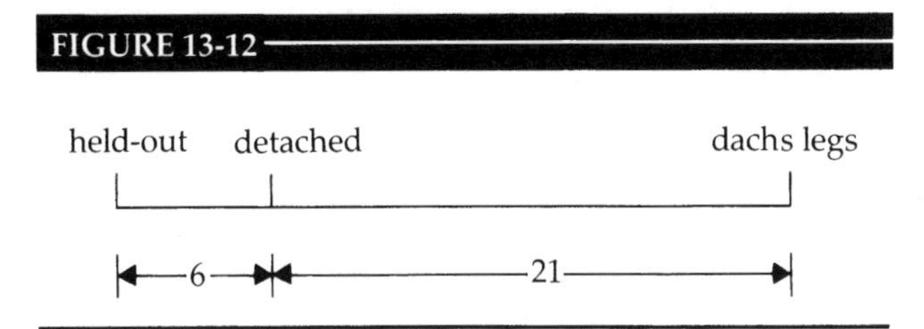

Through the use of additional testcrosses with loci in common, positions and distances for other loci on the chromosome can be determined. Eventually loci along the entire length of the chromosome can be mapped.

USING RECOMBINATION MAPS

Predicting Double-Crossover Frequencies

Recombination map distances give the probability of single crossovers occurring between adjacent loci and can be used to make certain useful predictions. For example, if crossover events in two chromosomal regions are independent, the expected frequency of double crossovers involving two adjacent regions could be calculated using the product law. As an illustration, consider the map in Figure 13-9, derived from the *Drosophila* testcross considered earlier in this chapter which involved the loci for vestigial wings (*vg*), cinnabar eye color (*cn*), and black body color (*b*) and produced a total of 3200 progeny.

A map distance of 9.5 units between the *vg* and *cn* loci tells us that crossing over occurs in the *vg*-to-*cn* region with a frequency of 9.5%, a value that can also be expressed as a probability of 0.095. A distance of nine units between the *cn* and *b* loci indicates a crossover frequency of 9.0% for the *cn*-to-*b* region, which can also be expressed as a probability of 0.09. The expected probability of a double crossover involving the two adjacent regions, that is, in the *vg*-to-*b* region, is given by the product of the probabilities of crossovers in the two adjacent regions. Multiplying these two probabilities gives $0.095 \times 0.09 = 0.0086$ and multiplying this probability by the total number of progeny, 3200, gives the expected number of double-crossover progeny: $0.0086 \times 3200 = 27.52$. The actual observed number of double-crossover progeny was 15, and since this reflects the outcome of a single testcross, the difference between the observed and expected values might well be attributed to chance. However geneticists, in repeating testcrosses, typically find that such discrepancies are real; in other words, fewer double-crossover events occur than would be expected. This observation leads to the conclusion that single-crossover events are not independent of each other. In other words, a single crossover occurring in one region can interfere with crossing over in a nearby region.

Coefficient of Coincidence and Interference

The ratio of observed to expected double-crossover frequencies, a value known as the **coefficient of coincidence,** provides an index of the expected double crossovers that actually occur.

$$\text{Coefficient of coincidence} = \frac{\text{Observed frequency of double crossovers}}{\text{Expected frequency of double crossovers}}$$

For example, the coefficient arising from the *Drosophila* testcross referred to above is

$$\text{Coefficient of coincidence} = \frac{\text{Observed doubles}}{\text{Expected doubles}} = \frac{15}{27.52} = 0.545 = 0.55.$$

This coefficient of coincidence indicates that only 55% of the expected double-crossover events actually occurred. The magnitude of the **interference,** symbolized by I, which produces this discrepancy is expressed by subtracting the coefficient of coincidence from 1:

$$\text{Interference} = 1 - \text{Coefficient of coincidence.}$$

FIGURE 13-13

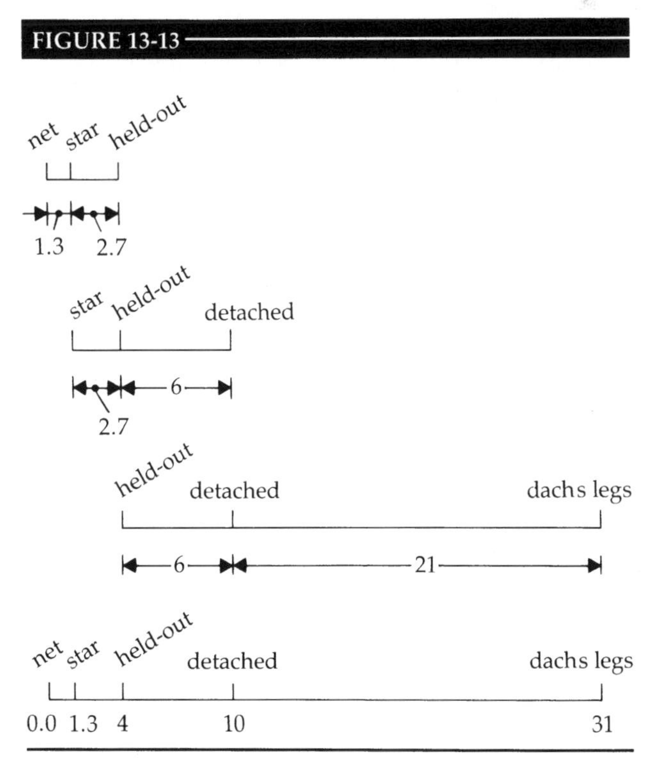

For the example we have been considering, the interference is 1 − 0.55 = 0.45. In other words, 45% of the expected double crossovers are *not* occurring. Interference values usually range between 0 and 1. When I = 0, the coefficient of coincidence is 1 and all the expected double crossovers occur. When I = 1, the coefficient of coincidence is 0, and interference is complete, with crossing over in one region completely blocking crossing over in the adjacent region. A eukaryotic exception to this positive interference occurs in the mold *Aspergillus* where interference values are negative for one chromosome section; here one crossover event seems to enhance the occurrence of other crossovers in adjacent sections.

High-Interference Regions

Experimental work indicates that interference generally increases (and the coefficient of coincidence correspondingly decreases) as the distance between loci diminishes. At distances greater than 50 map units, interference is very small. As the distance decreases below 50 units, interference progressively increases, operating most strongly for loci that are less than 10 to 15 map units apart. Interference is also high for loci that are found in sections of the chromosome adjacent to the centromere. The high levels of interference in both of these situations may be mechanical in origin, with strand flexibility insufficient to allow two crossover events to occur in short sections of a chromatid or in the regions adjacent to the centromere.

Predicting the Outcome of a Three-Point Testcross

If the coefficient of coincidence or the interference value is available for the portion of a recombination map that includes the loci to be studied in a three-point testcross, the outcome of the cross can be predicted. As an illustration of this, consider the portion of the third chromosome of *Drosophila melanogaster* which carries the loci for sepia eyes (*se*), dichaete bristles (*D*) (where the mutant allele which causes certain bristles to be missing is dominant to the wild-type allele for normal bristles), and curled wings (*cu*) as mapped in Figure 13-14. The coincidence coefficient for loci separated by these distances is approximately 0.25.

The testcross can be represented as + + +/*seDcu* (female) × *se*+*cu*/*se*+*cu* (male). A total of 1000 progeny are produced and you are asked to identify the various classes of progeny and predict their numbers. The procedure described as follows begins with consideration of the double crossovers.

FIGURE 13-14

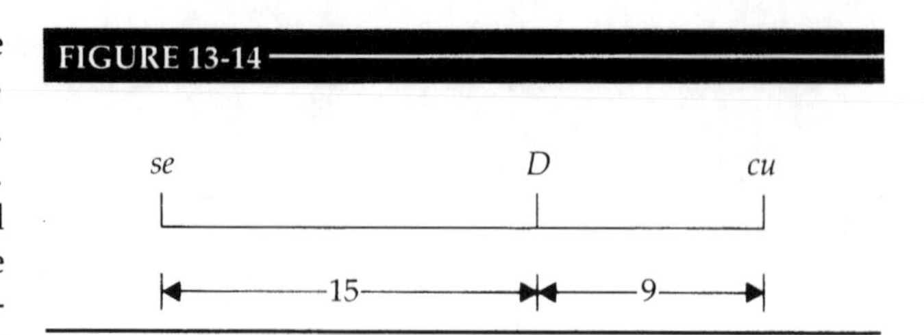

1. The expected frequency of double crossovers between the *se* and *cu* loci is determined by multiplying together the probability of crossovers in the *se*-to-*D* region and in the *D*-to-*cu* region. These probability values are obtained from the map. The expected frequency of doubles is (0.15)(0.09) = 0.0135.
2. The expected frequency of double crossovers is multiplied by the coefficient of coincidence to determine the frequency of the expected double-crossover events that would actually occur: (0.0135)(0.25) = 0.0034. Of 1000 progeny, approximately three (1000 × 0.0034) would be double-crossover recombinants.
3. Once the number of double crossovers is known, the single-crossover frequencies in the *se*-to-*D* and in the *D*-to-*cu* regions can be determined. Keep in mind that the probability of crossing over in each region that is obtained from the map distances reflects *all* the crossover events taking place between two loci, that is, it reflects both single- and double-crossover events. Subtracting the number of actual double crossovers from the total crossovers in a region gives the number of single crossovers in that region. For example, the total of all crossovers in the *se*-to-*D* region for 1000 progeny is 0.15 × 1000 = 150. Adjusting this value downward to compensate for the double crossovers, gives 150 − 3 = 147 of the 1000 progeny. In similar fashion, the total number of progeny with crossovers in the *D*-to-*cu* region is 0.09 × 1000 = 90. Reducing this by the number of crossovers in this region which occurred as part of double-crossover events gives us 90 − 3 = 87 progeny. Once the number of double crossovers (3) and the total number of single crossovers (147 + 87 = 234) have been determined, subtracting their total from 1000 gives the number of nonrecombinant-, or parental-, type progeny among the 1000 progeny: 1000 − (3 + 234) = 763.

SUMMARY

Three-point testcrosses are used to determine the relative order of, and the map distances between, three loci. Consideration of a third locus makes it possible to detect double-crossover events which would otherwise go undetected in two-point testcrosses and, thus, to obtain more accurate estimates of recombination frequencies and map distances. Three-point testcrosses with at least one common locus can be used to relate various sections of a chromosome to each other and to map a series of loci along a chromosome. Since recombination maps provide the probability of crossovers between adjacent loci, they can be used to predict the expected frequency of double crossovers involving two adjacent regions. If the coefficient of coincidence or the interference value is available for the section of a recombination map that includes three loci under consideration, the outcome of a three-point testcross involving these loci can be predicted. Summaries of the procedures for analyzing the outcome of a three-point testcross and for predicting testcross outcomes from recombination maps are included in this chapter.

For additional practice, a comprehensive set of problems for Chapters 1–13 is included in Appendix A.

PROBLEM SET

13-1. In a particular animal species, loci *s*, *t*, and *u* are linked, with *t* as the middle locus. Two testcrosses are carried out and each yields the same number of progeny. The first, a two-point testcross involving loci *s* and *u*, crosses an individual heterozygous at the *s* and *u* loci with an individual homozygous recessive at these two loci. The second is a three-point testcross between a triple heterozygote with the genotype *STU/stu* and an individual homozygous recessive at these three loci. The eight classes of progeny from the three-point testcross are as follows.

(1) *STU/stu* (33%)
(2) *stu/stu* (33%)
(3) *Stu/stu* (9%)
(4) *sTU/stu* (9%)
(5) *STu/stu* (6.5%)
(6) *stU/stu* (6.5%)
(7) *StU/stu* (1.5%)
(8) *sTu/stu* (1.5%)

a. Identify the progeny classes which received gametes carrying the parental, or noncrossover, linkage arrangements.
b. All of the progeny possess a chromosome bearing three recessive alleles. Where did these chromosomes come from?
c. Double-crossover progeny are produced from the three-point testcross. Are double-crossover progeny produced from the two-point testcross? Explain.
d. Would the map distance between the *s* and *u* loci based on the outcome of the two-point testcross be greater than, less than, or about the same as the *s*-to-*u* distance determined from the outcome of the three-point testcross? Explain.
e. Identify the chromosomal region or regions in which crossovers occurred to produce the progeny in classes 3 and 4.
f. Identify the chromosomal region or regions in which crossovers occurred to produce the progeny in classes 5 and 6.
g. Identify the chromosomal region or regions in which crossovers occurred to produce the progeny in classes 7 and 8.

13-2. Refer to the information given in problem 13-1.

a. Determine the number of recombination map units between loci *s* and *t* and between loci *t* and *u*.
b. What is the *s*-to-*u* map distance when determined from the three-point testcross data?
c. What is the *s*-to-*u* map distance when determined from the outcome of the two-point testcross?

13-3. The outcome of a three-point testcross involving linked loci depends on a number of factors, including the distance separating the loci, the sequence of the loci, and the linkage arrangement (coupling or repulsion) of the genes on the chromosomes of the triple-heterozygous parent. An individual heterozygous for three linked loci, *d*, *e*, and *f*, was crossed with an individual homozygous recessive at these three loci. The following three sets of progeny could arise from this testcross.

Outcome (a)

$\frac{D\ e\ f}{d\ e\ f}$ (36%) $\frac{D\ E\ F}{d\ e\ f}$ (7%) $\frac{D\ e\ F}{d\ e\ f}$ (3%) $\frac{D\ E\ f}{d\ e\ f}$ (1%)

$\frac{d\ E\ F}{d\ e\ f}$ (40%) $\frac{d\ e\ f}{d\ e\ f}$ (8%) $\frac{d\ E\ f}{d\ e\ f}$ (4%) $\frac{d\ e\ F}{d\ e\ f}$ (1%)

PROBLEM SET

Outcome (b)

$\frac{D\ E\ F}{d\ e\ f}$ (37%) $\frac{D\ e\ F}{d\ e\ f}$ (4%) $\frac{D\ E\ f}{d\ e\ f}$ (9%) $\frac{D\ e\ f}{d\ e\ f}$ (1%)

$\frac{d\ e\ f}{d\ e\ f}$ (35%) $\frac{d\ E\ f}{d\ e\ f}$ (5%) $\frac{d\ e\ F}{d\ e\ f}$ (8%) $\frac{d\ E\ F}{d\ e\ f}$ (1%)

Outcome (c)

$\frac{D\ e\ f}{d\ e\ f}$ (31%) $\frac{d\ e\ F}{d\ e\ f}$ (6%) $\frac{d\ e\ f}{d\ e\ f}$ (10%) $\frac{d\ E\ f}{d\ e\ f}$ (1%)

$\frac{d\ E\ F}{d\ e\ f}$ (34%) $\frac{D\ E\ f}{d\ e\ f}$ (7%) $\frac{D\ E\ F}{d\ e\ f}$ (10%) $\frac{D\ e\ F}{d\ e\ f}$ (1%)

For each outcome identify: (1) the progeny receiving parental- and double-crossover-type gametes from the triple-heterozygous parent, (2) the middle locus, and (3) whether the genes carried by the triple-heterozygous parent are in repulsion or in coupling.

13-4. An organism heterozygous for three loci is crossed with an individual homozygous for recessive alleles at the same three loci. No information is available as to whether these loci are linked or unlinked. The 1200 progeny fall into eight different phenotypic classes. Two of these classes contain about 375 progeny each, two others contain about 25 progeny each, and the remaining classes contain about 100 progeny each.

a. Would this outcome lead you to support the hypothesis that these loci are assorting independently? Explain.

b. Would this outcome lead you to support the hypothesis that these loci are linked? Explain.

c. What statistical test could be carried out to evaluate whether this ratio differs significantly from that expected under the hypothesis of independent assortment?

d. Without doing any calculations, what generalization can be drawn about the relative map distances between the three loci?

13-5. Among the loci carried on the third chromosome in the fruit fly *Drosophila melanogaster* are those determining the traits of antennae form (where the recessive allele, *th*, produces a thread-like tip to the antennae), wing form (where the recessive allele, *cu*, results in the wings being curled), and thorax appearance (where the recessive allele, *sr*, produces a striped thorax). The dominant allele at each of these loci produces the wild phenotype for each of these traits. Nothing is known about the sequence of these loci. A female fly, heterozygous for each of these loci, is mated with a male homozygous recessive for each of these loci. The 1000 progeny fall into the eight phenotypic classes listed in the following table.

PROBLEM SET

Phenotype			
Antennae	**Wing**	**Thorax**	**Frequency**
Wild	Wild	Wild	413
Thread	Wild	Wild	35
Wild	Wild	Striped	54
Wild	Curled	Wild	2
Thread	Curled	Wild	66
Wild	Curled	Striped	29
Thread	Curled	Striped	398
Thread	Wild	Striped	3
		Total:	1000

a. Identify the classes of progeny that received parental- or nonrecombinant-, type gametes from the triple-heterozygous parent.
b. Identify the classes that received double-crossover-type gametes from the triple-heterozygous parent.
c. Which locus is positioned between the other two?

13-6. Refer to the information given in problem 13-5.
a. Identify the two classes of progeny that arise as a result of a single crossover between the middle locus and the *th* locus, and determine the frequency of all crossovers in this region.
b. Identify the two classes of progeny that arise as a result of a single crossover between the middle locus and the *sr* locus, and determine the frequency of all crossovers in this region.
c. Sketch a map showing the sequence of these loci and the relative map distances separating them.

13-7. Three loci in the fruit fly *Drosophila melanogaster* are studied in a testcross. The loci control the traits of wing form (where the recessive allele, *c*, produces curved wings), eye color (where the recessive allele, *pr*, produces purple eyes), and body color (where the recessive allele, *b*, produces a black body). These loci are linked, with the purple-eye locus positioned between the other two. The dominant allele at each locus produces the wild phenotype for each of these traits. The testcross parents are a female fly heterozygous for each of these loci and a male fly homozygous for recessive alleles at these loci. The progeny fall into eight classes in the frequencies listed in the following table.

Phenotype			
Wing	**Eye**	**Body**	**Frequency**
Wild	Wild	Wild	4
Curved	Purple	Black	5
Curved	Wild	Wild	52
Wild	Purple	Black	60
Wild	Wild	Black	148
Curved	Purple	Wild	160
Curved	Wild	Black	529
Wild	Purple	Wild	509
		Total:	1467

a. Identify the two classes of progeny arising from parental-type gametes produced by the triple-heterozygous parent.

PROBLEM SET

b. Identify whether the genes of the triple-heterozygous parent are carried in the coupling or repulsion linkage arrangement.

c. (Verify that you have correctly answered 13-7b before answering this question.) Identify the two most common progeny classes for each of the alternative linkage arrangements possible for the triple heterozygote's genes.

13-8. Refer to the information given in problem 13-7. Prepare a map showing the sequence of the three loci and the distances separating them.

13-9. Three linked loci in corn, *Zea mays*, are studied in a testcross. These loci control the traits of endosperm shape—the endosperm is a portion of the seed containing food reserves—(where the recessive allele, *sh*, produces shrunken shape and the dominant wild-type allele, sh^+, produces full shape), endosperm composition (where the recessive allele, *wx*, produces a waxy endosperm and the dominant wild-type allele, wx^+, produces a starchy endosperm), and seed color (where the recessive allele, *c*, produces colorless seeds and the dominant wild-type allele, c^+, produces colored seeds). Nothing is known about the sequence of these loci on the chromosome. In the testcross, a plant heterozygous at each of these loci is crossed with a plant homozygous recessive for these loci. The 3400 progeny fall into the eight phenotypic classes listed in the following table.

Phenotype			
Endosperm shape	**Endosperm composition**	**Seed color**	**Frequency**
sh^+	wx^+	c	74
sh^+	wx^+	c^+	1210
sh	wx^+	c^+	5
sh	wx	c^+	62
sh	wx	c	1373
sh	wx^+	c	354
sh^+	wx	c	4
sh^+	wx	c^+	318
		Total:	3400

a. Identify the two parental classes, the two double-recombinant classes, and the middle locus.

b. Determine the frequency of crossing over in the two chromosomal regions on either side of the middle locus. (If a locus other than *wx* is in the middle, you may find it helpful, before continuing, to rewrite the sequence of loci for the eight progeny classes so they reflect the actual order.)

c. Sketch a map showing the sequence of loci and the distances between them.

13-10. a. From the crossover frequencies calculated in answering problem 13-9, determine the expected frequency of double crossovers between the two outer loci. What is the expcected number of double-crossover progeny?

b. Calculate the coefficient of coincidence and the interference value, and explain what these statistics mean.

PROBLEM SET

13-11. Three linked loci in the fruit fly *Drosophila melanogaster* are studied in a testcross. These loci control the traits of body color (where the recessive allele, *y*, produces a yellow body), wing venation (where the recessive allele, *cv*, produces crossveinless wings), and eye form, (where the recessive allele, *lz*, produces lozenge eyes). The dominant allele at each locus produces the wild phenotype for each of these traits. A testcross is carried out between a male homozygous recessive for these loci and a female heterozygous for the loci. The progeny fall into eight classes in the frequencies listed in the following table.

Phenotype			
Body color	**Eye shape**	**Wing type**	**Frequency**
y	lz	cv^+	81
y^+	lz^+	cv^+	87
y	lz^+	cv	8
y	lz^+	cv^+	463
y	lz	cv	77
y^+	lz	cv	488
y^+	lz	cv^+	7
y^+	lz^+	cv	89
		Total:	1300

a. Identify the nonrecombinant classes, the double-recombinant classes, and the middle locus.
b. Determine the frequency of crossing over in the two chromosomal regions on either side of the middle locus.
c. Sketch a map showing the sequence of loci and the relative distances between them.

13-12. a. From the crossover frequencies calculated in answering problem 13-11, determine the expected frequency of double crossovers between the outer loci. What is the expected number of double-crossover progeny?
b. Calculate the coefficient of coincidence and the interference value.

13-13. The portion of the chromosome map in the following figure shows three loci that are to be studied in a three-point testcross.

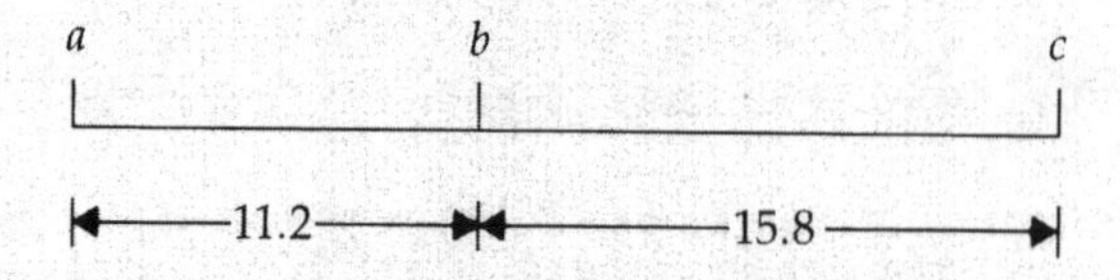

The coefficient of coincidence for loci this far apart is about 0.30. The testcross is represented as follows, where + is used to designate the wild-type allele at each locus.

$$+ + +/abc \text{ (female)} \times abc/abc \text{ (male)}$$

Assume that 1000 progeny are produced. Answer the following questions based on the predicted outcome of this cross.
a. What is the expected frequency of double crossovers?

PROBLEM SET

b. How many of the 1000 progeny would be double recombinants?
c. Predict the number of single crossovers in the *a*-to-*b* region.
d. Predict the number of single crossovers in the *b*-to-*c* region.
e. How many of the 1000 progeny would be expected to be parental types?

13-14. Each of the four maps shown in the following figure was prepared from the outcome of a different three-point testcross. Combine these four maps into a single map which includes all of these loci.

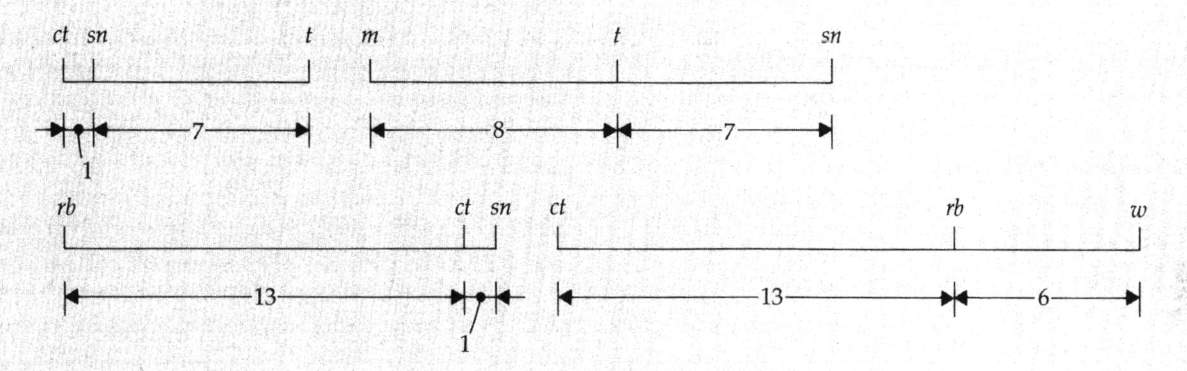

CHAPTER

14

Haploid Genetics: Tetrad Analysis I

INTRODUCTION

A number of eukaryotes including the mold *Neurospora*, the alga *Chlamydomonas*, and the yeast *Saccharomyces* are haploid for a significant portion of their life cycles. These organisms have several important features which make them useful subjects for studying genetic phenomena: (1) they can be cultured and crossed readily, and have a rapid life cycle, (2) their genotype can easily be inferred from their phenotype since each carries a single allele, and (3) their reproductive patterns make it possible to readily identify the four products arising from a particular nucleus that undergoes meiosis. These four products, collectively known as a **tetrad,** can be isolated, grown, and analyzed phenotypically and genotypically. (Note that the term "tetrad" as it is used here has a meaning different from that used to describe an arrangement of four chromatids during meiosis.) This chapter and the one that follows examine the role that tetrad analysis has played in broadening our understanding of the process of meiosis and the genetics of haploid organisms. Specifically, this chapter examines crosses that involve a single locus and two independently assorting loci. Tetrad analysis has also been used in the study of gene conversion which is discussed in Chapter 19.

ORDERED TETRADS IN *NEUROSPORA*

Neurospora crassa, better known as the common, or pink, bread mold, is a haploid ($n = 7$) multicellular fungus with a body made up of a mass of threads, or hyphae, each of which consists of a series of cells which remained joined end-to-end. Although reproduction is generally by means of asexual spores, sexual reproduction can occur when individuals of different mating types come into physical contact and nuclei of one mating type (male) are transferred to the hypha of the opposite type (female). Ultimately, pairs of these haploid nuclei fuse to form diploid **fusion nuclei** within specialized, elongated hyphae that develop into structures known as **asci** (singular: ascus). The diploid fusion nucleus of each ascus undergoes meiosis I to produce two nuclei, and meiosis II to form the four nuclei which comprise a tetrad. Each of these nuclei then replicates mitotically to form a total of eight nuclei, each of which serves as the nucleus of a haploid sexual spore, or **ascospore.** Figure 14-1 summarizes these events within a single ascus.

Upon maturing, the ascus wall dries out, breaks open, and releases the microscopic ascospores which are carried about by currents of air. Those landing on sites with favorable growth conditions grow mitotically to form new haploid mold colonies. When crosses be-

FIGURE 14-1

Diploid nucleus in the immature ascus undergoes meiosis-I, meiosis-II, and mitotic divisions to form eight haploid ascospores in the mature ascus.

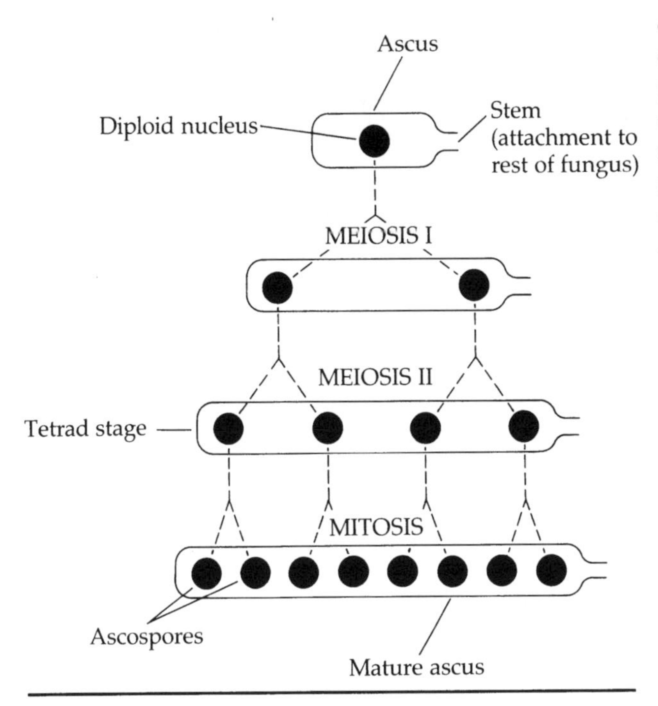

tween different strains of *Neurospora* are carried out in the laboratory, ascospores can be removed before the asci break open and germinated individually in the order in which they are removed from the ascus. (Note in Figure 14-1 that the stem end of each ascus provides an orientation to each ascus and is used by the geneticist as a reference point in the ordered removal of the ascospores.) The mold that develops may then be examined for the phenotypes of the traits under study, and the genetic makeup of each nucleus can be inferred.

A key factor that makes *Neurospora* particularly useful in genetic studies is that the eight nuclei within its asci are ordered, that is, they are always lined up in a manner which is related to the way in which they were produced. This ordering occurs because of the elongated nature of the ascus and the fact that the sets of spindles arising during the two meiotic and subsequent mitotic divisions do not spatially overlap with each other. Consequently, the history, or lineage, of each ascospore can be determined from its location within its ascus. In other words, the meiosis-I and meiosis-II nuclei from which each ascospore developed can be identified. For example (refer to Figure 14-1), the two ascospores closest to the stem end of the mature ascus (that is, the end on the right side of Figure 14-1) have developed from the meiosis-II nucleus on the far right, and that, in turn, arose from the meiosis-I nucleus on the right. This and other lineages are shown in Figure 14-1 by dotted lines.

Because the ascus contains the ordered products of a single meiosis, it is possible to determine, for example, the meiotic division during which allelic pairs segregate and whether or not crossing over occurs. If crossing over occurs, the ordered products allow us to tell when it occurs and to identify how many chromatids participate in crossover events. We can determine the frequency with which crossing over occurs and use the recombination frequency to estimate locus-to-centromere distances, to determine whether two loci are linked or assort independently, and to map linked genes. (A clear understanding of meiosis is essential to understand tetrad analysis. If necessary, review the description of meiosis in Chapter 1 or your text before continuing.)

CROSSES INVOLVING A SINGLE LOCUS

Detection of Crossing Over and M-1 and M-2 Segregation Patterns

First, we will consider how tetrad analysis can be used to determine whether or not crossing over has occurred during meiosis and to identify the stage during meiosis when alleles segregate. Assume that two alleles designated as A and a occur at a hypothetical locus. A cross is carried out between a strain carrying A and another strain carrying a to produce the diploid Aa fusion nucleus as follows.

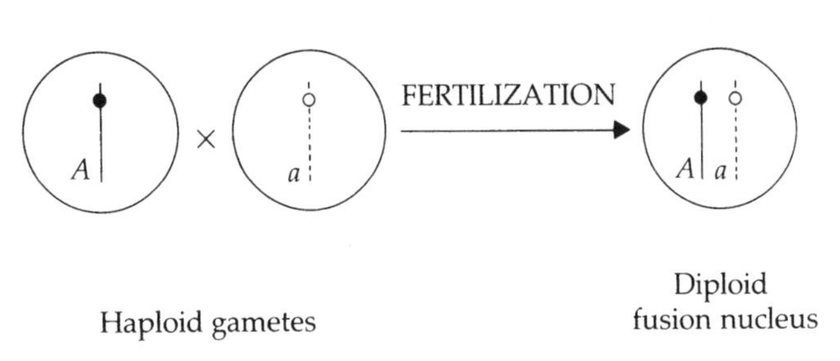

Complete the two schemes in Figure 14-2. Each scheme shows this diploid nucleus undergoing chromosomal replication, meiosis I, meiosis II, and the mitotic division. The scheme on the left (Figure 14-2a) shows no crossing over while that on the right (14-2b) involves crossing over between the a locus and the centromere.

Compare your completed schemes. During which meiotic division did the A and a alleles found in the diploid fusion nucleus segregate into separate nuclei

FIGURE 14-2

Diploid nucleus with genotype *Aa* undergoes chromosomal replication, meiosis-I, meiosis-II, and mitosis divisions.

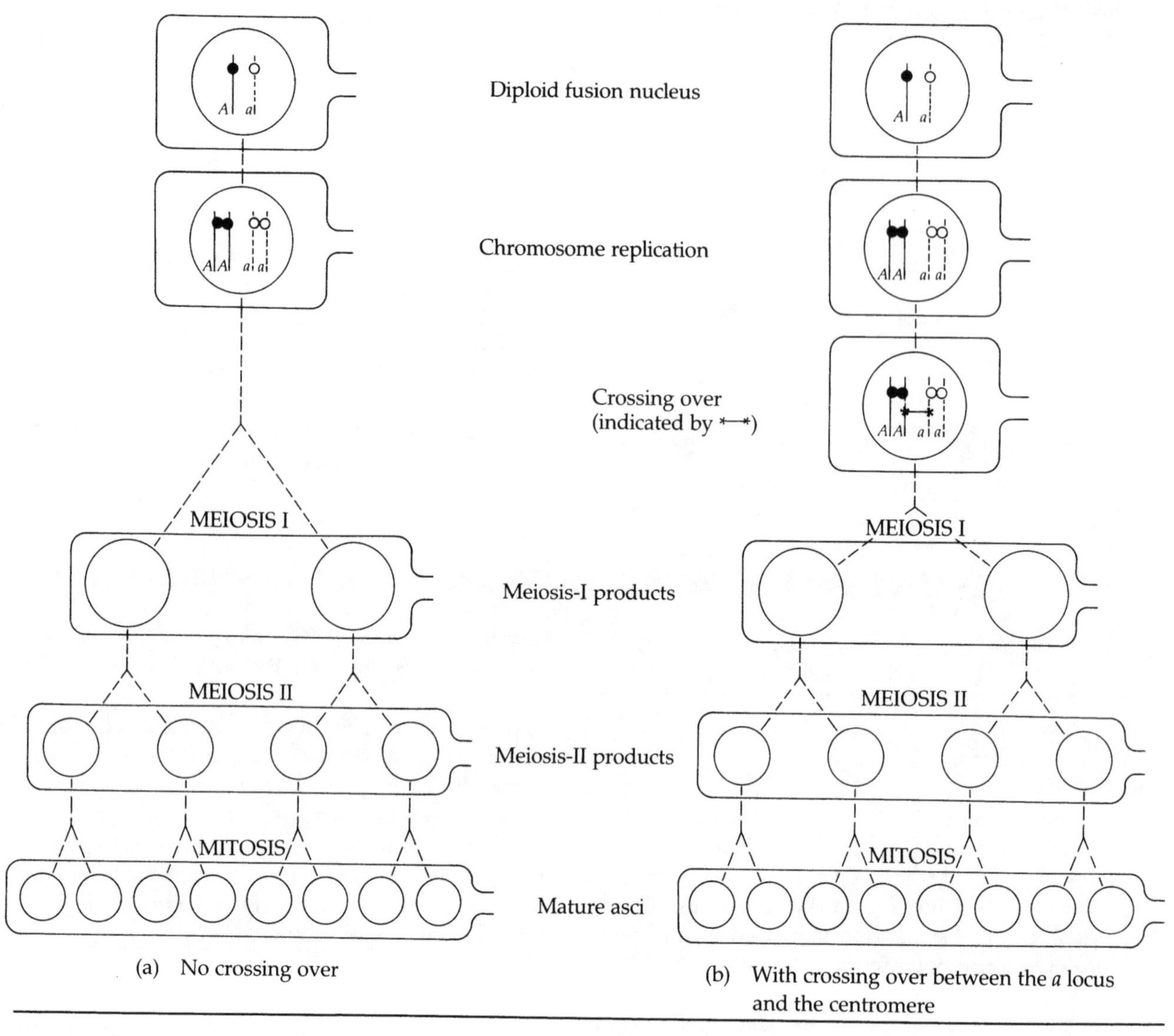

(a) No crossing over

(b) With crossing over between the *a* locus and the centromere

in the scheme with no crossing over? ________[1] in the scheme with crossing over? ________[2] Compare the arrangement of the alleles in the eight nuclei found in the ascus at the bottom of each scheme. (For example, do the different alleles alternate with each other? Are they grouped in pairs or in threes or fours?) What distribution pattern do the alleles show with no crossing over? ________________________________ ________[3] when crossing over occurs? ________ ________________________________[4] Is there a relationship between the occurrence of crossing over and the distribution pattern of alleles in the ascus? ________[5] What is that relationship? ________ ________________________________[6] If you had no information other than the distribution pattern shown by the alleles in the ascus, could you tell whether or not crossing over had taken place? ________[7] Explain. ________________________________[8]

To summarize, loci associated with the centromere segregate during meiosis I if there is no crossing over and during meiosis II if there is crossing over between the locus and the centromere. Ascospores

[1] During meiosis I. [2] During meiosis II. [3] A 4-4 pattern, that is, four spores in one half of the ascus carry one allele while the four spores in the other half carry the other allele: *A A A A a a a a*. [4] A 2-2-2-2 pattern, that is, two of the four nuclei in each half of the ascus carry one allele, and two carry the other allele: *A A a a A A a a*. [5] Yes. [6] Crossing over produces the meiosis-II segregation and the 2-2-2-2 pattern. [7] Yes. [8] The 4-4 pattern occurs with no crossing over and the 2-2-2-2 pattern with crossing over.

within a particular ascus will display one of two possible segregation patterns. (1) If there is no crossing over, the meiosis-I pattern, abbreviated as M-1, and also designated as 4-4 and as first-division segregation (FDS), will occur. (2) If there is crossing over between the locus and the centromere, the meiosis-II pattern, abbreviated as M-2, and also designated as 2-2-2-2 and as second-division segregation (SDS), will be found.

Random Variants of the M-1 and M-2 Patterns

Both the M-1 and M-2 patterns can occur in alternative ways. Refer to the scheme with no crossing over (Figure 14-2a). What is the alternative to the *A A A A a a a a* sequence? ________[9] Note that these two sequences can be distinguished only by making reference to the stem end of the ascus. What determines the particular sequence that is produced? (Clue: What would have to be modified in the diagram in order for the alternative sequence to be produced?)

____[10]

In the scheme shown in Figure 14-2a, either chromosome has an equal likelihood of being on the left and thus separating to the left. In other words, the two sequences are **random variants** of each other. Any particular ascus that was free of crossing over would,

FIGURE 14-3

Alternative arrangements of replicated chromosomes for single crossovers between the *a* locus and the centromere give rise to four random variants of the M-2 pattern.

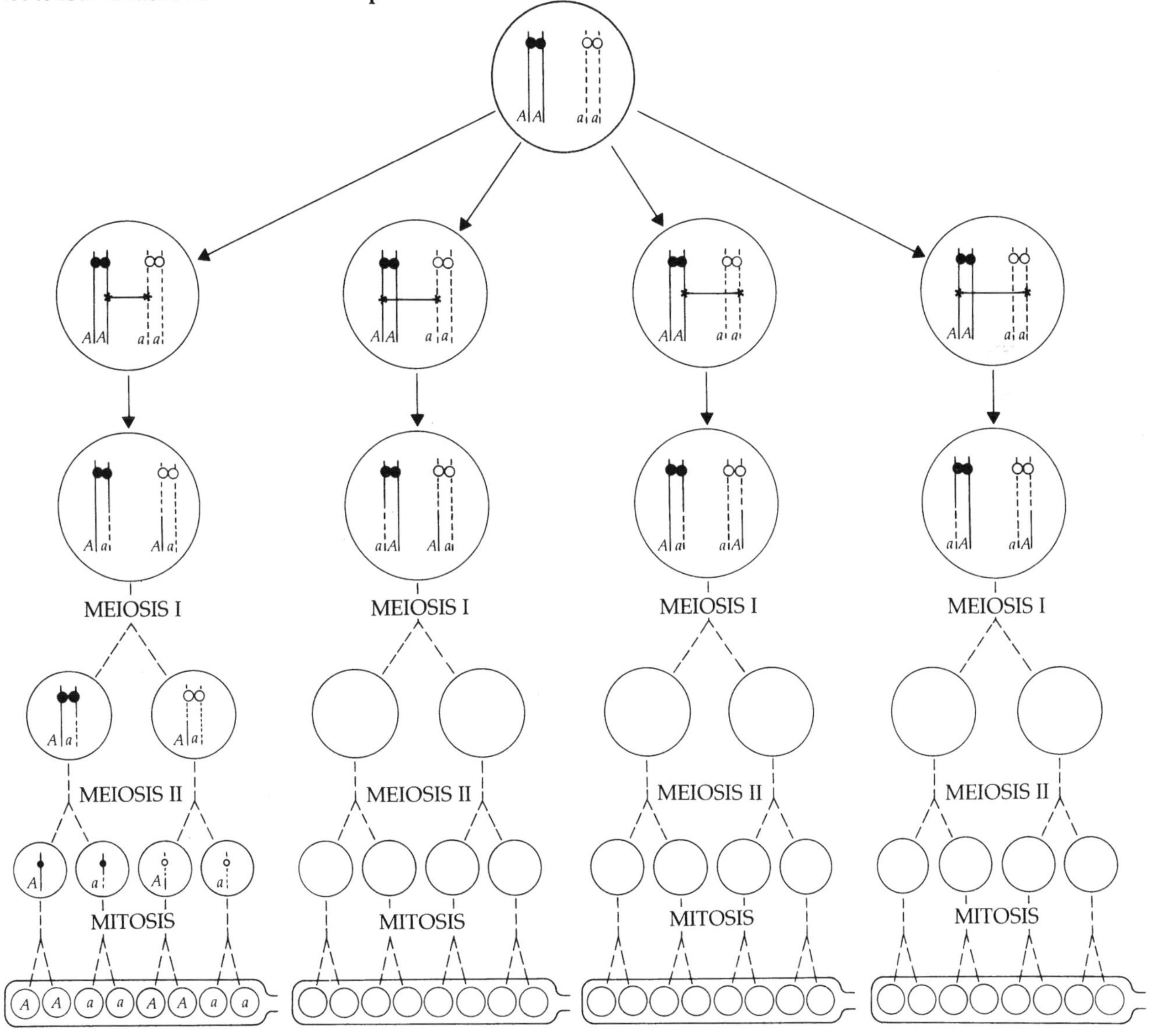

[9] The *a a a a A A A A* sequence. [10] The way in which the replicated chromosomes align during metaphase of meiosis I.

obviously, have just one of these sequences, but if a large number of such asci were examined, what would you expect to be the frequencies of the occurrences of each variant? ______________________________________ [11]

Similarly, the M-2 pattern produced in the scheme with crossing over (Figure 14-2b) also has random variants. In addition to *A A a a A A a a* (shown on the far left in Figure 14-3), three other sequences are possible. Identify them by completing Figure 14-3.[12] Again, it is the alignment of the chromosomes during meiosis I that determines the sequence produced in a particular ascus. One or another of these random variants would occur in an ascus after crossing over and, if a large number of asci are involved, the four sequences would occur among them in equal frequencies.

To summarize, six different arrangements of the *A* and *a* alleles are possible. Two are random variants of the M-1 pattern that occurs in the absence of crossing over and four are random variants of the M-2 pattern that arises when crossing over occurs between the locus and the centromere.

Estimating Centromere-to-Locus Distances

The centromere-to-locus distance can be estimated from the numbers of asci produced in a cross that show the M-1 and the M-2 patterns. For example, say that 250 asci are produced in the $A \times a$ cross whose outcome is described in Figure 14-2, with 200 showing the M-1 pattern and 50 showing the M-2 pattern. The relative number of asci showing the M-2 pattern reflects the frequency of crossing over that, in turn, is a function of the distance between the centromere and the locus being considered. Of the 250 asci examined, 50, or 20%, show the M-2 pattern. How many of the ascospores in each of these 50 asci carry recombinant-type strands? __________[13] How many of the ascospores in these asci carry parental-type strands? __________[14] Since only half the ascospores carry recombinations, the frequency of recombination is one-half of the frequency of asci with the M-2 pattern, that is, $1/2 \times 0.20 = 0.10$ or $0.10 \times 100 = 10\%$. Since 1% recombination equals one unit of map distance, the *a* locus is 10 map units from the centromere. This calculation is summarized as follows:

Centomere-to-locus distance

$$= \frac{(\text{Number of M-2 asci})/2}{\text{Total number of asci}} \times 100.$$

CROSSES INVOLVING TWO INDEPENDENTLY ASSORTING LOCI

Now consider a second locus, call it *b*, in addition to the *a* locus. These two loci may be carried on separate chromosomes and assort independently, or they may be linked. As you have just seen, when considered by themselves, the two alleles at the *a* locus give rise to a total of six different arrangements within the asci. When two loci are involved, as in the diploid *AaBb*, a total of $6^2 = 36$ different arrangements of alleles can occur, and these arrangements are produced whether the loci are linked or unlinked. Earlier, you verified that the six arrangements arising from the *Aa* diploid were random variants of two basic patterns. A similar inspection of the 36 possible arrangements produced by the *AaBb* diploid would show that they are random variants of seven basic ascospore patterns. We will take a look at these seven basic patterns and see how they arise when the loci assort independently.

Seven Basic Ascospore Patterns and Their Random Variants

When the loci are on separate chromosomes, the cross $AB \times ab$ and the diploid fusion nucleus that results can be represented as follows.

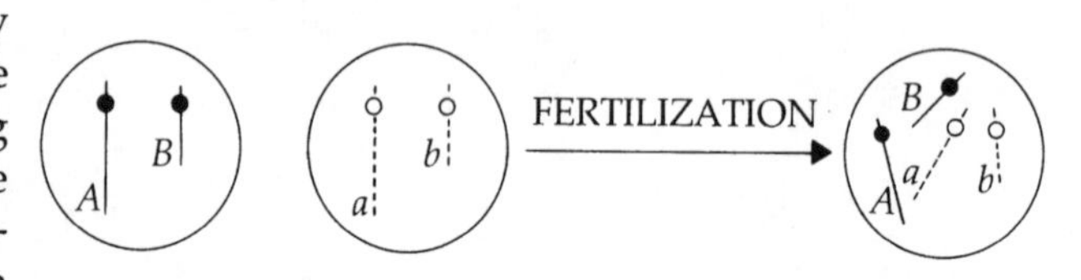

Patterns with No Crossing Over First we will consider the patterns that result when there is no crossing over during meiosis. Refer to Figure 14-4, which shows the meiosis-I and -II divisions as well as the mitotic division. Note that this figure shows two alternative, random arrangements of the chromosomes prior to the first meiotic division. Complete the two schemes in Figure 14-4, identify the genetic makeup of the ascospores, and complete the table at the bottom of the figure.

Each alternative chromosome arrangement in Figure 14-4 produces a different arrangement of ascospores. The scheme on the left produces the arrangement of *AB AB AB AB ab ab ab ab*. This asco-

[11] The variants would be expected in equal frequencies: [12] Beginning with the second ascus from the left, the sequences are *a a A A A A a a*, *A A a a a a A A*, and *a a A A a a A A*. [13] Half. [14] Half.

spore pattern is an example of a **ditype** tetrad, since it contains two different genotypes. It can be more precisely labeled as a **parental ditype,** abbreviated as PD, since both genotypes are parental types. Both loci in this parental-ditype ascospore pattern show the M-1 segregation pattern. Is there a second parental-ditype arrangement that could be produced in the scheme in Figure 14-4 on the left? ________[25] What is the arrangement? ______________________________[26] How could that arrangement arise? ______________________________[27]

FIGURE 14-4

Alternative arrangements of replicated chromosomes yield two ascospore patterns in the absence of crossing over.

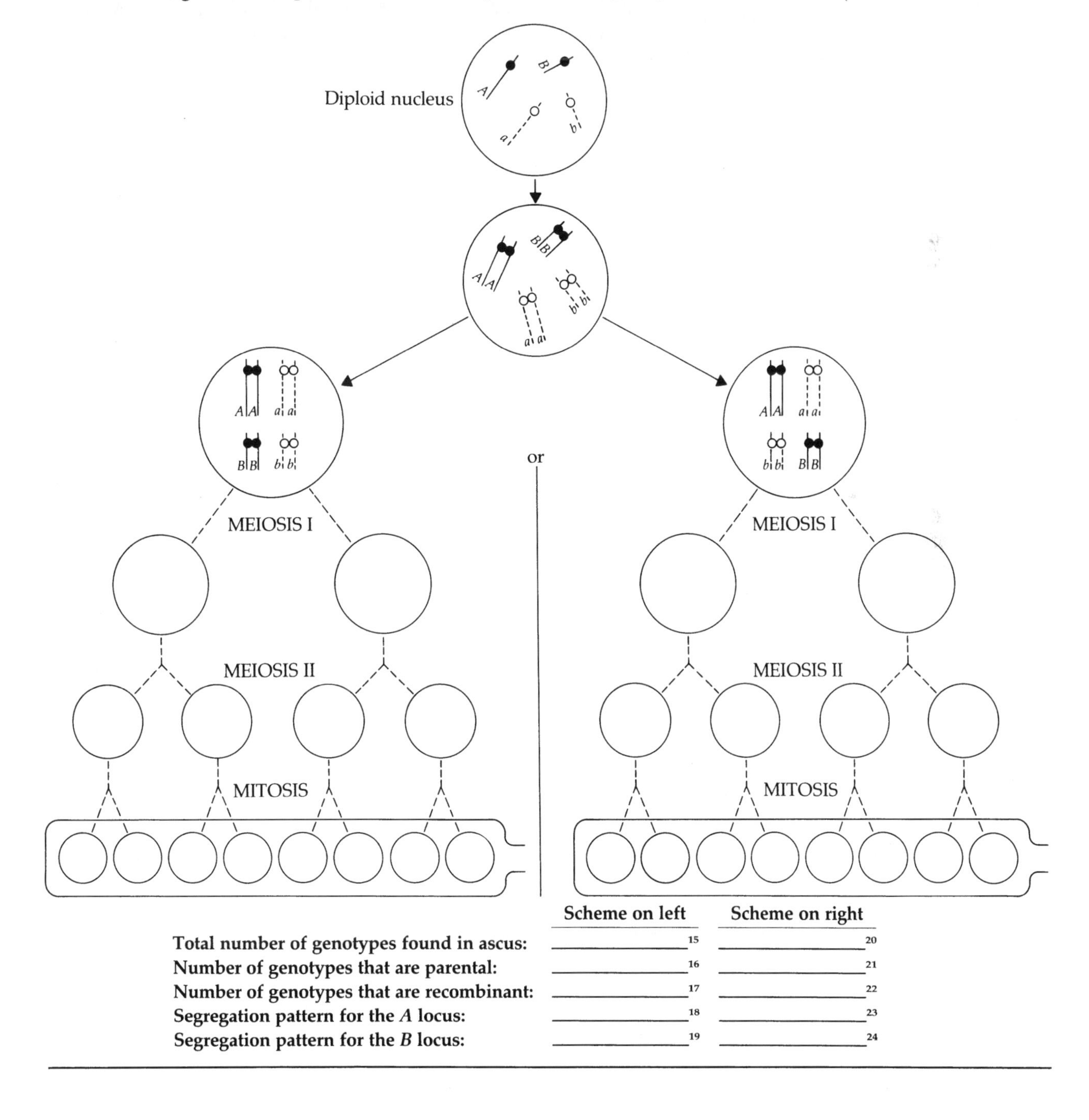

	Scheme on left	Scheme on right
Total number of genotypes found in ascus:	________[15]	________[20]
Number of genotypes that are parental:	________[16]	________[21]
Number of genotypes that are recombinant:	________[17]	________[22]
Segregation pattern for the *A* locus:	________[18]	________[23]
Segregation pattern for the *B* locus:	________[19]	________[24]

[15] Two. [16] Two. [17] Zero. [18] M-1. [19] M-1. [20] Two. [21] Zero. [22] Two. [23] M-1. [24] M-1. [25] Yes. [26] *ab ab ab ab AB AB AB AB.* [27] If an equally likely alternative chromosomal alignment occurs.

The scheme on the right in Figure 14-4 produces *Ab Ab Ab Ab aB aB aB AB*. With two genotypes present, this ascospore pattern is also an example of a ditype tetrad but is classified as a **nonparental ditype** tetrad, abbreviated as NPD, since both genotypes are nonparental types. Both loci in this ascospore pattern show the M-1 segregation pattern. Can a second nonparental ditype arrangement be produced in this scheme? ________[28] What is that arrangement? ____________________________[29] How could that arrangement arise? ____________________________[30]

To summarize, two basic ascospore patterns can occur in the absence of crossing over. One is a parental ditype tetrad with two random variants, and the other is a nonparental ditype tetrad with two random variants. What is the probability that any one of these four outcomes will occur in a particular ascus that is free of crossing over? ________[31] If 300 such asci were studied, what is the expected frequency for each outcome? ________[32] As will be discussed later, the 1:1 ratio of parental ditypes to nonparental ditypes confirms the fact that the genes involved are assorting independently.

FIGURE 14-5

Ascospore pattern produced following a single crossover in the *b*-to-centromere region.

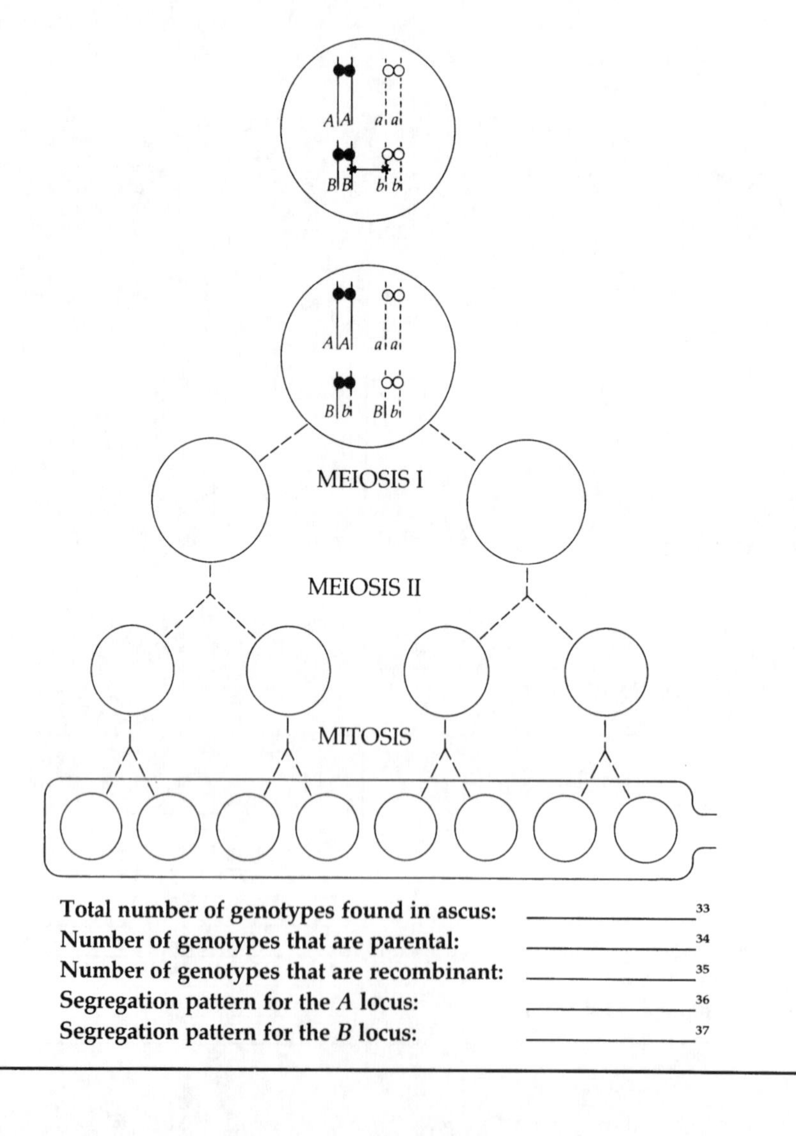

Total number of genotypes found in ascus: ________[33]
Number of genotypes that are parental: ________[34]
Number of genotypes that are recombinant: ________[35]
Segregation pattern for the *A* locus: ________[36]
Segregation pattern for the *B* locus: ________[37]

[28] Yes. [29] *aB aB aB aB Ab Ab Ab Ab.* [30] If an equally likely alternative chromosomal alignment occurred. [31] 1/4. [32] 300 × 1/4 = 75. [33] Four. [34] Two. [35] Two. [36] M-1. [37] M-2.

Patterns with a Single Crossover between One Locus and Its Centromere Now we consider the ascospore patterns that can arise when crossing over occurs. Assume that a single crossover occurs in the region between the *b* locus and its centromere. Complete Figure 14-5 by identifying the genetic makeup of the ascospores and completing the table at the bottom of the figure.

This type of ascospore pattern, with four different genotypes, two parental and two recombinants, is an example of a **tetratype** tetrad, abbreviated as T. The *a* and *b* loci show M-1 and M-2 segregation patterns, respectively. (Note that the M-2 pattern for the *b* locus would tell us, if we did not already know, that this locus had participated in crossing over.) Because alternative alignments of the chromosomes occur, this basic pattern can occur in a total of eight random variants that are listed in Table 14-1. The first arrangement listed in the table is the one you produced in Figure 14-5. For each of the other arrangements listed, designate, in the space provided, the tetrad type (parental ditype [PD], nonparental ditype [NPD], or tetratype [T]) and give the segregation pattern shown by each locus.

If you are puzzled as to how the arrangements listed in Table 14-1 arise, Figure 14-6 should be helpful. The eight arrangements are produced when the chromosome pairs are oriented as shown in nuclei 1 through 8, respectively, in Figure 14-6. Each arrangement listed in Table 14-1 has an equal probability of occurring after a single crossover between the *b* locus and its centromere.

Patterns with a Single Crossover between the Other Locus and Its Centromere Next, we will examine what happens following a single crossover between the *a* locus and its centromere, as shown in Figure 14-7. This chromosomal alignment coupled with a crossover in the *a*-to-centromere region could give rise to the following arrangement of ascospores within the ascus.

Ascospore:	1 & 2	3 & 4	5 & 6	7 & 8
	AB	*aB*	*Ab*	*ab*

What is the total number of genotypes found in this type of ascus? ________[45] How many of these genotypes are parental? ________[46] How many are recombinant? ________[47] What is the segregation pattern for the *a* locus? ________[48] What is the segregation pattern for the *b* locus? ________[49]

You saw that eight random variants could arise after a single crossover in the centromere-to-*b* locus region (see Table 14-1). The same number of variants can arise following a single crossover in the centromere-to-*a* locus region, as listed in Table 14-2. Verify that each variant is a tetratype tetrad showing M-2 segregation for the *a* locus and M-1 segregation for the *b* locus.

TABLE 14-1

Eight random variants of the ascospore pattern produced following a single crossover in the *b*-to-centromere region.

Class	Ascus Contents				Tetrad type	Segregation Pattern	
	1 & 2	3 & 4	5 & 6	7 & 8		*a* locus	*b* locus
1.	*AB*	*Ab*	*aB*	*ab*	T	M-1	M-2
2.	*Ab*	*AB*	*aB*	*ab*	________[38]		
3.	*AB*	*Ab*	*ab*	*aB*	________[39]		
4.	*Ab*	*AB*	*ab*	*aB*	________[40]		
5.	*aB*	*ab*	*AB*	*Ab*	________[41]		
6.	*ab*	*aB*	*AB*	*Ab*	________[42]		
7.	*aB*	*ab*	*Ab*	*AB*	________[43]		
8.	*ab*	*aB*	*Ab*	*AB*	________[44]		

[38] T, M-1, M-2. [39] T, M-1, M-2. [40] T, M-1, M-2. [41] T, M-1, M-2. [42] T, M-1, M-2. [43] T, M-1, M-2. [44] T, M-1, M-2. [45] Four. [46] Two. [47] Two. [48] M-2. [49] M-1.

FIGURE 14-6

Eight alternative arrangements of the replicated chromosomes that produce the eight tetratype random variants listed in Table 14-1.

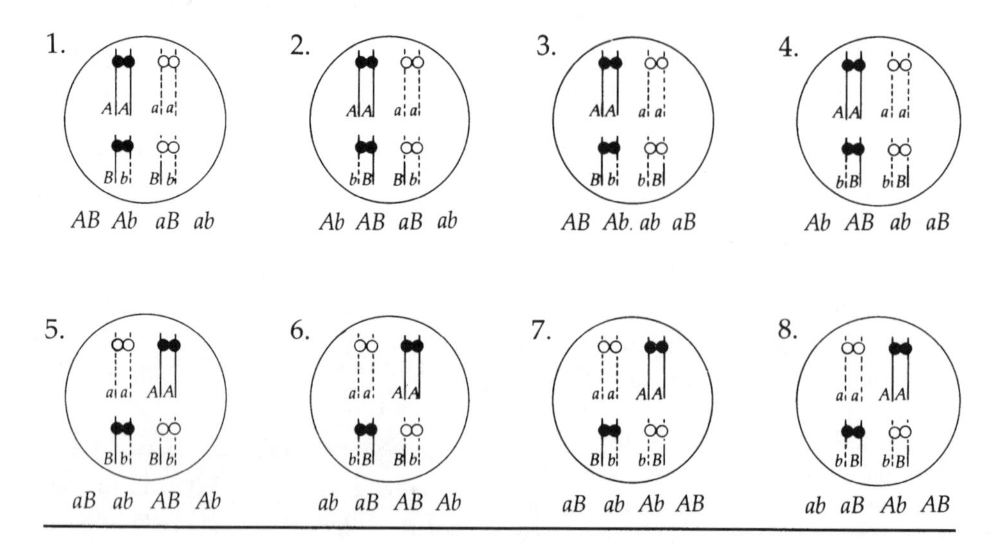

Patterns with Both Crossovers So far, we have looked at the ways in which four of the seven basic ascospore patterns are produced—two when there was no crossing over and two others when there was a single crossover. The three remaining basic patterns arise when *both* of the single crossovers we have considered occur simultaneously. This two-crossover situation is shown in Figure 14-8.

The three basic ascospore patterns that could arise are given in Table 14-3. The pattern produced within a particular ascus will, of course, depend on just how the chromosomes align during meiosis. Complete the table by indicating, for each class, its tetrad type (PD, NPD, or T) and the segregation pattern shown by each locus. Note that each of the three basic ascospore patterns can, of course, show random variants; the first two patterns listed show four variants each, and the remaining pattern shows eight.

Summary of Basic Ascospore Patterns We have just looked at the seven basic ascospore patterns and the way in which each is produced when the two loci involved assort independently.

1. Noncrossover parental ditype (PD).
2. Noncrossover nonparental ditype (NPD).
3. Single-crossover tetratype (T) (crossover involves one of two loci).
4. Single-crossover tetratype (T) (crossover involves the other of two loci).
5. Double-crossover parental ditype (PD).

TABLE 14-2

Eight random variants of the ascospore pattern produced following a single crossover in the *a*-to-centromere region.

Class	Ascus Contents			
	1 & 2	3 & 4	5 & 6	7 & 8
1.	*AB*	*aB*	*Ab*	*ab*
2.	*aB*	*AB*	*ab*	*Ab*
3.	*AB*	*aB*	*ab*	*Ab*
4.	*aB*	*AB*	*Ab*	*ab*
5.	*ab*	*Ab*	*aB*	*AB*
6.	*Ab*	*ab*	*AB*	*aB*
7.	*ab*	*Ab*	*AB*	*aB*
8.	*Ab*	*ab*	*aB*	*AB*

FIGURE 14-7

Diploid nucleus experiencing a crossover in the *a*-to-centromere region.

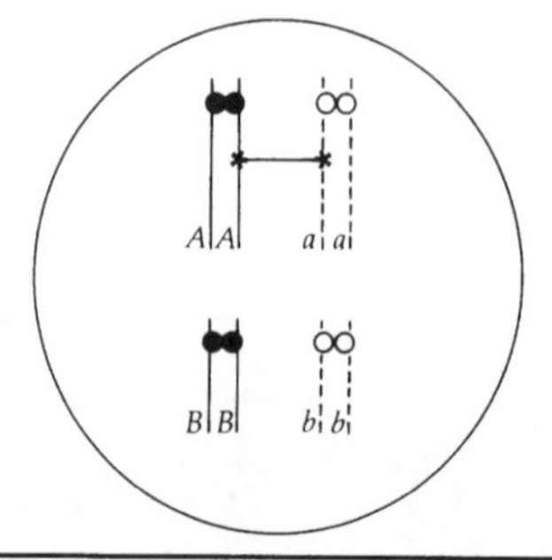

FIGURE 14-8

Diploid nucleus experiencing a crossover in the *a*-to-centromere region and another crossover in the *b*-to-centromere region.

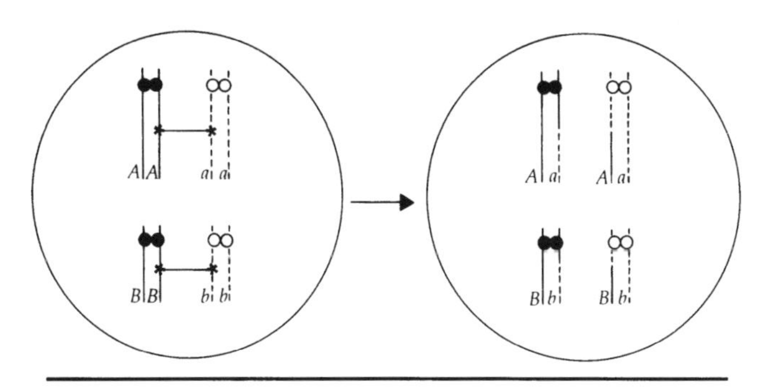

6. Double-crossover nonparental ditype (NPD).
7. Double-crossover tetratype (T).

Any particular ascus will, of course, show just one of these patterns, but if a large number of asci are formed from the diploid *AaBb* nucleus, which of the seven basic ascospore patterns would you expect to occur most frequently? ________[53] Explain why. ____________________________________[54] Which type or types would you expect to occur least frequently? ________[55] Explain why. ____________________________________[56] What type or types would occur with intermediate frequencies? ________[57] Explain why. ____________________________________[58]

Determining whether Loci Are Linked At this point it should be clear that when two loci are on different pairs of homologous chromosomes, random segregation results in approximately equal numbers of PD and NPD tetrads. The proportion of PD to NPD tetrads is important information when the outcome of a tetrad analysis is presented with no information about whether the two loci under consideration are linked. As you will see in the next chapter, when two loci are linked, the frequency of PD greatly exceeds the number of NPD. If the two types occur in approximately equal numbers, the loci are assorting independently. Whether the PD:NPD ratio approximates the 1:1 ratio expected with independently assorting genes can be statistically evaluated using the chi-square test discussed in Chapter 6.

Mapping Independently Assorting Loci through Tetrad Analysis

With an understanding of the seven basic ascospore patterns and the relative frequencies in which they occur when the two loci are on separate chromosomes, we are now ready to consider how tetrad analysis data are used to map such loci. We consider two loci in *Neurospora* each with two alleles. One locus controls growth form, with alleles for peach form, *pe*, and normal growth form, +. The other locus relates to nutrition, with an allele for wild-type nutrition, +, and one that imposes a requirement for the amino acid arginine, *arg*. The data in Table 14-4 show the classes of asci produced in a cross between a strain with peach growth form and a nutritional requirement for arginine and a strain with wild-type growth form and nutrition.

TABLE 14-3

Three basic ascospore patterns produced following a crossover in the *b*-to-centromere region and a crossover in the *a*-to-centromere region.

	Ascus Contents					Segregation Pattern	
Class	**1 & 2**	**3 & 4**	**5 & 6**	**7 & 8**	**Tetrad type**	***a* locus**	***b* locus**
1.	*AB*	*ab*	*AB*	*ab*	____	____	____[50]
2.	*Ab*	*aB*	*Ab*	*aB*	____	____	____[51]
3.	*AB*	*ab*	*Ab*	*aB*	____	____	____[52]

[50] PD, M-2, M-2. [51] NPD, M-2, M-2. [52] T, M-2, M-2. [53] The two patterns produced in the absence of crossing over, 1 and 2. [54] The noncrossover situation occurs more frequently than the relatively rare crossover situations. [55] The three patterns arising from two crossover events, 5, 6, and 7. [56] They depend on the simultaneous occurrence of two relatively rare and independent crossover events. [57] The two patterns produced by single crossovers, 3 and 4. [58] Single crossovers occur less frequently than noncrossover situations and more frequently than double crossovers.

TABLE 14-4

Outcome of *pearg* × + + cross in *Neurospora*.

Class	Ascus Contents				Frequency	Tetrad type	Segregation Pattern	
	1 & 2	3 & 4	5 & 6	7 & 8			*pe* locus	*arg* locus
1.	*pearg*	*pearg*	+ +	+ +	96	______59		
2.	*pe* +	*pe* +	+ *arg*	+ *arg*	103	______60		
3.	+ +	+ *arg*	*pe* +	*pearg*	29	______61		
4.	+ +	*pe* +	+ *arg*	*pearg*	69	______62		
5.	+ +	*pearg*	+ +	*pearg*	1	______63		
6.	+ *arg*	*pe* +	+ *arg*	*pe* +	1	______64		
7.	+ +	*pearg*	+ *arg*	*pe* +	1	______65		
				Total:	300			

We need to verify that the two loci are not linked and then determine the distance between each locus and its centromere.

Identifying Tetrad Types and Segregation Patterns For each ascus class given in Table 14-4, determine the total number of genotypes that are present, along with the number of these genotypes that are parental types and the number that are recombinant. Then, for each ascus class identify (1) whether it is a parental ditype (PD), nonparental ditype (NPD), or a tetratype (T) tetrad; and (2) the segregation pattern, M-1 or M-2, shown by each locus. Record this information in the appropriate columns in Table 14-4.

Verifying that Loci Are Not Linked In the next step, we need to confirm that the two loci are carried on different chromosomes. This can be done by comparing the number of PD with the number of NPD. Similar frequencies would indicate that the loci are carried on separate chromosomes, while a much higher frequency of PD relative to the number of NPD would indicate that the loci are found on the same chromosome. What does your comparison of the frequencies of classes 1 and 2 indicate? ________________________66 What does this lead you to conclude about the linkage of these two loci? ________________________67

Determining Locus-to-Centromere Distances Once it is established that the two loci are on separate chromosomes, we can determine locus-to-centromere distances from the frequency with which each locus participates in crossing over.

How can we identify the classes that have experienced a crossover involving the *pe* locus? ________________________68 Identify those classes by the numbers assigned to them in Table 14-4. __________69 What percentage of all the asci occur in these classes? __________70 How many of the ascospores in each of these asci are recombinants? ______71 What is the recombination frequency for the *pe* locus? __________72 How many map units separate the *pe* locus from its centromere? __________73

How can we identify the classes that have experienced a crossover involving the *arg* locus? ________________________74 Identify these classes. __________75 What percentage of all the asci produced occur in these classes? __________76 How many of the ascospores in each of these asci are recombinants? __________77 What is the recombination frequency for the *arg* locus? __________78 How many map units separate the *arg* locus from its centromere? __________79

[59] PD, M-1, M-1. [60] NPD, M-1, M-1. [61] T, M-1, M-2. [62] T, M-2, M-1. [63] PD, M-2, M-2. [64] NPD, M-2, M-2. [65] T, M-2, M-2. [66] The difference between 97 and 104 is small. [67] They are not linked but are carried on different chromosomes. [68] They show an M-2 segregation pattern for the *pe* locus. [69] 4, 5, 6, and 7. [70] The 69 + 1 + 1 + 1 = 72 asci make up 72/300 = 0.24, or 0.24 × 100 = 24%, of all asci. [71] Half. [72] 24% × 1/2 = 12%. [73] 12. [74] They show an M-2 segregation pattern for the *arg* locus. [75] 3, 5, 6, and 7. [76] The 29 + 1 + 1 + 1 = 32 asci make up 32/300 = 0.106, or 0.106 × 100 = 10.6%, of all asci. [77] Half. [78] 10.6% × 1/2 = 5.3%. [79] 5.3.

SUMMARY

In many haploid organisms, the tetrad, or four meiotic products, arising from a particular diploid fusion nucleus can be isolated. In the mold *Neurospora*, each of the nuclei in the tetrad experiences a mitotic division, and the eight nuclei that result develop into ascospores within a linear ascus. From the location of an ascospore within an ascus, the meiosis-I and meiosis-II nuclei giving rise to it can be identified. Individual ascospores can be removed, in order, from an ascus, cultured individually, and the genotype of each can be inferred from its phenotype. The ordered placement of the asci in *Neurospora* make it particularly useful in the study of a number of genetic phenomena.

In crosses involving a single locus, the pattern of genotypes shown by the ascospores within an ascus makes it possible to determine whether or not crossing over occurs and to identify the meiotic division during which two alleles segregate. The first-division, or M-1, allelic segregation pattern (showing the 4-4 distribution) occurs in the absence of crossing over, and the second-division, or M-2, segregation pattern (showing the 2-2-2-2 distribution) occurs with crossing over. Different alignments of the chromosomes during metaphase of meiosis I give rise to random variants of both the M-1 and M-2 segregation patterns. Locus-to-centromere distance can be estimated from the proportion of the total number of asci produced in a cross that shows the M-2 pattern. Since only half the ascospores in the M-2 asci carry recombinations, this proportion needs to be reduced by one-half to give the recombination frequency.

When two independently assorting loci are considered simultaneously, seven basic ascospore patterns arise: with no crossovers (2), with single crossovers (2), and with double crossovers (3). The patterns include a total of 36 random variants. From an understanding of the relative frequencies with which these seven basic ascospore patterns occur with independently assorting loci, tetrad analysis can be used to determine locus-to-centromere distances.

PROBLEM SET

14-1. In haploid organisms like *Neurospora* and baker's yeast, it is possible to determine the manner in which alleles segregate during meiosis by collecting the haploid products that develop from a single diploid fusion nucleus, culturing them individually, and identifying their genotypes. In most diploid organisms, however, this type of procedure is impossible. How is comparable information gathered for diploid organisms?

14-2. A mature *Neurospora* ascus such as in the following sketch contains eight ascospores. Ascospores 1 and 2 are genetically identical to each other as are 3 and 4, 5 and 6, and 7 and 8. Why are the ascospores making up each pair identical to each other?

① ② ③ ④ ⑤ ⑥ ⑦ ⑧

14-3. The following two types of *Neurospora* asci arise from the diploid fusion nuclei produced in the mating $T \times t = Tt$. Note that the alleles carried by the eight ascospores of each type of ascus show a different arrangement.

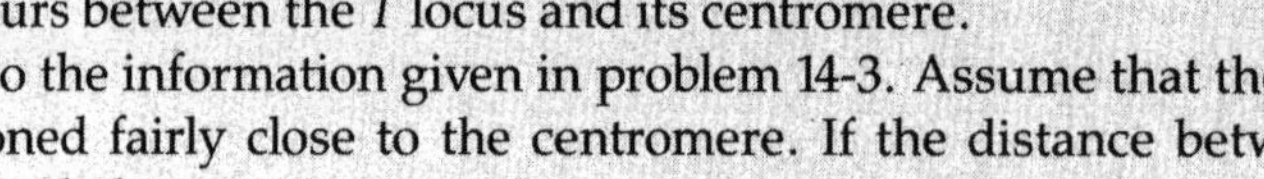

T T T T t t t t T T t t T T t t

a. For each type of ascus, identify the segregation pattern shown by the alleles and indicate whether allelic segregation occurs during meiosis I or meiosis II.

b. Identify the segregation pattern that arises whenever a single crossover occurs between the *T* locus and its centromere.

14-4. Refer to the information given in problem 14-3. Assume that the *T* locus is positioned fairly close to the centromere. If the distance between the *T* locus and the centromere were to increase, what would happen to the relative numbers of these two types of asci?

14-5. A strain of *Neurospora* carrying the allele for fluffy growth form, *fl*, is crossed with another strain carrying the allele for normal growth form, +. An analysis of ascospores in the asci that are produced indicates that they fall into the following six classes.

(1) *fl fl fl fl* + + + +
(2) + + + + *fl fl fl fl*
(3) + + *fl fl* + + *fl fl*
(4) *fl fl* + + + + *fl fl*
(5) + + *fl fl fl fl* + +
(6) *fl fl* + + *fl fl* + +

a. Identify the segregation pattern shown by the locus for each ascus class.

b. Which ascus classes arise in the absence of crossing over?

c. Which ascus classes are formed when crossing over occurs?

PROBLEM SET

14-6. Refer to the information given in problem 14-5.

a. Why does the cross produce more than one class of parental and more than one class of recombinant asci?

b. Would you expect that all six of these classes would arise from this mating in approximately equal frequencies? Explain.

14-7. Refer to the information given in problem 14-5.

a. The *fl* locus is separated from its centromere by a short distance. If this distance were greater, would your expectations about the frequencies of the six ascus classes change? Explain.

b. Explain why only half of the ascospores found in each of the asci in classes 3, 4, 5, and 6 are recombinants.

c. What class or classes of asci would be formed if a double crossover that involved the same two chromatid strands occurred in the centromere-to-*fl* region? What effect would the occurrence of this type of crossover have on the calculation of the centromere-to-*fl* distance?

14-8. A total of 340 asci produced in the cross described in problem 14-5 are examined, and the six classes of asci occur in the following frequencies: class 1, 151; class 2, 161; class 3, 8; class 4, 6; class 5, 7; class 6, 7. Determine the centromere-to-*fl* distance.

14-9. A strain of *Neurospora* carrying the *suc* allele, for succinate as a nutritional requirement, is crossed with a strain which carries the wild-type allele for this trait. A study of the ascospores within 410 of the asci that are formed indicates that 107 show an M-2 pattern of allelic distribution, while the remainder show the M-1 pattern. What is the centromere-to-*suc* distance?

14-10. The *leu-1* locus in *Neurospora* is carried on the third chromosome 10.6 map units from the centromere. A cross between a strain carrying the *leu-1* allele and a strain carrying the wild-type allele at this locus produces a total of 230 asci. Based on the map distance, how many of these asci would be expected to carry recombinations?

14-11. A cross is carried out between a wild-type strain of *Neurospora* and one that carries alleles for button growth form, *bn*, and for a histidine-5 nutritional requirement, *his-5*. An examination of ascospores in 620 asci indicates the following classes of asci.

	Ascus Contents				
Class	**1 & 2**	**3 & 4**	**5 & 6**	**7 & 8**	**Frequency**
1.	*bnhis-5*	*bnhis-5*	+ +	+ +	185
2.	+ *his-5*	+ *his-5*	*bn* +	*bn* +	198
3.	+ +	*bn* +	+ *his-5*	*bnhis-5*	49
4.	+ +	+ *his-5*	*bn* +	*bnhis-5*	185
5.	+ +	*bnhis-5*	+ +	*bnhis-5*	1
6.	+ *his-5*	*bn* +	+ *his-5*	*bn* +	1
7.	+ +	*bnhis-5*	+ *his-5*	*bn* +	1
				Total:	620

a. Identify each class as either a parental ditype, nonparental ditype, or tetratype tetrad and give the segregation pattern shown by each allele.

b. What data in the outcome of this mating indicates that these two loci are on separate chromosomes? Explain.

PROBLEM SET

14-12. Refer to the information given in problem 14-11.

a. Which class or classes of progeny experienced a crossover involving the *bn* locus?

b. What is the distance between the centromere and the *bn* locus?

c. Which class or classes of progeny experienced a crossover involving the *his-5* locus?

d. What is the distance between the centromere and the *his-5* locus?

14-13. A chi-square statistical test indicates that the difference between the numbers of parental and nonparental ditype tetrads produced in a cross between a *Neurospora* strain carrying alleles for two different nutritional differences and a strain carrying wild-type alleles at these loci is nonsignificant at the 0.01% probability level. Are these loci on the same or on separate chromosomes?

CHAPTER

15

Haploid Genetics: Tetrad Analysis II

INTRODUCTION

This chapter continues our consideration of the use of tetrad analysis in understanding the genetics of the fungus *Neurospora* and other haploid organisms. In Chapter 14 we considered crosses involving single loci and two independently assorting loci which provided an essential foundation for the material covered in this chapter. Here we consider crosses involving two linked loci as well as unordered tetrads and mass analysis of spores. Review Chapter 14, if necessary, before proceeding.

CROSSES INVOLVING TWO LINKED LOCI

Seven Basic Ascospore Patterns with Loci on the Same Chromosome Arm

We begin by considering two linked loci, say *c* and *d*, that are carried on the same arm of the chromosome, shown as follows.

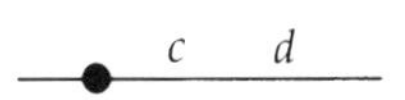

A mating between *Neurospora* strains *CD* and *cd* produces the following diploid *CcDd* fusion nucleus.

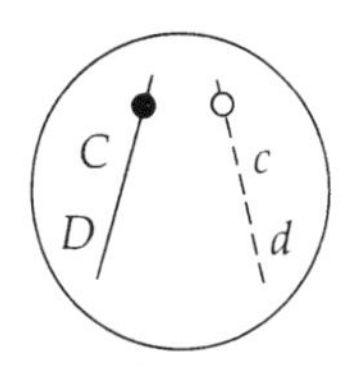

The two heterozygous loci in this fusion nucleus could give rise to 36 different ascospore combinations (depending on whether crossing over takes place and how it takes place). When the random variants are grouped, these combinations fall into seven basic ascospore patterns, just as we found for two unlinked loci (see Chapter 14). Before considering how tetrad analysis is used to map linked loci, we will examine how these seven basic ascospore patterns arise when the loci are carried on the same arm of the chromosome. These ascospore patterns will be considered in three groups, depending on the type of tetrad formed.

Patterns with No Crossing Over and with a Single Crossover between One Locus and the Centromere

First we will identify the ascospore patterns produced when there is no crossing over (Figure 15-1a) and when there is a single crossover between the centromere and the locus closest to the centromere (Figure 15-1b). Complete each scheme in Figure 15-1 to identify the ascospores formed.[1] Then fill in the table at the bottom of the figure. Both the noncrossover and the single-crossover situations produce parental ditype tetrads.

Note that in both the schemes in Figure 15-1, other arrangements of ascospores can occur within asci whenever equally likely alternative chromosomal alignments occur during meiosis. With no crossing over, there is an additional random variant (*cd cd cd cd CD CD CD CD*) and with one crossover there are three additional random variants (*cd cd CD CD CD CD cd cd*, *CD CD cd cd cd cd CD CD*, and *cd cd CD CD cd cd CD CD*). Each random variant is, of course, a parental ditype.

Patterns with Single, Two-Stranded Double, and Three-Stranded Double Crossovers Next we will compare the ascospore patterns arising from three other possible outcomes: (1) a single crossover between the two loci, (2) a double crossover involving two strands, with one crossover occurring between the centromere and the *c* locus and the second between the two loci, and (3) a double crossover involving three strands, with one crossover occurring between the centromere and the *c* locus and the second between the two loci. Each of these situations is shown in Figure 15-2. Complete this figure and identify the contents of each ascus.[12] Then complete the table at the bottom of the figure.

Each of these situations produces tetratype tetrads. Note that these crossovers can occur in ways other than those shown in the diagram. Each of these basic patterns shows seven additional random variants, each of which is a tetratype, for a total of eight random variants. These alternative arrangements of ascospores occur within asci whenever equally likely alternative chromosomal alignments occur during meiosis. With a single crossover between the two loci, the additional variants are *Cd Cd CD CD cD cD cd cd*, *CD CD Cd Cd cd cd cD cD*, *Cd Cd CD CD cd cd cD cD*, *cD cD cd cd CD CD Cd Cd*, *cd cd cD cD CD CD Cd Cd*, *cD cD cd cd Cd Cd CD CD*, and *cd cd cD cD Cd Cd CD CD*. With the two-stranded double crossover, the additional random variants are *cD cD CD CD Cd Cd cd cd*, *CD CD cD cD cd cd Cd Cd*, *cD cD CD CD cd cd Cd Cd*, *Cd Cd cd cd CD CD cD cD*, *cd cd Cd Cd CD CD cD cD*, *Cd Cd cd cd cD cD CD CD*, and *cd cd Cd Cd cD cD CD CD*. With the three-stranded double crossover, the random variants are *cd cd CD CD Cd Cd cD cD*, *CD CD cd cd cD cD Cd Cd*, *cd cd CD CD cD cD Cd Cd*, *Cd Cd cD cD CD CD cd cd*, *cD cD Cd Cd*

FIGURE 15-1

Ascospore patterns produced from a diploid *CcDd* fusion nucleus.

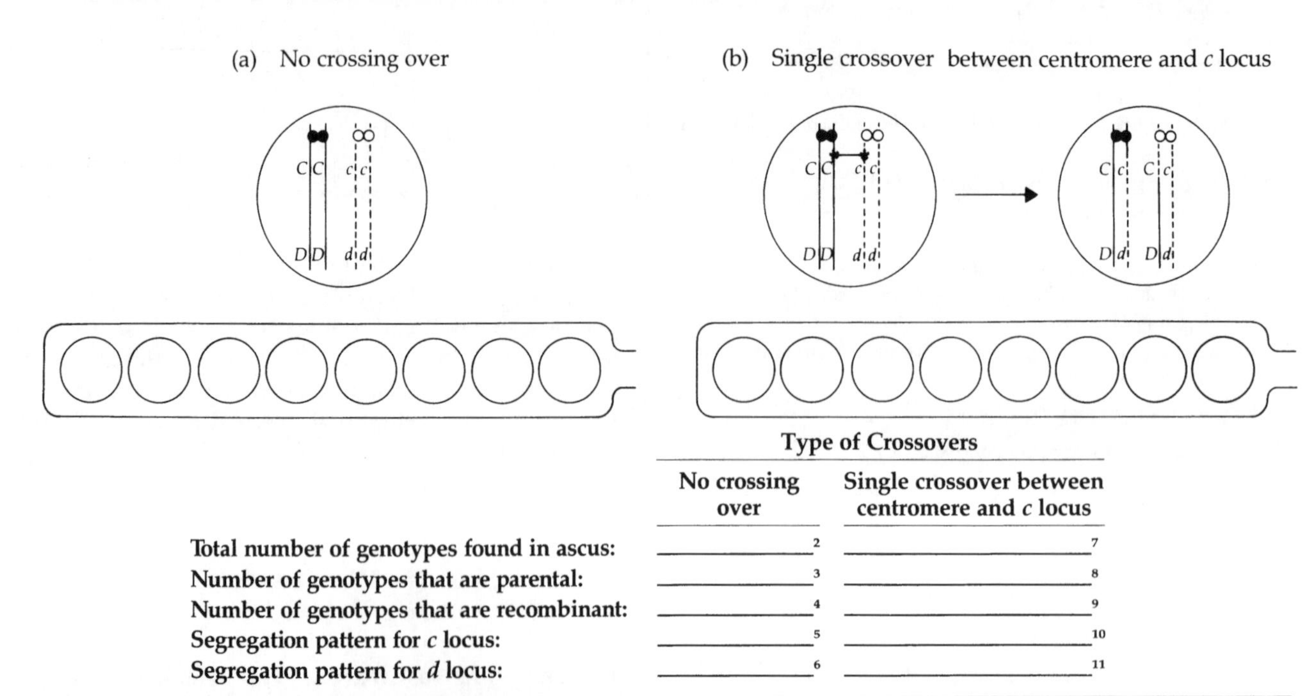

	Type of Crossovers	
	No crossing over	**Single crossover between centromere and *c* locus**
Total number of genotypes found in ascus:	______ [2]	______ [7]
Number of genotypes that are parental:	______ [3]	______ [8]
Number of genotypes that are recombinant:	______ [4]	______ [9]
Segregation pattern for *c* locus:	______ [5]	______ [10]
Segregation pattern for *d* locus:	______ [6]	______ [11]

[1] With no crossover, the ascus contains *CD CD CD CD cd cd cd cd*, and with the single crossover, *CD CD cd cd CD CD cd cd*. [2] Two. [3] Two. [4] Zero. [5] M-1. [6] M-1. [7] Two. [8] Two. [9] Zero. [10] M-2. [11] M-2.
[12] With the single crossover: *CD CD Cd Cd cD cD cd cd*; with the two-stranded double crossover: *CD CD cD cD Cd Cd cd cd*; and with the three-stranded double crossover: *CD CD cd cd Cd Cd cD cD*.

FIGURE 15-2

Additional ascospore patterns produced from a diploid *CcDd* fusion nucleus.

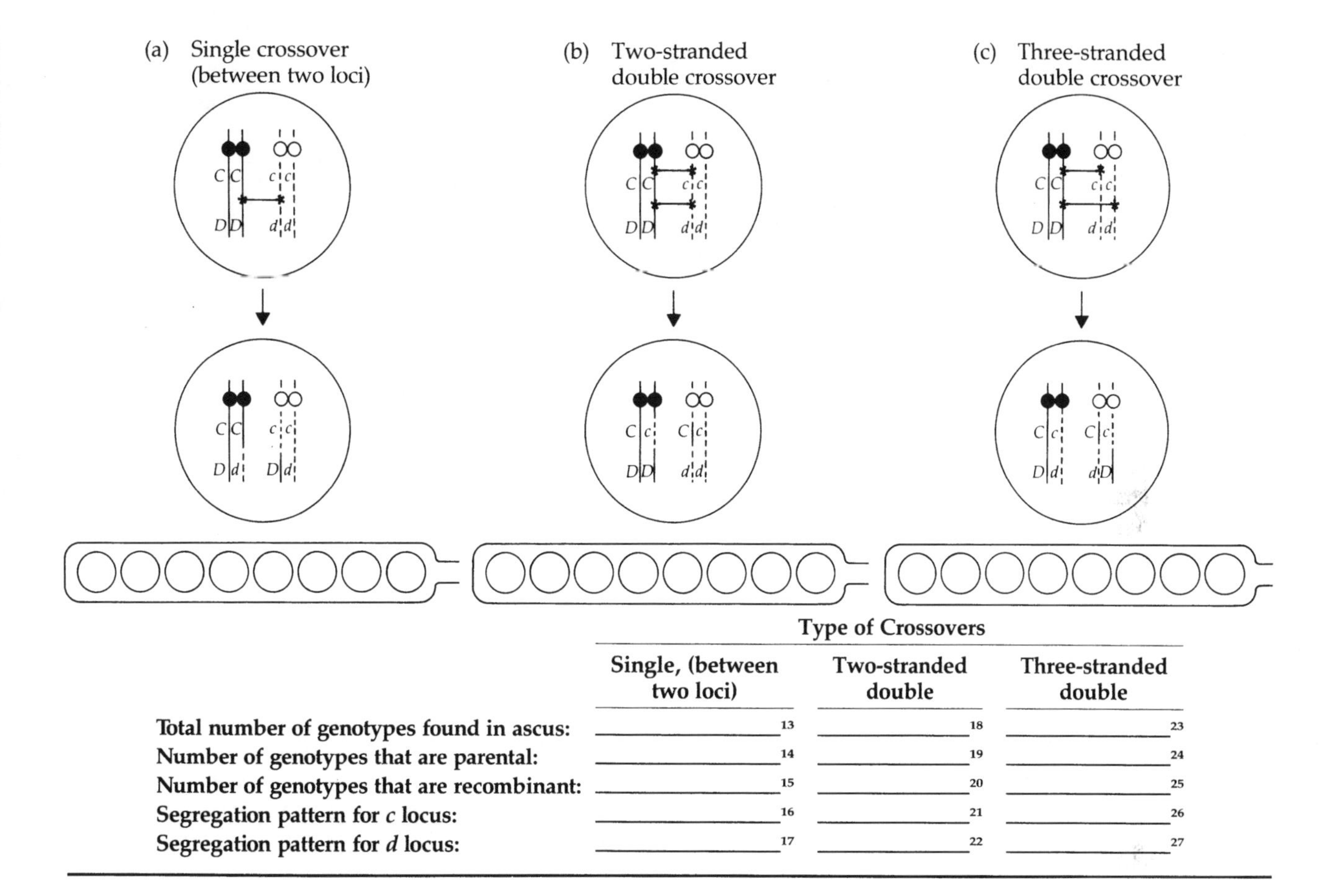

	Type of Crossovers		
	Single, (between two loci)	Two-stranded double	Three-stranded double
Total number of genotypes found in ascus:	______[13]	______[18]	______[23]
Number of genotypes that are parental:	______[14]	______[19]	______[24]
Number of genotypes that are recombinant:	______[15]	______[20]	______[25]
Segregation pattern for *c* locus:	______[16]	______[21]	______[26]
Segregation pattern for *d* locus:	______[17]	______[22]	______[27]

CD CD cd cd, Cd Cd cd cd cd cd CD CD, and *cD cD Cd Cd cd cd CD CD.* Again, each random variant in each situation is, of course, a tetratype.

Patterns with Four-Stranded Double and Four-Stranded Triple Crossovers The situations that produce the two remaining ascospore patterns are as follows: (1) a double crossover which involves all *four* strands—that is, two strands are involved in one crossover and the two other strands are involved in the second crossover, with both crossovers occurring in the region between the two loci; (2) a triple crossover which involves all *four* strands—two strands are involved in a single crossover (which occurs between the centromere and the *c* locus) and four strands are involved in a double crossover, with both crossovers occurring in the region between the two loci. These situations are shown in Figure 15-3. Complete this figure and identify the contents of each ascus.[28] Then complete the table at the bottom of the figure.

Both situations produce nonparental ditype tetrads. Note that in both the double- and triple-crossover situations, other arrangements of ascospores occur within asci whenever equally likely alternative chromosomal alignments occur during meiosis. With the double crossover, there is an additional random variant (*cD cD cD cD Cd Cd Cd Cd*), and with the triple crossover, there are three additional random variants (*cD cD Cd Cd Cd Cd cD cD, Cd Cd cD cD cD cD Cd Cd,* and *cD cD Cd Cd cD cD Cd Cd*). Each random variant in each situation is, of course, a nonparental ditype.

Summary of the Seven Basic Ascospore Patterns The seven basic ascospore patterns produced when the two loci are linked and carried on the same arm of the chromosome are

1. noncrossover parental ditype (PD),
2. single-crossover parental ditype (PD),

[13] Four. [14] Two. [15] Two. [16] M-1. [17] M-2. [18] Four. [19] Two. [20] Two. [21] M-2. [22] M-1. [23] Four. [24] Two. [25] Two. [26] M-2. [27] M-2.
[28] With the four-stranded double crossover: *Cd Cd Cd Cd cD cD cD cD* and with the four-stranded triple crossover: *Cd Cd cD cD Cd Cd cD cD*.

FIGURE 15-3

Additional ascospore patterns produced from a diploid *CcDd* fusion nucleus.

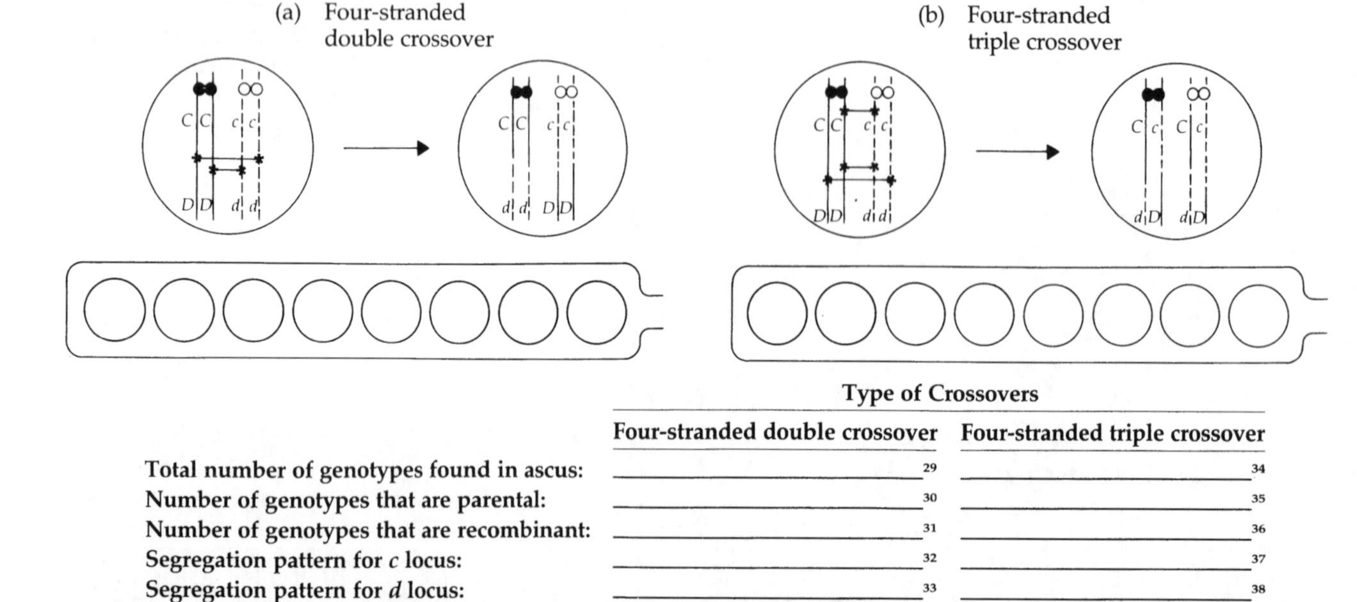

	Type of Crossovers	
	Four-stranded double crossover	**Four-stranded triple crossover**
Total number of genotypes found in ascus:	____ 29	____ 34
Number of genotypes that are parental:	____ 30	____ 35
Number of genotypes that are recombinant:	____ 31	____ 36
Segregation pattern for *c* locus:	____ 32	____ 37
Segregation pattern for *d* locus:	____ 33	____ 38

3. single-crossover tetratype (T),
4. two-stranded double crossover tetratype (T),
5. three-stranded double crossover tetratype (T),
6. four-stranded double crossover nonparental ditype (NPD),
7. four-stranded triple crossover nonparental ditype (NPD).

Relative Frequencies of Basic Ascospore Patterns Any particular ascus will, of course, show just one basic ascospore pattern, but if a large number of asci arose from diploid *CcDd* nuclei, which of the seven basic patterns would you expect to be most frequent? ________________ 39

Explain why. ________________ 40 Which of the seven patterns would you expect to be least frequent? ________________ 41 Explain why. ________________ 42

What patterns would occur with intermediate frequencies? ________________ 43 Explain why. ________________ 44

Detection of Linkage As noted in Chapter 14, a comparison of the frequencies of the PD and NPD tetrads produced in a cross can give an important clue as to whether two loci are linked. With two independently assorting loci, the PD and NPD tetrads are expected in equal frequencies. With linked loci, the number of PD tetrads is much larger than the number of NPD tetrads and the very different frequencies of these two types indicates that the two loci are linked.

Now that you have an understanding of how the seven basic ascospore patterns arise and the relative frequencies with which they are produced when two loci are carried on the same arm of a chromosome, we can consider an illustration of how two such loci are mapped.

Mapping Two Linked Loci on the Same Chromosome Arm

Two strains of *Neurospora*, one carrying alleles *C* and *D* and the other carrying *c* and *d*, are crossed. These two loci are known to be linked and to be carried on the same chromosome arm. Our task is to determine the distances between each locus and the centromere and between the two loci.

[29] Two. [30] Zero. [31] Two. [32] M-1. [33] M-1. [34] Two. [35] Zero. [36] Two. [37] M-2. [38] M-2.
[39] The single noncrossover pattern. [40] Crossing over is a relatively rare event. [41] The triple-crossover pattern.
[42] It depends on the simultaneous occurrence of three relatively rare and independent crossovers. [43] The two single-crossover patterns and the two double-crossover patterns. [44] Both types occur with a lower frequency than the noncrossovers and with a greater frequency than the triple crossovers, with the double-crossover frequencies less than the single-crossover frequencies.

TABLE 15-1

Outcome of a *CD* × *cd* cross in *Neurospora*.

Class	Ascus Contents				Frequency	Tetrad type	Segregation Pattern	
	1 & 2	3 & 4	5 & 6	7 & 8			*c* locus	*d* locus
1.	*CD*	*CD*	*cd*	*cd*	360	____ [45]		
2.	*Cd*	*Cd*	*cD*	*cD*	1	____ [46]		
3.	*Cd*	*CD*	*cD*	*cd*	42	____ [47]		
4.	*CD*	*cD*	*Cd*	*cd*	3	____ [48]		
5.	*CD*	*cd*	*CD*	*cd*	39	____ [49]		
6.	*Cd*	*cD*	*Cd*	*cD*	1	____ [50]		
7.	*Cd*	*cD*	*CD*	*cd*	4	____ [51]		
				Total:	450			

Identifying Tetrad Types and Segregation Patterns Analysis of the ascospores in a total of 450 asci indicates that seven classes of asci occur in the frequencies listed in Table 15-1. For each class, identify the tetrad as PD, NPD, or T and the segregation pattern shown by each allele as M-1 or M-2, and record this information in Table 15-1.

Verification of Linkage Although we have been told that these two loci are linked, how could this be verified? ____________ [52]

Determining Locus-to-Centromere Distances Now let us estimate the map distances. Which classes show the M-2 segregation pattern for the *c* locus? ______[53] What is the *c*-to-centromere distance? ________[54] Which classes show the M-2 segregation pattern for the *d* locus? ________[55] What is the *d*-to-centromere distance? ________[56] Since these two loci are on the same arm of the chromosome, our map would look as follows.

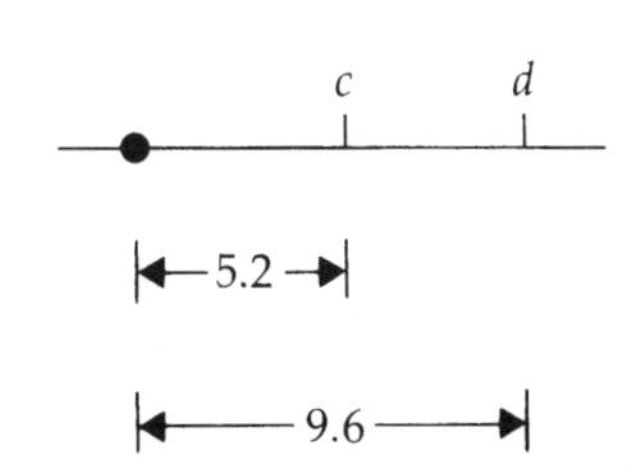

Subtracting the centromere-to-*c* distance from the centromere-to-*d* distance gives a distance of 9.6 − 5.2 = 4.4 map units between the two loci.

More Accurate Distances: Using Crossover Frequencies The distances just estimated are based on the number of asci that show the M-2 segregation pattern for the loci involved. There is a drawback to using this approach when the two loci are located on the same arm of the chromosome: some multiple crossovers involving the locus farthest from the centromere are overlooked. For example, consider tetrad class 4 produced in the *CD* × *cd* cross just considered (Table 15-1). Earlier, you demonstrated that this type of ascus arises following the two crossovers shown in Figure 15-2b. Although one of the two crossovers occurs in

[45] PD, M-1, M-1. [46] NPD, M-1, M-1. [47] T, M-1, M-2. [48] T, M-2, M-1. [49] PD, M-2, M-2. [50] NPD, M-2, M-2. [51] T, M-2, M-2. [52] By comparing the NPD and the PD frequencies. An NPD frequency significantly below the PD frequency indicates that the loci are linked. [53] 4, 5, 6, and 7. [54] The total number of asci in these classes is 3 + 39 + 1 + 4 = 47 for a frequency of 47/450 = 0.1044, or 0.1044 × 100 = 10.4%, of all asci. Since only half the ascospores in these asci carry recombinations involving the *c* locus, this percentage is reduced by half to give 10.4% × 1/2 = 5.2%, or 5.2 map units. [55] 3, 5, 6, and 7. [56] The total number of asci in these classes is 42 + 39 + 1 + 4 = 86 for a frequency of 86/450 = 0.191, or 0.191 × 100 = 19.1%. Reducing this by half gives 19.1% × 1/2 = 9.6%, or 9.6 map units.

the *c*-to-*d* region, the class of tetratypes arising from these two crossovers shows M-1 segregation for the *d* locus (implying, of course, that this locus has *not* participated in crossing over). What effect has overlooking this class had on our estimate of the *c*-to-*d* distance? ________________[57]

An approach which will allow us to detect more of the otherwise undetected multiple crossovers is to estimate distances on the basis of the number of single and double crossovers that occur in each region. Let us use this method to recalculate the distances we have just identified.

Table 15-2 summarizes, for each class of progeny from the *CD* × *cd* cross, the crossovers that occurred in the three chromosome regions with which we are concerned. (If necessary, verify this information by referring to the sketches you completed earlier in this chapter in Figures 15-1, 15-2, and 15-3.) Identify the classes that experience two crossovers in the *c*-to-*d* region. __________[58] How many of the ascospores in the asci of these classes carry recombinations? ______[59] Identify the classes experiencing single crossovers in the *c*-to-*d* region. __________[60] How many of the ascospores in asci of these classes carry recombinations? __________[61]

To calculate the distance between the *c* and *d* loci, we count all the asci in classes 2 and 6 (1 + 1 = 2) and half those in classes 3, 4, and 7 (42 + 3 + 4 = 49, 49 × 1/2 = 24.5). The frequency of the recombinants is 2 + 24.5 = 26.5, 26.5/450 = 0.059, and the number of map units separating the *c* and *d* loci is 0.059 × 100 = 5.9% = 5.9. Note that this more accurate estimate of the *c*-to-*d* distance differs from the 4.4 units arrived at earlier. The calculation you have just carried out for the gene-to-gene distance is summarized by the general expression

$$\text{Map distance} = \frac{(1/2)\text{T} + \text{NPD}}{\text{Total asci}} \times 100.$$

Basing your estimates on the number of crossovers, determine the centromere-to-*c* and the centromere-to-*d* distances. __________[62] The map, based on crossover frequencies, is as follows.

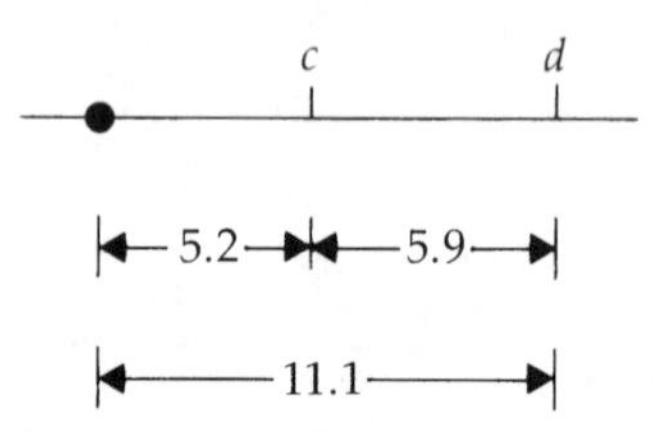

Determining the Arrangement of Loci: Same or Different Arms

In the mapping exercise just completed, we were told that the two linked loci were on the same chromosome arm. Normally, such information is not given and, as part of the challenge, you have to figure out whether the loci are on the same or on different arms of the chromosome. The outcome of a cross can give us this information, and answering the following question should give you an idea of how this is done.

TABLE 15-2

Summary of crossovers that occurred in the three chromosome regions for each class of progeny from the *CD* × *cd* cross.

Class	Tetrad type	Frequency	Crossover classifications	Number of Crossovers: Centromere-to-*c*	*c*-to-*d*	Centromere-to-*d*
1.	PD	360	noncrossover	0	0	0
2.	NPD	1	four-stranded double	0	2	2
3.	T	42	single	0	1	1
4.	T	3	two-stranded double	1	1	2
5.	PD	39	single	1	0	1
6.	NPD	1	four-stranded triple	1	2	3
7.	T	4	three-stranded double	1	1	2

[57] It underestimates the distance. [58] 2 and 6. [59] Since four chromatids of the replicated chromosome pair participated in crossing over, all the ascospores in these classes carry recombinations. [60] 3, 4, and 7. [61] Half. [62] In the centromere-to-*c* region, classes 4, 5, 6, and 7 each experience a single crossover, and thus half of the contents of each ascus carry recombinants: 3 + 39 + 1 + 4 = 47, 47 × 1/2 = 23.5. The frequency is 23.5/450 = 0.052 which is equivalent to 5.2%, indicating a distance of 5.2 map units. In the centromere-to-*d* region, classes 3 and 5 (42 + 39 = 81) experience one crossover and are thus half counted (81 × 1/2 = 40.5), while classes 2, 4, and 7 (1 + 3 + 4 = 8) experience two crossovers and are fully counted. Class 6 (1 ascus) experiences three crossovers and is thus counted $1\frac{1}{2}$ times ($1 \times 1\frac{1}{2} = 1.5$). The frequency for this total is 40.5 + 8 + 1.5 = 50, 50/450 = 0.111, which is equivalent to 11.1%, indicating a distance of 11.1 map units.

If two loci are on the same arm (for example, ——●—e—f——) and a crossover occurs in the centromere-to-*e* region, the *e* locus will be involved in a recombination. Will the *f* locus also be involved in this recombination? __________[63] If the loci are on different arms (for example, —f—●—e——) and a crossover occurs in the centromere-to-*e* region, will the *f* locus also be involved in the recombination? __________[64]

Once the centromere-to-locus distances are calculated and the locus closest to the centromere is identified, the linkage arrangement can be determined. Begin by adding up the number of asci in classes showing an M-2 pattern for the locus closest to the centromere. Then, from this group, determine the number of asci that *also* show the M-2 pattern for the other locus. If most of the asci showing the M-2 pattern for the closer locus also show the M-2 pattern for the other locus, the implication is that the loci are on the same chromosome arm. If, on the other hand, few of the asci show the M-2 pattern for the other locus, the implication is that the loci are on different arms.

Look at the outcome of the cross we have just considered (summarized in Table 15-1) to see if the data support the fact that the two loci are on the same chromosome arm. Which classes show the M-2 pattern for the locus closest to the centromere? __________[65] What is the number of asci in these classes? ______[66] How many of these classes with the M-2 pattern for the *c* locus *also* show the M-2 pattern for the *d* locus? __________[67] What is the number of asci in these classes? __________[68] Now compare the two asci frequencies. What do you conclude about the linkage arrangement? ______________________________ ____________[69]

So far in this chapter, we have considered the seven basic ascospore patterns that arise when two heterozygous loci are linked and carried on the same arm of the chromosome. We found that the seven patterns occurred in a total of 36 random variant arrangements. Although we will not take time to document this, it should be noted that things work out in the same way—that is, with 36 arrangements falling into seven basic patterns—when the two linked loci are carried on different chromosome arms.

A Final Example of *Neurospora* Tetrad Analysis

Let us take a look at a final example. A strain of *Neurospora* exhibiting yellow color, determined by the allele *ylo*, and a nutritional requirement for riboflavin, determined by the allele *rib*, is crossed with a strain carrying wild-type alleles at both of these loci.

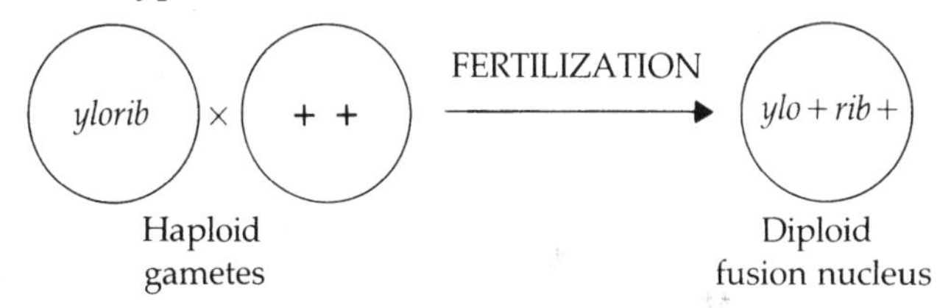

TABLE 15-3

Outcome of a *ylorib* × + + cross in *Neurospora*.

Class	Ascospores 1 & 2	3 & 4	5 & 6	7 & 8	Frequency	Tetrad type	Segregation Pattern *ylo* locus	*rib* locus
1.	+ +	+ +	*ylorib*	*ylorib*	499	____[70]		
2.	+ *rib*	+ *rib*	*ylo* +	*ylo* +	4	____[71]		
3.	+ *rib*	+ +	*ylo* +	*ylorib*	168	____[72]		
4.	+ +	*ylo* +	+ *rib*	*ylorib*	114	____[73]		
5.	+ +	*ylorib*	+ +	*ylorib*	6	____[74]		
6.	+ *rib*	*ylo* +	+ *rib*	*ylo* +	4	____[75]		
7.	+ *rib*	*ylo* +	+ +	*ylorib*	5	____[76]		
				Total:	800			

[63] Yes. We would most often expect that the *f* locus would also recombine. [64] No. *e*-locus recombinations would not be associated in high frequency with *f*-locus recombinations. [65] 4, 5, 6, and 7. [66] 3 + 39 + 1 + 4 = 47 asci. [67] Three of these four classes (5, 6, and 7). [68] 39 + 1 + 4 = 44 asci. [69] 44 of 47 asci (or 93.6%) showing a *c*-locus recombination also show a *d*-locus recombination; this verifies the information given to us earlier that the two loci are on the same arm. [70] PD, M-1, M-1. [71] NPD, M-1, M-1. [72] T, M-1, M-2. [73] T, M-2, M-1. [74] PD, M-2, M-2. [75] NPD, M-2, M-2. [76] T, M-2, M-2.

Analysis of ascospores from a total of 800 asci arising from the diploid fusion nuclei show that seven kinds of asci occur in the frequencies listed in Table 15-3. Our task is to map these loci.

Identifying Tetrad Types and Segregation Patterns Begin by completing Table 15-3. For each class, identify the tetrad type (PD, NPD, or T) and the segregation pattern shown by the alleles (M-1 or M-2).

Determining whether Loci Are Linked Next, determine whether the loci are on the same or different chromosomes. How is this done? ______________________________[77] How many of the 800 asci contain PD tetrads? __________[78] How many contain NPD tetrads? __________[79] What does this lead you to conclude about the linkage of the two loci? ______________________________[80]

Estimating Centromere-to-Locus Distances At this point we can determine the distance separating each locus from the centromere. The key to doing this is the frequency of asci showing the M-2 segregation pattern for each locus. What classes show the M-2 segregation pattern for the *ylo* locus? __________[81] What is the total number of asci found in these classes? __________[82] These asci make up what percentage of the 800 asci examined? __________[83] How many map units separate the *ylo* locus from the centromere? __________[84] Determine the *rib*-to-centromere distance. __________[85]

Determining Loci Placement: Same or Different Arms Next we need to determine whether the loci are on the same or different chromosome arms. How is this done? ______________________________[86] You already found that classes 4, 5, 6, and 7, containing 114 + 6 + 4 + 5 = 129 asci, show the M-2 pattern for the *ylo* locus. Which of these classes also show the M-2 pattern for the *rib* locus? __________[87] What percentage of the asci showing the M-2 pattern for *ylo* also show the M-2 pattern for the *rib* locus? __________[88] What do you conclude about the linkage arrangement? ______________________________[89]

Sketching the Map Now that the centromere-to-locus distances and the loci placement are known, sketch a map in the space provided that shows the loci.[90]

Summary of the Approach for Mapping Two Loci from Ordered Tetrads

To begin, you will need to determine whether the loci are on separate chromosomes or are linked, and if linked, whether they are on the same or different chromosome arms. You will also need to determine the distance in map units separating each locus from its centromere and, if they are linked, the distance between the two loci. A general approach for mapping two loci based on an analysis of the tetrads produced from a *Neurospora* cross is as follows.

1. Identify the number of loci involved in the cross, the genotypes of the two parental strains, and the genotype of the diploid fusion nucleus that is produced.
2. Examine the various kinds of asci arising from the cross and, for each, determine whether it contains a parental ditype (PD), nonparental ditype (NPD), or tetratype (T) tetrad, and identify the segregation pattern as M-1 or M-2 for each locus.
3. Determine whether the loci are linked by comparing the frequencies of the PD and NPD tetrads. If

[77] By comparing the frequencies of the PD and NPD tetrads. [78] 499 + 6 = 505. [79] 4 + 4 = 8.
[80] They are unlinked since the PD:NPD ratio of 505:8 is very different from a 1:1 ratio and would fail a chi-square test for a 1:1 ratio.
[81] 4, 5, 6, and 7. [82] 114 + 6 + 4 + 5 = 129. [83] 129/800 = 0.16, or 0.16 × 100 = 16%. [84] Since only half of the ascospores in each ascus have undergone crossing over, the 16% is reduced by half to 8%, giving a distance of eight map units. [85] Classes 3, 5, 6, and 7 show the M-2 pattern for the *rib* locus and contain 168 + 6 + 4 + 5 = 183 asci. These make up 183/800 = 0.2287, or 0.2287 × 100 = 22.9%, of the total number of asci. The map distance is 22.9% × 1/2 = 11.45 units. [86] By looking at the frequency of crossovers involving the locus closest to the centromere, *ylo*, that also involve the *rib* locus. [87] 5, 6, and 7. [88] Of the 129 asci showing the M-2 pattern for *ylo*, 6 + 4 + 5 = 15 of them also show it for *rib*, for a frequency of 15/129 = 0.116, or 0.116 × 100 = 11.6%.
[89] They are on different arms because so few of the tetrads showing M-2 for *ylo* also show M-2 for *rib*.
[90] The distance between the two loci is given by the sum of their locus-to-centromere distances: 8 + 11.6 = 19.6.

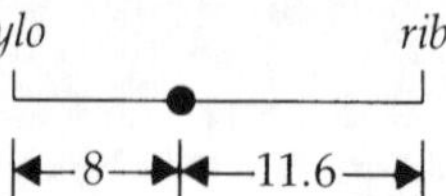

the loci assort independently, these two groups would be expected to occur in approximately equal frequencies. If linked, the NPD frequency would be expected to be significantly less than the frequency of PD.

4. Determine the distance in map units separating each locus from its centromere.
   a. If the loci are *unlinked,* sketch the chromosome that carries each locus and give the distance separating each locus from its centromere.
   b. If the loci are *linked,* first determine whether the loci are on the same or different arms of the chromosome. This is done as follows.
      (1) Identify the asci classes that show an M-2 segregation pattern for the locus closest to the centromere and determine the total number of asci in these classes.
      (2) Determine how many of these same asci also show an M-2 segregation pattern for the other locus. If most do, it indicates that a crossover involving the closer locus also moved the other locus, implying that the loci are on the same chromosome arm. If few of the asci show the M-2 pattern for the other locus, then, most of the crossovers moving the closer locus do not move the other locus, implying that the loci are on different arms of the chromosome. Sketch the chromosome that carries the loci and give the distance separating each locus from the centromere. (Note that if loci are on the same arm, distances are most accurately estimated by determining the number of single and double crossovers that occur in each interval. This approach will detect more of the multiple crossovers.)

UNORDERED TETRADS

In haploid organisms like the motile green alga, *Chlamydomonas,* and baker's yeast, *Saccharomyces cerevisiae,* the two meiotic divisions of a fusion nucleus are *not* followed by a mitotic division. The four meiotic products are found, as in *Neurospora,* in a single ascus. However the asci in these organisms are spherical and the tetrads are unordered. Consequently, the position of the spores in the ascus tells us nothing about the way in which they were formed during meiosis. Since we cannot identify the allelic segregation patterns, it is impossible to calculate locus-to-centromere distances. Nonetheless, analysis of such tetrads can be used to determine whether two loci are independently assorting or are linked; and, if they are linked, the distances between them can be calculated.

Consider the following example of a cross involving the union of haploid nuclei from the different mating types of yeast. One nucleus carries alleles *s* and *t* and the other carries the wild-type forms of these alleles, *S* and *T*. The union of these nuclei produces a diploid fusion nucleus of genotype *SsTt*. Regardless of whether the loci are linked or unlinked, meiosis will produce the three ascus types containing the spores listed as follows. Identify the tetrad type as PD, NPD, or T for each.

Spore-case contents	Tetrad type
st st ST ST	____________[91]
sT sT St St	____________[92]
st ST sT St	____________[93]

What percentage of the spores are recombinants in the PD tetrads? ____________[94] in the NPD tetrads? ________[95] in the T tetrads? ____________[96] If the loci are unlinked, what would you expect for the frequencies of the PD and NPD tetrads? ______________________________ ____________________[97] Explain why. ____________________ ______________________________[98] How would tetratype tetrads arise if the loci are unlinked? __________ ______________________________[99] If the loci are linked, what would you expect for the PD and NPD frequencies? ______________________________ ____________________[100] Explain why. ____________________ ______________________________[101] How would tetratype tetrads arise with linked loci? ______________ ______________________________[102]

In examining the results of tetrad analysis for a cross, what would indicate nonlinkage of the loci? ______________________________[103]

What would indicate linkage of the loci? __________ ______________________________[104] If the loci were linked, how would you determine the recombi-

[91] PD. [92] NPD. [93] T. [94] 0%. [95] 100%. [96] 50%. [97] They would be approximately equal. [98] Independent assortment. [99] From crossing over in the centromere-to-locus region of one of the pairs of chromosomes. [100] The frequency of PD would considerably exceed the NPD frequency. [101] The NPD tetrads arise from rare four-stranded double crossover events. [102] From single crossovers in the region between the two loci. [103] Equal numbers of PD and NPD would indicate independent assortment and, thus, nonlinkage. [104] A much higher frequency of PD relative to NPD would indicate linkage.

nation frequency? ______________________________

________________[105] Assume that the frequencies for the three tetrad types arising from this yeast cross are as follows: PD, 130; NPD, 7; and T, 30. Are the loci linked or unlinked? __________[106] What indicates this?

____[107]

Calculate the distance separating these two loci. ____[108]

MASS ANALYSIS OF SPORES

Even when carried out by an expert, analysis of individual spores in a hundred or more spore cases produced in a cross of haploid organisms is a time-consuming, tedious business. Sometimes, in the interest of efficiency, the spores arising from a cross are collected without regard to the particular spore case in which they developed and, in the case of species having ordered tetrads, without regard to the placement of the spores within the ascus. Once collected, the spores can be cultured and the phenotype and genotype for each identified. The relative frequencies of spores showing parental and recombinant phenotypes can be used to determine whether two loci are carried on separate chromosomes or are linked. If the loci are linked, the spore frequencies can be used to determine the map distance between them.

If two loci are being studied and a diploid zygote is heterozygous for the two loci, what ratio would you expect the spores to show for the parental and recombinant phenotypes if the loci are independently assorting. __________[109] If the loci are linked, how would this ratio be altered? ______________________________

________________[110] How could the recombination frequency by determined? ______________________________

______________________________[111]

SUMMARY

When two linked loci are considered simultaneously in haploid organisms, the basic ascospore patterns that arise with and without crossing over fall into seven types that comprise a total of 36 random variants. From an understanding of the relative frequencies with which these seven basic ascospore types occur when the loci are linked, tetrad analysis can be used to determine locus-to-centromere distances. Once these distances have been calculated, it is possible to determine whether the loci are on the same or different chromosome arms by noting the proportion of tetrad classes with the M-2 segregation pattern for the locus closest to the centromere that also show the M-2 pattern for the other locus. If the loci are located on the same chromosome arm, centromere-to-locus distances based on M-2 segregation patterns are underestimates because some crossovers go undetected. More crossovers are detected, and thus more accurate estimates are prepared, if distance calculations are based on crossover frequencies.

In many haploid organisms other than *Neurospora*, no mitotic division follows meiosis, and the four meiotic products arising from a fusion nucleus are unordered; that is, their position within the asci indicates nothing about how they are formed during meiosis. As a consequence, it is impossible to identify the allelic segregation patterns and to calculate locus-to-centromere distances. Nonetheless, tetrad analysis can indicate whether two loci are independently assorting or are linked; and if they are linked, distances between them can be determined.

[105] All the spores in the NPD asci and half the spores in the T asci are recombinants; adding the number of NPD asci to half the number of T asci and dividing by the total number of asci gives the recombination frequency. [106] Linked.

[107] The large difference in the PD and NPD frequencies.

[108] Map distance $= \frac{(1/2)\text{T} + \text{NPD}}{\text{Total asci}} \times 100 = \frac{(1/2)(30) + 7}{130 + 7 + 30} \times 100 = \frac{15 + 7}{167} \times 100 = \frac{22}{167} \times 100 = 13.2$ units. [109] 1:1:1:1.

[110] There would be an abundance of the two parental types and a deficiency of recombinants.

[111] By dividing the number of recombinant spores by the total number of spores. Note that since we are dealing with individual spores rather than asci, there is no need to correct the number of recombinants by multiplying by 1/2.

PROBLEM SET

15-1. *Neurospora* loci *a* and *b* are linked and positioned 10 and 12 map units, respectively, from the centromere. Following a cross between strains *AB* and *ab*, ascospores from 420 asci are individually cultured and analyzed; 98 show recombinations involving the *a* locus and, of these, 90 show recombinations involving the *b* locus. Determine whether these two loci are on the same or different chromosome arms and the number of map units separating them.

15-2. A cross is carried out between a strain of wild-type *Neurospora* and a strain carrying genes for a lysine nutritional requirement, *lys*-2, and a growth form known as spray, *sp*. An examination of the ascospores in 850 asci indicates that the asci fall into seven classes with the following frequencies.

	Ascus Contents				
Class	**1 & 2**	**3 & 4**	**5 & 6**	**7 & 8**	**Frequency**
1.	+ +	+ +	*lyssp*	*lyssp*	565
2.	+ *sp*	+ *sp*	*lys* +	*lys* +	2
3.	+ *sp*	+ +	*lys* +	*lyssp*	130
4.	+ +	*lys* +	+ *sp*	*lyssp*	8
5.	+ +	*lyssp*	+ +	*lyssp*	135
6.	+ *sp*	*lys* +	+ *sp*	*lys* +	1
7.	+ *sp*	*lys* +	+ +	*lyssp*	9
				Total:	850

a. For each class, determine the tetrad type and the segregation pattern shown by each locus.
b. Are these loci carried on the same or on different chromosomes?

15-3. Refer to the information given in problem 15-2.
a. Based on the classes showing the M-2 segregation pattern for the *lys* locus, what is the *lys*-to-centromere distance?
b. Based on the classes showing the M-2 segregation pattern for the *sp* locus, what is the *sp*-to-centromere distance?
c. Are these loci carried on the same or different arms of the chromosome?

15-4. Refer to the information given in problem 15-2.
a. Based on the locus-to-centromere distances you calculated, how many map units separate the two loci?
b. Identify the classes that experienced single and double crossovers in the *lys*-to-*sp* interval, and, based on the crossover frequencies, determine the number of map units separating the two loci.
c. Which estimate of the *lys*-to-*sp* distance, the one based on segregation patterns or on the number of crossovers, is more accurate? Explain.
d. Sketch a map showing the centromere and the two loci, and include the most reliable estimates of distance between these points.

15-5. A wild-type strain of *Neurospora* is crossed with a strain carrying genes for fissure growth form, *fi*, and for a cysteine nutritional requirement, *cys-10*. Examination of ascospores in 1100 asci indicates that the asci fall into seven classes with the frequencies listed as follows.

PROBLEM SET

Class	1 & 2	3 & 4	5 & 6	7 & 8	Frequency
	Ascus Contents				
1.	+ +	+ +	*ficys*	*ficys*	755
2.	+ *cys*	+ *cys*	*fi* +	*fi* +	3
3.	+ *cys*	+ +	*fi* +	*ficys*	209
4.	+ +	*fi* +	+ *cys*	*ficys*	11
5.	+ +	*ficys*	+ +	*ficys*	110
6.	+ *cys*	*fi* +	+ *cys*	*fi* +	2
7.	+ *cys*	*fi* +	+ +	*ficys*	10
				Total:	1100

a. Are these two loci linked or unlinked?
b. Based on segregation patterns, what are the centromere-to-locus distances for each locus?
c. Are the loci on the same or different chromosomal arms?

15-6. Refer to the information given in problem 15-5.
a. Based on crossover frequencies in the region between the two loci, estimate the map distance between the two loci.
b. Sketch a map showing the relative position of the centromere and the two loci, and indicate the estimated distances between these points.

15-7. A cross is carried out between wild-type *Neurospora* and a strain carrying genes for the crisp growth form, *cr*, and a nutritional requirement for arginine, *arg-1*. An examination of the ascospores in 620 asci indicates that seven classes of asci are formed in the frequencies listed as follows.

Class	1 & 2	3 & 4	5 & 6	7 & 8	Frequency
	Ascus Contents				
1.	+ +	+ +	*crarg*	*crarg*	406
2.	+ *arg*	+ *arg*	*cr* +	*cr* +	2
3.	+ *arg*	+ +	*cr* +	*crarg*	138
4.	+ +	*cr* +	+ *arg*	*crarg*	63
5.	+ +	*crarg*	+ +	*crarg*	5
6.	+ *arg*	*cr* +	+ *arg*	*cr* +	1
7.	+ *arg*	*cr* +	+ +	*crarg*	5
				Total:	620

a. Are the two loci linked or unlinked?
b. Based on segregation patterns, what are the locus-to-centromere distances for each locus?
c. Directly calculate the *cr*-to-*arg* distance.
d. Are the loci on the same or different chromosomal arms?
e. Sketch a map showing the relative positions of the two loci and the centromere, and indicate the distances between these points.

15-8. Take yourself back to the time when there was uncertainty as to just when crossing over took place during meiosis. One possibility was that it occurred before chromosomal replication when a homologous pair consists of two single-stranded chromosomes; another possibility was that it occurred after chromosomal replication when a homologous pair consists of four chromatids. Tetrad analysis in *Neurospora* can be used to identify when

PROBLEM SET

crossing over occurs during meiosis. Consider a cross between a wild-type strain of *Neurospora* and a strain carrying mutant alleles at two linked loci, *u* and *v*. In answering the following questions, make these assumptions:

(1) Assume that crossing over occurs before chromosomal replication;
(2) the *u* and *v* loci are carried on the same arm of the chromosome;
(3) crossing over occurs in some but not in all the cells undergoing meiosis; and
(4) the crossovers that do occur are single crossovers that involve breaks in the region between the two loci.

a. What general types of tetrads would result?
b. If the frequency of crossing over increased, what would happen to the relative frequencies of the types of tetrads produced?
c. Could tetratype tetrads be produced? Explain.

15-9. Refer to the information given in problem 15-8, but now assume that crossing over occurs after chromosomal replication.
a. What general types of tetrads would be formed?
b. If the frequency of crossing over increased, what would happen to the relative frequencies of the types of tetrads produced?
c. What would have to occur in order for NPD tetrads to be produced?
d. Does crossing over in *Neurospora* occur before or after chromosomal replication?

15-10. A cross is carried out between two strains of *Neurospora*, one carrying a mutation, designated as *a*, which imposes a requirement for a particular amino acid, and the other carrying a mutation, designated as *g*, which causes an unusual growth form. An examination of the ascospores in 437 asci indicates that seven classes of asci are formed in the following frequencies.

	Ascus Contents				
Class	**1 & 2**	**3 & 4**	**5 & 6**	**7 & 8**	**Frequency**
1.	*ag*	+ +	*ag*	+ +	1
2.	*a*+	+ +	*ag*	+*g*	37
3.	*ag*	*ag*	+ +	+ +	2
4.	*ag*	+ +	*a*+	+*g*	3
5.	*a*+	+*g*	*a*+	+*g*	3
6.	*a*+	*a*+	+*g*	+*g*	316
7.	*ag*	*a*+	+ +	+*g*	75
				Total:	437

Map these two loci.

15-11. The tetrads arising from meiosis in the flagellated green alga, *Chlamydomonas*, are unordered. A strain carrying genes for neamine resistance, *nr*, and paralyzed flagella, *pf*, is crossed with a wild strain producing a total of 900 asci which fall into the three following categories.

(1) 762 asci containing equal numbers of *nrpf* and + + spores.
(2) 132 asci containing equal numbers of *nrpf*, *nr* +, +*pf*, and + + spores.
(3) 6 asci containing equal numbers of *nr* + and +*pf* spores.

PROBLEM SET

a. Are the loci under consideration linked or unlinked?
b. If the loci are linked, determine the map distance separating them.

15-12. A strain of *Chlamydomonas* carrying the paralyzed-flagella allele, *pf*, is mated with a strain carrying a chloroplast ribosome–deficient allele, *cr*. The cross produces a total of 1100 asci in the three following categories.

(1) 519 asci containing equal numbers of *pf* + and + *cr* spores.
(2) 498 asci containing equal numbers of *pfcr* and + + spores.
(3) 83 asci containing equal numbers of *pf* +, + *cr*, *pfcr*, and + + spores.

a. Are these loci linked or unlinked?
b. If the loci are linked, determine the map distance separating them.

15-13. A strain of *Chlamydomonas* carrying alleles for long flagella, *lf*, and yellow color, *y*, is crossed with a wild-type strain. Spores are obtained from a large number of asci without regard to the particular ascus in which they were formed, and a mass culturing is carried out. Examination of the phenotypes of the algae arising from these spores indicates that the spores are of four types (+ +, *lfy*, *lf* +, and + *y*) which occur in roughly equal frequencies.
a. Are these two loci linked or unlinked?
b. If the loci are linked, determine the map distance separating them.

15-14. A strain of *Chlamydomonas* carrying the allele for paromomycin resistance, *pr*, is crossed with another strain carrying the allele for acetate dependence, *ac*. A total of 3600 spores is collected without regard to the ascus in which they were formed, and a mass culturing is carried out. Examination of the algae arising from these spores shows that the spores are of four types which occur in the following frequencies: *pr* +, 1602; + *ac*, 1710; *prac*, 130; and + +, 158.
a. Are these loci linked or unlinked?
b. If the loci are linked, determine the map distance separating them.

CHAPTER

16

Changes in Chromosome Structure and Number

INTRODUCTION

This chapter deals with chromosome mutations that involve changes in chromosome structure—referred to as **chromosome aberrations** or **anomalies**—and changes in chromosome number. These mutations are distinct from gene or point mutations which involve alterations in the structure of the nucleic acid comprising a gene.

CHROMOSOME ABERRATIONS

Alterations in chromosome structure fall into four categories: deficiencies, duplications, inversions, and translocations, each of which can be detected by conventional light microscopy. Note that similar alterations may also occur in prokaryotes but are not detected in this manner and are discussed elsewhere. While deficiencies and duplications involve the loss or gain of genetic material, inversions and translocations involve its rearrangement. Each aberration involves chromosomal breakage followed by the reattachment of chromosomal sections. If a chromosome carries an aberration and its homolog is normal, the individual possessing such a chromosome pair is referred to as heterozygous for the aberration. Each type of aberration usually gives rise to microscopically detectable abnormal patterns such as loops or crosses when a chromosome carrying the aberration synapses with a normal homolog during meiosis. One consequence that may arise from a chromosomal aberration is the altered phenotypic expression of certain genes. Since the expression of a gene may be influenced by the genes next to it on a chromosome (a phenomenon known as **position effect**), genes that end up with new neighbors may exhibit altered expressions.

Deficiencies

A chromosome carrying a **deficiency** has a missing segment. Such chromosomes generally arise when two breaks occur within a chromosome, the intervening segment is removed, and the two remaining chromosome segments rejoin. This is illustrated in the following representation where the letters denote a series of loci on a chromosome and the arrows point to the sites of the breaks. Breaks between *C* and *D* and between *F* and *G* release segment *DEF*.

```
   ↓   ↓
ABC DEF GHIJK  ⟶  ABC  GHIJK  +  DEF
               ⟶  ABCGHIJK  +  DEF
```

If the released segment lacks a centromere, it cannot attach to a spindle fiber and will not be transmitted to daughter nuclei; such fragments are generally enzymatically degraded or otherwise eliminated from the cell. During synapsis, when one member of the chromosome pair carries this deficiency, the portion of the normal chromosome corresponding to the deficiency lacks loci with which to synapse and consequently projects as an unpaired **loop** or buckle, as shown in the following diagram.

Unpaired loop: *E F D*

Normal homolog: *A B C G H I J K*
Homolog with deficiency: *a b c g h i j k*

The size of the loop provides an index of the size of the deficiency. Deficiencies are almost always harmful to the individual carrying them and if the loss of genetic information is too great, death generally results.

FIGURE 16-1

Unequal crossing over produces a duplication.

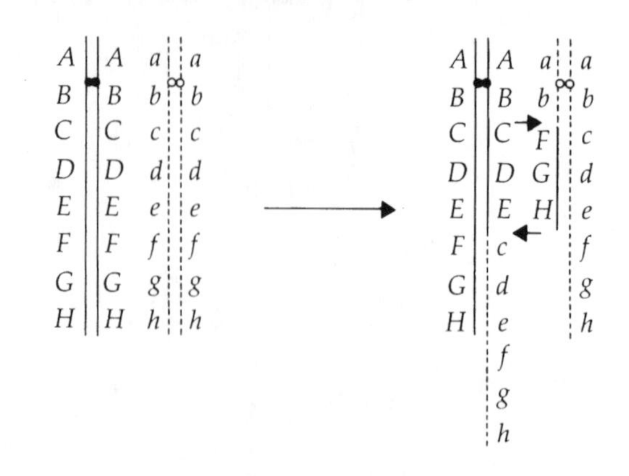

1 2 3 4
Replicated homologous pair prior to crossing over

1 2 3 4
Homologs following unequal crossing over

Duplications

A **duplication** exists when a segment of a chromosome is repeated, as in the following illustration for the segment carrying loci *EFG*.

ABCDEFGHI → *ABCDEFGEFGHI*

Duplications often arise through uneven breakage during crossing over, leaving one participating chromatid with two copies of a section, as shown in Figure 16-1 where the arrows indicate the break sites. Note that, in addition to the duplication, this uneven crossing over produces a second type of aberration.

Refer to Figure 16-1 as you answer the following questions. Which of the four product strands carries the duplication? ________[1] What loci are carried in duplicate? ________[2] What other type of chromosome aberration arises from this unequal crossing over? ________[3] Which loci are involved in this second aberration? ________[4] Which strand shows this other type of aberration? ________[5]

The synapsis of homologous chromosomes where one carries a duplication and the other is normal will often result in an unpaired loop forming at the site of one of the two regions carried in duplicate, as follows.

Normal homolog: *a b c d e f g h*
Homolog with duplication: *A B C D E F G H*

Unpaired loop: *C E D*

In some organisms, genes that code for proteins needed in large quantities may occur within cells in multiple copies. This **gene redundancy** may have arisen through duplication. In addition, duplications may play a role in evolution: if a chromosome carries two copies of a locus, one can mutate and specify a different phenotype without disrupting the original function of the gene.

Inversions

An **inversion** arises when two breaks occur in a chromosome. The released segment then turns around and becomes reinserted into the chromosome. This process is illustrated as follows, where the two breakage sites are indicated by arrows.

[1] Strand 2 carries the duplication. [2] Loci *CDE* are duplicated. [3] Deficiency. [4] Loci *CDE* are absent. [5] Strand 3.

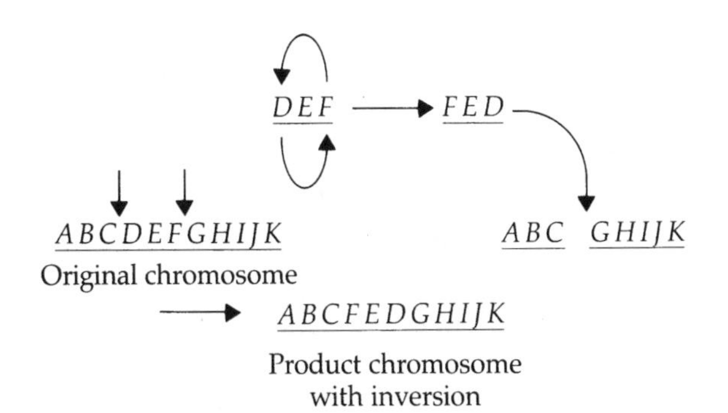

Compare the original and the product chromosomes. Do both carry the same loci? ________[6] What is different about the two chromosomes? ________________________________[7]

Inversions are classified into two categories. If the inverted section includes the centromere, it is a **pericentric inversion.** If the inverted section does not include the centromere, it is a **paracentric inversion.**

Crossing Over in Paracentric Inversions Figure 16-2a shows a replicated homologous pair where the chromosome on the right carries the paracentric inversion *fed*. Figure 16-2b shows these homologs synapsed, with a loop at the inversion site that involves all four strands of the tetrad. Because of this loop, single crossovers that occur within the inverted section, as illustrated in Figure 16-3, give rise to defective recombinant strands. In the space provided in Figure 16-3, sketch the four chromatids that result following a crossover in the region between loci *D* and *E*.

Referring to your drawing in Figure 16-3, list, in order, the loci carried by each of the two nonrecombinant chromatids. ________________________[8] List, in order, the loci carried by each of the two recombinant chromatids. ________________________[9] How many of the four product chromatids carry just the nine loci found on the chromosomes prior to crossing over? ________[10] How many chromatids have some loci duplicated and others missing? ________[11] Which chromatids have a single centromere? ________[12] Which chromatids are acentric? ________[13] Which chromatids are dicentric? ________[14]

FIGURE 16-2

Synapsis of homologs in a carrier of a paracentric inversion.

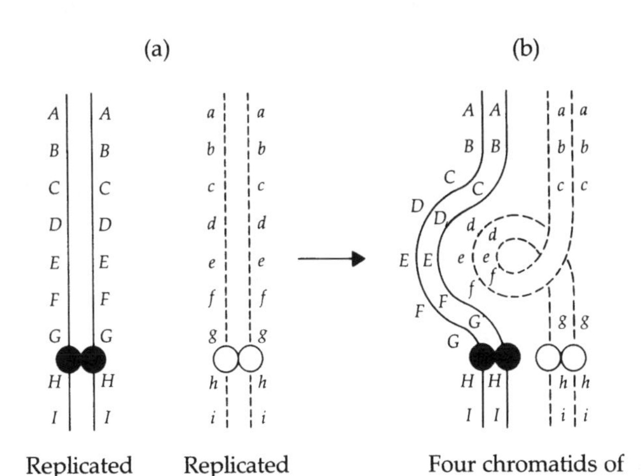

FIGURE 16-3

Crossing over in a paracentric inversion.

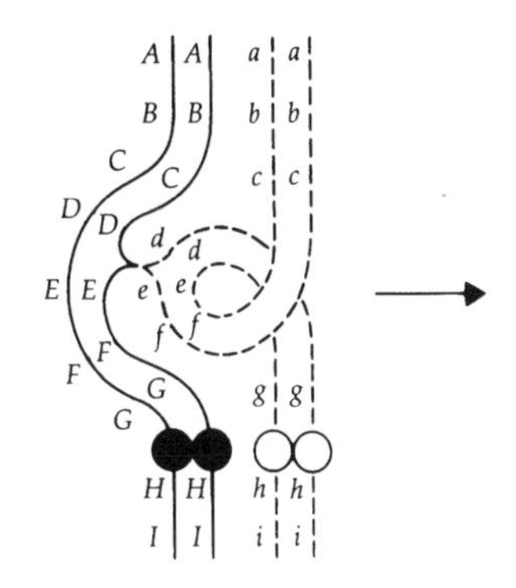

[6] Yes. [7] The order of loci is altered. [8] The nonrecombinant chromatids are genetically balanced; they carry the same loci, but in different orders: *A B C D E F G ● H I* and *a b c f e d g ● h i* [9] *A B C D e f c b a* and *i h ● g d E F G ● H I*. [10] Two. [11] Both recombinant chromatids have some loci duplicated and others missing. [12] Both nonrecombinant chromatids have a single centromere. [13] One recombinant chromatid, *A B C D e f c b a*, has no centromere (acentric). [14] One recombinant chromatid, *i h ● g d E F G ● H I*, has two centromeres (dicentric).

The irregularities with both the number of centromeres and of loci on the recombinant strands create problems. Without a centromere, a chromatid cannot join to a spindle fiber and thus cannot move to the poles of the spindle apparatus with the other chromosomes. A chromatid with two centromeres gets attached to spindle fibers from each pole during the first meiotic division and may become extended between the two poles forming a **bridge.** If the bridge fails to break, its chromosomal material fails to be incorporated into either of the product nuclei. If breakage does occur, it is often unequal, giving rise to further gene imbalance in the product nuclei.

Based on this, what general prediction can be made about the genetic makeup of the gametes arising from a primary sex cell carrying a paracentric inversion? ______________________________

________[15] If fertilized, these unbalanced gametes usually give rise to inviable zygotes. As a consequence, paracentric-inversion carriers produce fewer progeny than normal parents and are often designated as **semisterile.** Since they generally are able to transmit only the original chromosomal linkage groups to the next generation, inversion carriers usually produce no recombinant progeny.

Crossing Over in Pericentric Inversions Carriers of a pericentric inversion are also semisterile. Figure 16-4a shows a replicated homologous pair where the chromosome on the right carries the pericentric inversion *rqp*. To the right of the arrow (Figure 16-4b), these homologs are synapsed, showing a loop at the inversion site and a single crossover within the inversion between loci *P* and *Q*. Identify the number of centromeres and the loci carried by each of the four chromatids after crossing over. ______________________

______________________[16] The half of the gametes that receive the recombinant strands will have duplicated and missing loci, and following fertilization, usually will give rise to inviable zygotes.

FIGURE 16-4

Crossing over in a pericentric inversion.

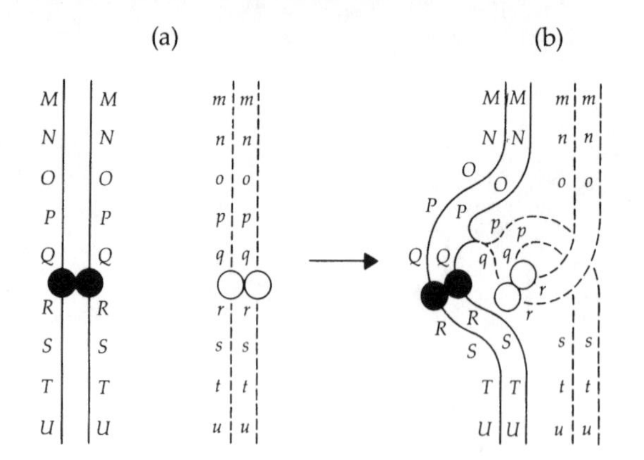

Replicated normal chromosome

Replicated inversion chromosome

Four chromatids of a tetrad at prophase of meiosis I, showing a crossover in a pericentric inversion

Consequences of Inversions For chromosomes carrying either paracentric or pericentric inversions, inversion heterozygotes generally transmit only the chromatids that do not participate in crossing over, that is, the parental types, to the next generation. Although crossing over does in fact occur, the absence of progeny recombinant for the loci found on the chromosome gives the *appearance* that the inversion suppresses crossing over. Because recombinants for the loci within the inversion do not survive, the genes within an inversion tend to be inherited together as a block or **supergene.** The development of a particular, favorable combination of genes within the inversion, that is, a combination that confers a selective advantage, will be preserved as a unit because of the inversion.

Inversions in *Drosophila* Inversions other than very small ones are not widespread in most plant and animal species. An exception to this occurs in many natural populations of *Drosophila* where inversions, often of considerable size, are very common. Since no crossing over occurs in males of this species, no unbalanced chromatids are formed by male inversion heterozygotes and their gametes carry only parental-type strands. Crossing over does occur in females, but because of the manner in which the meiotic products are positioned during meiosis, the two nuclei produced by an inversion heterozygote that carry the genetically unbalanced chromatids always end up in the polar bodies (tiny cells produced at meiosis that fail to develop into egg cells) which subsequently degenerate. The two remaining nuclei carry genetically balanced chromatids; one of these nuclei is also incorporated into a polar body (which also degenerates) and the other into the functional egg. Half the time the egg carries a chromosome with the inversion and half the time it carries a chromosome with the normal, noninverted sequence of loci. Since all eggs and all sperm of

[15] Half the gametes end up genetically unbalanced because of gene deficiencies. [16] All chromatids have one centromere; both recombinants, *M N O P q* ● *r o n m* and *U T S R* ● *Q p s t u,* have some loci duplicated and others missing, and the two nonrecombinants, *M N O P Q* ● *R S T U* and *m n o r* ● *q p s t u,* have all loci represented.

inversion heterozygotes carry genetically balanced copies of the chromosome with the inversion, there is no reduction in fertility of the inversion carriers and the inversion is readily transmitted to the next generation of the *Drosophila* population.

Translocations

A **translocation** involves the transfer of part of a chromosome to a new location in the genome where it becomes attached. An **intrachromosomal translocation** involves three breaks within the same chromosome and shifts a segment of the chromosome to a new site within that chromosome. The following representation shows a chromosome before and after an intrachromosomal translocation.

hijklmnopq	⟶	*himnjklopq*
Original chromosome		Product chromosome

What is the simplest scheme of events that could have occurred to produce the product chromosome?

______________________________[17]

Compare the original and the product chromosome. Is there a change in the number of genes present? ______[18]

An **interchromosomal translocation** involves shifting a segment from one chromosome to a **nonhomologous** chromosome. A **reciprocal translocation** involves a mutual exchange of segments between two nonhomologous chromosomes, as illustrated in Figure 16-5. Figure 16-5a shows two pairs of homologous chromosomes (1 and 2 comprise one pair and 3 and 4 comprise the other) and Figure 16-5b shows these two pairs following the reciprocal translocation. What is the minimal number of breaks required to produce this reciprocal translocation? ________[19] Where did these breaks occur? ______________________________

__________[20] What segments were transferred?

______________________________[21]

Has this reciprocal translocation altered the total number of genes present? ________[22]

FIGURE 16-5

Reciprocal translocation.

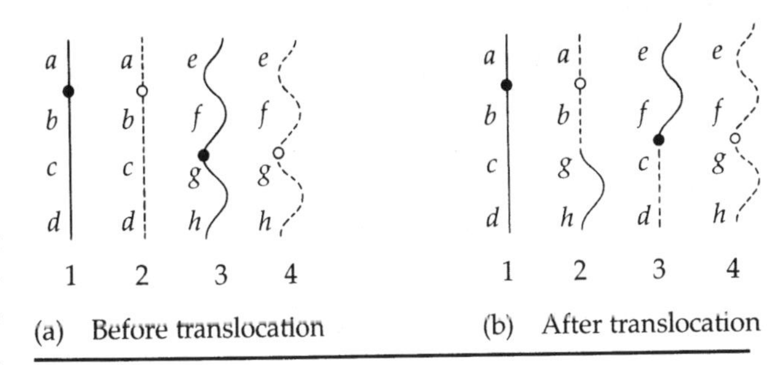

Consequences of Translocations Assume that the reciprocal translocation in Figure 16-5 occurred in the primary sex cells of an organism. During meiosis, chromosome pairs that are heterozygous for a mutual translocation generally achieve point-by-point pairing by assuming cross-shaped configurations during pachynema of meiotic prophase I. The problems associated with reciprocal translocations arise when the chromosomes separate during meiosis. Demonstrate this by identifying the various chromosome combinations possible in the gametes arising from these primary sex cells (remember that each gamete will get one member of each homologous pair which separates according to the law of independent assortment) and indicate whether each type of gamete carries a complete set of loci. ______________________________

______________[23]

The gametes carrying duplications and deficiencies will give rise to inviable or abnormal zygotes. As a consequence of this, a parent heterozygous for a reciprocal translocation is designated as semisterile since only about one half of its progeny survive. In addition to position effect, translocations also create new linkage relationships between genes.

Translocation Down Syndrome Down syndrome in humans is caused by an extra copy of chromosome 21. Although commonly produced through failure of the chromosome-21 pair to separate during meiosis (nondisjunction) in one parent, about 5% of the cases of this disorder arise through a simple nonreciprocal translocation where one copy of chromosome 21 (or most of it) becomes attached to a larger chromosome, often chromosome 14. Although individuals carrying this translocation will have 45 chromosomes, their phenotype will be normal since two copies of chromosome 21 are present. The carrier, however, runs the

[17] Breaks occurred between loci *i* and *j*, between *l* and *m*, and between *n* and *o*; the segment $\underline{m\ \ n}$ has been shifted to a new location.
[18] No. [19] Two. [20] One between *b* and *c* of chromosome 2 and the other between *f* and *g* of chromosome 3.
[21] Segment $\underline{c\ \ d}$ from chromosome 2 was transferred to chromosome 3 and segment $\underline{g\ \ h}$ from chromosome 3 was transferred to chromosome 2. [22] No. [23] Four combinations are possible: 1 and 4, full set of loci; 2 and 3, two translocations with full set of loci; 2 and 4, one translocation, one normal chromosome; incomplete set of loci with duplications and deficiencies; 1 and 3, one translocation, one normal chromosome, incomplete set of loci with duplications and deficiencies.

risk of producing a child affected with Down syndrome.

Meiosis in the carrier will involve problems with synapsis and with chromosome separation, and six types of gametes, with reference to chromosomes 14 and 21, can be formed. These are listed in the following table, where 14/21 represents the translocation product. Assume that a normal gamete carrying one 14 and one 21 unites with each of the gametes listed in the following table. Identify, in the space provided to the right of each type of gamete, the combinations of 14 and 21 that occur in the resulting zygotes.

	Chromosomes included in gametes			Chromosome combinations found in zygote
1.			21	____________[24]
2.	14			____________[25]
3.	14	14/21		____________[26]
4.		14/21	21	____________[27]
5.		14/21		____________[28]
6.	14		21	____________[29]

An extra copy of 14 or missing copies of 14 or 21 block the development of the zygote, thereby causing its death. Which of the zygote types listed in the preceding table carry lethal assemblages? ____________[30] Which type or types exhibit translocation Down syndrome? ____________[31] Which type or types carry the translocation (and have a normal phenotype)? ____________[32] Which type or types are normal? ____________[33]

CHANGES IN CHROMOSOME NUMBER

Each species has a standard number of chromosomes. In most sexually reproducing animals, for example, the chromosomes in most somatic cell nuclei generally occur in pairs, and the total number of chromosomes in each nucleus makes up the diploid ($2n$) number for the species. Spontaneous or induced alterations in chromosome number fall into two categories depending on whether individual chromosomes or whole haploid (n) sets are involved. **Aneuploidy** is the condition produced when the normal complement, or set of chromosomes, is altered by the addition or removal of individual chromosomes. **Euploidy** results when the chromosome number is increased through the addition of one or more complete haploid sets.

Aneuploidy

The more commonly encountered types of aneuploidy involve the loss or gain of one or two chromosomes. **Monosomics** are missing one member of a single homologous pair and are symbolized as $2n - 1$, **double monosomics** are missing one member of each of two pairs ($2n - 1 - 1$), and **nullisomics** are missing both members of a homologous pair ($2n - 2$). **Trisomics** have an extra copy of a homologous pair ($2n + 1$), **double trisomics** have an extra copy of each of two homologous pairs ($2n + 1 + 1$), and **tetrasomics** carry two extra copies of a single homologous pair ($2n + 2$).

Origin of Aneuploidy Aneuploids can often be traced back to the failure of homologs to separate during meiosis (nondisjunction). If this occurs in a primary sex cell that gives rise to four gametes and the nondisjunction affects a single homologous pair, two gametes will get an extra copy of the chromosome while the others lack the chromosome altogether. When these two types of abnormal gametes unite with normal gametes, what types of zygotes result? ____________[34]

Aneuploids can also arise through a simple translocation, when most of a chromosome attaches to a nonhomologous chromosome. Although the translocation reduces the chromosome number, most of the genetic material of the "missing" chromosome is still in the cell, joined to another chromosome. Translocation Down syndrome, described above, is an example of this type of aneuploidy.

Consequences of Aneuploidy Missing chromosomes are almost always fatal in diploid animals where double doses of many genes appear to be essential for normal development. The exception to this, occurring in a fair number of species, involves one of the sex chromosomes, as illustrated in those species where sex is determined by the XO method (see Chapter 10). In humans, no autosomal losses are tolerated, but the absence of a homolog for the X sex chromosome, designated as XO, results in a female afflicted with Turner syndrome. Such a female is sterile.

Extra copies of most chromosomes are often lethal in animals. Apparently a third dose of many genes drastically disrupts embryonic development. Some animal species may tolerate trisomy for some chromosomes; for example, extra copies of chromosomes 8, 13, 18, 21, X, and Y can occur in humans. It should be noted that with the exception of the X, these chromo-

[24] One 14, two 21. [25] Two 14, one 21. [26] Three 14, two 21. [27] Two 14, three 21. [28] Two 14, two 21. [29] Two 14, two 21.
[30] Types 1, 2, and 3. [31] Type 4. [32] Type 5. [33] Type 6. [34] Trisomic and monosomic zygotes are formed.

somes carry substantial blocks of genetically inactive material and this may explain why their trisomy does not cause death. The best known human trisomy involves chromosome 21 which causes Down syndrome. Other trisomies involving the sex chromosomes have been studied at some length and include XXY (sterile male, Klinefelter syndrome); XXX (phenotypically female) and XYY (phenotypically male).

Aneuploidy in plants carries much less serious consequences and may actually be of adaptive significance. For example, the thorn apple or jimsonweed, *Datura stramonium*, has a fertile trisomic for each of its 12 chromosome pairs; each has a different seed capsule and leaf morphology.

Euploidy

Euploidy results when the chromosomal complement is increased by one or more whole haploid sets. With euploids, the chromosome number is expressed in multiples of the haploid number. The general term **polyploidy** refers to situations where individuals carry more than two haploid sets. Over half of the species of higher plants are believed to be polyploids. The chromosome numbers of the species in many plant genera make up a series of multiples of the same basic number. For example, the genus *Triticum* which includes wheat and its relatives encompasses several groups of species with diploid numbers of 14, 28, and 42, all of which are multiples of 7. It is thought that the species with lower chromosome numbers gave rise to those with higher numbers through the duplication of entire chromosome sets. Polyploidy among animals is extremely rare.

There are two basic types of euploidy: (1) **autopolyploidy,** where the chromosome complement consists of identical sets, with the extra set or sets originating within the species, and (2) **allopolyploidy,** where the assemblage is comprised of different chromosomes sets, with the sets derived from different species.

Autopolyploidy Autopolyploids possess several copies of the same genome, with all the sets contributed from the same species. Autopolyploidy can arise in a number of different ways including the following.

1. Meiotic failure: The replicated chromosomes fail to separate during meiosis giving rise to diploid gametes. When united with a haploid gamete, a triploid ($3n$) zygote is produced which develops into a $3n$ individual. The union of two diploid gametes produces a tetraploid ($4n$) zygote.
2. Multiple fertilization: Fertilization of an egg by more than one sperm can result in a zygote with extra sets of chromosomes. For example, the union of a haploid egg with two haploid sperm will produce a $3n$ zygote.
3. Somatic doubling: Failure of chromosomal separation during mitosis can double the number of chromosomes in somatic cells, converting, for example, diploid cells to tetraploid. If these cells give rise to reproductive cells, as would be the case, for example, with the primary sex cells, diploid gametes would be formed through meiosis.
4. Artificial induction: The chemical colchicine blocks the formation of the spindle apparatus and thus disrupts chromosome separation. Somatic cells treated with this substance end up with twice the normal number of chromosomes. Similar results can be achieved through the use of thermal shock or chemicals that stimulate an extra round of chromosome replication. Artificial induction has been used to prepare polyploids in numerous plant species.

Gamete Formation and Reproduction in Autopolyploids Most triploids ($3n$) and other polyploids with odd numbers of chromosome sets ($5n$, $7n$, and so forth) usually exhibit greatly reduced fertility. This occurs because the synapsis of an uneven number of homologs creates some unusual configurations which lead to unbalanced gametes. The three homologs of any particular chromosome in a triploid, for example, may form a bivalent (composed of two synapsed homologs) and a univalent (composed of the remaining homolog). The replicated chromosomes making up the bivalent separate normally while the univalent goes to one of the two product nuclei. Alternatively, the three homologs may cluster to form a multivalent where synapsis is incomplete, with different portions of each replicated chromosome synapsed with different strands as shown in Figure 16-6.

Regardless of the synaptic configuration, the gametes or spores produced will have chromosomes missing or present in extra copies, and these serious imbalances often result in sterility. In tetraploids and other even-numbered polyploids, the reduction in fertility may be slight since each chromosome has a counterpart, and balanced gametes may be produced. If a self-fertilizing tetraploid plant, for example, produces diploid gametes, normal tetraploid individuals will result from their union.

Some species of autopolyploid plants are of interest to plant breeders since they generally have greater vigor and yield. Those plants exhibiting reduced fertility can be propagated through vegetative methods such as taking cuttings, grafting, and so on. In animals, autopolyploidy is restricted to a very limited number of species. This may be explained by the fact

FIGURE 16-6

Incomplete synapsis of three homologs.

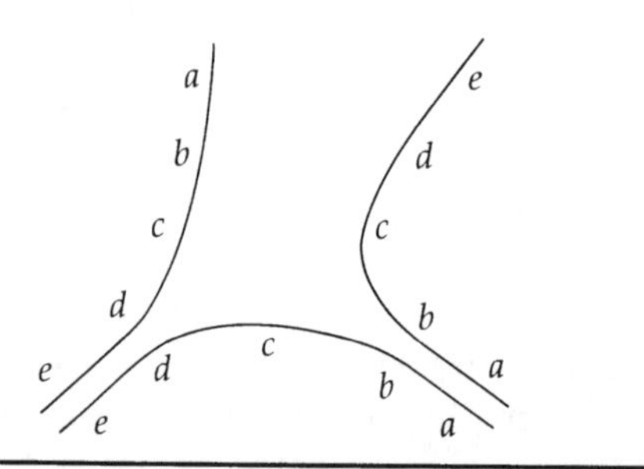

that sex determination in animals is chromosomally based and polyploidy disrupts this mechanism. This disruption does not occur in most plants since many are bisexual and lack differentiated sex chromosomes.

Allopolyploidy Allopolyploids may be formed when two species interbreed and contribute sets of chromosomes to a zygote. Normally, if a hybrid is formed, it is sterile since homologous pairing cannot be achieved during meiosis, and this leads to defective gametes. However, if the parents become tetraploids (through somatic doubling) *and* produce diploid gametes *and* these gametes unite, the $2n + 2n$ zygote formed may give rise to a fertile, self-perpetuating line which usually qualifies as a new species. The fertility exists since each chromosome now has a homologous counterpart, and standard synapsis and regular segregation can occur during meiosis. Allopolyploids may also arise through the union of unreduced, or $2n$, gametes produced by diploid members of two different species. In the $2n + 2n$ individual that results, each chromosome has a homolog, the gametes are balanced, and fertililty is the rule.

SUMMARY

This chapter examines alterations in chromosome structure and in chromosome number. Chromosome aberrations include deficiencies, duplications, inversions, and translocations and involve the breaking and reattachment of chromosomal segments. Both deficiencies and duplications can create serious developmental problems, while inversions and translocations may be less severe. Both inversions and translocations may alter gene expression through position effect as well as cause deficiencies and duplications in gametes, thereby producing semisterility. Inversions, in addition, appear to suppress crossing over, with inversion carriers tending to produce progeny that are nonrecombinant for the loci found within the inversion. Simple translocations involving the human chromosome 21 cause Down syndrome.

Aneuploids have individual chromosomes added to or removed from their chromosomal complement; such changes usually create a serious genetic imbalance which is often lethal. Euploids have their chromosome number increased by entire haploid sets. These sets may be copies of the same genome (a condition known as autopolyploidy, which usually arises through meiotic difficulties) or different genomes (a condition known as allopolyploidy, which arises when two species hybridize). The extra genetic material in odd-numbered polyploids can give rise to developmental problems and meiotic difficulties that lead to unbalanced gametes. Polyploids are usually sterile or semisterile. If allopolyploids arise through the hybridization of two tetraploid species, the genomes from each species are contributed in pairs. Because of this, meiosis is normal, the gametes are balanced, and a fertile hybrid is usually produced.

PROBLEM SET

16-1. Identify the chromosome aberration or aberrations that
- **a.** increase the genetic material on a particular chromosome.
- **b.** decrease the genetic material on a particular chromosome.
- **c.** change the position of genetic material within a chromosome without altering the amount of genetic material carried by the chromosome.
- **d.** shift genetic material from one chromosome to a nonhomologous chromosome.

16-2. Dogs have a diploid ($2n$) number of 76. Identify the number of chromosomes that would be expected in a dog that is
- **a.** a monosomic.
- **b.** a trisomic.
- **c.** an inversion carrier.
- **d.** a translocation carrier.
- **e.** a deficiency carrier.
- **f.** a nullisomic.
- **g.** a tetraploid.

16-3. Cri-du-chat (cat-cry) syndrome in humans, characterized by severe mental retardation and early death, is caused by a deficiency in the fifth chromosome. Make a sketch to show how a pair of fifth chromosomes would appear during synapsis when one carries this deficiency and its homolog is normal. Label the normal chromosome and the one carrying the deficiency.

16-4. Assume that the loci in a particular region of a chromosome have been assembled through standard crossing over into a particular, effective group in which the genes work well together. If a deficiency eliminates this region from one of the chromosomes of a homologous pair, will crossing over be able to disrupt this favorable assemblage of loci? Explain.

16-5. The two kinds of polypeptide chains (alpha and beta) that make up the human hemoglobin molecule are very similar in the number of amino acids they have (141 and 146, respectively) and carry identical amino acids at about 45% of their sites. A different gene codes for each polypeptide. Speculate on how the genes coding for these two polypeptides might have arisen in the evolutionary past.

16-6.
- **a.** How many chromosomes are found in the diploid nuclei of a human afflicted with a standard Down syndrome? with translocation Down syndrome?
- **b.** How many chromosomes are found in the diploid nuclei of each of the parents of an individual affected with translocation Down syndrome?
- **c.** What proportion of the zygotes produced by a chromosome 14/21 translocation carrier and a normal individual would be expected to fail to develop (because of problems with chromosomes 14 or 21)?
- **d.** What is the probability that a child produced by the couple described in 16-6c will have Down syndrome? will have the normal chromosome makeup, that is, will not be a translocation carrier? will be a translocation carrier?

16-7. Refer to Figure 16-5b which shows the products of a reciprocal translocation between two nonhomologous chromosomes. Make a sketch showing the configuration assumed by these chromosomes during synapsis in an individual heterozygous for this translocation.

16-8. Assume that nondisjunction of the X chromosome occurs during the first meiotic division of egg production in a human female and that the egg is

PROBLEM SET

fertilized by a normal sperm cell. Identify all of the sex-chromosome combinations that might occur in a zygote.

16-9. Assume that each of two loci in the fruit fly *Drosophila melanogaster* controls a different trait. The gene at each locus has two alleles, one of which is completely dominant to its recessive counterpart. Both loci are carried on the second chromosome, although nothing is known about their relative positions on that chromosome. A fly that was expected to be heterozygous at each of these loci (and therefore exhibit the dominant phenotype for each trait) expresses the recessive allele at each locus. A cytological examination of synapsed chromosomes from this fruit fly shows that one of the chromosomes making up the second pair exhibits a small loop.

a. Explain what may have happened to allow the expression of the recessive alleles at these two loci.

b. Why did the loop form?

c. What could be tentatively concluded about the location of these two loci on the chromosome?

d. What general conclusion could be tentatively made about the distance separating these two loci?

16-10. Identify two likely ways in which a human with a sex-chromosome makeup of XXXY could arise.

16-11. Most of the somatic cells in a human male, who develops from a normal XY zygote, carry XY chromosomes. However, some of his somatic cells are XO. How can this difference in chromosome makeup be explained?

16-12. In 1928, the Russian geneticist Karpechenko crossed a radish ($2n = 18$) with a cabbage ($2n = 18$) and the progeny were sterile. Through somatic doubling, some plants with a chromosome number of 36 were formed. Crossing them produced fertile offspring.

a. Speculate on the reason why the F_1 plants were sterile.

b. Why were the plants with a chromosome number of 36 most likely able to produce fertile offspring?

c. When the plants with a chromosome number of 36 were crossed with either of the original parents, sterile offspring were produced. Speculate on the reason why these offspring were sterile.

16-13. Speculate on why polyploidy is so rare among animals while it is very common among plants.

16-14. a. Is there really a suppression of crossing over within the inversion carried by inversion heterozygotes? Explain.

b. If there were no crossing over within the inversion, could recombinants for loci outside the inversion be detected? Explain.

c. Is there an apparent suppression of crossing over in chromosomes homozygous for an inversion? Explain.

16-15. A student researcher studying a group of laboratory animals detects both semisterility and alterations in the expression of several genes because of position effect. She is uncertain of the cause and seeks your assistance.

a. What would you suggest to explain her observations?

b. What additional things would you suggest she do to more precisely identify the cause?

16-16. Differentiate between autopolyploidy and allopolyploidy on the basis of the source of the extra set or sets of chromosomes necessary to produce these conditions.

PROBLEM SET

16-17. Through the use of thermal shock, a pair of plants is induced to produce polyploid progeny. The crossing of two of these progeny carrying odd-numbered sets of chromosomes produces no progeny while crosses between even-numbered polyploids yield some progeny. Explain this discrepancy.

16-18. **a.** Differentiate between paracentric and pericentric inversions.

b. Do heterozygotes for each type of inversion give rise to recombinant chromatids with abnormal numbers of centromeres? Explain.

c. Do heterozygotes for each type of inversion exhibit the apparent suppression of crossing over? Explain.

16-19. Assume that crossing over has occurred within the inversion carried by a sexually reproducing inversion heterozygote. Microscopic slides have been prepared showing primary sex cells (the cells that give rise to gametes) from this organism during anaphase of the first meiotic division. Could microscopic examination of these slides allow you to determine whether the inversion in question is paracentric or pericentric? Explain.

16-20. Unbalanced gametes are produced by sexually reproducing animals heterozygous for a reciprocal translocation. They are also produced by animals heterozygous for an inversion. These unbalanced gametes arise in different ways for each type of aberration. In general terms, describe the mechanism responsible for the production of the unbalanced gametes for each type of aberration.

CHAPTER

17

Quantitative Inheritance, Statistics, and Heritability

INTRODUCTION

Phenotypic differences for many of the traits considered in earlier chapters arise because of genotypic differences at a single locus. These phenotypic differences are often clear cut, with organisms falling into distinct phenotypic classes (for example, tall and short pea plants; A, B, AB, and O blood types in humans; and so on). Since the phenotypic variation shown by a population for such traits is **discontinuous** and since the number of different genotypes for the trait is limited, genetic analysis of single-locus traits is often relatively straightforward.

In contrast, many of the traits of practical interest to plant and animal breeders (for example, milk yield from dairy cows, weight of hogs, and egg production in chickens) are controlled by **multiple genes,** or **polygenes.** Such traits are designated as **quantitative,** and each of the several to many loci involved make a cumulative, or **additive,** contribution to the expression of the trait. In addition, the environment often exerts a major effect on the phenotypic expression of these traits. As a consequence, quantitative traits usually lack the clear-cut phenotypic expression shown by single-locus traits, and the phenotypes shown by members of a population grade into each other, showing **continuous** variation. Variation of this type is seen, for example, with the trait of height in humans which ranges between short and tall extremes. Quantitative traits are more difficult to study because of the large number of possible genotypes and the difficulty of associating specific genotypes with particular phenotypes.

This chapter looks at quantitative traits; some of the statistics used to summarize their expression; and one of their attributes, heritability, which assesses the extent to which the phenotypic variation shown by a trait can be attributed to genetic variation. The chapter begins by considering a series of examples where increasing numbers of loci interact to determine the phenotypic expression of a trait.

QUANTITATIVE TRAITS

An Overly Simple Example: A Trait Determined by Two Loci

Assume that kernel color in wheat is determined by two independently assorting loci, *A* and *B*, each of which has two alleles. Alleles *A* and *B* each add an equal quantity of red pigment to the phenotype while alleles *a* and *b* contribute no pigment. A plant of genotype *aabb* makes no red pigment and thus has white kernels. Plants with one or more pigment alleles will have red kernels, with the color intensity increasing as the number of these alleles increases.

What kernel color occurs in the F_1 when pure-breeding lines of white, with genotype *aabb*, and dark red, with genotype *AABB*, are crossed? ________[1] The outcome of interbreeding the F_1, symbolized as *AaBb* × *AaBb*, is shown in the following Punnett square (boxes within the square are numbered for reference purposes).

	AB	*Ab*	*aB*	*ab*
AB	*AABB* (1)	*AABb* (2)	*AaBB* (3)	*AaBb* (4)
Ab	*AABb* (5)	*AAbb* (6)	*AaBb* (7)	*Aabb* (8)
aB	*AaBB* (9)	*AaBb* (10)	*aaBB* (11)	*aaBb* (12)
ab	*AaBb* (13)	*Aabb* (14)	*aaBb* (15)	*aabb* (16)

From the number of pigment alleles carried by each of the 16 genotypes, determine the phenotypes shown by the F_2 generation. What proportion of the F_2 progeny show the white phenotype? ________[2] the darkest shade of red? ________[3] How many different phenotypes occur in the F_2? ________[4] What is the F_2 phenotypic ratio? ________[5]

Because of variation in environmental factors, such as exposure to sun or rainfall, some variation would be expected within each of the five phenotypic classes. For example, not all of the dark-red kernels would show the same exact shade of red. However, even with this variation, discontinuities remain between the classes and the progeny can be sorted into five distinct groups whose frequency distribution can be shown graphically in the frequency histogram in Figure 17-1. This histogram shows the greatest number of individuals at the midpoint of the phenotypic range and tapers on either side to low frequencies at the two phenotypic extremes.

FIGURE 17-1

Histogram showing the percentage of F_2 individuals with different kernel-color phenotypes when the trait is controlled by two gene pairs.

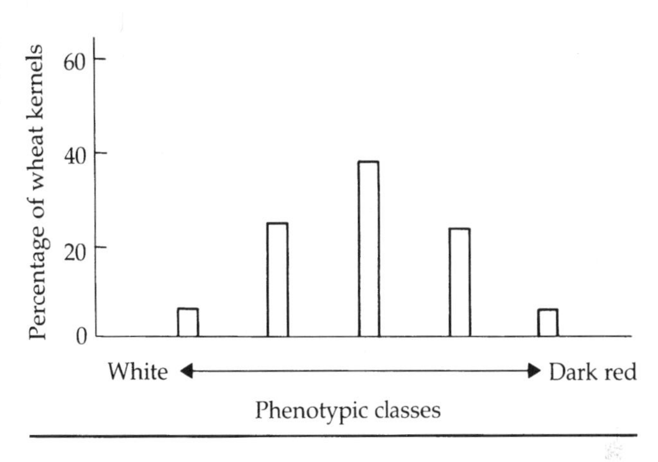

A Trait Determined by Three Loci

In another strain of wheat, kernel color is determined by three independently assorting loci, *A*, *B*, and *C*, each of which has two alleles. Alleles *A*, *B*, and *C* each add an equal quantity of red pigment to the phenotype while alleles *a*, *b*, and *c* contribute no pigment.

What would be the genotype for a pure-breeding line with white kernels? ________[6] for a pure-breeding line with dark red kernels? ________[7] What would be the genotype and phenotype of F_1 plants produced by crossing members of these two pure-breeding lines? ________[8] In the interbreeding of F_1 plants, represented as *AaBbCc* × *AaBbCc*, how many different phenotypic classes would be expected? ________[9] How many different shades of red would be represented among the F_2 progeny? ________[10] What proportion of the F_2 progeny would show the white phenotype? ________[11] What proportion of the F_2 progeny would show the darkest shade of red? ________[12] What is the F_2 phenotypic ratio? ________[13]

As with our earlier example, environmental factors would produce some variation within each of the seven phenotypic classes. Despite this, the progeny could most likely be sorted into distinct classes as shown in the frequency histogram in Figure 17-2.

Generalizing for Traits Determined by More Than Three Loci

Compare the frequency histogram in Figure 17-2 with that in Figure 17-1. How do the gaps or discontinuities between the phenotypic classes shown on these two

[1] Medium red (genotype *AaBb*). [2] 1/16. [3] 1/16. [4] Five phenotypes: dark red (four pigment alleles [box 1]); medium dark red (three pigment alleles [boxes 2, 3, 5, 9]); medium red (two pigment alleles [boxes 4, 6, 7, 10, 11, 13]); light red (one pigment allele [boxes 8, 12, 14, 15]); white: (no pigment alleles [box 16]). [5] 1:4:6:4:1. [6] *aabbcc*. [7] *AABBCC*. [8] *AaBbCc*; medium-red.
[9] Seven, determined by the number of pigment alleles which would range from 0 to 6. [10] Six. [11] 1/64 (genotype *aabbcc*).
[12] 1/64 (genotype *AABBCC*). [13] 1:6:15:20:15:6:1.

FIGURE 17-2

Histogram showing the percentage of F_2 individuals with different kernel-color phenotypes when the trait is controlled by three genes.

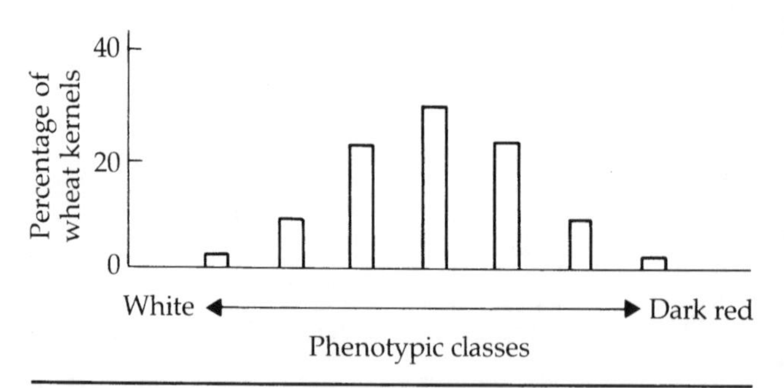

histograms compare? ______________________________[14] Predict what would happen to the discontinuities as the number of loci involved in additively determining a trait increases beyond three. ______________________________[15] As the number of loci determining a trait increases, will it always be possible to distinguish each of the increasing number of phenotypic classes? __________[16]

A phenotypic distribution arising from the crossing of individuals heterozygous for all the several loci determining a polygenic trait is shown in Figure 17-3. The number of phenotypes is large enough so that the histogram shows no discontinuity. Such histograms are typically seen when the trait is determined by approximately five or more loci. Connecting the tops of the frequency bars of the histogram traces out a symmetrical **normal,** or **bell-shaped,** curve which represents the so-called **normal frequency distribution.**

Generalized Expressions Useful with Quantitative Traits

Estimating the Number of Loci Involved It may be possible to estimate the number of loci involved in determining a quantitative trait. The clue to this, as the following discussion will show, lies with the fraction of F_2 individuals that show one or the other phenotypic extreme.

This time, assume that kernel color in yet another strain of wheat is determined by a single pair of alleles, where one allele contributes to pigment production and its counterpart adds nothing to pigment production. Pure-breeding white and dark red plants are crossed to produce the F_1 which is self-fertilized to form the F_2. What proportion of the F_2 progeny shows a phenotype as extreme as either of the original parents? __________[17] What proportion of the F_2 shows a phenotype as extreme as either of the original parents of the double-locus trait in the example at the beginning of the chapter? __________[18] What proportion of the F_2 shows a phenotype as extreme as either of the original parents of the triple-locus trait in the preceding example? __________[19]

This regular decrease from 1/4 to 1/16 to 1/64 leads us to the generalized expression for the fraction of individuals in the F_2 showing one of the phenotypic extremes exhibited by the original pure-breeding parents. The expression is $1/(4^n)$, where n is the number of loci at which the F_1 parents are heterozygous. This expression can be used to estimate the number of allelic pairs involved in determining a trait. For example, if 1/1024 of a particular F_2 generation shows a phenotype as extreme as that exhibited by one of the original parents, how many heterozygous loci for this trait were carried by the F_1 individuals? __________[20]

This generalized expression can be used only when the phenotypic classes are sufficiently distinct to allow counting of the members of an extreme phenotypic class. Also, as the number of additive loci determining a trait grows, it is necessary to have an increasingly larger sample of F_2 progeny in order to get a reliable estimate of the number of individuals showing the phenotypic extremes.

FIGURE 17-3

Frequency histogram for a phenotype showing a continuous distribution. Connecting the tops of the frequency bars fits a normal curve to the distribution.

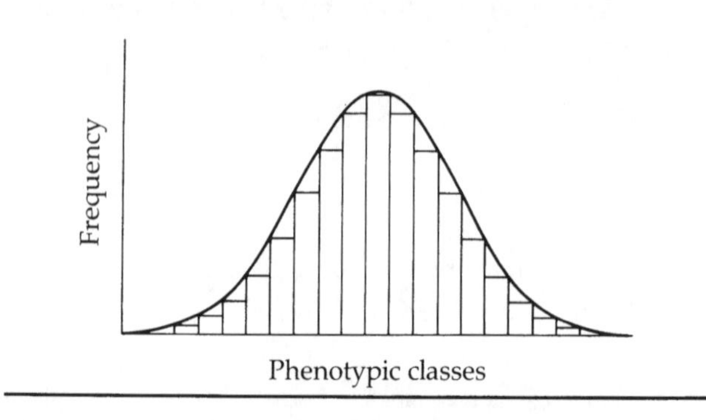

[14] Smaller for the trait determined by three loci. [15] The discontinuities get progressively smaller and the distribution becomes smoother. [16] No. It will be increasingly more difficult and then impossible. [17] 1/4 white, 1/4 dark red. [18] 1/16 white, 1/16 dark red. [19] 1/64 white, 1/64 dark red. [20] Since $1024 = 4^5$, there are five pairs of alleles.

Determining the Number of F_2 Phenotypes and Genotypes The number of phenotypes for a quantitative trait expected in an F_2 generation is given by the expression $2n + 1$, where n is the number of gene pairs determining the trait. For example, with three pairs of genes, $2(3) + 1 = 7$ phenotypes would be expected.

The number of genotypes in the F_2 is given by the expression 3^n, where n is the number of gene pairs determining the trait. For example, three pairs of genes would give $3^3 = 27$ genotypic classes. Determine the number of F_2 phenotypes and genotypes when the F_1 parents are heterozygous at four loci. ________[21]

These expressions can also be used to identify the number of gene pairs involved in determining a polygenic trait if the number of F_2 genotypic or phenotypic classes is known. For example, assume that a polygenic trait shows 11 phenotypic classes in an F_2 generation. Setting the expression $2n + 1$ equal to 11 and solving for n gives the number of gene pairs as follows.

$$2n + 1 = 11$$
$$2n = 11 - 1 = 10$$
$$n = 10/2 = 5 \text{ gene pairs}$$

Determining F_2 Phenotypic Ratios The phenotypic ratio for an F_2 generation is given by the coefficients associated with the terms in an expansion of the binomial $(a + b)^{2n}$, where n is the number of gene pairs determining the trait. As we saw in an earlier example, the phenotypic ratio 1:4:6:4:1 arises with two pairs of genes. These numbers are the coefficients for the five expressions obtained by expanding the binomial $(a + b)^4$ which gives $\mathbf{1}a^4 + \mathbf{4}a^3b + \mathbf{6}a^2b^2 + \mathbf{4}ab^3 + \mathbf{1}b^4$.

STATISTICS

For traits showing discontinuous variation, the clear-cut distinctions between the phenotypic classes make it easy to describe how the trait appears in a particular population. For example, in a population of pea plants, half might be tall and the other half short. For traits showing continuous variation, the lack of clear-cut distinctions between the various phenotypic classes makes it more difficult to characterize a population. This difficulty is overcome through the use of some basic statistics.

Sampling

The phenotypes shown within a population for many quantitative traits produce a distribution in the form of a normal, or bell-shaped, curve. In most instances the measurements for a particular quantitative trait are obtained from a limited number of individuals or a sample of the population because it is often impractical or impossible to gather information from all of the members of the population. For example, the outcome of one or a few crosses of a particular type provides a sample of the results that could be obtained if all possible crosses of that type had been carried out within a population. If the sample is random and representative of the entire population, its distribution will approximate that of the entire population. Often geneticists wish to compare the expression of a trait in two samples to determine if they are the same or they may wish to judge how reliably a sample reflects the population from which it was drawn before making inferences about the population.

Certain statistical measures which summarize key features of a distribution are routinely used to describe samples and to compare sample and population distributions. The more important of these are the mean, variance, standard deviation, and standard error of the mean. The values obtained from a sample are designated as **statistics** and symbolized by English letters while the values for a population, often estimated from sample values, are known as **parameters** and symbolized by Greek letters.

The Mean

The mean identifies the center or central tendency of the distribution. The **mean,** symbolized as $\bar{x}$ and read as "x bar," is the arithmetic average of all the values (each represented by x) making up the distribution and is calculated according to the expression

$$\bar{x} = \frac{\sum x}{n}$$

where n represents the number of values and Σ designates "taking the sum of" whatever value follows it, so that Σx represents the sum of all the values of x. This calculation is often made easier by setting up a **frequency distribution,** where the number of individuals showing each of the values in the distribution is determined. The mean is obtained by multiplying the value for each class, v, by its frequency, f, and these (v)(f) values are added together for all of the classes and divided by the total number of individuals.

A frequency distribution giving the height in inches for 100 college-age women is shown in Table

[21] Phenotypes: $2(4) + 1 = 9$; genotypes: $3^4 = 81$.

TABLE 17-1

Frequency distribution of height in inches for 100 women.

Class value, v

(height):	50	51	52	53	54	55	56	57	58	59	60	61	62	63	64	65	66
Frequency, f:	1	0	3	1	4	6	9	16	19	15	11	7	5	1	1	0	1
(v)(f) =	50	0	156	53	216	330	504	912	1102	885	660	427	310	63	64	0	66

17-1. Determine the mean height for this sample by completing the following calculation.

$$\text{Mean} = \frac{\text{Sum of (v)(f) values}}{\text{Total number of values}} = \underline{\qquad\qquad}$$ [22]

Measures of Variability

The graph in Figure 17-4 shows two normal distributions with the same mean. How do these distributions differ? ________________________________

________[23] The different curve widths reflect different ranges of values on each side of the mean or different degrees of **variability.** As the two curves in Figure 17-4 indicate, the mean by itself is inadequate to describe a distribution and needs to be supplemented with a measure of variability showing the extent to which the values making up the distribution differ from the mean. Two common indices of variability are the variance and its square root, the standard deviation.

Variance The **variance** of a sample, symbolized as V or s^2, is the average squared difference between each value in the distribution and the distribution mean. It is obtained by (1) determining the difference between each measurement (x) and the mean ($\bar{x}$), (2) squaring each difference, (3) summing these differences and dividing by the degrees of freedom, which is 1 less than the total sample size and is symbolized as $n - 1$. This procedure is summarized in the expression

$$V = \frac{\sum (x - \bar{x})^2}{n - 1}.$$

Note that variance is expressed in square units because of the way in which it is calculated.

The calculation of the variance for the sample of heights of college-age women is set up in Table 17-2. Each of the first eleven class values has been subtracted from the mean (57.98 inches) and the difference $(x - \bar{x})$ has been squared and then weighted by multiplying by the class frequency. Carry out these calculations for the remaining six class values and determine the variance in the space available in Table 17-2.

The variance directly reflects the spread of the distribution; the larger its value, the wider the distribution. All of the various factors that contribute to the phenotypic variation in a population, including genetic differences and environmental factors, add together to make the population variance. Once the total population variance is known, an estimate of the variability contributed by a specific factor can be subtracted to indicate the variability due to other causes.

Standard Deviation The **standard deviation** for a sample, symbolized as s or SD, is the square root of the variance: $SD = \sqrt{V}$. The standard deviation for the distribution we have been considering is $SD = \sqrt{7.01} = 2.65$ inches. Like the variance, the standard deviation directly reflects the spread of a distribution: the larger its value, the greater the variability. For a normal or bell-shaped distribution, 68.3% of the values in the distribution fall within one standard deviation above or below the mean (often written as mean ± 1 SD), 95.5% lie within two standard deviations, and 99.7% fall within three standard deviations of the mean. This relationship is summarized on the bell-shaped distribution shown in Figure 17-5.

If we infer that the mean height (57.98 inches) and standard deviation (2.65 inches) for our sample of col-

FIGURE 17-4

Two normal distributions with the same mean.

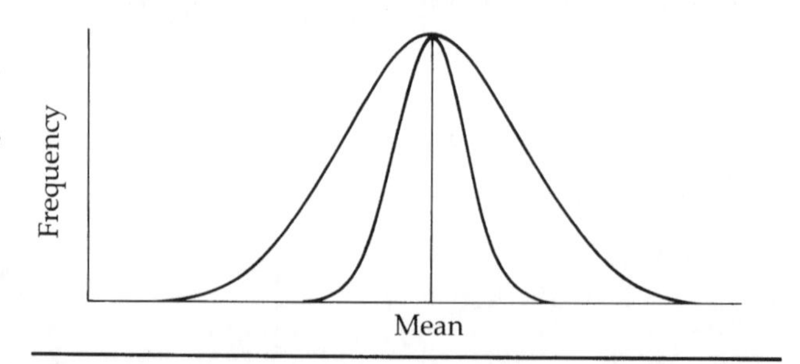

[22] Mean = 5798/100 = 57.98 inches. [23] They have very different widths.

TABLE 17-2

Calculation of the variance for the frequency distribution in Table 17-1.

Class value	Frequency (f)	(x − x̄)	(x − x̄)²	(f)(x − x̄)²
50	1	−7.98	63.68	63.68
51	0	−6.98	48.72	0
52	3	−5.98	35.76	107.28
53	1	−4.98	24.80	24.80
54	4	−3.98	15.84	63.36
55	6	−2.98	8.88	53.28
56	9	−1.98	3.92	35.28
57	16	−0.98	0.96	15.37
58	19	0.02	0.0004	0.01
59	15	1.02	1.04	15.61
60	11	2.02	4.08	44.88
61	7	______[24]	______[25]	______[26]
62	5	______[27]	______[28]	______[29]
63	1	______[30]	______[31]	______[32]
64	1	______[33]	______[34]	______[35]
65	0	______[36]	______[37]	______[38]
66	1	______[39]	______[40]	______[41]
			Total:	______[42]
			Variance =	______[43]

lege-age women reflect the true average height and standard deviation for all college-age women, within what range would the height of 68.3% of the population of college-age women be expected to fall? ______[44] Within what range would 95.5% of the population fall? ______[45] Within what range would 99.7% of the population fall? ______[46]

Standard Error of the Mean Often it is useful to know how well a particular sample reflects the entire population or, put in terms of our example, how reliably the mean height for a sample of 100 college-age women reflects the mean height of the population of all college-age women. Another measure of variability, the **standard error,** SE, is useful for evaluating this and is calculated from the standard deviation of a sample according to the following equation:

$$SE = \frac{SD}{\sqrt{n}}$$

where SD is the standard deviation and n is the sample size. Determine the standard error for our sample of heights. ______[47]

FIGURE 17-5

Normal distribution showing the percentage of distribution included between ±1SD, ±2SD, and ±3SD.

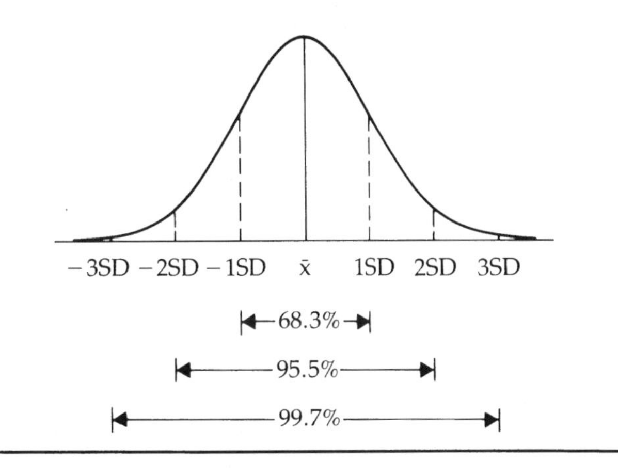

[24] 3.02. [25] 9.12. [26] 63.84. [27] 4.02. [28] 16.16. [29] 80.80. [30] 5.02. [31] 25.20. [32] 25.20 [33] 6.02. [34] 36.24. [35] 36.24. [36] 7.02. [37] 49.28. [38] 0. [39] 8.02. [40] 64.32. [41] 64.32. [42] 693.95. [43] 7.01 sq. inches. [44] 57.98 ± 2.65 inches, or between 55.33 and 60.63 inches. [45] 57.98 ± (2)(2.65) = 57.98 ± 5.30 inches, or between 52.68 and 63.28 inches. [46] 57.98 ± (3)(2.65) = 57.98 ± 7.95 inches, or between 50.03 and 65.93 inches. [47] SE = $2.65/\sqrt{100}$ = 2.65/10 = 0.265 inches.

The standard error represents the standard deviation of the large set of means that would be obtained by measuring the height of a great many samples of college-age women. These sample means would vary from each other and collectively would trace out a bell-shaped curve. The standard error gives an idea of how close the mean of our single sample lies to the true mean for this distribution of sample means. Thus, the smaller the standard error for a sample, the more reliable it is in reflecting the larger population from which it was taken. In other words, the standard error indicates the reliability of the sample mean as an index of the true mean height of the population of all college-age women. Often a sample mean is given with its standard error and is written as "mean ± SE."

Refer to the formula for the calculation of the standard error as you answer the following questions. What effect will an increase in sample size have on the size of the standard error? ______________________

______________________[48] What effect does an increase in sample size have on estimating the true mean? ______________________

________[49] What does this imply regarding the sample size necessary to reliably estimate population parameters? ______________________

________[50]

The standard error of a sample mean allows us to do two things: (1) set a range of values, or a confidence interval, within which the true mean would be expected to lie and (2) predict the probability of the true mean falling within this range. Since the standard error represents a standard deviation for the normal distribution of all sample means, we could predict that the true mean is expected to fall within one standard error of the sample mean 68.3% of the time, within two standard errors 95.5% of the time, and within three standard errors 99.7% of the time.

For our example, the SE of 0.265 inches for the sample mean of 57.98 inches tells us that the true mean for the entire population of college-age women has a 68.3% chance of falling in the range of 57.98 ± 0.265 inches, that is, between 57.72 and 58.25 inches.

HERITABILITY

For many quantitative traits, environmental as well as genetic factors produce the phenotypic variability seen in a population. This can be expressed as $V_p = V_g + V_e$, where V_p, V_g, and V_e, represent the phenotypic, genetic, and environmental variance, respectively. Genetic variance contributes to phenotypic variance in three key ways. Some genes, as discussed earlier in this chapter, cause additive effects where substituting one allele for another in one or more gene pairs results in a different phenotype for a trait. Other genes cause dominant effects where one copy of an allele (the dominant one) at a locus results in the appearance of a different phenotype for a trait. Yet other genes cause epistatic effects where the phenotype produced by one gene pair is controlled by a gene pair at a different locus.

Heritability, H, expresses the proportion of the total phenotypic variability due to genetic factors. The heritability for certain traits is of considerable interest to plant and animal breeders who seek to improve the quality of their stock by selecting the individuals showing the strongest expression of a desirable trait to serve as the parents of the next generation. This selection is most effective with traits showing a high level of heritability.

Broad and Narrow Heritability

If heritability is estimated from the total genotypic variance, V_g, which takes all categories of gene action (that is, additive as well as dominant and epistatic effects) into account, it is known as **broad heritability** (H_B). It is the ratio of the total genetic variance to the total phenotypic variance, expressed symbolically as $H_B = V_g/V_p$, and measures the extent to which phenotypic variation in a population arises because of all genetic differences.

If heritability is estimated solely on the variation caused by additive genes, V_a, it is known as **narrow heritability** (H_N). Narrow heritability is the ratio of additive genetic variance to the total phenotypic variance and is expressed symbolically as $H_N = V_a/V_p$. Since H_N is based on the variance contributed by additive genes, it is far more useful to plant and animal breeders than is a measure of broad heritability since the traits with which breeders are concerned are generally controlled by additive genes.

Estimating Narrow Heritability

There are two ways of estimating narrow heritability. One approach is to measure the phenotypic change that arises in the progeny following human selection of the parents. This is calculated in the following way.

[48] An increase in sample size would cause the standard deviation to be divided by a larger number which will decrease the standard error. [49] Increased reliability. [50] Larger samples are more reliable.

1. The mean phenotype, that is, the **average yield,** Y, for a quantifiable trait is determined for the generation from which the parents are selected.
2. After reproduction, the average yield is determined for the offspring, Y_o, and the difference between the average yields of the parental and progeny generations, designated as the **gain,** is determined.
3. The gain is divided by the difference between the average yield of the selected parents, Y_p, and the average yield in the parental generation, Y, which gives an index of the amount of selection practiced.

This procedure is summarized in the following equation.

$$\text{Heritability} = \frac{\text{Gain because of selection}}{\text{Degree of selection}} = \frac{Y_o - Y}{Y_p - Y}$$

Heritability assessed in this way is designated as **realized heritability** since it is determined after the breeding has been carried out. Traits with realized heritability values greater than 0.5 are generally considered to have high heritability, those with values between 0.2 and 0.5 have medium heritability, and those with values less than 0.2 have low heritability.

As an example, consider wool length in sheep. Wool length is a quantitative trait and one that is of key interest to breeders since longer strands of wool bring a higher market price. The average length of a strand of wool in a particular generation is 1.1 inches. The sheep selected from this generation to be the parents of the next generation have an average wool length of 1.6 inches. Their offspring show a wool length of 1.4 inches. Determine the realized heritability for this trait. ________[51]

A second approach to estimating heritability is to compare the phenotypic variability of populations with similar and diverse genotypes. Grown under the same range of environmental conditions, individuals with similar genotypes (such as inbred strains or close relatives) would show less variability for traits with high heritability than individuals with less similar genotypes. This approach is discussed next, in connection with human-twin studies.

Heritability in Humans: Twin Studies

Heritability for human quantitative traits can be estimated through a comparison of pairs of twins. **Monozygotic,** or **identical, twins** arise from the same zygote through the separation of embryonic cells after the start of cleavage and are genetically identical; their phenotypic variability can be attributed to environmental variation. The degree to which traits are environmentally influenced can be assessed through comparison of pairs of identical twins raised together with pairs that are reared apart. **Dizygotic,** or **fraternal, twins,** in contrast, arise from two different zygotes and show no more and no less genetic similarity than would be expected for any two siblings. Within pairs of dizygotic twins, both genetic and environmental factors contribute to the phenotypic differences. Comparisons with regard to weight and height, for example, indicate that identical twins reared together show marked similarities; identical twins reared apart have these similarities slightly reduced; and fraternal twins show marked differences. This outcome for weight and height supports a strong genetic basis for these traits.

Another Measure of Heritability: Concordance of Twin Traits

A trait exhibited by both twins of a pair or by neither is said to be **concordant** while a **discordant** trait is expressed by one twin but not by the other. **Concordance** is the percentage of twin pairs who are concordant relative to the total number of twin pairs surveyed. For example, a study showed that both members of 24 of 50 pairs of identical twins were affected with diabetes mellitus for a concordance of 24/50 = 0.48, or 48%. A survey of fraternal twins showed that 23 of 230 pairs were concordant for diabetes mellitus giving a concordance of 23/230 = 0.10, or 10%. Comparing the concordance of identical twins with that of fraternal twins provides a measure of heritability for the trait. If identical twins show a concordance significantly greater than that of fraternal twins, the trait has a strong genetic basis. For example, the preceding data support the conclusion that there is a genetic basis for the variation in diabetes mellitus.

SUMMARY

Quantitative traits are controlled by multiple genes, or polygenes, with each of the several to many loci making an additive and usually equal contribution to the expression of the trait. These traits can be measured and exhibit continuous variation within a population.

[51] The gain because of selection is 1.4 − 1.1 = 0.3, and the degree of selection is 1.6 − 1.1 = 0.5. Realized heritability is 0.3/0.5 = 0.6.

Measurements are generally carried out on a sample of individuals from the population and are statistically summarized to provide the average value (mean) and a measure of the variability (variance and standard deviation). These statistics provide a way of comparing trait expression in samples and populations. The standard error of the mean makes it possible to assess how well a sample mean reflects the mean of the larger population from which the sample was drawn.

The phenotypic variability associated with a quantitative trait is produced by both genetic and environmental factors. Heritability is the proportion of that variability that is due to genetic factors. Traits with high heritability feature prominently in the selection carried out by plant and animal breeders. Heritability can be estimated by measuring the change produced by selective breeding (realized heritability) or by comparing the phenotypic variability in groups with similar and diverse genotypes, as is done for human traits through comparative studies involving twins.

PROBLEM SET

17-1. Two pure-breeding lines of corn with short and long seed length, respectively, are crossed to produce an F_1 which is allowed to self-fertilize to produce an F_2. The distribution of seed lengths for the F_2 is shown in part (a) of the following figure. This crossing procedure is repeated with two other pure-breeding lines of corn exhibiting short and tall plant height, respectively, and their F_2 produces the distribution of heights as shown in part (b) of the following figure.

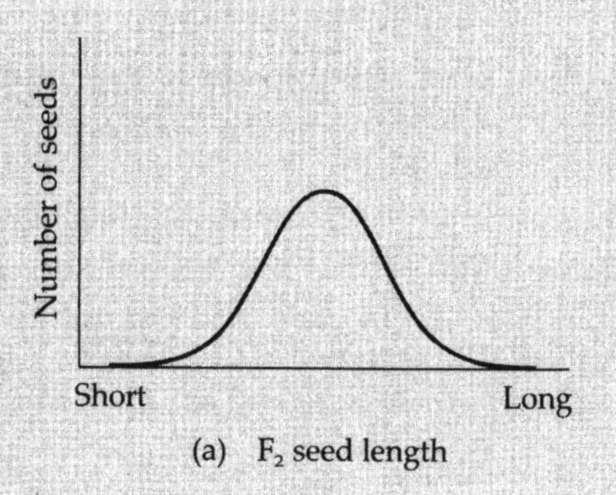

(a) F_2 seed length

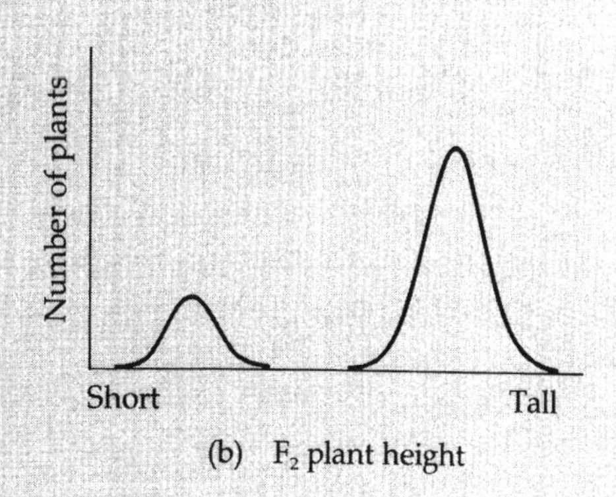

(b) F_2 plant height

a. Which of these traits is likely to be quantitative? Explain the basis for your decisions.
b. What can be concluded about the genetic basis for each of these traits? Explain your conclusions.

17-2. Refer to the information given in problem 17-1. With regard to the trait of plant height, the members within each F_2 phenotypic class have the same genotype. In light of this, why is it that the plants within each group show a range of heights rather than exactly the same height?

17-3. Sketch the shape of the frequency distribution of phenotypes expected in the F_1 for each of the corn plant traits described in problem 17-1.

17-4. Assume that skin pigmentation in humans is a quantitative trait controlled by two loci, *A* and *B*. Alleles *A* and *B* each make an equal and additive contribution to pigment production while alleles *a* and *b* make no contribution to pigment production.
a. How many skin-color phenotypes are possible?
b. Identify the genotype or genotypes possible for a black man, for a white woman, and for their children.
c. Assume that a person with intermediate pigmentation is heterozygous at both pigment loci. What genotypes and phenotypes would be expected among the progeny if this person marries a black individual?
d. Could two parents of intermediate color produce black children among their progeny? Explain.

17-5. The F_2 phenotypic ratio for a particular quantitative trait is 1:6:15:20:15:6:1. How many pairs of alleles are involved in determining this trait?

17-6. Assume that varieties of wheat are available in which kernel color is known to be determined by three loci, designated *A*, *B*, and *C*. Assume that each locus has two alleles. Alleles *A*, *B*, and *C* each add an equal quantity of red pigment to the phenotype while alleles *a*, *b*, and *c* contribute no pigment. Pure-breeding strains with white and dark-red kernels are crossed to produce an F_1 which is in turn selfed to form an F_2. Pairs of F_2 plants are then crossed to produce the F_3 phenotypes in the following list. For each

PROBLEM SET

F_3 phenotype, identify, to the extent possible, the genetic makeup of the F_2 parents.

a. All white.

b. 63/64 showing varying shades of red, and 1/64 white.

c. 15/16 showing varying shades of red, and 1/16 white.

17-7. Assume that tail length in white rats is a quantitative trait. In one pure-breeding line the tail averages 7 inches and in another it averages 4 inches. Progeny arising from the interbreeding of these two lines have tails which average $5\frac{1}{2}$ inches. When members of the F_1 interbreed, the 496 offspring have tails which also average $5\frac{1}{2}$ inches in length, but which vary much more than those of their parents, with a range between 4 and 7 inches. Two members of the F_2 have 7-inch tails.

a. Based on these data, what can be tentatively concluded about the number of pairs of genes involved in determining tail length?

b. Why would the conclusion be tentative?

c. How many members of the F_2 would be expected to have 4-inch tails?

17-8. Assuming an equal contribution from each effective allele, how much does each allele contribute to the trait described in problem 17-7?

17-9. The length of the corolla (a structure within the flower) of the tobacco plant, *Nicotiana*, is believed to be a quantitative trait. A cross is carried out between two plant varieties each of which expresses a different phenotypic extreme for this trait and is pure breeding. The corollas of the F_1 show a range of intermediate lengths while those of the F_2 show somewhat greater variation in lengths. Although over 1000 F_2 plants are examined, none has a corolla as long or as short as those found in the original parents. Based on these results, what, in general, can be said regarding the number of loci involved in determining this trait?

17-10. Assume that six allelic pairs make equal and additive contributions to the trait described in problem 17-9. What is the number of phenotypic and genotypic classes expected in the F_2?

17-11. The following figure shows the frequency distribution curves obtained when three different populations are examined for the same trait. Compare these and indicate, by number, the distributions with the

a. same variance.

b. same mean.

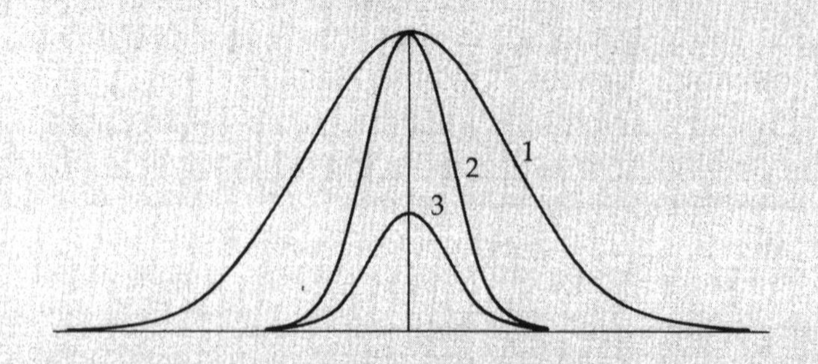

17-12. A sample of 21 adults of a particular species of fish are obtained from a freshwater pond and their lengths, rounded to the nearest centimeter, are as follows: 23, 18, 25, 26, 23, 21, 25, 23, 20, 27, 28, 24, 23, 22, 22, 21, 24, 22, 22, 23, and 24.

PROBLEM SET

a. Prepare a frequency distribution for these data.
b. Determine the mean.
c. Determine the variance.
d. Determine the standard deviation.

17-13. Distinguish between a sample's standard deviation and its standard error.

17-14. The distribution of ear lengths for a sample of 60 adult rabbits approximates a normal curve. The mean ear length is 5.27 centimeters with a variance of 1.21 centimeters.
a. Determine the standard deviation for this sample.
b. Within what range do 68% of the sample values fall?
c. Within what range do 99% of the sample values fall?
d. Determine the standard error for this sample.

17-15. Daily egg production is monitored over a 24-hour period for a sample of 100 female fruit flies who are of the same age. The mean egg production is 37.6 eggs per fly and the standard error of this mean is 0.7 eggs.
a. What effect would a larger sample have on the standard error?
b. What is the 95.5% confidence interval for this mean?
c. How often will a researcher be correct if she assumes that the true population mean lies within the 95.5% confidence interval?
d. Determine the standard deviation for this sample.

17-16. Distinguish between
a. monozygotic and dizygotic twins.
b. concordance and discordance.

17-17. **a.** Distinguish between broad heritability and narrow heritability.
b. Which of these two estimates of heritability is of greatest interest to plant and animal breeders? Explain why.

17-18. Egg weight in chickens is a quantitative trait. The average egg weight in a particular generation is 25.7 grams. The hens selected from this generation by the chicken breeder to give rise to the next generation produce eggs with an average weight of 29.1 grams, and their offspring produce eggs with an average weight of 28.0 grams.
a. Determine the realized heritability of this trait.
b. Is this trait likely to have a significant genetic basis? Explain.

17-19. A study indicates that litter size in swine has a realized heritability of 0.11. A generation of swine shows an average litter size of 9.2, and the individuals selected from this generation to be the parents of the next generation have an average litter size of 11.1. Predict the average litter size that will be expected in the next generation.

17-20. Statistics published by the U.S. Public Health Service based on a study of mental retardation in twins shows concordance levels of 67% and 0% for identical and fraternal twins, respectively.
a. If 45 pairs of identical twins were examined in this study, how many pairs showed both members exhibiting mental retardation?
b. Based on this study, what conclusion can be drawn about the heritability of mental retardation?

17-21. The number of tomatoes produced per plant was determined for the F_2 generation of a cross between two varieties of tomato plants. This tomato production exhibited a normal distribution with a mean of 27 and a variance of 9. What percentage of the F_2 plants would be expected to produce
a. between 24 and 30 tomatoes?
b. more than 30 tomatoes?

PROBLEM SET

c. between 27 and 33 tomatoes?
d. fewer than 21 tomatoes?

17-22. A study of development time in a sample of a pure-breeding variety of self-fertilizing pea plants indicates that the time from planting to maturity follows a normal distribution with a mean of 77 days and a standard deviation of 2.39 days.

a. How much of the phenotypic variance shown for this trait is due to genetics? Explain.
b. What is the environmental variance for this trait?
c. What is the heritability for this trait?
d. Assume that a plant breeder selects plants that mature in 69 days to be the parents of the next generation. Estimate the time required for the next generation to mature.

17-23. Pea plants of the variety described in problem 17-22 are crossed with pea plants from another pure-breeding line. The time from planting to maturity of the F_1 and F_2 progeny is determined. The variances for this trait shown by the F_1 and F_2 are 4.85 and 9.67 days, respectively.

a. What are the environmental and genotypic variances for this trait?
b. What is the broad heritability for this trait?

CHAPTER

18

Nucleic Acid Structure

INTRODUCTION

Deoxyribonucleic acid (DNA) and ribonucleic acid (RNA) are polymeric molecules that contain and transmit genetic information. To understand the structure, and, in turn, the function of these molecules, we first need to consider their building blocks, then the way in which these building blocks are assembled into strands, and finally the manner in which these strands make up the functional molecules.

NUCLEIC ACID BUILDING BLOCKS

DNA and RNA are made up of **nucleotide** building blocks known as **deoxyribonucleotides** and **ribonucleotides,** respectively. Each nucleotide consists of three kinds of covalently bonded subunits: (1) a five-carbon, or pentose, sugar, (2) a nitrogen-containing base (either a purine or pyrimidine), and (3) one, two, or three phosphate groups.

Five-Carbon Sugars: Ribose and Deoxyribose

Ribonucleotides contain **ribose sugar** while deoxyribonucleotides have **deoxyribose sugar.** The structural formulas for these two sugars are shown in Figure 18-1. Note that numbers with prime symbols (1′ through 5′) are assigned to specific carbons to allow us to distinguish between them. Compare the structural formulas of the two sugars; what difference do you notice?

____ [1]

The missing oxygen in the deoxyribose sugar explains the prefix "deoxy-."

Nitrogenous Bases: Purines and Pyrmidines

The bases that are present in nucleotides are grouped into two classes: **purines** and **pyrimidines.** The struc-

FIGURE 18-1

Ribose and deoxyribose sugars.

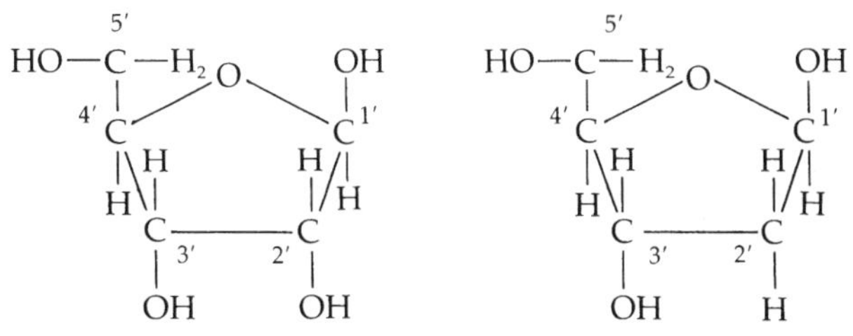

Ribose sugar: 5′ HO—C—H_2, O, 4′ C, 1′ C (OH, H), 3′ C (H, OH), 2′ C (H, OH)

Deoxyribose sugar: 5′ HO—C—H_2, O, 4′ C, 1′ C (OH, H), 3′ C (H, OH), 2′ C (H, H)

Ribose sugar

Deoxyribose sugar

tural formulas for two purines, **adenine** (A) and **guanine** (G), and for three pyrimidines, **cytosine** (C), **thymine** (T), and **uracil** (U), are shown in Figure 18-2. Note that numbers without prime symbols are assigned to each nitrogen and carbon atom in the ring structures of purines and pyrimidines. Compare the structural formulas of the two purines. In general terms, how are they similar? ______________________[2] How do they differ? ______________________[3]

Compare the structural formulas of the pyrimidines. In general terms, how are they similar? ______________________[4] How do they differ? ______________________[5] The purines adenine and guanine and the pyrimidine cytosine occur in both DNA and RNA. The pyrimidine thymine is found only in DNA while the pyrimidine uracil usually occurs only in RNA.

Phosphate Group

The phosphate group is the same in DNA and RNA and is shown in Figure 18-3. A single nucleotide may contain one, two, or three phosphate groups.

FIGURE 18-3

Phosphate group.

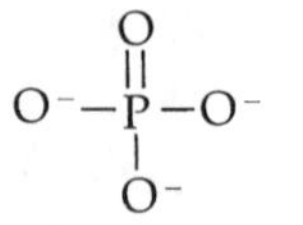

Nucleotide Structure

A nitrogenous base bonded to the 1′ carbon of a sugar forms a molecule called a **nucleoside;** the bonding of one or more phosphate groups, most commonly to the 5′ carbon of the sugar, converts the nucleoside into a nucleotide. Both nucleosides and nucleotides are named according to the base they contain. Single nucleotides that exist free in the cell are usually triphosphorylated (that is, with three phosphate groups joined to the sugar) and it is in this form that nucleotides are supplied for use in nucleic acid synthesis. Figure 18-4 shows a triphosphorylated molecule of deoxyribonucleotide containing thymine (thymidine triphosphate, TTP).

FIGURE 18-2

Structural formulas for purines and pyrimidines.

Purines:

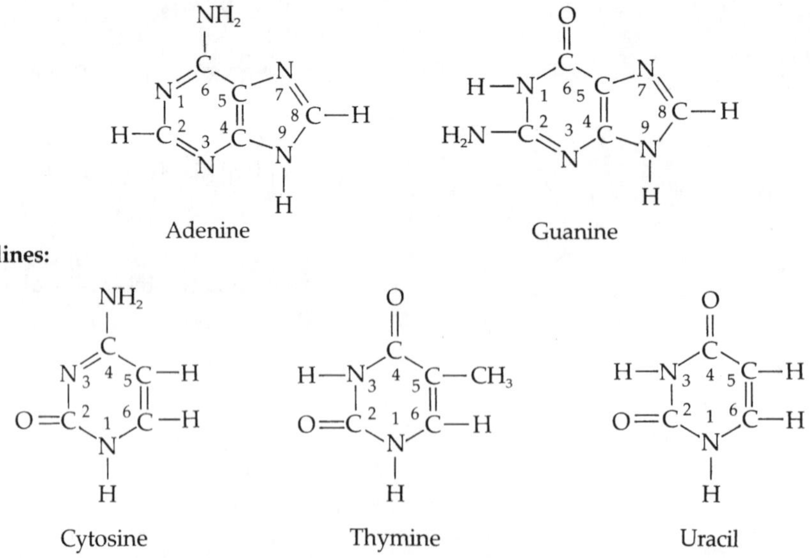

Adenine　　Guanine

Pyrimidines:

Cytosine　　Thymine　　Uracil

[1] The 2′ carbon of the ribose sugar has an -OH group attached; in the deoxyribose sugar this carbon carries an -H. [2] Both purines have double-ringed skeletons. [3] In the side groups that are attached to the skeleton. [4] The pyrimidines have single-ringed skeletons. [5] In the side groups that are attached to the skeleton.

FIGURE 18-4

Thymidine triphosphate (TTP).

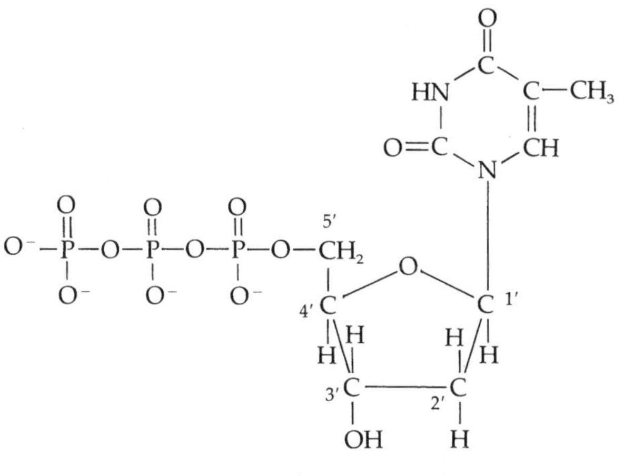

Thymidine triphosphate (TTP)

JOINING NUCLEOTIDES INTO A STRAND

Phosphodiester Bond

The joining of two triphosphorylated nucleotides involves releasing two phosphate groups from one nucleotide and establishing a covalent bond between its remaining phosphate group (attached to the 5′ carbon of its sugar) and the hydroxyl group (-OH) of the 3′ carbon of the sugar of the other nucleotide. The 5′–3′ phosphate linkage is referred to as a **phosphodiester bond.** It is designated as "-diester" because the joining phosphate group carries an ester linkage (-O-) on either side. Figure 18-5 shows two deoxyribonucleotides joined into a **dinucleotide.** The bonding of a third nucleotide through its 5′ phosphate to the 3′ hydroxyl of the bottom nucleotide shown in Figure 18-5 would produce a **trinucleotide,** and so on. Each nucleotide becomes incorporated in monophosphorylated form with the liberation of two of its phosphate groups. A large number of nucleotides assembled through this 5′-to-3′ growth make up a **polynucleotide strand** that may contain hundreds or thousands of nucleotides.

Abbreviated Structural Formulas for Strands

For convenience, the structural formulas of nucleic acid strands are often abbreviated. A variety of systems are used, two of which are shown in Figure 18-6. In Figure 18-6a, a phosphate group is represented by an encircled P, a sugar by a pentagon, and a base by its letter abbreviation. With the more abbreviated scheme shown in Figure 18-6b, a horizontal line is used to represent the sugar. Both schemes in Figure 18-6 show the same five-nucleotide DNA strand.

Strand Polarity

Examine the nucleotides at both ends of the polynucleotide strand shown in Figure 18-6. What is the terminal group at the upper end of the strand? (Specifically, what is joined to the 5′ carbon of the uppermost nucleotide?) ________[6] What is the terminal group at the lower end of the strand? (Specifically, what is joined to the 3′ carbon of the lowermost nucleotide?) ________[7] Are the two ends of the strand different? ________[8] Because its ends are different, the strand is said to exhibit **polarity.** Polarity is a feature of all nucleic acid strands.

DNA STRUCTURE

Double-Stranded Molecules

In some viruses, the DNA molecule consists of a single strand, but in all cells the DNA is double-stranded.

FIGURE 18-5

An adenine and a cytosine joined by a 5′-3′ phosphodiester bond.

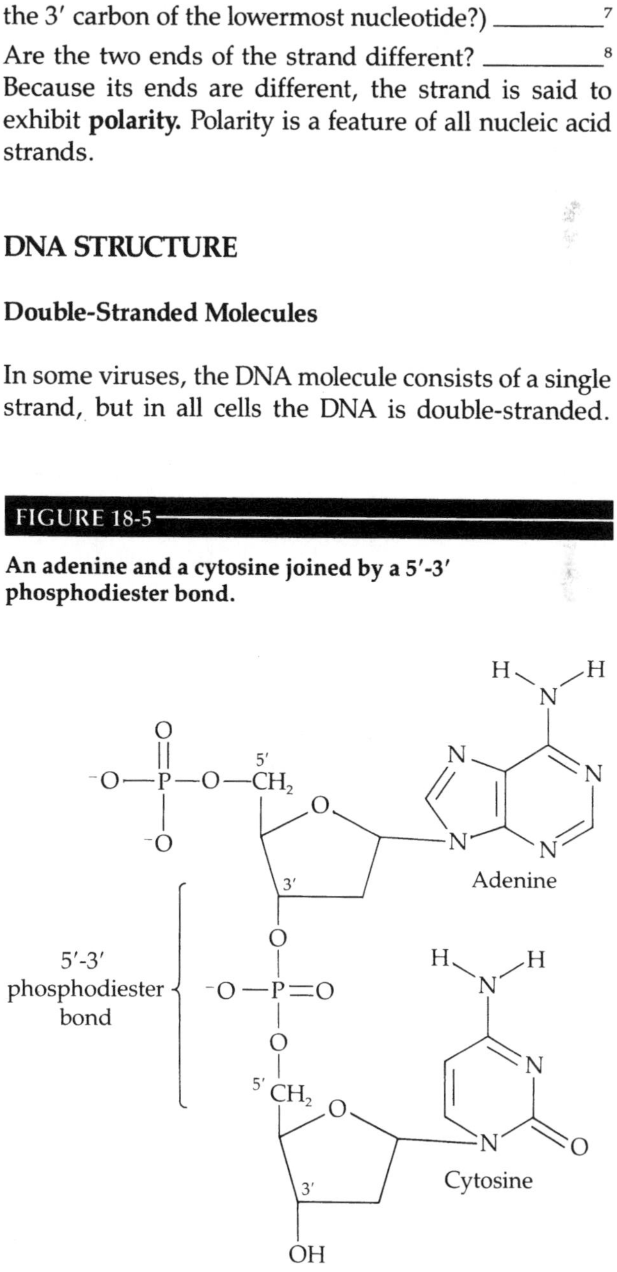

[6] A 5′ phosphate. [7] A 3′ hydroxyl. [8] Yes.

FIGURE 18-6

Two representations of the same single-stranded DNA molecule.

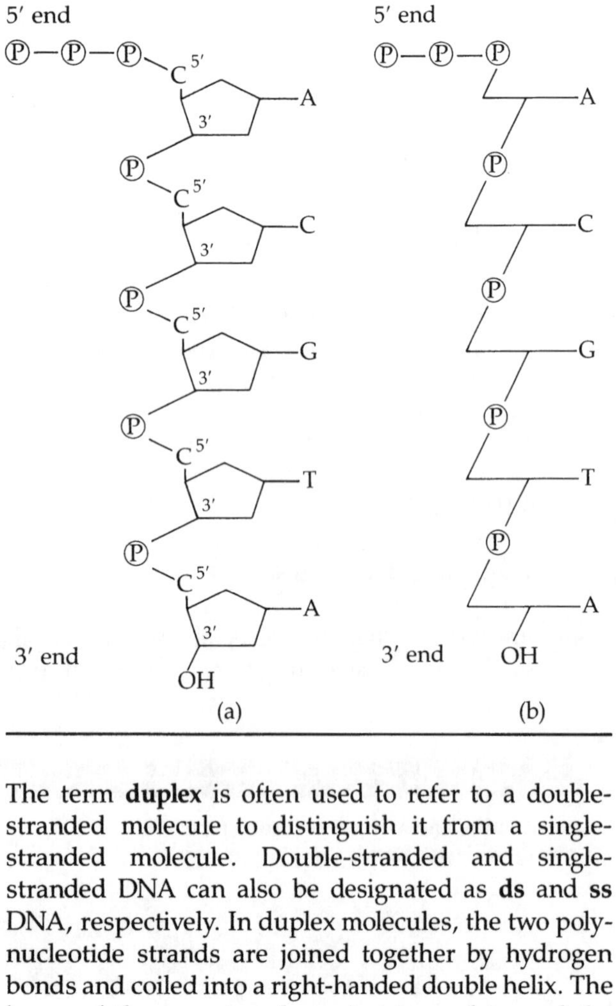

(a) (b)

The term **duplex** is often used to refer to a double-stranded molecule to distinguish it from a single-stranded molecule. Double-stranded and single-stranded DNA can also be designated as **ds** and **ss** DNA, respectively. In duplex molecules, the two polynucleotide strands are joined together by hydrogen bonds and coiled into a right-handed double helix. The bases of the two strands project inward toward the central axis of the molecule and the alternating phosphate and sugar units of each strand's backbone make up the outer portion of the duplex molecule, as shown in the two representations in Figure 18-7.

The form of the duplex molecule has been likened to a ladder, where the two uprights consist of alternating sugar and phosphate units and each rung is made up of two bases. Each polynucleotide strand contributes one upright and half of each rung. To complete the analogy, if the bottom of the ladder is then held stationary and the top is twisted in a right-hand direction, the uprights take on the double-helix configuration exhibited by the molecule. A duplex molecule has a uniform diameter of about 2.0 nanometers (1 nanometer is one-billionth of a meter) which provides sufficient space between the uprights to accommodate rungs consisting of a purine and a pyrimidine. The double ring of a purine next to the single ring of a pyrimidine fills this space perfectly. What would be the effect if a rung were made up of two pyrimidines?

______________________________ [9]

What would be the effect if a rung were made up of two purines? ______________________________ [10]

Specificity of Base Pairing: Hydrogen Bonding

Base pairing normally occurs only between purines and pyrimidines that can form the same number of **hydrogen bonds.** Hydrogen bonds are very weak attractions that get established between a weakly electropositive hydrogen atom (covalently bonded to a nitrogen or oxygen atom) and a nearby, weakly electronegative oxygen or nitrogen atom. Adenine and thymine have the potential to form two hydrogen bonds and they pair to make up one type of rung. Similarly, guanine and cytosine have the capacity to form three hydrogen bonds and they pair to form the second type of rung found in duplex DNA. The molecular structures of both types of rungs are given in Figure 18-8, which shows a two-rung portion of a DNA duplex. DNA rungs consisting of AC and GT pairs do not normally occur. Although hydrogen bonds are much weaker than covalent bonds, the collective effect of the large number of hydrogen bonds that occurs in a duplex DNA molecule confers great stability upon the molecule under the physiological conditions of living cells.

Ten nucleotide pairs or rungs are found in each complete turn of the double-helix molecule. With the base pairs evenly spaced every 0.34 nanometers, a complete turn of the molecule occurs every 3.4 nanometers. The outer surface of the duplex molecule is contoured by two grooves; the deeper is known as the major groove and the shallower, the minor groove.

Complementary, Antiparallel Strands

A consequence of the highly specific base pairing is that the two strands making up a duplex molecule **complement** each other. This means that if we know the base composition of one strand, the specificity of pairing allows us to identify the bases in the other strand. The strands also exhibit opposite polarity: one strand is oriented in the 5′-to-3′ direction while the other strand is oriented in the 3′-to-5′ direction. This is shown in both Figures 18-7 and 18-8. This **antiparallel** orientation of the two strands aligns the bases so that the appropriate hydrogen bonds can form.

[9] The space would not be sufficiently filled. [10] The space would be overfilled, causing the duplex molecule to bulge.

FIGURE 18-7

Two ways of representing double-stranded DNA molecules. Hydrogen bonds are shown as dotted lines.

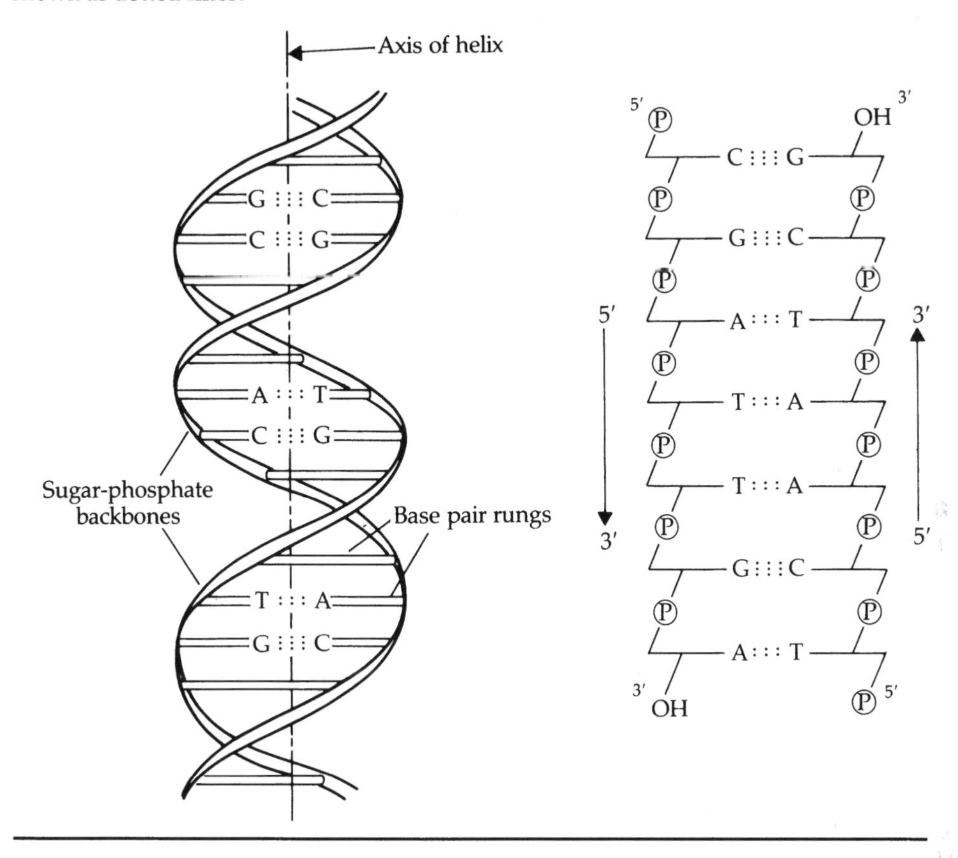

Features of the Watson-Crick Model

The double-helix model for DNA, proposed by J. Watson and F. Crick in 1953, provided explanations as to how the genetic material could replicate, store information, and mutate.

Replication begins with the breaking of the hydrogen bonds, the unwinding of the double helix, and the separation of the two strands. Each strand then serves as a template to guide the formation of a new complementary strand using nucleotides that are present in the cell. The specificity of base pairing normally ensures precise and accurate replication; the two double-helix molecules that result are identical to each other as well as to the original parent molecule. One strand of each new duplex molecule is a strand from the parental molecule and the other is newly synthesized. This pattern of replication is referred to as **semiconservative,** since half of the parental duplex is preserved, or conserved, in each of the new double-helix molecules.

Information is stored in the duplex molecule in the sequence of base pairs. There is no restriction on the length or sequence of base pairs making up the molecule.

A change, or mutation, in the sequence of bases in a portion of a DNA molecule alters its information and can change the phenotype of the organism. Such a change in base sequence could arise occasionally because of a mispairing of bases during DNA replication.

Other Types of DNA

DNA exhibiting the features of the Watson-Crick model is often designated as a **B form** DNA. High-resolution x-ray diffraction studies on natural and synthetic molecules indicate that DNA sometimes occurs in configurations different from those of the Watson-Crick model. The more common of these is designated as **Z form** DNA and, unlike the B form, its helical structure is left-handed, its sugar-phosphate backbones exhibit a zigzag appearance (hence the name "Z form"), and its base pairs are positioned away from the center of the molecule and more toward the outside. Some DNA molecules are able to change back and forth be-

FIGURE 18-8

Molecular structure of two types of rungs found in duplex DNA.

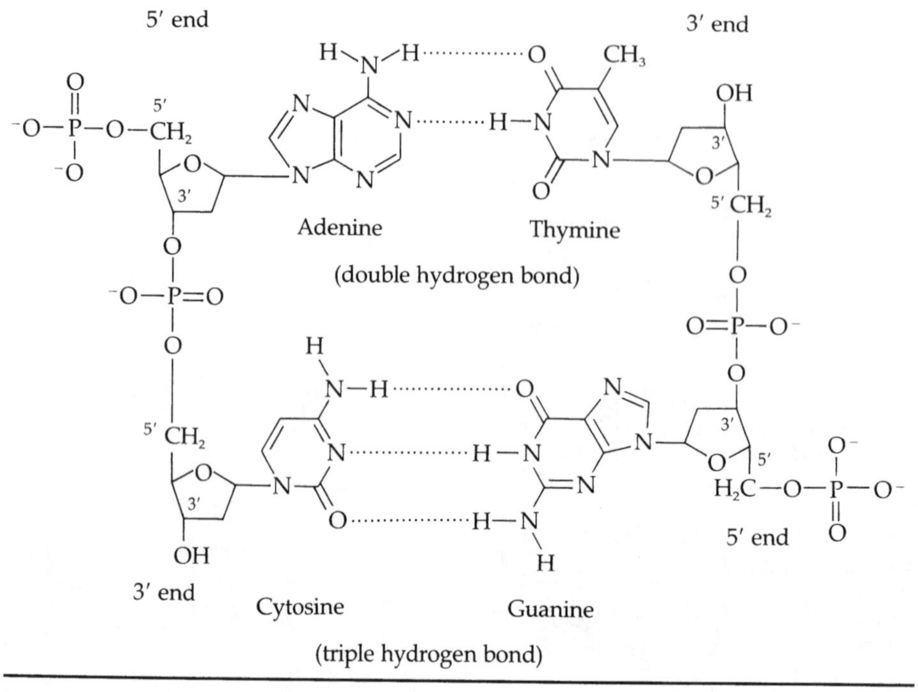

tween the Z and B forms; this ability may possibly serve to control gene expression. A few other forms of DNA, all right-handed helices, have been characterized as well.

RNA STRUCTURE

Ribonucleotides differ from deoxyribonucleotides in that they contain ribose rather than deoxyribose sugar. In addition, the base thymine does not occur in ribonucleotides but is replaced by a similar pyrimidine, uracil. Ribonucleotides are assembled into polynucleotide strands similar to those found in DNA. Whereas DNA usually exists as a double-stranded molecule, RNA usually occurs in cells in single-stranded forms. These single-stranded molecules rarely exist in linear form. Since bases within one section of a strand complement bases in another section of the strand, hydrogen bonds may form between the bases in these sections. This produces folded regions (an RNA molecule may contain more than one such region) in which the two complementary sections run antiparallel to each other and assume the form of a double helix. This is illustrated in Figure 18-9.

Usually the molecular configuration assumed by the RNA is essential to its functional role. RNA occurs in several functionally different types within the cell, including **messenger RNA** (mRNA), **transfer RNA** (tRNA), and **ribosomal RNA** (rRNA). These types are discussed in greater detail in later chapters. Although all cells and many viruses encode genetic information in DNA some viruses use RNA as their genetic material, which occurs as either single or double-stranded molecules.

SUMMARY

Nucleotides, the building blocks of DNA and RNA molecules, each consist of a five-carbon sugar, a nitrogenous base (either a purine or a pyrimidine), and a phosphate group. Deoxyribonucleotides contain deoxyribose sugar and a purine (adenine and guanine) or a pyrimidine (thymine or cytosine). Nucleotides are joined by covalent phosphodiester bonds into polynucleotide strands. Most DNA molecules consist of two such strands held together by hydrogen bonds and coiled into a right-handed double helix. The bases of the two strands project inward toward the central axis

FIGURE 18-9

Folded region in an RNA strand.

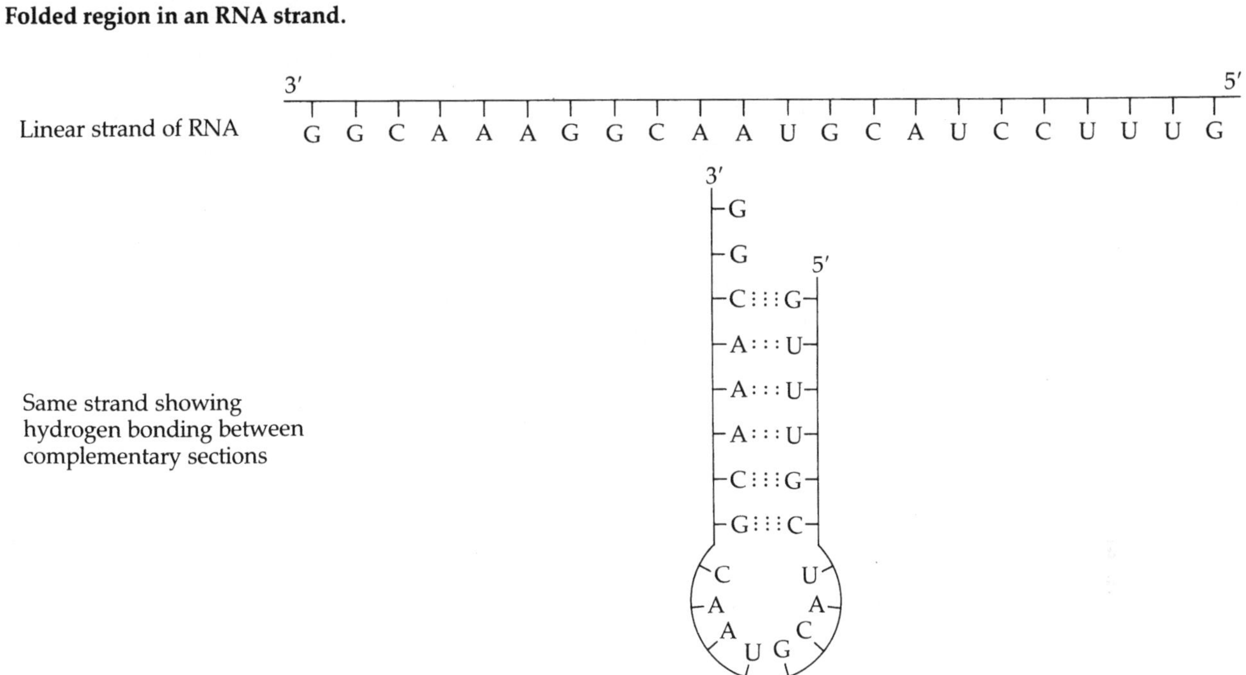

of the molecule, and the alternating sugar and phosphate units of each strand's backbone make up the outer portions of the duplex molecule. The two backbones run in opposite directions relative to each other, and the purine and pyrimidine making up each rung exhibit complementary hydrogen bonding with each other. The Watson-Crick model (B form configuration) provides explanations of how the genetic material replicates, stores information, and mutates. Other known configurations for DNA molecules include the Z form.

Ribonucleotides differ from deoxyribonucleotides in that they contain ribose rather than deoxyribose sugar. In addition, the base thymine does not occur in ribonucleotides but is replaced by a similar pyrimidine, uracil. Most RNA molecules are single-stranded and contain folded regions in which two complementary sections of the strand run antiparallel to each other and assume the form of a double helix. RNA serves as the genetic material in some viruses and plays key roles in protein synthesis in all organisms.

PROBLEM SET

18-1. Write the DNA base sequences that complement each of the following DNA strands.

a. 5′ AATCGCCCATTGCAGTTC 3′

b. 5′ CGATTGGCTTA 3′

18-2. The application of heat (about 90°C) to double-stranded DNA causes the hydrogen bonds holding the strands together to break down and the strands to unwind and separate, a process known as denaturation. The greater the number of hydrogen bonds in a duplex, the greater the amount of thermal energy required for the denaturation. Bases within single strands absorb more ultraviolet radiation than do those in double-stranded molecules. Thus, if ultraviolet absorption is monitored as heat is applied, the extent of denaturation can be measured. The temperature at which half of the hydrogen bonds in a sample of duplex DNA have denatured is called the melting point, or t_m. Listed below are the base compositions of three different duplex DNA molecules.

(1) 5′ ATTCTAACTTTGAT 3′
3′ TAAGATTGAAACTA 5′

(2) 5′ CGCACTGGCTAACC 3′
3′ GCGTGACCGATTGG 5′

(3) 5′ CGCATTATTGTCAA 3′
3′ GCGTAATAACAGTT 5′

a. Order these three molecules in terms of their melting points from highest to lowest.

b. State the relationship between the percentage of AT base pairs in a double-stranded DNA molecule and the melting point of that molecule.

18-3. An analysis of the base composition of duplex DNA molecules from two bacterial cells indicates that the four kinds of bases occur with identical frequencies in the DNA from each cell. Does this indicate that the bacteria have identical DNA molecules?

18-4. An analysis of a nucleic acid molecule indicates that the base thymine is present. No other information is available about the structure or composition of the molecule.

a. What kind of sugar would you expect to find in the nucleotides of this molecule?

b. What other kinds of bases might be found in this molecule?

c. If it is established that thymine makes up 13% of the bases in the molecule and that the molecule is double-stranded, what are the percentages of the other bases?

d. If it is established that thymine makes up 13% of the bases in the molecule and that the molecule is single-stranded, can you predict the percentages in which the other bases would occur? Explain.

18-5. An analysis of a nucleic acid molecule indicates that the base uracil is present. No other information is available about the structure or composition of the molecule.

a. What kind of sugar would you expect to find in the nucleotides of this molecule?

b. What other kinds of bases might be found in this molecule?

c. If it is established that uracil makes up 23% of the bases in the molecule and that the molecule is double-stranded, what are the percentages of the other bases?

PROBLEM SET

d. If it is established that uracil makes up 23% of the bases in the molecule and that the molecule is single-stranded, can you predict the percentages in which the other bases would occur? Explain.

18-6. Two duplex DNA molecules, each consisting of 25 base pairs, have the following base sequences.

5′ ATATCGGGCTATGAGTACAGTAATG 3′
3′ TATAGCCCGATACTCATGTCATTAC 5′

5′ AGCATTTCTGACTAATGTACACACC 3′
3′ TCGTAAAGACTGATTACATGTGTGG 5′

a. Calculate the A-to-T ratio for each molecule.
b. Calculate the G-to-C ratio for each molecule.
c. Calculate the (A + T) to (G + C) ratio for each molecule.
d. The molecules obviously have different sequences, but do they have the same overall nucleotide composition? Which of the three ratios that you have just calculated will allow you to answer this question quickly without having to count the number of times each kind of base occurs?

18-7. Nucleic acid molecules have been isolated from seven types of viruses. The base composition of these molecules has been analyzed and the findings are presented in the following table. In some instances only a limited amount of information is available regarding the composition of the nucleic acid. Data are given in terms of the percentage of total base composition.

	Virus Type						
Base	1	2	3	4	5	6	7
A	12	–	20	18	–	26	23
T	12	18	0	26	–	0	18
C	38	–	30	26	–	24	28
G	38	–	30	30	–	26	–
U	0	–	20	0	17	24	–

Identify the type of nucleic acid found in each of the seven viruses. Also indicate, if possible, whether each molecule is single-stranded or double-stranded. In each case, indicate why you classified the molecule as you did.

18-8. Refer to the information in problem 18-7. Detailed study of the nucleic acid found in virus 1 indicates that the molecule contains 4200 base pairs.
a. How many nucleotides are present in the molecule?
b. How many sugar molecules are present in the nucleic acid molecule?
c. How many complete turns would occur in the configuration that this molecule assumes?
d. What is the length of this molecule?

18-9. Refer to the information in problem 18-7. Detailed study of the nucleic acid molecules present in viruses 3 and 4 indicates that the total number of nucleotides present in each molecule is the same. Would these molecules have the same or different lengths? Explain.

18-10. The relationship between DNA weight and number of nucleotides is as follows: 1 gram of nucleic acid contains 2.0×10^{21} nucleotides. The relation-

PROBLEM SET

ship between the number of nucleotide pairs and length of a duplex DNA molecule is as follows: 2.9×10^6 nucleotide pairs equals 1 millimeter of duplex DNA. The weight of DNA carried in a haploid cell is called the C-value, and in humans this amounts to approximately 3.2×10^{-12} grams.

a. Determine the number of nucleotides present in this amount of human DNA.

b. Identify the number of nucleotide pairs present in this DNA.

c. What is the total length of the DNA in this human haploid cell?

18-11. Refer to the information given in problem 18-10. An *E. coli* cell contains a single genome consisting of 4.7×10^{-15} grams of duplex DNA.

a. Determine the number of nucleotides present in the DNA in this *E. coli* cell.

b. Identify the number of nucleotide pairs present in this DNA.

c. What is the length of the DNA in this *E. coli* cell?

CHAPTER

19

DNA Replication and Recombination

INTRODUCTION

Precise replication of the genetic material is essential if genetic information possessed by one generation is to be accurately transmitted to the next. Watson and Crick, as described in Chapter 18, proposed that DNA replication involves the unwinding of the double helix and the separation of the two strands, with each strand then serving as a template for the formation of a new complementary strand. The specificity of base pairing normally ensures accurate replication, and the two daughter double-helix molecules produced are identical to each other as well as to the original parent molecule. One strand of each new duplex molecule is a strand from the parent molecule and the other is newly synthesized. This pattern of replication is referred to as **semiconservative,** since half of the parental duplex is preserved, or conserved, in each of the new double-helix molecules.

Two additional alternative models of DNA replication were proposed by other researchers in the mid-1950s. In one, the strands of the parental duplex separate and serve as templates for the synthesis of new complementary strands just as was proposed under the semiconservative model. Upon the completion of replication, however, the two new strands separate from their templates and join together to make a duplex consisting solely of newly synthesized DNA. At the same time, the two template strands rejoin to form a duplex consisting of the two strands of the parental duplex. This type of replication is referred to as **conservative,** since the parental duplex is conserved as one of the two product duplexes.

In the third proposed type of replication, the strands of the parental duplex separate and break up lengthwise into short segments. Each segment then replicates. The short double-stranded segments that result then rejoin to make up two duplexes, with each duplex strand consisting of segments of newly synthesized DNA and segments of the original parental strands. This type of replication is referred to as **dispersive,** since the nucleotides making up the parental duplex are dispersed in each of the daughter duplexes.

THE NATURE OF DNA REPLICATION: THE MESELSON-STAHL EXPERIMENT

Important experimental evidence for the nature of DNA replication was provided in 1958 by M. Meselson and F. Stahl. They grew *Escherichia coli* bacteria on food (medium) containing ammonium chloride (NH_4Cl) as the sole nitrogen source. The nitrogen in this ammonium chloride was ^{15}N, an isotope known as heavy nitrogen, that contains one more neutron than is found in the natural, common form of nitrogen, ^{14}N. The nitrogen in the medium is used by *E. coli* to synthesize an array of nitrogen-containing compounds, including the nitrogenous bases required for DNA replication. Eventually, after many generations of growth, all of the nitrogen-containing compounds in the bacteria, including DNA, contain ^{15}N.

The type of nitrogen in DNA molecules can be determined by extracting the DNA from the bacteria and subjecting it to a procedure known as **density-gradient equilibrium centrifugation.** This procedure involves adding the DNA to a tube of concentrated cesium chloride solution. High-speed centrifugation establishes a linear density gradient of cesium chloride molecules, with the density increasing from the top to the bottom of the tube. During centrifugation the DNA molecules move until they reach a point where their density equals the density of the cesium chloride. DNA containing ^{15}N bases has a greater density and collects to form a band deeper in the centrifuge tube than does DNA containing ^{14}N. When complexed with a stain that fluoresces under ultraviolet light, the DNA can be located by exposing the centrifuge tube to ultraviolet light.

After many generations of growth on an ^{15}N medium, samples of *E. coli* from this medium were transferred to a medium containing ^{14}N. These bacteria, designated as belonging to generation 0, were allowed to grow for two generations. DNA duplexes extracted from samples of bacteria taken from generations 0, 1, and 2 were separately subjected to density-gradient equilibrium centrifugation to determine their densities. The generation-0 duplex can be represented as

FIGURE 19-1

Generation-0 DNA.

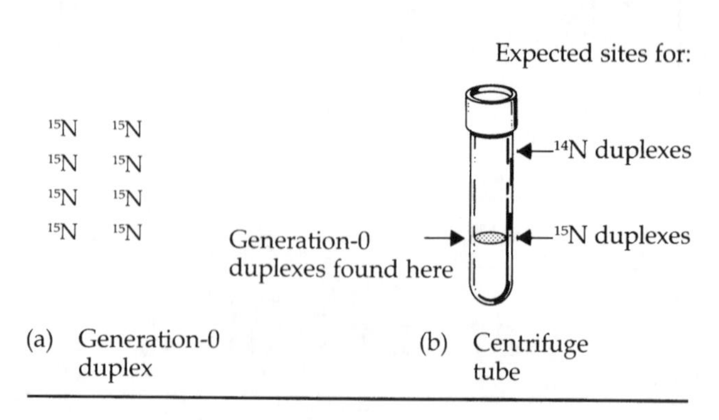

shown in Figure 19-1a. Upon centrifugation, the duplexes extracted from generation-0 bacteria collect in the centrifuge tube at the site predicted for duplexes containing only ^{15}N, as shown in Figure 19-1b.

Now let us make some predictions regarding the generation-1 DNA based on the proposed alternative mechanisms for DNA replication. First, assume that replication is semiconservative. In the space provided in Figure 19-2b, sketch the two generation-1 duplexes produced by the replication of the generation-0 duplex shown in Figure 19-2a. What is the distribution of ^{14}N and ^{15}N in these daughter duplexes? ______________ ______________________________[1] In the centrifuge tube in Figure 19-2c, sketch the location of the band or bands formed by the centrifugation of the generation-1 DNA duplexes.[2]

FIGURE 19-2

Semiconservative replication of generation-0 DNA.

^{15}N ^{15}N
^{15}N ^{15}N
^{15}N ^{15}N
^{15}N ^{15}N

Replication on an ^{14}N medium →

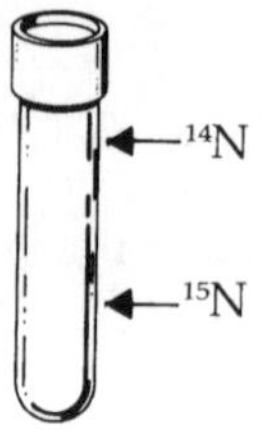

^{14}N
^{15}N

(a) Generation-0 duplex

(b) Generation-1 duplexes

(c) Centrifuge tube

[1] One strand of each duplex is newly synthesized and contains just ^{14}N bases; the other strand is an ^{15}N-containing strand from the parental duplex. [2] Since all the duplexes produced have a density halfway between the densities expected for duplexes containing just ^{15}N and those containing just ^{14}N, your band should be halfway between the ^{15}N and ^{14}N duplex sites.

FIGURE 19-3

Conservative replication of generation-0 DNA.

^{15}N ^{15}N
^{15}N ^{15}N
^{15}N ^{15}N
^{15}N ^{15}N

Replication on an ^{14}N medium

$\leftarrow ^{14}N$
$\leftarrow ^{15}N$

(a) Generation-0 duplex (b) Generation-1 duplexes (c) Centrifuge tube

Now assume that replication is conservative. In the space provided in Figure 19-3b, sketch the two duplexes produced by the replication of the generation-0 duplex shown in Figure 19-3a. What is the distribution of ^{14}N and ^{15}N in the daughter duplexes? ______[3] In the centrifuge tube in Figure 19-3c, sketch in the band or bands formed by centrifugation of the generation-1 DNA duplexes.[4]

Now assume that replication is dispersive. In the space provided in Figure 19-4b, sketch the two duplexes produced by the replication of the generation-0 duplex shown in Figure 19-4a. What is the distribution of ^{14}N and ^{15}N in the daughter duplexes? ______[5] In the centrifuge tube in Figure 19-4c, sketch in the band or bands formed by the centrifugation of the generation-1 DNA duplexes.[6]

When Meselson and Stahl analyzed the generation-1 DNA, they found that *all* the daughter duplexes had densities halfway between those expected for duplexes containing just ^{15}N and those containing just ^{14}N. Compare these results with your predictions above. Do these results allow us to identify the correct method of DNA replication in *E. coli*? ______[7] Explain. ______[8] Do these results make it possible to eliminate

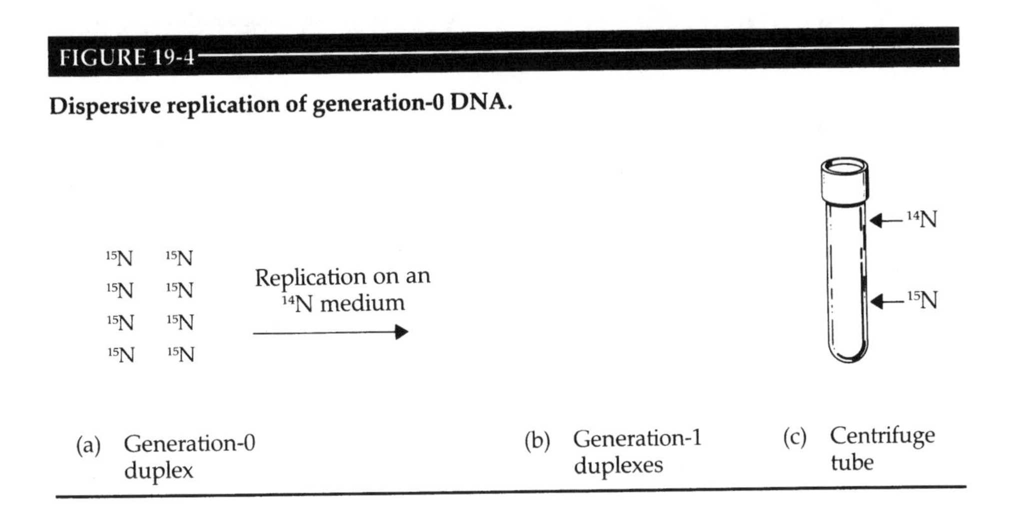

FIGURE 19-4

Dispersive replication of generation-0 DNA.

[3] Both strands of one duplex are newly synthesized and contain just ^{14}N bases; both strands of the other duplex are the ^{15}N-containing strands of the parental duplex. [4] Two bands result: the ^{14}N duplexes form a band at the site expected for ^{14}N duplexes and ^{15}N duplexes form a band at the site expected for ^{15}N duplexes. [5] Both strands of each duplex consist of a combination of ^{14}N and ^{15}N bases. [6] Since both duplexes have a mixture of ^{14}N and ^{15}N bases, their densities would be about halfway between the densities expected for duplexes containing just ^{15}N and those containing just ^{14}N, and they would form a band about halfway between the ^{14}N and ^{15}N duplex sites. [7] No. [8] There is ambiguity since the results are consistent with either semiconservative or dispersive replication.

FIGURE 19-5

Semiconservative replication of generation-1 DNA.

^{15}N ^{14}N
^{15}N ^{14}N
^{15}N ^{14}N
^{15}N ^{14}N

Replication on an ^{14}N medium →

← ^{14}N
← ^{15}N

(a) Generation-1 duplex (b) Generation-2 duplexes (c) Centrifuge tube

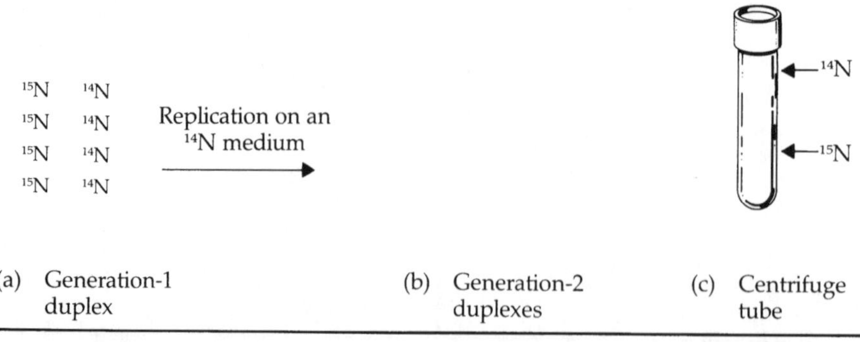

one of the three proposed methods of replication? ________[9] Explain. ______________________________[10]

Having eliminated conservative replication from further consideration, Meselson and Stahl went on to analyze DNA from generation-2 bacteria. Again, before we consider the outcome of their study, we will predict the type of DNA duplexes formed, first by assuming that replication is semiconservative and then by assuming it is dispersive. In the space provided in Figure 19-5b, sketch the generation-2 duplexes produced by the semiconservative replication of the generation-1 duplex shown in Figure 19-5a. What is the distribution of ^{14}N and ^{15}N in the daughter duplexes? ______________________________[11] In the centrifuge tube in Figure 19-5c, sketch the band or bands formed by the centrifugation of the generation-2 DNA duplexes.[12]

Now assume that replication is dispersive. In the space provided in Figure 19-6b, sketch the daughter duplexes produced by the replication of the generation-1 duplex shown in Figure 19-6a. What is the distribution of ^{14}N and ^{15}N in the daughter duplexes? ______________________________[13] In the centrifuge tube in Figure 19-6c, sketch the band or bands formed by the centrifugation of the generation-2 DNA.[14]

FIGURE 19-6

Dispersive replication of generation-1 DNA.

^{15}N ^{14}N
^{14}N ^{15}N
^{14}N ^{15}N
^{15}N ^{14}N

Replication on an ^{14}N medium →

← ^{14}N
← ^{15}N

(a) Generation-1 duplex (b) Generation-2 duplexes (c) Centrifuge tube

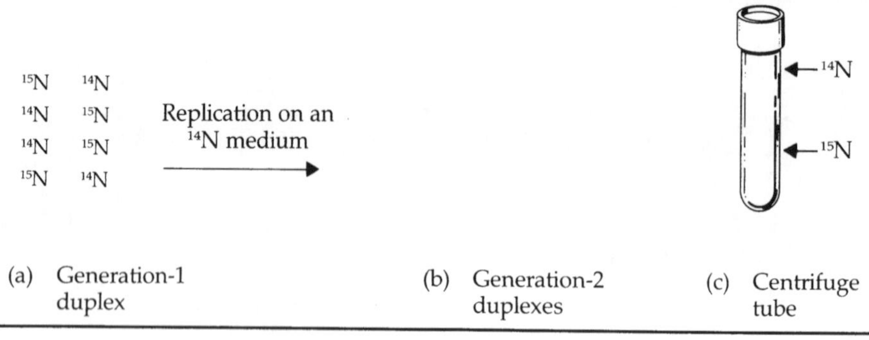

[9] Yes. [10] The results are not consistent with conservative replication which would have generated two types of duplexes, one containing just ^{14}N and the other containing just ^{15}N. [11] Two kinds of duplexes are formed. The ^{15}N strand from the generation-1 duplex is paired with a newly synthesized ^{14}N strand and its ^{14}N counterpart is paired with a newly synthesized ^{14}N strand. [12] Half the duplexes have a density halfway between those expected for duplexes containing just ^{15}N and those containing just ^{14}N, and they will form a band halfway between the expected sites for ^{15}N and ^{14}N duplexes. The other half of the duplexes containing just ^{14}N will form a band at the expected site for ^{14}N duplexes. [13] Both strands of each duplex would consist of newly synthesized sections containing ^{14}N bases and sections containing ^{15}N bases derived from the parental duplex, with the ^{14}N bases making up approximately 3/4 of each duplex. [14] Since both duplexes would have a mixture of ^{14}N and ^{15}N bases, their densities would be intermediate and they would form a band between the expected sites for duplexes comprised of ^{14}N and ^{15}N bases. Since the ^{14}N bases predominate in these duplexes, this band would be closer to the expected site for ^{14}N duplexes.

When Meselson and Stahl analyzed the generation-2 DNA, they found daughter duplexes of two densities. The band for one type was halfway between the sites expected for duplexes with just ^{15}N and duplexes with just ^{14}N, and the band for the other type was located at the site expected for ^{14}N duplexes. Compare these results with your predictions. Do they indicate the method of DNA replication in *E. coli*? ____[15] Explain. ____________________

____[16]

Based on their experimental results, Meselson and Stahl concluded that DNA replication in *E. coli* is semiconservative. Keep in mind that although we have considered individual DNA molecules in discussing their work, their studies dealt with large populations of DNA molecules. Experiments carried out by other researchers on other organisms, including some eukaryotes, provide additional support for the semiconservative nature of DNA replication.

STUDYING DNA REPLICATION USING RADIOACTIVE TRACERS

Studies using radioactive elements, or tracers, have provided important details about DNA replication in both prokaryotic and eukaryotic cells. For example, supplying one of the precursors of DNA, thymidine (the deoxyribonucleoside of thymine), labeled with the radioactive isotope tritium, ^{3}H, to a cell that is about to undergo DNA replication makes it possible to monitor the progress of DNA replication. Any newly synthesized DNA can be identified because the presence of tritiated thymidine makes it radioactive. This radioactivity may be detected through a technique known as **autoradiography.** This involves covering the DNA with a special photographic film for a period of time during which electrons given off by the tritiated thymidine hit the film directly above the radioactive DNA. The sites hit by electrons show up as dark spots when the film is developed and examined under a microscope. J. Cairns used this technique to confirm that replication of an individual *E. coli* chromosome was semiconservative (and at the same time demonstrated that the chromosome was circular). Additional autoradiography studies have confirmed that the DNA replication preceding eukaryotic mitosis and meiosis is also semiconservative.

Autoradiography has also demonstrated that DNA replication consistently begins at sites within chromosomes designated as **origins.** The circular prokaryotic chromosome has a single origin whereas eukaryotic chromosomes have many origins spaced at intervals of 30,000 to 100,000 base pairs, or 30 to 100 kilobases, where 1 kilobase equals 1000 base pairs. The DNA replicated from any particular origin is known as a **replicon.** What amount of the chromosome comprises the replicon in bacteria? ________[17] What amount of the chromosome comprises the replicon in eukaryotes? ________[18]

Autoradiography also indicates that with the separation of the two strands of the duplex at an origin, there are usually two Y-shaped zones of replication, known as **replication forks,** which form on either side of each origin and move out along the chromosome in opposite directions. Figure 19-7a shows a schematic representation of a portion of a DNA duplex molecule containing an origin. Figure 19-7b shows the two strands of the duplex beginning to separate at the origin and the two replication forks that result.

In addition, autoradiography has revealed two further points about DNA replication. (1) Replication is **bidirectional** since it occurs at both replication forks as they move away from each other; put another way, replication proceeds in both directions from the origin along the original duplex molecule. (Some exceptions occur; for example, mitochondrial DNA replication is unidirectional.) (2) Replication occurs simultaneously along both duplex strands at each replication fork with both newly synthesized strands showing overall growth in the same direction. A major challenge to

FIGURE 19-7

Portion of a DNA duplex that shows (a) an origin and (b) the two strands of the duplex beginning to separate at the origin to form two replication forks.

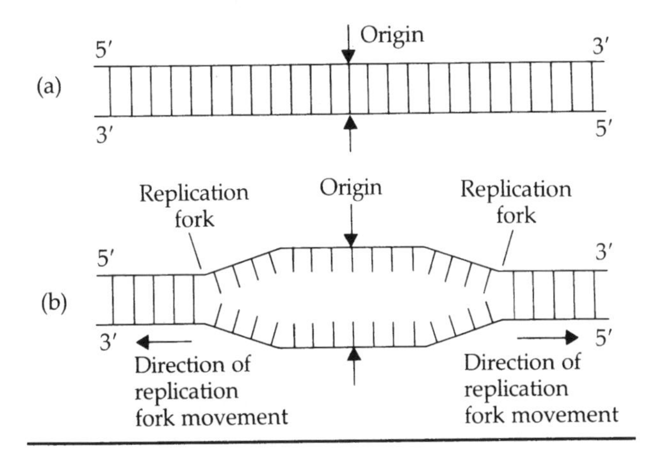

[15] Yes. [16] The results are consistent with semiconservative replication; the presence of the ^{14}N duplexes would not occur with dispersive replication, since all new DNA would be expected to contain some ^{15}N bases present in the parent duplex.
[17] The entire chromosome. [18] A portion of the chromosome, often between 30 to 100 kilobases in length.

molecular geneticists and others was to explain how these events observed through autoradiography occurred at the molecular level.

GENERAL FEATURES OF DNA REPLICATION

Many researchers working for over thirty-five years have contributed to our present-day understanding of the complexities of semiconservative replication. A summary of the key events in this process follows. Much of this information has been gathered by studying *E. coli*.

DNA Polymerase III

DNA synthesis in *E. coli* is catalyzed by the enzyme **DNA polymerase III** (so designated because it was the third kind of polymerase enzyme to be isolated from *E. coli*). This enzyme catalyzes the addition of nucleotides, one at a time, to the terminal nucleotide at the 3′ end of a growing DNA strand during DNA replication. Nucleotides serving as the raw material for this process are synthesized in the cell in triphosphorylated form. As each nucleotide is added, two of its phosphate groups are released as its 5′ carbon becomes covalently bonded to a hydroxyl group of the 3′ carbon of the strand's terminal nucleotide. Note that the polymerase III enzyme can add nucleotides only to the 3′ end of an *already existing* strand of nucleotides.

There are two key facts about the functioning of this enzyme which must be considered in producing any model for DNA replication.

1. The enzyme cannot begin the assemblage of a DNA strand from scratch; that is, it cannot lay down the first deoxyribonucleotide to which another nucleotide could be added to begin the assemblage of a new DNA strand.
2. Since the enzyme can catalyze the addition of nucleotides only to the 3′ end of a strand, replication proceeds only in a single direction, from 5′ to 3′.

Initiation Sites

As previously mentioned, DNA synthesis is initiated at specific origins. In prokaryotes, origins sometimes include **palindromic regions** where the base composition of one strand reads the same as the complementary strand read in the same direction. This is illustrated as follows for a section of duplex DNA where the upper strand, read for example from 5′ to 3′, reads the same as the complementary strand from 5′ to 3′.

5′ . . . AAAAGCTTTT. . . 3′
3′ . . . TTTTCGAAAA. . . 5′

Unwinding and Separation of Duplex Strands

The DNA duplex unwinds and the strands separate in both directions from the origin. The unwinding involves the breaking of hydrogen bonds holding the two strands together, requires energy supplied by adenosine triphosphate (ATP), and is catalyzed, in *E. coli*, by the enzyme **DNA helicase.** This unwinding exposes the bases so each strand may serve as a template. Once the unwinding occurs, molecules of another protein, **single-stranded DNA-binding (SSB) protein,** associate with the individual strands to maintain their separation. The unwinding by DNA helicase creates a twist tension which produces supercoiling (the coiling of coils) beyond the replication fork in the regions of the duplex that have yet to unwind. This tension is relieved when the enzyme **DNA gyrase** creates temporary nicks in both strands of the duplex in the region under stress. This nicking forms a **swivel point** which permits the tension to be relieved by allowing the DNA molecule to untwist in the direction opposite to which it is supercoiled. After the untwisting, the DNA gyrase reestablishes the phosphodiester bonds broken by the nicking, thereby restoring the duplex to its original configuration.

Synthesis Initiation: RNA Primers

The initiation of DNA synthesis requires the synthesis of short (10- to 50-nucleotide) segments of RNA known as **primer RNA.** The synthesis of these RNA primers occurs in a 5′-to-3′ direction (that is, with new ribonucleotides added to the 3′ end of the RNA segment) and is catalyzed by an enzyme known as **primase.** Each primer is assembled in complementary fashion, one ribonucleotide at a time, and becomes hydrogen bonded to the section of the DNA strand serving as its template. These RNA primers are essential to the initiation of DNA synthesis since they provide a free 3′ end to which the DNA polymerase III enzyme can join the initial deoxyribonucleotide of each new DNA strand. We mentioned earlier that DNA synthesis takes place along both duplex strands at each replication fork. Since replication occurs by somewhat different mechanisms along each strand, we will consider each mechanism separately. Note, however, that in a cell, both mechanisms operate simultaneously at each replication fork.

Continuous Synthesis of the Leading Strands

DNA synthesis along one of the strands at each replication fork occurs by **continuous synthesis.** This process begins with RNA primer formation. Figure 19-8 shows two RNA primers synthesized in the vicinity of the origin; each primer extends along a short portion of one of the separated strands of the original duplex that served as its template. Since the strands making up a duplex run antiparallel to each other, what can be said about the direction in which the two RNA primers shown in Figure 19-8 run relative to each other on their template strands? ______________________________ ______________[19]

Once primers are formed, DNA polymerase III begins catalyzing the addition of deoxyribonucleotides, one at a time, to the 3′ end of each primer. The two new DNA strands synthesized in this manner are known as **leading strands.** Sketch them in Figure 19-8 and place an arrowhead on each to indicate its direction of growth.[20] Note that each newly synthesized strand runs antiparallel to its template strand. Through the continuous addition of deoxyribonucleotides, these leading DNA strands continue to lengthen as the duplex continues to unwind and as the replication forks move farther apart.

Discontinuous Synthesis of the Lagging Strands

Continuous synthesis explains the formation of DNA along only one of the two template strands at each replication fork. Since the two template strands are antiparallel and since DNA polymerase III synthesizes DNA only in the 5′-to-3′ direction, synthesis along the other template strand *cannot* occur through the continuous process we have just described; some other mechanism must be operating. That other mechanism, known as **discontinuous synthesis,** involves the synthesis of a series of relatively short segments (1000 to 2000 nucleotides in *E. coli*, 100 to 200 nucleotides in eukaryotic forms) which are then joined together into a so-called **lagging strand.**

Multiple Primers and Okazaki Fragments The synthesis of each short DNA segment is initiated on a separate RNA primer formed at regular intervals along this portion of the 5′-to-3′ DNA template strand. These primers, like the single primers for the leading strands considered earlier, are synthesized from the 5′ to the 3′ end. Once formed, the DNA polymerase catalyzes, in step-wise fashion, the addition of deoxyribonucleotides to the primer's 3′ end, resulting in DNA synthesis in the standard 5′-to-3′ direction. This DNA synthesis goes on until the neighboring RNA primer is encountered. These short segments of DNA, named after their discoverer, are known as **Okazaki fragments.** Figure 19-9 shows an origin with its two replication forks. The two leading strands and the single RNA segment that primed each are shown with arrowheads indicating the direction of synthesis. In addition, a few RNA primers are shown along the sections of the original strands that served as their templates. In Figure 19-9, sketch the DNA fragments that will grow from

FIGURE 19-8

The beginning of continuous DNA synthesis. An RNA primer has been synthesized at the origin on each strand of the original duplex.

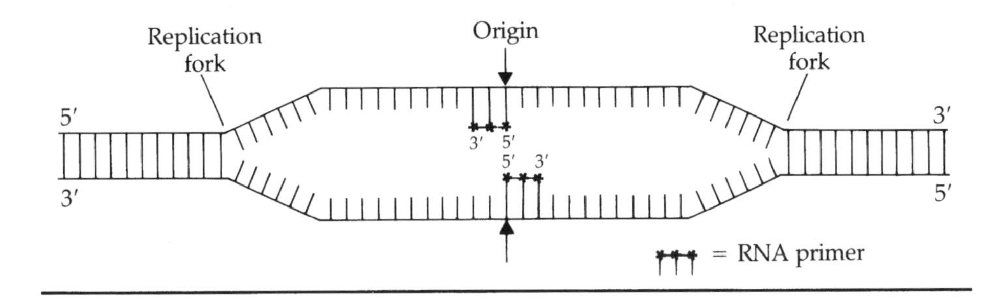

[19] Since both primers are synthesized through the addition of nucleotides at their 3′ end, they run in opposite directions relative to each other. [20] Both strands are growing from 5′ to 3′. The arrowheads on your upper and lower strands should point toward the left and right, respectively. These arrowheads indicate that the leading strands grow in opposite directions relative to each other because their template strands are antiparallel.

FIGURE 19-9

Two leading strands with their RNA primers are shown along with the RNA primers for the lagging strands.

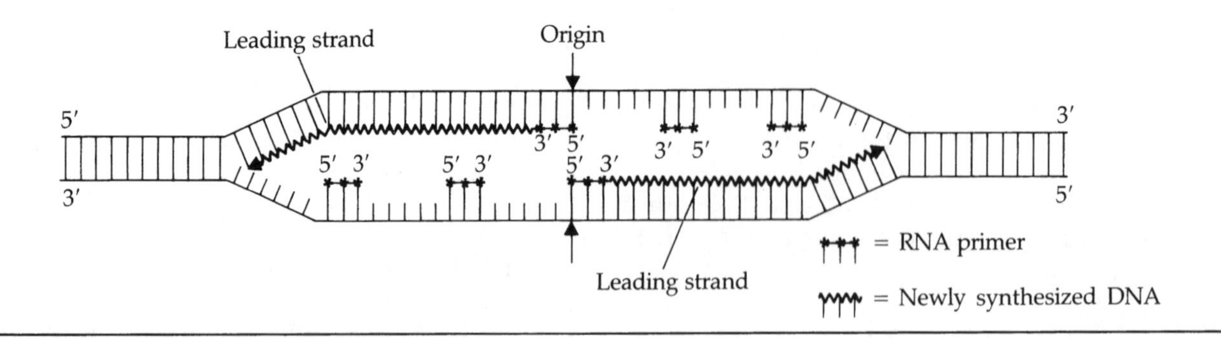

each of these primers, placing an arrowhead on each to indicate the direction of its synthesis.[21]

Assembling the DNA Fragments into Strands Next, these DNA fragments are joined into lagging strands. Speculate on what three key events must occur now in order for this to happen.

1. ______________________________ [22]
2. ______________________________ [23]
3. ______________________________ [24]

The manner in which these three events occur is detailed in the following paragraphs.

Once a series of DNA fragments has been synthesized to make up the discontinuous strands, the RNA primer attached to each fragment is enzymatically removed. Figure 19-10a includes an RNA primer–DNA fragment complex on its DNA template. The removal starts at the 5′ end of the RNA primer where ribonucleotides are cleaved off, one at a time.

The removal of each ribonucleotide exposes a base in the DNA template strand that now pairs with a complementary deoxyribonucleotide triphosphate as shown in Figure 19-10b. As each new deoxyribonucleotide moves into place, it forms a phosphodiester linkage to the nucleotide at the 3′ end of the adjacent DNA segment as shown in Figure 19-10c. This removal of ribonucleotides and their replacement with deoxyribonucleotides goes on until the entire RNA primer segment is replaced with DNA. In *E. coli*, this process is catalyzed by **DNA polymerase I,** or **Kornberg polymerase,** which has a dual functional role. In removing the ribonucleotides from the RNA primer, this enzyme displays a 5′-to-3′ exonuclease capacity (an **exonuclease** is an enzyme removing nucleotides from the end of a strand); and in filling in the space by adding deoxyribonucleotides to the 3′ end of the adjacent DNA segment, it exhibits its polymerase role. (Studies on the DNA polymerase enzymes isolated from eukaryotic cells indicate that they lack the 5′-to-3′ exonuclease activity of DNA polymerase I. Consequently, the removal of the RNA primers in eukaryotic cells is most likely carried out by a ribonuclease.)

Once the RNA primers have been removed, the one remaining task is the joining together of the neighboring DNA fragments into a strand. This is accomplished by the formation of 5′-to-3′ phosphodiester bonds between the terminal nucleotides of adjacent DNA fragments, a process catalyzed by an enzyme known as **DNA ligase** and shown in Figure 19-10d. Figure 19-11, which shows just the left-hand replication fork from the single origin we have been considering, summarizes these events as discontinuous replication proceeds along a DNA template strand. As the DNA fragments are assembled into the growing lagging strand, the overall direction of this strand's growth is toward the replication fork. Since the leading strand also grows toward the replication fork, we now have an explanation for the autoradiograph results mentioned earlier, confirming that both of the new DNA strands grow along their templates toward their replication fork.

Assume that the entire segment of the original DNA duplex (shown initially in Figure 19-7a) gets to the point where it is fully separated and each of the original strands has a newly synthesized complement

[21] Your DNA fragments should be attached to the 3′ end of each RNA fragment with the arrowheads indicating 5′-to-3′ growth of each fragment. Note that the newly synthesized fragments run antiparallel to their template strands.

[22] The RNA primers must be removed. [23] The gaps this creates must be filled with DNA nucleotides.

[24] The DNA fragments must be joined together, end-to-end, to make each strand.

as shown in Figure 19-12. Consider each newly synthesized strand in its entirety. How has each been assembled: continuously or discontinuously? Explain. [25]

____

Summary of Events in DNA Replication

DNA replication is initiated at specialized sites within the duplex molecules; these sites are characterized by unique sequences of base pairs and are designated as origins. The two strands of the duplex unwind and separate from each other in the region of the origin. Two replication forks develop at each origin and move away from each other in opposite directions as the two strands of the duplex unwind. At each fork, both strands of the original duplex serve as templates to guide the synthesis of new complementary strands. Complementary strands grow from the origin toward the forks. Toward any particular fork, synthesis occurs continuously along one strand of the original DNA duplex molecule and discontinuously along the other strand.

Short RNA primers, formed as complements to sections of the template strands, are essential for DNA synthesis since they provide a free 3′ end to which the DNA polymerase III enzyme attaches the initial deoxy-

FIGURE 19-10

Series of steps involved in RNA primer removal.

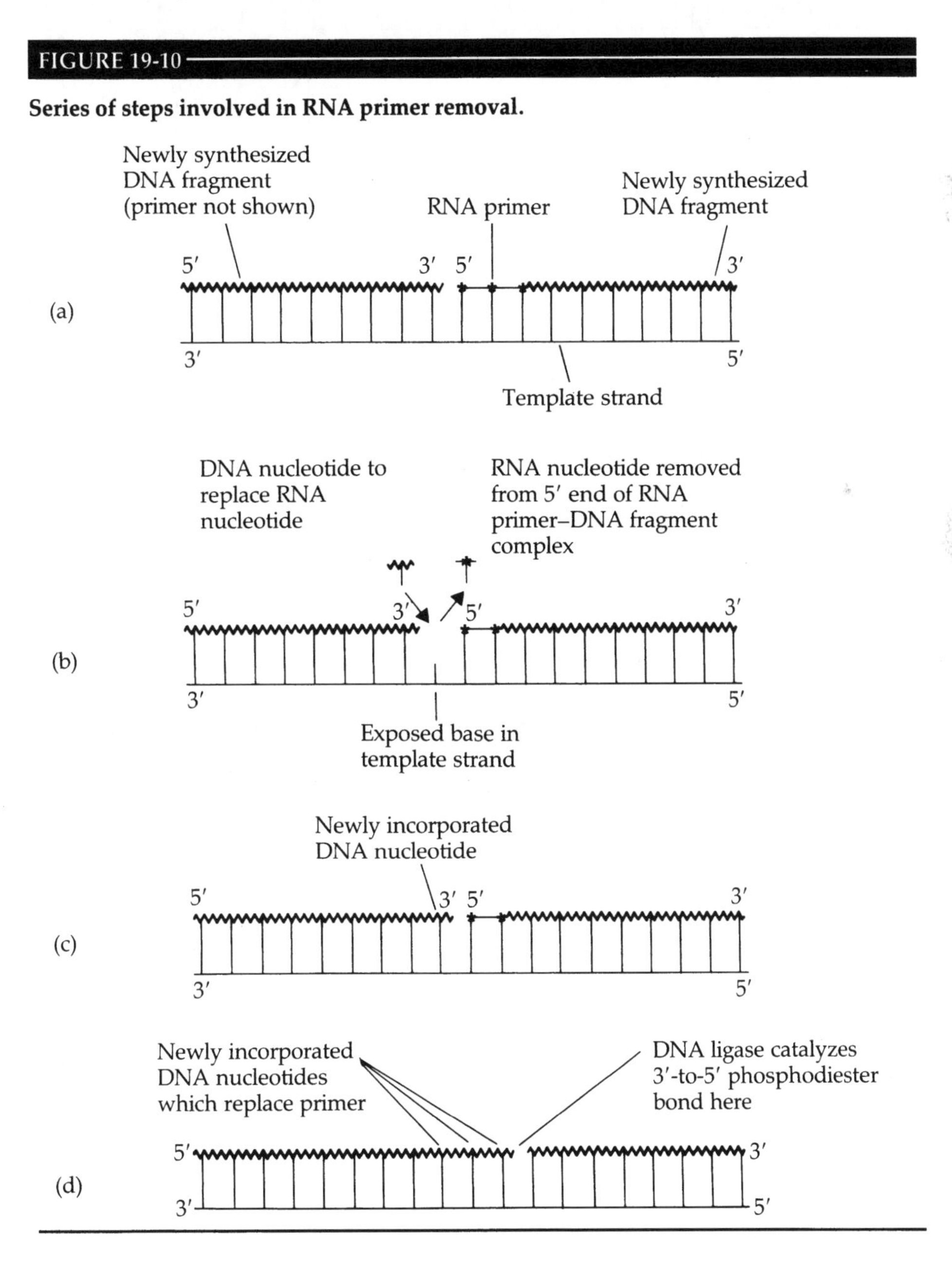

[25] Half of each strand has been assembled continuously and half discontinuously.

FIGURE 19-11

Summary of events involved in discontinuous replication. Shown is the left-hand replication fork from the single origin we have been considering.

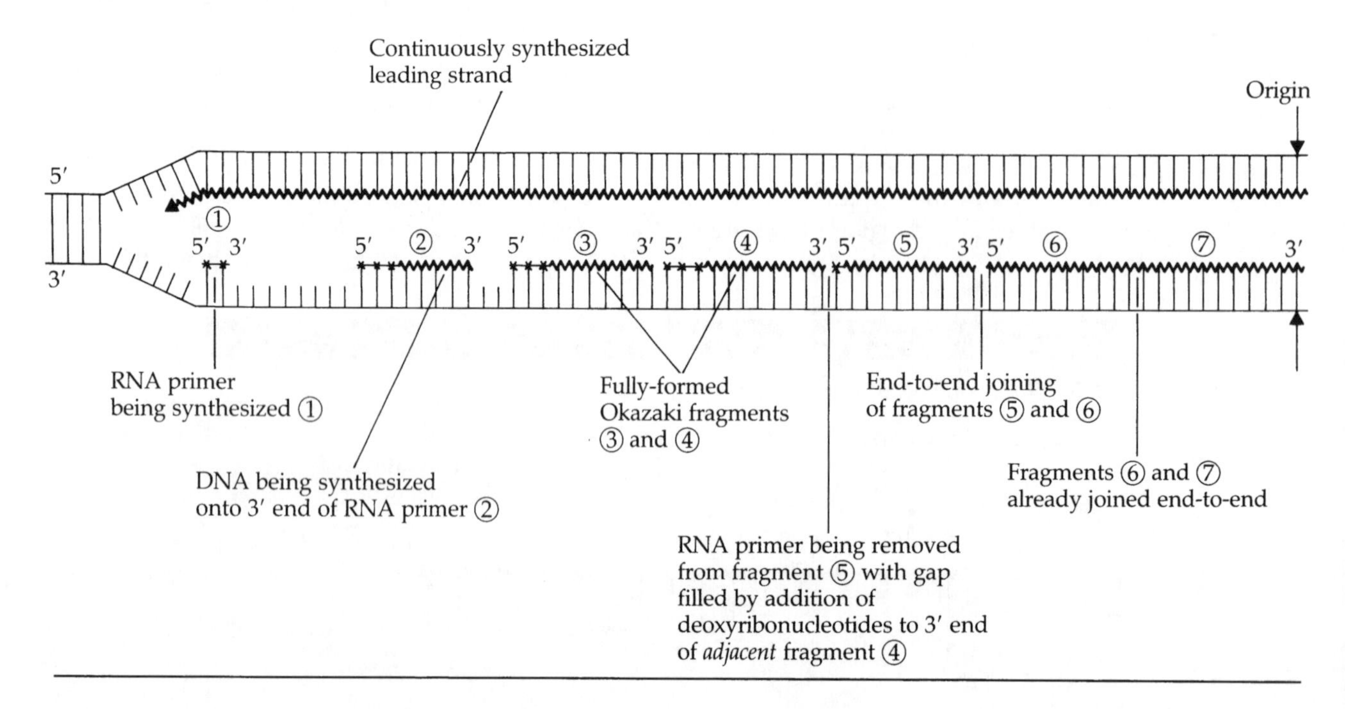

ribonucleotide to initiate DNA synthesis. Synthesis of the continuous, or leading, strands requires a single RNA primer that is synthesized at the origin and to which the continuous addition of deoxyribonucleotides forms a single lengthening strand.

Formation of the discontinuous strand involves several to many RNA primers (with the exact number depending on the strand length), the first of which is synthesized at the origin, with the rest formed along the DNA template strand at regular intervals. Each of these primers initiates the formation of an Okazaki fragment that will be incorporated into the lagging strand. To assemble these segments, the RNA primer must be removed from each and the gap must be filled in with DNA nucleotides (which join to the 3′ end of the neighboring segment by phosphodiester bonds). Then the adjacent DNA segments are linked by phosphodiester bonds.

Fate of the Replication Forks

As replication proceeds from the single origin in a circular bacterial chromosome, what eventually happens to the two replication forks that form at that origin and move apart from each other? ____________________ [26] As replication proceeds from an origin in a linear eukaryotic chromosome, what will eventually happen to the two replication forks originating at that origin? ____________________ [27]

FIGURE 19-12

Each of the original duplex strands has a newly synthesized complement.

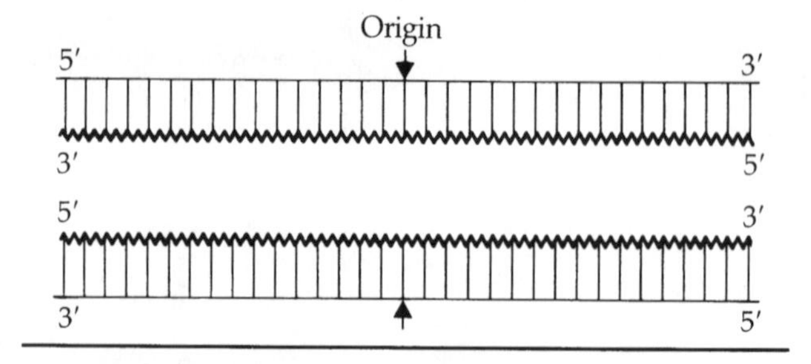

[26] They will move around the chromosome in opposite directions and eventually meet. [27] Eukaryotic chromosomes have several to many origins; the two replication forks arising from an origin will move away from each other until each meets with a replication fork moving toward it from a neighboring origin.

FIGURE 19-13

Theta form of a replicating prokaryotic chromosome.

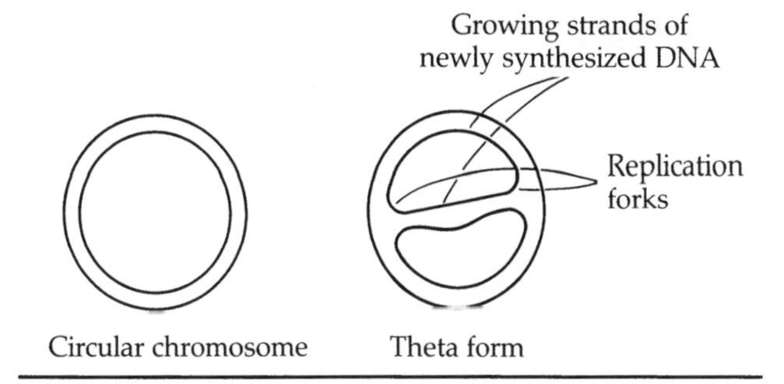

CONFIGURATIONS OF REPLICATING CHROMOSOMES

Observations of replicating chromosomes using the electron microscope and photographic plates exposed during autoradiographic-tracer studies show that chromosomes may exhibit characteristic configurations during DNA replication. Replicating circular chromosomes of prokaryotes, for example, take on a resemblance to the Greek letter theta (θ) as the two replication forks move around the chromosome in opposite directions and are referred to as **theta forms.** An interpretation of a theta form, with the replication fork sites noted, is shown in Figure 19-13.

As the two strands of a linear chromosome separate on both sides of an origin and as the replication forks move apart, the expansion created in the duplex is referred to as a **bubble,** or an **eye.** The number of eyes exhibited by a linear chromosome during its replication, of course, indicates the number of origins on the chromosome: linear chromosomes of viruses exhibit a single bubble, while those of eukaryotic organisms show many bubbles. Figure 19-14 shows a linear chromosome with three origins in bubble, or eye, form, with the sites of the replication forks indicated.

FIGURE 19-14

Section of a eukaryotic chromosome showing three bubbles. Arrows indicate sites of replication forks.

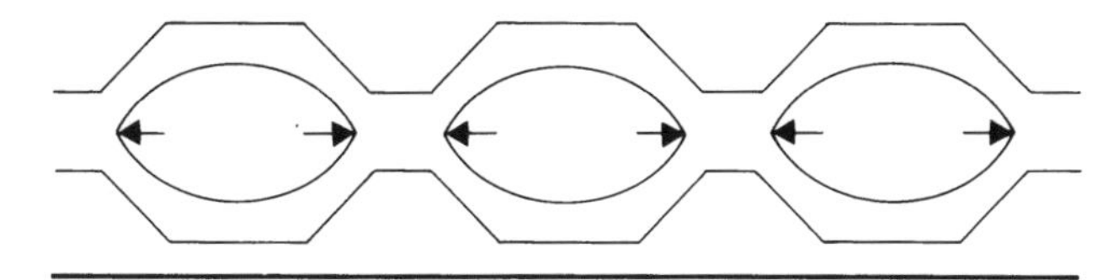

Note that both the theta and bubble forms are temporary configurations assumed by chromosomes as they replicate. They disappear when the replication is completed and do not reappear until the next round of chromosome replication.

ROLLING-CIRCLE DNA REPLICATION

The chromosomes of certain types of bacteria during conjugation (a process involving the one-way transfer of genetic material; see Chapter 22) and of certain kinds of viruses, such as the *E. coli* bacteriophages λ and $\phi\chi174$, carry out DNA replication by a method called **rolling-circle replication.** The process always begins with a circular duplex molecule. In viruses possessing a single-stranded chromosome, this circular duplex molecule is formed by the synthesis of a complementary strand for the chromosome. Depending on the particular type of virus or bacterium, rolling-circle replication results in the production of either circular or linear copies of the chromosome. The basic events of DNA replication that we have just described (that is, the growth of DNA through the addition of nucleotides, one at a time, to a preexisting 3′ end, with continuous replication along one duplex template strand and discontinuous replication along the other) occur during rolling-circle replication.

General Features

The key steps in the process are as follows. Rolling-circle replication is initiated when an endonuclease enzyme breaks a phosphodiester linkage at the origin in one of the two strands of the circular duplex DNA molecule, as shown in Figure 19-15. The strand that is broken, or nicked, is designated as the positive (+) strand and, following the break, it has two free ends. How is it possible to distinguish between these two ends? ______________________ [28]

Starting at the nicked strand's 5′ end, the hydrogen bonds holding the two duplex strands together begin to break, and the positive strand, led by its 5′ end, begins to peel off the unbroken negative (−) strand of the duplex. The intact portion of the duplex rotates or rolls on its axis (hence the term "rolling circle") as the positive strand, led by its 5′ end, continues to separate from the negative strand. The peeled-off portion of the positive strand is sometimes referred to as the "tail," and the resemblance of the circle and tail to the Greek letter sigma (σ) gives origin to the term **sigma mode** to describe the rolling-circle method

[28] One end is 3′ hydroxyl, and the other, 5′ phosphate.

FIGURE 19-15

Representation of a circular duplex DNA molecule comprising a bacterial chromosome. One strand of the duplex has been nicked.

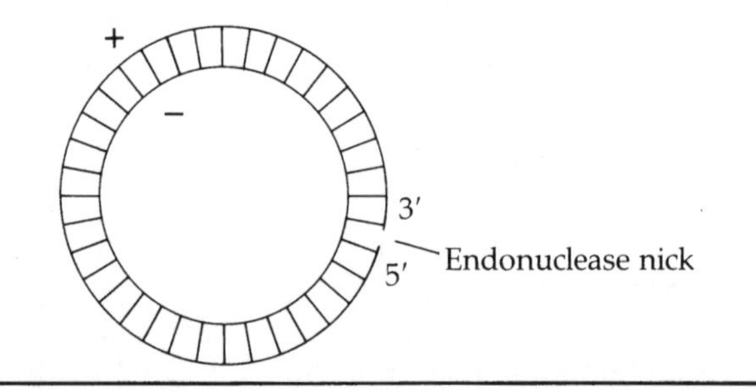

of replication. The peeling of the tail exposes portions of both the negative and positive strands as shown in Figure 19-16.

Once exposed, both the positive and negative strands can serve as templates, and DNA replication begins. The intact negative strand serves as a template for the addition of new deoxyribonucleotides to the 3′ end of the positive strand. Do you think an RNA primer is required to initiate this DNA synthesis? ______[29] Explain. ______________________________ ________[30] Do you think the synthesis at this 3′ end will be continuous or discontinuous? __________[31] Will the synthesis occurring at the 3′ end give rise to a new positive or a new negative strand? __________[32]

The exposed portion of the original positive strand, the tail, also serves as a template. As with DNA synthesis along the negative template, a 3′ end is essential for initiating DNA synthesis along this template. Based on our earlier discussions in this chapter, where do you think this 3′ end will come from? ______ ______________________________[33] As the positive strand continues to peel off and as more and more of it is exposed, are additional RNA primers required to initiate and sustain replication along this template? _________[34] Explain. _______________ ____________________________[35] Is the DNA replication along this template continuous or discontinuous? _________[36] Will the strand synthesized on this template be positive or negative? __________[37]

Synthesis of both new DNA strands continues as the original duplex rotates and more of the original positive strand unpeels as shown in Figure 19-17. As the discontinuous replication proceeds along the positive strand template, RNA primers and short DNA fragments are formed at regular intervals along its length. As described earlier, to join these into a single strand, the RNA primers are removed one ribonucleotide at a time, the gaps are filled with deoxyribonucleotides, and the DNA fragments are joined together.

Eventually, the entire length of the original positive strand unpeels from the original negative strand, as shown in Figure 19-18. At this point, what is joined end-to-end with the original positive strand and what does it complement? ______________________ _________________[38] At this point, what complements the original positive strand? _____________ ______________________________[39] What will be re-

FIGURE 19-16

Peeling of the "tail" of the positive strand exposes portions of both the negative and positive strands.

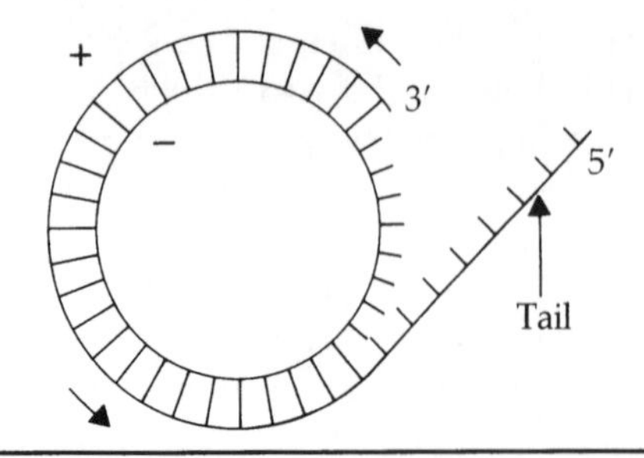

FIGURE 19-17

Both continuous and discontinuous synthesis occur as the original positive strand unpeels.

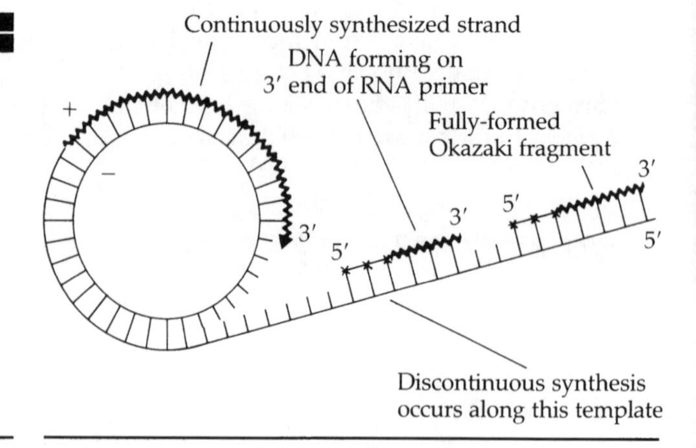

[29] No. [30] A 3′ hydroxyl end already exists as the 3′ end of the positive strand. [31] Continuous. [32] Positive strand.
[33] Synthesis of an RNA primer. [34] Yes. [35] Additional primers are required since the template strand runs from 5′ to 3′.
[36] Discontinuous. [37] Negative. [38] A newly synthesized positive strand that complements the original negative strand.
[39] A newly synthesized negative strand.

FIGURE 19-18

Once all of the original positive strand has unpeeled, both of the original strands have synthesized new complementary counterparts.

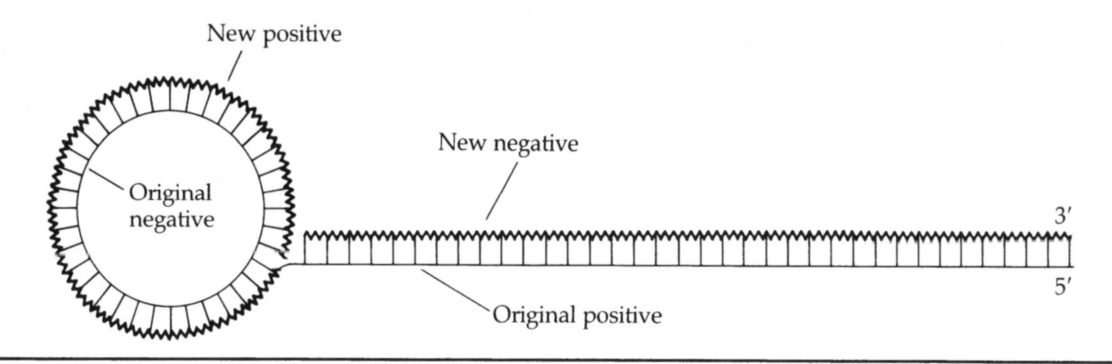

quired to separate the original positive strand and its new complement from the original negative strand and its new complement? ______________________________[40] This separation is brought about by a nuclease enzyme and the two duplex molecules that result are shown in Figure 19-19.

Immediately after separation occurs, what is the configuration of the duplex molecule containing the original negative strand? __________[41] What is the configuration of the duplex containing the original positive strand? __________[42] What would be required to convert the linear duplex into circular form? ______________________________[43] The enzyme responsible for the linear-to-circular conversion is DNA ligase. Is the rolling-circle pattern consistent with the semiconservative scheme of replication? __________[44] Explain. ______________________________[45]

Variations on Rolling-Circle Replication

The general scheme of rolling-circle replication just considered can vary depending on the type of bacterium or virus involved.

λ Bacteriophage Variation is shown, for example, by λ, a bacteriophage of *E. coli.* In this virus, the duplex DNA chromosome always exists in linear form, but once injected into a host cell, it may assume a circular configuration. Replication always proceeds from this circular form and occurs by the rolling-circle method. However, the process does not stop with the formation of a single duplicate chromosome, but continues until

FIGURE 19-19

Upon the completion of replication, the two daughter duplexes separate. The linear duplex then assumes a circular configuration.

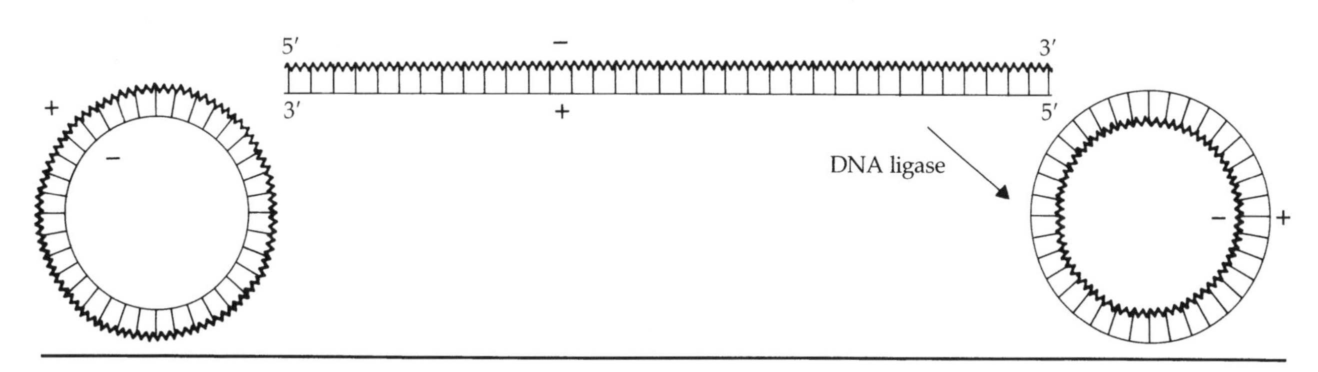

[40] An enzyme to break the phosphodiester bond linking the 3′ terminal nucleotide of the original positive strand to the first nucleotide at the 5′ end of the newly synthesized positive strand. [41] Circular. [42] Linear. [43] Joining the two ends of the duplex. [44] Yes.
[45] Since one strand of each daughter duplex is one of the strands of the original duplex molecule and the other is newly synthesized, the pattern is semiconservative.

one linear string comprising several chromosomes has been produced. Several copies result because the negative strand of the original duplex keeps on turning, and with each turn, it repeats its template role and guides the continuous synthesis of another complementary positive strand. As the tail peels off (growing longer with each turn), it serves as a template for the discontinuous synthesis of a complementary series of negative strands joined end to end. What results is a linear duplex molecule that contains several copies of the original duplex chromosome. A linear molecule such as this, consisting of a series of repeating subunits, is called a **concatemer.** It is subsequently subdivided by a nuclease enzyme into the viral chromosomes which are then incorporated, in linear configuration, into protein coats.

$\phi\chi$174 Bacteriophage A different variation is shown by $\phi\chi$174, another *E. coli* bacteriophage that has a chromosome consisting of a single-stranded (positive) circular molecule of DNA. When injected into a host cell, this strand serves as a template to guide the formation of a complementary negative strand that remains hydrogen-bonded to it to make up a duplex molecule. This duplex, known as the **replicative form,** or **RF,** undergoes rolling-circle replication to make additional copies of itself. When a number of RFs are present in the host cell, they carry out rolling-circle replication but in a modified way. As the 5′ of the positive strand peels off from an RF, a viral protein (Rep A) binds to it, thereby preventing it from serving as a template for the formation of a complementary negative strand. Because of this modification, only positive strands are now synthesized. At the end of one cycle, the displaced positive strand is cut from the tail and circularized by Rep A. Continuation of this cycle supplies many chromosome copies.

Bacterial Conjugation In the rolling-circle replication of a bacterial chromosome during conjugation, the 5′ end peeling off the circular duplex chromosome of the donor cell moves through a cytoplasmic connection known as the conjugation tube into the recipient cell. The continuous replication guided by the negative strand of the original duplex takes place in the donor cell while the discontinuous replication guided by the positive strand template occurs in the recipient cell after the positive strand has been transferred. Conjugation is described in much greater detail in Chapter 22.

CROSSING OVER AT THE MOLECULAR LEVEL: THE HOLLIDAY MODEL OF RECOMBINATION

This model, proposed initially by R. Holliday in 1964 and subsequently modified, presents the molecular details of how crossing over may occur between a pair of homologous chromosomes. As we consider this model, keep in mind the following points that were discussed in Chapter 1: (1) crossing over occurs after chromosome replication and while the replicated homologs are synapsed during prophase of meiosis I; (2) each replicated chromosome consists of two sister chromatids, and the four chromatids of a synapsed pair of homologous chromosomes make up a configuration known as a tetrad; (3) each crossover occurs between two nonsister chromatids of a tetrad; and (4) each chromatid contains a duplex DNA molecule.

Prior to replication, a homologous pair of chromosomes can be represented as shown in Figure 19-20a. Note that one chromosome carries linked genes *A* and *B* while the other carries their counterpart alleles, *a* and *b*. Since each homolog is derived from a different parent, one is depicted using a solid line and the other using a dashed line. The DNA duplex of each chromosome is represented in Figure 19-20b, where details of the centromere have been omitted.

FIGURE 19-20

Schematic representation of a homologous pair of chromosomes and the DNA duplexes they contain.

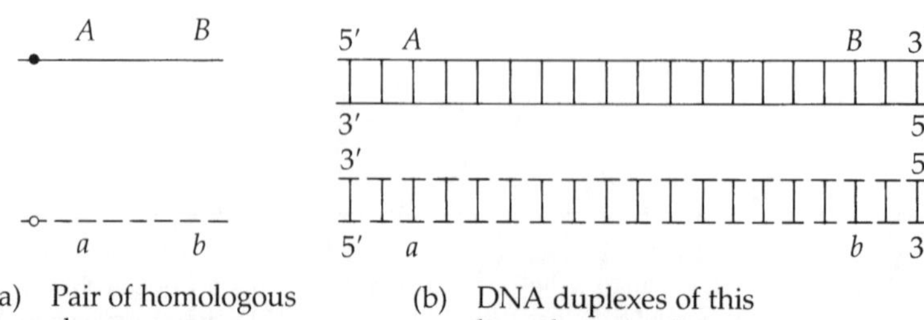

FIGURE 19-21

Replicated pair of homologous chromosomes and the DNA duplexes they contain.

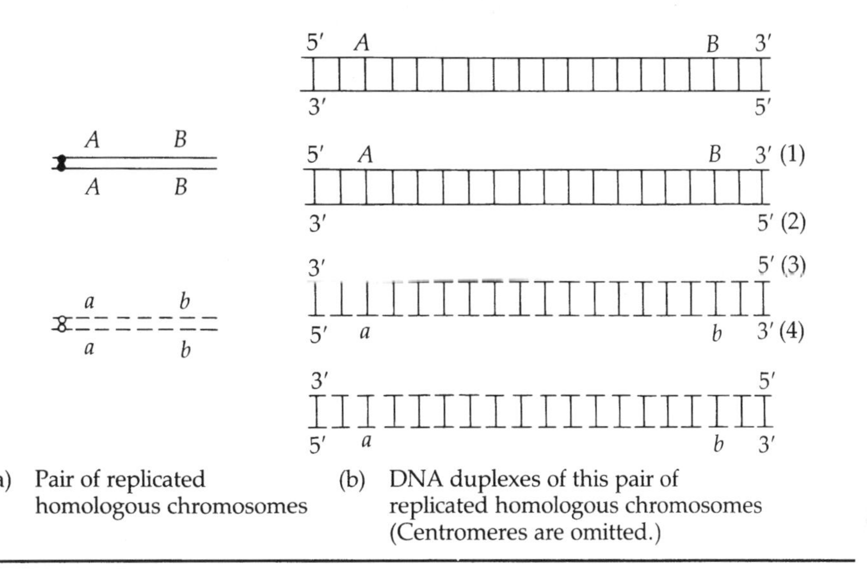

(a) Pair of replicated homologous chromosomes (b) DNA duplexes of this pair of replicated homologous chromosomes (Centromeres are omitted.)

Following replication and synapsis, the homologous pair of chromosomes can be represented as shown in Figure 19-21a; each replicated chromosome consists of two sister chromatids. Each chromatid contains a duplex DNA molecule, as shown in Figure 19-21b.

Assume that a single crossover occurs between the two inner nonsister chromatids of this tetrad. To illustrate the Holliday model, we need to consider the DNA duplexes found within these chromatids. The DNA strands making up these two duplexes are numbered 1 through 4 on the far right in Figure 19-21. Complementary strands 1 and 2 make up the duplex molecule of one chromatid and complementary strands 3 and 4 make up the duplex molecule of the nonsister chromatid. The key events in the Holliday model are described in the following paragraphs. Note that the diagrams illustrating these events will show just the four DNA strands of the two nonsister chromatids that participate in crossing over.

1. An endonuclease enzyme produces a nick in the sugar-phosphate backbone of one DNA strand of each duplex molecule. These nicks occur at equivalent sites in strands with the same polarity and are shown in strands 2 and 3 in Figure 19-22.

2. On one side of each nick, a part of the nicked strand separates, by breaking hydrogen bonds and unwinding, from the equivalent complementary section of it uncut partner strand. This is shown in Figure 19-23, where part of strand 2 separates from strand 1 and part of strand 3 separates from strand 4.

3. The free or displaced portion of each cut strand now pairs up with the corresponding portion of the nonnicked, intact strand of the opposite duplex. In Figure 19-24, the displaced portion of strand 2 pairs up with the corresponding portion of strand 4 and the displaced portion of strand 3 pairs up with the corresponding portion of strand 1. This complementary base pairing between strands from different parental duplexes generates sections of what is known as **heteroduplex DNA** within each duplex (circled in Figure 19-24) and establishes a cross bridge between the two duplexes.

4. DNA ligase establishes covalent bonds between the strand ends as shown in Figure 19-25. This bonding joins DNA from two different sources to generate two recombinant DNA strands. These two strands

FIGURE 19-22

DNA duplexes from two inner nonsister chromatids of a tetrad. The inner strand of each has been nicked by an endonuclease.

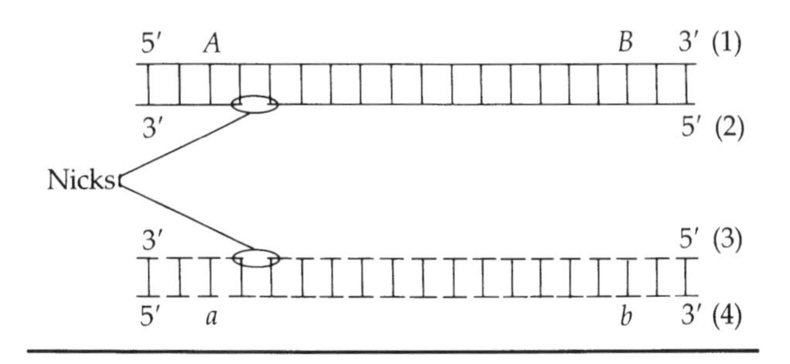

FIGURE 19-23

A part of each nicked strand separates from the equivalent complementary section of its uncut partner strand.

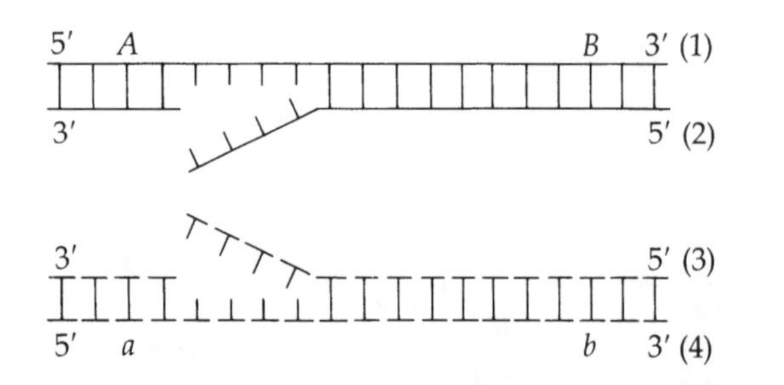

bridge the two duplexes in a configuration known as a **Holliday structure,** or **chi form.**

5. The cross bridge can move along the two duplexes in a process called **branch migration.** During this process an additional portion of each recombinant strand breaks the hydrogen bonds linking it to its original complementary strand and reestablishes these bonds with a portion of the nonnicked strand of the nonsister chromatid. This lengthens the portion of each recombinant strand that is transferred from one duplex to another, thereby lengthening the section of heteroduplex DNA within each duplex, as shown in Figure 19-26.

6. The two duplexes spread apart from each other to form the extended configuration shown in Figure 19-27.

FIGURE 19-24

The displaced portion of each cut strand pairs up with the corresponding portion of the intact strand of the opposite duplex to produce a heteroduplex DNA region within each duplex.

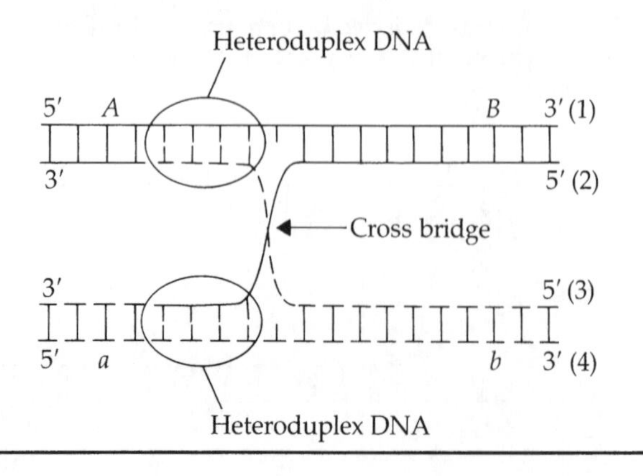

FIGURE 19-25

DNA ligase establishes covalent bonds between the strand ends (as shown) to generate two recombinant DNA strands.

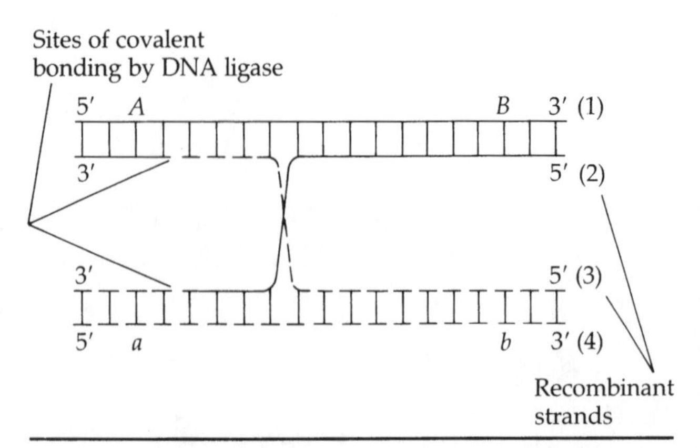

7. Half the time the extended configuration shown in Figure 19-27 undergoes no change and half the time it **isomerizes** to generate an alternative form. The series of steps involved in this isomerization are as follows. (1) First, the bottom half of the configuration (that is, the portion below the two crossed strands) rotates 180 degrees relative to the top half to produce the configuration shown in Figure 19-28. (2) Next, the entire structure rotates 90 degrees counterclockwise in the plane of the paper to produce the configuration shown in Figure 19-29. (3) Finally, there is a 180-degree rotation of the top half of the configuration relative to the bottom half to produce the configuration shown in Figure 19-30. The net effect of this series of conformational changes is to reposition the DNA strands. The

FIGURE 19-26

The cross bridge can move along the two duplexes (branch migration) lengthening the sections of heteroduplex DNA within each duplex.

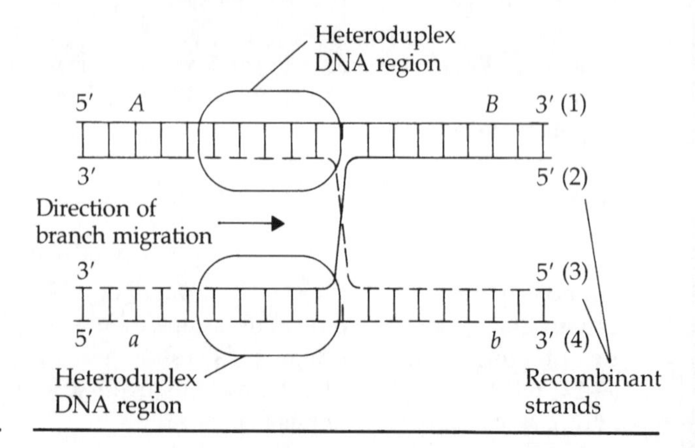

FIGURE 19-27

The two duplexes spread apart from each other to form an extended configuration.

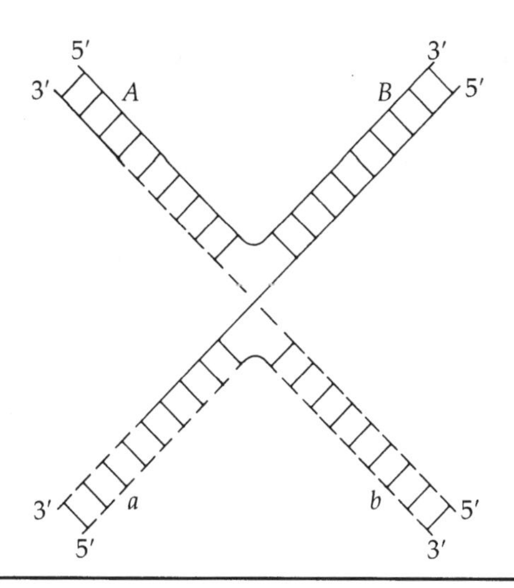

two recombinant strands that initially bridged the duplexes become outside strands and the two nonrecombinant strands that were initially the outside strands become the bridging strands, as shown in Figure 19-31.

8. Next, the two duplexes are separated by an endonuclease that cuts the two bridging strands where they cross each other. The nature of the duplexes that result (Figure 19-32) depends on the configuration possessed by the joined duplexes prior to their separation.

FIGURE 19-28

The first step in isomerization. The bottom half of the configuration shown in Figure 19–27 rotates 180 degrees relative to the top half to produce this configuration.

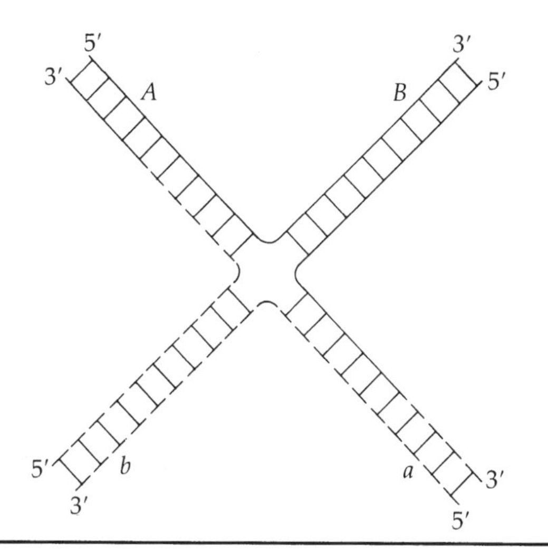

FIGURE 19-29

The entire structure shown in Figure 19-28 rotates 90 degrees counterclockwise in the plane of the paper to produce this configuration.

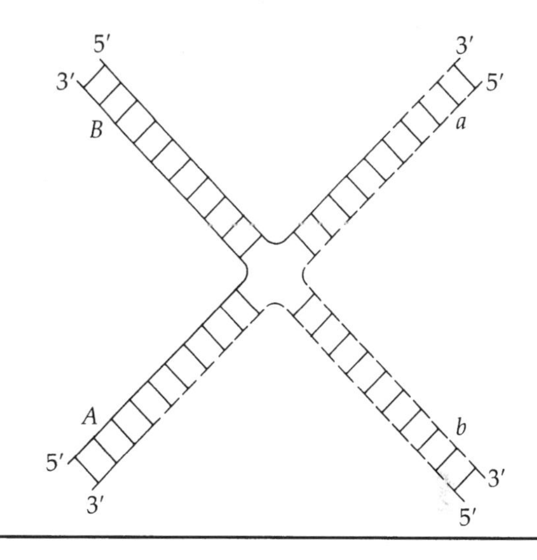

If the joined duplexes show the original configuration (where the two recombinant strands bridge the duplexes as shown in Figure 19-26), they produce the set of duplexes shown in Figure 19-32a. Here, following the ligasing of the nicks, each duplex ends up with one recombinant strand and one nonrecombinant strand and carries a section of heteroduplex DNA. The arms flanking the heteroduplex region show the same arrangement of the genes under consideration that were present in the original duplex molecules. Thus, each chromatid that is formed is nonrecombinant with

FIGURE 19-30

A 180-degree rotation of the top half relative to the bottom of Figure 19-29 produces this configuration.

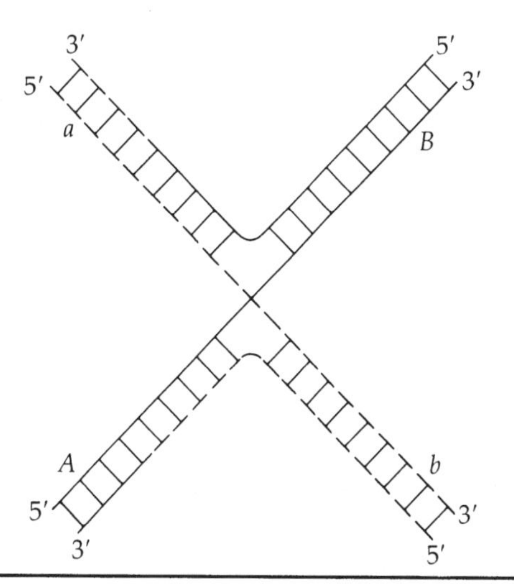

FIGURE 19-31

The series of conformational changes reposition the DNA strands so the recombinant strands are the outside strands and the nonrecombinant strands are the bridging strands.

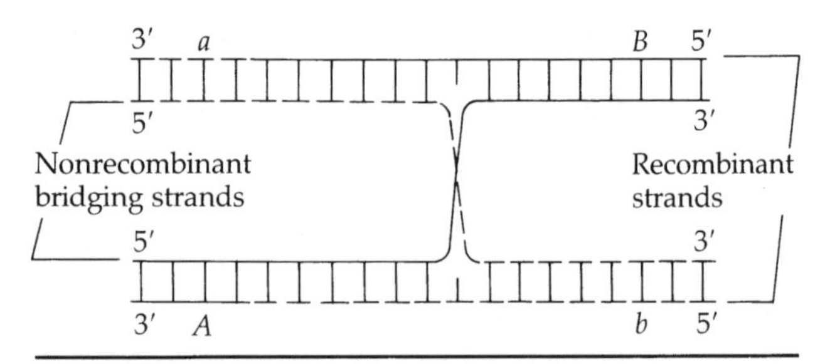

regard to these genes and shows the same *AB* and *ab* flanking-gene combinations exhibited by the parents.

If the joined duplexes show the isomerized configuration (where the two nonrecombinant strands bridge duplexes as shown in Figure 19-31), they produce the set of duplexes shown in Figure 19-32b. Here, following ligasing of the nicks, each duplex consists of two recombinant strands and has a region of heteroduplex DNA. The arms flanking this heteroduplex region show arrangements of genes differing from those of the original duplex molecules. Thus, each chromatid that is formed is recombinant with regard to these genes, showing either the *Ab* or *aB* combination.

These two outcomes are equally likely and thus half of all crossovers generate recombinations of the flanking genes. (Note that with each outcome, crossing over results in at least one heteroduplex region within each duplex.)

Base Mismatches in Heteroduplex DNA and Mismatch Repair

The two strands making up the heteroduplex DNA regions formed during recombination may not complement each other exactly since they are derived from different parental duplexes. In other words, the heteroduplex region may contain **base-pairing mismatches**—for example, A might be paired with G rather than with its normal complement, T. Such mismatches may be repaired by a nuclease enzyme that recognizes the mismatch and removes one base of the mismatched pair and a few of its neighboring nucleotides to create a gap in one DNA strand of the heteroduplex. Then DNA polymerase guides the placement of nucleotides to fill the gap, using the section of the intact strand opposite the gap as a template.

Gene Conversion

Chapters 14 and 15 describe how the inheritance in haploid fungi such as *Neurospora* is studied through tetrad analysis. As discussed there, ascospore segregation patterns are used to determine the distance between a gene and its centromere and between two linked genes. You will recall that two linked genes, for example, c and d, studied in the cross $c^+d^+ \times cd$, segregate into tetrads of three types: parental ditypes (produced in the highest frequency), tetratypes (intermediate frequency), and nonparental ditypes (lowest frequency) (see Chapter 15). Regardless of its type, each tetrad usually contains four copies of each of the four alleles under consideration. However, when the alleles under consideration are very closely linked, segregation may produce unexpected patterns. For example, if the c and d loci are very closely linked, the cross $c^+d^+ \times cd$ will generate the usual tetrads with four copies of each allele, but in addition, some tetrads will show unusual numbers of these alleles. For example, the ascospore content of some tetrads may be as follows: two c^+d^+, two c^+d, two cd, two cd. Here, c^+ and c occur in the expected equal numbers (4:4) but d^+ and d occur in unequal numbers, in a ratio of 2:6. Since two of the expected copies of d^+ appear to have been converted to d, the term **gene conversion** is used to describe this phenomenon.

Mismatch Repair and Gene Conversion

Gene conversion can be explained by the mismatch repair that occurs in heteroduplex regions formed according to the Holliday model during crossing over.

FIGURE 19-32

The product duplexes. The two outcomes are equally likely.

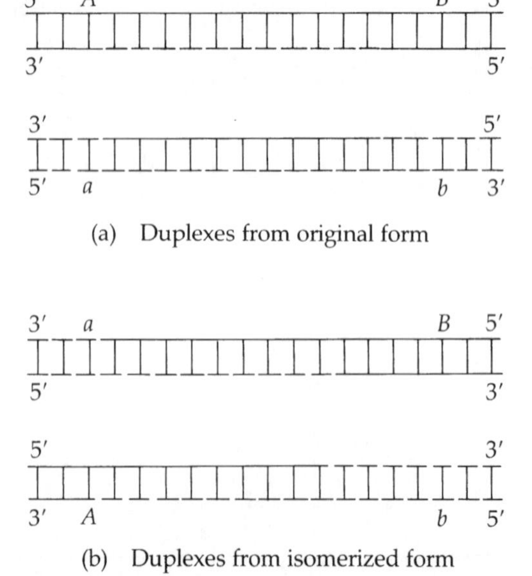

(a) Duplexes from original form

(b) Duplexes from isomerized form

FIGURE 19-33

Four chromatids comprising a tetrad. The *c* and *d* loci are very closely linked.

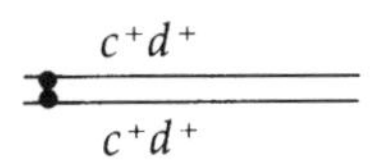

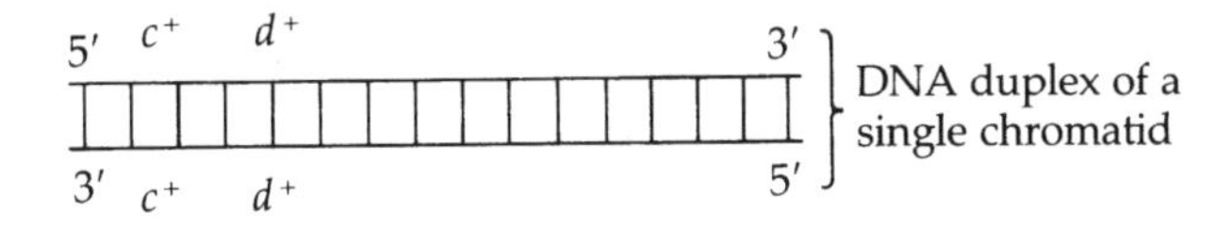

Figure 19-33 shows the four chromatids comprising a tetrad with the two upper sister chromatids carrying genes c^+ and d^+ and the two lower sister chromatids carrying their counterpart alleles *c* and *d*.

Figure 19-34 shows the four duplex DNA molecules found in these four chromatids following a single crossover that involved the two inner chromatids. This crossover produced a heteroduplex DNA region in the duplexes of each of the two inner chromatids. Note that one strand in each heteroduplex region carries a single-stranded nucleotide sequence for allele d^+ paired with a single-stranded nucleotide sequence for allele *d*.

FIGURE 19-34

Four duplex molecules found in four chromatids shown in Figure 19-33 following a single crossover between the two inner chromatids.

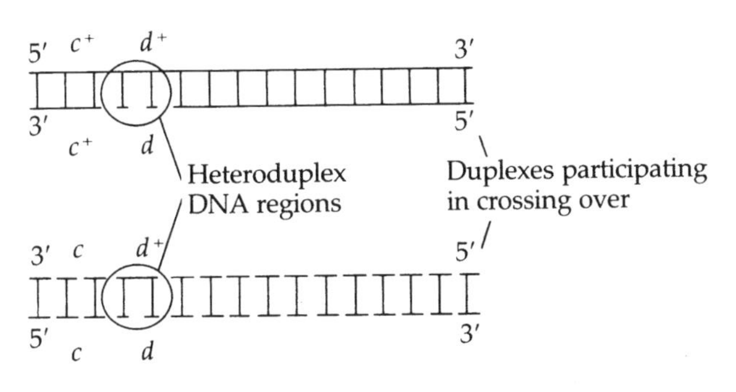

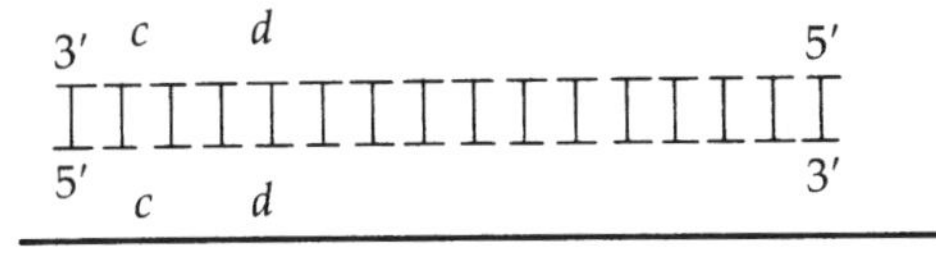

FIGURE 19-35

A section has been excised from one strand of each heteroduplex.

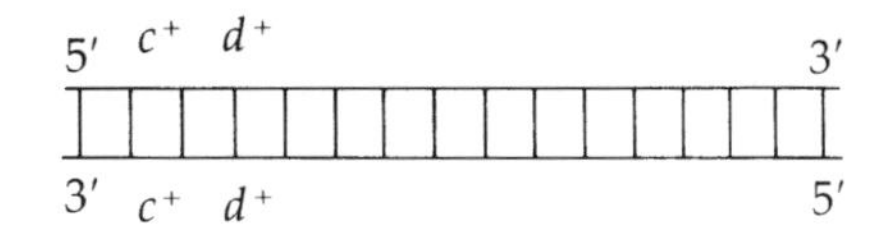

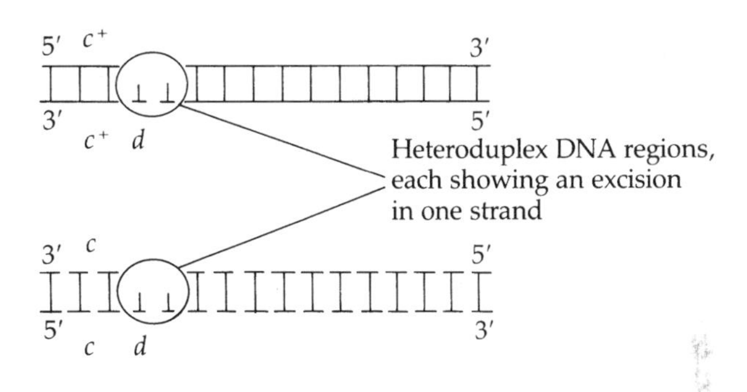

If the heteroduplex DNA contains mismatched bases, mismatch repair may occur. This is initiated with the excision of a portion of one strand of the heteroduplex region at the site of the *d* locus. However, the mismatch repair may excise either of the bases involved in the mismatch and thus, depending on which base is removed, a segment may be excised from either heteroduplex strand. Figure 19-35 shows that a section has been excised from one strand of both heteroduplexes. In this case, the single-stranded nucleotide sequence for allele d^+ was removed from each heteroduplex. In other cases, the sequence for allele *d* could have been excised from both heteroduplexes. In yet other instances, the sequence for allele d^+ could have been removed from one heteroduplex with the sequence for allele *d* removed from the other. The gaps created by excision are then filled in by DNA synthesis, with the intact strand of each heteroduplex region serving as a template. The duplexes that result after this gap filling are shown in Figure 19-36a. This process caused a gene conversion in the upper heteroduplex where d^+ was converted to *d*. There was no gene conversion in the lower heteroduplex (the repair restored the original *d* allele).

Following the completion of meiosis and the subsequent mitotic division that characterizes ascospore production in *Neurospora*, the spore case or ascus produced from the nucleus containing the four duplexes

FIGURE 19-36

The four duplexes give rise to the ascospores shown in the ascus.

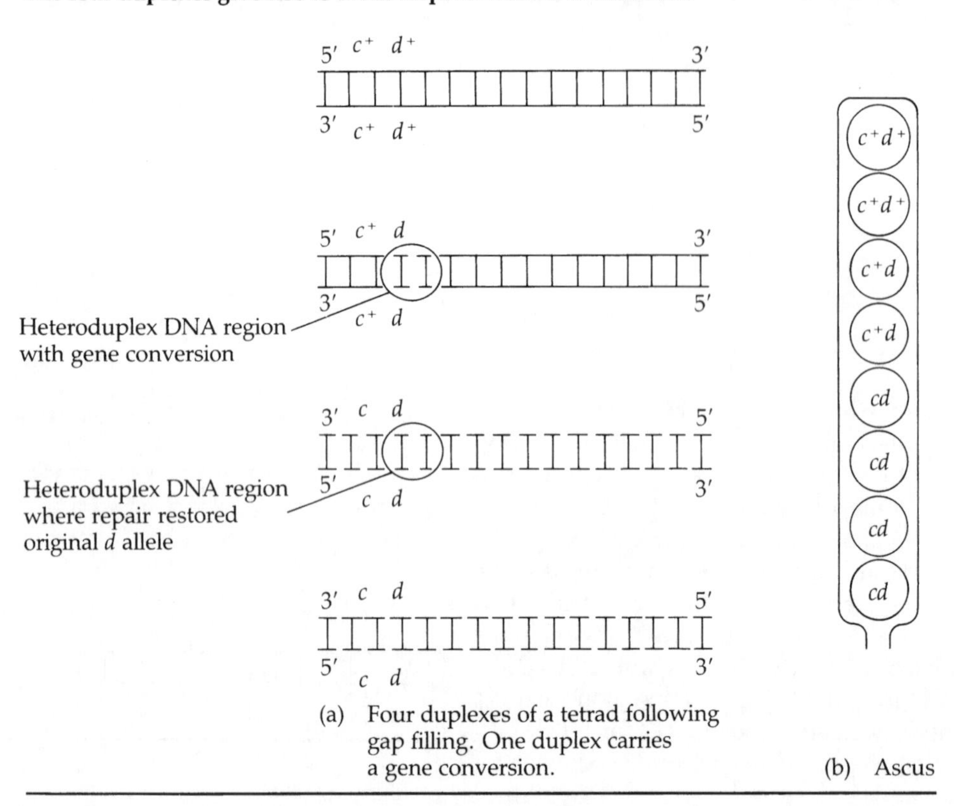

(a) Four duplexes of a tetrad following gap filling. One duplex carries a gene conversion.

(b) Ascus

shown in Figure 19-36a would contain the following ascospores: two c^+d^+, two c^+d, two cd, two cd, as shown in Figure 19-36b. The ascus contains the expected number of c genes, with c^+ and c in a 4:4 ratio, but an unexpected number of d genes, with d^+ and d in a 2:6 ratio.

Other types of tetrads with unusual ratios of genes may result from the $c^+d^+ \times cd$ cross. To illustrate this, what ratios of these two genes would occur in the ascus if the heteroduplex segments excised from both duplexes were opposite those removed in the scheme we have just considered? ______________________ ______________________[46]

SUMMARY

Based on experimental results, Meselson and Stahl concluded that DNA replication in *E. coli* is semiconservative as proposed by Watson and Crick. Experiments carried out by other researchers on other organisms, including some eukaryotes, provide additional support for the semiconservative nature of DNA replication.

DNA replication is initiated when the strands of a duplex unwind and separate from each other at one or more origin sites. Two replication forks develop at each origin and move apart as the two strands of the duplex continue to unwind. At each fork, both strands of the original duplex serve as templates to guide the synthesis of new complementary strands. This synthesis occurs continuously along one template strand and discontinuously along the other template strand. Short RNA primers, formed as complements to sections of the template strands, are essential for DNA synthesis since they provide a 3′ hydroxyl end to which the DNA polymerase III enzyme attaches the initial deoxyribonucleotide.

Synthesis of the continuous, or leading, strand requires a single RNA primer that is synthesized at the origin and to which the continuous addition of deoxyribonucleotides forms a single, lengthening DNA strand. Formation of the discontinuous, or lagging,

[46] Excision would remove the section carrying d from the lower strand of the upper heteroduplex-carrying duplex and the section carrying d from the lower strand of the lower heteroduplex-carrying duplex. Following gap filling, each strand of each duplex would carry d^+, giving rise to an ascus containing two c^+d^+, two c^+d^+, two cd^+, two cd. The c^+ and c alleles would be in a 4:4 ratio and d^+ and d would be in a 6:2 ratio.

strand involves several to many RNA primers (with the exact number depending on the strand length), the first of which is synthesized at the origin, with the rest formed at regular intervals along the template strand. Each of these primers initiates the formation of an Okazaki fragment that becomes incorporated into the lagging strand. To assemble these fragments, the RNA primer must be removed from each, the gap filled in with DNA nucleotides, and the adjacent DNA segments linked by phosphodiester bonds.

Autoradiographic studies of replicating chromosomes have revealed that the circular chromosomes of numerous prokaryotes assume theta forms as the two replication forks move around the chromosome in opposite directions. The strands of a linear chromosome separate on both sides of each origin and as the replication forks move apart, the expansion creates a bubble or eye. The chromosomes of certain types of bacteria during conjugation and of certain kinds of viruses carry out DNA replication by the rolling-circle method: one strand of a circular duplex molecule becomes nicked and peels away, serving as the template for the formation of a discontinuously synthesized counterpart, while the other original strand guides the formation of a complement by continuous synthesis.

The Holliday model presents the molecular details of how crossing over may occur between a pair of homologous chromosomes. Recombination results in the formation of at least one heteroduplex region in each of the DNA duplexes involved in the crossover. Since the two strands making up each heteroduplex region are derived from different parental duplexes, they may contain base mismatches. Mismatch repair may lead to gene conversion which, in fungi like *Neurospora,* gives rise to tetrads with unusual ratios of one or more of the genes under study. Gene conversion represents an avenue for altering the genome.

PROBLEM SET

19-1. As described in this chapter, Meselson and Stahl grew ^{15}N-labeled *E. coli* on a medium containing ^{14}N and extracted DNA from samples of cells following replication. After subjecting the DNA produced in the second round of replication (that is, the DNA derived from generation-2 bacteria) to density-gradient equilibrium centrifugation, they found two types of duplexes. One type's density was halfway between that expected for ^{14}N and ^{15}N duplexes and the other type's density was that expected for ^{14}N duplexes. Heating a double-helix DNA molecule breaks the hydrogen bonds holding the complementary base pairs together, causing the two strands to separate, or denature. Once separated, the strands can be subjected to density-gradient centrifugation. Assume that a sample of the duplexes extracted from Meselson and Stahl's generation-2 *E. coli.* are heat denatured and subjected to density-gradient centrifugation. Predict the banding pattern shown by the DNA strands following centrifugation.

19-2. **a.** If conservative replication occurred in *E. coli*, what banding pattern would have been produced in the experiment described in problem 19-1?

b. If dispersive replication occurred in *E. coli*, what banding pattern would have been produced in the experiment described in problem 19-1?

19-3. After bacteria are grown for many generations on a medium containing normal, nonradioactive phosphorus (P), some are transferred to a medium where all the phosphorous is in the form of its radioactive isotope, ^{32}P. The bacteria use this ^{32}P to synthesize (among other things) the deoxyribonucleotides they need for DNA replication. The bacteria grow on this ^{32}P medium for three generations and a sample of bacteria is removed at the end of each round of replication. DNA duplexes from these samples are isolated and analyzed for ^{32}P content.

a. For each of the three samples of bacteria, estimate the proportion of duplexes that would have (1) all of their nucleotides containing ^{32}P, (2) some of their nucleotides containing ^{32}P and some containing normal P, and (3) all of their nucleotides containing normal P.

b. For the duplexes that have some nucleotides containing ^{32}P and some containing normal P, estimate the proportion of ^{32}P nucleotides in these duplexes.

19-4. Assume that the conditions of the experiment in problem 19-3 are the same except that following the second round of replication, the bacteria are transferred back to a medium where all the phosphorus is nonradioactive P. Describe the distribution of radioactive and nonradioactive P in the duplexes in the sample taken after the next (third) round of replication. In what proportion would the various kinds of duplexes occur in this sample?

19-5. Assume that the DNA replication in the *E. coli* described in problem 19-3 occurred according to the conservative model. What would you expect regarding the proportion of duplexes, in each of the three samples, that would have (1) all of their nucleotides containing ^{32}P, (2) some nucleotides with ^{32}P and some with nonradioactive P, and (3) all nucleotides containing nonradioactive P?

19-6. The following enzymes play key roles in DNA replication: DNA polymerase I, DNA polymerase III, RNA primase, and DNA ligase.

a. Identify the function of each enzyme.

PROBLEM SET

b. Beginning with the initiation of DNA replication, list these enzymes in the order in which they function; that is, identify which would be required first, second, and so on.

c. Some of these enzymes are used much more frequently during discontinuous replication than during continuous replication. Identify these enzymes and explain why this is so.

19-7. Among the different kinds of protein molecules identified as playing important roles at DNA replication forks are single-stranded DNA-binding protein, DNA gyrase, and DNA helicase.

a. Identify the function of each of these proteins.

b. List these proteins in the order in which they prepare the replication fork prior to DNA replication.

c. Is there any difference in the frequency with which these proteins function during discontinuous replication compared with continuous replication? Explain.

19-8. Distinguish between a template and a primer. Why are both essential during DNA synthesis?

19-9. According to autoradiographic studies on replicating *E. coli* chromosomes, the synthesis of new DNA strands at a replication fork occurs along both strands of the original duplex with the overall growth of the new strands occurring in the same direction. Given that the template strands are antiparallel and that the enzyme guiding this synthesis, polymerase III, catalyzes the addition of nucleotides only in the 5′-to-3′ direction, how does synthesis occur simultaneously along both strands?

19-10. The following figure shows a replication fork in a segment of duplex DNA, with the upper template strand running from left to right in the 3′ to 5′ direction.

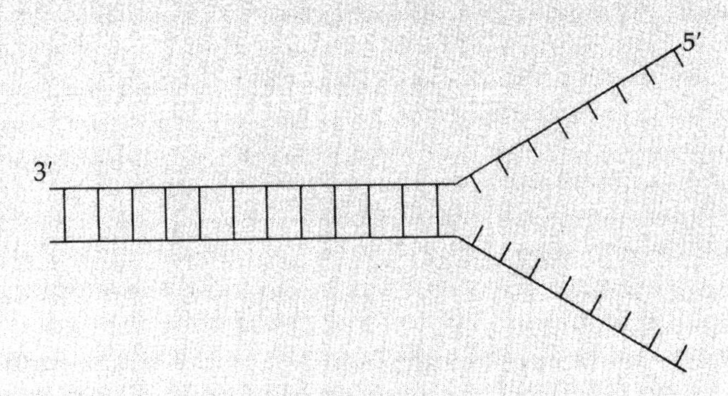

a. In what direction (left or right) does the replication fork move as DNA synthesis proceeds?

b. Along which template strand (upper or lower) does continuous synthesis occur?

c. Along which template strand are numerous Okazaki fragments formed?

d. Along which template strand are numerous RNA primers synthesized?

e. Along which template strand is the leading strand formed?

19-11. In the fruit fly *Drosophila melanogaster,* DNA replication at a single replication fork occurs at a rate of about 2600 nucleotide pairs per minute. The DNA molecule occurring in one of the largest chromosomes of this species has been estimated to contain 6×10^7 nucleotide pairs.

PROBLEM SET

a. If the replication of this molecule was initiated at a single origin in the middle of the chromosome, estimate the time, in days, required for complete replication of the chromosome.

b. Estimates based on living cells indicate that this chromosome replicates in about four minutes. Assuming that the origins are spaced equally along the DNA, how many of them would be required to completely replicate this chromosome in four minutes?

19-12. Supplying a radioactively labeled precursor of DNA, tritiated thymidine, to bacteria undergoing DNA replication results in its incorporation into the newly synthesized DNA. During the first 15 seconds or so the label is found in short fragments of DNA which are about 1000 to 2000 nucleotides long.

a. As the interval of exposure to the tritiated thymidine increases, what would you predict about the length of the DNA fragments containing the label; that is, would they be longer, shorter, or about the same length as those detected during the first 15 seconds? Explain.

b. Would it be possible to detect short fragments of labeled DNA throughout the interval of DNA synthesis?

c. Would your answer to 19-12b differ if DNA replication were continuous along both strands of the duplex bacterial chromosome? Explain.

d. What is the advantage of using thymidine in a study of this sort, rather than a radioactive form of one of the other deoxyribonucleotides?

19-13. Assume that a strain of *E. coli* carries a mutation causing a deficiency of the enzyme DNA ligase and that this mutation expresses itself only at temperatures up to 30°C. Above 30°, normal amounts of DNA ligase are formed. Assume that bacteria of this strain are grown at a temperature below 30° and supplied with tritiated thymidine (see Problem 19-12) just prior to DNA replication. Halfway through replication, the DNA from these cells is isolated and studied. Characterize the DNA molecules you would expect to find with regard to relative length and the presence or absence of the labeled thymidine.

19-14. A duplex DNA molecule makes a complete turn every 3.4 nanometers, or 0.0034 micrometers. The circular duplex chromosome of *E. coli* is about 1300 micrometers long. This chromosome has a single origin which gives rise to two replication forks, with the replication being bidirectional. Traveling in opposite directions and moving at the same rate, these forks require, at 37°C, about 40 minutes to move around the entire chromosome. Assume that two swivel points, one beyond each replication fork, relieve the twist tension by allowing the DNA duplex to untwist as the fork travels. How many revolutions per minute would each swivel point experience during the replication of the chromosome? (Assume that one revolution is required for each helical turn.)

19-15. J. Cairns carried out an experiment in which *E. coli* were grown on a medium containing tritiated thymidine for one 30-minute generation period. Autoradiography of samples of these cells indicated that this procedure caused one strand of the new chromosomes to be labeled. The cells were left in the same medium for another 30 minutes during which samples of cells were withdrawn at regular intervals. Autoradiographs of cells from several of these samples showed chromosomes in theta forms.

a. Using a dashed line to represent a labeled DNA strand and a solid line to represent an unlabeled DNA strand, sketch a chromosome produced

PROBLEM SET

at the end of the first 30-minute interval. In a separate sketch, use a dotted line to indicate how the autoradiograph produced from your chromosome would appear. Assume that each radioactive strand appears as a single strand of dots in the autoradiograph.

b. Using a dashed line to represent a labeled DNA strand and a solid line to represent an unlabeled DNA strand, sketch a theta-form chromosome like the ones Cairns may have observed during the second 30-minute interval of his study. In a separate sketch, use dotted lines to indicate how the autoradiograph produced from the theta-form chromosome would appear.

19-16. The following figure represents a circular chromosome of duplex DNA that is about to undergo rolling-circle replication. The outer (+) strand has been nicked and a section of it has separated from the intact (−) strand to form a short tail.

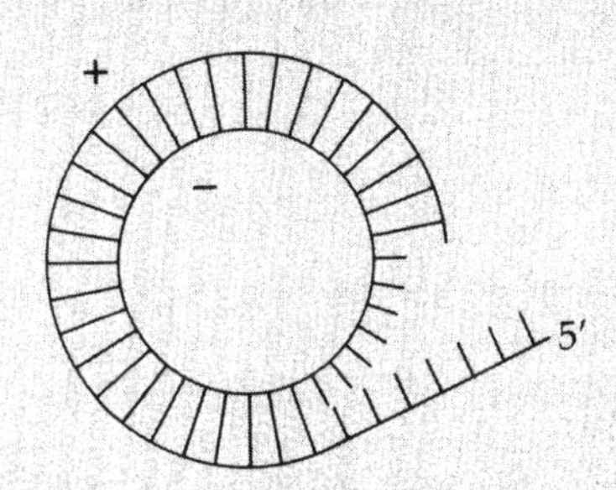

a. Will the chromosome rotate clockwise or counterclockwise as replication proceeds?

b. Which of the original strands, positive or negative, will serve as a template for discontinuous synthesis?

c. Along which template strand will Okazaki fragments form?

d. Is an RNA primer required to initiate the synthesis of the new continuously synthesized strand? Explain.

19-17. Gene conversion has been studied in a variety of fungal species, including the smut fungus *Ustilago*. R. Holliday isolated a strain of this fungus that was deficient in a nuclease enzyme that, along with DNA polymerase, brings about the repair of base-pairing mismatches in heteroduplex DNA. What prediction would you make about the frequency of gene conversion in this strain relative to the wild-type strain?

19-18. Assume that in the *Neurospora* cross $a^+s^+ \times as$ recombination produces a heteroduplex section that consists of one strand of s^+ and the other strand of s in two of the four duplexes in a nucleus. These strands of s^+ and s differ in a single nucleotide. Mismatch repair occurs in one duplex and restores it to its original state in this region, but fails to occur in the heteroduplex region of the other duplex. Keeping in mind that the four meiotic products undergo mitosis to form the ascospores, what gene ratios would you expect in the ascus that results?

19-19. Through crosses that simultaneously study the inheritance of four linked genes in yeasts, for example $s^+t^+u^+v^+ \times stuv$, the two intermediate genes are often found to be simultaneously converted with a relatively high frequency. What feature of the Holliday model explains this phenomenon?

CHAPTER

20

The Basis of Prokaryotic Inheritance

INTRODUCTION

As discussed in Chapter 1, prokaryotic organisms include the blue-green algae and bacteria, and are characterized by the absence of membrane-bound organelles such as a nucleus. Most of their genetic material is in the form of circular chromosomes containing a double-stranded molecule of DNA. Recombination in bacteria depends upon the transfer of genetic material from one cell to another, with the donated DNA usually replacing homologous sections of the recipient's chromosome. In nature, this transfer is achieved in one of three ways: conjugation, transduction, or transformation. **Conjugation** involves direct cell-to-cell contact with genetic material transferred from a donor bacterium to a recipient cell. During **transformation,** DNA derived from ruptured bacteria is picked up from the environment by a recipient cell. In **transduction,** a viral particle, having "accidentally" packaged bacterial DNA into its protein head during its infective cycle in one cell, serves as the agent for transferring the DNA to another bacterium. Regardless of the method, once bacterial DNA has been transferred, it may undergo recombination in the recipient cell. This apparently involves a pairing of the linear piece of transferred DNA with homologous sections of the recipient's chromosome. Then, through crossing over, genes of the donated DNA are incorporated into the chromosome while the original genes are excised. How many crossovers are required to insert a linear segment of DNA into the circular chromosome? ____[1] What effect would a single crossover have on the circular configuration of the bacterial chromosome? ________________________________[2]

In order for recombination to be useful in mapping, its occurrence has to be detectable. How do geneticists usually know that an organism has acquired an allele through recombination? ________________________________[3] If a gene occurs in a population in just one form, that is, if only a single allele is present, would it be possible to detect the participation of the gene in recombination? ________[4] Explain. ________________________________[5] If a bacterium acquired, through recombination, an allele that it did not previously possess, and the new allele failed to alter the phenotype of the bacterium, could the recombinations be readily detected? ________[6] Explain. ________________________________[7] Identify two requirements that must be met before the participation of a particular allele in recombinations can be detected. ________________________________[8] Once recombinations are detected, the frequency with which they occur between two loci provides an index of the chromosomal distance separating the two loci.

Before getting to the details of bacterial genetics, it is essential to have some familiarity with the methods of culturing bacteria, the bacterial traits that are commonly studied, and the techniques for identifying or selecting mutant forms from large populations of bacteria.

CULTURING BACTERIA

Bacteria are readily grown in the laboratory on an artificial nutritional **medium** in liquid form (as a **broth**) or in semisolid, gel-like form (when **agar** is used as a base). When warmed, agar-based medium converts to a liquid form making it possible to pour it into test tubes or lid-covered shallow plates called petri dishes or plates. Broth cultures are generally set up in test tubes. Various substances—for example, metabolites—can readily be added to either type of medium to tailor it to the nutritional needs of specific types of bacteria. Provided with a suitable medium and a favorable temperature, a small number of bacteria will produce an enormous number of descendent cells in a few hours. On an agar medium, the descendents of a single cell will, during overnight culture, form a visible mass known as a **colony** or **clone.** When a small quantity (approximately two drops) of an overnight liquid culture containing an appropriate concentration of bacteria (around 10^8 cells) is spread over the surface of agar medium (or added to the agar medium before it cools), the colonies that develop within a few hours enlarge and grow together forming a continuous layer of cells known as a **lawn.**

BACTERIAL PHENOTYPES

Many mutations, whether spontaneous or induced, alter the phenotype of wild-type bacteria. Sometimes individual genes, or markers, studied in recombination experiments determine aspects of **colony morphology** such as size, shape, texture, or color. These markers are easy to study since a large number of colonies can be screened at a glance to detect any that exhibit a different appearance. Other genetic markers frequently studied are those affecting nonmorphological aspects of the bacterial phenotype, such as **nutritional requirements.**

Conditional Mutations

Mutations referred to as **conditional** affect the phenotype of bacteria only under certain environmental conditions designated as **restrictive.** Under other, **permissive,** environmental conditions, conditional mutations fail to alter the phenotype and organisms carrying them exhibit the normal, wild-type phenotype. We will consider some examples of conditional mutations shortly.

Nutritional Requirements: Prototrophs and Auxotrophs

Wild-type bacteria of some species including the much studied *Escherichia coli* have the remarkable ability to grow on a **minimal medium,** a very simple medium consisting of an organic compound (often the simple sugar glucose, or sometimes an amino acid) which serves as a source of energy and carbon, a few inorganic salts, and water. (Some bacterial species require a few other supplements as well.)

From these few substances, wild strains, designated as **prototrophs,** can synthesize all the building blocks required to make all of the organic molecules necessary for their survival, growth, and reproduction. These synthetic reactions can occur because the bacteria possess genes that code for all of the large array of enzymes that catalyze these reactions. Often the synthesis of a particular building block involves a stepwise series of reactions, where the product of one reaction serves as a reactant in the next. Since each reaction is catalyzed by a different enzyme, several genes and the enzymes they code for are essential to complete the biochemical pathway. The following illustrates such a scheme, where you may assume that the formation of the final product results in a wild-type phenotype for a particular trait and the absence of the final product results in a mutant phenotype.

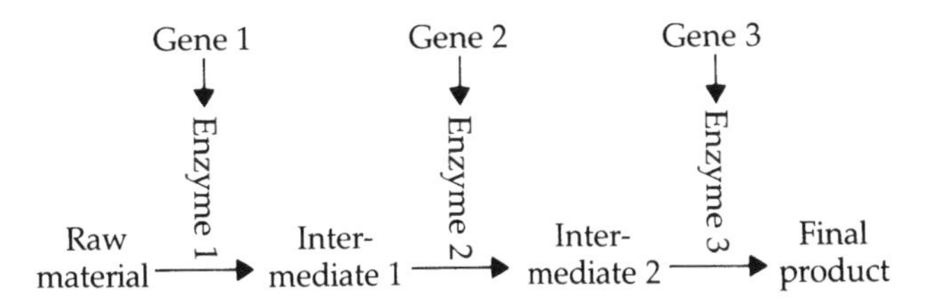

[1] Two, or another even number. [2] It would lose its circularity and become linear. [3] Through an inherited change in its phenotype. [4] No. [5] The organisms participating in the recombination would end up with the same type of allele and thus have the same phenotype that they started with. [6] No. [7] If the phenotype is the same after recombination as it was before, the recombination would ordinarily go undetected. [8] The gene must exist in alternative forms (that is, a wild-type allele and at least one mutant of that allele) and the receipt of an alternative allele must result in a detectable change in the phenotype of the recipient cell.

What would be the consequence for the final product if a mutation occurred in any of the three genes in the pathway shown here? ____________________ ____________________[9] From the standpoint of the final product, would a mutation in any of these genes result in a different phenotype? __________[10] Explain. ____[11]

Possession of a mutant allele that in its wild-type form codes for the production of an essential enzyme is **conditionally lethal** because the bacterium will be incapable of synthesizing a necessary substance and will be unable to grow unless the substance is added to the medium. For example, if that enzyme happens to catalyze the formation of an amino acid, the absence of that amino acid will block the synthesis of all enzymatic and structural polypeptides of which it is a part. Bacteria possessing such nutritional mutations are called **auxotrophs.** How could an auxotroph be made to grow? ____________________ __________[12] The nutritional supplement required for the growth of an auxotroph gives an indication of the metabolic reaction disrupted by the auxotroph's mutation.

Genotype and Phenotype Nomenclature and Symbols

In bacteria, a locus and its alleles are generally designated by an abbreviation, often consisting of three lowercase, italic letters. For loci that control the production of a particular growth substance, the abbreviation is derived from the name of that substance. For example, a locus controlling threonine production is designated by the abbreviation *thr*. Alleles at this locus are symbolized using the same abbreviation with the addition of a superscript minus sign to designate the mutant allele and a superscript plus sign to indicate the normal or wild-type allele. Thus, the mutant and wild-type alleles at the *thr* locus are symbolized as thr^- and thr^+, respectively. The phenotype associated with a mutation is often designated by the same abbreviation as the allele with the first letter capitalized. This abbreviation is not italicized and may or may not be followed by the minus superscript. Thus, threonine-deficient cells could be designated as threonine$^-$, Thr, or Thr$^-$

IDENTIFICATION AND ISOLATION OF NUTRITIONAL MUTANTS

Selective Media and Replica Plating

Often in genetic studies it is necessary to screen large numbers of bacteria in order to identify and isolate those that are auxotrophic for a particular metabolite. A specific auxotroph can be identified and isolated by using a **selective medium** that permits it to survive and reproduce while simultaneously preventing the growth of bacteria lacking the auxotroph's mutation. The technique of **replica plating,** developed in 1952 by E. Lederberg and J. Lederberg, is often used in connection with selective media to screen for auxotrophs.

To illustrate how a selective medium and replica plating are used, assume that a broth culture contains prototrophs and several different types of auxotrophs. Furthermore, assume that the broth is a **complete nutrient medium** which contains, preformed, all the metabolites such as amino acids, vitamins, and so on, essential for bacterial growth and reproduction. (Often the complex mixture of nutrients found in substances like yeast extract, blood, or beef heart infusion serve as the basis for a complete medium.) Such a medium allows both prototrophs and auxotrophs to grow and reproduce.

A diluted sample of bacterial cells from this broth culture is spread over the surface of a complete nutrient agar medium in a petri dish. If incubated overnight, usually at 37°C, individual bacteria will reproduce and form colonies visible to the unaided eye. (Note that the sample of bacteria used to prepare, or inoculate, the plate is diluted sufficiently to insure that each of these colonies arises from a single cell.) Each colony will consist of the genetically identical progeny of either a prototrophic or auxotrophic bacterium. What would determine the pattern or positions of the colonies on the surface of the agar following incubation? ____________________ __________[13]

Bacteria from each of the colonies formed on this plate can be collectively transferred to the agar surface of other petri dishes by replica plating. To do this, a piece of sterile velvet cloth or other absorbent material is placed over the end of a cylinder whose diameter is

[9] It would usually disrupt the biochemical pathway and prevent the production of the final product. [10] No. [11] It would generally produce the same mutant phenotype since the final product could not be made. [12] Supplementing its medium with the particular substance that it is incapable of synthesizing. [13] The pattern is determined by the distribution of the individual cells on the agar during inoculation.

slightly less than that of a petri dish. The cylinder is then gently pressed against the surface of the agar in the original, or master, petri plate and some bacteria from each colony transfer to the velvet. The velvet can now be used the way an inked rubber stamp is used to print an ink pattern onto paper: the velvet is pressed to the agar surface of one or more sterile petri dishes. With each application, some bacteria from each of the original colonies are left behind in the same relative positions shown by colonies in the master petri dish. This procedure is illustrated in Figure 20-1.

Assume that the petri dish to which bacteria are transferred contains a minimal medium. Which bacteria in our original sample (prototrophs or auxotrophs) will be able to reproduce and form colonies on the minimal medium? ________[14] Why would the nongrowing bacteria be unable to reproduce? ______ ____[15] How could the pattern of colonies on the minimal-medium plate be used to identify colonies of auxotrophs on the master plate? ________________________________ ____________[16]

The procedure described so far allows us to identify all the master-plate colonies that consist of auxotrophic bacteria. The following illustrates the procedure for identifying specific auxotrophs. Assume that one of the auxotrophs present in our mixed, broth culture requires the amino acid leucine in order to grow, and we wish to identify Leu⁻ master-plate colonies. Replica plating from the master plate to a sterile plate containing a minimal medium supplemented with leucine will allow us to do this. Will the Leu⁻ cells be able to reproduce on this medium? _____[17] Explain why or why not. ______________________ ____________________[18] Following overnight incubation, how would the pattern of colonies that develop on this minimal-medium + leucine plate be expected to differ from the pattern on the minimal-medium plate? ________________________________ ______[19] What type of bacteria would make up these colonies? _________[20] If desired, samples of the leucine auxotrophs could be removed from these colonies for further study. Repeating this procedure using additional agar plates each of which contains a minimal

FIGURE 20-1

Replica-plating procedure.

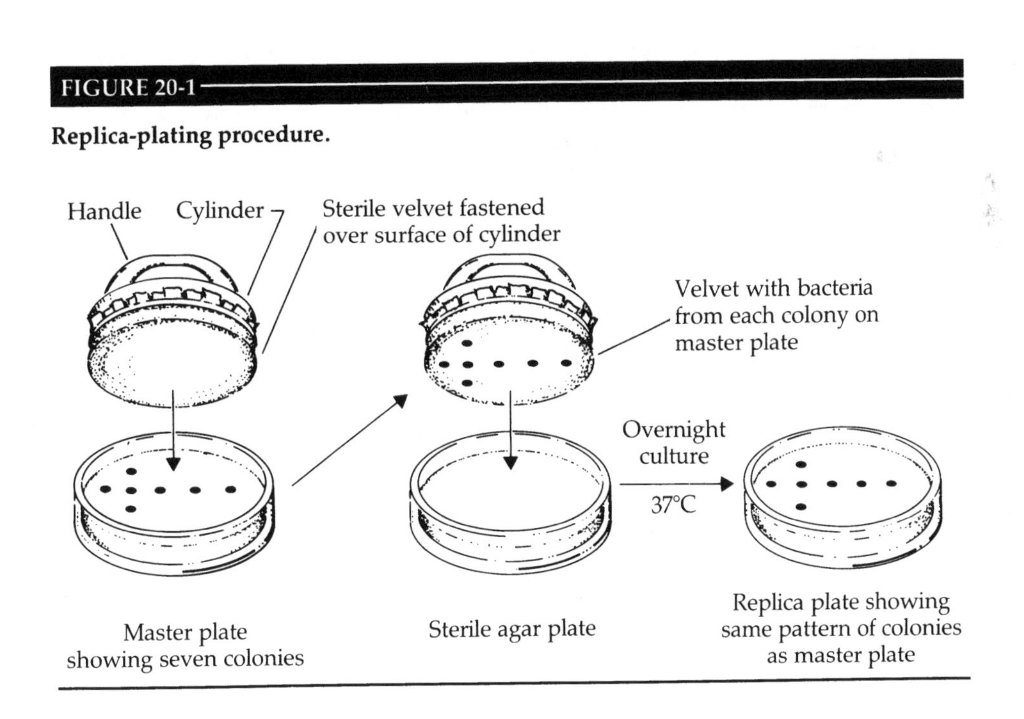

[14] Prototrophs, since they can synthesize all essential nutrients from the minimal medium. [15] The auxotrophs are unable to synthesize essential substances from the minimal medium. [16] By comparing the patterns on the two plates: colonies present on the master plate but absent from the minimal-medium plate would most likely consist of auxotrophs. [17] Yes. [18] The leucine supplement in the medium provides the leucine that the Leu⁻ cells are unable to synthesize. [19] One or more additional colonies could be present. [20] Any colonies found on the minimal-medium + leucine plate that were not on the minimal-medium plate should consist of Leu⁻ cells.

medium supplemented with a different growth factor (for example, an amino acid or a vitamin) makes it possible to specifically identify an array of auxotrophs.

Enriching for Auxotrophs

Often samples of bacteria contain enormous quantities of wild-type prototrophic cells and a very limited number of mutant auxotroph cells. Screening such a sample to identify and isolate the mutants can be facilitated by an **enrichment procedure** which increases the number of mutant cells relative to the number of wild-type cells.

With *E. coli* and certain other bacteria, the incorporation of penicillin into the minimal medium makes it possible to carry out this enrichment. The basis for this procedure rests with the fact that penicillin-sensitive cells are killed in the presence of this antibiotic when they are actively growing: the penicillin blocks the formation of the cell wall and, as a consequence, the cell breaks open. To demonstrate how this procedure works, assume that all the bacteria in a mixed culture of prototrophs and auxotrophs are penicillin-sensitive and that we wish to enrich for the auxotrophs. A sample from this culture is used to inoculate a plate containing a minimal agar medium to which penicillin has been added. During overnight incubation of this plate, which cells, prototrophic or auxotrophic, would be able to reproduce and form colonies? __________[21] Which cells would be killed by the penicillin? __________[22] Which cells would survive on the minimal medium? __________[23] The auxotrophic cells could be cultured for further study by removing the penicillin (by either washing it away or breaking it down by applying the enzyme penicillinase) and transferring them to a complete medium. Any colonies that develop should consist of auxotrophs.

To identify specific types of auxotrophs, those developing on the complete medium can be replica plated onto an array of agar plates each of which contains minimal medium supplemented with a different growth factor. Comparing the colony patterns that develop on these replica plates makes it possible to identify specific auxotrophs: colonies found on a particular plate that are absent from the other plates should consist of auxotrophs deficient for the particular nutrient used to supplement the minimal medium in that plate. For example, an auxotroph that lacks the ability to synthesize the amino acid methionine will be able to form colonies only on a minimal medium supplemented with methionine. Colonies that develop on this plate and that are absent from the other plates in the study should consist of methionine-deficient auxotrophs.

Another approach to identifying auxotrophs involves adding a sample of bacteria that may contain some auxotrophs to a petri dish containing agar minimal medium supplemented with trace amounts of a complex mixture of nutrients. After sufficient time for growth, auxotrophs are identified as producing tiny colonies while prototrophs produce large colonies. Samples of bacteria from the auxotrophic colonies can then be cultured for further study if desired.

ADDITIONAL TYPES OF MUTATIONS

Nutritional Mutations Affecting Degradation Reactions

In addition to the synthetic, or **anabolic,** biochemical reactions that we have discussed, bacteria also carry out degradation, or **catabolic,** reactions. For example, many bacteria have the potential to use carbon sources that are chemically more complex than simple sugars. Before these compounds can be used by the cell, however, they must be broken, or digested, into simpler, smaller molecules. For example, lactose, a 12-carbon sugar, can be used only after it is broken down into two 6-carbon monosaccharide entities, glucose and galactose.

As with synthesis reactions, each degradation reaction is catalyzed by an enzyme. For example, lactose is a type of disaccharide sugar designated as a galactoside, and the enzyme catalyzing its breakdown is known as β-galactosidase. A mutation in the gene that codes for β-galactosidase will prevent a cell from making this enzyme and therefore from using lactose as a carbon source. Such a cell is designated as lactose$^-$ or Lac$^-$. If lactose happens to be the only carbon source available, this cell is in trouble; it can survive and reproduce only if an alternative carbon source is provided in its medium.

Mutations Affecting Sensitivity to Bacteriophages or Antibiotics

Another category of frequently studied bacterial traits is that relating to sensitivity or resistance to bacteriophages, or phages (viruses that attack bacteria), or to antibiotics. Bacteriophages show specificity as to the

[21] Prototrophs. [22] Prototrophs. [23] Auxotrophs, although not able to reproduce, would survive.

type of host cell they infect. The capacity of the phage to recognize a specific receptor molecule (either a protein or a carbohydrate) on the surface of the particular type of host bacterium enables the virus to attach and to initiate its infective cycle. A mutation in a bacterial gene that codes for the receptor molecule recognized by the bacteriophage could alter the receptor sufficiently to prevent viral recognition, and thereby confer resistance to the virus. Within a large bacterial population consisting mostly of wild-type cells, what procedure could be used to identify cells possessing the mutant allele for viral resistance? ______________ ______________[24]

Antibiotics interfere with vital activities of bacterial cells. A number of antibiotics such as tetracycline, streptomycin, and erythromycin prevent protein synthesis, thereby preventing cell growth. Others, like bacitracin and penicillin, disrupt cell-wall synthesis and thus kill any actively growing cells (resting or nondividing cells are not killed by this type of antibiotic). Cells that survive and replicate in the presence of an antibiotic are said to be resistant to that antibiotic and can readily be isolated and used for further study. Bacterial resistance to antibiotics or phages is customarily designated by affixing a superscript "s" or "r," designating sensitivity or resistance respectively, to the abbreviation of the antibiotic or the designation for the virus. For example, an *E. coli* strain labeled Pen^s, λ^r would be sensitive to penicillin and resistant to the phage λ.

Temperature-Sensitive Mutations

Temperature-sensitive, or ts, mutations comprise another frequently studied group. These mutations are conditional since they express themselves only in a certain, restrictive, range of temperatures. At permissive temperatures, bacteria with ts mutations show the wild-type phenotype. For example, a certain strain of *E. coli* carrying a conditionally lethal ts mutation thrives at 37°C but is unable to reproduce at temperatures above 45°C. Underlying such temperature sensitivity is the fact that the product of a ts gene apparently functions normally at the permissive temperatures, but becomes unstable at the restrictive temperatures. Exposing a ts mutant strain to restrictive temperatures at various points during its development often makes it possible to identify the specific time interval during which the gene with the temperature-sensitive allele is operating.

SUMMARY

This chapter provides basic information about methods of culturing bacteria, reviews several categories of bacterial mutations that are commonly used as markers in genetic studies, and discusses the use of selective media and replica plating to screen large populations of bacteria for mutant cells.

[24] Infect the bacteria with the bacteriophage and then culture them on agar; most of the colonies that develop will be resistant to that type of virus.

PROBLEM SET

20-1. Bacteria are generally cultured either in a broth or on an agar medium. Distinguish between these two types of media.

20-2. **a.** The terms "colony" and "lawn" are often used to describe bacterial growth on the surface of agar-based media. Distinguish between these two terms.
b. How is a lawn produced?

20-3. Assume that a sample of bacteria, obtained from a liquid culture, is to be used to inoculate a sterile petri dish containing nutrient agar. What should be done before inoculation to insure that each colony develops from a single bacterium?

20-4. **a.** Distinguish between a minimal medium and a complete medium.
b. Which of these two types of media would allow prototrophic bacteria to reproduce?
c. Which of these two types of media would allow auxotrophic bacteria to reproduce?

20-5. A comparison of bacteria that can live and reproduce on a minimal medium with those requiring a complete medium indicates that both types have the same requirements for amino acids and vitamins. Bacteria living and reproducing on a complete medium acquire their amino acids and vitamins directly from the medium. Where or how do the bacteria that can live and reproduce on a minimal medium acquire their amino acids and vitamins?

20-6. Can auxotrophic bacteria be grown on a medium other than one that is complete? Explain.

20-7. Assume that a single prototrophic bacterium is used to inoculate a sterile test tube containing a complete nutrient broth. The tube is incubated overnight and a large population of cells results. A study of this population indicates the presence of a few auxotrophic cells.
a. What is the most reasonable explanation for the origin of these auxotrophic cells?
b. Why were the auxotrophic cells able to survive in this medium?
c. If it were possible to identify all the enzymes that could be produced by the prototrophic bacteria and compare them with the enzymes that could be synthesized by the auxotrophic cells, what would you expect to find?

20-8. Describe a way in which the auxotrophic cells referred to in problem 20-7 could be isolated from the largely prototrophic population.

20-9. **a.** Assume that the auxotrophic bacteria described in problem 20-7 were all incapable of synthesizing the same amino acid. Describe a procedure that could be used to identify the amino acid that the auxotrophs were unable to make.
b. How could these auxotrophs be cultured for further study?

20-10. A sample of bacteria is removed from a broth culture, diluted to an appropriate level, and used to inoculate a complete-medium agar plate. Overnight growth of the bacteria on this master plate produces the colony pattern shown in the following figure, where each of the 14 colonies has been numbered for reference. This master plate is replica plated onto four sterile plates, each of which contains a different type of agar. Plate 1 contains a minimal-medium agar and each of the other plates has a minimal medium supplemented with a single amino acid: proline has been added to plate 2, aspartic acid to plate 3, and valine to plate 4. Following

PROBLEM SET

overnight culture, these plates exhibit the colony patterns shown in the figure.

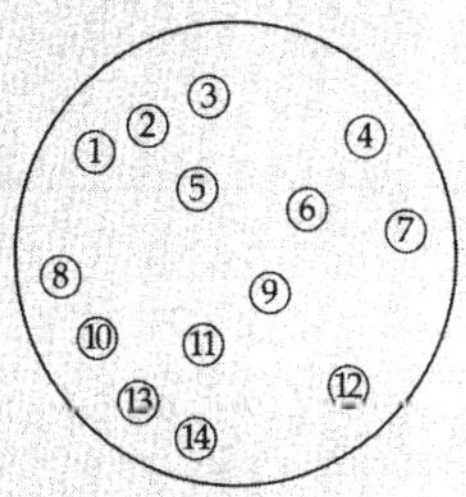

Master plate

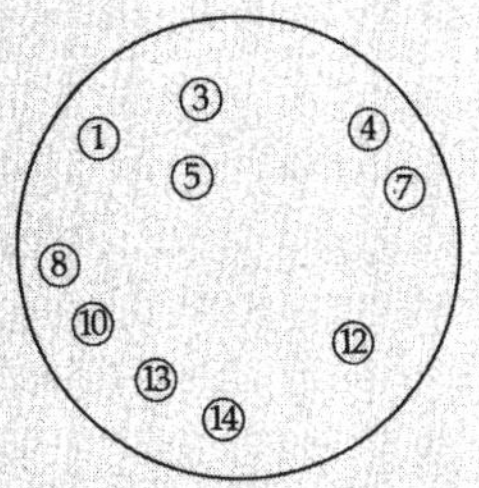

Plate 1: minimal medium

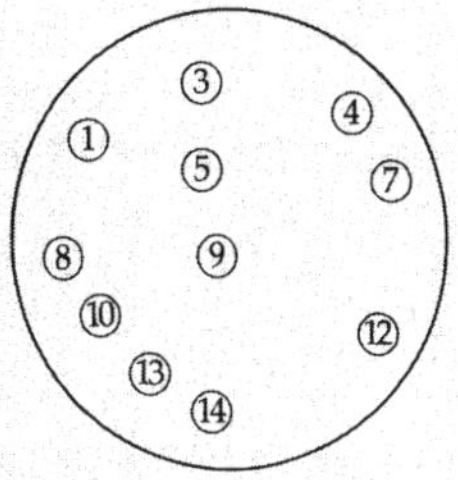

Plate 2: minimal medium + proline

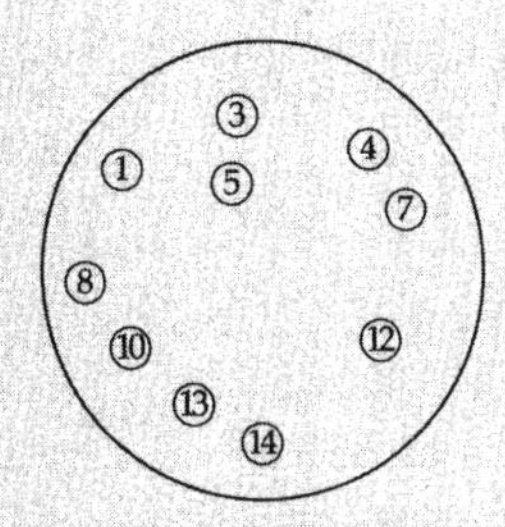

Plate 3: minimal medium + aspartic acid

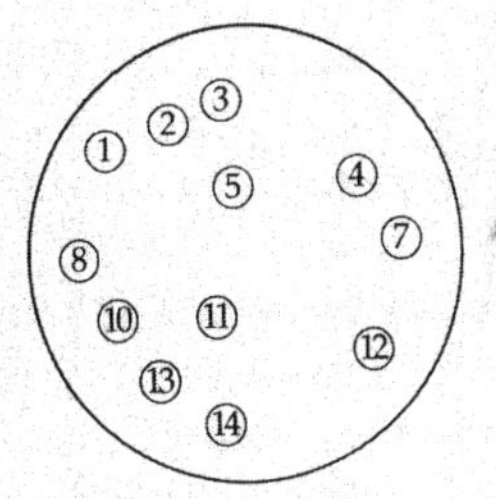

Plate 4: minimal medium + valine

a. Some of the colonies growing on the master plate failed to grow on plate 1 containing the minimal medium. How many failed to grow and, in general, why did they fail to grow?

b. How many of the bacteria used to inoculate the master plate were prevented from growing on the minimal medium by the inability to synthesize proline? Explain.

c. How many of the bacteria used to inoculate the master plate were unable to synthesize valine? Explain.

d. How many of the bacteria used to inoculate the master plate were able to synthesize aspartic acid? Explain.

20-11. a. Based on the information supplied in problem 20-10, what tentative conclusion could be made about the nature of the nutritional deficiency shown by the remaining bacterial type which failed to grow on the minimal-medium plate, that is, the bacteria of colony 6?

b. Further studies indicate that the bacteria making up colony 6 are able to grow on minimal medium supplemented with both valine and proline. How would this information cause you to modify the conclusion you drew in response to 20-11a?

PROBLEM SET

c. Speculate on the number of nutritional mutations carried by the bacteria of colony 6.

20-12. The synthesis of a particular essential amino acid is known to be a two-step process that requires two different enzymes, as follows.

$$\text{Raw material} \xrightarrow{\text{Enzyme A}} \text{Intermediate} \xrightarrow{\text{Enzyme B}} \text{Amino acid}$$

A research laboratory has cultured stocks of wild-type bacteria and two mutant strains, one of which is deficient for enzyme A and the other for enzyme B. Due to an oversight, the tubes in which these three types of bacteria are cultured were not labeled. In an attempt to identify them, the three strains are cultured under the following conditions. (Note that the raw material necessary for the formation of the amino acid under study is found in the minimal medium.)

Strain	Minimal medium	Minimal medium + intermediate
1	No growth	No growth
2	No growth	Growth
3	Growth	Growth

a. Identify which strain is wild-type, which is deficient for enzyme A, and which is deficient for enzyme B.

b. Assume that a fourth strain is available that is unable to synthesize either enzyme A or enzyme B. What results would be expected if it were cultured on the minimal medium and on the minimal medium supplemented with the intermediate? Could this double-deficient strain be unambiguously distinguished from the three other strains?

20-13. Bacteria carrying a mutant allele that blocks the production of an enzyme essential for the synthesis of the amino acid phenylalanine are designated as Phe^-. Assume you are given a mixed liquid culture of wild-type (Phe^+) and Phe^- bacteria. Outline the key steps that you would follow in order to determine the relative numbers of Phe^- bacteria in the mixed culture.

20-14. Bacteria carrying a mutant allele that blocks the production of an enzyme essential for the synthesis of the amino acid tyrosine are designated as Tyr^-: A diluted sample from a mixed culture containing penicillin-sensitive wild-type (Tyr^+) and Tyr^- cells is spread on a penicillin-containing medium deficient in tyrosine.

a. Identify the type or types of bacterial cells that will be able to survive on this medium. Explain.

b. What must be done so that the type or types of cells that survive can reproduce?

CHAPTER

21

Transformation and Mapping the Bacterial Chromosome

INTRODUCTION

Transformation, a process that occurs in a rather limited number of bacterial species, involves the transfer of genetic information from a **donor** bacterium to a **recipient** bacterium through the uptake of extracellular, or **exogenous,** fragments of DNA. These fragments, released upon the lysis, or breaking up, of donor cells, are acquired by recipient cells from the medium in which they live. The precise events that occur during the uptake of DNA vary, depending on the species and the nature of the recipient's cell wall, specifically, on whether the recipient is Gram-positive or Gram-negative. In Gram-positive bacteria, one of the strands of the duplex DNA is degraded during uptake so that only one strand of each fragment enters the cell. With Gram-negative bacteria, the exogenous DNA is taken into the recipient in double-stranded form and remains double-stranded until uptake is complete, when one strand is enzymatically degraded. During the final phase of transformation, the single-stranded donor fragment becomes incorporated into one of the two strands of the recipient's chromosome through pairing with the homologous region and crossing over. This displaces a portion of a strand of the original chromosome which subsequently degrades.

Following recombination, the recipient's chromosome consists of an original DNA strand and a recombinant strand which includes the donated fragment, and is known as a **heteroduplex.** When this chromosome replicates, each strand, of course, forms a new complement; one daughter chromosome is identical to the original, pretransformation, chromosome and the other carries the genes acquired through transformation in a segment identical to the original double-stranded donor DNA fragment. When the cell containing these two chromosomes divides, one daughter cell will have the same genome (genetic complement) as the original recipient cells and the other cell will carry the recombination produced through transformation.

FEATURES OF TRANSFORMATION

The following features apply to transformations in most transformable species of bacteria, including the *Bacillus*, *Streptococcus* (formerly *Diplococcus*), and *Hemophilus* groups.

1. Bacteria are physiologically receptive to transformation, that is, they are competent, for a very limited portion of their growth cycle when they may take up fragments of DNA.
2. Uptake by the recipient cells requires that the transforming pieces be double-stranded and rela-

tively large (usually in the range of 10,000 to 20,000 nucleotide base pairs).

3. A direct relationship exists between the number of **transformants,** or bacteria that have been transformed, and the number of molecules of foreign DNA in the external medium up to a concentration of about 10 molecules of DNA per recipient cell. Higher concentrations of DNA produce no further increase in the number of transformants, and the number of transformed cells levels off.

In designing transformation experiments, the donoı and recipient strains must carry different alleles at the loci under study (this is essential if the recombination is to be detected). The DNA extracted from donor cells generally is sheared, or randomly broken into pieces, and added to a suspension of recipient cells. Transformants can be identified since they now express the donor gene or genes.

TWO-POINT TRANSFORMATION STUDIES

Two-point studies involve two strains of bacteria that carry contrasting alleles for two traits. For example, say the recipient strain, carrying mutations at loci controlling the synthesis of the amino acids tryptophan and tyrosine, has a genotype of $trp^- tyr^-$. Donor DNA is obtained from a wild strain with genotype $trp^+ tyr^+$ and is supplied to the recipient cells, which are then tested for nutritional competency for each amino acid. How many genotypes, with respect to these two traits, would occur among the cells exposed to the DNA? ________[1] What would these genotypes be? ______________________________________[2] The transformants are identified using appropriate selective media and the relative number of each type is counted and used to assess the degree of linkage between the two loci. Before considering this, however, we will look at how each of the three transformant classes produced in this cross arises.

Crossing Over and the Production of Transformants

The incorporation of a fragment of donor DNA into the recipient's chromosome requires an even number of crossovers. What would be the consequence of an odd number of crossovers? ______________________________________[3] In the following diagrams you will be able to show how the three types of transformants arise when the recipient's chromosome is $trp^- tyr^-$ and the donor DNA fragment is $trp^+ tyr^+$. On the left of each diagram you will find a portion of the recipient's chromosome and the donor DNA fragment. Note that the regions of the chromosome are labeled for reference: zone 1 is to the left of the *trp* locus, zone 2 is between the two loci, and zone 3 is to the right of the *tyr* locus. Determine the smallest number of crossovers necessary to produce each of the transformants shown on the right and draw an "X" at the site of each crossover.

Production of single transformant $trp^+ tyr^-$ [4]

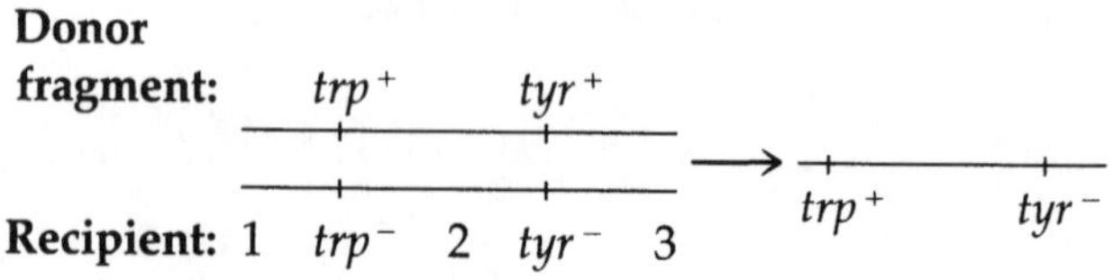

Donor fragment: trp^+ tyr^+

Recipient: 1 trp^- 2 tyr^- 3 → trp^+ tyr^-

Production of single transformant $trp^- tyr^+$ [5]

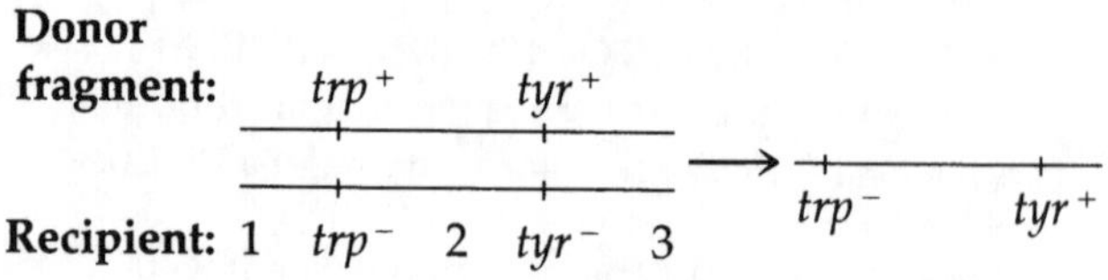

Donor fragment: trp^+ tyr^+

Recipient: 1 trp^- 2 tyr^- 3 → trp^- tyr^+

Production of double transformant $trp^+ tyr^+$ [6]

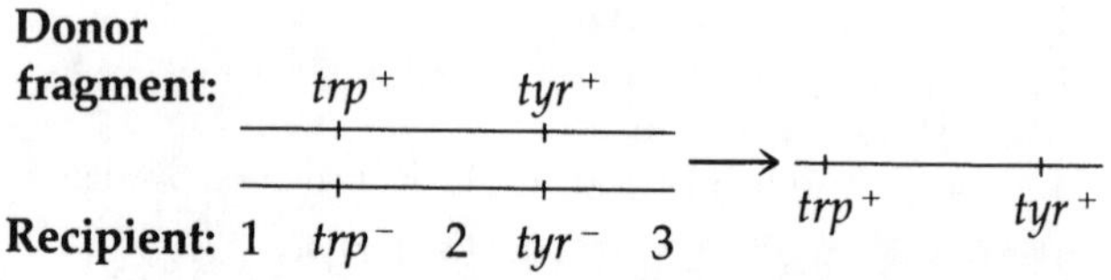

Donor fragment: trp^+ tyr^+

Recipient: 1 trp^- 2 tyr^- 3 → trp^+ tyr^+

Identify another way in which the double transformant could arise. ______________________________________[7] If the double transformant had been produced in this manner, how many crossovers would have been required to integrate the DNA? _____[8] This alternative way of producing the double transformant is represented as follows.

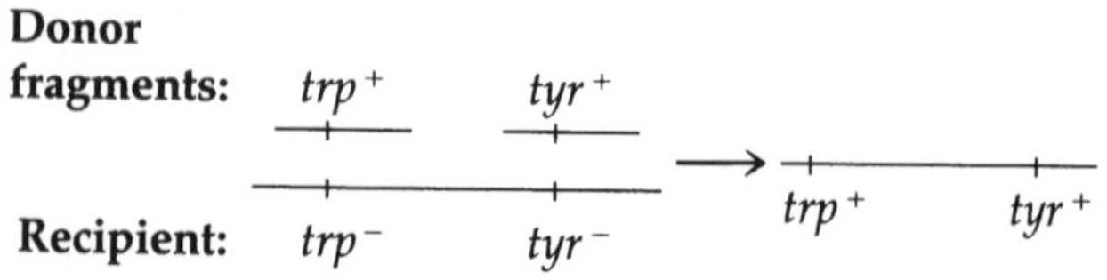

Donor fragments: trp^+ tyr^+

Recipient: trp^- tyr^- → trp^+ tyr^+

[1] Four. [2] $trp^- tyr^-$ (nontransformants), $trp^+ tyr^-$ and $trp^- tyr^+$ (single transformants), and $trp^+ tyr^+$, (double transformants).
[3] The donor fragment would attach, but the chromosome would become linear; this would prevent its replication. [4] Crossovers in zones 1 and 2. [5] Crossovers in zones 2 and 3. [6] Crossovers in zones 1 and 3. [7] Through the uptake and incorporation of two pieces of DNA, one carrying trp^+ and the other, tyr^+. [8] Four, two to integrate each fragment.

Formation of Double Transformants and the Distance between Loci

The manner in which double transformants, or **co-transformants,** arise can provide information about the distance separating the loci. If the distance between two loci exceeds the length of an average donor fragment, how would double transformants arise? ____[9] Alternatively, if the distance between two loci is less than the length of an average donor fragment, what is the most likely way of producing double transformants? ____[10]

The number of transformation events can usually be determined from the relative frequency with which the double transformants are produced in a transformation experiment. How is the probability of the simultaneous occurrence of two independent transformation events determined? ____[11] In general, how does this probability value compare with the probability of a single transformation event? ____[12] If two loci are relatively far apart on the chromosome, what prediction could be made about the relative numbers of single and double transformants? ____[13] If two loci are closely linked, what prediction could be made about the relative numbers of single and double transformants? ____[14]

To summarize, if two loci are closely linked, double transformants can arise through a single transformation event. If the loci are more distant, two independent transformation events are required. Thus, the frequency of double transformants is very low for distant loci and much higher for closely linked genes.

Interpreting a Pair of Transformation Experiments

Consider the outcome of a pair of separate transformation experiments carried out on *Bacillus subtilus* (by E. W. Nester, M. Schafer, and J. Lederberg, 1963, *Genetics* 48:529–51), each involving recipient cells of genotype *trp*$^-$*tyr*$^-$. In the first experiment, the donor DNA is a mixture of equal parts derived from two strains with genotypes *trp*$^+$*tyr*$^-$ and *trp*$^-$*tyr*$^+$. Examine the results of this experiment given in Table 21–1 and, in the space provided within the table, identify each class of transformants as a single or double type.

Note that the single- and double-recombinant classes in experiment 1 do not occur in equal or nearly equal frequencies. Recall that, in eukaryotes, reciprocal recombinant classes are generally found in equal frequencies because they originate through the same crossover event. Here, however, each class arises through a different double crossover. In general, how does the number of double transformants compare with the frequencies of the single transformants? ____[18]

What does the double-transformant frequency suggest about the way in which this transformant class arose? ____[19]

Given the nature of the donor DNA in this experiment, how *must* these double transformants have arisen? ____[20]

What does the outcome of experiment 1 tell us about the distance between these two loci? ____[21]

In the second experiment in this study, all the donor DNA supplied to the *trp*$^-$*tyr*$^-$ cells came from

TABLE 21-1

Experiment 1: Outcome of transformation when the recipient's genotype is *trp*$^-$*tyr*$^-$.

Donor DNA	Transformant genotype	Transformant type	Frequency
Mixture of *trp*$^+$*tyr*$^-$ and *trp*$^-$*tyr*$^+$	*trp*$^+$*tyr*$^-$	____________[15]	190
	trp$^-$*tyr*$^+$	____________[16]	256
	trp$^+$*tyr*$^+$	____________[17]	2

Frequencies from E. W. Nester, M. Schafer, and J. Lederberg, 1963, Gene linkage in DNA transfer: A cluster of genes concerned with aromatic biosynthesis in *Bacillus subtilus, Genetics* 48:529–51.

[9] Through two independent transformation events, each of which involves a separate fragment of donor DNA. [10] Through a single transformation event. [11] From the product of their individual probabilities. [12] It is very small relative to the probability of a single transformation. [13] Since double transformants could arise only through two simultaneous transformations, they would occur in a very low frequency relative to the single transformants. [14] Since double transformants would arise most commonly through single transformations, the number of double transformants would be equivalent to the number of single transformants. [15] Single. [16] Single. [17] Double. [18] It is very low relative to either of the single-transformant frequencies. [19] It arose through two transformation events, each involving a separate piece of DNA. [20] Since each type of donor DNA carries a different wild-type allele, there must have been two simultaneous transformation events. [21] Nothing; given the makeup of the donor DNA, the double transformants will always be produced with a very low frequency.

cells with genotype trp^+tyr^+ and produced the outcome given in Table 21-2. In the space provided within this table, identify each transformant class as a single or double type. How does the number of double transformants produced in experiment 2 compare with the number of single transformants? ______________________________[25] What does this suggest about the way in which the double transformants arose in this experiment? ______________________________[26] What does the outcome of experiment 2 imply about the distance between these two loci? ______________________________[27]

If the two loci just considered had been separated by a considerably greater distance so that they were never found together on a piece of transforming DNA, would the number of double transformants in experiment 2 change relative to the number of single transformants? __________[28] Explain. ______________________________[29]

Transformation as a Function of DNA Concentration

Transformation studies designed to compare the relative numbers of single and double transformants produced at two or more different DNA concentrations can provide very reliable information about the degree of linkage between two loci. As mentioned earlier, the number of bacteria transformed is directly related to the number of molecules of foreign DNA in the external medium up to a concentration of about 10 molecules of DNA per recipient cell.

Assume that there are two transformation setups which differ in the concentration of donor DNA. In one, concentration is such that the probability of a single transformation event is 0.01. The second has a tenfold reduction in DNA concentration such that the probability of a single transformation event is reduced tenfold to 0.001. What is the probability of two simultaneous transformations in the first setup? ______[30] What is the probability of two simultaneous transformations in the second setup? __________[31] How much does the tenfold reduction in DNA concentration reduce the probability of two simultaneous transformations? __________[32] Two facts emerge from this illustration: (1) the probability of a single transformation event declines at the same rate as the decline of the concentration of transforming DNA, while (2) the probability of two simultaneous transformation events declines at a rate much greater than the decline of the concentration of transforming DNA.

This rate difference makes it possible to determine the degree of linkage for two loci from changes that occur in the relative numbers of single and double transformants produced as the concentration of donor DNA declines. If, as the concentration of DNA declines, the numbers of single and double transformants both decline at about the same rate, how are double transformants likely to be formed? ______________________________[33] What does this tell us about the distance separating these two loci? ______________________________[34]

If, as the concentration of DNA declines, the number of double transformants declines much more rapidly than the number of single transformants, how are double transformants likely to be formed? ______________________________[35] What does this tell us about the distance separating the two loci? ______________________________[36]

In this type of study, the key to determining whether two loci are closely linked is to compare the rates of decline that occur in the number of double and single transformants as the concentration of transforming DNA is reduced.

TABLE 21-2

Experiment 2: Outcome of transformation when the recipient's genotype is trp^-tyr^-.

Donor DNA	Transformant genotype	Transformant type	Frequency
	trp^+tyr^-	__________[22]	196
trp^+tyr^+	trp^-tyr^+	__________[23]	328
	trp^+tyr^+	__________[24]	367

Frequencies from E. W. Nester, M. Schafer, and J. Lederberg, 1963, Gene linkage in DNA transfer: A cluster of genes concerned with aromatic biosynthesis in *Bacillus subtilus*, *Genetics* 48:529–51.

[22] Single. [23] Single. [24] Double. [25] The number is comparable to the single-transformant frequencies.
[26] They arose through a single transformation event. [27] They are linked closely enough to be transformed on the same fragment of DNA. [28] Yes. [29] It would be very low since each double transformant would require two separate transformation events.
[30] $0.01 \times 0.01 = 0.0001$. [31] $0.001 \times 0.001 = 0.000001$. [32] 100-fold. [33] Through the uptake of single pieces of DNA.
[34] They are close enough to be transformed on a single piece of DNA. [35] By the simultaneous uptake of two separate pieces of DNA.
[36] They are not closely linked.

Interpreting Transformation Studies Conducted at Different Concentrations

Table 21-3 (based on research data from S. H. Goodgal, 1961, *Journal of General Physiology* 45:211) gives the outcomes of two transformation studies, each involving a different pair of loci in *Hemophilis influenzae*. For each study, we need to determine whether the loci involved are closely linked or relatively far apart. Note that Table 21-3 gives results from each study at three different concentrations of transforming DNA.

Look at the results of study 1 in Table 21-3. First, consider the reduction in DNA concentration in relation to the reduction in the number of single transformants. What effect does a reduction to 50% of the original DNA concentration, that is, from 0.0050 to 0.0025, have on the number of single transformants? ______________________________ [37]

What effect does a reduction of the concentration of transforming DNA to 10% of the original concentration, that is, from 0.0050 to 0.0005, have on the number of single transformants? ______________________________ [38] Now consider the reduction in DNA concentration in relation to the number of double transformants in study 1. What effect does a 50% reduction in the concentration of DNA have on the number of double transformants? ______________________________ [39] What effect does a reduction of the DNA concentration to 10% of the original concentration have on the number of double recombinants? ______________________________ [40] Based on a comparison of the declines in the numbers of single and double transformants, what can be concluded about the distance separating the two loci involved in study 1? ______________________________ [41]

Now examine the results of study 2 in Table 21-3. What effect does a reduction in the amount of transforming DNA to 50% of the original concentration have on the number of single recombinants? ______________________________ [42] What effect does a reduction in the amount of transforming DNA to 10% of the original concentration have on the number of single recombinants? ______________________________ [43] What effect does a 50% reduction of the concentration of DNA have on the number of double recombinants? ______________________________ [44] What effect does a reduction of the concentration of DNA to 10% of the original concentration have on the number of double recombinants? ______________________________ [45] Based on the decline in the number of single and double transformants, what can be concluded about the distance separating the two loci involved in study 2? ______________________________ [46]

TABLE 21-3

The outcomes of two transformation studies, each involving a different pair of loci in *Hemophilis influenzae*.

	Concentration of transforming DNA in medium (μg DNA/ml)	Relative Number or Transformants	
		Single	Double
Study 1	0.0050	90	7.0
	0.0025	50	1.0
	0.0005	10	0.1
Study 2	0.0050	91	90.0
	0.0025	49	49.0
	0.0005	9	8.0

Calculation of Approximate Distance between Closely Linked Loci

If two loci are closely linked, the map distance between them can be estimated from the frequency with which they are cotransformed. This estimate is obtained by selecting among the recipients for either of the donor genes by replica plating onto the appropriate media and then determining the number of those cells that have also been transformed for the other donor gene. For example, in a study with donor DNA from a strain

[37] It reduces the number from 90 to 50, or to 50/90 = 0.556, or 0.556 × 100 = 55.6%, of the original number.
[38] It reduces the number from 90 to 10, or to 10/90 = 0.111, or 0.111 × 100 = 11.1%, of the original number.
[39] It reduces the number from 7.0 to 1.0, or to 1.0/7.0 = 0.1429, or 0.1429 × 100 = 14.29%, of the original number.
[40] It reduces the number from 7.0 to 0.1, a reduction to 0.1/7.0 = 0.0143, or 0.0143 × 100 = 1.43%, of the original number.
[41] The loci are relatively far apart since the number of double transformants declines at a much faster rate than the number of single transformants. [42] It reduces the number from 91 to 49, or to 49/91 = 0.538, or 0.538 × 100 = 53.8%, of the original number.
[43] It reduces the number from 91 to 9, or to 9/91 = 0.989, or 0.989 × 100 = 9.89%, of the original number.
[44] It reduces the number from 90.0 to 49.0, or to 49.0/90.0 = 0.544, or 0.544 × 100 = 54.4%, of the original number.
[45] It reduces the number from 90.0 to 8.0, or to 8.0/90.0 = 0.0889, or 0.0889 × 100 = 8.89%, of the original number.
[46] They are close enough to be transmitted on the same piece of DNA since the numbers of double transformants and single transformants decline at about the same rate.

with genotype $trp^+ tyr^+$ and recipient cells of genotype $trp^- tyr^-$, an initial selection for the trp^+ found 281 cells transformed for trp^+ and, of these, 190 were cotransformed for tyr^+. From these relative numbers, the frequency of cotransformation of the *tyr* locus with the *trp* locus can be calculated to give an estimate of the distance between the two loci. This frequency is given by the proportion of *trp-tyr* cotransformants relative to the total number transformed for the *trp* locus, as summarized in the following equation.

$$\text{Cotransformation frequency} = \frac{\text{Number of cotransformants}}{\text{Total number of transformants for } trp}$$

What is the distance between the two loci? ______[47] What is the relationship between the cotransformation frequency and the distance separating two loci?

______________________________[48]

Cotransformation frequencies range between 0 (0%) and 1 (100%). What does a cotransformation value close to 1 imply about the distance separating two loci?

______________________________[49]

What does a value close to 0 imply about the distance between two loci? ______________________________

______________[50]

Ordering Loci from Two-Point Transformation Studies

Two closely linked loci can be ordered relative to other loci through additional transformational studies that involve either of the loci and at least one additional locus. For example, if one study indicates that loci *s* and *t* have a high frequency of cotransformation and are thus closely linked, and if a second study shows *t* is closely linked to *u*, and if a third indicates that *s* and *u* are not transformed together, what is the order for the three loci? __________[51]

THREE-POINT TRANSFORMATION STUDIES

Three-point transformation studies, much like three-point crosses in eukaryotes, are used to determine the order of loci and to estimate the distances between them. Assume that donor DNA from $trp^+ his^+ tyr^+$ bacteria is supplied to $trp^- his^- tyr^-$ cells. (The *his* locus controls the synthesis of the amino acid histidine.) At this point, nothing is known about the order of loci; the order in which they are written here may or may not be correct. Following transformation, the vast majority of recipient cells remain untransformed while others are found to be transformed to one of the seven transformant classes listed in Table 21-4. One of these transformant classes arose through a quadruple crossover and the rest from double crossovers. In general, how would the expected frequency of the quadruple crossover class compare with those of the classes arising from a double crossover? ______________________________

______________________________[52] Identify the quadruple-crossover class. __________[53]

Identifying the Middle Locus

Once the quadruple-crossover class is known, the middle locus can be identified by comparing its genotype with either of the parental types. If compared with the recipient parent's genotype, the locus carrying the same allele in both the recipient parent and in the quadruple-crossover type is the middle locus. Compare the following genotypes and identify the middle locus.

TABLE 21-4

Outcome of transforming experiment when donor DNA from $trp^+ his^+ tyr^+$ bacteria is supplied to $trp^- his^- tyr^-$ cells.

Class	Transformant genotype	Frequency
1.	$trp^+ his^+ tyr^+$	4676
2.	$trp^- his^+ tyr^+$	1505
3.	$trp^- his^- tyr^+$	267
4.	$trp^- his^+ tyr^-$	100
5.	$trp^+ his^- tyr^-$	1054
6.	$trp^+ his^- tyr^+$	24
7.	$trp^+ his^+ tyr^-$	484
	Total:	8110

[47] There are 190 cotransformants and 281 *trp* transformants. The cotransformation frequency is 190/281 = 0.676, or 0.676 × 100 = 67.6%. [48] It is an inverse relationship: as the cotransformation frequency increases, the distance between the loci decreases. [49] The genes are transformed together with a very high frequency and thus are very closely linked. [50] The loci are very rarely transformed together and thus separated by a distance greater than the length of an average transforming fragment. [51] Since *s* and *u* are not transformed together, they are not close to each other; the order is *s-t-u*. [52] It would be much smaller since the frequency of a quadruple crossover equals the product of the probabilities of two double crossovers. [53] It is the least frequent class, 6.

Recipient-parental type: $trp^{-}his^{-}tyr^{-}$

Quadruple-crossover type: $trp^{+}his^{-}tyr^{+}$

Middle locus: ________[54]

If compared with the genotype of the donor parent, the loci carrying the same alleles in both the donor and the quadruple-crossover class flank the middle locus. Compare the following genotypes and identify the middle locus.

Donor-parental type: $trp^{+}his^{+}tyr^{+}$

Quadruple-crossover type: $trp^{+}his^{-}tyr^{+}$

Middle locus: ________[55]

Since *his* is the middle locus, the order in which the loci are written in the seven transformant classes in Table 21-4 reflects the correct order.

Diagraming the Quadruple Crossover

Now that the middle locus has been identified, it is worth taking a moment to illustrate how the quadruple-crossover event generates the rare quadruple-crossover recombinants. The following diagram shows the donor DNA fragment and a portion of the recipient's chromosome. The regions of the chromosome are labeled for reference: zone 1 is to the left of the *trp* locus, zone 2 is between *trp* and *his*, zone 3 is between *his* and *tyr*, and zone 4 is to the right of the *tyr* locus. Draw an "X" at the site of each crossover necessary to produce the transformant shown on the right.[56]

Donor: trp^{+} his^{+} tyr^{+} → trp^{+} his^{-} tyr^{+}

Recipient: 1 trp^{-} 2 his^{-} 3 tyr^{-} 4

Calculating Cotransformation Frequencies

Once the order of loci is established, the cotransformation frequencies can be calculated to provide an estimate of the distances separating the loci. The best approach to this is to inspect the loci two at a time for each of the classes; while this is done, the third locus is temporarily disregarded. For example, to determine the cotransformation of the *his* locus with the *trp* locus, the *his* and *trp* loci would be considered, while ignoring the *tyr* locus. With regard to these two loci, each class is first examined to identify those that are transformed for *trp*, and then these classes are examined to identify those that are also transformed for *his*. Refer to Table 21-4 and begin by classifying the *trp* and *his* loci combinations in each of the seven classes of transformants. Which classes are transformed for the *trp* locus? ________[57] Which classes are cotransformed for the *trp* and *his* loci? ________[58]

Cotransformation frequency for trp and his: The cotransformation frequency for *trp* and *his* is determined by adding together the number of bacteria cotransformed for these two loci and expressing that number as a proportion of the total number of *trp* transformants.

$$\text{Cotransformation frequency} = \frac{\text{Number of cotransformants for } trp \text{ and } his}{\text{Total number of transformants for } trp}$$

Calculate this cotransformation frequency. ________[59]

Cotransformation frequency for trp and tyr: Repeat this procedure by referring to Table 21-4 and identifying each class that is transformed for *trp*, and from that group, the classes that are also transformed for *tyr*. Which classes are transformed for the *trp* locus? ________[60] Which classes are cotransformed for the *trp* and *tyr* loci? ________[61] Calculate the cotransformation frequency for the *trp* and *tyr* loci. ________[62]

Cotransformation frequency for his and tyr: Repeat this procedure by referring to Table 21-4 and identifying each class that is transformed for *his*, and from these classes identify those that are also transformed for *tyr*. What classes are transformed for the *his* locus? ________[63] What classes are cotransformed for the *his* and *tyr* loci? ________[64] Calculate the cotransformation frequency for the *his* and *tyr* loci. ________[65]

Cotransformation Frequency Map

This cotransformation study results in the following map, where the numbers reflect the cotransformation frequencies given in percentages. Remember that the cotransformation frequency is inversely related to the distance between the loci under study; that is, the

[54] The *his* locus carries the same allele, his^{-}, in both types and is thus the middle locus. [55] The *trp* and *tyr* loci carry the same allele in both types and are thus the loci flanking *his*. [56] A crossover in each of the four zones generates the $trp^{+}his^{-}tyr^{+}$ recombinant. [57] 1, 5, 6, and 7. [58] 1 and 7. [59] The cotransformant classes (1 and 7) contain a total of $4676 + 484 = 5160$ bacteria. The classes transformed for *trp* (1, 5, 6, and 7) contain $4676 + 1054 + 24 + 484 = 6238$ bacteria. The cotransformation frequency is $5160/6238 = 0.827$, which is equivalent to $0.827 \times 100 = 82.7\%$. [60] 1, 5, 6, and 7. [61] 1 and 6. [62] The cotransformation frequency is $(4676 + 24)/6238 = 4700/6238 = 0.753$, or $0.753 \times 100 = 75.3\%$. [63] 1, 2, 4, and 7. [64] 1 and 2. [65] The cotransformation frequency is $(4676 + 1505)/(4676 + 1505 + 100 + 484) = 6181/6765 = 0.914$, or $0.914 \times 100 = 91.4\%$.

higher the percentage of cotransformation, the closer the loci.

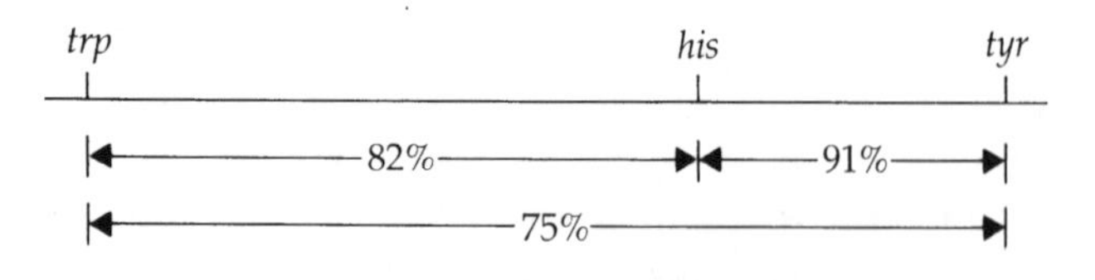

LIMITATIONS OF TRANSFORMATION STUDIES

Transformation studies have some limitations when it comes to bacterial chromosome mapping—chief among them is the fact that a very limited number of bacterial species readily undergo transformation. In species where transformation studies can be undertaken, the information derived from them, used in conjunction with other techniques (which will be discussed in the following chapters), has made a significant contribution to our understanding of the bacterial genome.

SUMMARY

Transformation involves the uptake, by the recipient, of extracellular pieces of donor DNA followed by recombination. Transformation is studied by using donor and recipient strains carrying contrasting alleles at the loci being considered. If the loci are closely linked, both double and single transformants can arise through single-transformation events involving single pieces of transforming DNA. Under these circumstances, double transformants are relatively common, with their frequency rising as the distance between the loci is diminished.

If the loci are not closely linked, double transformants arise only through the simultaneous occurrence of two separate transformation events, each involving a separate piece of transforming DNA. Since the two transformation events are independent, their simultaneous occurrence is very rare and double transformants occur with a very low frequency relative to that of the single-transformant classes.

Additional information about the degree of linkage between two loci can be supplied by comparing the relative numbers of single and double transformants produced at different concentrations of transforming DNA. As the concentration of transforming DNA declines, the reduction in the number of double transformants is much greater than the reduction of single transformants when loci are not closely linked.

The distance between two closely linked loci can be estimated from the frequency with which they are cotransformed. The analysis of three-point transformation experiments involves identifying the middle locus to establish the order of three loci and determining cotransformation frequencies to estimate the distances between them.

PROBLEM SET

21-1. Bacteria that are capable of being transformed may take up pieces of eukaryotic DNA just as they take up pieces of DNA derived from bacterial cells of the same species. The eukaryotic DNA very rarely becomes incorporated into the bacterial chromosome, while the bacterial DNA is usually incorporated. Explain why.

21-2. The results of a series of transformation experiments involving a particular bacterial species indicate that two loci carried on the bacterial chromosome are cotransformed only with great rarity. The DNA fragments used in these transforming experiments are approximately 20,000 base pairs in length. Based on this information, can anything be concluded about the distance separating these two loci?

21-3. The following sketch represents a portion of a bacterial chromosome showing the sites of two adjacent loci occupied by genes a^- and b^-. The portion of the chromosome to the left of locus a is designated as zone 1, that between the two loci as zone 2, and that to the right of locus b as zone 3.

a^- b^-

1 2 3

Next are shown three donor DNA fragments derived from bacteria of genotype a^+b^+. For each fragment, indicate the zone or zones in which crossing over must occur in order for transformation to give the a^-b^- recipients the post-transformation genotype that is listed.

	Donor fragment	Corresponding portion of recipient's chromosome following transformation
a.	a^+ b^+	a^+ b^-
b.	a^+ b^+	a^+ b^+
c.	a^+ b^+	a^- b^+

21-4. Assume that the linkage between two loci is tight enough so that a double transformant for these two loci can arise through the uptake of a single piece of transforming DNA. Is it possible for double transformants for these two loci to arise in any other way? Explain.

21-5. Three loci, designated as e, f, and g, are known to be adjacent to each other on a bacterial chromosome with the f locus in the middle position. The distances between e and f and between f and g are such that a transforming fragment can include loci e and f, or loci f and g, but not all three. Two transforming experiments are carried out. Experiment 1 involves transforming DNA derived from e^+f^+ bacteria and e^-f^- recipients. Experiment 2 involves transforming DNA derived from f^+g^+ bacteria and f^-g^- recipients. The three types of transformants produced in each experiment and their frequencies are as follows.

PROBLEM SET

Experiment 1		Experiment 2	
Transformants	Frequency	Transformants	Frequency
e^-f^+	132	f^-g^+	65
e^+f^-	158	f^+g^-	45
e^+f^+	106	f^+g^+	177
Total:	396	Total:	287

a. Calculate the frequency with which the *f* locus cotransforms with the *e* locus.
b. Calculate the frequency with which the *g* locus cotransforms with the *f* locus.
c. Based on the cotransformation studies, how does the distance separating *e* and *f* compare with the distance separating the *f* and *g* loci?

21-6. The transformability of two loci is studied at three different concentrations of transforming DNA and the relative number of single and double transformants is determined at each concentration. These results are summarized in the following table.

Concentration of transforming DNA in medium (μg DNA/ml)	Relative Number of Transformants	
	Single	Double
0.0040	100	10
0.0020	56	6
0.0010	30	2

a. What conclusion can be drawn regarding the manner in which the double transformants were produced; that is, did they arise from one or two transformation events? Explain.
b. Based on these data, are the two loci considered here closely linked or relatively far apart on the chromosome?

21-7. The transformability of two loci is studied at three different concentrations of transforming DNA, and the relative number of single and double transformants is determined at each concentration. These results are summarized in the following table.

Concentration of transforming DNA in medium (μg DNA/ml)	Relative Number of Transformants	
	Single	Double
0.0030	81	75
0.0010	38	8
0.0003	19	0.8

a. What conclusion can be drawn regarding the manner in which the double transformants were produced; that is, did they arise from one or two transformation events? Explain.
b. Are the two loci considered here closely linked or relatively far apart on the chromosome?

PROBLEM SET

21-8. Two loci, designated *c* and *d*, each controlling the synthesis of a different amino acid, designated as C and D, respectively, are studied in transformation experiments. The recipient strain, carrying mutant alleles at both of these loci, has genotype c^-d^-. In the first experiment, the donor DNA is obtained from two bacterial strains with genotypes c^+d^- and c^-d^+, respectively. A mixture of equal amounts of DNA from these two donor types is added to the medium in which cells of the recipient strain are growing. Following this, three classes of transformants are found in these frequencies: c^+d^-, 224; c^-d^+, 163; and c^+d^+, 3; for a total of 390.

a. If you were given a mixture of the three transformant types arising from this experiment, how would you isolate the bacteria of each type?

b. What is the only way in which a double transformant (genotype c^+d^+) could be produced in this experiment?

21-9. Refer to the information given in problem 21-8. A second experiment is carried out in which the donor DNA is derived from c^+d^+ bacteria. When this is supplied to c^-d^- bacteria, three classes of transformants are found in these frequencies: c^+d^-, 175; c^-d^+, 293; and c^+d^+, 332; for a total of 800.

a. What do the data from this second experiment lead you to conclude about the manner in which double transformants arose during the second experiment?

b. What do you notice about the frequency of double transformants in the second experiment relative to the number produced in the first experiment (described in problem 21-8)?

c. What does the difference in double-transformant frequencies between these two experiments suggest about the distance between the two loci?

d. If the double transformants produced in the second experiment arose in the same manner as those produced in the first experiment, would their frequency be different? Explain.

21-10. A wild strain of bacteria with the genotype $s^+t^+u^+$ serves as the source of donor DNA for a transformation study. The recipient strain is $s^-t^-u^-$. Assume that the three loci are far enough apart so that it is impossible for any two of these loci to be included in the same piece of transforming DNA.

a. What general type or types of transformants would you expect to be most common?

b. Generally, how would the frequency of transformants that had experienced recombination at two of the three loci compare with the frequency of those receiving a single locus from the donor DNA?

21-11. Refer to the information given in problem 21-10, but now assume that the *t* and *u* loci are linked closely enough to be included in a single piece of transforming DNA, while the *s* locus is located too far away to be included in a piece of transforming DNA carrying both loci *t* and *u*.

a. What general type or types of transformants would you expect to be most common?

b. Generally, how would the frequency of $s^-t^+u^+$ transformants compare with the frequency of $s^+t^-u^-$ transformants?

21-12. Two loci in the bacterium *Bacillus subtilus* were studied in a transformation experiment. The recipient strain carried mutant alleles at these two loci (n^-o^-). The donor DNA was derived from a strain that had wild-type

PROBLEM SET

alleles at these loci (n^+o^+). The transformed classes were then isolated and their frequencies were n^+o^-, 200; n^-o^+, 259; and n^+o^+, 283; for a total of 742. What is the frequency with which the *o* locus is cotransformed with the *n* locus?

21-13. Assume that a transformation study is carried out with a bacterial species involving the loci *met*, *ile*, and *thr*, which control, respectively, the production of amino acids methionine, isoleucine, and threonine. Donor DNA from donor cells of genotype $met^+ile^+thr^+$ is supplied to recipient cells of genotype $met^-ile^-thr^-$. The order of these loci on the chromosome is unknown, and they may or may not be positioned as they are written here. Identification and counting of the transformants at the end of the experiment shows that there are seven recombinant classes which occur in the following frequencies.

Class	Transformant genotype	Frequency
1.	$met^-ile^+thr^-$	417
2.	$met^+ile^-thr^+$	65
3.	$met^+ile^+thr^+$	8012
4.	$met^-ile^+thr^+$	2800
5.	$met^-ile^-thr^+$	815
6.	$met^+ile^+thr^-$	1002
7.	$met^+ile^-thr^-$	2106
	Total:	15,217

a. Identify each of the transformant classes as either a single, double, or triple recombinant.

b. The three loci considered here can fall into one of two general categories: they can be tightly linked or not tightly linked. Based on just an inspection of the given data, how would you categorize these loci? Explain.

c. Verify your answer to 21-13b before proceeding. If these loci were in the category other than the one you placed them in in answering 21-13b, state, in general terms, how the frequency of the class containing transformants $met^+ile^+thr^+$ would be different?

d. Identify the recombinant class that arose through a quadruple crossover.

e. Which of the three loci is betwen the other two?

21-14. Refer to the information given in problem 21-13.

a. Determine the cotransformation frequencies of *ile* with *met*, of *thr* with *met*, and of *ile* with *thr*.

b. Sketch a map showing the relative positions of the three loci and the cotransformation frequencies separating them.

21-15. Assume that a transformation study is carried out with a bacterial species involving the loci *arg*, *cys*, and *leu*, which control, respectively, the production of amino acids arginine, cysteine, and leucine. Donor DNA from donor cells of genotype $arg^+cys^+leu^+$ is supplied to recipient cells of genotype $arg^-cys^-leu^-$. The order of these loci on the chromosome is unknown, and they may or may not be positioned as they are written here. Identification and counting of the transformants at the end of the experiment shows that there are seven recombinant classes which occur in the following frequencies. Which of these loci are tightly linked?

PROBLEM SET

Class	Transformant genotype	Frequency
1.	$arg^+ cys^+ leu^-$	78
2.	$arg^+ cys^- leu^+$	61
3.	$arg^- cys^- leu^+$	2938
4.	$arg^+ cys^+ leu^+$	6
5.	$arg^- cys^+ leu^+$	49
6.	$arg^+ cys^- leu^-$	3357
7.	$arg^- cys^+ leu^-$	3118
	Total:	9607

21-16. Assume that a transformation study is carried out with a bacterial species involving the loci *aro*, *his*, and *try*, which control, respectively, the production of aromatic amino acids, histidine, and tryptophan. Donor DNA from donor cells of genotype $aro^+ his^+ try^+$ is supplied to recipient cells of genotype $aro^- his^- try^-$. The order of these loci on the chromosome is unknown, and they may or may not be positioned as they are written here. Identification and counting of the transformants at the end of the experiment shows that there are seven classes which occur in the following frequencies.

Class	Transformant genotype	Frequency
1.	$aro^+ his^+ try^-$	148
2.	$aro^+ his^- try^+$	2285
3.	$aro^- his^- try^+$	197
4.	$aro^+ his^+ try^+$	5860
5.	$aro^- his^+ try^+$	11
6.	$aro^+ his^- try^-$	111
7.	$aro^- his^+ try^-$	878
	Total:	9490

a. Identify each of the transformant classes as either a single, double, or triple transformant.
b. Identify the transformant class that arose through a quadruple crossover.
c. Which of the three loci is between the other two?

CHAPTER

22

Conjugation and Mapping the Bacterial Chromosome

INTRODUCTION

Conjugation involves the transfer of genetic material from one bacterium, the donor, to another, the recipient, through a protoplasmic connection. Once transferred, the genetic material may be incorporated into the recipient's chromosome through crossing over. The homologous sections of the recipient's chromosome displaced by this crossing over are generally lost during subsequent cell divisions. Although there are several kinds of conjugation systems, this chapter focuses on the one found in *E. coli*.

MATING TYPES

An *E. coli* bacterium can act as a donor if it possesses a genetic entity referred to as the **F factor,** also known as the **F element** or the **fertility factor,** a self-replicating, circular unit of double-helix DNA which contains about 2% of the DNA found in the bacterial chromosome. Cells with an F factor are designated as F^+ and those without as F^-. Only F^- cells can serve as genetic recipients. In F^+ cells, the F factor usually exists as an autonomous unit separate from the main bacterial chromosome. However, occasionally the F factor integrates, through crossing over, into the main bacterial chromosome (see Figure 22-1) to produce a specialized type of donor or F^+ cell designated as **Hfr.** The integration of the F factor is facilitated by short sequences of DNA known as **insertion sequences** (IS) which are carried on the F factor and on the *E. coli* chromosome. (Insertion sequences are discussed further in Chapter 29.) Genetic units like the F factor, which can be either autonomous and replicate independently of the bacterial chromosome or integrated into the chromosome and replicate with it, are designated as **episomes.**

An autonomous F factor can replicate independently of the donor's chromosome and be transferred to an F^- cell. This transfer is facilitated by a specialized surface component produced by a donor cell that allows it to make physical contact with an F^- cell. Following this contact, a protoplasmic connection is established between the donor and recipient cells. The transfer involves rolling-circle DNA replication which is described in Chapter 19. One strand of the F factor is enzymatically nicked, and the 5′ end that is created peels off as the nicked strand unrolls from the intact strand of the F factor. As this unrolling proceeds, this strand passes from the donor into the F^- cell where it serves as a template for the synthesis of a new, complementary, counterpart strand. When this transfer is finished and the synthesis is complete, the ends of this linear double-stranded molecule are covalently joined, giving the F factor its circular form. Meanwhile, back

FIGURE 22-1

Integration of an F factor into the main bacterial chromosome of an F^+ cell to form an Hfr cell.

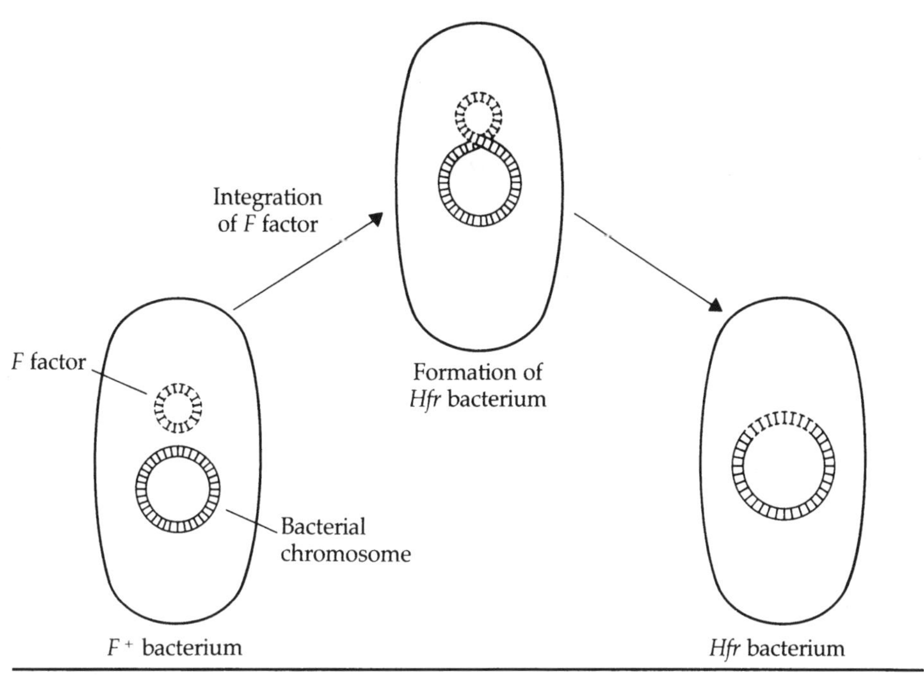

From *Concepts of Genetics*, Second Edition, by William S. Klug and Michael R. Cummings. Copyright © 1986, 1983 Scott, Foresman and Company.

in the donor cell, a replacement for the transferred strand is synthesized with the original, unnicked strand serving as its template. Both the donor and recipient cells end up with an intact F factor. The donor cell remains F^+ and can again serve as a donor; the recipient is now F^+ and can serve as a donor. These events are summarized in Figure 22-2.

Hfr CELLS

The replication and transfer of an autonomous F factor does not by itself transfer any of the genetic material of the bacterial chromosome. However, this may occur when the donor's F factor is integrated into its chromosome. An integrated F factor acts like a part of the chromosome: it replicates with the chromosome and gets transmitted with it during cell reproduction. Bacteria with integrated F factors are designated as Hfr because of the *h*igh-*f*requency of *r*ecombination that occurs when they participate in conjugation. Like an F^+ cell, an Hfr bacterium can establish contact with an F^- cell and transfer genetic material to it by rolling-circle replication. A nick occurs within one strand of the Hfr cell's integrated F factor and the 5′ end of that strand peels off as it unwinds and passes through the connection. However, in this case, the transferred portion of the F factor is continuous with a strand of the chromosome, which is also transferred. As this combined F factor–chromosome strand passes into the recipient, it serves as a template for the synthesis of a complementary strand. The transfer occurs at a relatively constant rate, with about 100 minutes required to complete the transfer of the entire chromosome strand and the balance of the F factor. During the transfer, the intact F factor–chromosome strand in the Hfr donor serves as a template for the synthesis of a new complementary counterpart.

Note again that the nick that initiated the transfer occurs within the F factor and that only a part of the F factor leads the chromosomal strand into the recipient cell. If the recipient receives the "leading" portion of the F factor *and* the entire chromosomal strand *and* the balance or "trailing" portion of the F factor, the recipient is converted from F^- to Hfr. Such conversions are relatively rare, however, since most conjugating pairs spontaneously separate before the transfer is complete. Usually the "leading" section of the F factor and only a portion of the chromosome strand are transferred. Since the speed of transfer is relatively con-

FIGURE 22-2

Transfer of an F factor from an F^+ cell to an F^- cell, converting the F^- cell to F^+.

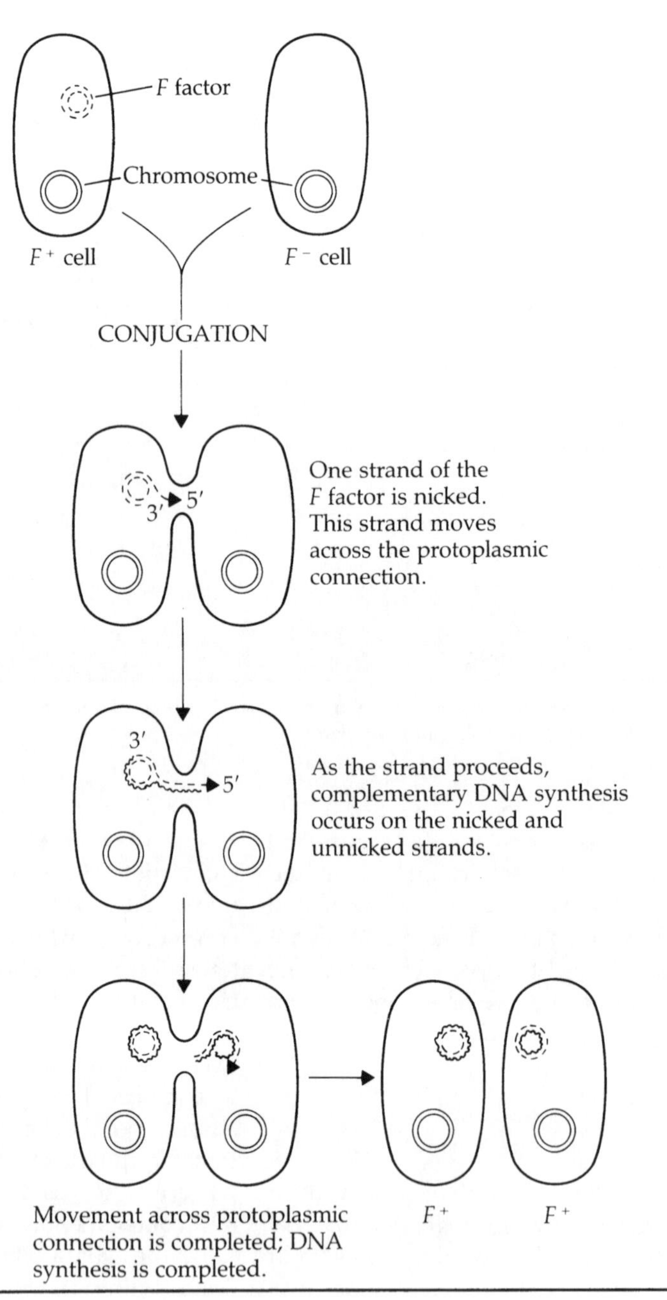

From *Concepts of Genetics*, Second Edition, by William S. Klug and Michael R. Cummings. Copyright © 1986, 1983 Scott, Foresman and Company.

stant, what determines the amount of chromosomal material transferred? ______________________________ ______________________[1] If the transfer is disrupted prior to completion, what will be the mating type of the recipient? __________[2]

Regardless of the amount of the Hfr chromosome transferred, once it is inside the recipient it may pair with homologous regions of the recipient's chromosome and undergo recombination. The reciprocal exchange that occurs incorporates donor genes into the

[1] The time the cells are in contact. [2] F^-, since it lacks a complete copy of the F factor.

recipient's chromosome. The free DNA that remains following recombination is subsequently lost from the cell. During the interval when both donor and recipient genes are present, the cell is diploid for those genes and is referred to as a **partial zygote** or **merozygote.** The events involved in the incomplete transfer of the F factor–chromosome strand from an Hfr cell to an F^- cell are summarized in Figure 22-3.

CONJUGATION STUDIES

Conjugation is studied through matings between Hfr donors and F^- recipients. If the two strains exhibit contrasting traits at some loci, then it is possible to determine which of these loci get transferred and recombined because each locus transferred will alter the recipient's phenotype in a specific way. Furthermore,

FIGURE 22-3

Incomplete transfer of the F factor–chromosome strand from an Hfr cell to an F^- cell. (a) Protoplasmic connection is established between Hfr donor and F^- recipient cells. (b) Transfer of Hfr chromosome is underway, led by the leading portion of the F factor. DNA replication occurs to form complements for transferred and intact strands of the Hfr chromosome. Conjugation usually terminates before the entire Hfr strand is transferred. (c) Conjugation is completed with only the *A* and *B* loci transferred. Transferred Hfr genes may recombine with the F^- chromosome.

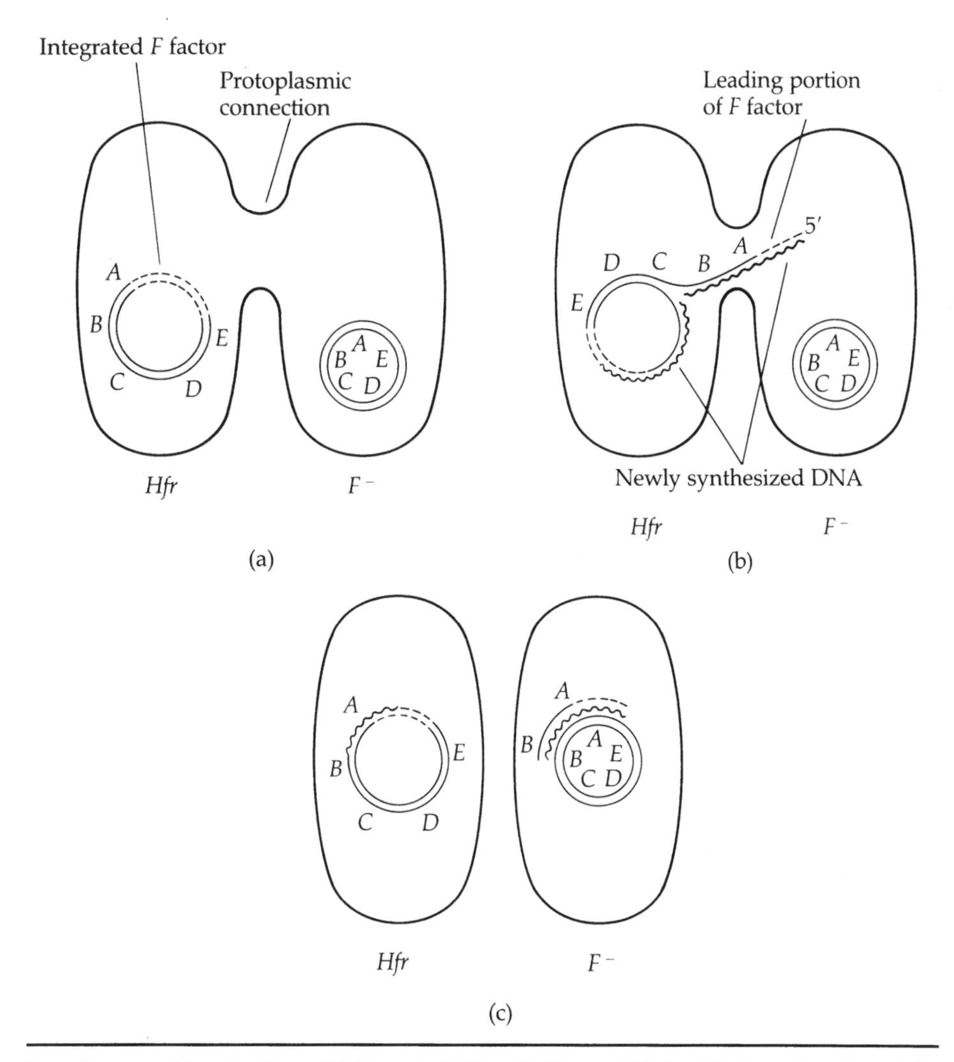

From *Concepts of Genetics*, Second Edition, by William S. Klug and Michael R. Cummings.

since the transfer of the genetic material by a particular strain of Hfr cells always occurs in the same direction and is a sequential process, it is possible to determine the order of gene transfer. Genes adjacent to the leading portion of the F factor are transferred first. The further from the leading portion chromosomal genes are, the longer the cells must be in contact to transfer the genes.

Two forms of chromosomal maps can be prepared from conjugation studies—one expressing distance in terms of minutes required for transfer and the other, in terms of recombination percentages. Each type of map is produced from a somewhat different type of conjugation study.

Interrupted-Mating Studies

Interrupted-mating studies, initially carried out by F. Jacob and E. Wollman in the 1950s, are designed to deliberately interrupt the Hfr × F^- matings after specific intervals of time. These experiments are carried out by mixing bacteria of an Hfr strain with a population of F^- cells. After allowing sufficient time for the donor and recipient cells to establish contact, samples of cells are removed at regular intervals and placed in a kitchen blender which separates the conjugating cells without damaging the bacteria, thereby preventing any subsequent transfer of genetic material. Following blender treatment, the bacterial sample consists of a mixture of Hfr cells, F^- cells, and, importantly (from the standpoint of mapping), some F^- cells that have acquired one or more of the donor's genes through conjugation and subsequent recombination. Since only these recombinant cells are useful in preparing a map, a selective medium is used that allows only the recombinant cells to grow. Recipient cells conjugating for short intervals would show recombination at a limited number of loci. Where would these loci be positioned relative to the leading portion of the F factor?

____[3]

As the conjugation interval increases, what happens to the number of loci participating in recombination?

____[4]

Parental Strains After a conjugation study is completed, it is necessary to distinguish among Hfr cells, F^- cells that have not participated in conjugation, and F^- cells that have acquired DNA from Hfr donor cells. The Hfr and F^- bacteria used in conjugation studies are usually chosen to facilitate this distinction and often exhibit the following characteristics. Hfr donor cells (1) are all from the same strain so they carry the F factor inserted at the same site (origin) and with the same orientation; (2) carry wild-type alleles at the loci being mapped; and (3) possess an allele for sensitivity to either an antibiotic such as streptomycin or to a particular bacteriophage, with this allele positioned far enough along the chromosome, relative to the leading portion of the F factor, so that it is one of the last loci to be transferred. F^- recipient cells (1) carry mutant alleles at the loci being mapped; and (2) carry an allele for resistance to the antibiotic or bacteriophage to which the Hfr strain is sensitive.

Detecting the Transfer of a Single Locus Assume we wish to determine the time required for the transfer of a single locus that controls, say, the production of the amino acid methionine. The Hfr parental strain is nutritionally competent and streptomycin-sensitive, with genotype met^+str^s. What genotype should the F^- parental strain for this cross exhibit? ____________[5] What type of medium would support the growth of this F^- strain? ____

________[6]

The study is carried out as follows: Hfr and F^- cells are cultured together at 37°C and samples are withdrawn at regular intervals. These samples, containing both conjugating and nonconjugating cells, are promptly treated in a kitchen blender to separate the cells and prevent further transfer of genetic material. What types of cells would you expect to find in the blender-treated samples taken before the *met* locus is transferred? ______________________________

______________[7] What types of cells would you expect to find in samples taken after the *met* locus is transferred and before the *str* locus is transferred?

____[8]

What type of medium could be used to select for F^- recombinants? ______________________________

______________[9] Indicate whether each of the cell types listed as follows would grow when placed on the methionine-deficient, streptomycin-containing medium, and explain why growth would or would not occur.

[3] They would be nearest to it. [4] The number of participating loci would increase as loci progressively further from the F factor are transferred. [5] met^-str^r. [6] One containing methionine with or without streptomycin. [7] Hfr, met^+str^s; and F^-, met^-str^r.
[8] Hfr, met^+str^s; F^-, met^-str^r; and F^-, met^+str^r (recombinant.) [9] One lacking methionine and containing streptomycin.

Hfr, $met^+ str^s$: ______________[10]

F^-, $met^- str^r$: ______________[11]

F^-, $met^+ str^r$: ______________[12]

Following overnight culture on this medium, each F^- recombinant would grow into a colony visible to the unaided eye.

Assume that samples are taken from the conjugating mixture at two-minute intervals where 0 minutes marks the start of conjugation. Samples withdrawn at the end of 2, 4, 6, 8, 10, and 12 minutes show no recombinant cells. Samples at 14 minutes and thereafter contain numerous met^+ recombinants. What conclusion would you draw about the time of transfer? ______________________________[13] If the map distance is expressed in units of minutes and if the F-factor insertion point is designated as 0 minutes, how far is the *met* locus from the F-factor insertion point? __________[14]

Studies Involving Two or More Loci Conjugation studies are usually designed so that two or more loci can be mapped simultaneously. Assume that an F^- parental strain is like the one just considered, but also lacks the capacity to synthesize the amino acid threonine and has genotype $met^- thr^- str^r$. If the goal is to map both the *met* and the *thr* loci, what genotype should a parental Hfr strain possess? __________[15] What type of medium should the blender-treated samples be placed upon in order to select for F^- cells recombinant for both loci? ______________________________[16] Assume that samples are taken from the conjugating mixture at two-minute intervals. If this type of recombinant is found only in samples at 22 minutes and thereafter, what could be concluded about the time required for the transfer of both loci? ______________________________[17] The previous experiment placed the *met* locus between 12 and 14 map minutes, and this experiment indicates that the *thr* locus is passed to the F^- cell at between 20 and 22 minutes.

A similar study with an F^-, $met^- thr^- leu^-$ strain might indicate that recombinant F^- cells capable of growing on medium deficient in methionine, threonine, and leucine are found in samples taken at 26 minutes and thereafter. Based on the outcome of these three studies, what is the order of these three alleles in the Hfr donor strain relative to the insertion point, or origin, of the F factor? __________[18] Approximately what distance separates the *met* and the *thr* loci? ______[19] the *thr* and *leu* loci? __________[20]

Spontaneous-Interruption Studies

The interrupted-mating technique just described deliberately disrupts the mating process at regular intervals. A second type of conjugation study relies on the fact that conjugating cells separate spontaneously as a consequence of chance and thermal agitation. Initially, when F^- and Hfr cells are mixed together, a great many conjugating pairs are formed. As time passes, more and more of the pairs spontaneously separate (the probability of this separation remains constant per unit time) until a point is reached where very few, if any, of the conjugating pairs remain in contact. An illustration of this type of study is as follows. Assume that the *pro, leu, trp,* and *arg* loci are to be mapped. The F^- parental strain carries mutant alleles for these loci and is streptomycin-resistant, with genotype $pro^- leu^- trp^- arg^- str^r$. The Hfr strain carries wild-type alleles at these loci and is streptomycin-sensitive.

The Hfr and the F^- strains are mixed together and cultured for 100 minutes or longer—an interval sufficient to allow the transfer of the entire strand of the Hfr chromosome (for any cells that happen to stay in contact that long). At the end of the period, samples containing equal numbers of cells are spread on a series of agar plates, each containing a medium designed to select for F^- cells recombinant for a different donor allele. What, for example, would be the critical features of a medium designed to select for F^-, pro^+ cells? ______________________________[21] Culturing samples on selective media for $str^r leu^+$,

[10] No growth because of their streptomycin sensitivity. [11] No growth because of the absence of methionine.
[12] Growth since they are streptomycin-resistant and can synthesize methionine. [13] Between 12 and 14 minutes after initiation.
[14] Between 12 and 14 map minutes. [15] $met^+ thr^+ str^s$. [16] It should lack methionine and threonine, and contain streptomycin; growth of F^- and Hfr parental strains would be blocked. [17] Both loci were transferred by 22 minutes. [18] *met-thr-leu*.
[19] This distance is the time difference between when $met^+ thr^-$ and $met^+ thr^+$ recombinants are found: $22 - 14 = 8$ map minutes.
[20] $(26 - 22) = 4$ map minutes. [21] It must contain streptomycin, which will eliminate the Hfr cells, and be free of proline but contain all of the other necessary nutrients to grow mutant cells, which will eliminate nonrecombinant F^- cells.

$str^r trp^+$, and $str^r arg^+$ would be done in a similar way to the $str^r pro^+$ selection. How could the first transferred locus be identified? ______ [22]

Say that the media designed to select for F^-, pro^+ recombinants has the largest number of colonies, indicating that the *pro* locus is the first transferred. After the first transferred locus is identified, the next step is to determine the percentage of F^-, pro^+ cells that are recombinant for each of the other genes being mapped. One approach to gathering this data is to replica plate the F^-, pro^+ recombinant colonies onto a series of plates, each of which is designed to select for one of the other genes being studied. Once this has been done, it is an easy matter to calculate the percentage of the F^-, pro^+ cells that have also received each of the other genes. One might find, for example, that of the pro^+ recombinants, 93% were leu^+, 86% were trp^+, and 45% were arg^+. What is the relationship between these recombination percentages and the distance of the loci from the first transferred locus? ______ [23]

The percentages indicate the relative position of these genes on the chromosome, with distances expressed in terms of recombination percentages.

Both the interrupted-mating and spontaneous-interruption approaches to mapping provide a good approximation of the relative positions of genes separated by distances equivalent to three or more map minutes. Experimental limitations preclude reliable estimates for loci that are much closer than this. Such loci may be more accurately mapped through transformation or transduction studies, which are discussed in Chapters 21 and 23, respectively.

F-FACTOR INSERTION SITES AND ORIENTATIONS

Hfr strains can differ in the F-factor insertion sites and in the orientation of the F factors they carry. This can be verified by comparing the outcomes of a series of interrupted-mating studies. We will consider a series that involves five different Hfr strains. Each Hfr strain is identical with regard to eight wild-type genes, say $a^+b^+c^+d^+e^+f^+g^+h^+$, whose sequence is initially unknown. Each of these Hfr strains is separately mated with F^- cells of genotype $a^-b^-c^-d^-e^-f^-g^-h^-$. The order of transfer of loci, detected through interrupted-mating studies, is given in the following table. Look closely at the gene sequences for each Hfr strain. Note that the position designated as O represents the **origin of transfer** or the site where transfer is initiated; and remember that since each sequence is carried on a circular chromosome, the first and last loci to be transferred flank the F factor carried by the chromosome.

Hfr strain	Order of gene transfer to F^-
1	O *bfdaeghc*
2	O *ghcbfdae*
3	O *daeghcbf*
4	O *fbchgead*
5	O *bchgeadf*

At first glance it may appear that each strain has a different sequence of loci. However, look at the neighbors of a particular locus in all of these Hfr strains. Repeat this with another locus. What do you notice? ______ [24]

Compare the loci on either side of the origin in strains 1, 2, and 5. Are they the same or do they differ? ______ [25] What does this tell you about the F-factor insertion sites? ______ [26] Compare the loci on either side of the origin for strains 3 and 4. Do they carry the F factor at the same site? ______ [27] In what way *do* strains 3 and 4 differ? ______ [28] What does this indicate about the F factor? ______ [29]

FIGURE 22-4

Eight loci on a bacterial chromosome.

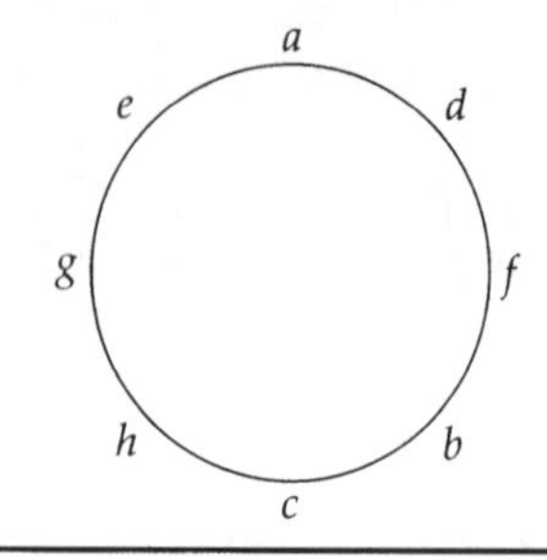

[22] From the medium supporting the greatest number of colonies, since recombinants carrying the first transferred locus would be more common than recombinants for any other locus. [23] The lower the percentage, the farther the locus is from the first transferred locus.
[24] A particular locus is always flanked by the same two loci; taken collectively, this indicates that all of the strains have the same sequence.
[25] They differ from one strain to another. [26] They vary. [27] Yes. [28] The direction in which the loci are transferred is reversed.
[29] Its orientation differs with respect to the direction of transfer.

As you have just shown, the differences in these five strains can be explained either by different insertion sites or orientations of the F factor. The map in Figure 22-4 shows the eight loci carried by these five Hfr strains. For each strain, indicate on Figure 22-4 the F-factor insertion site and the direction of transfer as either clockwise or counterclockwise.[30] The F factor's ability to insert at different sites and with different orientations is apparently due to the fact that the insertion sequences that facilitate F-factor insertion can occur at different locations and with different orientations in the *E. coli* chromosome.

F-FACTOR EXCISION AND SEXDUCTION

The F factor incorporated into the chromosome of an Hfr bacterium can spontaneously remove itself from the chromosome, thereby converting the cells to F^+. Such cells are designated as **F^+ revertants.** The excision of the F factor is apparently accomplished by a reversal of the crossover process that inserted the F factor initially. Occasionally, an error occurs during excision and a limited portion of the bacterial chromosome from either one side or the other of the F-factor insertion site is removed along with the F factor. Such a modified F factor, designated as an **F′** (F prime) **factor,** can be transmitted to an F^- cell. The transfer of one or more chromosomal genes by the F′ is called **sexduction** or **F-duction.** Following such a transfer, the recipient cell is now diploid for the chromosomal gene or genes carried by the F′ factor. For example, conjugation between a cell with an F′ factor that carries the allele for galactose fermentation (gal^+) and an F^-, gal^- cell that lacks the ability to ferment galactose can convert the recipient to an F^+, gal^+ status. The recipient now carries both the gal^+ and gal^- alleles and is a partial diploid for at least a short time. As you will see, sexduction is similar to the process of specialized transduction which is discussed in Chapter 23.

SUMMARY

The mating type of *E. coli* cells depends on the presence or absence of a small unit of DNA known as the fertility, or F, factor. An F^- cell lacks the F factor and can serve as a genetic recipient. Both F^+ and Hfr (*h*igh-*f*requency *r*ecombination) cells are genetic donors. An F^+ cell has an autonomous F factor while an Hfr cell has an F factor integrated into its chromosome.

During conjugation, an F^+ cell transfers one strand of its F factor to an F^- cell where it serves as a template for the synthesis of a complementary strand. When the recipient cell has a complete copy of the F factor, the cell is converted to F^+ status. Synthesis of a complement to the F-factor strand that remains behind in the donor F^+ cell allows the donor cell to retain its F^+ status. No chromosomal genes are involved in this type of transfer.

An Hfr cell can transfer a copy of its chromosome to an F^- cell, beginning at a site within the F factor known as the origin of transfer. This transfer occurs at a relatively constant rate, and the longer a pair of Hfr and F^- cells remain in conjugation, the greater the number of loci transferred. The time required for the transfer of a particular donor gene serves as an index of its distance from the origin and provides the basis for chromosomal mapping through interrupted-mating studies.

An F factor can be incorporated into the bacterial chromosome at a number of different sites. The insertion site determines the origin of transfer for Hfr genes, and the orientation of the F factor defines the direction of transfer.

Occasionally, when the F factor of an Hfr bacterium excises from the chromosome, it takes a limited portion of the bacterial chromosome with it. This modified F factor and the chromosomal genes it carries can be transmitted to an F^- cell in the process of sexduction.

[30] Strain 1, between *c* and *b*, counterclockwise; 2, between *e* and *g*, counterclockwise; 3, between *d* and *f*, counterclockwise; 4, between *d* and *f*, clockwise; 5, between *f* and *b*, clockwise.

PROBLEM SET

22-1. The Hfr donor strains used in conjugation studies are generally sensitive to an antibiotic or to infection by a bacteriophage while the F^- recipients are resistant to the antibiotic or phage. Why is this Hfr sensitivity important to the design of the study?

22-2. After an interval of conjugation, the Hfr donors in the conjugation study are eliminated from the sample of bacteria, leaving a mixture of F^- and F^- recombinant cells. How are the F^- and F^- recombinants usually distinguished from each other?

22-3. Hfr strains chosen for use in conjugation studies have the locus that controls antibiotic or phage sensitivity at a site very distant from the F-factor insertion site. Why is such a location for the "sensitivity" locus critical to the design of the study?

22-4. Briefly explain how each of the following conversions occurs.

a. F^- to F^+.
b. F^+ to Hfr.
c. F^- to Hfr.
d. Hfr to F^+.
e. Hfr to F'.

22-5. In $F^+ \times F^-$ matings, the recipient is converted to F^+ status. Yet in most Hfr $\times$ F^- matings the recipient remains F^-. Explain why this is so.

22-6. Explain why Hfr $\times$ F^- matings result in the transfer of chromosomal genes while $F^+ \times F^-$ matings do not.

22-7. In *E. coli*, the *arg* and *his* loci control the ability to synthesize the amino acids arginine and histidine, respectively. The *gal* locus controls the ability to utilize the sugar galactose as a carbon source. An Hfr strain, sensitive to infection by the bacteriophage T2, carries wild-type alleles at these loci: $arg^+gal^+his^+$. These Hfr bacteria are mated with a strain of F^- cells that is resistant to the T2 phage and that has mutant alleles at these three loci: $arg^-gal^-his^-$. Studies indicate that this Hfr strain transmits the T2-sensitivity locus very late in conjugation. Samples of cells are removed from the conjugating mixture at regular intervals and shaken strongly enough to separate conjugating pairs. Subsamples of cells are placed on each of three different types of agar-based media, each of which is designed to select for one of the three donor genes, and, following this, T2 phages are added to each petri dish.

a. Indicate the key features of the type of media used to select for cells recombinant for arg^+, for gal^+, and for his^+.

b. The following data give the number of bacterial colonies found growing on the three selective media from samples taken at intervals of five minutes after the start of conjugation.

Medium selecting for	Number of Colonies at Samples Times of (minutes of conjugation)											
	0	5	10	15	20	25	30	35	40	45	50	55
arg^+	0	0	0	0	0	0	0	0	0	0	0	0
gal^+	0	0	0	120	150	150	150	150	150	150	150	150
his^+	0	0	0	0	0	0	0	0	0	130	150	150

From these data, determine the order of the three loci on the chromosome relative to the insertion point of the F factor and the distances separating these sites.

PROBLEM SET

c. When the F^- recipient used in the study is mated in a separate experiment with a different Hfr, $arg^+ gal^+ his^+ T2^s$ strain, the three loci are transmitted in the same order as seen in 22-7b. However, the *gal* and *his* loci are transmitted at about 30 and 60 minutes, respectively, after the start of conjugation. How could these later transmission times be explained?

22-8. An interrupted-mating experiment is carried out between an Hfr strain that carries wild-type alleles for four different nutritional genes ($r^+ s^+ t^+ u^+$) and is streptomycin-sensitive (str^s). This strain transmits the r^+ gene shortly after the start of conjugation and transmits the str^s gene very late in the process. The recipient strain is $f^- r^- s^- t^- u^- str^r$. Following the mixing of the two strains, samples are removed every five minutes and $r^+ str^r$ recipients are selected for. Through the use of appropriate media and replica plating, these recipient cells are then scored for the presence of the other donor genes. The data giving the percentage of r^+ recipient cells that possess the u^+ in the samples taken at five-minute intervals during the 55 minutes of conjugation are as follows.

	Minutes of Conjugation										
	0	**5**	**10**	**15**	**20**	**25**	**30**	**35**	**40**	**45**	**50**
Percentage of r^+ cells with u^+:	0	0	15	80	90	90	90	90	90	90	90

These data are plotted on the following graph.

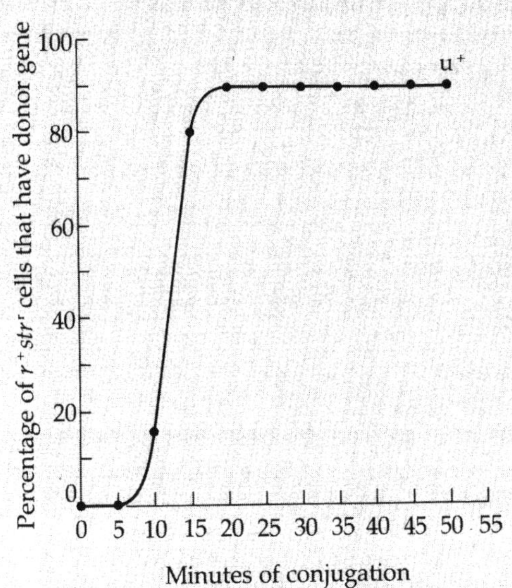

a. Identify the interval during which r^+ was transferred.
b. Identify the interval during which u^+ was transferred.
c. Why does the increase in the number of cells recombinant for u^+ occur over a 15-minute interval rather than all at once?
d. Why does the percentage of cells with u^+ plateau at 90% rather than continue to increase?

22-9. Refer to the basic information given in problem 22-8. The data giving the percentage of r^+ recipient cells that possess the other donor genes that are

PROBLEM SET

studied in this experiment (that is, s^+ and t^+) are given in the following table. Plot these data on the graph in problem 22-8 before you answer the questions that follow.

	Minutes of Conjugation										
	0	**5**	**10**	**15**	**20**	**25**	**30**	**35**	**40**	**45**	**50**
Percentage of r^+ cells with s^+:	0	0	0	0	0	0	7	20	27	30	30
Percentage of r^+ cells with t^+:	0	0	0	0	3	20	40	50	50	50	50

a. Are the lower levels of recombination exhibited by the s^+ and t^+ genes due to insufficient time for transfer of the genes by the donor cells? Explain.
b. What is the order of the *u*, *s*, and *t* loci on the bacterial chromosome?
c. Estimate the distances in map minutes between these three loci.

22-10. Five bacterial loci, each controlling the production of a different amino acid, are to be studied in an interrupted-mating experiment. The relative order of these loci on the chromosome is known to be *trp-his-gly-met-arg* but nothing is known about the insertion point of the F factor in the Hfr parental strain. The Hfr donor strain is $trp^+his^+gly^+met^-arg^+$ streptomycin-sensitive. The F^- recipient strain is $trp^-his^-gly^-met^+arg^-$ streptomycin-resistant. Earlier interrupted-mating studies with this particular Hfr strain indicate that (1) the first of these loci gets transferred 10 minutes after the start of conjugation, (2) the remaining four loci are transferred at successive 10-minute intervals, and (3) the gene for streptomycin sensitivity is one of the last to be transmitted (that is, it is transmitted near the end of the 100 minutes required for complete chromosome transfer). A conjugating mixture of Hfr and F^- cells is prepared and samples of cells are taken from it at 5, 15, 25, 35, 45, 55, 65, 75, and 85 minutes after cell mixing. Following cell separation, these samples are separately placed on a minimal medium containing streptomycin. Of these samples, only the one taken at 45 minutes of contact shows a large number of cells capable of growth on this medium.
a. What is the genotype of the cells in the 45-minute sample that are capable of growing on the minimal medium containing streptomycin?
b. Sketch the Hfr strain chromosome showing the six loci, the distances between them, the relative position of the F-factor insertion site, and the direction of transfer.
c. Explain why some of the bacteria from samples taken after 45 minutes are not able to grow on the streptomycin-containing minimal medium.

22-11. Conjugational matings are set up between F^- recipient cells and five different strains of Hfr bacteria. The loci studied in each cross and their order of transfer are indicated in the following table. Note that the leftmost letter written for each strain represents the first locus transferred.

Hfr strain	Order of gene transfer
1	*vncysu*
2	*odpem*
3	*bodp*
4	*ycnvwme*
5	*subod*

PROBLEM SET

a. Why does the first gene transferred differ with each Hfr strain?
b. Set up a chromosome map giving the sequence of all the loci studied in these matings.
c. Indicate on this map the origin, or insertion point, of the F factor for each of the five Hfr strains as well as the direction of transfer.
d. Do all the Hfr strains transfer in the same direction? If not, how is this explained?

22-12. a. Assume that it is necessary to obtain recombinant cells that are Hfr from the cells arising in a cross involving Hfr strain 4 in problem 22-11. Which gene of the seven studied would you select for (by using the appropriate selective medium) in order to find the greatest number of Hfr recombinant cells? Explain.
b. Would all of the recombinant cells carrying the gene you designated in your answer to 22-12a be Hfr cells? Explain.

22-13. A strain of Hfr *E. coli* cells is combined with a population of F^- cells for 100 minutes at 37°C. Seven loci are studied in this cross. The Hfr cells carry wild-type alleles for these loci and have genotype $l^+a^+g^+t^+d^+m^+n^+$. The F^- strain carries mutant alleles at these same loci: $l^-a^-g^-t^-d^-m^-n^-$. Locus *l* controls the production of an essential amino acid and is known to be transferred very early by this Hfr strain. Nothing is known about the sequence or transfer time of the other six loci. The Hfr strain is sensitive to streptomycin, and the F^- strain is resistant. Following the 100-minute interval, bacteria are placed on a medium lacking amino acid *l* and containing streptomycin. Replica plating of the colonies developing on the minimal medium onto appropriate media indicates that 37% are a^+, 87% are g^+, 23% are t^+, 54% are d^+, 3% are m^+, and 49% are n^+.
a. How many of the colonies developing on the medium lacking amino acid *l* and containing streptomycin have developed from Hfr parental cells? Explain.
b. Would nonrecombinant F^- cells be able to grow on the streptomycin medium lacking amino acid *l*?
c. What is the order of transfer for the seven loci and the approximate distances separating them?

22-14. A mating is set up between bacteria that are Hfr, $s^+t^+u^+$ streptomycin-sensitive and F^-, $s^-t^-u^-$ streptomycin-resistant. The *s* locus is known to be transferred very early by this Hfr strain. The two strains of bacteria are allowed to remain in contact for 110 minutes, an interval sufficient to allow complete transfer of the Hfr chromosome to an F^- cell. Recombinant cells that are s^+ are selected using an appropriate medium. Colonies arising from these s^+ recombinants are replica plated onto a series of media designed to select for the presence of the other two wild-type alleles. Among the total of 422 recombinant cells, four genotypes are found in the following frequencies: $s^+t^-u^-$, 310; $s^+t^-u^+$, 100; $s^+t^+u^-$, 2; and $s^+t^+u^+$, 10.
a. What is the relative order of these three loci on the chromosome?
b. How do you explain the small number of $s^+t^+u^-$ individuals among the recombinants?

22-15. Conjugation studies indicate that the *met* and *pro* loci are very close to each other, and that both are about six map minutes form the *pur* locus. It is impossible to order the *met* and *pro* loci relative to the *pur* locus by inter-

PROBLEM SET

rupted-mating techniques and either *pro-met-pur* or *met-pro-pur* sequences are possible.

a. Assume that the order is *pro-met-pur*. Two matings, Hfr, $pro^+ met^+ pur^-$ × F^-, $pro^- met^- pur^+$, and its reciprocal, Hfr, $pro^- met^- pur^+$ × F^-, $pro^+ met^+ pur^-$, are carried out. For each cross, sketch the crossovers required to produce recombinants carrying wild-type alleles at all three loci.

b. Assume that the order is *met-pro-pur*. Two matings, Hfr, $met^+ pro^- pur^+$ × F^-, $met^- pro^+ pur^-$, and the reciprocal, Hfr, $met^- pro^+ pur^-$ × F^-, $met^+ pro^- pur^+$, are carried out. For each cross, sketch the crossovers required to produce recombinants carrying wild-type alleles at all three loci.

c. The outcome of the reciprocal crosses indicates that each cross produces approximately equal numbers of recombinants with wild-type alleles at all three loci. What is the order of the three loci? Explain.

CHAPTER

23

Transduction and Mapping the Bacterial Chromosome

INTRODUCTION

Transduction is a process that results in the transfer of genetic material from one bacterial cell, the donor, to another, the recipient, with a bacteriophage, or phage, serving as the carrier or **vector.** Transduction provides a common and accurate method of mapping loci that are very close to each other.

BACTERIOPHAGE LIFE CYCLE AND TRANSDUCING PARTICLES

Since bacteriophages, typified by the *E. coli* phage P1, and the *Salmonella typhimurium* phage P22, are the agents of transduction, a brief summary of their life cycle is in order. Each phage consists of a chromosome surrounded by a protein coat. The infection begins when a phage attaches to the surface of a bacterial **host** and injects its chromosome into the cell. The bacterial genetic material stops functioning and the metabolic machinery of the cell, guided by viral genes, manufactures many copies of the viral chromosome and components of the protein coat which are then assembled into intact viruses. Once in a while this assembly goes amiss and a piece of the now disintegrating bacterial chromosome, comparable in size to the viral chromosome, gets incorporated into a protein coat. The **transducing particles,** or **transducing phages,** resulting from this mispackaging are released into the environment along with normal viruses when the host cell **lyses** (breaks open).

Transducing particles, like normal viruses, are infective—they can inject the bacterial DNA they carry into another bacterium. Since transducing phages carry no viral genes, this injection is accomplished without harm to the new host. Transduction occurs when this injected "foreign" DNA lines up with the homologous region of the host's chromosome and, through crossing over and covalent bonding, replaces a section of it. Since any part of the bacterial chromosome, provided it is of the appropriate size, can be mispackaged into a transducing particle and bring about transduction, the process just described is known as **generalized transduction. Specialized transduction,** which is restricted to the transfer of very specific portions of the bacterial chromosome, is discussed at the end of this chapter.

COTRANSDUCTION

Loci that are transduced together are said to be **cotransduced.** In theory, there are two possible ways in which the cotransduction of two loci could occur. Two separate transducing particles, each carrying a different locus, could simultaneously infect the same host cell, or alternatively, the cell could be infected by a single transducing phage carrying both loci. In general, transducing particles are produced with a frequency of one in approximately one million (10^6) viruses. With a probability value this small, the likelihood of a cell being simultaneously transduced by two viral particles is negligible. As a consequence, virtually all cotransductions can be assumed to arise from the uptake of the DNA fragment carried by a single viral particle.

What limits the amount of bacterial genetic material that can be transferred through transduction?

____[1]

In general, the length of a viral chromosome is about 0.5 to 1% of the bacterial genome. If two loci are never cotransduced, what general conclusion can be drawn about the distance separating the two loci? ________

____[2] What can be concluded about the distance separating two loci that are cotransduced? ________________

________________[3] What relationship exists between the frequency with which two loci are cotransduced and the distance separating the loci on a transducing fragment? ________________

________________[4]

Two- and three-point transduction studies are used to determine the order of loci on the bacterial chromosome and to estimate the distances separating them. These studies involve infecting donor bacteria with bacteriophages, collecting the **lysate** (the material released upon lysis), and supplying it to recipient bacteria. Transduction can be detected because the donor and recipient strains chosen for use in the study differ in the alleles they carry at the loci under consideration. Transductants, although arising with a very low frequency, can readily be distinguished through the use of media designed to select for recipient cells that have received the donor allele or alleles.

TWO-POINT TRANSDUCTION STUDIES

Assume we wish to determine the distance separating two *E. coli* loci that are known to be cotransduced at least some of the time. The loci, designated as *pro* and *met*, control the synthesis of the amino acids proline and methionine, respectively. The donor strain, carrying wild-type alleles at these loci (genotype pro^+met^+), is exposed to phages of an appropriate type. Following infection, the phage **lysate,** containing both normal and transducing phages, is used to infect a population of auxotrophic pro^-met^- recipient bacteria. Following this infection, how many genotypes with respect to these two traits would be expected among the recipient bacteria? ________[5] What would these genotypes be?

____[6]

Each of these transductant classes arises through a double crossover as follows. Each crossover is indicated by an "X."

Single transductant, pro^+met^-

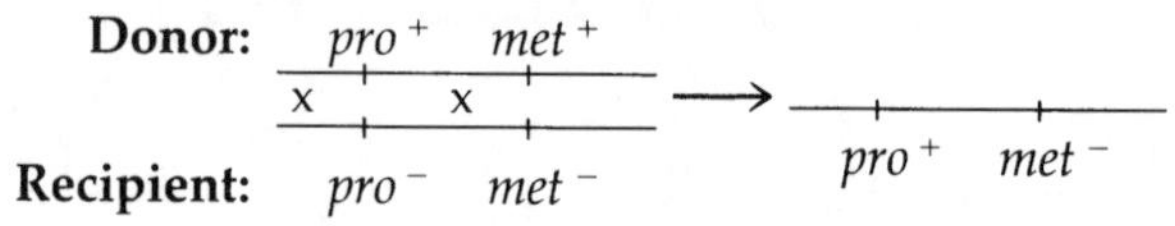

Donor: pro^+ met^+
X X → pro^+ met^-
Recipient: pro^- met^-

Single transductant, pro^-met^+

Donor: pro^+ met^+
X X → pro^- met^+
Recipient: pro^- met^-

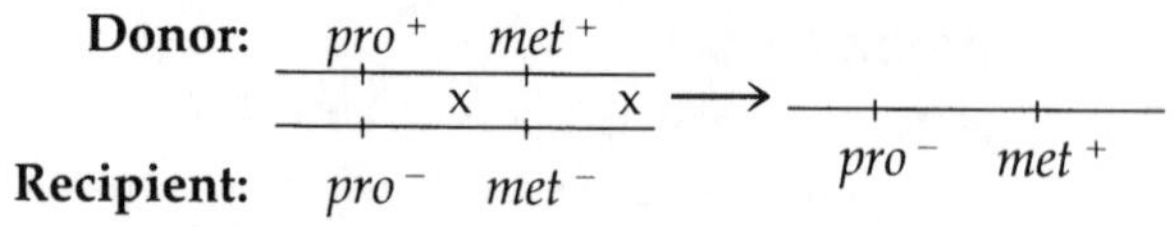

Double transductant, pro^+met^+

Donor: pro^+ met^+
X X → pro^+ met^+
Recipient: pro^- met^-

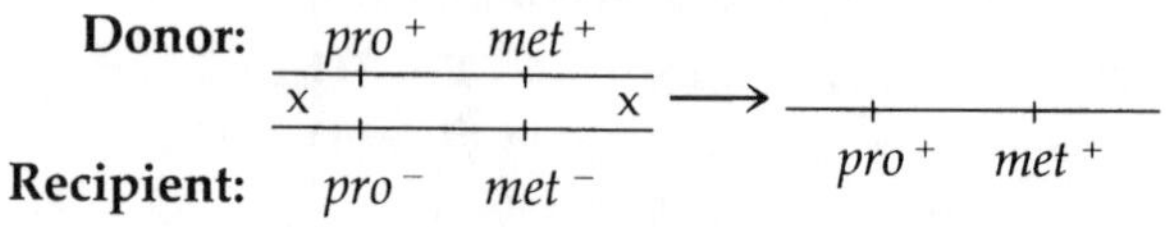

As was pointed out, the likelihood of the double transductant arising through the uptake of two separate pieces of DNA and a quadruple crossover is exceedingly rare.

Identification of the transductants is initiated by selecting for either of the donor genes by replica plating onto appropriate selective media. Assume that selection is carried out for cells that are pro^+. What type of media could be used to select for pro^+ transductants? ________________

________[7] Then the proportion of these pro^+ cells that are also met^+ is determined by replica plating onto an appropriate selective medium. What type of media could be used to select for these pro^+met^+ transduc-

[1] The amount of DNA that can be packaged in the phage particle. [2] It exceeds the length of the DNA packaged in the phage.
[3] It is within the length of the DNA packaged in the phage. [4] The higher the cotransduction frequency, the closer the loci. [5] Four.
[6] Untransduced bacteria (pro^-met^-), single transductants (pro^+met^- and pro^-met^+), and double transductants (pro^+met^+).
[7] A minimal medium supplemented with methionine will allow both pro^+met^- and pro^+met^+ cells to grow.

tants? ______________________

______[8] Assume that the outcome of this procedure is as follows.

Genotype	Type	Frequency
pro⁺—	Transduced for *pro*	818
pro^+met^+	Transduced for *pro* and *met*	681

The relative frequency with which *met* is cotransduced with *pro* provides an estimate of the distance between the two loci. This is obtained by calculating the ratio of the number of *pro-met* transductants relative to the total number of bacteria transformed for *pro*, as summarized in the following equation.

$$\text{Cotransduction frequency} = \frac{\text{Number of cotransductants}}{\text{Total number of } pro \text{ transductants}}$$

Calculate the cotransduction frequency for these two loci. ______[9]

Cotransduction Frequencies and the Distances between Loci

Cotransduction frequencies serve as a measure of the relative proximity of two loci and range between 0 (0%) and 1 (100%). As the cotransduction frequency for two loci approaches 1, what could be concluded about the distance separating the two loci? ______________________[10] As the cotransduction frequency for two loci approaches 0, what could be concluded about the distance separating the loci? ______________________[11] If the cotransduction frequency for two loci is 0, what would be the conclusion? ______________________[12] To summarize, there is an inverse relationship between the cotransduction frequency and the distance separating two loci: as the cotransduction frequency increases, the distance between the loci decreases.

Deducing Gene Order

From the relative frequencies with which various genes are cotransduced, the relative positions of the loci on the chromosome can often be deduced. To illustrate this, assume that a pair of two-point transduction studies indicate that the *cys* locus is transduced with the *arg* locus 45% of the time and with the *trp* locus 3% of the time. From these data, what general conclusion can be drawn about the *cys*-to-*arg* distance relative to the *cys*-to-*trp* distance? ______________________[13] Based on these data, identify the two possible orders for these loci. ______________________[14] Another pair of two-point transduction experiments indicate that the *trp* locus is cotransduced with *cys* 4% of the time and with *arg* 0% of the time. What do these data indicate about the *trp*-to-*cys* and the *trp*-to-*arg* distances? ______________________[15] What is the correct order for these three loci? ______[16]

Deducing the Size of a Transducing Fragment

Data of the sort just considered can give an idea of the size of the transducing fragment that can fit into the transducing phage used as the vector. Do the data indicate that it is possible for both the *trp* and *cys* loci to be included in the same phage? ______[17] Is it possible for both the *trp* and *arg* loci to be included in the same phage? ______[18] What can be concluded about the size of the DNA fragment that can fit into the phage? ______________________[19]

THREE-POINT TRANSDUCTION STUDIES

Two-point transduction studies may prove inconclusive in establishing the order for loci that are very closely linked, since the cotransduction frequencies are very high and very similar. In many instances, however, such loci can be ordered and the distance

[8] A minimal medium. [9] 681/818 = 0.833, or 0.833 × 100 = 83.3%. [10] The distance grows shorter. [11] The distance increases, approaching the length of the DNA segment that can be packaged in the phage. [12] They are never transduced together and thus are separated by a distance greater than the length of the DNA found in the phage. [13] Since *cys* has a higher cotransduction frequency with *arg* than with *trp*, *cys* must be more closely linked with *arg* than with *trp*. [14] *arg-cys-trp* or *cys-arg-trp*. [15] The low frequency with which *trp* and *cys* are cotransduced suggests a relatively substantial distance between these loci. *Trp* and *arg*, however, have not been cotransduced at all, indicating that the distance between them, relative to that between *trp* and *cys*, is even greater. [16] *arg-cys-trp*. [17] Yes, since they are cotransduced. [18] No, since the *arg* locus is never cotransduced with *trp*. [19] Its length must be somewhere between the *trp*-to-*cys* and the *trp*-to-*arg* distance.

separating them can be estimated through three-point transduction studies. We will consider an example that involves three loci, designated as g, h, and i, that are known to be cotransduced at least some of the time. Each of these loci controls the production of a different amino acid and the mutant allele at each locus confers a deficiency for that amino acid. Donor bacteria carrying wild-type alleles at these three loci, $g^+h^+i^+$, are exposed to phages to form transducing particles, which, in turn, are used to infect recipient bacteria of genotype $g^-h^-i^-$. Following this infection, procedures are carried out to identify the bacterial transductants. This is initiated by selecting for bacteria that have received one of the genes being studied (any one of the three loci can be used at this stage). Cells carrying this gene are then tested by replica plating onto the appropriate selective media to determine how many of them have received one or both of the other genes. For example, g^+ cells identified in the initial round of selection are subsequently tested to see if they are h^+ or h^-, and also tested to see if they are i^+ or i^-. Once this is done, a count is obtained for each transductant class. Assume that our experiment indicates that the transductants receiving g^+ fall into four classes with the genotypes and frequencies as follows: (1) $g^+h^-i^-$, 350; (2) $g^+h^+i^-$, 80; (3) $g^+h^-i^+$, 1; and (4) $g^+h^+i^+$, 60. Note that the order in which the loci are listed is arbitrarily chosen and may or may not represent the correct order.

Determining Gene Order from Three-Point Data

Regardless of the order, three of these transductant classes are produced through double crossovers and the fourth through a quadruple crossover. How will the frequency of the quadruple-crossover class compare with those of the double-crossover classes?

______________________________[20]

Which of the four transductant classes listed above is the quadruple-transductant class? __________[21]

Once the quadruple-transductant class is identified, the locus in the middle position can be determined. The $g^+h^-i^+$ class arises as the quadruple-crossover class only when a particular locus occupies the middle position. The following sketches show homologous sections of donor and recipient chromosomes, with each carrying a different locus in the middle position. Subject each to a quadruple crossover, using an "X" to designate the site of each crossover, and sketch the recombined bacterial chromosome that results.

Middle locus	Donor and recipient DNA	Recombined bacterial DNA
h	$g^+ \quad h^+ \quad i^+$ / $g^- \quad h^- \quad i^-$	__________[22]
i	$h^+ \quad i^+ \quad g^+$ / $h^- \quad i^- \quad g^-$	__________[23]
g	$h^+ \quad g^+ \quad i^+$ / $h^- \quad g^- \quad i^-$	__________[24]

Which of the sequences just listed gives rise to the quadruple-crossover class produced in this experiment? ______[25] What is the middle locus? ______[26] The order in which the loci were written initially is, in fact, the actual order.

There is a short-cut for determining the middle locus that eliminates having to sketch quadruple crossovers for all possible orders. It involves comparing the alleles found at each locus in either of the parental types with those found in the quadruple-crossover type. For example, compare the recipient parent ($g^-h^-i^-$) and the quadruple-crossover type ($g^+h^-i^+$). Are the alleles occupying the g locus the same? _____[27] Are those occupying the i locus the same? __________[28] What about those at the h locus? __________[29] The locus that carries the same allele in both the recipient and quadruple-crossover genotypes is the middle locus. (Alternatively, the genotypes of the quadruple-crossover type and the donor parent can be compared. Here the loci that carry the same alleles in both genotypes are the loci that flank the middle locus.) To summarize, the least frequent class produced in a three-point transduction study arises through a quadruple crossover. The locus carrying the same allele in both the quadruple crossover and recipient genotypes is the middle locus.

Cotransduction Frequencies from Three-Point Transduction Data

Three-point transduction data can provide us with an index of the distances separating the loci. The data from our experiment, once again, are as follows: (1) $g^+h^-i^-$, 350; (2) $g^+h^+i^-$, 80; (3) $g^+h^-i^+$, 1; and (4) $g^+h^+i^+$, 60, for a total of 491. We must determine the frequency with which the gene we initially selected

[20] The quadruple-crossover class, since it involves the independent incorporation of two fragments of DNA, will have the lowest frequency. [21] Class 3, $g^+h^-i^+$. [22] $g^+h^-i^+$. [23] $h^+i^-g^+$. [24] $h^+g^-i^+$. [25] g-h-i. [26] h. [27] No. [28] No. [29] Yes.

for, in this case g^+, is transduced with each of the other two genes, that is, with h^+ and then with i^+.

What is the total number of bacteria transduced for both g^+ and h^+? ________[30] The proportion of g^+h^+ cotransductants to the total number of g^+ transductants gives the cotransduction frequency for the g and h loci.

$$\text{Cotransduction frequency for } g \text{ and } h = \frac{\text{Number of } g^+h^+ \text{ transductants}}{\text{Total number of } g^+ \text{ transductants}}$$

Calculate this frequency. ________[31] What is the total number of bacteria that are transduced for both g^+ and i^+? ________[32] The proportion of g^+i^+ cotransductants to the total number of g^+ transductants gives the frequency of cotransduction for the g and i loci. Calculate this frequency. ________[33] Which of the cotransduction frequencies just calculated represents the greatest actual distance? ________[34]

Note that the data considered in this example, that is, bacteria initially selected for transduction at the g locus, represent just a portion of the results that could be obtained from this three-point transduction cross. Initial selections could be carried out to detect bacteria transduced at the h locus or at the i locus, with subsequent determination of the frequency with which each of these is cotransduced with the other two loci. Although all of this data was not presented, that which was available was sufficient to order the three loci and provide an index of the distances separating them. Generalized transduction analysis has been useful in detailing the chromosome maps produced from the interrupted-mating technique with conjugating bacteria.

SPECIALIZED TRANSDUCTION

The process of specialized transduction is considerably less important in bacterial chromosome mapping than is generalized transduction. Only a small, select portion of the bacterial chromosome can be transferred in this process. The viruses involved in transferring the bacterial DNA in specialized transduction belong to a group known as the **temperate phages,** typified by the lambda (λ) phage of *E. coli.*

Temperate Phage Life Cycle

Temperate phages may behave in one of two ways upon injecting their chromosome into a bacterium. Sometimes the metabolic machinery of the cell is taken over immediately and used to make viral particles in the standard infective cycle. Other times however, the viral chromosome may become integrated into the bacterial chromosome. The integrated phage is known as a **prophage,** and the bacterium into whose chromosome the phage is integrated is referred to as **lysogenic.** A specific type of temperate phage consistently integrates into the bacterial chromosome at the same position (because of correspondence between a short sequence of nucleotides in the viral chromosome and in the bacterial chromosome).

While in the integrated, or **temperate,** state, the viral genes behave just like a part of the bacterial chromosome: when the bacterial chromosome replicates, the viral genes replicate along with the bacterial genes and the integrated virus is transmitted in the daughter chromosomes to all descendent cells. But the viral genetic material may excise from the bacterial chromosome at any time through a reversal of its insertion process. Upon excision, the viral genes become functional and the infective cycle is initiated.

Defective Viral Particles

Occasionally, the excision of the integrated viral chromosome is imprecise: a portion of the viral chromosome remains behind with the bacterial chromosome and a bit of the bacterial chromosome next to the viral insertion point is removed. When this happens, the excised piece of DNA is a hybrid that carries most but not all of the viral genes, along with a very limited amount of bacterial genetic material. If this DNA gets packaged into a protein coat, the resultant **defective particle** may be able to infect a new host cell and produce progeny phages. In other cases, the missing phage genes may prevent the defective particle from infecting a host cell and reproducing. This limitation may be overcome if the defective particles are mixed with normal phages and supplied to the host. Any bacterium that is infected by both types of phage now contains a complete set of viral genes (since the normal or "helper" virus has supplied those that are missing from the defective phage). The infective cycle can occur in these bacteria with the production of both normal and defective phages.

[30] Classes 2 and 4, containing a total of 80 + 60 = 140 progeny. [31] 140/491 = 0.285.
[32] Classes 3 and 4, containing a total of 1 + 60 = 61 progeny. [33] 61/491 = 0.124.
[34] That for the g and i loci; the lower the cotransduction frequency, the greater the distance between the two loci.

Specialized Transduction in *E. coli*

Specialized transduction in the K12 strain of *E. coli* with the λ phage serving as the transferral agent has been studied extensively. The phage consistently inserts itself between the *gal* locus, controlling galactose metabolism, and the *bio* locus, controlling synthesis of the vitamin biotin, and consequently, these are the only loci that the defective phage, designated as λ*d*, can transduce. Figure 23-1a shows a λ phage integrated into the *E. coli* chromosome. Figures 23-1b and c show the viral and bacterial chromosomes that result following normal excision and imprecise excision, respectively.

Defective transducing particles carrying one or the other of the *E. coli* genes on either side of the insertion site are designated as λ*dgal*$^+$ or λ*dbio*$^+$. When λ*dgal*$^+$ phages, for example, infect K12 *gal*$^-$ cells, the bacteria become partial heterozygotes for the *gal* locus, that is, *gal*$^+$/*gal*$^-$, and now exhibit a *gal*$^+$ phenotype. Since specialized transducing particles only transport bacterial loci that are immediately adjacent to the phage insertion point, the recombinants arising from the process are of limited use in mapping the bacterial chromosome.

FIGURE 23-1

Integration and excision of a λ phage from an *E. coli* chromosome.

(a) *E. coli* chromosome with integrated λ phage

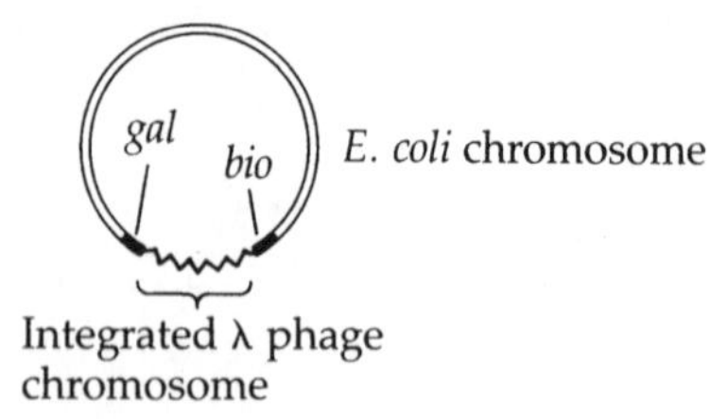

(b) Products of normal excision

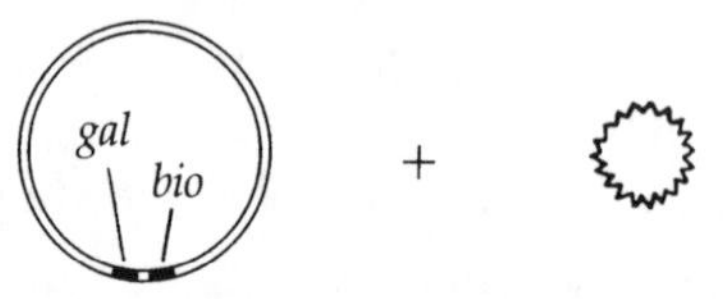

(c) Products of imprecise excision

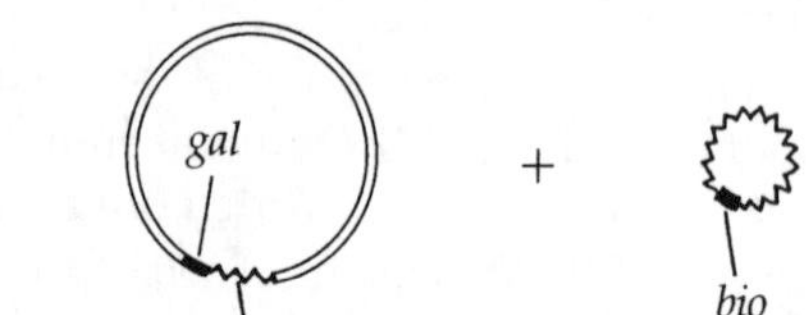

SUMMARY

Occasionally, a fragment of DNA from a bacterial host may be incorporated into a protein coat during the infective cycle of a bacteriophage thus forming a transducing phage. Transduction involves the transfer of this bacterial DNA to a recipient cell by the transducing phage and its incorporation into the recipient's chromosome. Transduction is studied by using donor strains with wild-type alleles and recipient strains with mutant alleles at the loci under consideration, and by using selective media to identify recipients that end up with donor alleles. Since a very limited quantity of bacterial DNA can fit into a phage coat, loci that are transduced by the same phage must be closely linked; the higher the frequency of cotransduction, the tighter the linkage.

Specialized transduction involves temperate phages which integrate into the bacterial chromosome. Through imperfect excision, such a phage may remove a very limited portion of bacterial DNA which then may be transferred to another bacterium. Since only the genes adjacent to the phage's specific insertion site can be transferred in this process, its usefulness in mapping is limited.

PROBLEM SET

23-1. Assume that the transduction of the *arg* and *met* loci is studied in the bacterium *Salmonella typhimurium* through the use of an appropriate bacteriophage. The DNA carried by the transducing phages is derived from bacteria of genotype arg^+met^+ and the recipient cells have genotype arg^-met^-. This two-factor study gives rise to three transductant classes that are produced in the following frequencies.

Genotypes	Frequency
arg^+met^-	99
arg^-met^+	185
arg^+met^+	1289

a. Identify each class as either a single or double transductant.
b. Calculate the frequency with which met^+ is cotransduced with arg^+.

23-2. A transduction study involving two loci shows that they are never cotransduced. Does this provide any information that might be useful in preparing a map?

23-3. **a.** A series of two-point transduction studies involving a particular bacterial species indicates that locus *s* is frequently cotransduced with locus *u*, that locus *t* is commonly cotransduced with locus *u*, and that locus *s* is rarely transduced with *t*. In what order would you position these loci?
b. Additional studies indicate that locus *v* is frequently transduced with locus *s*, but not with either of the other loci being considered here. Where would you position locus *v* in the order of loci you worked out in answering 23-3a?

23-4. A series of three cotransduction studies involving three *E. coli* loci, *u*, *v*, and *w*, is carried out using the bacteriophage known as P1. The bacterial DNA carried by the transducing phages is $u^+v^+w^+$ and the genotype of the recipient cells is $u^-v^-w^-$. In the first study, recombinant cells are selected for the presence of the u^+ allele and of these, 76% are found to be w^+ and 3% are v^+. In the second study, recombinant cells are selected for the presence of the v^+ gene and of these, 6% are u^+ and 0% are w^+. In the third study, recombinant cells that are both u^+ and v^+ are selected, and of these none has the w^+ gene.
a. Solely on the basis of the first study, what can be concluded about the relative positions of the loci and the estimated distances between them? Sketch the map or maps consistent with the results of this study.
b. Does the outcome of the second study make it possible to narrow the conclusion you drew from the first study? Explain. Indicate the map or maps consistent with the results of both the first and second studies.
c. Does the outcome of the third study allow us to say anything about the size of the DNA fragment that can be incorporated into the P1 phage?

23-5. A cotransduction study involving five bacterial loci, *a*, *b*, *c*, *d*, and *e*, is carried out. Transductants are initially selected on the basis of the *a* gene. The frequency with which each of the other four genes is cotransduced with *a* is as follows: *a* with *b*, 47%; *a* with *c*, 56%; *a* with *d*, 63%; and *a* with *e*, 64%.
a. Based on a quick inspection of these cotransduction frequencies what general statement can be made about the degree of linkage between these loci?

PROBLEM SET

b. Assuming the differences in these cotransduction frequencies to be statistically significant, what do they suggest about the order of these five loci?

c. The relative positions of two of these loci cannot be reliably stated. Identify these two loci and explain why this is true.

d. What type of procedure would you suggest be carried out to resolve the ambiguity in ordering these two loci?

23-6. Bacteria of genotype $l^-m^-n^-$ were infected with transducing particles obtained following phage infection of bacteria with the genotype $l^+m^+n^+$.

a. If the middle locus is *m*, identify the genotype arising from a quadruple crossover.

b. If the middle locus is *n*, identify the genotype that would be expected to have the lowest frequency.

c. If a quadruple crossover produces the genotype $l^-m^+n^+$, which locus is the middle locus?

23-7. A three-factor transduction study of the *a*, *d*, and *e* loci referred to in problem 23-5 is carried out. Bacteria of the genotype $a^-d^-e^-$ are infected with transducing particles obtained following phage infection of bacteria with the genotype $a^+d^+e^+$. Using appropriate selection techniques, cells transduced for the *a* locus are identified and then tested further to determine how many of them have been transduced for one or both of the other loci. The genotypes of the transductants and their frequencies are as follows.

Class	Genotype	Frequency
1.	$a^+d^+e^+$	161
2.	$a^+d^+e^-$	6
3.	$a^+d^-e^+$	120
4.	$a^+d^-e^-$	820

Recall from problem 23-5 that the *d* and *e* loci are very close to each other, but we do not know the correct order for these loci, that is, either *d* or *e* could be in the middle position.

a. Do these data make it possible for us to identify the middle locus? If so, what is that locus?

b. Calculate the cotransduction frequency for *a* and *d*.

c. Calculate the cotransduction frequency for *a* and *e*.

23-8. Bacteria of the genotype $i^+j^-k^+$ are infected with transducing particles obtained following phage infection of bacteria with the genotype $i^-j^+k^-$. Assume that cells transduced at the *i* locus can be selectively identified. These cells are then tested to determine how many of them have been transduced for one or both of the other loci. The four classes of transductants, their genotypes, and their frequencies are as follows.

Class	Genotype	Frequency
1.	$i^-j^+k^+$	24
2.	$i^-j^+k^-$	213
3.	$i^-j^-k^+$	687
4.	$i^-j^-k^-$	1810

PROBLEM SET

Note that the correct order of the loci is unknown; the order written here is arbitrarily chosen and may or may not reflect the correct sequence.

a. How many crossovers would be necessary to produce each of the four transductant genotypes if the actual sequence is the one written above, that is, *i-j-k*? Make a sketch to show each, if necessary.

b. How many crossovers would be necessary to produce each of the four transductant genotypes if the actual gene order has *k* as the middle locus, that is, *i-k-j*?

c. How many crossovers would be necessary to produce each type of transductant if the actual gene order has *i* as the middle locus, that is, *j-i-k*?

d. Of the two general types of crossover events that you have described in answering 23-8a, b, and c, which type would you expect to occur with the lowest frequency? Explain.

23-9. Refer to the information given in problem 23-8.

a. Based on the data presented in the problem, why is the order *i-j-k* not the correct one?

b. Of the two remaining orders, *i-k-j* and *j-i-k*, which is most likely to be the actual order? Explain.

23-10. Assume that the reciprocal of the cross described in problem 23-8 is carried out by infecting $i^-j^+k^-$ cells with transducing phages grown on $i^+j^-k^+$ host bacteria. Among the transductants are found 50 cells of the genotype $i^+j^-k^-$ and 847 cells of the type $i^-j^-k^+$. Do these data provide further support for the correctness of the sequence of loci arrived at in answering problem 23-9? Explain.

23-11. Earlier in this chapter we considered a set of data from a three-point transduction study. The study involved bacteria carrying wild-type alleles at three loci, $g^+h^+i^+$, which were exposed to phages to form transducing particles. These particles were then used to infect bacteria of the $g^-h^-i^-$ genotype. Transductants for the *g* locus were tested for transduction at the *h* locus and also at the *i* locus. The transductants receiving g^+ fell into four classes and the genotypes and frequencies for these are as follows.

Class	Genotype	Frequency
1.	$g^+h^-i^-$	350
2.	$g^+h^+i^-$	80
3.	$g^+h^-i^+$	1
4.	$g^+h^+i^+$	60

a. Do these results cover all possible ways in which the g^+ locus is transduced with i^+ and with h^+?

b. Can these results be used to calculate the total frequency with which *h* and *i* are cotransduced? Explain.

c. What modification would you make in the experimental procedure in order to get a complete set of data for determining cotransductance of *h* with *i*?

23-12. A researcher is attempting to establish the order of two bacterial loci, *b* and *c*, relative to a third locus, *a*. Preliminary studies indicate that two orders are possible: either the *b* locus is in the middle (giving an order of *a-b-c*) or the *c* locus is in the middle (giving an order of *a-c-b*). In an attempt to

PROBLEM SET

correctly order these loci, a pair of reciprocal transduction crosses are carried out as follows:

(1) single-mutant donor × double-mutant recipient
(2) double-mutant donor × single-mutant recipient.

If the correct order is *a-b-c*, these crosses can be written as

(1) $a^+b^+c^- \times a^-b^-c^+$
(2) $a^-b^-c^+ \times a^+b^+c^-$.

If the correct order is *a-c-b*, these crosses can be written as

(1) $a^+c^-b^+ \times a^-c^+b^-$
(2) $a^-c^+b^- \times a^+c^-b^+$.

The crosses are carried out and, using the appropriate selective media and replica plating, the genotypes of all the transductants formed in each cross are tallied. Due to an oversight, the only results recorded are the number of bacteria produced in each cross that carry the wild-type alleles at all three loci, that is, those that are $a^+b^+c^+$ (or $a^+c^+b^+$, if *c* is the middle locus).

a. If the correct order is *a-b-c*, what type of crossover (double or quadruple) is required in cross 1 to produce the transductants carrying wild-type alleles at all three loci?

b. If the correct order is *a-b-c*, what type of crossover (double or quadruple) is required in cross 2 to produce the transductants carrying wild-type alleles at all three loci?

c. If the correct order is *a-b-c*, what prediction would you make about the relative numbers of wild-type transductants arising from crosses 1 and 2?

d. If the correct order is *a-c-b*, what type of crossover (double or quadruple) is required in cross 1 to produce the transductants carrying wild-type alleles at all three loci?

e. If the correct order is *a-c-b*, what type of crossover (double or quadruple) is required in cross 2 to produce the transductants carrying wild-type alleles at all three loci?

f. If the correct order is *a-c-b*, what prediction would you make about the relative number of wild-type transductants arising from cross 1 and 2?

23-13. Refer to the information given in problem 23-12. The actual numbers of transductants carrying wild-type alleles at all three loci are as follows: cross 1 produces 47 and cross 2 produces 539. On the basis of these data, what is the correct order of the loci?

CHAPTER 24

Viral (Bacteriophage) Genetics

INTRODUCTION

Viruses lack cellular structure and are consequently classified separately from prokaryotes and eukaryotes. Each virus consists of two components: a chromosome (which, depending on the type of virus, may consist of either DNA or RNA) and a protein coat. Viral genes carry instructions for the synthesis of both viral components, but since the virus lacks an energy source and raw materials for producing these components, viral reproduction can occur only within a living cell, or host. Mutant variants of viral genes make it possible to study inheritance and recombination, and to prepare chromosome maps. Much of what we know about viral genetics comes from the study of bacteria-attacking viruses known as **bacteriophages** ("bacteria eaters"), or **phages.**

THE BACTERIOPHAGE CYCLE OF INFECTION

The phage infective cycle, as shown by the T4 phage of *E. coli,* begins when a virus attaches to the wall of a bacterial host and injects its viral chromosome into the cell. Bacterial genes stop functioning and the metabolic machinery of the cell, following instructions encoded in the viral chromosome, is harnessed to turn out many copies of the viral chromosome and protein coat. Near the end of this infective cycle, each viral chromosome is packaged into a protein coat to make an intact viral particle. More than 250 phages can be formed in a single host cell. At this point the bacterial cell wall disintegrates, the cell breaks open, or **lyses,** and releases the phages which now can go on to infect other bacteria. This type of life cycle which kills the host cell is referred to as the **lytic infection cycle.**

Bacteriophage Phenotypes

The extremely small size of phages makes direct observation difficult, and consequently, the phenotypic traits commonly studied are those manifested when the virus interacts with its host. For example, the trait known as **host range** determines the type or types of host cells that a virus can infect. A mutation in this gene broadens the range of host cells that a virus can infect. Another trait, **plaque morphology,** relates to the appearance of the clear area or **plaque** that arises when phages infect some of the bacteria grown in a dense, opaque layer, or lawn, on the nutrient agar in a petri dish. One aspect of plaque morphology under genetic control is the clarity of the plaque. If the viruses kill all of the bacteria within the boundary of a plaque, the zone will be clear; however, if only some are destroyed, the plaque will appear cloudy or turbid. Other features of plaque morphology may also be under genetic control: wild-type plaques are smaller and have "soft" or fuzzy edges while the rapid-lysis mutation results in larger plaques with "hard," well-defined borders.

Other types of mutations that have been isolated and studied in bacteriophages include conditional, temperature-sensitive mutations. These mutations block the reproductive ability of the virus at higher temperatures, for example, 42°C, but allow normal reproduction at cooler temperatures such as 25°C. Consequently, the virus exhibits the normal, wild-type, phenotype at cooler, permissive, temperatures and the mutant phenotype at higher, restrictive, temperatures.

FIGURE 24-1

A bacterial cell infected by two different phages.

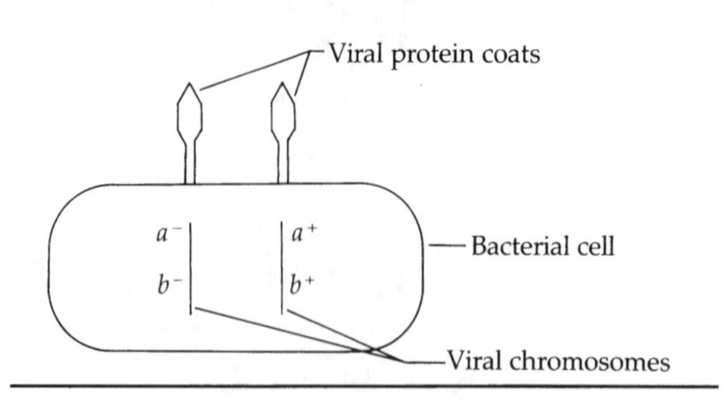

Bacteriophage Recombination

Recombination of viral chromosomes can occur whenever a host cell is infected by two or more phages that reproduce in the cell. Before the newly reproduced chromosomes have been packaged into viral coats, there is an opportunity for crossing over to take place between them. Figure 24-1 shows a bacterial cell which has been infected by two phage strains: one carries mutant alleles at two loci (genotype a^-b^-) while the other has wild-type alleles at these loci (genotype a^+b^+). Assume that a limited number of chromosomes of the two types experience a single crossover in the region between the two loci. What would be the genotypes of the viruses released upon lysis of the host bacterium? ________[1] Which genotypes would you expect to be relatively abundant? ________[2] Which genotypes would be produced in relatively low frequencies? ________[3] Upon what would the frequencies of the recombinant types depend? ________ ________[4]

TWO-POINT CROSSES

Recombination studies in viruses involve infecting host bacteria with phage strains that carry different alleles at the loci under consideration and determining the proportion of recombinants among the progeny phages. Since recombination involving a large number of viruses produces reciprocal recombinants in nearly equal numbers, the basic principles involved in mapping are similar to those used in eukaryotic mapping. A typical recombination study involving two loci in the T2 virus of *E. coli* is described next. One locus influences host range while the other controls the speed of lysis. The presence of a wild-type host-range allele, h^+, allows the T2 virus to grow on *E. coli* strain B. A mutant form of this allele, h^-, broadens the infective capabilities of the virus to include the B/2 strain of *E. coli*. A wild-type allele, r^+, at the rapid-lysis locus results in the production of small plaques with fuzzy edges while a mutant allele, r^-, speeds up the rate of lysis, causing larger plaques with smooth edges. Our task is to find out the frequency with which recombination occurs between these two loci.

The key steps in the experiment are as follows. A mixture of both h^+r^- and h^-r^+ viruses is added to a liquid culture of B bacteria. High concentrations of phages (5 to 10 per bacterium) assure that at least some bacteria will be simultaneously infected by both types of phage. (The phage concentration in an experiment such as this is usually described by the term **multiplicity of infection,** which is the number of bacteriophage particles available to infect host cells divided by the number of bacteria available for infection.) Recombination can occur among the many copies of the two kinds of viral chromosomes produced during the infection before the chromosomes are packaged in protein coats. Upon lysis of the host bacteria, the liberated progeny phages are collected and a suitable dilution of the virus is added to melted agar containing about 10^8 B and B/2 bacteria. The mixture is then poured on an agar plate and 12 to 18 hours later the plaques are counted and classified. The dilution of the viruses is such that each plaque that develops is assumed to be produced by the descendants of a single phage. In light of this, what will counting the number of plaques that show a particular trait tell us? ________ ________[5] The characteristics of these plaques will indicate the genotype of the viruses that are produced.

[1] a^-b^-, a^+b^+, a^-b^+, and a^+b^-. [2] Parental types: a^-b^- and a^+b^+. [3] Recombinants: a^-b^+ and a^+b^-.
[4] The frequency with which crossing over occurred between the two parental types, which is related to the distance separating the loci.
[5] The number of viruses showing each trait.

The progeny viruses that have not participated in recombination will, of course, have the h^+r^- and h^-r^+ parental genotypes. With h^+r^- phages, which strain or strains of *E. coli* would be expected to lyse? ______[6] Would the plaques be clear or cloudy? ________[7] Would the plaques be large with well-defined borders or small with fuzzy borders? ________[8] With the h^-r^+ phages, what type or types of bacteria would be expected to lyse? ________[9] Would the plaques be clear or cloudy? ________[10] Would the plaques be large with well defined borders or small with fuzzy borders? ________[11] Additional plaques, produced through the infective activities of the recombinant viruses (genotypes h^+r^+ and h^-r^-), will differ in appearance from those produced by the parental type progeny. What type of plaque would be produced by h^+r^+ viruses? ________[12] By the h^-r^- viruses? ________[13]

Following incubation of the progeny viruses on the mixed lawn of the two *E. coli* strains, a count of each of the four possible plaque types is made. The number of recombinant plaques relative to the total number of plaques gives us the recombination frequency.

$$\text{Recombination frequency} = \frac{\text{Number of recombinant plaques}}{\text{Total number of plaques}}$$

The recombination frequency, in turn, provides an indication of the distance between the two loci under consideration. For example, say the progeny virus genotypes (as deduced from the appearance of the plaques) and their frequencies are as shown in Table 24-1. Calculate the recombination frequency for these two loci. ________[14] The recombination frequency is 24/176 = 0.136. Sketch a map in the space provided showing these two loci and the distance between them.[15]

TABLE 24-1

Outcome of the $h^+r^- \times h^-r^+$ bacteriophage cross.

	Genotype	Frequency
Parental types	h^+r^-	80
	h^-r^+	72
Recombinant types	h^+r^+	11
	h^-r^-	13
	Total:	176

THREE-POINT CROSSES

Three-point crosses are used to establish the sequence of loci and the distances between them. Assume that three bacteriophage loci, designated as *a*, *b*, and *c*, are to be mapped. The genotypes of the parental viruses are $a^-b^-c^-$ and $a^+b^+c^+$ and both types are supplied in appropriate concentrations to host bacterial cells. Following lysis, the progeny phages are diluted to a suitable concentration, added to melted agar containing approximately 10^8 host bacteria per milliliter, and poured onto an agar plate. Following incubation, examination of the plaques indicates eight classes of progeny viruses. The genotypes (as deduced from the appearance of the plaques) and frequencies of these classes are listed in Table 24-2. Note that the sequence in which the three loci are written is arbitrary and any of the loci could occupy the middle position.

TABLE 24-2

Outcome of the $a^+b^+c^+ \times a^-b^-c^-$ bacteriophage cross.

Class	Genotype	Frequency	Percentage
1.	$a^-b^-c^-$	2831	44.3
2.	$a^+b^+c^+$	3011	47.1
3.	$a^-b^+c^+$	89	1.4
4.	$a^+b^-c^-$	73	1.1
5.	$a^-b^-c^+$	154	2.4
6.	$a^+b^+c^-$	178	2.8
7.	$a^-b^+c^-$	22	0.3
8.	$a^+b^-c^+$	29	0.5
	Total:	6387	

[6] B strain. [7] Cloudy, since B/2 bacteria will still be alive in that plaque zone. [8] The r^- allele produces rapid lysis which causes large plaques with well-defined borders. [9] The B and B/2 strains. [10] Clear, since both types of bacteria are killed. [11] The r^+ allele reduces the speed of lysis which results in small plaques with fuzzy borders. [12] Cloudy and small with fuzzy borders. [13] Clear and large with well-defined borders. [14] The number of recombinant plaques is 11 + 13 = 24 and the total number of plaques is 176. [15] Expressed in terms of percent, the recombination frequency is 0.136 × 100 = 13.6%. The two loci are separated by 13.6 map units.

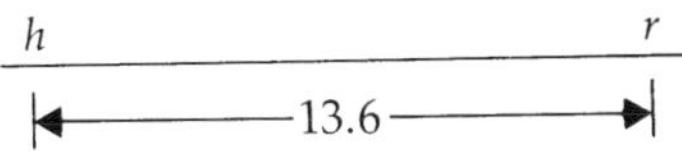

Identifying the Middle Locus

The analysis is carried out much as it would be with the progeny of eukaryotic three-point cross. First parental and double-crossover classes are identified and the middle locus designated. The two least common classes (7 and 8) are the double-crossover types, while the two most common classes (1 and 2) are the parental or noncrossover products. As you realize from earlier discussions of eukaryotic three-point crosses (see Chapter 13), a double crossover appears to change the position of the middle locus. Compare the genotypes of the double-crossover products with those of the parental types and identify the middle locus. ________[16]

Determining Distances between Loci

The next step in the analysis is to establish the *a*-to-*b* and the *b*-to-*c* distances. This is done by determining the number of crossovers occurring in each of these two regions. First, consider the *a*-to-*b* region. Which of the single-crossover classes have experienced a crossover in the *a*-to-*b* region? ________[17] Are there other classes which have experienced crossovers in this region? ______[18] If so, identify these classes. ______[19] Similarly, crossovers in the *b*-to-*c* region have taken place with classes 5 and 6 as well as in each of the double-crossover classes. The frequencies with which these crossovers occur are summarized in Table 24-3. Dividing each of these totals by the total number of progeny (6387) gives the crossover frequency for each chromosome region. Calculate the recombination frequency for the *a*-to-*b* region. ________[20] Calculate the recombination frequency for the *b*-to-*c* region. ________[21] Sketch a map in the space provided showing the sequence of loci and the distances separating them.[22]

TABLE 24-3

Crossover frequencies in the three-point bacteriophage cross.

Class	Number of Crossovers: Between *a* and *b*	Number of Crossovers: Between *b* and *c*
3 (single crossover)	89	
4 (single crossover)	73	
5 (single crossover)		154
6 (single crossover)		178
7 (double crossover)	22	22
8 (double crossover)	29	29
Totals:	213	383

Note again that the procedures that we used for mapping from the outcome of a three-point cross are fully detailed in Chapter 13; if necessary, refer to that chapter.

MULTIPLE RECOMBINATION IN VIRUSES

In mapping viral loci, the frequencies of recombinants arising from viral two- and three-point crosses have been used like those from eukaryotic recombination studies. However, the situation in eukaryotic forms, where recombination involves a single round of reciprocal exchange between a pair of homologous chromosomes, differs greatly from the exchange pattern that occurs in viruses. A host cell that has been infected by two genetically different viruses will make a few hundred copies of each type of viral chromosome. Half the time, chance tells us that the interacting chromosomes will be identical to each other, that is, derived from the same parental virus, and consequently, the crossover products will be identical to the parental chromosomes. Only when the interacting chromosomes are different, that is, derived from different parental viruses, are detectable recombinants produced. In addition, more than one mating can occur. Chromosomes can interact repeatedly (through random pairing, crossing over, and recombining) prior to their packaging in protein coats. Thus, each recombinant chromosome detected in a three-point cross may

[16] Interchanging the *b* alleles would convert the double-crossover genotypes into the parental genotypes; thus *b* is the middle locus.
[17] 3 and 4. [18] Yes. [19] They have occurred in each of the double-crossover classes, 7 and 8. [20] 213/6387 = 0.0333.
[21] 383/6387 = 0.0600. [22] The recombination frequency of 0.033 is equivalent to 3.3%, so there are 3.3 map units separating the *a* and *b* loci. The recombination frequency of 0.0599 is equivalent to 5.99%, so there are 6.0 map units between the *b* and *c* loci.

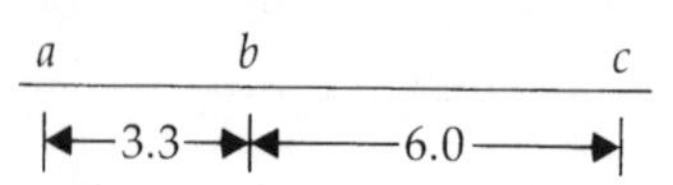

well be the product of several rounds of such interaction.

To illustrate multiple recombinations, consider a host bacterium that becomes infected by three genetically different viruses, each carrying a mutant allele at a different locus: $a^+b^+c^-$, $a^-b^+c^+$, and $a^+b^-c^+$. Among the recombinants are viruses of the type $a^-b^-c^-$. Generally speaking, what is the most direct way in which recombinants of this type could arise? [23]

The recombination history of the $a^-b^-c^-$ recombinant might, for example, involve the following scheme: chromosome $a^+b^+c^-$ crosses over with chromosome $a^+b^-c^+$ in the b-to-c region to produce an $a^+b^-c^-$ recombinant. This newly formed $a^+b^-c^-$ recombinant could then cross over with the $a^-b^+c^+$ in the a-to-b region to give rise to the $a^-b^-c^-$ recombinant virus released by the host cell. These crossovers are shown in Figure 24-2. Although this two-recombination scheme represents a direct way of producing the $a^-b^-c^-$ recombinant genotype, it is by no means the only way this recombinant could arise.

Two important considerations arise from the occurrence of multiple recombination in viruses. (1) Multiple recombinations produce frequencies of double recombinant classes that are higher than expected. This excess of double recombinants is attributed to **negative interference,** a situation where the occurrence of a recombination event appears to increase the probability that another recombination event will occur in a neighboring region. This contrasts with the **positive interference** seen in many eukaryotic organisms where, for example, the double crossovers arising in a three-point cross often are produced in frequencies below those expected because a crossover in one region blocks crossing over in an adjacent region. (2) Although each particular recombination event gives rise to two strands that are reciprocals of each other, there is nothing to guarantee that the recombinants ultimately released from the host cell will have reciprocal recombinants present in equal or roughly equal frequencies. Indeed, studies carried out on the viral recombinants released by a single bacterial cell infected with two genetically different viruses show that reciprocal recombinants do not occur in anything close to equal frequencies. This however does not pose a problem because, when mapping studies are carried out, the progeny phages come not from a single bacterium but from a very large population of host cells. In short, viral recombination can be approached as a population phenomenon. When an enormous number of viruses are involved, reciprocal recombinants, like those produced in eukaryotic three-point crosses, are found in equal or nearly-equal numbers.

FIGURE 24-2

A direct scheme for a multiple recombination in a virus.

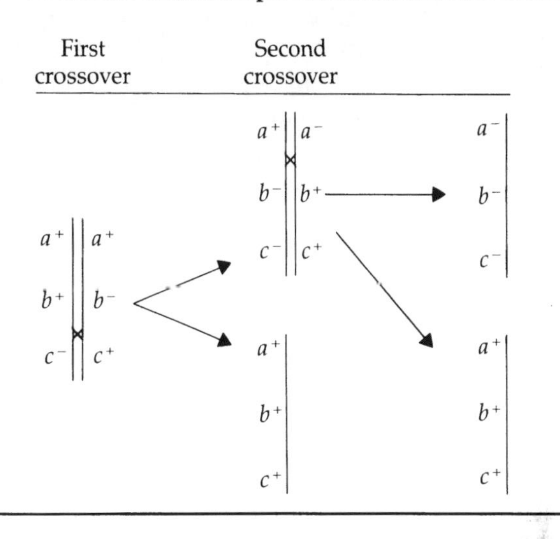

SUMMARY

Phage recombination occurs when a host cell is simultaneously infected by two (or more) genetically different phage strains. Commonly studied traits are those that arise from the interaction of the virus with its host and are detectable with the unaided eye; these include plaque morphology and host range. Since recombination involving a large number of viruses produces reciprocal recombinants in nearly equal numbers, the basic principles involved in mapping are similar to those used in eukaryotic mapping. The progeny phages released after a double infection are spread on a host lawn. Following incubation, the lawn is examined to determine the number of parental- and recombinant-type plaques. The recombination frequency provides an index of the map distance separating the loci. Three-point crosses are used to establish the sequence of loci as well as the distance between them. The chromosomes of recombinant viruses may participate in several rounds of genetic exchange during chromosomal replication. As a consequence, the reciprocal recombinants released from a single bacterial cell do not occur in equal frequencies. This poses no problem as long as the progeny phages used in recombination studies come from a large population of host cells. Under these circumstances, where an enormous number of viruses are involved, reciprocal recombinants are found in equal or nearly equal numbers.

[23] One of the parental viral chromosomes participates in at least two different recombinations, each supplying it with one of the two mutant alleles it lacks.

PROBLEM SET

24-1. What is a plaque and how does it arise?

24-2. Why is it necessary that the bacteriophages used in viral recombination studies be supplied to the bacterial host cells in very high concentration?

24-3. In carrying out recombination studies involving bacteriophages, it is necessary that the lysate released as the bacterial hosts are destroyed, be diluted to a concentration appropriate for plating the phages on a bacterial lawn. What is meant by a "concentration appropriate for plating"?

24-4. The following sketch shows two bacteriophage chromosomes that have been injected into a bacterial host. One chromosome carries mutant alleles a^- and b^-, while the other carries wild-type alleles, a^+ and b^+.

a^- b^-

a^+ b^+

Identify the viral genotypes that would arise following

a. a double crossover, with both crossovers occurring in the region between the two loci.

b. a double crossover, with one crossover in the region between the two loci and one in the region to the right of the *b* locus.

c. a double crossover, one in the region to the left of the *a* locus and one in the region to the right of the *b* locus.

24-5. Two traits influencing plaque morphology are studied in a type of virus that has *E. coli* as its host. A concentrated mixture of viruses of genotypes a^+b^+ and a^-b^- are added to a culture of *E. coli* of appropriate density (containing about 5×10^8 bacteria per milliliter). Following infection and lysis, samples of progeny viruses are cultured. Inspection and counting of the plaques that develop indicate the following viral genotypes and frequencies: a^+b^+, 1086; a^-b^+, 38; a^+b^-, 43; and a^-b^-, 1190.

a. Determine the percentage of recombination that has taken place.

b. What map distance separates the *a* and *b* loci?

24-6. The data in problem 24-5 show that the numbers of recombinant progeny (38 and 43) are approximately equal. Through the use of appropriate dilutions, it is possible to isolate the viruses released upon the lysis of a single bacterial cell. Do you think that the general pattern of recombinant-progeny frequencies seen in the data from problem 24-5 would also apply to the viral progeny released from a single bacterium? Explain.

24-7. Two loci, *m* and *r*, have been mapped in a particular type of virus and are separated by 12.8 map units. A concentrated mixture of viruses with genotypes m^+r^+ and m^-r^- is supplied to a concentrated culture of the appropriate bacterial host. Culturing a sample of the progeny viruses produces 1600 plaques. Estimate the number of these plaques that would show traits characteristic of the recombinant types (that is, m^+r^- and m^-r^+).

24-8. Assume that a viral geneticist has discovered a mutant bacteriophage strain that produces tiny plaques when plated onto a lawn of the host

PROBLEM SET

bacterium *E. coli*. The wild strain of this phage produces large plaques. When wild and mutant strains are simultaneously supplied at a high multiplicity of infection to *E. coli* cells, a total of 1858 plaques are counted, of which 892 are the large wild-type and 966 are the tiny mutant type. The data is recorded by a technician, who concludes that a mutation at a single locus is responsible for the mutant phenotype. A second technician, in reviewing the data, decides that mutant alleles at two loci interact to produce the mutant phenotype. Which conclusion is correct? Explain.

24-9. Two-point crosses are carried out using a particular *E. coli* bacteriophage. Mutant alleles at the two loci under study, *s* and *c*, influence plaque morphology and produce small and clear plaques, respectively. Wild-type viruses produce plaques that are turbid and large. One cross is set up so that the two mutant alleles are carried in coupling, that is, on the same chromosome: $s^+c^+ \times s^-c^-$. The other cross is set up so that the two mutant alleles are carried in repulsion, that is, on different chromosomes: $s^-c^+ \times s^+c^-$.

- **a.** Would you expect the types of recombinants produced from these two crosses to be the same or different? Explain.
- **b.** Would you expect the recombination frequencies calculated from the outcomes of each of these crosses to be similar or different? Explain.

24-10. Two plaque-morphology loci, *b* and *d*, of a bacteriophage are studied in a two-point testcross. The progeny phages of the mating $b^-d^+ \times b^+d^-$ produce a total of 2188 plaques, broken down as follows: b^-d^+, 1051; b^+d^-, 1109; b^+d^+, 16; and b^-d^-, 12. Another cross involving the *b* locus and another plaque-morphology locus, *e*, is set up as follows: $b^+e^+ \times b^-e^-$. The progeny phages of this cross produce a total of 2412 plaques as follows: b^+e^+, 1079; b^-e^-, 1111; b^-e^+, 103; and b^+e^-, 119.

- **a.** Calculate the recombination frequency for the *b* and *d* loci and determine the distance separating them.
- **b.** Calculate the recombination frequency for the *b* and *e* loci and determine the distance separating them.
- **c.** Sketch the map or maps consistent with the data from these two crosses that show the relative positions of and the distances separating these loci.

24-11. Additional two-point crosses carried out using the bacteriophage described in problem 24-10 provide the following additional percentages of recombination between pairs of loci: *a* and *d*, 2.9%; *c* and *a*, 4.8%; *c* and *e*, 0.4%; *a* and *b*, 4.0%; and *d* and *e*, 7.9%. Combine this information with that obtained from problem 24-10 and set up a map showing the relative positions of these five loci (*a*, *b*, *c*, *d*, and *e*).

PROBLEM SET

24-12. Three bacteriophage loci are to be mapped. Genes at each locus influence a plaque trait. Parental phages of the genotypes $s^- t^- u^-$ and $s^+ t^+ u^+$ are supplied to a population of the appropriate host bacterium. Culturing samples of progeny phages yields the results that follow. Note that nothing is known about the sequence of the loci and that the order in which they are written here is arbitrary. Determine the sequence of these three loci and the map distances between them.

Class	Genotype	Number of plaques
1.	$s^- t^- u^-$	2847
2.	$s^+ t^+ u^+$	2631
3.	$s^- t^+ u^+$	360
4.	$s^+ t^- u^-$	321
5.	$s^- t^- u^+$	683
6.	$s^+ t^+ u^-$	729
7.	$s^- t^+ u^-$	81
8.	$s^+ t^- u^+$	72
	Total:	7724

CHAPTER

25

Complementation Testing and Fine-Structure Mapping of the Gene

INTRODUCTION

Most mapping efforts focus on determining the sequence of genes along the chromosome and the distances between them. More precise analytical techniques, developed largely through studies with bacteriophages, are used to map the sites of mutation and recombination *within* individual genes. As with intergenic mapping, **intragenic,** or **fine-structure, mapping** depends on detecting recombinants. However, because of the closeness of intragenic sites, the frequency with which intragenic recombinants arise is very low. This means that enormous numbers of progeny have to be grown and screened to find the recombinants—a tedious and time-consuming project. Bacteriophages, because of their ability to quickly form large populations that can be screened with ease, are ideal subjects for this type of study which can provide important information on the genetic and physical organization of the phage genome.

T4 PHAGES AND *rII* MUTATIONS

The pioneering studies of intragenic mapping were carried out by S. Benzer in the 1950s using the T4 bacteriophage of *E. coli*. Wild-type T4 phages grow on the wild-type strain B of *E. coli* and produce small plaques with fuzzy margins. One class of T4 mutant phages carry mutations, designated as rapid, or *r*, which produce a speeded-up lysis of *E. coli* B cells and result in large plaques with clear margins. Using conventional intergenic mapping procedures, the *r* mutations were mapped to several locations in the phage chromosome with many located at a site designated as the ***rII* region.**

In addition to speed of lysis, *rII* mutations also influence the trait of host range. Because of this, the mutations produce a trait that can be classified as conditionally lethal, that is, one which is lethal under certain, restrictive, environmental conditions and viable under other, permissive environmental conditions. Viruses carrying these mutations are incapable of growing on a strain of *E. coli* designated as K12(λ), a strain that readily hosts wild-type T4 viruses. (Note that the K12 (λ) strain carries the genetic material of the λ phage inserted into its circular chromosome.) In this case, the K12(λ) strain which is lethal to the *rII* mutants is their **restrictive host,** while strain B serves as their **permissive host.** Summarize the growth capabilities of the wild T4 and *rII* T4 phages by completing the following table. In each case indicate whether or not there is viral reproduction, and if reproduction occurs, indicate the margin type and size of the plaques that are formed.

		Host	
		E. coli B	*E. coli* K12(λ)
Virus	wild T4	________[1]	________[2]
	rII T4	________[3]	________[4]

COMPLEMENTATION TESTS

The large number of independently arising *rII* mutants mapped to the *rII* region raises a question regarding the number of genes in that region. All the *rII* mutants might be produced by different alleles of a single gene or more than one gene could be involved, with some of the mutants caused by alleles of one gene while others were due to alleles of other genes. A **complementation test** can be used to identify the number of loci that are independently capable of giving rise to the same mutant phenotype. The following illustration presents the basis for this test.

Assume that among the substances necessary for the multiplication of a virus are two different protein products, A and B, produced by genes *a* and *b*, respectively. A virus with wild-type alleles at both loci will make both proteins and will be able to reproduce. In contrast, a virus carrying a mutant allele at one or both of these loci will lack one or both of these proteins and will be unable to reproduce. Assume that two different mutant viruses infect the same restrictive host cell of strain K12(λ): one virus carries a mutant allele at locus *a* and a wild-type allele at locus *b* (genotype a^-b^+) and the other carries a wild-type allele at locus *a* and a mutant allele at locus *b* (genotype a^+b^-). The arrangement of alleles on these two viral chromosomes, with each carrying a different mutation, is designated as the ***trans*** **configuration.** If both of the mutations were carried on a single chromosome, they would show the ***cis*** **configuration.** Another name for a complementation test, ***cis-trans*** **test,** is derived from these two terms. Which of the two protein products (A or B) will the a^-b^+ virus be able to form? ________[5] Which of the two protein products will the a^+b^- virus be able to produce? ________[6] Will both protein products be available in the doubly infected host cell? ________[7] Will both strains of virus be able to multiply in the same restrictive host cell? ________[8] In this example, the genetic makeup of one virus complements, or completes, the makeup of the other. Each type of virus supplies the gene for the production of a different substance essential for viral multiplication. Each type of virus is able to reproduce in K12(λ) because of the presence of the other.

Now assume that the two coinfecting viruses carry the same mutation with, say, genotype a^-b^+. Will these viruses be able to make a normal A protein? ________[9] Will they be able to make a normal B protein? ________[10] Will they be able to multiply in the K12(λ) host cell? ________[11] Do these two viruses complement each other? ________[12] To summarize, when the mutant allele carried by each virus belongs to a different locus, intergenic complementation takes place and reproduction occurs. In contrast, when each virus carries a mutant allele belonging to the same locus, no complementation occurs and there is no reproduction.

In the two cases just considered, information about the number of loci producing the mutant phenotype was provided to us. If this information were not available and we wished to determine the number of loci involved, a complementation test could be carried out. Assume that two viruses, each with a different *rII* mutation, coinfect type K12(λ) bacteria. (Remember that bacteria of this type do not support the growth of either mutant.) If the bacteria is lysed, releasing many progeny phages, would you conclude that complementation occurred? ________[13] Are the *rII* mutations carried by these viruses due to a single gene or to different genes? ________[14] If another pair of coinfecting *rII* mutants fails to lyse the K12(λ) cells and no progeny viruses were formed, would you conclude that complementation occurred? ________[15] What could you infer about the mutations carried by the viruses? ________________[16]

Through pairwise matings, Benzer found that virtually all of the *rII* mutations fell into one or the other of two genes or **cistrons,** designated as *A* and *B*, in the *rII* region of the T4 chromosome. The term "cistron" (derived from the *cis* and *trans* configurations) can be used interchangeably with the term gene to designate each of the two units of function in the *rII* region. Each gene or cistron carries out a different task, coding for a different polypeptide product essential for phage reproduction.

[1] Reproduction; fuzzy-edged, small plaques. [2] Reproduction; fuzzy-edged, small plaques. [3] Reproduction; clear-edged, large plaques. [4] No reproduction. [5] Protein B. [6] Protein A. [7] Yes. [8] Yes. [9] No. [10] Yes. [11] No. [12] No. The absence of reproduction indicates that the viruses do not complement. [13] Yes. Reproduction indicates complementation. [14] Each is a variant of a different gene. [15] No, as indicated by the absence of viral reproduction. [16] They are variants of a single gene.

INTRAGENIC MAPPING

Once a number of mutant alleles has been mapped to the same locus, pairs of viral strains carrying these mutations can be crossed in the permissive host, recombinant information can be gathered, and the sites of the mutations *within* the gene can be mapped. Whereas intergenic mapping generally depends upon the frequency of recombination between two genes, intragenic mapping depends upon the frequency of recombination between two sites within a gene.

Procedure for Intragenic Mapping

The procedure for determining the recombination frequency between two mutation sites within a gene is as follows. Assume that there are two different strains of *rIIB* mutants, designated as $rIIB^1$ and $rIIB^2$, which carry point mutations *r1* and *r2*, respectively. (Note that a point mutation, discussed in detail in Chapter 28, involves a change in a single base pair within the double-helix DNA molecule of the viral chromosome.) As the following sketch indicates, each mutant carries its mutation at a different site within the *rIIB* gene.

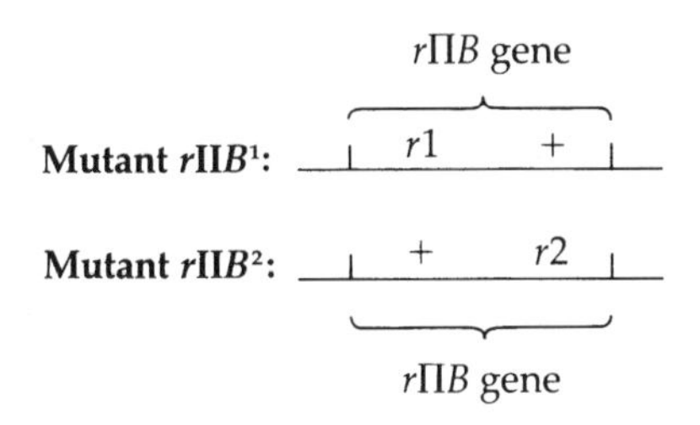

The site occupied by the mutation in each DNA molecule carries the normal (+) nucleotide pair in the counterpart DNA molecule.

Viruses of each type simultaneously infect *E. coli* strain B host cells. Each type of mutant can grow by itself in strain B cells; when the two mutants coinfect the same B cell, the many copies of both viral genomes produced facilitate recombination. Following the replication of the viral chromosomes, assume that crossing over takes place between the two types of chromosomes in the region between the mutation sites. What types of recombinants would result from this crossing over? ________[17] What determines the frequency with which the recombinants are formed? ____[18] What other types of viruses would be released upon lysis of the host cells? ________[19]

Determining Recombination Frequencies

Two equal-sized samples of the phages released from the *E. coli* B cells are collected and diluted as appropriate. One sample is added to a lawn of *E. coli* B cells. Which of the four types of viruses released by the host cells, that is, + +, *r1r2*, *r1* +, and + *r2*, can grow on this lawn? ________[20] The second sample is added to a lawn of *E. coli* K12(λ) cells. Which of the four types of progeny phages would grow here? ________[21] (Note that because all four types of viruses grow on the B strain bacteria, the phage sample used there is usually diluted so that it is 100 times less concentrated than the sample added to the K12(λ) cells.) How is the total number of viral particles added to the *E. coli* B determined? ________________________[22] How is the total number of viral particles added to the K12(λ) lawn determined? ________________________[23]

Does the screening procedure using the K12(λ) lawn detect all of the recombinants in the sample of viruses? ________[24] Explain. ________________________[25] How can the total number of recombinants added to the K12(λ) plate be estimated? ________________________[26] How is the proportion of recombinants among all the viral progeny determined? ________________________[27] If there were three plaques on the K12(λ) plate and a total of 30 plaques on the B plate, and the sample supplied to the B plate was 100 times less concentrated than that supplied to the K12(λ) plate, what is the recombination frequency? ________[28] What is the map distance

[17] Wild-type (+ +) and double mutant (*r1r2*). [18] The crossing over frequency, which is a function of the distance separating the two mutational sites; the greater the distance, the more likely a recombination event will occur. [19] The two parental, that is, single-mutant types. [20] All four types. [21] Only the wild-type recombinants. [22] By counting the total number of plaques, since each plaque is derived from a single phage particle. [23] By multiplying the number of plaques on the B strain lawn by the dilution factor. [24] No. [25] The double-mutant recombinant type does not grow here and could not be counted. [26] By multiplying the number of wild-type recombinants by 2, since the wild-type and double-mutant recombinants would be expected in equal frequencies. [27] By dividing the total number of recombinant plaques, obtained from the K12(λ) plate, by the total number of plaques found on the strain B plate multiplied by the dilution factor. [28] The number of wild-type recombinants in the sample is 3 and the total number of recombinants is estimated as twice this, or 6. Multiplying the total number of plaques on the B plate (30) by 100, the dilution factor, gives a total of $30 \times 100 = 3000$ viruses present in each of the initial samples. The recombination frequency is $6/3000 = 0.002$.

separating the two mutant sites? ________[29] The procedure just reviewed is sufficiently sensitive to pick up one wild-type recombinant virus among 10^8 and makes it possible to screen an enormous number of viruses in a few hours.

DELETION MAPPING

As more *rII* mutants were discovered by Benzer, the process of mapping the mutation site of each by mating it with all the known mutants became increasingly tedious. The potential for speeding up this process came with the discovery of mutants that are missing one or more nucleotide pairs from the *rII* region. However, before these **deletion mutants** could be used in mapping studies, the size of the deletion each carried and its position within the *rII* region (that is, whether it fell within gene *A* or *B*, or overlapped into both) had to be determined. This information was gathered by separately mating each deletion mutant with a number of viruses carrying known (that is, already mapped) *rII* point mutations. As the following example shows, the outcome of each cross indicates whether the site of the point mutation falls within or outside of the deletion.

Using Known *rII* Mutations to Map Deletion Mutations

Assume that viruses carrying an unmapped deletion, and others carrying a mapped point mutation, coinfect an *E. coli* strain B cell. Represented in the following drawing are the *rII* regions of each type of virus. One DNA molecule bears a deletion (shown as an open area) while the other carries a point mutation (shown as an asterisk). In the space to the right, sketch the recombinant molecules that would arise following a crossover between these two DNA molecules at a point between the site of the *rII* point mutation and the *rII* deletion mutation.[30]

rII region of point mutant: ___*_______

rII region of deletion mutant: _________( )_

Crossing over →

Recombinant molecules

How could the wild-type progeny phages be detected?
____________________________________[31]

As you have just demonstrated, wild-type recombinants will be produced if the two mutations are located in different sections of the *rII* region. If no wild-type recombinant progeny are produced, what conclusion could be drawn regarding the relative location of the two mutation sites? ____________________
____________________[32]

Mating the deletion mutant with a series of viruses carrying different point mutations provides an idea of the size and location of the deletion. The following sketch shows the sites of eight known point mutations, numbered 1 through 8, which are spaced throughout the *rII* region.

1 2 3 4 5 6 7 8

rII region

Viruses carrying each of these *rII* point mutations are separately crossed with the unknown deletion mutant, using *E. coli* B host cells. The progeny of each of these matings are screened for the presence of wild-type recombinants. Assume that wild-type recombinant progeny are produced only from crosses involving point mutants *rII-6*, *rII-7*, and *rII-8*; crosses involving the other point mutations produce no wild-type phages. What does the presence of wild-type recombinants among the progeny of the *rII-6*, *rII-7*, and *rII-8* matings tell us about the location of the deletion relative to the positions of these mutations? ____________
____________________________[33] What does the absence of wild-type progeny from the other matings tell us about the location of the deletion? ________
____________________________________[34]

[29] Multiplying the recombination frequency by 100 gives 0.2%, or 0.2 map units. [30] One recombinant, carrying both the *rII* and the deletion mutation, is a double mutant; the other, consisting of normal DNA, is a wild-type. [31] By placing the progeny viruses on a lawn of *E. coli* K12(λ) where only wild-type phages would grow. [32] They are in the same section; that is, the site of the point mutation falls within the deleted zone. [33] The deletion does not include sites *6*, *7*, or *8*. [34] The deleted DNA covers the region defined by mutants *1*, *2*, *3*, *4*, and *5*.

Consider another deletion mutation (call it *D-2*) that has arisen independently of the one just assigned to the *1–5* zone (which we will designate as *D-1*). Mutants carrying *D-2* are mated separately with the same eight *rII* mutants used in mapping *D-1*. Crosses involving *rII* mutants *1, 2,* and *3* produce wild-type recombinants while the others yield no such recombinants. What can you conclude about the location of deletion *D-2*? ______________________

______________ [35] Shade in the deleted areas of mutations *D-1* and *D-2* on the following sketches of the *rII* region.[36]

Mutation *D-1*: 1 2 3 4 5 6 7 8

Mutation *D-2*: 1 2 3 4 5 6 7 8

*r*II region

Using Reference Deletion Markers to Map Point Mutations

Once the size and location of a number of deletion mutations is known, they can serve as reference markers to map unknown *rII* point mutations. Assume that a newly discovered *rII* point mutation designated as *rII-x* is to be mapped. Our reference markers are deletion mutations *D-1* and *D-2* that were mapped previously. Mutant *rII-x* is separately mated with *D-1* and with *D-2* viruses and the progeny from each mating are screened for wild-type recombinants. The following three outcomes could occur.

1. Possible outcome 1: The mating of *rII-x* with *D-1* produces no wild-type recombinants, and the mating with *D-2* produces wild-type recombinants. From the mating with *D-1*, what could be concluded about the location of the *rII-x* mutation?

   ______________________________________ [37]

   From the mating with *D-2*, what could be concluded about the location of the *rII-x* mutation?

   ______________________________________ [38]

   Taking the outcome of these two crosses together, what would you conclude about the location of *rII-x*? ______________________

   ____________ [39] The location of *rII-x* is as follows.

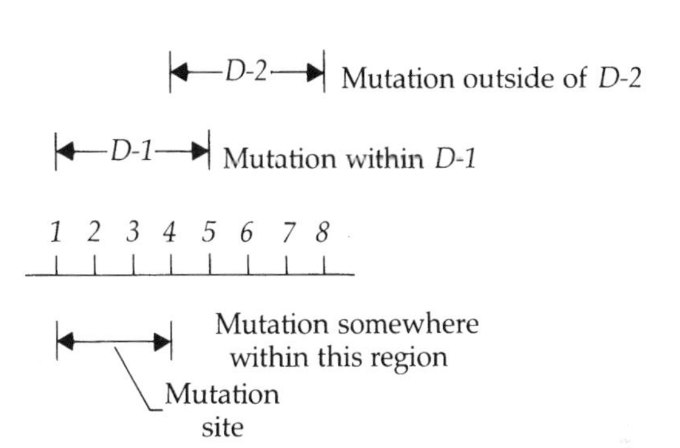

2. Possible outcome 2: Wild-type recombinants result from the mating of *rII-x* mutants with *D-1* but not with *D-2*. From these results, what would you conclude about the location of *rII-x*? ____________

   ______________________________________ [40]

3. Possible outcome 3: Each cross fails to produce wild-type recombinants. What could be concluded about the mutation site? ______________________

   ______________________________ [41]

More Precisely Locating a Deletion

Once a point mutation has been mapped to a general region, it can be more precisely located within that region through additional matings with deletion mutations previously mapped to that region. How would these deletions generally compare in size with those used to map a point mutation to a general region?

______________________________________ [42]

To illustrate this, assume that the *rII-x* mutation lies within the *4–5* zone, which is shown, as follows, as enlarged and arbitrarily subdivided into 10 subzones.

[35] The presence of wild-type recombinants with point mutations *1, 2,* and *3* indicates that the deleted DNA in *D-2* does not include these sites; the absence of recombinants with the other mutants indicates that the deletion must span the 4–8 region of the gene.
[36] The shading should block out positions *1–5* inclusive for *D-1* and positions *4–8* inclusive for *D-2*. Notice that the deletions overlap at positions *4* and *5*. [37] The absence of wild-type recombinants indicates that the point-mutation site is within the area of *D-1*.
[38] The occurrence of wild-type recombinants indicates that the mutant site is outside the *D-2* deletion. [39] Since it falls within the *D-1* zone (that is, sites *1–5*, inclusive) and lies outside the *D-2* zone (that is, sites *4–8*, inclusive) the point mutation must lie within the portion of *D-1* that does not overlap with *D-2* (that is, sites *1–3*, inclusive). [40] The wild-type recombinants from the *D-1* mating indicate that the point mutation is outside the *D-1* area, that is, outside of sites *1–5*, inclusive. The absence of wild-type recombinants from the *D-2* mating places the point mutation within the *D-2* deletion, that is, between sites *4–8*, inclusive. The mutation is located between sites *5* and *8*. [41] The absence of wild-type recombinants with both *D-1* and *D-2* indicates that the site of *rII-x* lies within an area common to both deletions. Only sites *4* through *5* occur within both deletions, and thus the *rII-x* site must be located in the *4–5* zone. [42] They would be smaller and thus permit finer mapping.

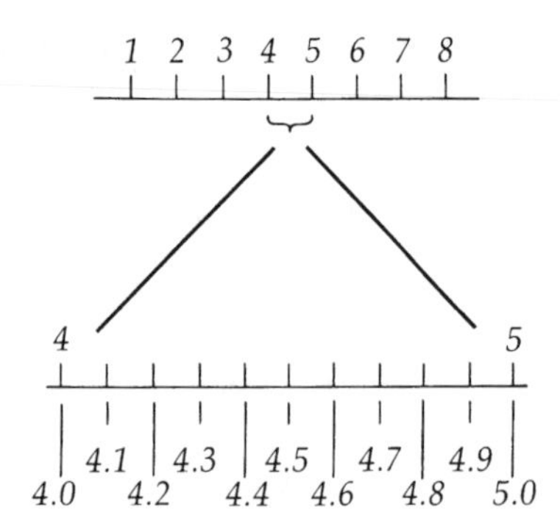

Four reference deletion markers have been mapped in this *4–5* zone. The following table gives the range of each and tells whether wild-type recombinants arise when viruses carrying the deletion are crossed with *rII-x*.

Deletion mutation	Range of deletion (inclusive)	Wild-type recombinant progeny in mating with *rII-x*
D-a	*4.7–5.0*	Yes
D-b	*4.1–4.6*	No
D-c	*4.5–5.0*	Yes
D-d	*4.1–4.4*	Yes

In the spaces provided, indicate what can be concluded about the location of *rII-x* from the outcome of each mating.

Recombinants with *D-a*: ________________ [43]

No recombinants with *D-b*: ________________ [44]

Recombinants with *D-c*: ________________ [45]

Recombinants with *D-d*: ________________ [46]

Where is the *rII-x* site? ________________ [47]

The two successive mapping steps that we have worked through with *rII-x* (first placing it in the *4–5* zone and then narrowing it to the *4.4–4.5* subzone) provide an idea of how deletion mapping works. Each step, of course, depends on the availability of a number of deletion mutations whose location and size have been previously established. Each step further narrows the possible region within which the point mutation is located. The next step in mapping *rII-x*, if the process were to be carried further, would be to mate *rII-x* viruses with a series of deletion mutations that had been mapped to regions within the *4.4–4.5* subzone. Once the mutation has been situated within the smallest region for which reference deletion markers are available, the precise location of *rII-x* within that subzone could be determined from the relative recombinant frequencies produced from crosses between viruses carrying *rII-x* and point mutations that had been previously mapped to this same subsection of the gene.

SUMMARY

Complementation or *cis-trans* testing is used to determine the number of loci independently capable of giving rise to the same mutant phenotype. If phage multiplication occurs following the coinfection of the restrictive host (in this case, *E. coli* K12(λ) strain) by two T4 *rII* mutant strains, the mutations producing the mutants are carried at separate complementary loci. If no multiplication occurs, the mutations are carried at the same locus and do not complement each other.

Mapping mutation sites within a gene depends upon the frequency of recombination between two sites. The *rII* point mutations are mapped by coinfecting the permissive host (in this case, *E. coli* strain B) with viral strains carrying different *rII* mutations. Both mutants can reproduce in this host and once their chromosomes are replicated, recombination can occur. Samples of progeny phages are grown on the *E. coli* B strain to provide an estimate of the total number of progeny and on the K12(λ) strain to detect wild-type recombinants. The recombination frequency is obtained by expressing the total number of recombinants (obtained by doubling the number of wild-type recombinants) as a proportion of the total number of progeny.

Deletion mutations are mapped by crossing phages carrying them with phages carrying previously mapped point mutations using *E. coli* strain B hosts. The production of wild-type phages (detected using the restrictive host, K12(λ)) indicates that the deletion does not include the site of the point mutation. The absence of wild-type phages shows that the deletion includes the site of the point mutation. A series of known (that is, already mapped) reference deletions can be used to map newly arisen point mutations. Viruses carrying the point mutations are individually mated with a series of reference deletions and the progeny from each are screened for wild-type recombinants using K12(λ) host cells. If wild-type recombinants are found, the point-mutation site lies outside the deletion; if none is found, the point mutation lies within the deletion.

[43] *rII-x* is outside *4.7–5.0*. [44] It is within *4.1–4.6*. [45] It is outside *4.5–5.0*. [46] It is outside *4.1–4.4*. [47] It falls between *4.4* and *4.5*.

PROBLEM SET

25-1. Assume that two independently arising *rIIB* mutants of the T4 virus have recently been discovered and there is uncertainty as to whether the mutation carried by each is located at the same or at different sites within the gene. Viruses from each strain are used to coinfect *E. coli* B cells.

a. What is the purpose of coinfecting the *E. coli* B cells?

b. A suitably diluted sample of the lysate from the *E. coli* B cells is combined with melted agar containing *E. coli* K12(λ) and poured onto an agar plate. Four plaques develop during incubation. What type or types of progeny phages would form these plaques?

c. Identify another way in which wild-type viruses could arise in this procedure and design a control experiment which would detect these cells.

d. Based on the information presented in 25-1b, what can be concluded about the relative positions of the two *rII* mutations in the gene?

e. Another lysate sample of the same size as that mentioned in 25-1b is used to prepare a lawn with *E. coli* B, and 1110 plaques develop. What type or types of progeny phages would form plaques on this lawn?

f. What is the recombination frequency for these two mutant loci?

g. How many map units separate the two loci?

25-2. Five independently arising, newly discovered *rII* mutants of T4 are used in a series of pairwise matings carried out using *E. coli* K12(λ) host cells. The progeny phages from these matings are summarized in the following matrix. A plus sign indicates the production of progeny phages, while a minus sign denotes their absence. For example, the plus sign at the intersection of mutant 2 with mutant 3 indicates that the mating of 2 and 3 produces progeny phages.

		rII Mutant					
		1	*2*	*3*	*4*	*5*	*6*
rII Mutant	*1*	–	–	+	+	–	–
	2		–	+	+	–	–
	3			–	–	+	+
	4				–	+	+
	5					–	–
	6						–

a. Based on these data, how many genes are present in the *rII* region? Explain.

b. Identify the mutations which belong to each of the genes represented here.

25-3. Another set of T4 *rII* mutants different from that described in problem 25-2 is studied in pairwise matings using *E. coli* K12(λ) as the host. No mating produced progeny. What can be concluded about the mutations carried by these viruses?

25-4. Each of six mutant strains of a little-studied bacteriophage show the same phenotype. Research indicates that the mutation carried by each of these strains is located in the same general region of the viral chromosome. A series of pairwise matings between these six mutant types is carried out using an appropriate restrictive bacterial host. The outcome of each mating

PROBLEM SET

is summarized in the following matrix where a plus sign indicates the production of progeny phages and a minus sign indicates no progeny.

		Mutant					
		A	*B*	*C*	*D*	*E*	*F*
Mutant	*A*	–	+	+	–	+	–
	B		–	+	+	–	+
	C			–	+	+	+
	D				–	+	–
	E					–	+
	F						–

a. How many genes are found in the chromosomal region containing these mutations?

b. Identify the mutations occurring in each of these genes.

25-5. A pair of T4 *rIIB* mutants are used to coinfect B strain *E. coli*. No plaques are found when an appropriate dilution of lysate is placed on a lawn of K12(λ). What can be concluded about the site of the mutation carried by each of these *rIIB* mutants?

25-6. The DNA making up the T4 chromosome consists of about 200,000 base pairs, and the circular map of this chromosome involves about 1500 map units. Two strains of *rIIB* mutants are mated, and wild-type recombinants are found with a frequency of 0.0002.

a. How many map units separate these two mutation sites?

b. How many base pairs are found in a unit of map distance?

c. How many base pairs separate the two mutation sites?

25-7. Examination of a laboratory notebook indicates that two newly discovered *rII* mutants, temporarily designated as carrying mutations *s* and *t*, have been studied. Two strains, with genotypes s^+ and ^+t, respectively, were used to coinfect *E. coli* host cells, and progeny phages were formed. Because of an omission in the laboratory notebook, it is unclear whether the *E. coli* hosts belonged to strain B or to strain K12(λ).

a. If the *E. coli* were strain B cells, would complementation or recombination be responsible for the formation of the progeny phages? What would be the genotype or genotypes of these phages?

b. If the *E. coli* were strain K12(λ) cells, would complementation or recombination be responsible for the formation of these phages? What would be the genotype or genotypes of the progeny phages?

c. The researcher responsible for this work later recalls that the K12(λ) strain was used. Are the two mutations variants of the same gene?

25-8. A complementation test is carried out with two *rII* mutants with *E. coli* K12(λ) as the host. If complementation occurs, phages are normally formed at a rate of approximately 250 per infected host cell. With this particular pair of mutants, phages are formed at a rate of 0.1 per infected host cell.

a. Do the mutant genes carried by these viral strains occupy the same locus? Explain.

b. What genotype or genotypes do the progeny phages have.

c. Identify two alternate ways in which the progeny phages could have been formed.

PROBLEM SET

25-9. The following sketch shows the *rIIB* gene in the bacteriophage T4 subdivided into 10 zones. Seven mutations that involve deletions of portions of the *B* gene have been mapped to the zones as shown.

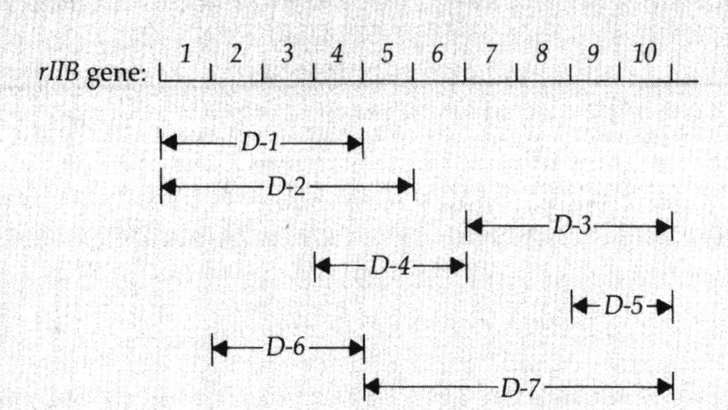

Viruses carrying certain of these mutations are mated in permissive host strain B bacteria. For each cross, indicate whether wild-type recombinant viruses would be expected among the progeny. Explain each answer.

(1) *D-1* × *D-2*
(2) *D-2* × *D-3*
(3) *D-4* × *D-1*
(4) *D-2* × *D-7*
(5) *D-4* × *D-5*
(6) *D-6* × *D-2*

25-10. Refer to the information in problem 25-9. Viruses carrying a point mutation known to be located in zone 2 of the *rIIB* gene are separately mated in permissive host strain B bacteria with stocks of viruses carrying each of the seven deletion mutations. Identify the crosses that would give rise to wild-type progeny phages.

25-11. Individual crosses in permissive host strain B bacteria between a phage strain carrying a newly arisen unmapped point mutation and viruses carrying each of the seven deletion mutations shown in problem 25-9 indicate that some wild-type recombinants are produced with deletion mutants *D-1, D-2, D-3, D-5,* and *D-6*. No such recombinants arise with *D-4* and *D-7.* As far as the data allow, identify the zone within which this point mutation is located.

25-12. The sites of three point mutations, designated as *6-2, 6-3,* and *6-4,* are known to be in zone 6. When viruses carrying the point mutation mapped in problem 25-11 (designate it as *6-1*) are separately crossed with viruses carrying point mutations *6-2, 6-3,* and *6-4,* a relatively high level of wild-type recombinants is produced in the cross involving mutant *6-4*; a low level is formed in the cross with mutant *6-2*; and an intermediate number results from the cross with mutant *6-3*.

a. What can be said about the position of the point mutation mapped in 25-11, that is, *6-1*, relative to point mutations *6-2, 6-3,* and *6-4*?

b. Could the sequence of these point mutation sites be determined from these data?

25-13. In the bacteriophage T4, five different independently arising *rII* point mutations (designated as *F, G, H, I,* and *J*) produce no recombinants when

PROBLEM SET

mated separately in permissive host bacteria with phages carrying a deletion of the entire *B* cistron of the *rII* gene (note that this deletion of the *B* cistron does not extend into the *A* cistron). What does the absence of recombinant progeny indicate about the location of these point mutations?

25-14. The following sketch shows the *rIIB* gene in the bacteriophage T4 subdivided into 15 zones. Five deletion mutant strains, designated as *D-11, D-12, D-13, D-14,* and *D-15,* have been mapped to the *rIIB* gene; the zones missing from each deletion are indicated.

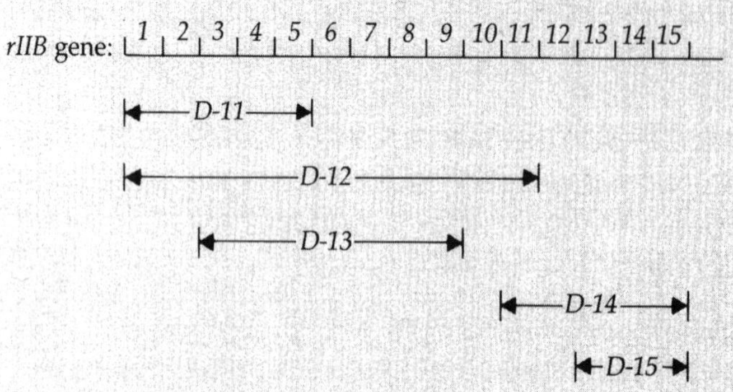

rIIB gene: 1 2 3 4 5 6 7 8 9 10 11 12 13 14 15

D-11

D-12

D-13

D-14

D-15

A series of pairwise matings is set up (in permissive host strain B bacteria) between each of the five deletion mutations and each of the five *rII* point mutations, *F, G, H, I,* and *J,* referred to in problem 25-13. The outcome of these matings is summarized in the following table.

	Deletion Mutation				
Point mutation	***D-11***	***D-12***	***D-13***	***D-14***	***D-15***
F	+	+	+	−	+
G	−	−	+	+	+
H	+	−	+	+	+
I	+	−	−	+	+
J	+	−	−	+	+

Identify the zone or zones of the *B* gene containing the sites of point mutations *F, G, H, I,* and *J*. Be as specific as the data allow.

25-15. Refer to the information given in problem 25-14. Assuming that point mutations *I* and *J* are not at the same site, what procedure could be used to distinguish between them?

CHAPTER

26

Gene Expression: Transcription and Translation

INTRODUCTION

Genetic information is stored in sequences of bases in DNA. Some DNA guides the formation of **messenger RNA (mRNA)** which goes on to specify the amino acid sequence for polypeptide molecules which go on to assume structural or enzymatic roles in the cell. Other DNA guides the synthesis of **transfer RNA (tRNA)** and **ribosomal RNA (rRNA)** which also have key roles in polypeptide synthesis. Transfer RNA molecules complex with individual amino acids and guide their placement within a growing polypeptide strand. Ribosomal RNA molecules complex with an array of protein molecules to form ribosomal subunits that join together and become functional ribosomes—the central processing units for polypeptide synthesis.

Each type of RNA is synthesized through complementary base pairing with a DNA template. This DNA-to-RNA transfer rewrites the information, but still uses the nucleotide language. The information stored in mRNA is transferred one step further when it guides the assemblage of a polypeptide molecule. This transfer involves rewriting the instructions in a different language that uses amino acids. The term **transcription** means to make a copy of something and the term **translation** means to rewrite something in a different language. Which of these terms describes the information transfer that occurs in the synthesis of any of the three types of RNA from DNA templates? ____[1] Which term describes the information transfer that occurs in the synthesis of polypeptides from mRNA? ________[2] The relationship between transcription and translation is summarized as follows.

```
         TRANSCRIPTION       TRANSLATION
DNA ─────────────► RNA ─────────────► Polypeptide
```

This relationship, called the **central dogma of molecular biology,** was formulated by Francis Crick. In this chapter we will examine transcription and translation. Since both of these processes have been extensively studied in bacteria, especially in *E. coli,* our discussion will focus on prokaryotic organisms with some mention of eukaryotic forms.

TRANSCRIPTION

Units of Transcription

A **unit of transcription** is defined as a portion of a double-stranded DNA molecule carrying information

[1] Transcription. [2] Translation.

that is incorporated into a single molecule of RNA. In eukaryotic organisms, the gene is the unit of transcription. In prokaryotic forms, a unit of transcription may be one gene (or cistron) and is known as a **monocistron,** or it may consist of two or more genes and is referred to as a **polycistron.** Genes that encode for polypeptides are known as **structural genes** and are transcribed into mRNA which serves to guide polypeptide formation. Other genes are transcribed into tRNA or rRNA. Only one of the two DNA strands, designated as the **sense strand,** serves as a template to guide RNA synthesis; its complementary counterpart is known as the **antisense strand.** Some genes or group of genes use one duplex strand as the sense strand while other genes use the other strand of the same duplex in that capacity.

Successful initiation and termination of transcription depend on DNA nucleotide sequences known as **promoters** and **terminators.** The promoter sequence precedes the unit of transcription while the terminator is part of the unit of transcription and is located at its end. These sequences are illustrated in Figure 26-1 for a unit of transcription where the 3′-to-5′ template strand is the sense strand.

The promoter serves as the site where RNA polymerase attaches to the sense strand prior to beginning synthesis. The terminator signals the end of synthesis and causes the RNA polymerase to detach from the sense strand and release the newly synthesized RNA molecule. For convenience in discussion, the promoter is often described as being "upstream" from the transcribed section while the terminator is located "downstream." In addition, the first DNA nucleotide transcribed into RNA is often designated as +1, and thus the nucleotides upstream are designated with minus signs (−1, −2, and so on) while those downstream are given plus signs (+2, +3, and so on).

FIGURE 26-1

Unit of transcription with promoter and terminator regions.

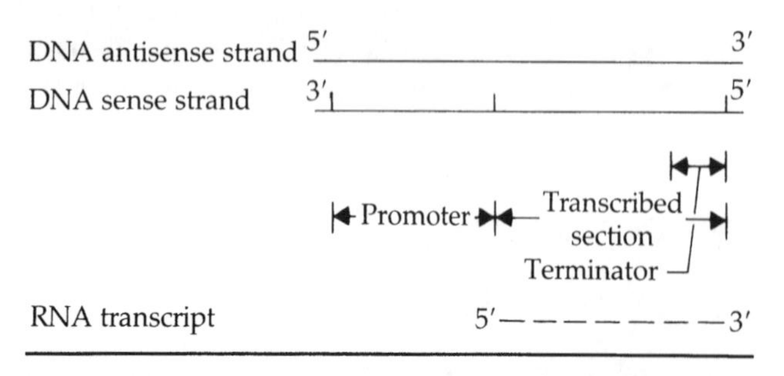

Prokaryotic RNA Polymerase

The transcription of RNA molecules in prokaryotes is catalyzed by a single enzyme, **RNA polymerase.** Because transcription requires a DNA template, this enzyme is sometimes referred to by the more specific term, **DNA-dependent RNA polymerase.** In addition to this enzyme and the DNA template, the process requires Mg^{++} and supplies of all four ribonucleotides in triphosphorylated form (ATP, GTP, CTP, and UTP). (Remember that thymine nucleotides are replaced by uracil nucleotides in RNA.) RNA polymerase initiates RNA formation from scratch (that is, it requires no primer RNA) and proceeds to add ribonucleotides one at a time, as dictated by the DNA template strand, to the 3′ end of the growing RNA strand.

Core Enzyme DNA-dependent RNA polymerase enzymes have been isolated from a variety of prokaryotes, and in most cases, the molecules have been shown to be complexes of several polypeptide subunits. In *E. coli*, for example, there are three kinds of basic subunits. Two of these polypeptides (designated as beta and beta prime) are present in single copies, and the third subunit (known as alpha) is present in two copies, making a four-polypeptide unit known as the **core.**

Sigma Factor Another polypeptide, the **sigma factor,** must complex temporarily with the core in order to initiate normal transcription. Try to determine the role played by the core enzyme and by the sigma subunit from the following information. When the core enzyme and the sigma factor are added to *in vitro* RNA-synthesizing experimental setups containing all the other necessary ingredients, RNA synthesis is initiated only at the sites at which synthesis is normally initiated *in vivo.* When the core polymerase is added without the sigma factor, RNA synthesis still occurs, but it is initiated at just about any point on either of the DNA strands. What conclusion can be drawn from this information regarding the role of the core enzyme?

____[3]

What conclusion can be drawn from this information regarding the role of the sigma factor? ____________

____________________________[4] Once transcription is underway and ten or so ribonucleotides

[3] The core enzyme catalyzes the polymerization of RNA and initiates transcription at most any point on either the sense or antisense strands.

[4] When the sigma factor is present, RNA synthesis begins at the normal and appropriate sites along the sense strand, implying that the sigma factor is essential for RNA polymerase to recognize the promoter sites.

have been covalently joined, the sigma subunit dissociates from the core and the balance of the RNA transcript is synthesized by the core enzyme. The released sigma factor can then reattach to another molecule of core polymerase.

Rho Factor The termination of RNA synthesis for some prokaryotic units of transcription requires an additional polypeptide called the **rho,** or **termination, factor.** In these cases, the rho factor, which apparently complexes with the core enzyme, is essential for RNA polymerase to recognize the termination signals present in the DNA. The rho factor dissociates from the polymerase when RNA transcription is complete. Once released, the rho polypeptide is free to be used again. For other prokaryotic units of transcription, the RNA polymerase enzyme is able to recognize the termination signal without the rho factor.

Eukaryotic RNA Polymerases

Much less is understood about eukaryotic RNA polymerases. Three types are known and all are complex molecules made up of aggregations of polypeptide subunits of differing size. RNA polymerase I occurs in the nucleolus and synthesizes most types of rRNA. RNA polymerase II, located in the nucleoplasm, is responsible for the formation of mRNA. RNA polymerase III, also located in the nucleoplasm, synthesizes tRNA and one type of small rRNA.

Prokaryotic Promoter Sequences

Sequencing the nucleotides of promoters from different prokaryotic species and from different transcriptional units within the same species indicates considerable variation in length and sequence. Despite this, all promoters seem to have some similar sections. One such section consists of seven nucleotide pairs, most of which are made up of adenine and thymine. This section, known as the **Pribnow box** (named after the scientist who first described it), is usually centered around the nucleotide pair in position -10 and thus has one of its ends just six or so nucleotides away from the first nucleotide transcribed. Can you see any advantage to this DNA section consisting primarily of AT pairs? ______________________________

______[5] Another section showing similarity is found in the -35 region and is also rich in AT pairs. An RNA polymerase molecule that binds to both these regions is properly positioned to initiate RNA synthesis.

Prokaryotic Terminator Sequences

Terminator sequences are at the end of the transcriptional unit and, as mentioned above, are of two types depending on whether their recognition requires that the special terminator protein, rho, be complexed with the core enzyme.

Rho-independent Terminators **Simple,** or **rho-independent, terminators** can be recognized by the core enzyme without rho. They consist of signals, within the DNA of the sense strand, that include a poly-A sequence occurring at the very end of the sense strand of the unit of transcription. Also, upstream from this, these terminators contain an **inverted-repeat sequence** with a high proportion of GC pairs.

An inverted-repeat sequence consists of a series of nucleotide pairs in a portion of a duplex DNA molecule that is inverted and repeated in a neighboring section of the same molecule. For example, assume the sequence

5′ . . . CGACCG . . . 3′
3′ . . . GCTGGC . . . 5′

occurs in one portion of a DNA duplex and is inverted and repeated in a nearby section of the duplex to produce this inverted-repeat sequence.

5′ . . . CGACCG . . . CGGTCG . . . 3′
3′ . . . GCTGGC . . . GCCAGC . . . 5′

If the lower, or 3′-to-5′ strand, is part of the sense strand of a unit of transcription, what is the base sequence of its RNA transcript? __________[6] These two sections of the RNA molecule complement each other, and when the RNA strand folds back on itself, they become hydrogen-bonded together to form the **hairpin,** or **stem-and-loop configuration** shown in Figure 26-2.

Once the hairpin has developed in the transcript, it plays a key role in the termination of transcription by apparently causing a change in the conformation of the RNA polymerase molecule downstream from it. This alters the enzyme's functional properties so that both the enzyme and the RNA transcript dissociate from the template strand. The poly-U sequence at the end of the transcript which is weakly bound to the poly-A sequence at the tail end of the transcribed DNA may facilitate this release of the polymerase from the DNA.

[5] AT pairs have fewer hydrogen bonds than CG pairs, and fewer hydrogen bonds in that region may facilitate the separation of the two DNA strands. [6] 5′ . . . CGACCG . . . CGGUCG . . . 3′.

FIGURE 26-2

Hairpin, or stem-and-loop, configuration of a portion of an RNA transcript.

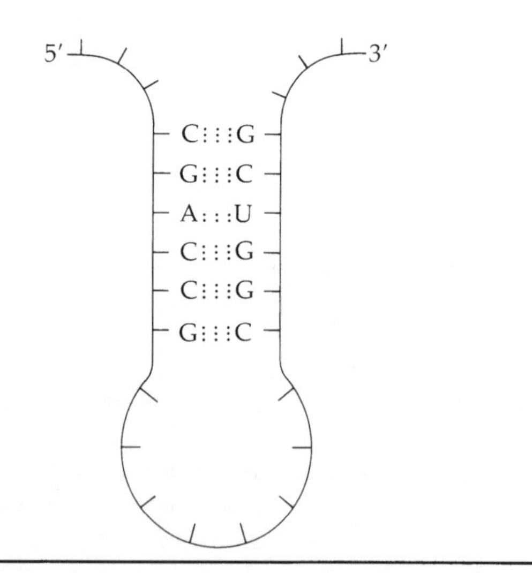

Rho-dependent Terminators The recognition of **rho-dependent terminators** requires that rho be complexed with the core enzyme. These terminators lack the poly-A series but do possess inverted repeats. These repeats, however, are not as rich in CG pairs as are those of rho-independent terminators.

Eukaryotic Promoter and Terminator Sequences

Eukaryotic promoters for RNA polymerase II possess two similar sections. One of these, the **Goldberg-Hogness box** (named after the scientists who first described it), is located around nucleotide pair -30. Since it is rich in AT pairs, it is also known as a **TATA box.** The second sequence, located around nucleotide pair -80, is known as a **CAAT box.** Both of these sections are believed to serve as a recognition site for the attachment of the RNA polymerase II to the promoter.

Promoters for RNA polymerase III also include TATA and CAAT boxes but these boxes are located downstream from the site of transcription initiation, within the unit of transcription. Studies of RNA polymerase I promoters from a variety of eukaryotes indicates little similarity among them. Terminator sequences from eukaryotic cells are under study, but little information is available at this time.

Key Steps in Transcription

Binding of RNA Polymerase to the Template RNA polymerase, consisting of the core enzyme complexed with a sigma factor, recognizes and binds with the promoter. This binding causes the breaking or denaturation of hydrogen bonds in the region of the Pribnow box and results in the separation of the two strands of the DNA duplex.

Synthesis Initiation The RNA polymerase moves along the promoter to the first nucleotides of the transcription unit, causing the breaking of more hydrogen bonds and further separation of the duplex strands. The enzyme then begins RNA synthesis through the stepwise pairing of complementary nucleotides. The triphosphorylated ribonucleotide complementing the first transcribed DNA nucleotide is positioned opposite the DNA nucleotide. As the next triphosphorylated nucleotide is positioned, RNA polymerase catalyzes the formation of a covalent phosphodiester bond joining the second ribonucleotide to the 3′ hydroxyl of the first with the release of two phosphate groups from the second nucleotide.

Strand Elongation As the enzyme continues to move along the sense strand, additional ribonucleotides are positioned opposite the template bases and the formation of phosphodiester bonds between the ribonucleotides continues. (In *E. coli* at 37°C, 40 to 50 ribonucleotides can be added per minute to a growing RNA strand.) Each new ribonucleotide is added to the 3′ hydroxyl of the most recently incorporated ribonucleotide. The growing transcript, of course, exhibits an orientation opposite to that of template DNA strand. In what direction is the RNA polymerase moving along a 3′-to-5′ DNA sense strand? __________[7] With regard to this DNA sense strand, is transcription initiated toward its 3′ or 5′ end? __________[8] Is the RNA strand growing from 5′ to 3′ or from 3′ to 5′? __________[9]

In the separate transcriptions of two genes carried on the same DNA duplex, assume that a different duplex strand serves as sense strand for each gene. Given the antiparallel nature of a DNA duplex, what can be said about the relative directions of RNA transcript synthesis from these two genes? ____[10] As the enzyme continues to move along the sense strand, the DNA duplex continues to open ahead of the enzyme. After the RNA strand has grown to a certain size, it begins to separate from the DNA template strand, starting at its 5′ end. As this occurs, the two strands of the DNA duplex reestablish hydrogen bonds with each other.

[7] It moves along the sense strand in the 3′-to-5′ direction. [8] Towards its 3′ end. [9] From 5′ to 3′. [10] Each strand grows in the 5′-to-3′ direction; the RNA is synthesized along the two genes in opposite directions relative to each other.

Synthesis Termination As already described, the termination of RNA synthesis with rho-independent terminators is signaled by an inverted-repeat sequence followed by a poly-A series at the end of the unit of transcription. As the inverted-repeat section is transcribed, base complementarity within the RNA causes a hairpin structure to develop. Transcription of the poly-A series generates a poly-U series in the RNA following the hairpin structure. The hairpin causes the RNA polymerase molecule downstream from it to change its configuration and the polymerase stops adding nucleotides to the RNA transcript, dissociating from the DNA. Rho-dependent terminators also feature an inverted-repeat sequence but require a rho factor for termination. In these cases, the rho factor dissociates when the enzyme is released from the template.

To summarize, identify four key things that RNA polymerase must do in order to guide the formation of an RNA transcript from the sense strand of a double-stranded unit of transcription.[11]

1. ______________________________

2. ______________________________

3. ______________________________

4. ______________________________

POST-TRANSCRIPTION MODIFICATION OF RNA

The transcription process just considered produces all the types of RNA found in the cell. Before messenger, ribosomal, and transfer RNA molecules can assume their very different functional roles, they usually undergo some post-transcription modification in which a ribonuclease enzyme and an RNA-splicing enzyme play prominent roles. We will take a look at the major modifications carried out on the primary transcripts of each of the three types of RNA to produce mature functional molecules in prokaryotes and eukaryotes.

Messenger RNA

Prokaryotic mRNA There is virtually no post-transcription modification of prokaryotic mRNA. This may be due to the fact that the translation of mRNA is often initiated before the transcription of the mRNA strand is completed, as described later in this chapter.

Eukaryotic mRNA Eukaryotic mRNA molecules undergo considerable post-transcription modification before they are translated into polypeptides. The initial mRNA transcript, known as **precursor mRNA (pre-mRNA)** or **heterogeneous nuclear RNA (hnRNA),** experiences modifications involving each end of the transcript and internal sections, as described next.

1. Capping of 5′ end: A methylated form of a guanine nucleotide, known as 7-methylguanosine (m^7G), is added enzymatically to the 5′ end of the transcript. This m^7G plus the original 5′ nucleotide make up what is referred to as the **cap.** After the mRNA is transported into the cytoplasm, the cap interacts with a **cap-binding protein** that assures proper orientation of the mRNA strand relative to the ribosome during translation.
2. Addition of a poly-A tail to the 3′ end: A portion of the transcript's 3′ end is removed to create a site to which a series of up to 200 adenine nucleotides known as the poly-A tail is added enzymatically. This polyadenylation may contribute to the stability of the mRNA.
3. Removal of noncoding internal sections: This modification occurs because most eukaryotic genes (and the mRNA transcribed from them) include noncoding regions, known as **intervening regions,** or **introns,** that are not translated into amino acids. These introns lie between sections that are translated, known as **expressed regions,** or **exons,** that collectively code for the gene's polypeptide product. A few to as many as 100 introns, varying in length from a few to 1000 or so nucleotide pairs, may be present in these **interrupted,** or **split genes.** Following transcription, the sections transcribed from the introns are removed and the remaining sections derived from the exons are spliced together to form a strand of mRNA containing the coding sequence of nucleotides. These modifications are summarized in Figure 26-3.

Sequencing the nucleotides at the RNA–intron-exon boundaries indicates some similarity that is summarized by the GU-AG rule: each intron has the sequence GU at its 5′ end and the sequence AG at its 3′ end. These doublets may serve as signal sites for the enzymes catalyzing the very precise cutting and splicing that occur here. Almost every type of eukaryotic nucleus contains an additional type of RNA designated as **small nuclear RNA (snRNA).** Studies on a limited number of these snRNA molecules indicate that portions of them complement the sequences found at the intron-exon boundaries and that they play a role in the splicing process.

[11] 1. Recognize and bind to the promoter region; 2. Break the hydrogen bonds joining the two DNA strands; 3. Polymerize ribonucleotides; 4. Recognize the terminator.

FIGURE 26-3

The formation of mature eukaryotic mRNA from an interrupted gene.

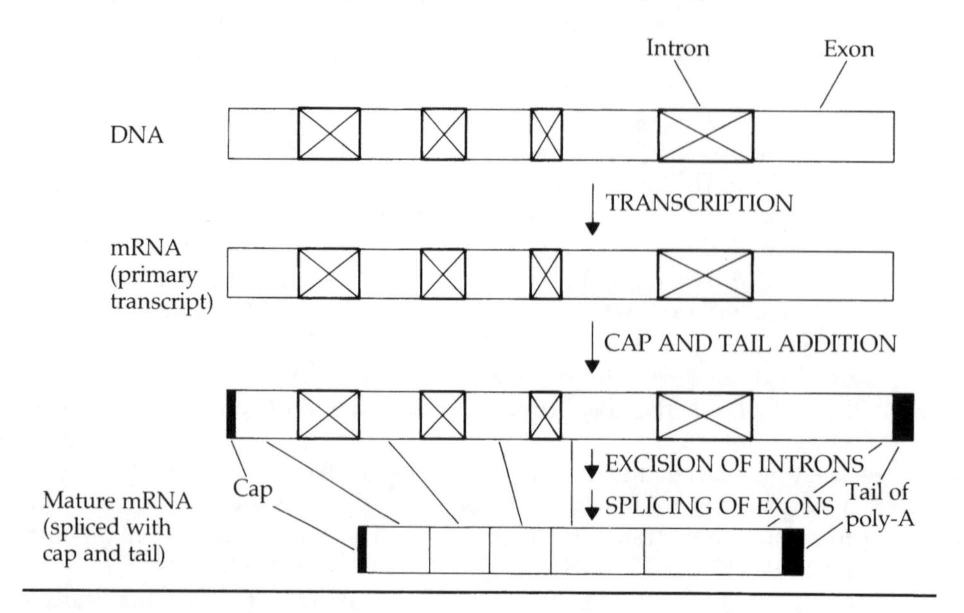

The mRNA that results after all three types of modification (capping, tailing, and intron removal) has occurred, designated as **mature mRNA,** is then transported to the cytoplasm for translation by the ribosomes.

Transfer RNA

The modification of the tRNA primary transcripts shows some similarities in prokaryotes and eukaryotes. For example, the primary transcript in both types of organisms may carry more than a single copy of a tRNA molecule. In some instances, one primary transcript may carry different tRNA molecules; in other instances the transcript may carry rRNA molecules in addition to tRNA molecules. Furthermore, after such transcripts have been reduced in size and cut up, some of the standard tRNA bases are converted to unusual forms before the tRNA molecules become functional. For example, adenine can be modified to the base inosine, I, through the deamination of its sixth carbon. Other unusual bases found in tRNA molecules include ribothymidine and pseudouridine.

Prokaryotic tRNA In prokaryotes, a precursor tRNA molecule experiences the following events.

1. The removal of a large section of nucleotides from its 3′ end, cutting the transcript back to a particular CCA sequence which then becomes the 3′ terminus of the mature tRNA molecule.
2. The removal of a series of nucleotides from the 5′ end.
3. Modification of some of the standard bases to unusual forms.

Eukaryotic tRNA In eukaryotes, precursor tRNA molecules tend to be considerably longer than the mature tRNA molecules. Post-transcription modification involves the following events.

1. The removal of a series of nucleotides from the 5′ end.
2. The addition of a three nucleotide sequence (5′-CCA-3′) to the 3′ end.
3. If intervening sequences are present, they are removed and the exons are spliced together.
4. Modification of some of the standard bases to unusual forms.

Mature tRNA In their mature form, tRNA molecules in both prokaryotes and eukaryotes consist of between approximately 75 and 90 nucleotides and show a number of similarities. Each tRNA strand folds back on itself and, because of intramolecular complementarity, takes on a secondary structure that can be represented in two dimensions by a cloverleaf form, as shown in Figure 26-4 for a generalized tRNA molecule.

Examine Figure 26-4 and note the following features. Intramolecular hydrogen bonding of complementary sections produces a double-stranded configuration in the parts of the molecule making up

FIGURE 26-4

Schematic diagram of a tRNA molecule shown in the two-dimensional cloverleaf configuration.

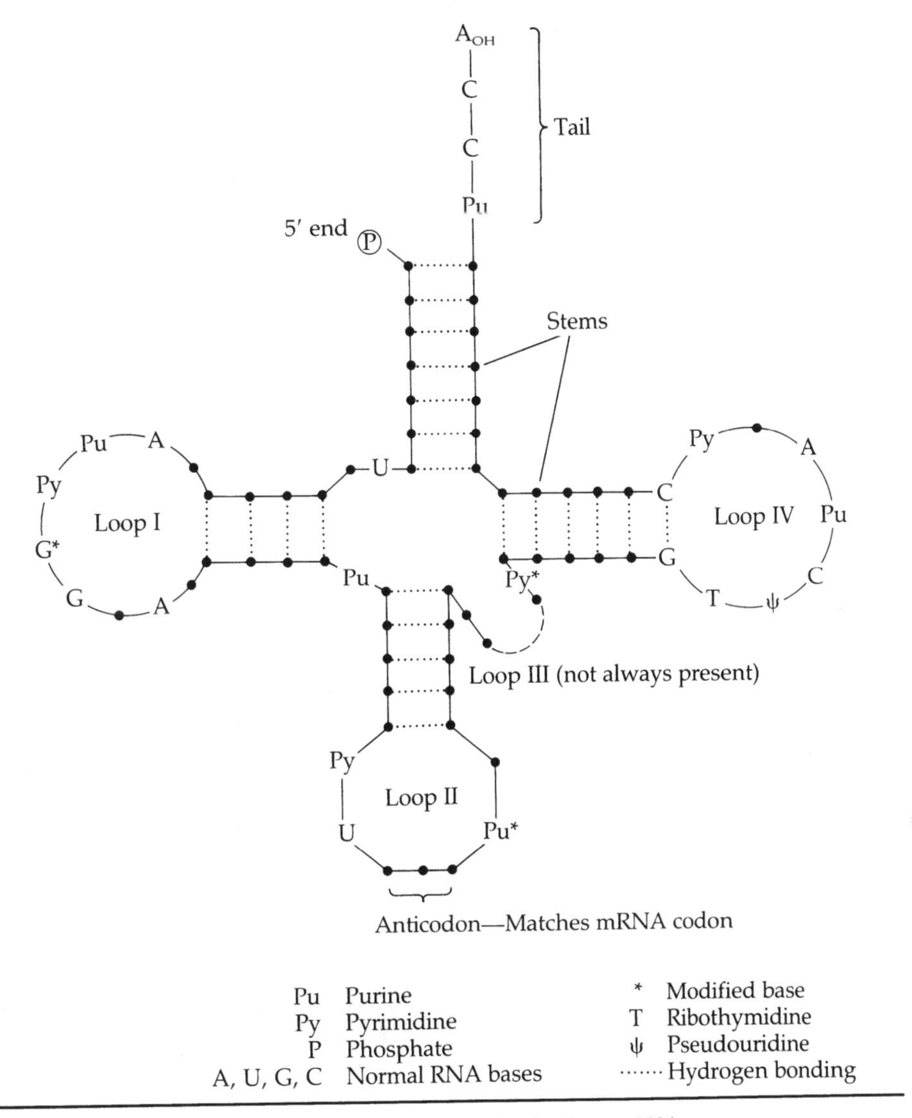

Peter J. Russell, *Genetics*, First Edition. Scott, Foresman/Little, Brown, 1986.

the stem parts of the cloverleaf. Noncomplementary sections (which exist, in part, because of the unusual bases formed during post-transcriptional modification) manifest themselves as the single-stranded loops making up the leaf portion of the cloverleaf. As will be discussed, the 3′ end of each type of tRNA molecule can join, in the presence of a specific enzyme, with a particular type of amino acid. The three successive unpaired bases making up the **anticodon** are prominently displayed on the loop section of the anticodon arm of the cloverleaf. Eventually, the tRNA anticodon pairs with a complementary codon of an mRNA molecule. This pairing positions the amino acid carried by the tRNA molecule at the appropriate site in the polypeptide strand for which the mRNA codes. Both prokaryotic and eukaryotic cells contain between 50 and 60 different types of tRNA molecules.

Ribosomal RNA

Before considering the post-transcription modification of rRNA, we need to make a brief digression to consider a commonly used method of characterizing rRNA molecules. This is based upon the rate at which the molecules settle when subjected to density-gradient equilibrium centrifugation. A molecule's rate of sedimentation is given as a **sedimentation coefficient,** expressed in terms of **Svedberg,** or **S, units,** and is influenced by its size, density, and weight. The larger a molecule's sedimentation coefficient, the faster its rate of settlement during centrifugation. Sedimentation coefficients are also used to characterize particles such as ribosomes and ribosome subunits.

The post-transcription processing of ribosomal RNA shows some similarity in prokaryotes and eukaryotes. In both, a single primary transcript is cut up to form functional rRNA molecules of different sizes. This cutting is accomplished by enzymatic removal of nonessential or spacer sections of the primary transcript. This is followed by further trimming to produce the functional rRNA molecules. (Note that most of the nonessential sections removed from the transcript are not introns: nonessential sections occur *between* the rRNA molecules; introns occur *within* genes.)

Prokaryotic rRNA One molecule of each of three kinds of rRNA is found in a functional prokaryotic ribosome. These types, designated as 5S, 16S, and 23S, are derived from a single 30S primary transcript. This transcript is cleaved into fragments during its transcription that are further trimmed to give rise to these three types of rRNA. This rRNA then interacts with approximately 50 kinds of proteins to form small (30S) and large (50S) ribosomal subunits. The union of a small and a large ribosomal subunit produces a functional ribosome (70S). (Note that S values are not strictly additive.) Can you see any advantage to having each of these three types of rRNA derived from a single precursor transcript? ______________________

______________________ [12]

Eukaryotic rRNA Eukaryotic ribosomes contain four types of rRNA molecules designated as 5S, 5.8S, 18S, and 28S. These molecules are produced from two kinds of primary transcripts. One kind arises from a ribosomal-DNA transcriptional unit encoding for the 18S, 5.8S, and 28S types of rRNA. Once formed, this transcript is processed enzymatically by removal of a segment from its 5′ end and by cutting most of what remains into one copy of each of these three types of rRNA. The second kind of primary transcript arises from a separate transcriptional unit (located in the nucleolus) that encodes for the 5S rRNA molecules. These four types of rRNA interact with a large array of proteins to form the small (40S) and large (60S) eukaryotic ribosomal subunits. The combining of a small and a large ribosomal subunit produces a functional ribosome (80S).

TRANSLATION

Before looking at the details of polypeptide formation, we need to briefly review some key points about amino acids and the way in which they are assembled into polypeptides.

Amino Acid and Polypeptide Structure

There are twenty different **amino acids** that serve as polypeptide building blocks. These amino acids have certain structural features in common. All carry a carbon atom, designated as the alpha carbon, which is bonded to a carboxyl group (-COOH), an amino ($-NH_2$) group, and a residue (-R) group. In all the amino acids except proline, the fourth bond of the alpha carbon is bound to a hydrogen atom. The properties of the different amino acids arise from differences in their -R groups. The generalized formula for an amino acid is shown in Figure 26-5.

The linkage that joins amino acids together is the **peptide bond.** This bond links the alpha amino group of one amino acid to the alpha carboxyl group of another, with the release of a molecule of water. Figure 26-6 shows the formation of a peptide bond between two generalized amino acids. Joining amino acids together produces a **peptide.** Two linked amino acids

FIGURE 26-5

Generalized structural formula for an amino acid.

Radical group (varies from one amino acid to next)

R

α-carbon atom

C

NH_2 H COOH

Amino group

Carboxyl group

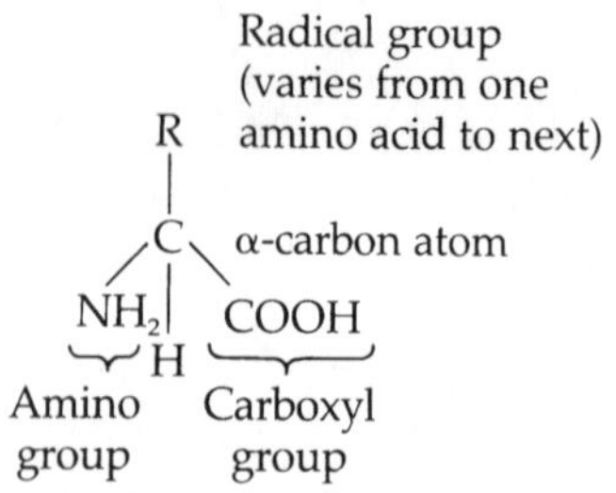

[12] It assures that the three types of rRNA will be produced in equal quantities, thereby assuring that all three will be available for ribosome production.

FIGURE 26-6

Formation of a peptide bond between two amino acids.

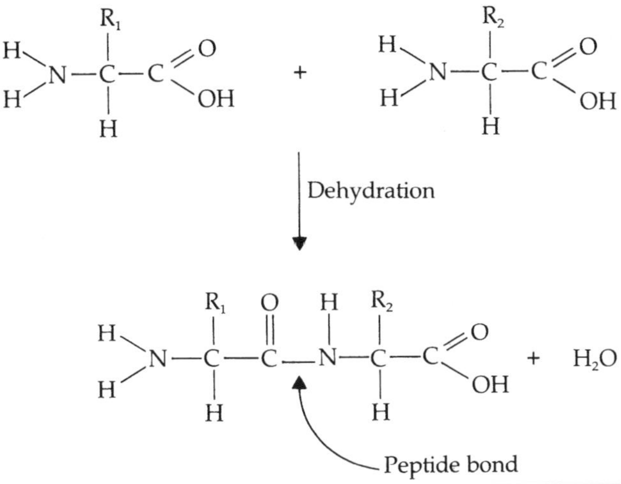

Dehydration

$+ H_2O$

Peptide bond

make up a dipeptide, three joined make up a tripeptide, and so on. The joining of many amino acids makes up a **polypeptide.** An assemblage of amino acids, regardless of its length, displays polarity: one end bears a chemically unbonded amino group (designated as the N-terminal end) while the other end carries a chemically unbonded carboxyl group (C-terminal end). The interaction of the -R groups of the amino acids in a polypeptide determines the chemical properties and three-dimensional configuration of the molecule. The structural configuration, in turn, defines the functional role of the molecule.

The three-dimensional configuration of a polypeptide is the product of either three or four levels of organization. The linear sequence of amino acids, joined by covalent peptide bonds, makes up the polypeptide's **primary structure (1°).** Interactions, usually between the -R groups of the amino acids and generally of a noncovalent nature (for example, hydrogen bonds, hydrophobic interactions, and ionic bonds), confer additional orders of polypeptide structure. Hydrogen-bond stabilized interaction between neighboring amino acids imposes a repeating configurational pattern or **secondary structure (2°)** that commonly takes the form of either an alpha helix or a beta pleated sheet. Covalent and noncovalent interactions between -R groups of amino acids in different sections of the polypeptide can cause the secondary configuration to bend and fold back on itself to produce the **tertiary structure (3°)** for a polypeptide. A fourth order of structure comes into play if two or more polypeptides must join together before the protein complex that results can become functional. This **quaternary structure (4°)** refers to the final structure assumed by the complex of bonded polypeptides. Hemoglobin is an example of a protein with a quaternary structure. A normal hemoglobin molecule must consist of two copies of an alpha polypeptide complexed with two copies of a beta polypeptide (along with four iron-containing heme groups) in order to function in oxygen transport. Note that the ability of a polypeptide to assume any of its particular configurations above the primary level depends on its primary structure.

Components of Polypeptide Synthesis

Next we look at the three components that feature prominently in polypeptide synthesis: amino acid–tRNA complexes, the ribosomes, and mRNA. We then consider how these components interact to produce polypeptides.

Amino Acids, tRNA, and Aminoacyl-tRNA Complexes Polypeptide synthesis begins with a mature mRNA molecule carrying the instructions for assembling a particular polypeptide and ends with the completed polypeptide—a covalently bonded series of amino acids of the right kind and in the right numbers, assembled in the correct sequence. What happens in between involves the very precise interaction of many different kinds of molecules that are available in the cell, including numerous enzymes. Of central importance are the twenty kinds of amino acids and the various kinds of tRNA molecules.

Two things must happen to an amino acid before it can be incorporated into a polypeptide molecule. First it must be activated by reacting with ATP so that it has sufficient energy to participate in the polypeptide-synthesis process (that is, to form a peptide bond). Then it must be joined with an appropriate tRNA molecule so that it can be inserted at the correct point into a growing polypeptide. Both of these reactions are catalyzed by an enzyme known as **aminoacyl-tRNA synthetase.** At least one specific type of aminoacyl-tRNA synthetase exists for each of the twenty amino acids. In the first reaction, the **activation reaction,** the enzyme catalyzes the breakdown of ATP to AMP (adenosine monophosphate) which then becomes joined by a high-energy bond to the amino acid, with the release of pyrophosphate. This is summarized as follows, where ~ represents a high-energy bond.

$$\text{Amino acid} + \text{ATP} \xrightarrow{\text{Aminoacyl-tRNA synthetase}} \text{Amino acid}{\sim}\text{AMP} + \underbrace{\text{P}{\sim}\text{P}}_{\text{(Pyrophosphate)}}$$

The second reaction follows on the heels of the first and occurs while the amino acid~AMP complex is still linked to the aminoacyl-tRNA synthetase enzyme. The amino acid~AMP complex reacts with an appropriate tRNA molecule to produce the amino acid–tRNA complex, with the release of the AMP, as follows.

$$\text{Amino acid}\sim\text{AMP} + \text{tRNA} \xrightarrow{\text{Aminoacyl-tRNA synthetase}} \text{Amino acid}\sim\text{tRNA} + \text{AMP}$$

Most types of amino acids can complex with more than one type of tRNA molecule. As discussed in Chapter 27, since each mRNA codon is made up of three nucleotides and since there are four kinds of nucleotides, a total of $4^3 = 64$ different codons can occur in an mRNA molecule. This means that a total of 64 kinds of tRNA molecules, each with a different anticodon, is possible.

The amino acid–tRNA complex, or **aminoacyl—tRNA complex,** is now ready to participate in polypeptide synthesis. The tRNA anticodon ensures that the amino acid is positioned at an appropriate site within a polypeptide, and the high-energy bond allows the amino acid to form a covalent bond joining it to a growing polypeptide chain. (Note: The various types of tRNA molecules are identified by the amino acids they carry. A superscript abbreviation for the amino acid is used to designate a particular type of tRNA molecule. For example, tRNA^{ala}, designates the tRNA molecule that carries the amino acid alanine. Aminoacyl-tRNA molecules are designated by placing the abbreviation for the amino acid in front of the symbol for the tRNA that carries it. For example, ala-tRNA^{ala} represents the alanine–tRNA complex.)

How many active sites would you expect a molecule of aminoacyl-tRNA synthetase to possess? _____[13] What type of molecule would be accommodated by each of these active sites? ______________________

______________________[14] As already noted, each of the twenty biologically important amino acids has its own type of aminoacyl-tRNA synthetase. In comparing the configurations of these different kinds of aminoacyl-tRNA synthetase molecules, which active sites would show variation from one kind of synthetase to another? ______________________

__________[15] Most of the amino acids can form aminoacyl-tRNA complexes with more than just a single type of tRNA molecule. What does this indicate about the tRNA active site of at least some of the aminoacyl synthetase enzymes? ______________________

______________________[16]

Ribosomes Ribosomes are the cellular organelles that serve as the processing units for polypeptide synthesis. A functional ribosome consists of a large and a small ribosomal subunit and is responsible for bringing together a strand of mRNA and an array of aminoacyl-tRNA complexes. It possesses a pair of adjacent sites that bind aminoacyl-tRNA complexes, one known as the **A,** or **aminoacyl, site,** and the other as the **P,** or **peptidyl, site.** Once the ribosome is attached to a strand of mRNA, it passes along the strand sequentially exposing codons so that they pair, in complementary fashion, with the anticodons of the appropriate aminoacyl-tRNA complexes. The temporary stability conferred by the ribosome on the hydrogen bonds of the codon-anticodon linkages enables the amino acids carried by the aminoacyl-tRNA complexes to become covalently bonded to the growing polypeptide chain.

Mature mRNA: Leader, Translated, and Trailer Sequences Mature mRNA molecules that are ready to guide polypeptide synthesis consist of some regions that will be translated and others that will not. One nontranslated region, known as the **leader sequence,** or **5′ noncoding sequence,** consists of a series of nucleotides between the 5′ end of the mRNA strand and the first codon to be translated. The number of nucleotides present in the leader sequence varies with the transcriptional unit and ranges from 20 to more than 600. In prokaryotes and in prokaryotic viruses, a portion of the leader sequence known as the **ribosome-binding sequence,** or the **Shine-Dalgarno sequence,** (named after its discoverers) plays a key role in binding the mRNA to the ribosome. This consists of around five or six bases that complement a series of bases at the 3′ end of the 16S rRNA found in the small ribosome subunit. This base pairing results in proper orientation of the mRNA strand and the ribosome, and ensures recognition of the codon that initiates translation. The cap-binding protein associated with the 5′ end of the mRNA strand is also believed to act as a ribosome-binding signal. (The existence of comparable ribosome-binding sequences in eukaryotic mRNA has yet to be demonstrated.)

Immediately adjacent to the leader is the portion of the mRNA transcript that is to be translated. Monocistronic transcripts, guiding the formation of single

[13] Three. [14] tRNA, an amino acid, and ATP. [15] Amino acid and tRNA sites would show variation while the ATP site would not.
[16] They must have sufficient "flexibility" so that more than one kind of tRNA molecule can be accommodated.

polypeptides, have a single section to be translated, whereas polycistronic transcripts have two or more. With polycistronic transcripts, the translated sections are separated by nontranslated **spacer** sections. Translated sections are often followed by another nontranslated section known as the **trailer sequence,** or **3′ noncoding sequence,** found between the final codon to be translated, that is, the terminator codon, and the 3′ end of the mRNA transcript.

Key Events in Translation

Now we are ready to consider how the ribosomes, mRNA, and amino acid–tRNA complexes interact to carry out polypeptide synthesis. The process can be divided into four stages: formation of an initiation complex, amino acid transfer and peptide bond formation, translocation, and termination.

Formation of an Initiation Complex Polypeptide synthesis begins with the formation of an **initiation complex.** At least three types of special protein molecules known as **initiation factors** and designated as IF-1, IF-2, and IF-3 facilitate the formation of this complex. The complex consists of a small ribosomal subunit, a strand of mRNA, and an initiator aminoacyl-tRNA assemblage. The energy for assembling the initiation complex comes from the breakdown of a molecule of guanosine triphosphate (GTP) into guanine diphosphate (GDP) and inorganic phosphate (P_i); the GTP carries high energy bonds and is broken down much like ATP.

The first translated codon, or **initiation codon,** of a strand of mRNA is almost always AUG although there are certain mRNA strands that initiate with GUG. Regardless of its makeup, the initiation codon always pairs with a special initiator tRNA molecule that bears the anticodon UAC and carries the amino acid methionine. The fact that the U of the anticodon 3′-UAC-5′ can pair with either the A or G at the 5′ ends of the initiator codons (5′-AUG-3′ or 5′-GUG-3′) is due to some play or **wobble** that exists when hydrogen bonds are formed between these bases. (Such wobble occurs with some frequency in codon-anticodon interactions, but with noninitiator codons, the wobble occurs at the base position located at the opposite (3′) end of the codon. Wobble is discussed further in Chapter 27.) The formation of an initiation complex is summarized in Figure 26-7.

FIGURE 26-7

Formation of an initiation complex.

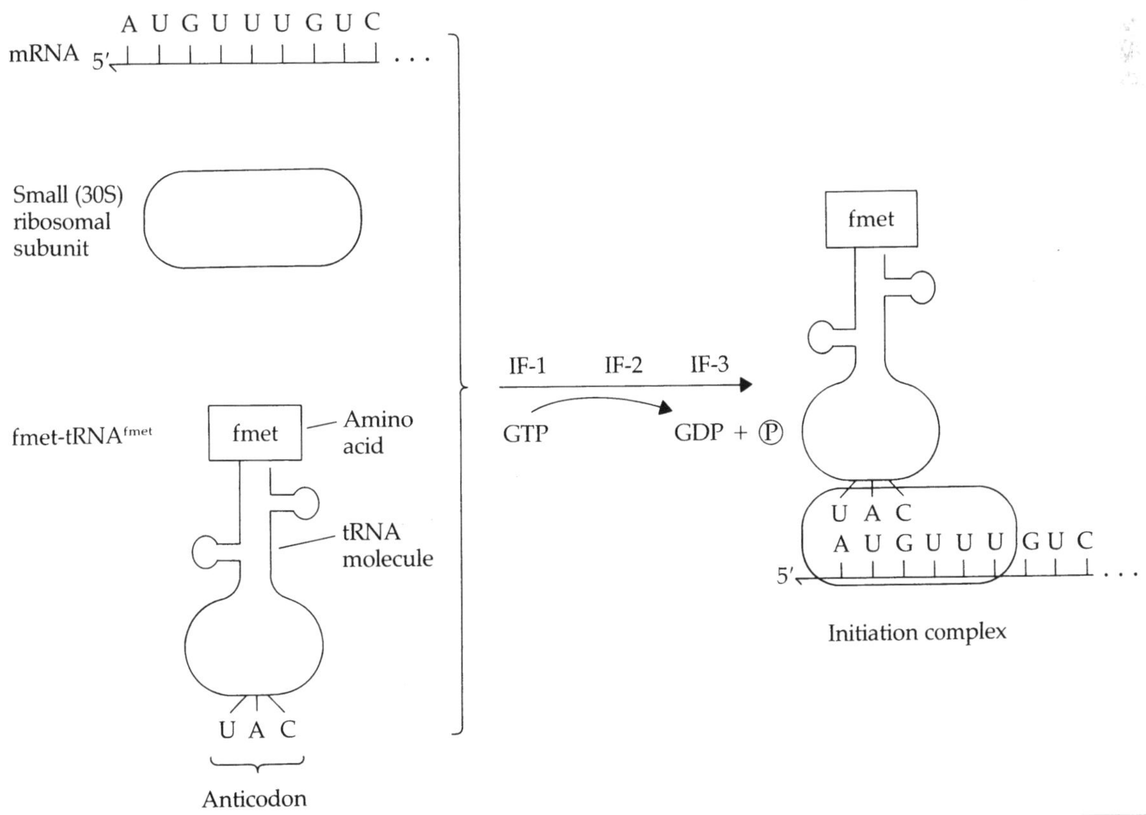

In prokaryotes (and in eukaryotic chloroplasts and mitochondria) the methionine carried by the initiator tRNA has been modified by adding a formyl group (-CHO) to the nitrogen of its amino group. This modified methionine is referred to as **N-formylmethionine** and the tRNA carrying it is designated as $tRNA^{fmet}$. The addition of the formyl group blocks the amino portion of the methionine from participating in peptide bond formation. With its amino terminal blocked, is it possible for the methionine to participate in peptide bond formation? ________[17] Explain. ________ ________________________________[18] If it is to participate in polypeptide synthesis, is there a restriction on the site or sites that can be occupied by N-formylmethionine? ________[19] Explain. ________ ________________________________[20]

There are three additional points regarding N-formylmethionine and $tRNA^{fmet}$.

1. The formyl group of the N-formylmethionine is removed from all prokaryotic polypeptides before their translation is completed. With more than half of the polypeptides, the methionine that remains is removed after translation is completed and before the polypeptides assume functional roles in the cell.

FIGURE 26-8

(a) Large (50S) ribosomal subunit joins with an initiation complex. P site is filled and A site is about to be filled by incoming amino acid–tRNA complexes. (b) Both P and A sites are filled.

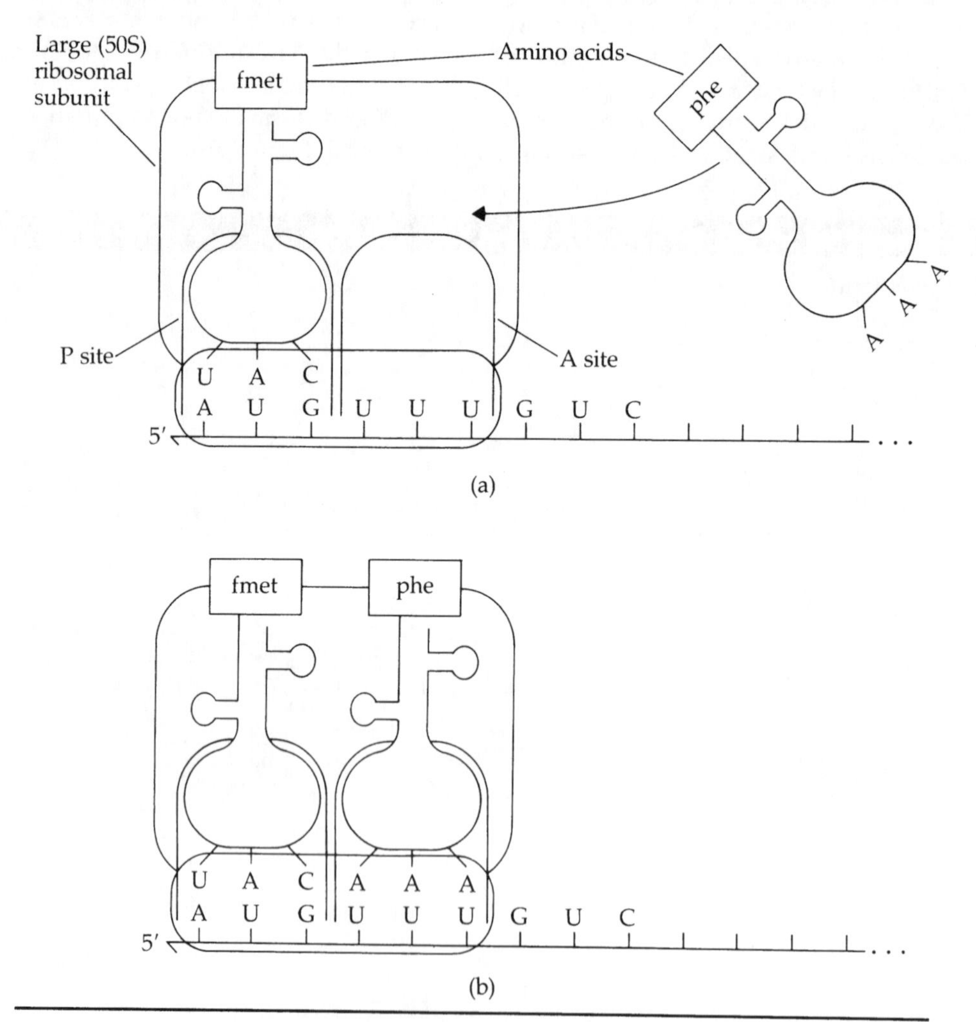

[17] Yes. [18] It can participate only by joining its carboxyl group to the amino terminal of another amino acid. [19] Yes.
[20] N-formylmethionine can occur only as the first amino acid at the N-terminal end of a polypeptide; it cannot occur at an internal position nor can it occur at the C-terminus of a polypeptide.

2. Methionine-coding AUG codons that occur within the mRNA strand do not pair with N-formylmethionine-tRNA^{fmet}. As we will consider shortly, before an amino acid-tRNA complex can pair with an internal codon, it must interact with proteins known as elongation factors. N-formylmethionine-tRNA^{fmet} cannot interact with these elongation factors and thus cannot pair with AUG or GUG codons when they occur within an mRNA strand. Internal AUG codons interact with a different type of tRNA designated as tRNA^{met} that complexes with normal methionine. Internal GUG codons pair with tRNA molecules bearing the anticodon CAC and carrying the amino acid valine.

3. In eukaryotes, the methionine carried by the initiator tRNA, designated as tRNA$_i^{met}$, is not formylated. Its role as the first amino acid–tRNA complex to be involved in polypeptide formation is assured by the fact that this is the only amino acid–tRNA complex able to interact with the initiation factors.

Formation of a Functional Ribosome Once the initiation complex is formed, it is united, using energy supplied by the breakdown of GTP, with a large ribosomal subunit to form a fully functional ribosome. The joining of the two ribosomal subunits makes it possible for both the A and P binding sites to begin accepting aminoacyl-tRNA complexes. Each of these sites is lined up with a single codon of the mRNA strand. The P site is aligned with the AUG (or GUG) initiator codon and is occupied by the initiator tRNA^{fmet}. The A site is lined up with the second mRNA codon to be translated and will become filled by the amino acid–tRNA complex carrying the complementary anticodon (see Figure 26-8a). However, before this amino acid–tRNA complex (or any other) can bind to the A site, it must interact with two protein elongation factors designated as EF-T_s and EF-T_u. Following this interaction, the binding to the A site occurs, using energy supplied by the breakdown of a molecule of GTP. The complete ribosome with its P and A sites occupied is shown in Figure 26-8b.

Amino Acid Transfer and Formation of the First Peptide Bond The next event in polypeptide synthesis involves the formation of a peptide bond between the two amino acids carried by the tRNA molecules in the P and A sites, a process catalyzed by a ribosome-bound enzyme called **peptidyl transferase.** The peptide bond forms when the amino acid carried by the tRNA molecule in the P site is transferred to the amino end of the amino acid carried by the tRNA molecule in the A site. The tRNA in the P site is now without an amino acid and it is now released from the ribosome, emptying the P site. The tRNA occupying the A site has two amino acids, that is, a dipeptide, joined to it as shown in Figure 26-9.

Translocation As soon as a peptide bond forms between the amino acids and the tRNA molecule is released from the P site, the ribosome moves relative to the mRNA molecule. This movement, known as **translocation,** requires the breakdown of one molecule of GTP and advances the ribosome one codon along the

FIGURE 26-9

A peptide bond is formed between the amino acids carried by the adjacent amino acid–tRNA complexes. The dipeptide is attached to the amino acid–tRNA complex in the A site and the tRNA molecule in the P site is released.

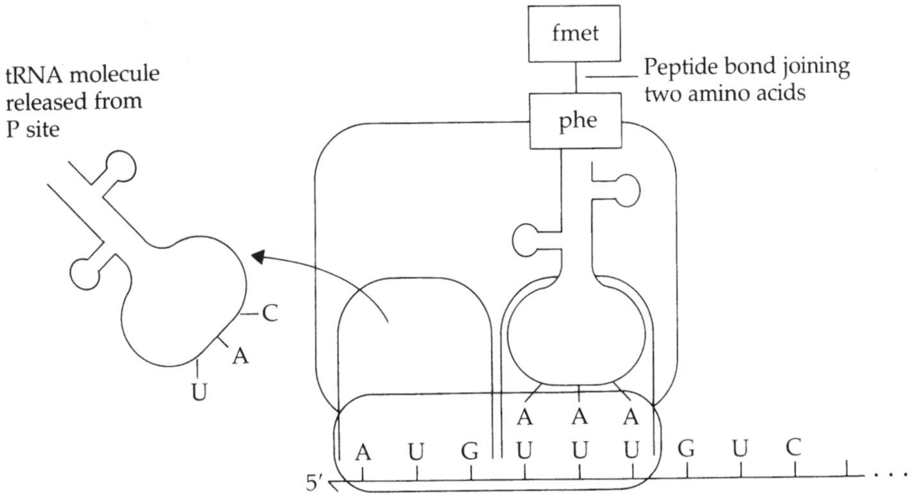

mRNA strand. Examine Figure 26-10 which shows the ribosome that we have been considering after it has undergone translocation. In what direction, relative to the mRNA strand, has the ribosome moved? _____[21] How far, in terms of ribonucleotides and in terms of codons, has the ribosome moved? __________[22] Prior to translocation, the dipeptide-tRNA complex occupied the A site. In what site is this complex found following translocation? __________[23] Note that the A site is now empty. Which codon is now lined up with the A site? __________[24] What do you think will happen next? ______________________________[25] What determines the particular amino acid–tRNA complex that occupies the A site? ______________________________[26] Where does this amino acid–tRNA complex come from? ______________________________[27]

Figure 26-11 shows the translocated ribosome with the P site filled by the dipeptide-tRNA complex and the A site now filled by an amino acid–tRNA complex with an anticodon that complements the third codon of the mRNA strand. The stage is now set for the next major event in the synthesis of our polypeptide. What do you think that event will be? ______________________________[28] Following this, how many amino acids have been joined together? __________[29] Which ribosome site now holds the tRNA molecule carrying the tripeptide? _____[30] Once the peptide bond is established to form the tripeptide, what do you think happens to the tRNA molecule in the P site? ______________________________[31] What will happen to this tRNA after its release? ______________________________[32] What happens to the ribosome when the tRNA molecule is released from the P site? ______________________________[33] In which site will the tripeptide-tRNA complex be found following this translocation? __________[34]

List the four major events that next take place to continue the synthesis of our peptide.[35]

1. ______________________________
2. ______________________________

FIGURE 26-10

Translocation advances the ribosome one codon along the mRNA strand and positions the dipeptide-tRNA complex in the P site.

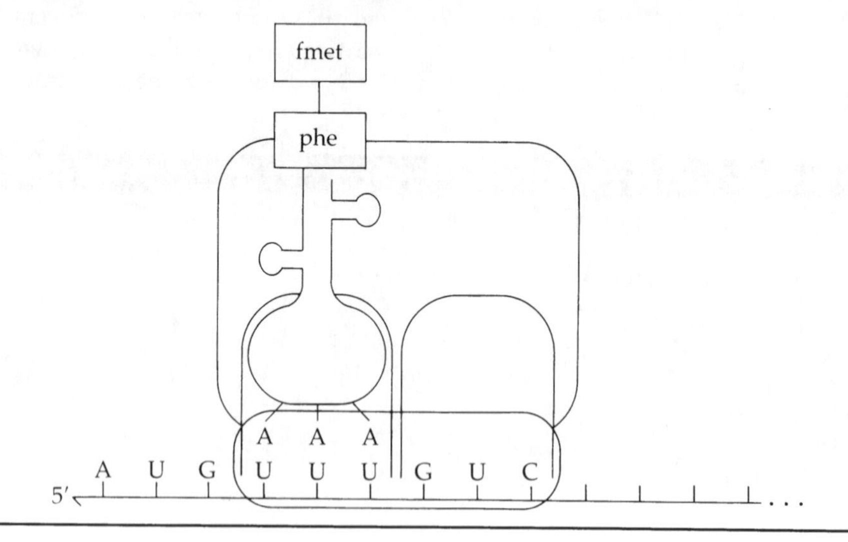

[21] Along the mRNA strand in the 5′-to-3′ direction. [22] Three nucleotides or one codon. [23] The P site. [24] The third codon, GUC. [25] The A site gets filled by an amino acid–tRNA complex with the anticodon CAG. [26] Codon-anticodon interaction; the amino acid–tRNA complex must have an anticodon that complements the codon opposite the A site. [27] Supplies of all the amino acid–tRNA complexes are present in the cell's cytoplasm. [28] The transfer and peptide bonding of the dipeptide carried by the tRNA in the P site to the amino acid carried by the tRNA in the A site. [29] Three, making up a tripeptide. [30] The A site. [31] It is released. [32] It returns to the cytoplasmic pool where it is free to again interact with the appropriate aminoacyl-tRNA synthetase enzyme and to link with another molecule of its amino acid. [33] It undergoes translocation, moving one codon along the mRNA molecule. [34] The P site.
[35] 1. The now vacant A site becomes occupied by the appropriate amino acid–tRNA complex. 2. Peptide bonds form. 3. The tRNA in the P site is released. 4. Translocation occurs.

FIGURE 26-11

The A site is now filled with an amino acid–tRNA complex with an anticodon that complements the third codon of the mRNA strand.

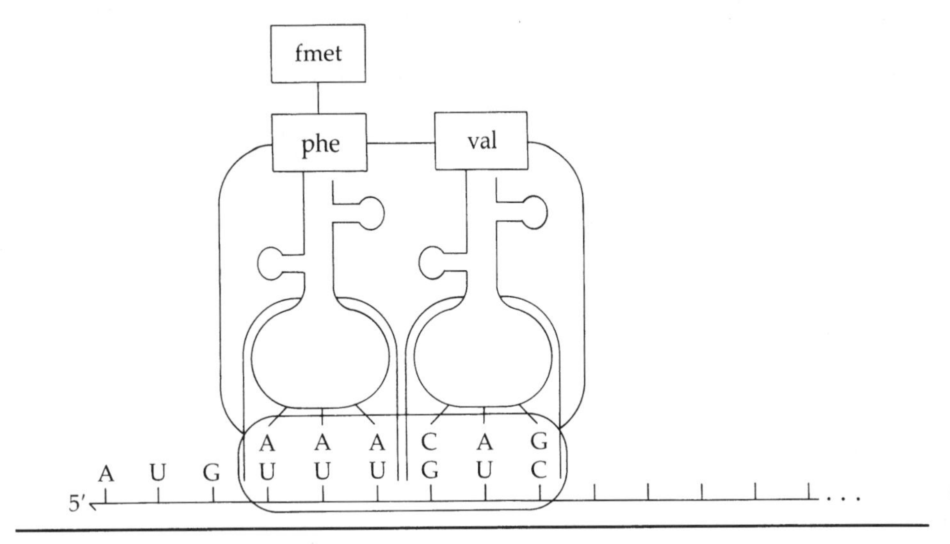

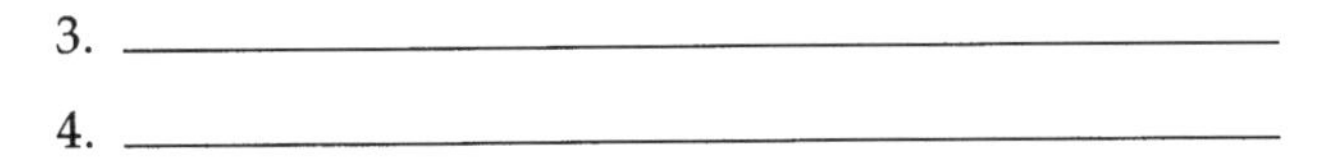
3.
4.

By now you can see that there is a pattern to the events occurring here—a pattern repeated over and over until the ribosome has moved the length of the mRNA strand and encounters a terminator, or stop, codon.

Termination of Polypeptide Formation Polypeptide synthesis terminates when a translocation occurs that lines up the A site of a ribosome with a **terminator codon,** either UAA, UAG, or UGA. These codons are not recognized by any of the normal aminoacyl-tRNA complexes, and when one becomes lined up with the A site, it interacts with a protein **release factor** which apparently blocks off the A site. Once this happens, three events occur: (1) the polypeptide is cleaved from the tRNA molecule occupying the P site, (2) the tRNA molecule in the P site is released, and (3) the ribosome is released from the mRNA strand and then dissociates into large and small subunits. The polypeptide goes on to take up its functional role in the cell and the ribosomal subunits are available to serve again as components of functional ribosomes.

Polysomes

Our look at polypeptide synthesis has focused on the activities of a single ribosome as it translates the message encoded in an mRNA strand. In actuality, several ribosomes can simultaneously translate a strand of mRNA. After a ribosome has moved 25 or so codons along an mRNA strand, the 5′ end of the mRNA can again participate in the formation of an initiation complex. As a consequence, a strand of mRNA may have a number of ribosomes spaced out along it at approximately equal intervals. The assemblage made up of a strand of mRNA, along with the ribosomes that are translating it and the growing polypeptides they carry, is known as a **polyribosome** or **polysome.** In moving closer to the 3′ end of an mRNA strand that is read simultaneously by a number of ribosomes, what would you expect regarding the length of the growing polypeptides associated with these ribosomes? ________
________[36]

Linkage of Transcription and Translation in Prokaryotes

In prokaryotes, the processes of transcription and translation may be coupled. As a strand of mRNA is being synthesized (in a 5′-to-3′ direction), its 5′ end becomes free of the DNA template strand. This free end can then join with a small ribosome subunit and an initiator aminoacyl-tRNA to form an initiation complex. When this initiation complex joins with a large ribosome subunit, translation of the mRNA strand begins. So before the synthesis of an mRNA molecule is completed, ribosomes may have already started translating at its 5′ end. Can you see an advantage to this

[36] They would be successively longer.

coupling of the processes of transcription and translation? ________[37] Explain. ____________________

____________________[38] Can this linkage of transcription and translation occur in eukaryotic cells? ________[39] Explain. ____________________

____________________[40] In some instances, prokaryotic mRNA molecules that are experiencing translation while still being transcribed have been found to experience simultaneous degradation in a 5′-to-3′ direction. Could further translation be initiated once this degradation starts? ________[41] Explain. ____________________

____________________[42]

SUMMARY

Genetic information is stored in sequences of bases in DNA. Transcription, catalyzed by RNA polymerase and using one duplex strand as a template, forms mRNA, tRNA, and rRNA. Successful initiation and termination of transcription depend on nucleotide promoter and terminator sequences associated with the unit of transcription. Transcription begins when RNA polymerase (core enzyme + sigma factor) recognizes and binds with the promoter, initiating the separation of the two duplex strands. The RNA polymerase carries out a stepwise pairing of nucleotides in the transcription unit with complementary ribonucleotides and catalyzes the formation of phosphodiester bonds joining the ribonucleotides, with the strand growing in a 5′-to-3′ direction. Terminator signals at the end of the unit of transcription cause the RNA polymerase to stop adding ribonucleotides and to dissociate from the DNA template, releasing the RNA transcript. Complexing of the rho factor with RNA polymerase may be required for recognition of the terminator. Before RNA molecules assume their functional roles, their primary transcripts usually undergo some post-transcriptional modification to produce mature functional molecules. These alterations may include the removal of some terminal and internal sections and the addition of others. A ribonuclease enzyme and an RNA-splicing enzyme play prominent roles in this modification.

Polypeptide synthesis begins with a mature mRNA molecule carrying the instructions for assembling a particular polypeptide and ends with a completed polypeptide—a covalently bonded series of amino acids. Before an amino acid can be incorporated it must be activated by reacting with ATP so that it has sufficient energy to form a peptide bond. Then it must be joined with an appropriate tRNA molecule so that it can be inserted at the correct point into a growing polypeptide. Both of these reactions are catalyzed by aminoacyl-tRNA synthetase. A ribosome serves as the processing unit for polypeptide synthesis by bringing together the mRNA strand and an array of aminoacyl-tRNA complexes. The ribosome pairs mRNA codons with complementary anticodons of aminoacyl-tRNA complexes. This enables the amino acids carried by the aminoacyl-tRNA complexes to become covalently bonded into a growing polypeptide chain. Synthesis terminates when a translocation aligns the ribosome with a terminator codon. This releases the completed polypeptide from the ribosome and releases the ribosome from the mRNA strand. In prokaryotes, transcription and translation may be coupled; as a strand of mRNA is synthesized, its 5′ end may form an initiation complex and translation begins. This makes it possible for prokaryotic forms to rapidly produce the gene products of activated genes.

[37] Yes. [38] This makes it possible for prokaryotic forms to promptly produce the gene products of activated genes. [39] No. [40] The nuclear transcription site is separated from the cytoplasmic translation site by the nuclear membrane; only after the mRNA moves through the nuclear membrane to the cytoplasm can there be translation. [41] No. [42] Once the degradation is initiated, the Shine-Dalgarno sequence is no longer there to guide the formation of the initiation complex.

PROBLEM SET

26-1. **a.** Define what is meant by a unit of transcription.

b. What is the relationship between the sense strand and the antisense strand of a unit of transcription?

c. What is the function of the sense strand of a unit of transcription?

d. What is the function of the antisense strand of a unit of transcription?

26-2. **a.** Where, relative to a transcriptional unit, is its promoter sequence located?

b. What function is served by a promoter sequence?

c. Is the promoter sequence transcribed?

d. The promoter region of a "wild-type" eukaryotic gene includes the sequence TATAAAAAA. *In vitro* mutagenic procedures make it possible to substitute a G nucleotide for the interior T nucleotide so that this section of the promoter reads TAGAAAAAA. A study comparing the transcription frequencies for the gene with the wild-type form of the promoter and the gene with the mutant promoter is undertaken in separate but identical *in vitro* RNA synthesizing setups. The results indicate that the transcription rate is significantly reduced for the gene with the mutant promoter. Based on this outcome, what conclusion can you draw regarding the functional significance of this portion of the promoter region?

26-3. One point of uncertainty that existed in 1960, as scientists began to gather information about polypeptide synthesis, related to the way in which the amino acid–tRNA complexes interacted with the mRNA strand that was guiding the process. It was unclear whether it was the tRNA molecule or the amino acid portion of the amino acid–tRNA complex that recognized the appropriate codon within the mRNA molecule. F. Chapeville and several colleagues reported in 1962 that they were able to prepare quantities of the $tRNA^{cys}$ molecule that carried the amino acid cysteine labeled with radioactive ^{14}C. Through the use of an appropriate catalyst, these researchers were able to modify the ^{14}C cys-$tRNA^{cys}$ complexes by converting the cysteine attached to the $tRNA^{cys}$ molecule into the amino acid ^{14}C alanine to produce ^{14}C ala-$tRNA^{cys}$ complexes. Each of these two types of amino acid–tRNA complexes was then supplied to a separate *in vitro* polypeptide synthesizing system containing quantities of the same artificial mRNA strand that coded for a polypeptide containing cysteine but not alanine.

a. Based on what you know about polypeptide formation, what would you predict about the makeup of the polypeptide synthesized in the *in vitro* setup supplied with the ^{14}C cys-$tRNA^{cys}$ and why?

b. What results would you predict for the *in vitro* setup supplied with the ^{14}C ala-$tRNA^{cys}$ and why?

c. Assume that a third *in vitro* polypeptide synthesizing setup, identical to those described in 26-3a and b, is supplied with quantities of a $tRNA^{ala}$ molecule complexed with ^{14}C-labeled alanine. What results would you predict for this setup and why?

d. What purpose does the third experiment (26-3c) serve?

e. Why do you think the cysteine and alanine amino acids used in these experiments are labeled?

f. Does this series of experiments answer the question about whether it is the tRNA molecule or the amino acid portion of the amino acid–tRNA

PROBLEM SET

complex that recognizes the codon of the mRNA? Explain. (Clue: think about how the outcome of these experiments would differ if the mRNA had interacted with the amino acid part of the amino acid–tRNA complexes.)

26-4. Most of the mRNA synthesized in bacteria has a half life of one or two minutes.

- **a.** Speculate on an advantage that this short life span provides to the bacteria.
- **b.** Is there any disadvantage associated with the short life span of this mRNA?
- **c.** What prediction would you make about the life spans of rRNA and tRNA in bacteria?

26-5. A professor is overheard making the following statement: ". . . tRNA molecules, because of their small size, must exhibit very little in the way of specificity." Do you agree or disagree with this statement? Explain.

26-6. A number of antibiotics have been identified that block the process of polypeptide synthesis. Several of these are listed along with the component of the polypeptide-synthesizing system that they affect. For each compound, indicate the stage of polypeptide synthesis (that is, initiation-complex formation, chain initiation, chain elongation, or chain termination) that is disrupted or blocked by the antibiotic.

- **a.** Sparsomycin inhibits the action of the enzyme peptidyl transferase.
- **b.** Puromycin acts like an amino acid–tRNA complex. It can occupy the A site and receive the growing amino acid chain from a tRNA molecule, but it lacks a terminal carboxyl group.
- **c.** Tetracycline binds to the small ribosome subunit and blocks the A site.
- **d.** Aurintricarboxylic acid prevents the joining of mRNA molecules to the ribosomes.

26-7. Do all of the aminoacyl-tRNA complexes that participate in the formation of a polypeptide do so by first binding at the A site of the ribosome? Explain.

26-8. The primary transcripts of mRNA molecules are often much longer than the mature mRNA molecules that guide polypeptide synthesis in eukaryotic cells. For example, the initial mRNA transcript for a beta-globin gene contains about 1500 nucleotides while the mature mRNA molecule that directs the synthesis of this polypeptide contains about 800 nucleotides. In contrast, the primary transcripts of prokaryotic mRNA molecules and the mRNA molecules that direct polypeptide synthesis are just about the same length. Account for this difference between eukaryotic and prokaryotic mRNAs.

26-9. In addition to the difference cited in problem 26-8, describe three additional significant ways in which the synthesis, or structural or functional features of eukaryotic mRNA differ from those of prokaryotic mRNA.

26-10.
- **a.** What effect would a mutation in an intron section of a gene have on the expression of the gene? Explain.
- **b.** Identify a circumstance under which a mutation in an intron could significantly alter the expression of a gene.

26-11. Primary transcripts of prokaryotic tRNA and rRNA undergo post-transcriptional modification which includes the removal of segments of nucleotides. How do these removed segments differ from the introns removed from primary transcripts of some eukaryotic RNA?

PROBLEM SET

26-12. Assume that the transcription and translation of a polypeptide that is routinely synthesized in reptiles and secreted by a certain type of glandular cell has been studied.

a. The polypeptide's unit of transcription consists of 5318 base pairs. How many bases will occur in the primary mRNA transcript produced from this transcription unit?

b. During its post-transcriptional modification, the length of this mRNA transcript is increased by several hundred nucleotides. What is added to contribute to this increase in length?

c. At the end of its post-transcriptional modification, the mRNA, now designated as mature mRNA, consists of 1222 nucleotides. What has been responsible for decreasing the length of the transcript?

d. The polypeptide consists of 262 amino acids. How many of the nucleotides present in the mature mRNA strand are required to code for these amino acids?

e. Identify the sections of the mature mRNA strand that make up the remaining, that is, noncoding, portions of the mature mRNA.

26-13. Polycistronic mRNA strands carry information for the synthesis of several polypeptides. Often these polypeptides are involved in the same metabolic pathway. For example, a polycistronic mRNA in *E. coli* codes for ten different enzymes, all of which play a role in the synthesis of the amino acid histidine. What is the advantage to this arrangement in which all the genes involved in a metabolic pathway are transcribed together?

26-14. Assume you are able to observe RNA synthesis at numerous points along the length of the DNA duplex molecule present in a eukaryotic chromosome.

a. In what direction is each RNA strand being synthesized (5′ to 3′ or 3′ to 5′)? Explain.

b. Would you expect to see strands of RNA synthesized in one or in both directions relative to the DNA strands? Explain.

26-15. Several DNA or mRNA sequences that play key roles in polypeptide synthesis are listed. For each, indicate whether it occurs in DNA or RNA, in prokaryotes or in eukaryotes, indicate the region where it is found (for example, promoter, terminator, leader, and so forth) and the role that it is believed to play.

a. Inverted-repeat sequence followed by a series of thymine nucleotides.

b. -35 region.

c. Pribnow box (-10 region).

d. Cap.

e. Goldbert-Hogness box (also known as the TATA box) in the -30 region.

f. Ribosome-binding sequence (also known as the Shine-Dalgarno sequence).

26-16. a. Compare the consequences of an error made in (1) the synthesis of a molecule of a polypeptide, (2) the transcription of a molecule of mRNA, and (3) the replication of the sense strand of a transcriptional unit that gives rise to a molecule of mRNA.

b. Which of these three types of errors is potentially heritable?

26-17. Determine whether each of the following statements is true or false. If a statement is false, explain why.

a. The first amino acid to be incorporated into polypeptides synthesized in *E. coli* is a modified form of methionine known as formylmethionine.

PROBLEM SET

b. The amino acid found at the N terminus of functional polypeptides in *E. coli* is always the modified form of methionine, formylmethionine.
c. The initiation codon in *E. coli* is always AUG.
d. All AUG codons, regardless of their position in the coding section of an mRNA strand, always code for exactly the same amino acid.
e. It is believed that methionine is always the first amino acid to be incorporated into polypeptides synthesized in eukaryotes.

26-18. The following protein molecules participate in either transcription or translation in *E. coli*. For each, identify the process in which it participates and state the role it plays in that process.
a. Sigma factor.
b. Peptidyl transferase.
c. Initiation factors.
d. Rho factor.
e. Elongation factors.
f. Release factor.

26-19. Determine whether each of the following statements is true or false. If a statement is false, explain why.
a. A single type of aminoacyl-tRNA synthetase catalyzes the joining of all the various types of tRNA molecules to their respective amino acids.
b. An amino acid is referred to as activated if it has already complexed with the appropriate tRNA molecule.
c. A peptide bond joins the amino group of one amino acid to the amino group of another amino acid.
d. The primary structure of a polypeptide consists of a linear series of amino acids joined by covalent peptide bonds.
e. Protein molecules with three orders of structure, that is, primary, secondary, and tertiary, consist of single polypeptides, whereas proteins exhibiting quaternary structure are always made up of two or more polypeptides.

26-20. The joining of amino acids to the growing polypeptide during polypeptide synthesis involves a cycle of events that gets repeated with each additional amino acid. Identify the key events involved in this cycle.

26-21. Determine whether each of the following statements is true or false. If a statement is false, explain why.
a. The components of the initiation complex in *E. coli* are an mRNA strand, a small (30S) ribosome subunit, and a tRNA complex that carries N-formylmethionine.
b. The number of kinds of tRNA molecules found in prokaryotes and eukaryotes is equal to the number of different types of amino acids.
c. Polypeptide synthesis terminates when a terminator codon becomes aligned with the A site of the ribosome and interacts with a protein release factor.
d. A molecule of prokaryotic mRNA with a number of ribosomes bound to it is known as a polycistron.
e. The breakdown of a molecule of GTP to GDP and P_i is necessary each time an incoming aminoacyl-tRNA molecule binds to the A site of a ribosome.

CHAPTER

The Genetic Code

INTRODUCTION

The linear sequence of nucleotides in a DNA segment (and in the messenger RNA subsequently made from it) determines the sequence of amino acid building blocks that make up the DNA's polypeptide product. The rules that govern this transfer of information from nucleic acid to polypeptide and the system of signals that conveys the information make up the **genetic code.** Breaking the code involved many geneticists working over a period of nearly a decade. Two major tasks were involved in deciphering the code. First, the number of nucleotides making up the code word, or codon, for a single amino acid had to be identified. Then the kinds and sequence of nucleotides making up the codon or codons for each amino acid had to be determined.

THE NUMBER OF NUCLEOTIDES IN A CODON

Polypeptide molecules are made up of 20 kinds of amino acid building blocks. As you know, nucleic acid molecules are made of just four kinds of nucleotides. What is the greatest number of codons possible in a system involving code words of one nucleotide? ____[1] What is the greatest number of amino acids that this singlet type of system could code for? ________[2] What is the maximum number of code words possible if a codon consisted of two nucleotides? ________[3] What is the greatest number of amino acids that a doublet codon system could code for? ________[4] What is the maximum number of code words that could exist with a triplet codon system? ________[5] What is the simplest system adequate to code for the 20 biologically important amino acids? ________[6] What assumption do you suspect that most molecular geneticists made about the number of nucleotides making up a code word as they began the work of breaking the code? ______________________ ________________[7]

BREAKING THE CODE

The Use of RNA

Determining the makeup of the code words was done using RNA rather than DNA. During the 1960s when

[1] $4^1 = 4$. [2] 4. [3] $4^2 = 16$. [4] 16. [5] $4^3 = 64$. [6] Triplet code.
[7] Most assumed that the code words were triplets.

this work was carried out, an enzyme was available that allowed geneticists to assemble ribonucleotides into molecules of RNA in the laboratory. At that time, no such enzyme had yet been isolated which would make possible a similar procedure for DNA. Being restricted to defining the relationship between amino acids and their RNA code words rather than their DNA code words posed no problem, however. Remember that in the cell, the information encoded in DNA's nucleotide sequence is transferred to RNA which, in turn, serves to guide polypeptide synthesis. The DNA-to-RNA transfer occurs as the RNA is synthesized as a complement to the DNA template strand. Because of this complementation, once the makeup of an RNA code word is known, the makeup of the DNA code word guiding its synthesis can readily be determined.

Basic Experimental Approach

The procedures used in many experiments providing information about the genetic code were basically the same and involved four steps.

1. Carry out *in vitro* synthesis of RNA. This was done by combining ribonucleotides in diphosphorylated form with the enzyme polynucleotide phosphorylase which catalyzed the joining of these nucleotides into polyribonucleotide strands. Since researchers controlled the kinds of ribonucleotides and their relative concentrations, deductions could be made about the way in which these building blocks assemble into RNA molecules and, in turn, about the makeup of the triplet code words within these artificial molecules.
2. Supply the RNA to an *in vitro* polypeptide-synthesizing system where it acts like mRNA and guides the formation of polypeptide molecules. Such a system contains ribosomes, a mixture of all of the kinds of aminoacyl-tRNA molecules, all of the enzymes and cofactors necessary for translation, and some GTP as an energy source; all of these raw materials are obtained from *E. coli*.
3. Collect the polypeptide molecules that are formed and analyze their amino acid composition.
4. Relate the amino acid makeup of the polypeptide molecules to the nucleotide triplets present in the RNA. The DNA triplets complementing the RNA triplets can then be deduced.

Major Types of Experiments Used

Next we look at some of the major kinds of experiments used to elucidate the genetic code and their contributions to the breaking of the code. These experiments will show that, as more sophisticated biochemical techniques were developed, geneticists were able to more precisely define the composition, and eventually the length, of the RNA that was formed *in vitro*. With this more precisely tailored RNA, the deductions made about the genetic code became more specific and our knowledge of the code more extensive. In considering each experiment, we will begin with a description of the type of RNA that was synthesized and supplied to the *in vitro* polypeptide-synthesizing system.

Monopolymers of RNA This RNA is made with a single kind of nucleotide. For example, when uracil ribonucleotides are supplied to an RNA-synthesizing system, the monopolymer polyuracil is formed. When this, in turn, is supplied to the polypeptide-synthesizing system, a polypeptide made solely of phenylalanine is formed. Based on this outcome, what would you conclude about the RNA code word for the amino acid phenylalanine? ________[8] What is the complementary DNA triplet? ________[9] Is there a limit to the usefulness of this type of procedure for elucidating the genetic code? ________[10] Explain. ________________________________[11] Using this procedure, AAA and CCC were found to code for lysine and proline, respectively. Strands of poly-G could not be used to guide polypeptide synthesis since they joined to each other by hydrogen bonding.

Random Copolymers of RNA This type of RNA is made from two kinds of ribonucleotides supplied in known proportions. The RNA molecules produced contain both kinds of nucleotides joined randomly in proportion to the concentration in which they were supplied to the RNA-synthesizing system. The basic approach is to supply the same two ribonucleotides to the RNA-synthesizing system in a series of different concentrations. For example, A and C ribonucleotides are supplied in a 1-to-1 ratio in one study, in a 1-to-5 ratio in another, and in a 5-to-1 ratio in a third study. Polypeptides translated from these RNA molecules were analyzed, and the differences in their amino acid composition were related to differences in the relative amounts of different nucleotides present in the RNA.

[8] UUU. [9] AAA. [10] Yes. [11] No more than four code words, UUU, AAA, CCC, and GGG, could be identified using this approach.

To illustrate this, first consider a study in which A and C ribonucleotides are supplied in a 1-to-1 ratio. From these concentrations, we can predict the kinds of triplets that would be in the RNA and their expected frequencies. What is the probability of an A nucleotide occurring in a triplet? ________[12] What is the probability of a C nucleotide occurring in a triplet? ____[13] Complete the following table to give the eight kinds of triplets found in the RNA and the expected frequency of each[14].

Triplet containing	Possible sequences	Expected frequencies
3A	AAA	1/2 × 1/2 × 1/2 = 1/8
2A, 1C	AAC	________
	ACA	________
	CAA	________
1A, 2C	________	________
	________	________
	________	________
3C	________	________

When this RNA was supplied to the polypeptide-synthesizing system, six amino acids—asparagine, glutamine, histidine, lysine, proline, and threonine—were incorporated into the polypeptides formed. The first four of these each made up about 1/8 of the polypeptides and each of the last two made up about 1/4. Since there are eight kinds of triplets, we could speculate that each of the four amino acids occurring in the polypeptide with a frequency of 1/8 is coded for by one of the eight triplets and that each of the two amino acids occurring with a frequency of 1/4 is coded for by two triplets.

These results, of course, do not tell us which triplet codes for which amino acid. Some clues for getting closer to that information can be obtained by repeating the experiment using the same ribonucleotides in different relative concentrations. For example, A and C could be supplied in a 1-to-5 ratio, respectively, so that 1/6 = 0.1667, or 0.1667 × 100 = 16.67%, of the mixture is A and 5/6 = 0.8333, or 0.8333 × 100 = 83.33%, of the mixture is C. Would the kinds of triplets within the RNA made from this mixture be identical to those produced when the two nucleotides were in a 1-to-1 ratio? ________[15] What would differ? ________[16] What is the probability of an A nucleotide being incorporated into a triplet? ________[17] of a C nucleotide being incorporated? ________[18] Determine the expected percentage for each of the triplets listed in the following table.

Triplet containing	Possible sequences	Expected frequencies
3A	AAA	________[19]
2A, 1C	AAC	________[20]
	ACA	________[21]
	CAA	________[22]
1A, 2C	ACC	________[23]
	CAC	________[24]
	CCA	________[25]
3C	CCC	________[26]

When this type of RNA was supplied to the *in vitro* polypeptide-synthesizing system, the resultant polypeptide contained the same amino acids present when the A to C ratio was 1-to-1, but in different concentrations, as follows: proline, 64.5%; histidine, 14.8%; threonine, 13.5%; asparagine, 3.2%; glutamine, 3.2%; lysine, 0.6%.

Now let us see if we can correlate the frequencies of these amino acids in the polypeptide with the frequencies of the various codons in the RNA and make some tentative assignments of codons to amino acids. Keep in mind that there may be discrepancies between the observed and expected frequencies, and that an assignment based on the incorporation of a small amount of an amino acid is less reliable than one based upon the incorporation of a larger amount. As discussed earlier in the chapter, from experiments with RNA made up of a single type of nucleotide, the code words CCC and AAA were assigned to proline and lysine, respectively. The results here further support these assignments: AAA is the least common triplet (0.46%) and lysine is the least frequent amino acid

[12] 1/2. [13] 1/2. [14] The 1A, 2C combinations are ACC, CAC, and CCA. Each of the eight triplets has an expected frequency of 1/2 × 1/2 × 1/2 = 1/8. [15] Yes. [16] Their frequencies. [17] 1/6. [18] 5/6. [19] 1/6 × 1/6 × 1/6 = 1/216 = 0.0046, 0.0046 × 100 = 0.46%. [20] 5/6 × 1/6 × 1/6 = 5/216 = 0.0231, 0.0231 × 100 = 2.31%. [21] 2.31%. [22] 2.31%. [23] 5/6 × 5/6 × 1/6 = 25/216 = 0.1157, 0.1157 × 100 = 11.57%. [24] 11.57%. [25] 11.57%. [26] 5/6 × 5/6 × 5/6 = 125/216 = 0.5787, 0.5787 × 100 = 57.87%.

(0.6%), while CCC is the most frequent triplet (57.8%) and proline is the most common amino acid (64.5%). Next consider the three codons containing 2A, 1C and the three others containing 1A, 2C. Compare the relative frequencies of codons with 2A, 1C and codons with 1A, 2C. What do you notice? ________________________[27] Now compare the frequencies of asparagine and glutamine incorporation relative to the frequencies of histidine and threonine. What do you notice? ________________________[28] What is the relationship between the frequency with which a codon occurs in the artificial RNA and the frequency with which the amino acid that it codes for occurs in the polypeptide? ________________________[29] What tentative conclusion would you make regarding the type of codon coding for asparagine and for glutamine? ________[30] What tentative conclusion would you make regarding the type of codon coding for histidine and for threonine? ________[31]

At this point we have established tentative assignments for six codons. One codon of the 1A, 2C type and one of the 2A, 1C type remain to be assigned. Go back and calculate the difference between the percentage of proline incorporated into the polypeptide and the percentage value for the CCC codon. ________[32] How could this difference be explained? ________________________[33] What type of codon might this be? ________[34] The remaining unassigned codon is of the 2A, 1C type. Which of our amino acids might this codon code for? ________[35]

Now consider the outcome of a third experiment in which the relative amounts of A and C are reversed from those in the experiment we have just considered, that is, the A and C are supplied in a 5-to-1 concentration ratio. Using RNA synthesized at this ratio, what would you expect regarding the relative frequencies of polypeptide incorporation for proline and lysine? ________________________[36] What would you expect about the relative amounts of incorporation of histidine and threonine versus the incorporation of asparagine and glutamine? ________________________[37] The outcome of this 5A-to-1C experiment in terms of the percentage of amino acids incorporated into the polypeptide is as follows: proline, 3.7%; histidine, 3.2%; threonine, 13.9%; asparagine, 12.8%; glutamine, 12.8%; lysine, 53.5%.

Compare these results with the predictions you just made about the expected frequencies of amino acid incorporation. Are there any surprises? ________[38] Explain. ________________________[39] How can this "surprise" be accounted for? (Clue: Remember that in assigning codons following the 1A-to-5C experiment, there was a codon of the 2A, 1C type for which a tentative assignment was a toss up: it may have coded for histidine or for threonine.) ________________________[40] To which of these two amino acids, histidine or threonine, might this codon now be tentatively assigned? ________[41]

Experiments of the type we have just considered using random copolymers have a major drawback—they cannot give conclusive evidence for linking a specific codon to a particular amino acid. For example, even if all the evidence supports the idea that the code word for glutamine is of the 2A, 1C type, we have no way of telling whether that codon is AAC, ACA, or CAA. In short, this type of experiment can give information about the base composition of a codon but not the base sequence.

Regular Copolymers of RNA This RNA is made with two kinds of nucleotides bonded together to form a regular copolymer, where the two types of nucleotides alternate with each other. For example, C and A ribonucleotides supplied to the appropriate RNA-synthesizing system could be made to produce the regular copolymer . . . ACACACACACAC. . . . Supplying this type of RNA to the polypeptide-synthesizing system produces a polypeptide (actually a copeptide) which consists of alternating threonine and histidine nucleotides: . . .-thr-his-thr-his-thr-his-. . . . How many trip-

[27] The relative frequences of 1A, 2C codons are five times greater than those of the 2A, 1C codons. [28] The relative amounts of histidine and threonine are a little more than four times those for asparagine and glutamine. [29] There is a direct correlation between the expected frequency of a codon and the relative amount of the amino acid it codes for. [30] 2A, 1C. [31] 1A, 2C.
[32] The difference is about 64.5% − 57.9% = 6.6%. [33] The difference suggests than another proline codon exists. [34] A 1A, 2C type.
[35] Either threonine or histidine—but the experiment results make it impossible to suggest one or the other.
[36] Since the frequency of the AAA codon is relatively high, the incorporation of lysine would be high, and since the frequency of the CCC codon is relatively low, the incorporation of proline would be low. [37] Histidine and threonine incorporation frequencies would be relatively low and those for asparagine and glutamine would be relative high. [38] Yes. [39] The predicted amount of histidine and threonine were relatively low and these results show that while the concentration of histidine is relatively low, that for threonine is relatively high. [40] The expected frequency of the 2A, 1C codon was relatively low with the 1A-to-5C concentration ratio and relatively high with the 5A-to-1C ratio; if this codon does in fact code for threonine, its relative abundance in the 5A-to-1C RNA would result in a higher incorporation of threonine. [41] Threonine.

let sequences occur in the regular RNA copolymer that guides the formation of this copeptide? ________[42] What are these triplet sequences? ________[43] Can you tell from the outcome of *this* study which of these triplets codes for threonine? ________[44] for histidine? ________[45]

Think about the results gathered in the previous experiments involving random copolymers that supported the idea that both histidine and threonine have a codon of the 1A, 2C type (that is, CCA, CAC, or ACC). We also speculated that threonine might be coded for by an additional codon of the 2A, 1C type (that is, CAA, ACA, or AAC). To how many codons does the outcome of the present study allow us to narrow this list of six codons? ________[46] What are those codons? ________[47] How many of these can actually code for histidine? ________[48] What codon could be tentatively assigned to histidine? ________[49] What must the other codon code for? ________[50]

Regular Tri- and Tetrapolymers of RNA This type of RNA is made from two or three kinds of ribonucleotides joined to form regular tri- and tetrapolymers. For example, A and C ribonucleotides can be made to combine into RNA with the sequence . . . ACCACCACCACCACC. . . . Assuming that the reading of this RNA can start at different points along the strand, how many kinds of triplets occur within this sequence? ________[51] Identify these triplets. ________[52] This RNA translates into three kinds of monopeptides: polythreonine, polyproline, and polyhistidine. If you start the reading of this RNA with any of the A bases, is there any variation in the codon that is encountered as you move along the strand? ________[53] if you start with a C immediately to the right of an A? ________[54] if you start with a C immediately to the left of an A? ________[55] Why are three different kinds of monopeptides formed rather than one kind of polypeptide containing these three amino acids? ________________[56] Note that in these *in vitro* polypeptide synthesis setups, magnesium is supplied in very high concentration, and because of that, translation can be initiated at any point along the RNA strand and reading occurs in all possible frames.

This experiment provides further support for a conclusion reached in the studies discussed earlier involving RNA formed from a mixture of A and C in varying concentration ratios. That conclusion was that threonine, proline, and histidine were each coded for by a triplet of the 2C, 1A type. Also earlier, in working with the regular copolymer . . . ACACACACAC . . . , evidence suggested that CAC coded for histidine. In light of that information, which of the three triplets that we are now concerned with, ACC, CCA, or CAC, can be assigned to histidine? ________[57] Of the two remaining codons, ACC and CCA, one must code for proline and the other for threonine. However, neither the outcome of the current study nor any of the ones that we have considered earlier allow us to make assignments for these codons.

These assignments can be made by carrying out another experiment using RNA assembled from A, C, and U ribonucleotides joined into the sequence . . . CCAUCCAUCCAUCCAU. . . . How many different kinds of triplets are possible with this type of RNA? ________[58] Identify them. ________[59] This RNA guides the translation of a polypeptide with repeating sequences of proline, histidine, isoleucine, and serine. Our task, again, is to figure out which amino acids (proline and threonine) go with which codons (ACC and CCA). Which of these two codons we are trying to assign appears in this . . . CCAUCCAUCCAU . . . RNA? ________[60] From the preceding study using the . . . CCACCACCACCA . . . RNA, what are the two amino acids that this codon could possibly code for? ________[61] Which one of those amino acids is incorporated into a polypeptide translated by the . . . CCAUCCAUCCAU . . . RNA? ________[62] Which amino acid is coded for by the CCA codon? ________[63] Which codon is left to be assigned to threonine? ________[64]

So far we have considered the outcomes of experiments using four different categories of artificially synthesized RNA. These experiments, taken together, are useful in identifying the triplets that code for only some of the 20 amino acids. It was not until a technique was available which allowed the synthesis of a fifth category of RNA—triribonucleotides—that the code was completely broken.

Customized Trinucleotides of RNA This RNA is made from two or three kinds of ribonucleotides joined

[42] Two. [43] CAC and ACA. [44] We have no way of telling from this study which triplet codes for which amino acid. [45] We have no way of telling. [46] Two. [47] CAC and ACA. [48] One. [49] It must be the codon from the 1A, 2C group; that is, CAC. [50] Threonine. [51] Three. [52] ACC, CCA, and CAC. [53] No. [54] No. [55] No. [56] The kind of monopeptide produced depends on the starting point; with three possibilities, three kinds of monopeptides can be produced. [57] CAC. [58] Four. [59] CCA, CAU, AUC, and UCC. [60] CCA. [61] CCA *could* code for proline *or* threonine. [62] Proline. [63] CCA. [64] ACC.

into trinucleotides with known sequences that can be considered as customized individual codons. Supplying a batch of identical trinucleotides to the polypeptide-synthesizing system resulted not in the formation of polypeptides (because the RNA is too short), but rather in the bonding of the trinucleotide to one kind of amino acid transfer RNA complex. Once the amino acid was identified, there was unambiguous evidence for linking the amino acid and the triplet of the trinucleotide. For example, the triribonucleotide GUG binds to an amino acid–tRNA complex carrying valine. Thus, the codon GUG codes for valine. Repeating this with most of the possible codons made it possible to completely break the genetic code.

THE CODE AND ITS FEATURES

Of the 64 possible code words, 61 were found to code for amino acids. The remaining three did not bind with any of the amino acid–tRNA complexes and because of that were referred to as **nonsense codons.** These nonsense codons were soon found to play a key role in polypeptide synthesis by serving as punctuation marks: they signal the termination of polypeptide synthesis.

The Genetic Code Table

Table 27-1 gives the genetic code. It contains the mRNA codons and the amino acid or punctuation that each calls for. The codons are written with their 5′ ends on the left and are read in a 5′-to-3′ direction. How many codons are there for serine? ________[65] for arginine? ________[66] for leucine? ________[67] Identify the amino acids that each have four codons. ________[68] that each have three codons. ________[69] How many amino acids are coded for by two codons? ________[70] Identify any amino acids that are coded for by just a single codon. ________[71]

Consistency and Degeneracy of the Code

Two features of the code should emerge from your examination of Table 27-1. The first is that the code is **consistent and reliable**—any codon coding for an amino acid codes consistently for that amino acid. The second feature is that since most amino acids are coded for by more than a single codon, the code is **degenerate.** Two factors contribute to this degeneracy. One is that a particular amino acid can complex with more than one kind of tRNA molecule. Serine, for example, can complex with six different kinds of tRNA

TABLE 27-1

The genetic code.

First nucleotide	Second nucleotide: U		C		A		G		Third nucleotide
U	5′ UUU 3′	phe	UCU	ser	UAU	tyr	UGU	cys	U
	UUC	phe	UCC	ser	UAC	tyr	UGC	cys	C
	UUA	leu	UCA	ser	UAA	stop	UGA	stop	A
	UUG	leu	UCG	ser	UAG	stop	UGG	trp	G
C	CUU	leu	CCU	pro	CAU	his	CGU	arg	U
	CUC	leu	CCC	pro	CAC	his	CGC	arg	C
	CUA	leu	CCA	pro	CAA	gln	CGA	arg	A
	CUG	leu	CCG	pro	CAG	gln	CGG	arg	G
A	AUU	ile	ACU	thr	AAU	asn	AGU	ser	U
	AUC	ile	ACC	thr	AAC	asn	AGC	ser	C
	AUA	ile	ACA	thr	AAA	lys	AGA	arg	A
	AUG	met and fmet	ACG	thr	AAG	lys	AGG	arg	G
G	GUU	val	GCU	ala	GAU	asp	GGU	gly	U
	GUC	val	GCC	ala	GAC	asp	GGC	gly	C
	GUA	val	GCA	ala	GAA	glu	GGA	gly	A
	GUG	val	GCG	ala	GAG	glu	GGG	gly	G

[65] Six. [66] Six. [67] Six. [68] Proline, valine, threonine, glycine, and alanine. [69] Isoleucine. [70] Nine.
[71] Tryptophane and methionine.

molecules. Each kind of tRNA has its own particular anticodon and because of that, serine is coded for by six different codons, each complementing a different anticodon. Molecules of tRNA that accept the same amino acid but have different anticodons are known as **isoacceptor tRNA** molecules.

Another factor contributing to the degeneracy is that a particular amino acid–tRNA complex may have an anticodon capable of pairing with more than a single kind of codon. This phenomenon can occur as the bases of the codon and anticodon are pairing up. This pairing occurs along the codon in the 5′-to-3′ direction and along the anticodon in the 3′-to-5′ direction, and the bases of each of the first two pairs of codon-anticodon bases must be complementary with normal hydrogen bonding between them. However, the base at the 5′ end of the anticodon may exhibit some flexibility or leeway in terms of the type of hydrogen bonding it carries out. Stated another way, there may be a breakdown in specificity of base pairing: the base in the third position of the anticodon may sometimes form hydrogen bonds with a base other than its standard complement. The pairing leeway at this site is referred to as **wobble** and its extent varies depending on the base occupying the third position of the anticodon. A and C show no leeway at all, that is, they can pair only with their usual counterparts. G and U however, in addition to their normal complements, can pair with U and G, respectively. In addition to these standard RNA bases, some types of tRNA molecules are known to contain a fifth type of base, the purine inosine, I. Inosine is similar to guanine and arises through enzymatic modification that occurs after the tRNA molecules have been transcribed. Inosine normally pairs with cytosine but, when it occurs in the third position of the anticodon, it can also pair with A or U.

Adaptive Features of the Code Minimizing the Effects of Mutation

The evolution of the genetic code may have been guided by selective pressures serving to minimize the effects of mutations. Let us take a closer look at two features of the code which could do this.

Similarity of Synonymous Codons First we consider the makeup of **synonymous codons,** that is, codons that code for the same amino acid. For example, compare the four triplets coding for proline. What similarity do you observe in these codons? ________[72] Are the differences restricted to a particular position within the codons? ________[73] What is that position? ________[74] Now compare the four triplets coding for threonine. What similarity do you observe in these codons? ________[75] Are the differences restricted to a particular position within the codons? ________[76] What is that position? ________[77] Examine other synonymous codons. What trend do you observe regarding similarities and differences in the triplets specifying a particular amino acid? ________________________________[78] What are the exceptions to this general pattern? ________________________________[79] These three exceptions have six codons apiece: four codons for each of these amino acids begin with the same two bases while the other two codons begin with a different pair of bases. Within each of these subgroups of codons, however, the variation is restricted to the third position.

Take a moment to consider the effects of a mutation causing a change in the base in the third position of a triplet that, say, codes for proline. For example, if CCU were to change to CCA, what amino acid would it then call for? ________[80] What effect would a change to CCC or to CCG have? ________[81] What effect would a change in the third base of a codon calling for threonine have on the amino acid placed in a growing polypeptide chain? ________[82] What generalization could you make regarding the effect on an amino acid arising from a change in the third base of one of the amino acid's synonymous codons? ________________________________[83] Why do you think such mutations are referred to as silent mutations? ________________________________[84] Coding for serine, leucine, and arginine shows even greater flexibility in terms of diminishing the effects of mutation, since some alterations in the first base position as well as any alteration in the third base position can occur without consequence.

[72] All begin with CC. [73] Yes. [74] The third position. [75] All begin with AC. [76] Yes. [77] The third position.
[78] With a few exceptions, all the synonymous codons for a particular amino acid have the same first two bases, with the variation restricted to the third position. [79] Codons for serine, leucine, and arginine. [80] Proline. [81] Proline would still be specified.
[82] No effect—threonine would still be specified. [83] Such a change produces no change in the amino acid called for and thus has no effect on the polypeptide product. [84] Because the amino acid coded for is not changed and the mutation has no effect on the encoded polypeptide's sequence.

Similarity of Codons for Similar Amino Acids

The code also diminishes the effects of mutations because amino acids with similar structural and/or chemical properties have similar codons. Consequently, a mutation producing a single base change may result in the substitution of an amino acid that is structurally or chemically similar to the amino acid originally specified. For example, the amino acids glutamic acid and aspartic acid both carry negatively charged side groups at the pH levels that exist in most living cells. Compare the codons that code for these two amino acids. What similarities and differences do you observe? ____

________[85] What effect would result from changing a GAG codon to GAC? ________________________

____________________[86]

Often, the middle positions are similar in the codons for amino acids with similar chemical properties. For example, amino acids with strongly polar side groups, such as histidine, lysine, and arginine, are coded for by triplets carrying purines in the middle position. If that purine is replaced with another, a different amino acid may be substituted but it will usually be one that is strongly polar. Similarly, amino acids that are strongly nonpolar, such as leucine, valine, and proline, all have pyrimidines as the middle base of the triplet. A mutation resulting in the replacement of one pyrimidine with another will still, most likely, position a strongly nonpolar amino acid in the polypeptide. Generally speaking, only when a pyrimidine is replaced with a purine or vice versa is a markedly different type of amino acid placed in the polypeptide.

Other Features of the Code

Absence of Spacers Codons adjoin each other, and there are no spacers or "commas" between codons—the base at the 3′ end of one codon is immediately adjacent to the base at the 5′ end of the next codon. The bases are read continuously and without omissions. For example, the sequence AAGUUG specifies two amino acids, lysine and leucine, in triplet, nonoverlapping code.

Initiator Codons Two of the 64 codons, AUG and GUG, may serve as **initiator codons** for polypeptide synthesis. One of these is always the first codon read during translation and codes for the first amino acid incorporated into a polypeptide. These initiator codons, in determining the starting point for polypeptide synthesis, also serve as the reference point for establishing the reading frame for the translation of the mRNA strand.

Nonoverlapping Reading Adjacent codons are read in nonoverlapping fashion. The only known exceptions to this occur in a small number of viruses (for example, the *E. coli* phage, $\phi\chi$174) where overlapping reading occurs.

Terminator Codons Three of the 64 codons, UGA, UAA, and UAG, (also known as the **opal, ochre,** and **amber** codons, respectively) are **terminator codons.** They fail to complex with any of the amino acid–tRNA units and instead signal the termination of polypeptide synthesis, causing the ribosome-mRNA assemblage to release its polypeptide product.

Universality The evidence that is currently available suggests that the code is the same in virtually all organisms, that is, that the genetic code is essentially universal. A key exception to this is found in the translation of mitochrondria mRNA codons of several species, including humans and yeasts, but the number of genetic code differences involved here is very limited. In addition, the chloroplast genetic code differs slightly from the standard code and some nuclear encoded genes in the protozoan *Tetrahymena* use an alternative genetic code. (Note that mitochrondria and chloroplasts possess limited quantities of DNA along with some tRNA molecules and ribosomes, and have the capacity to synthesize a few of the polypeptides necessary for their functioning.)

Colinearity of Nucleic Acids and Their Polypeptide Products

The linear sequence of nucleotides in a nucleic acid molecule stores the genetic information. Through transcription and translation this information guides the assemblage of a linear sequence of amino acids into a polypeptide product. In theory, if the genetic code works and if we know the sequence of nucleotides in a gene, we can identify, because of complementarity, the nucleotides in the mRNA transcribed from the gene and can, in turn, predict the sequence of amino acids in the polypeptide product of that gene. In laboratory studies, the sequence of bases occurring in the DNA of certain viruses has been determined and the sequence of amino acids in the polypeptide products of this DNA have been identified. The amino acid se-

[85] Each is specified by two different codons; all four begin with GA and have variation restricted to the third position.
[86] Substitution of glutamic acid for aspartic acid.

quence predicted from the DNA nucleotide sequence and that actually observed in the polypeptide were found to be identical. This linear correspondence between the nucleic acid triplets and the amino acids in the polypeptide is known as **colinearity** and provides *in vivo* evidence of the validity of the genetic code.

SUMMARY

The genetic code was deciphered using *in vitro* synthesized RNA. By controlling the kinds and relative concentrations of ribonucleotides supplied to the synthesizing systems, deductions could be made about the way in which these building blocks assembled into RNA molecules and, in turn, about the makeup of the triplet code words within these molecules. When supplied to a cell-free, polypeptide-synthesizing system (including ribosomes, a mixture of all kinds of aminoacyl-tRNA molecules, and all enzymes and cofactors necessary for translation), this artificial RNA acted like mRNA and guided the formation of polypeptide molecules. Relating the amino acid composition of these polypeptides to the nucleotide triplets present in the RNA made it possible to break the code and to deduce the DNA triplets complementing the RNA code words.

The genetic code is triplet, with three adjacent mRNA nucleotides making up a unique code word or codon that specifies a particular amino acid. There are 64 possible codons and the code is degenerate, with 18 of the 20 biologically important amino acids coded for by more than one codon. Two factors contribute to this degeneracy. Most amino acids can be accepted by more than one type of tRNA molecule, each with a different anticodon. In addition, a particular amino acid–tRNA complex may have an anticodon capable of pairing with more than a single codon because of pairing leeway or wobble at the third anticodon position. The code is free of spacers, has certain codons that specify initiation and others that specify termination, is read in nonoverlapping fashion, and is virtually universal. The evolution of the genetic code may have been guided by selected pressures serving to minimize the effects of mutations; the similarity of synonymous codons and of codons for similar amino acids are features of the code that contribute to this. The colinearity of nucleotide triplets in DNA and the amino acids in the polypeptide products of this DNA provide *in vivo* evidence of the validity of the genetic code.

PROBLEM SET

27-1. Prior to the beginning of laboratory studies elucidating the genetic code, why did geneticists feel that the code words consisted of three nucleotides rather than a larger number, say, four or five?

27-2. Artificially made RNA can be prepared using one, two, three, or all four kinds of ribonucleotides. These various types of RNA are individually supplied to *in vitro* polypeptide-synthesizing systems and the polypeptide products are analyzed. The results of these various experiments indicate that when the RNA is made up of a single kind of ribonucleotide, a single amino acid is incorporated into the polypeptide chain, and when the RNA is made from various combinations of two or of three kinds of ribonucleotides, other amino acids are incorporated. No amino acid requires the presence of all four nucleotides for its incorporation. What, if anything, can be concluded from these results about the size of the code words? Do these results allow us to designate the number of nucleotides in a code word?

27-3. What is meant when the genetic code is described as degenerate? If the code lacked this feature, what relationship would we find between the number of amino acids coded for and the number of codons?

27-4. What are synonymous codons? Would you expect to find them in a code that was not degenerate?

27-5. **a.** How many codons are changed by a mutation resulting in the substitution of a single base for another in a strand of mRNA read in standard, nonoverlapping fashion?

b. How many amino acids have the potential to be affected?

27-6. Assume that the strand of mRNA referred to in problem 27-5 is read in overlapping fashion, where the overlap involves two bases (that is, where the second and third bases of the codon read become the first and second bases of the next codon to be read and so on) and that the base substitution occurs near the middle of the strand.

a. How many codons are changed by the single-base substitution?

b. How many amino acids have the potential to be affected?

27-7. The polyribonucleotide sequence AAACCC is supplied to an *in vitro* polypeptide-synthesizing system.

a. How many amino acids would this sequence code for when read in standard, nonoverlapping fashion starting with the first A on the left? Identify those amino acids.

b. How many amino acids would this code for when read in overlapping fashion starting with the first A on the left? Identify those amino acids.

c. If the code were read in overlapping fashion, could the amino acid specified by AAA ever occur next to the amino acid called for by CCC? Explain.

27-8. Sequencing studies on polypeptides tell us the kinds of amino acids present and their order of arrangement. Such studies on a wide array of polypeptides indicate that there is no restriction on the kinds of amino acids that can occur next to any one of the amino acids; that is, each of the twenty amino acids can occur next to any of the other amino acids. Explain how this supports the idea of the genetic code being read in nonoverlapping fashion?

27-9. Genetic material in a very limited number of viruses is read in overlapping fashion. Cite a major advantage and a major disadvantage of this type of reading.

PROBLEM SET

27-10. Assume that you are at work in a laboratory, helping to elucidate the genetic code. You have demonstrated experimentally that triplets UUU and GGG code for phenylalinine and glycine, respectively. A coworker prepares a synthetic, random RNA copolymer using U and G, that, in turn, is supplied to a cell-free, *in vitro* polypeptide-synthesizing system. An analysis is performed on the polypeptide molecules resulting from the translation of this RNA. Weeks later you are reviewing the procedure and outcome of this experiment in your colleague's data book, and you notice that no record was made of the relative amounts of U and G incorporated into the synthetic RNA. In fact, the only information recorded is that phenylalanine made up 8/27 (= 0.296, or 29.6%) of the amino acids incorporated into the polypeptide.

a. What codons could occur in the RNA?

b. How many of those codons code for phenylalanine?

c. What were the relative concentrations in which U and G were supplied to the RNA synthesizing system?

27-11. A space probe, designed to land on a distant planet, is equipped to obtain samples of material, identify them according to general molecular type, and analyze their chemical makeup. Results obtained from samples from one planet indicate that the nucleic acid molecules contain just three types of nucleotides. Polypeptide molecules from this planet are found to be made of 26 different kinds of amino acids.

a. What is the minimal number of bases required in a code word in order to code for the 26 kinds of amino acids?

b. If, in addition, the code were to include two different terminator codons, could the codons be of the size designated in your answer to 27-11a? Explain.

27-12. Another space probe (see problem 27-11) sent to a different planet reports nucleic acids composed of just two kinds of nucleotides and polypeptides made up of 12 types of amino acids. Nucleic acid molecules studied are found to contain either 156 or 272 nucleotides.

a. What is the minimal number of bases required in a code word in order to code for the 12 types of amino acids and a single terminator codon?

b. Assume that the code words are read in nonoverlapping fashion and that the translation mechanism reads only in one direction along a nucleic acid strand. What is the maximum number of amino acids that could be incorporated into the polypeptides synthesized from single-stranded nucleic acid molecules that contain 156 nucleotides?

c. What is the maximum number of amino acids that could be incorporated into the polypeptides synthesized from double-stranded nucleic acid molecules that contain 272 nucleotides if only one strand is read?

27-13. A tRNA molecule carries the anticodon 3′ AUG 5′.

a. Identify the codon complementing this anticodon.

b. Identify the amino acid that would complex with this tRNA molecule.

c. Because of wobble, this kind of anticodon may be able to match up with additional codons. Identify the additional codons and the amino acids specified.

d. Does wobble alter in any way the amino acid associated with a particular tRNA molecule? Does it alter the amino acid specified by a particular code word? Exactly what does wobble alter?

PROBLEM SET

27-14. Assume this strand of mRNA is available for translation within an *E. coli* cell: 5′ AUGCUAUACCUCCUUUAUCUGUGA 3′.

a. How many amino acid molecules will be in the polypeptide synthesized from this mRNA strand?

b. How many kinds of amino acids will be in the polypeptide?

c. Give the sequence of amino acids that will be in the polypeptide.

d. Assume that there is no wobble. How many kinds of tRNA would be necessary to translate this mRNA strand into a polypeptide?

e. Assume that wobble occurs. What is the minimal number of kinds of tRNA molecules that would be necessary to translate this mRNA strand?

27-15. *In vitro* studies show that the codon GUG codes for valine. *In vivo* studies on prokaryotic organisms verify that this is the case when the codon is in an internal position within the mRNA strand. However, this codon can occupy the initiator site and, when there, acts like the AUG initiator codon in that it causes the placement of (N-formyl) methionine at the beginning of the polypeptide strand. Explain how this can occur.

27-16. A group of compounds known as the acridine dyes can cause the insertion of an extra nucleotide or the loss of a nucleotide during DNA replication. Either type of alteration, or mutation, causes the frame in which the nucleotides of mRNA are read to be shifted, resulting in garbled translation from the site of the alteration onward. In studying the phenotypic consequences of such mutations in the T4 phage of *E. coli*, it was found that the effect of a mutation deleting a nucleotide could generally be overcome (that is, the wild-type phenotype could be restored) if a second mutation inserted a nucleotide at a site fairly close to the deletion. Alternatively, if the initial mutation added a nucleotide, the wild-type phenotype could usually be restored if the second mutation produced a nearby deletion.

a. Explain why the second mutations generally restored the original, wild-type phenotype.

b. Do you think the entire polypeptide produced upon the occurrence of the second mutation would be identical to the polypeptide produced in the absence of both mutations? Explain.

c. What would you expect about the relationship between the number of nucleotides separating the first mutation site from the second mutation site and the likelihood of the polypeptide being restored to its original function following the second mutation?

27-17. a. Refer to the material presented in problem 27-16. A viral strain with two nucleotide additions or a strain with two nucleotide deletions exhibits a mutant phenotype. Yet viral strains carrying either three deletions or three additions display the normal, wild-type phenotype. Propose a simple and reasonable explanation for this difference.

b. Would the viral strains carrying either three deletions or three additions and exhibiting the wild-type phenotype produce exactly the same polypeptide as the original, premutation wild strain? Explain.

27-18. What does the outcome of the work with the viral mutants described in problems 27-16 and 27-17 indicate about the nature of the genetic code?

CHAPTER

Mutation

INTRODUCTION

A mutation refers to a change in the genetic material. This chapter will focus on changes that occur at the level of the gene and result in an altered nucleotide sequence. Gene mutations that involve a single base pair, often referred to as point mutations, fall into two general categories: those arising from the substitution of one nucleotide for another and those arising from the deletion or the addition of a nucleotide. Chromosome mutations involving changes in chromosome structure or number are discussed in Chapter 16.

NUCLEOTIDE SUBSTITUTIONS

Origin and Transmission of Substitutions

We begin by considering nucleotide substitutions. Such mutations could originate, for example, when a base in one strand of a duplex DNA molecule experiences a temporary change in its hydrogen-bonding capabilities (how these capabilities are altered will be discussed later in this chapter). Using the following series of figures, we will follow two rounds of DNA replication to see how a substitution arises and gets transmitted from one generation to the next. The following section of the DNA duplex shows normal base pairing.

G C
C G
T A
A T

Prior to replication, the two strands of this duplex separate to serve as templates for the formation of complementary counterpart strands. Assume that the adenine in the strand on the right has its hydrogen-bonding capabilities altered so that it pairs with C during replication.

G C
C G
T **A** ← Nucleotide with altered hydrogen bonding
A T

The two daughter duplexes produced in this replication are as follows.

G	C	G	C	
C	G	C	G	
T	A	**C**	**A**	← Pairing error produces this base pair
A	T	A	T	

The daughter duplex on right carries a single base substitution with a C replacing the T with which the A would normally pair. When this duplex replicates, it gives rise to a duplex carrying a base-pair substitution (on the left in the following sketch) and a normal duplex (on the right). Note that the production of this normal duplex assumes that the A nucleotide that had its hydrogen-bonding capacity altered now exhibits its normal hydrogen bonding and pairs with its normal partner, T.

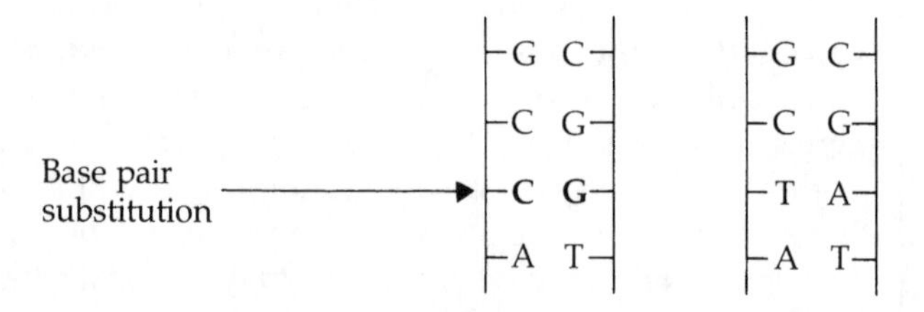

How many of the two daughter cells produced following this most recent round of DNA replication would get the base-pair substitution mutation? ________[1] To summarize, because of altered hydrogen-bonding capabilities of an A nucleotide in the original duplex, the nucleotide C substituted for T during a round of replication. This substitution, following a second round of replication, resulted in a duplex molecule with a CG pair replacing the TA pair of the original duplex.

To consider another example, assume that during the replication of a duplex, a particular G has its hydrogen-bonding abilities altered so that it pairs with a T rather than its usual C partner. When the duplex carrying this base substitution replicates, what will occupy the site of the original GC base pair in the product duplex carrying the base-pair substitution? ________[2]

Base-pair substitutions, such as the two that we have just considered, that replace bases with others of the same type (that is, replace a purine, A or G, with a purine and replace the complementary pyrimidine, T or C, with a pyrimidine), are referred to as **transitions.** Transitions preserve the purine-pyrimidine arrangement of the original duplex: the duplex strand originally carrying the purine at the site under consideration still has a purine at that site, and the strand originally with the pyrimidine still carries a pyrimidine. How many different transition mutations are possible? ________[3] Identify them. ________________[4] In other base-pair substitutions, the purine is replaced by a pyrimidine and the pyrimidine by a purine in what is known as a **transversion.** This reverses the purine-pyrimidine arrangement of the original molecule at the site under consideration so that the strand originally carrying the pyrimidine now has the purine and vice versa. How many different transversion mutations are possible? ________[5] Identify them. ________________[6]

Consequences of Substitutions

Consider the effect a substitution has when a DNA strand carrying a substitution serves as the template for the transcription of mRNA. How many codons in the mRNA strand would be altered by a single DNA base substitution? ________[7]

Missense Mutations Under what circumstances would the altered codon code for an amino acid different from the one coded for by the original unaltered codon? ________________[8] What effect would the substitution of one amino acid for another have on the polypeptide? ________________[9] Which do you think would have the greater effect on a polypeptide: the substitution of a chemically similar amino acid *or* of a chemically dissimilar amino acid? ________[10] the substitution of an amino acid within the region of a polypeptide carrying out its biological activity (that is, its active site) *or* at a site in the polypeptide remote from that region? ________[11] A substitution of the sort that changes a DNA triplet so that it codes for a different amino acid is known as a **missense mutation.**

[1] Since only one of the two duplexes carries the base-pair substitution, only one of the daughter cells will receive the mutation.
[2] The substituted T will pair with an A, to produce an AT pair. [3] Four. [4] AT to GC, TA to CG, GC to AT, and CG to TA. [5] Eight.
[6] AT to TA, TA to AT, AT to CG, CG to AT, GC to TA, TA to GC, CG to GC, and GC to CG. [7] One. [8] If it was nonsynonymous with the original codon, the translated amino acid would change. [9] It could vary from almost nonexistent to drastic, depending on the properties of the substituted amino acid and where the substitution occurs. [10] Dissimilar. [11] Within the active site.

Silent Mutations Under what circumstances would the altered codon code for the same amino acid as the original codon? ______________________________ ____________[12] If it coded for the same amino acid, would the mutation have any effect on the polypeptide? __________[13] A substitution that changes a codon to another coding for the same amino acid is designated as a **silent mutation.**

Nonsense Mutations Under what circumstances would the altered codon bring about the premature termination of polypeptide synthesis? ____________ ______________________________[14] What effect would this have on the functional properties of the polypeptide translated from the gene? ____________ ______________________________[15] A base substitution that changes a triplet coding for an amino acid into one coding for a terminator signal is a **nonsense mutation.**

NUCLEOTIDE ADDITIONS OR LOSSES: FRAMESHIFT MUTATIONS

The second major class of point mutations is that involving the loss or addition of DNA nucleotides. Since mRNA nucleotides are read in sequence and in groups of three, a mutation of this type throws off the reading frame from the point of the mutation onward through all the nucleotides subsequently read. Because of this, this type of mutation is referred to as a **frameshift mutation.** The following DNA strand (where vertical lines separate the triplets) transcribes into an mRNA strand containing 11 codons as shown. The initiator codon, AUG, establishes or defines the reading frame and the amino acids translated from this series of codons are given below the mRNA strand.

DNA:5′ TAC|GCA|GGG|AAA|TAT|TTG|CCT|CGT|TGA|TTC|ATT 3′
mRNA:3′ AUG|CGU|CCC|UUU|AUA|AAC|GGA|GCA|ACU|AAG|UAA 5′
Amino acids: met arg pro phe ile asn gly ala thr lys (stop)

Now let us examine the effect of a nucleotide deletion on the mRNA strand, its reading frame, and the polypeptide produced from it. Assume that the loss of a G nucleotide from the third DNA triplet from the 5′ end causes the middle nucleotide in the third mRNA codon from the left (CCC) to be deleted as follows.

Deletion site
↓
DNA: 5′ TAC|GCA|G G|AAA|TAT|TTG|CCT|CGT|TGA|TTC|ATT 3′
mRNA: 3′ AUGCGUC CUUUAUAAACGGAGCAACUAAGUAA 5′
Amino acids: met arg

Does the deletion alter the way in which the first six mRNA nucleotides are read as codons? __________[16] Are the first two amino acids in the polypeptide altered by this mutation? __________[17] What does the base composition of the third codon read? ________[18] What amino acid does this third triplet code for (refer to the genetic code in Table 27-1 for this information)? __________[19] Continue to identify the codons (fourth, fifth, and so on) read by the altered reading frame by drawing brackets around them. Identify the amino acids that will be coded for. ______________________ ______________________[20] How many amino acids would occur in the polypeptide product of this altered gene? __________[21] Do you think it likely that this polypeptide would be able to function in the same manner as the original polypeptide? __________[22] Specifically, what caused the difference in overall length of the polypeptide produced from the mutant gene and that produced from the original gene? __________[23].

As this example shows, the altered reading of the codons caused by a deletion mutation changes the amino acids from the site of the deletion onward, with the possibility that the polypeptide product will be prematurely shortened by the creation of a new, "early" termination signal. In other cases, another type of punctuation problem could arise: a reading frameshift could cause the original terminator codon to be overlooked, with translation extending beyond the point where it would normally end. Although our example involved a deletion, you should realize that similar consequences could arise from the addition of a nucleotide to a gene.

Before going further, take a moment to compare substitution and reading frameshift mutations involving single base pairs and the consequences arising from them in the following table.

[12] If it were synonymous with the original codon. [13] No. [14] Changing a sense codon to a terminator codon.
[15] The prematurely shortened polypeptide would probably be nonfunctional. [16] No. [17] No. [18] CCU. [19] Proline.
[20] The fourth is UUA (leucine), and the fifth is UAA (chain termination). [21] 4. [22] No, there is little chance of this.
[23] Creation of a new terminator codon.

	Substitution	Frameshift
Effect on DNA of gene:	______[24]	______[27]
Effect on mRNA transcribed from gene:	______[25]	______[28]
Effect on polypeptide translated from mRNA:	______[26]	______[29]

Although both substitution and frameshift mutations can have drastic consequences for the gene's polypeptide product (and, therefore for the phenotype of the organism carrying the mutation), which type carries a considerably greater likelihood of this happening? ______[30] Explain. ______[31] A substitution mutation affects a single codon. Under what circumstance would a frameshift mutation alter the reading of just a single codon? ______[32]

REVERSING MUTATIONS

In altering a gene, a missense, nonsense, or frameshift mutation alters the polypeptide product and, in turn, may change the phenotype of the organism carrying the mutation. Occasionally, the phenotypic consequences of these mutations can be reversed by a second point mutation that restores the original wild-type phenotype or something very similar to it. The **reversing mutation** may restore the original phenotype by changing the same nucleotide altered by the first mutation, another nucleotide in the same gene, or a nucleotide in a different gene.

Altering the Nucleotide Changed by the Original Mutation

True Reversions Let us begin by considering the reversing mutations known as **true reversions** or **true back-mutations** that change the nucleotide altered by the original mutation back to its original state. The following table shows a DNA triplet of a gene altered by two successive mutations, with the second restoring the original DNA triplet. Note that the nucleotide altered by each mutation is in boldface.

	DNA triplet	Complementing mRNA codon	Amino acid coded for
Original condition:	GAC	CUG	Leucine
After first mutation:	**C**AC	GUG	Valine
After second mutation:	**G**AC	CUG	Leucine

Compare the DNA triplet before the first mutation and after the second; at how many points do they differ? ______[33] Compare the mRNA codons transcribed before the first mutation and after the second; at how many points do they differ? ______[34] Would the polypeptide produced from the gene before the first mutation differ from that produced after the second mutation? ______[35] In this example, the second, or reversing, mutation qualifies as a true reversion or true back-mutation since it restored the gene sequence (and in turn, the polypeptide and the phenotype) back to the original conditions.

Operational Reversions A second type of reversion known as an **operational reversion** changes the same nucleotide altered by the original mutation and produces a codon synonymous with the original codon. What would have happened to the polypeptide if the second mutation shown in the previous table had converted the first nucleotide of the DNA triplet to A rather than to G (refer to the genetic code in Table 27-1)? ______[36] How can an operational reversion be distinguished from a true reversion? ______[37]

Pseudoreversion A third type of reversion known as a **pseudoreversion** changes the nucleotide altered by

[24] Replaces one base pair with another. [25] Replaces one ribonucleotide with another. [26] It varies: no change; single amino acid substitution; or premature termination. [27] Deletes or adds a base pair. [28] Deletes or adds ribonucleotides. [29] Changes amino acids from site of mutation onward and may shorten the polypeptide. [30] Frameshift. [31] Depending on where it begins, a frameshift usually alters the reading of several to many codons and usually has drastic consequences; the consequences of a single altered codon can be less severe. [32] If the nucleotide deletion or addition causing the frameshift mutation occurred in the final triplet of the DNA of the gene. [33] None. [34] None. [35] No. [36] The triplet would be AAC which transcribes into the codon UUG which codes for the amino acid leucine: the polypeptide would be identical to the original one. [37] By comparing the DNA sequences of the genes: an operational reversion differs from the wild-type by one nucleotide in the same triplet altered by the original mutation, whereas a true reversion is identical to the wild-type. With both mutations, the phenotypes and polypeptide products would be the same.

the original mutation to produce a codon that codes for an amino acid different from the original amino acid, but very similar to it in chemical properties. The polypeptide produced after this type of reversion differs from the original in a single amino acid. The chemical similarity of the new amino acid to the amino acid it replaces allows the new polypeptide to function much like the original which, in turn, restores the phenotype to its original or near original condition. Since the resulting phenotype looks like the original but is based on a slightly different polypeptide, it is referred to as a **pseudowild-type.**

An example of a pseudoreversion is presented in the following table.

	DNA triplet	Complementing mRNA codon	Amino acid coded for
Original condition:	TCT	AGA	Arginine
After first mutation:	T**G**T	A**C**A	Threonine
After second mutation:	T**T**T	A**A**A	Lysine

The amino acid originally coded for, arginine, and the amino acid coded for after the second mutation, lysine, are chemically similar since each is a basic amino acid, whereas the amino acid intermediately coded for, threonine, is neutral (that is, neither acidic nor basic). The substitution of one basic amino acid for another might well have little effect on the biological properties of the polypeptide and could restore the original phenotype. Compared to the original polypeptide, how does the polypeptide produced after a true reversion or after an operational reversion differ from that produced following a pseudoreversion? ______________________________ [38]

Altering a Nucleotide Other Than That Changed by the Original Mutation: Suppressor Point Mutations

So far we have considered reversing mutations that alter the same nucleotide changed by the initial mutation. Another type of reversing mutation changes a nucleotide other than the one altered by the first mutation. Reversing mutations of this type fall under the heading of **suppressor mutations** and they may alter (1) a nucleotide in the codon carrying the initial mutation, (2) a nucleotide in a different codon of the gene carrying the initial mutation, or (3) a nucleotide in a gene different from the one carrying the initial mutation. The first two reversing mutations are designated as **intragenic** suppressor mutations since they alter nucleotides in the same gene, while the third qualifies as an **intergenic** suppressor mutation since it changes a nucleotide in a second gene. Since both types of intragenic suppressors occur at a different site in the same gene, they are also known as **second-site reversions.** Compared with the original DNA, what difference would you expect to find in the DNA following a suppressor mutation and following a true back-mutation? ______________________________ [39]

Intragenic Suppressor Mutations: Altering a Nucleotide in the Same Triplet An intragenic suppressor mutation that changes a base in the codon carrying the initial mutation is illustrated in the following table.

	DNA triplet	Complementing mRNA codon	Amino acid coded for
Original condition:	GAG	CUC	Leucine
After first mutation:	**A**AG	UUC	Phenylalanine
After second mutation:	AA**C**	UUG	Leucine

Compare the DNA triplet sequences before the first mutation and after the second; at how many points do they differ? __________[40] Compare the sequences of mRNA produced before the first mutation and after the second; at how many points do they differ? _____[41] Would the polypeptide produced before the first mutation differ from that produced after the second mutation? __________[42]

This type of suppressor mutation, that is, one affecting the same codon, can also correct frameshift mutations that arise when a nucleotide becomes deleted or an extra nucleotide is added. For example, if a second mutation inserts an additional nucleotide into the same codon that earlier experienced a nucleotide deletion, what will happen to the reading frame? ______________________________ [43]

[38] The polypeptide produced after a true or operational reversion is identical to the original polypeptide; the polypeptide produced after a pseudoreversion differs from the original by a single amino acid. [39] Following a suppressor mutation, DNA shows altered nucleotides at two points; following a true back-mutation, the DNA is identical to the original DNA. [40] Two points. [41] Two points. [42] No. [43] It restores the original reading frame.

mutation differ from the wild-type polypeptide? ____________[44]

Describe the suppressor mutation required to suppress an initial mutation that inserted an *additional* nucleotide into a codon? ____________[45] How will the polypeptide produced after the suppressor mutation differ from the wild-type polypeptide? ____________[46]

Intragenic Suppressor Mutations: Altering a Nucleotide in a Different Triplet A second category of suppressor mutations involves those occurring in a triplet of the gene different from the triplet carrying the original mutation. Most often, this type of mutation suppresses a frameshift mutation. For example, just as a nucleotide loss can be suppressed by a nucleotide addition in the same triplet, such a loss can also be suppressed by adding a nucleotide at a subsequent point in the gene. When this happens, the reading frame, shifted by the codon carrying the initial loss, remains altered until the nucleotide inserted as a result of the second mutation is encountered; from that point onward, the reading frame is restored to its normal pattern. In general terms, describe the polypeptide produced following the second mutation? ____________[47] Upon what will the size of the garbled section depend? ____________[48]

In order for the wild-type phenotype or something similar to be restored, this garbled zone would, of course, have to have a minimal effect on the functional properties of the polypeptide.

This type of mutation is illustrated in the following sequence of sketches each of which shows a portion of a wild-type gene (with vertical lines separating the triplets), the complementary mRNA strand (with arcs designating its reading frame), and the polypeptide it codes for. An initial mutation results in a nucleotide deletion that, in turn, causes a frameshift and garbling of the polypeptide. A second mutation adds a nucleotide that, in turn, limits the extent of the frameshift.

Prior to Mutation

Portion of DNA of wild-type gene: ... CAA|GTT|ATA|GAC|GTA|TTT|GCA ...

Portion of complementary mRNA: ... GUUCAAUAUCUGCAUAAACGU ...

Translated portion of polypeptide: ... val gln tyr leu his lys arg ...

Initial Mutation: Loss of Nucleotide (T) from Indicated Site

DNA after initial mutation: ... CAAGT•ATAGACGTATTTGCA ... (T removed)

mRNA after initial mutation: ... GUUCA UAUCUGCAUAAACGU ...

Polypeptide: . . . val his ile cys ile asn val . . .

Garbled sequence →

Second Mutation: Addition of Nucleotide (G) at Indicated Site

DNA after second mutation: ... CAAGT ATAGAGCGTATTTGCA ...

Complementary mRNA after second mutation: ... GUUCA UAUCUCGCAUAAACGU ...

Polypeptide: . . . val his ile ser his lys arg . . .

Garbled sequence

How many of the polypeptide's amino acids are altered by the original mutation? ____________[49] How many are altered following the second mutation? ____________[50]

Our example has focused on a nucleotide loss created by an initial mutation that is suppressed by a second mutation adding a nucleotide in another portion of the gene. You should realize that in similar fashion, a mutation involving nucleotide gain can be suppressed by a second mutation removing a nucleotide from another portion of the gene.

Intergenic Suppression The final category of suppressor mutations, known as intergenic suppressors, alters a nucleotide in a gene different from the one altered by the first mutation. The gene changed by the second mutation is referred to as the **suppressor gene.**

[44] It would differ in no more than a single amino acid. [45] The deletion of a nucleotide will restore the original reading frame.
[46] It would differ in no more than a single amino acid. [47] It will carry a garbled section since the reading frame is still altered between the site of the initial loss and the site of the nucleotide addition; the garbled section will have normal sections on either side of it.
[48] The number of nucleotides between the sites of the original mutation and the subsequent mutation.
[49] All beyond the site corresponding to the mutation. [50] Three.

Suppressor mutations in genes coding for specific transfer RNA molecules have been much studied and generally cause a change in the anticodon of the tRNA molecule without altering the ability of the tRNA molecule to bind with its amino acid. If the tRNA molecule's anticodon now complements the mRNA codon altered by the original mutation, suppression of the first mutation may occur.

For example, assume that an initial mutation in a gene coding for a polypeptide converts a sense codon to a nonsense codon. What would be the effect of this mutation on the polypeptide product? ____[51] Now assume that a second mutation occurs in the same cell in a gene coding for a particular type of tRNA that results in an alteration of the anticodon of the tRNA molecule so that it can now pair with this nonsense codon. What effect would this have on the synthesis of the polypeptide? ____[52] Compare the polypeptide produced with and without the suppressor gene; how would they differ? ____[53]

In the situation just described, would the mutation affecting the tRNA anticodon always result in the suppression of the original mutation? __________[54] Explain why or why not. ____[55] If it did not, would the mutation in the tRNA gene qualify as a suppressor mutation? __________[56]

To consider another example, assume that a mutation in a triplet of a structural gene results in the substitution of an acidic amino acid for the original basic amino acid at a particular site in the polypeptide product. This substitution changes the functional properties of the polypeptide which, in turn, alters the phenotype of the organism. In the same cell, a gene coding for a type of tRNA molecule carrying an amino acid chemically similar to the original amino acid subsequently experiences a mutation that allows the anticodon of this tRNA molecule to complement the missense codon in the gene affected by the first mutation. Specifically, what would be the effect of this second mutation on the polypeptide product? ____[57] Generally speaking, how would the structure and biological activity of the polypeptide product produced prior to and after the occurrence of the second mutation differ? ____[58]

In the example just considered, would a mutation altering the anticodon of any kind of tRNA, so that it could pair with the missense codon, always result in the suppression of the original mutation? __________[59] Explain why or why not. ____[60] If it did not, would it be considered a supressor mutation? __________[61]

Another type of suppressor tRNA mutation gives rise to tRNA molecules with an abnormal number of anticodon nucleotides and may suppress certain frameshift mutations. As you know, normal tRNA anticodons consist of three nucleotides and read three mRNA nucleotides at a time during translation. A frameshift mutation caused by the addition of a single nucleotide could be suppressed by a tRNA molecule, for example, with a four-nucleotide anticodon. The anticodon would read four nucleotides together, three from a standard codon plus an extra one, thereby restoring the original reading frame. Such suppressor mutant tRNA molecules are known to occur in the bacterium *Salmonella typhimurium*.

It is important to note that most genes coding for tRNA molecules occur in cells in multiple copies. If there was only one copy, a suppressor mutation producing an altered anticodon for the tRNA molecule coded for by the gene could have lethal consequences. This is because the tRNA molecule would not be able to place its amino acid at the correct locations along the mRNA strand during polypeptide synthesis. Polypeptides requiring the amino acid carried by this tRNA would be abnormal and cause cell death. However, with the gene present in multiple copies, those lacking the suppressor mutation would produce tRNA molecules with the correct anticodon, ensuring that normal polypeptide synthesis would occur.

[51] Premature termination of polypeptide synthesis. [52] The premature termination would be overcome: an amino acid would be inserted at that codon and polypeptide synthesis would continue to the end of the gene. [53] Without the suppressor mutation, the polypeptide would be prematurely shortened; with the suppressor mutation, the polypeptide would be of normal length and would carry a single altered amino acid. [54] No. [55] The amino acid inserted by the mutant tRNA might be so different from the original amino acid that the polypeptide could not resume its original function. [56] No. [57] The amino acid inserted by the original missense mutation would be replaced by a different amino acid with properties similar to those of the original amino acid. [58] Prior to the suppressor mutation, the missense mutation would sufficiently alter the polypeptide so that its biological activity would be changed. The tRNA altered by the second mutation would insert an acidic amino acid at that site, and because of its chemical similarity to the original amino acid, the acidic amino acid may restore the original biological activity and the original phenotype, or something very close to it. [59] No. [60] The newly inserted amino acid might be so different from the original amino acid that the polypeptide could not resume its original function. [61] No.

CAUSES AND REPAIR OF MUTATIONS

Tautomeric Base Shifts

The various types of mutations we have considered occur spontaneously. In many instances, they arise because the configuration of the purines and pyrimidine bases of DNA molecules are not perfectly stable. Once in a while, a hydrogen atom within a nitrogenous base can spontaneously shift from one position to another and cause a change in the three-dimensional shape of the molecule. Configurational shifts of this sort are known as **tautomeric shifts** and the alternate forms of each base are referred to as **tautomeric forms** or **tautomers.** Occurring infrequently, the altered base configurations are short lived and readily reversible. Despite this, the altered base configurations are of great significance because while they last, they alter the hydrogen-bonding capacity of the base and can result in abnormal base pairings.

Figure 28-1 illustrates such a hydrogen-atom shift that occurs in the base adenine: the hydrogen atom is normally part of an amino group (-NH_2) but can shift to a nitrogen atom found in the ring, thus converting the amino group into an imino group, -NH. A base carrying the amino group is said to be in its amino form and a base carrying the imino group, in its imino form. The amino configuration (Figure 28-1a) is the form in which adenine almost always exists in living cells. The hydrogen-atom shift produces the imino form (Figure 28-1b). Note that when the hydrogen atom shifts to the ring nitrogen, the double bond normally positioned between the 6C and the ring nitrogen shifts away from the ring nitrogen to the imino group. This hydrogen-atom shift can occur in an adenine already incorporated into a DNA strand or in an adenine of a nucleotide about to be incorporated into a growing strand.

FIGURE 28-1

Tautomeric forms of adenine.

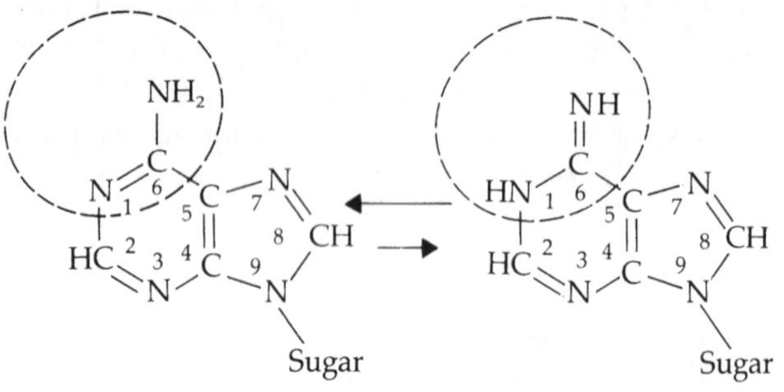

(a) Adenine: common amino form

(b) Adenine: rare imino form

FIGURE 28-2

Mispairing resulting from a rare tautomeric form of adenine.

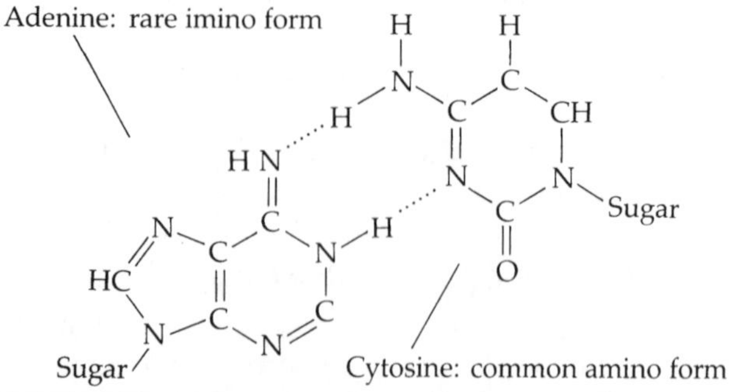

Adenine: rare imino form

Cytosine: common amino form

In the common amino configuration, adenine exhibits its standard, complementary hydrogen-bond formation with thymine. However, in its imino form, it acts like guanine and forms hydrogen bonds with amino cytosine, as shown in Figure 28-2.

Consequences of Base Mispairing

Let us examine the effects of this inappropriate pairing. Assume a duplex DNA molecule experiences a configurational shift in one of its adenine bases just as the duplex is about to replicate. A portion of this duplex is shown in Figure 28-3, and the shift occurs in the adenine in the strand on the left which, in its rare imino form, is shown as A*. Indicate in the space provided in the figure the bases that will pair with the bases of its two strands as the duplex replicates. How do the daughter duplexes compare with the original duplex? ____________________________[62]

Since the rare forms of bases tend to be unstable, they often shift back to their common form by the time another round of replication occurs. Assume that before the strands of the daughter duplex carrying the A*C base pair in Figure 28-3 separate to serve as templates for a second round of DNA replication, the imino configuration of adenine changes back to the amino form as shown in Figure 28-4. Indicate, in the space provided, the bases that will pair with the bases of the two separated strands of this duplex during this second round of replication. How do the two duplexes

[62] One is identical; the other carries an A*C pair in place of the AT pair of the original.

FIGURE 28-3

Consequences for the first-generation daughter duplexes of a configurational shift that converts the amino adenine in the left strand of the original duplex to its rare imino form.

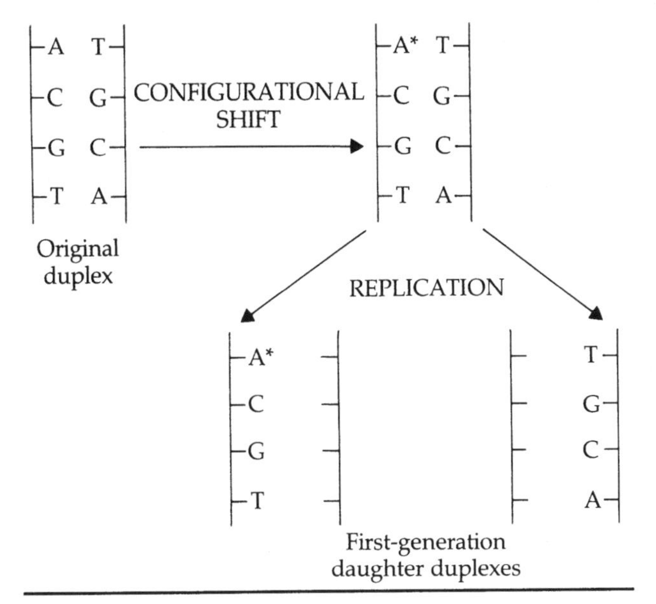

produced by this replication compare with the *original* duplex shown in Figure 28-3? ________________ [63]

How would the presence of the base-pair substitution in a gene that serves as a template for mRNA transcription alter the mRNA molecule produced? ________________ [64]

How many mRNA codons would be altered? ________ [65]

Now we will consider replication of a duplex DNA molecule where one of the triphospho-deoxyribonucleotides available for incorporation into newly synthesized DNA carries the rare imino form of adenine (A*). A portion of a duplex undergoing replication and corresponding sections of the two daughter duplexes produced from it are shown in Figure 28-5. As the daughter duplex on the right shows, the nucleotide with the imino form of adenine forms hydrogen bonds with the cytosine in a strand derived from the original duplex. Henceforth in our discussion, we will consider just this daughter duplex with this A*C base pair.

Assume that before the strands of this daughter duplex separate to serve as templates for a second round of replication, the imino form of adenine changes back to the amino form (A) as shown in Figure 28-6. In the space provided, indicate the bases that will pair with the bases of the two separated strands during this second round of DNA replication. How do the two duplexes produced by this second round of replication compare with the original duplex (Figure 28-5)? ________________ [66]

How would the presence of this base-pair substitution in a gene that serves as a template for mRNA transcription alter the mRNA transcript? ________________ [67] How many mRNA codons would be altered? ________ [68]

Cytosine can undergo a shift similar to the one we have considered for adenine, with the movement of a hydrogen atom away from an amino group converting the base to its imino form. While in its imino form, cytosine can form hydrogen bonds with adenine, which can, following a round of replication, give rise to a base-pair substitution in one of the duplexes formed.

Both guanine and thymine can undergo configurational shifts that differ from the amino-imino conversion that we have just considered for adenine and cytosine. For these bases, a hydrogen atom normally bonded to a ring nitrogen can shift to a keto group

FIGURE 28-4

Consequences for the second-generation daughter duplexes of a configurational shift that converts the imino adenine in the left strand of the first-generation duplex to its amino form.

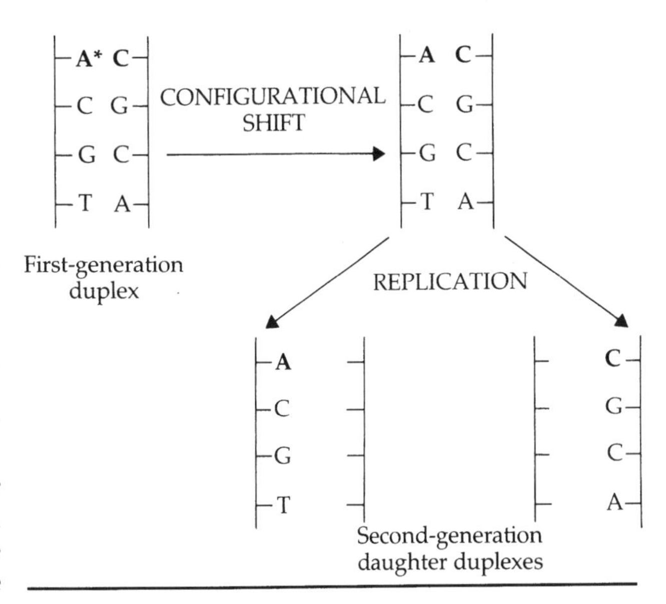

[63] At the site under consideration, one carries the AT pair of the original molecule and the other carries a GC pair substitution.
[64] A different ribonucleotide would be inserted at the site corresponding to the base substitution in the DNA. [65] One.
[66] At the site under consideration, one carries the GC pair found in the original molecule and the other carries an AT pair.
[67] A different ribonucleotide would be inserted at the site corresponding to the base substitution in the DNA. [68] One.

FIGURE 28-5

Replication of the original duplex incorporates imino adenine into one of the daughter duplexes.

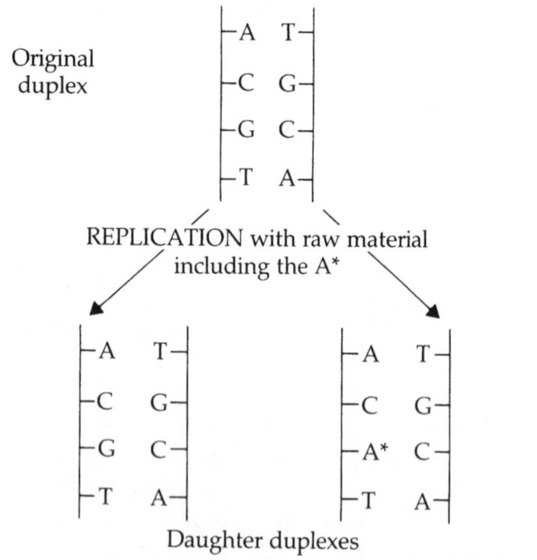

Original duplex

A T
C G
G C
T A

REPLICATION with raw material including the A*

A T
C G
G C
T A

A T
C G
A* C
T A

Daughter duplexes

FIGURE 28-6

Replication of the daughter duplex following the conversion of its imino adenine to amino adenine.

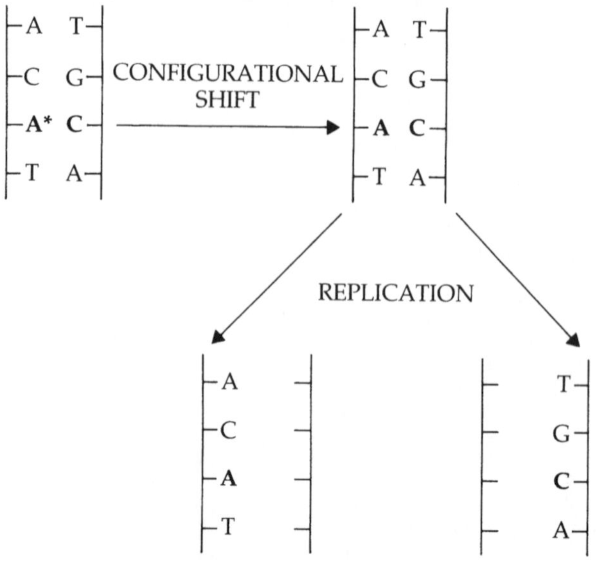

A T
C G
A* **C**
T A

CONFIGURATIONAL SHIFT

A T
C G
A **C**
T A

REPLICATION

A
C
A
T

T
G
C
A

(C=O), converting it to an enol group (C=OH). A base carrying the keto group is said to be in its **keto form** and a base carrying the enol group, in its **enol form.** Figure 28-7 shows both these forms for thymine. The keto configuration (Figure 28-7a) is the form in which thymine almost always occurs in living cells. The hydrogen atom shift produces the enol form (Figure 28-7b). Note that when the hydrogen atom shifts away from the ring nitrogen, the double bond normally found between the 6C and the oxygen shifts so that it is now between the 6C and the ring nitrogen.

In the common keto form, thymine exhibits its standard complementary hydrogen-bond formation with adenine. However, in its enol form, it has an altered hydrogen-bonding capacity allowing it to form hydrogen bonds with the common, or keto, form of guanine, as shown in Figure 28-8.

Guanine can undergo a similar shift, converting from its common keto form to its rare enol form in which it forms hydrogen bonds with the common, or keto, form of thymine.

For each of the bases we have considered, the conversion to its rare tautomere causes the base that it would normally complement to be replaced by a base of the same type. Specifically, a purine gets replaced by a purine or a pyrimidine by a pyrimidine. Consequently, the mutations arising from tautomeric shifts can be classified as transitions.

In addition to the many spontaneous mutations due to tautomeric shifts in the four naturally occurring DNA bases, other mutations arise because the replication enzymes occasionally guide the wrong base into position during DNA synthesis. The mispairings that result can lead to base-pair substitutions like those we have just considered that arise from tautomeric shifts.

Mutagenic Agents and Repair

The rate at which mutations occur spontaneously can be accelerated by exposure to ultraviolet and ionizing radiation and to certain types of chemicals. We will take a brief look at the more important of these mutagenic agents.

Ionizing Radiation Ionizing radiation includes x-rays, gamma rays, and cosmic rays. When this radiation

FIGURE 28-7

Tautomeric forms of thymine.

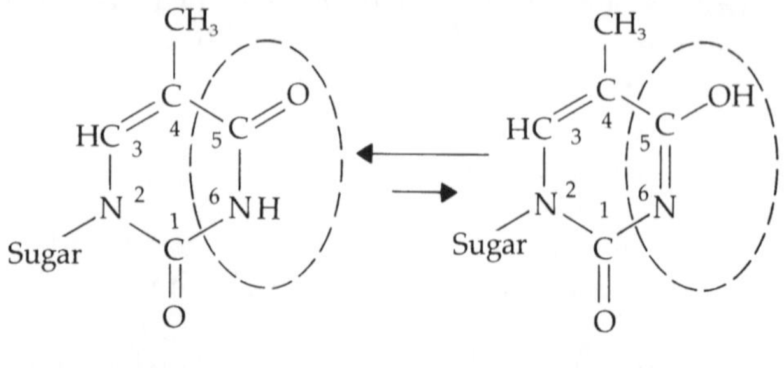

(a) Thymine: common keto form

(b) Thymine: rare enol form

FIGURE 28-8

Mispairing resulting from a rare tautomeric form of thymine.

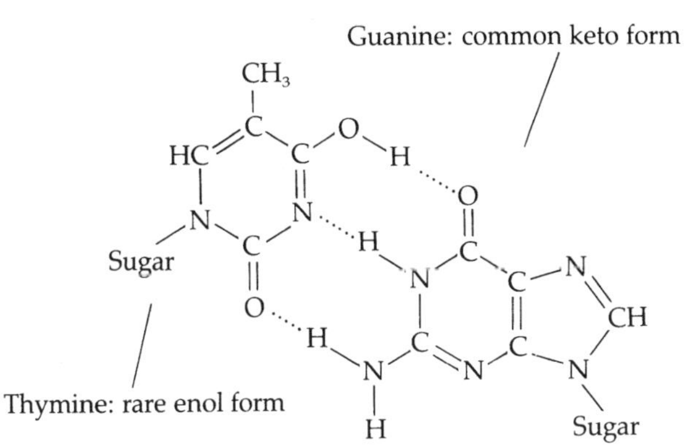

Guanine: common keto form

Thymine: rare enol form

passes through living cells, it collides with atoms, releasing electrons and creating positively charged ions. These ions can alter the nitrogenous bases of the DNA and change their hydrogen-bonding properties so that, during subsequent DNA replication, base substitutions occur in the synthesized DNA strands. At higher doses, ionizing radiation may further alter DNA strands by breaking their sugar-phosphate backbones.

Ultraviolet Radiation The wavelengths of ultraviolet light have too little energy to cause ionization, but they are readily absorbed by the DNA bases, especially the pyrimidines. One effect of ultraviolet radiation is to cause the formation of stable carbon-to-carbon covalent linkages between adjacent pyrimidines in the same DNA strand. These abnormal bonds are most commonly formed between two thymines to form what is known as a **thymine dimer,** but dimers of two cytosines and mixed dimers of cytosine and thymine also occur. Dimer formation causes a distorting bulge to develop in the duplex at the dimer site which may have lethal consequences unless it is repaired. Strands with dimers are not able to serve as normal templates and complementing strands of either DNA or RNA synthesized on them have gaps at the sites corresponding to those occupied by the dimers.

Dimer Removal: Mutation Repair A number of repair mechanisms remove dimers or counter their effects.

Photoreactivation or light repair: DNA molecules with dimers can be converted back to their original dimer-free state by exposure to the blue wavelengths (320–370 nanometers) of visible light. A light-sensitive enzyme, **photolyase,** catalyzes this repair. Apparently the enzyme recognizes the distortion created by the dimer, attaches at that site, and, when activated by blue light, breaks the bond joining the two bases of the dimer. Photoreactivation has been found in a wide range of prokaryotes and eukaryotes.

Excision repair: A more elaborate mechanism for removing dimers is known as **excision repair** or, since it requires no light, as **dark repair.** This process involves an array of enzymes and removes a short DNA segment containing the dimer and replaces it with a newly synthesized segment. The key steps that achieve this repair in *E. coli* are as follows.

1. An endonuclease detects the dimer distortion in the DNA duplex, attaches at that site, and nicks, or breaks, the sugar-phosphate backbone of the strand carrying the dimer on the 5′ side of the dimer (Figure 28-9).
2. A 5′ exonuclease, beginning at the 5′ end created by the nick, works in stepwise fashion to remove the dimer and some additional nucleotides as it moves along the strand in a 5′-to-3′ direction (Figure 28-10).
3. DNA polymerase I, beginning at the 3′ end created by the nick and working in stepwise fashion, adds nucleotides as it moves along the strand in a 5′-to-3′ direction. This process uses the exposed complementary portion of the intact DNA strand as a template and fills in the gap created by the 5′ exonuclease (Figure 28-11).
4. DNA ligase now catalyzes the formation of a phosphodiester bond between the 3′ end of the newly synthesized segment and the neighboring 5′ end of the original section of this strand. This establishes the continuity of the strand that once carried the dimer and completes the repair process (Figure 28-12).

Postreplication repair: An additional repair mechanism occurring in some bacteria including *E. coli* takes place after replication of dimer-carrying duplexes and is thus known as **postreplication repair.** To consider one

FIGURE 28-9

An endonuclease nicks the sugar-phosphate backbone of the strand carrying the dimer.

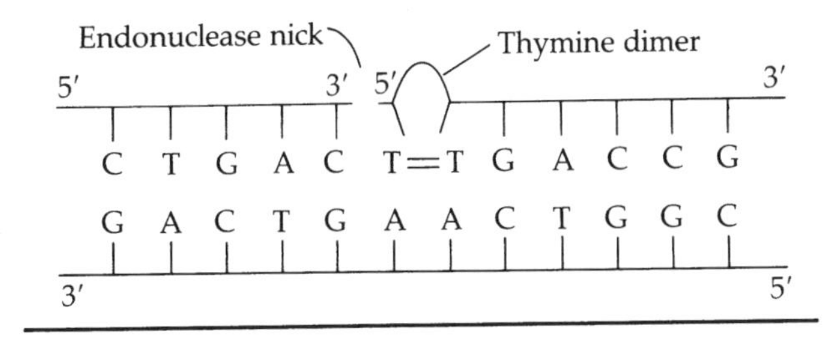

FIGURE 28-10

A 5′ exonuclease, beginning at the 5′ end created by the nick, removes the dimer and some additional nucleotide.

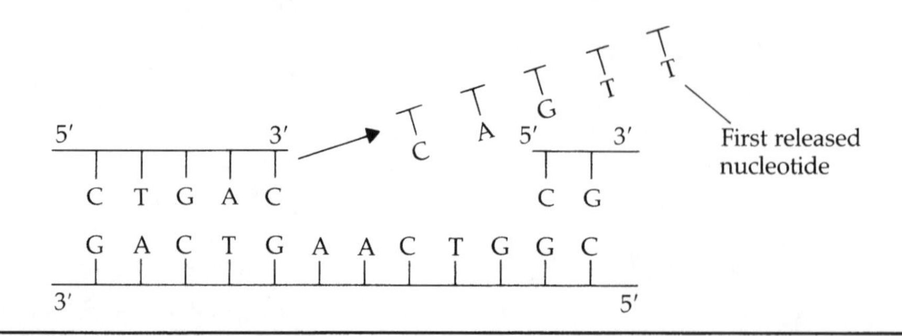

model of how this mechanism may work, assume that a DNA duplex, exposed to ultraviolet light, carries a thymine dimer in one of its strands, as shown in Figure 28-13. When this duplex replicates, both of its strands, of course, separate to serve as templates for the synthesis of complementary strands. Once replication is underway, DNA polymerase is unable to function when it encounters the distorted site of the dimer and skips over the site before resuming replication. When replication is complete, one of the product duplexes (consisting of strands 1 and 2 in Figure 28-14) is defective; newly synthesized strand 2 has a gap, referred to as a **postreplication gap,** opposite the dimer on its template strand 1. The other duplex, consisting of strands 3 and 4, is normal. The gap created because of the dimer may be filled in by a recombination mechanism. One model for how the key events in this process may occur involves the following steps.

1. An endonuclease nicks strand 4 which was derived from the parent duplex and is now part of the intact daughter duplex (Figure 28-15). Strand 4 was the complementary partner of the dimer-carrying strand, 1, in the parent duplex.
2. Moving in the 5′-to-3′ direction away from the nick in strand 4, hydrogen bonds joining bases of the nicked strand to those in its complementary partner, strand 3, are enzymatically broken for a distance along the duplex. A segment of the nicked strand 4 is now transferred to the gap-carrying strand 2. This occurs after the 5′ end created by the nick in strand 4 attaches to gap-carrying strand 2 at the 3′ end adjacent to the gap (Figure 28-16).
3. Following excision of the transferred segment from strand 4, the segment patches the gap opposite the dimer. This transfer creates a gap in the strand contributing the transferred segment (strand 4) that is subsequently filled in by repair synthesis. This synthesis restores the duplex donating the segment (consisting of strands 3 and 4) to its original state (Figure 28-17).
4. The dimer in strand 1 (Figure 28-11) may be removed from the defective duplex by the excision repair system. If it is not removed, the strand containing it will again, in the next round of replication, guide the formation of a complementary strand containing a gap.

FIGURE 28-11

DNA polymerase I fills the gap created by the 5′ exonuclease with newly synthesized DNA.

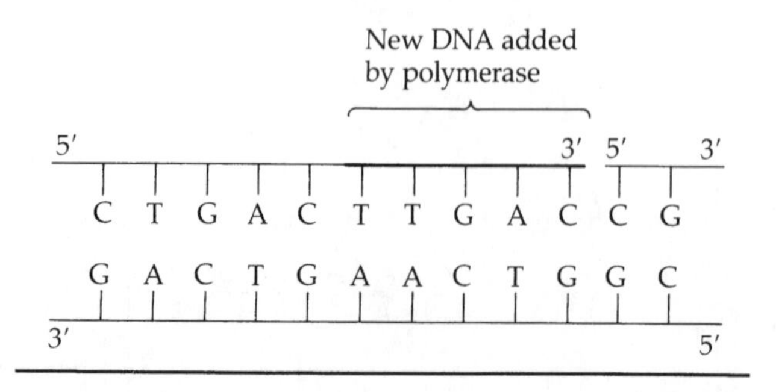

FIGURE 28-12

DNA ligase establishes continuity of the sugar-phosphate backbone of the strand that once carried the dimer, thereby completing the repair process.

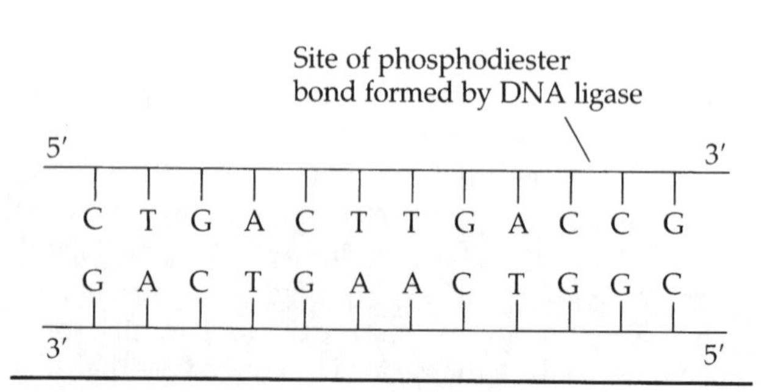

FIGURE 28-13

One strand of the DNA duplex carries a thymine dimer.

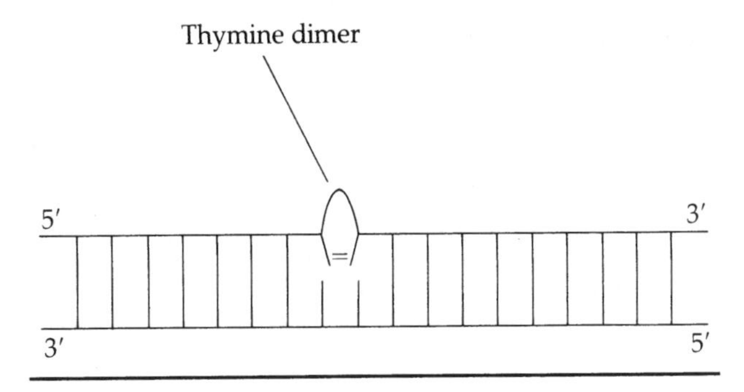

A comparison of mutation rates following ultraviolet exposure among several strains of *E. coli*, each defective for a different repair mechanism, indicates that those with a reduced ability to carry out recombination (because of a deficiency of a protein, rec A, essential to the process) have relatively high mutation rates. What does this indicate about the relative importance of postreplication repair in *E. coli*? ________[69]

SOS repair: There are other types of postreplication repair including one named after the international distress signal, **SOS repair.** This repair which has been much studied in *E. coli* appears to be a final, desperate response that is invoked to repair gaps, including those caused by dimers, in cells experiencing considerable genetic damage. You will recall that DNA polymerase III carries out an editing role by virtue of its 3′ exonuclease ability; it proofreads newly added nucleotides at the 3′ end of a growing DNA strand. Normally, any that are mispaired with the corresponding base in the template strand are immediately removed and replaced with the correct nucleotide. During SOS repair, however, this proofreading function is relaxed and errors generated during synthesis are not corrected, leading to repair that is "error prone." With ultraviolet-induced damage, this system appears to be activated only in emergency situations when DNA is too extensively damaged to be repaired by excision repair. A key role in turning on this system is played by the protein recA which is also involved in recombination. SOS repair allows a cell that would otherwise have lethal gaps in its DNA to form intact duplexes that carry mutations. Sometimes these mutations, like the gaps, bring about the death of the cell but in other cases they may be tolerated, allowing the cell to survive.

FIGURE 28-14

Replication of the duplex shown in Figure 28-13 produces two daughter duplexes, one of which carries the dimer (strand 1) and a postreplication gap (strand 2) and is defective. The other daughter duplex (strands 3 and 4) is normal.

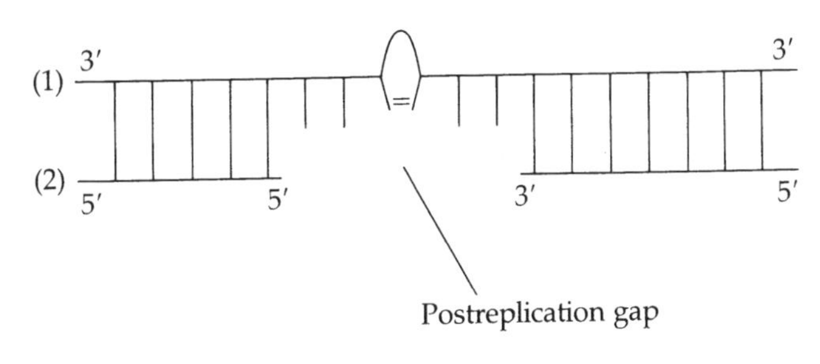

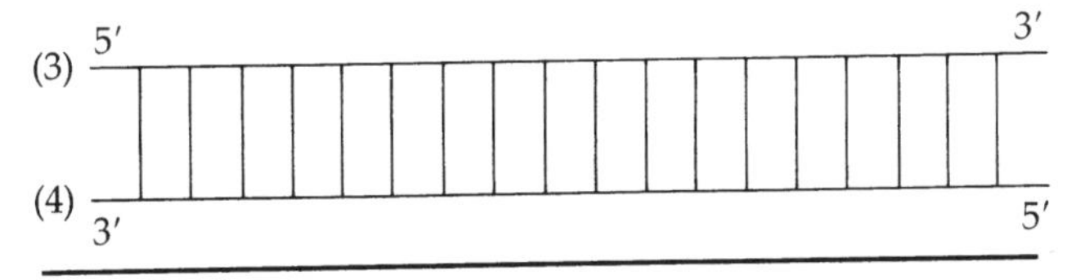

FIGURE 28-15

An endonuclease nicks strand 4.

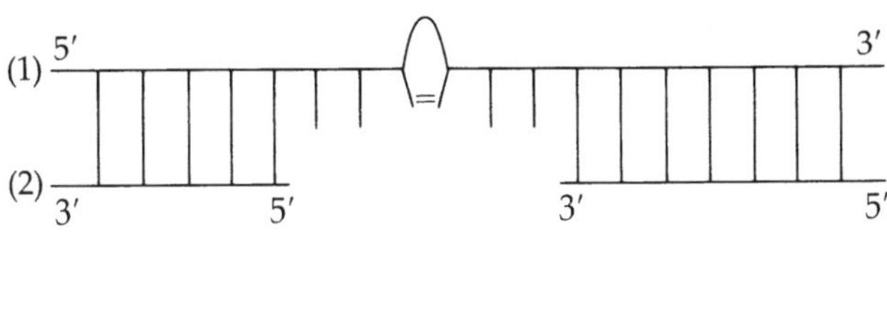

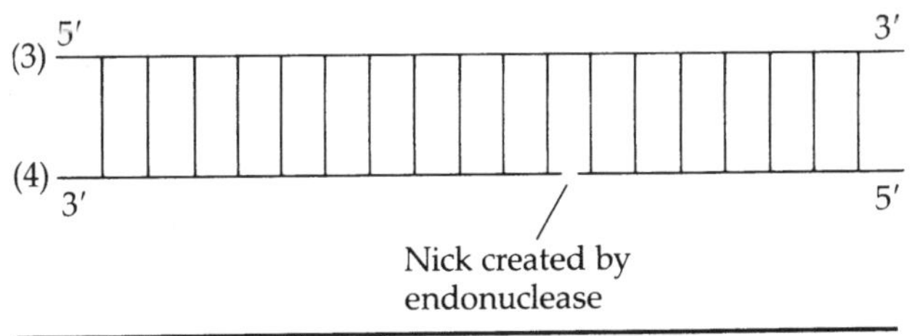

Mutagenic Chemicals

Among the chemicals known to increase mutation rates, three general types are widely used in laboratory studies and operate in reasonably well-understood ways. They include base analogs, alkylating agents, and acridine dyes.

Base Analogs Base analogs are purines and pyrimidines other than the naturally occurring bases that are so similar structurally and functionally to the natural bases that they get metabolized and incorporated into

[69] Since mutation rates are high in its absence, it must be an important repair mechanism.

FIGURE 28-16

A segment of strand 4 is transferred to the gap-carrying strand 2.

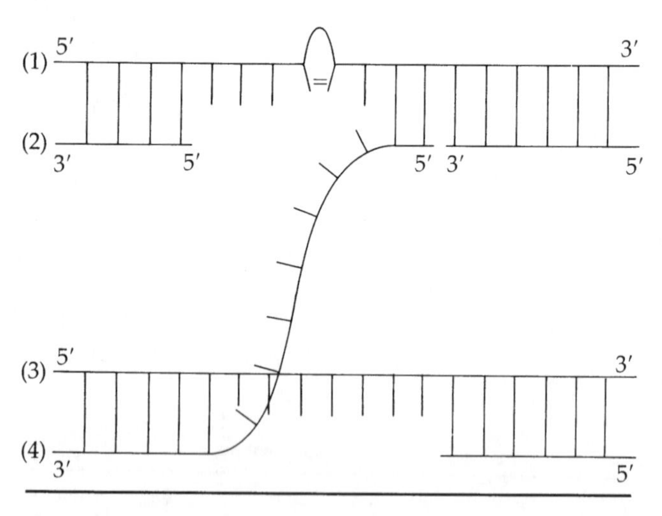

DNA strands during replication. This can lead to problems because the base analogs are configurationally less stable than the naturally occurring bases and therefore are more likely to experience tautomeric shifts. For example, the base analog 5-bromouracil (5-BU) in its common keto form acts like thymine and can substitute for it, forming hydrogen bonds with adenine. In its rarer enol form, however, 5-BU acts like cytosine and pairs with guanine. Mispairings that occur during a round of DNA replication result in the occurrence of a base-pair substitution during a second round of DNA replication. For example, assume that a molecule of the common form of 5-BU is available for incorporation into the daughter duplexes formed by the replication of the following original duplex. Identify the complement synthesized for each of the parental duplex strands in the space provided.[70]

FIGURE 28-17

The transferred segment patches the gap opposite the dimer in strand 2, and the gap created by the transfer in strand 4 is filled in by repair synthesis. The dimer may or may not be removed from strand 1 by the excision-repair process.

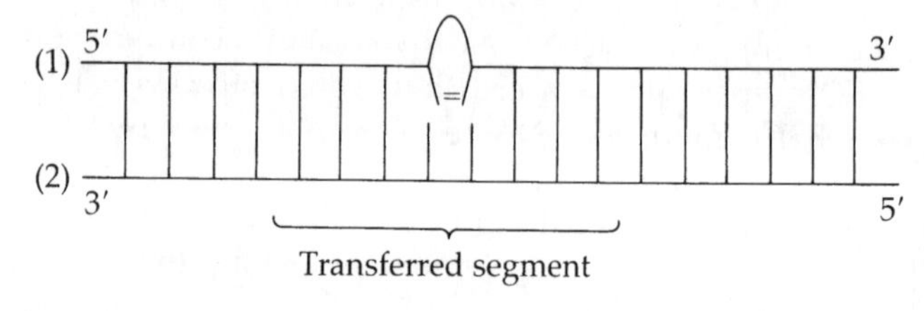

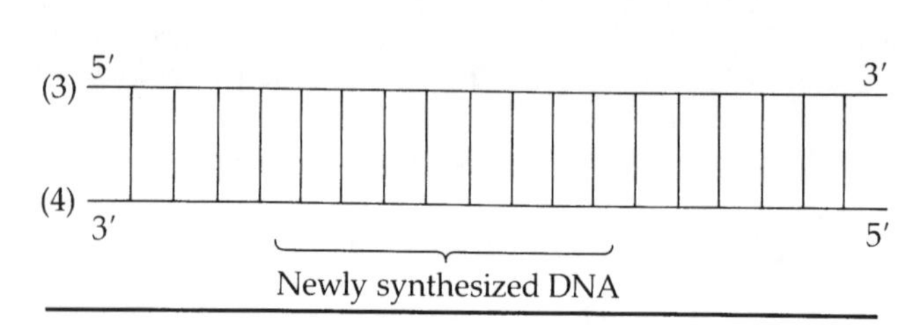

C G
T A
G C
REPLICATION
C
T
G
G
A
C
Original duplex
Daughter duplexes

Now assume that just prior to a second round of replication, the BU in the daughter duplex on the right undergoes a tautomeric shift to its rare enol form (BU*). What base pairs will make up the duplex containing the BU* following this second round of replication? __________[71] Following a third round of replication, what base pair substitutes, in one of the product duplexes, for the TA pair of the original duplex molecule? __________[72]

Now we will consider the consequences of incorporating 5-BU in its rare form (BU*) into newly synthesized DNA. Assume that a molecule of the rare form is available for incorporation into the daughter duplexes formed by the replication of the original duplex in the following sketch. Identify the complement synthesized for each of the strands in this parental duplex in the space provided.[73]

A T
T A
G C
REPLICATION
A
T
G
T
A
C
Original duplex
Daughter duplexes

[70] The daughter duplex on the left consists of base pairs CG, TA, and GC, from top to bottom; and the duplex on the right consists of CG, BUA, and GC. [71] The strand containing the BU* consists of C, BU*, G. With its complement it makes up a duplex consisting of base pairs CG, BU*G, and GC. [72] A CG pair. [73] The daughter duplex on the left consists of base pairs AT, TA, and GBU*, from top to bottom; and the duplex on the right consists of AT, TA, and GC.

Assume that prior to a second round of replication, the BU* in the daughter duplex undergoes a tautomeric shift to its common form (BU). What base pairs make up the duplex containing the BU following this second round of replication? ________[74] Following a third round of replication, what base pair substitutes in one of the product duplexes for the GC pair of the original duplex molecule? ________[75]

Alkylating Agents Alkylating agents such as ethylmethane sulfonate (EMS) and nitrogen mustards are chemicals that transfer alkyl groups such as methyl ($-CH_3$) and ethyl ($-CH_3CH_2$) groups to sites on DNA bases. This alters the capacity of the bases to form hydrogen bonds and, in turn, their base-pairing capabilities. For example, both guanine and thymine can be altered so that they pair with each other.

To illustrate the consequences of this, consider a particular AT base pair in a duplex molecule in a cell in which the bases have been exposed to an alkylating agent. When this duplex replicates, assume that the T of this base pair mispairs with G. Following a second round of replication, what will replace the AT pair found in the original duplex? ________[76] Now assume that the base pair under consideration is CG and that the G of this pair mispairs with T. Following a second round of replication, what will replace the CG pair found in the original duplex? ________[77] As you have just demonstrated, these mispairings, followed by replication, lead to transitions.

The alkylation of guanine can also lead to **depurination:** the sugar-base linkage within a guanine nucleotide breaks, releasing the base from the DNA, as follows.

A T → A T
C G —DEPURINATION→ C
T A → T A

Original duplex

Generally, a repair mechanism removes the remainder of the guanine nucleotide and replaces it with another guanine nucleotide. However, if the repair mechanism is faulty, or error prone, the removed nucleotide can be replaced by any of the nucleotides, and both transitions and transversions can result. What base pair will replace the CG pair of the original duplex after a round of replication if the G is replaced by A? ________[78] if the G is replaced by T? ________[79] if the G is replaced by C? ________[80] Classify each of these base-pair substitutions as a transition or a transversion. ________[81] Alkylating agents are particularly effective mutagens in eukaryotes.

Acridine or Intercalating Dyes Molecules of the acridine dyes such as proflavin or ethidium bromide are able to insert, or **intercalate,** themselves into a duplex DNA molecule between the adjacent base of one or both strands. A dye molecule inserted in a DNA strand serving as a template causes an extra nucleotide to be incorporated in the newly synthesized strand opposite the dye molecule. When that strand subsequently serves as a template, a nucleotide will be incorporated into the newly synthesized strand opposite this extra template nucleotide to generate a duplex with a base-pair addition. This addition results in a frameshift mutation that, when translated, generally causes a significant alteration in the polypeptide product of the gene.

Assume that during the replication of another duplex molecule, an acridine dye molecule inserts itself into one of the newly synthesized DNA strands in place of a nucleotide. Assume that the dye molecule then removes itself before the new strand serves as a template in a subsequent round of replication. What difference would be noted between the duplex formed by this template and the original duplex molecule? ________[82] In general terms, what are the consequences of this type of mutation? ________[83] How could a mutation of this sort be reverted? ________[84]

SUMMARY

This chapter considers two basic categories of gene mutations: (1) substitutions and (2) additions or dele-

[74] The strand containing BU consists of T, A, and BU. With its complement, it makes up a duplex consisting of base pairs AT, TA, and ABU. [75] An AT pair substitutes for the original molecule's GC pair. [76] GC. [77] AT. [78] G replaced by A leads to a TA base substitution. [79] G replaced by T leads to an AT base substitution. [80] G replaced by C leads to a GC base substitution. [81] Transition, transversion, and transversion, respectively. [82] The duplex formed by this template would carry a base-pair deletion. [83] Frameshift mutation. [84] Treatment with the same or similar mutagen could generate a base-pair insertion that would restore the original reading frame; the resultant polypeptide might have the same biological properties as the original polypeptide.

How will the polypeptide produced after this second tions of individual nucleotides (or pairs of nucleotides) in the genetic material. A nucleotide substitution may change a triplet so that it guides the synthesis of a codon that codes for the same amino acid (synonymous or silent mutation), a different amino acid (missense mutation), or termination (nonsense mutation). Substitution may drastically alter the functional properties of the product polypeptide or, possibly, not change it at all. The addition or loss of a nucleotide produces a frameshift mutation that throws off the reading frame from the point of the mutation onward to the end of the mRNA strand and results in a garbled and usually functionless polypeptide. In addition, the altered reading frame may introduce new punctuation or cause the original punctuation to be overlooked.

Nonsense, missense, and frameshift mutations may be reversed by the occurrence of a second mutation known as a reversion. True back-mutations, operational reversions, and pseudoreversions are reversions that change the same nucleotide altered by the initial mutation. A reversing mutation altering a nucleotide other than the one changed by the original mutation is called a suppressor mutation. A suppressor mutation may be intragenic and alter a nucleotide in the triplet changed by the original mutation or another triplet in the gene carrying the original mutation. Other suppressor mutations are intergenic and alter a triplet in a gene other than the one carrying the original mutation, often a gene coding for a tRNA molecule.

Mutations arise in a variety of ways. A tautomeric shift in a nitrogenous base alters its hydrogen-bonding ability and results in abnormal base pairing during replication. With a second round of replication, the abnormally paired base causes a new duplex to carry a base-pair substitution. Other mispairings occur because replication enzymes occasionally guide the wrong bases into position during DNA synthesis. Ionizing radiation can lead to dimer formation which carries lethal consequences unless repaired. Repair mechanisms play a vital role in reducing the mutation level in living cells. Several repair mechanisms, including photoreactivation, excision repair, postreplication repair, and SOS repair are discussed. Mutation rates can also be increased by certain chemicals including base analogs, alkylating agents, and acridine dyes.

PROBLEM SET

28-1. **a.** What is the effect of the addition of a single nucleotide to a single triplet within a functional genetic unit of DNA that is transcribed into mRNA?
b. If additional nucleotides could be inserted into this DNA, what is the minimum number needed to restore the reading frame?
c. Will the restoration of the reading frame restore the polypeptide product of this DNA to its original composition and structure? Explain.

28-2. Assume that in a eukaryotic organism a nucleotide was deleted from a strand of mRNA during transcription. Does this deletion qualify as a mutation?

28-3. A base substitution changes a DNA triplet from GTA to GTT.
a. Specify the kind of mutation in the GTT triplet required to produce a true reversion.
b. Specify the type of mutation in the GTT triplet required to produce an operational reversion.
c. Distinguish between the polypeptide produced following the true reversion and that produced following the operational reversion.

28-4. Human hemoglobin is a complex protein molecule made up of four polypeptides joined to an iron-containing heme group. In normal human adult hemoglobin, hemoglobin A, or HbA, two kinds of polypeptides designated as alpha and beta are found. Two identical alpha and two identical beta chains plus the heme group make up each molecule of hemoglobin A. Hemoglobin S, or HbS, is a hemoglobin variant occurring in individuals affected with the heritable disorder sickle cell anemia. A comparison of the amino acid sequences of the polypeptides in hemoglobins A and S indicates that the alpha chains in the two molecules are identical, but that the beta chains differ in a single interior amino acid.
a. How many kinds of genes are necessary for the synthesis of hemoglobin S?
b. Assuming that the gene for the beta chain of HbS arose from the gene for the beta chain of HbA through a single mutation, what general type of mutation was most likely responsible?
c. The difference between the beta chains of hemoglobin S and A is restricted to the sixth amino acid from the amino ($-NH_2$) end of the polypeptide. In hemoglobin S, the amino acid at that site is valine and in hemoglobin A it is glutamic acid. From your knowledge of the genetic code (and with reference to the genetic code in Table 27-1), what can you deduce about the specific (and simplest) base alteration in the beta polypeptide gene that would produce this amino acid substitution.

28-5. Refer to the basic information given in problem 28-4. Hemoglobin S is only one of many variant human hemoglobins that has been studied. Another is hemoglobin C, which, like HbS, differs from normal hemoglobin (HbA) in a single amino acid substitution in the sixth position from the amino end of the beta polypeptide, where HbC carries lysine. If we assume that HbC arose from HbA through a single mutation, what is the simplest nucleotide change that could have produced this amino acid substitution?

28-6. Refer to the basic information given in problems 28-4 and 28-5. The alpha and beta chains of an additional variant human hemoglobin designated as HbCS are compared with the alpha and beta chains of normal human hemoglobin, HbA. The alpha chains from HbCS and HbA have the same number of amino acids. The beta chain of HbCS has an additional 31 amino acids attached to one end which are not found in the beta polypeptide of

PROBLEM SET

HbA. What is the simplest type of mutation that could be responsible for the presence of these extra 31 amino acids?

28-7. Assume a DNA triplet reading CCC becomes altered by the insertion of an additional C nucleotide. What is the effect of this mutation on the translation of the mRNA transcribed from the gene containing this triplet?

28-8. An analysis of tRNA molecules in bacterial cells carrying the mutation described in problem 28-8 indicates that a type of tRNA that complexes with the amino acid glycine carries an additional nucleotide in its anticodon. Sequencing the nucleotides in this tRNA indicates that its anticodon is CCCC. Studies indicate that this particular type of tRNA causes ribosomes to advance four nucleotides along the mRNA strand rather than the usual three. Specify two ways in which the translation of the mRNA transcribed from the mutant gene described in problem 28-8 will be influenced by the presence of the glycine-tRNA that carries the CCCC anticodon.

28-9. Specifically, what type of mutation produced the suppressor gene described in problem 28-8?

28-10. A gene in a haploid organism produces an mRNA transcript containing a section with three adjacent codons that are identical and that code for leucine. Sequencing mRNA from the same gene following exposure of the organism to a mutagenic chemical indicates that codons at these three sites now code for tryptophan, serine, and leucine. Assume that the mutations that occurred were single-base substitutions. What was the makeup of the codons present in the original mRNA?

28-11. Assume that the single amino acid changes that are listed occur in specific polypeptides. For each, determine the minimal number of codon nucleotides that would need to be changed, specify what the codon changes would be, indicate the change that would be necessary in the DNA strand serving as the template for transcription of the mRNA involved, and identify each mutation as either a transition or a transversion.

a. Lysine to arginine.
b. Leucine to proline.
c. Arginine to tryptophan.

28-12. After a strain of a haploid organism showing the wild-type phenotype is grown in laboratory culture for several generations, it is noted that a few individuals exhibit a mutant phenotype. Some of these mutant forms are isolated and grown in pure culture. After several generations some of their descendents show the original wild-type phenotype. The restoration of the wild-type phenotype might be due to either a true back-mutation or a suppressor mutation.

a. What is the difference between a true back-mutation and a suppressor mutation?
b. Explain how the outcome of a cross between mutant organisms showing the restored original phenotype and organisms of the original wild-type strain would allow you to identify whether the original phenotype was restored by a true back-mutation or by a suppressor mutation.

28-13. Assume that in *E. coli* a base-substitution mutation occurs in the DNA of a gene coding for a polypeptide. The mutation causes the sense codon, GGA, located in the interior of the gene's mRNA transcript, to change to UGA. Another base substitution occurs in a gene transcribing a type of tRNA that complexes with the amino acid tryptophan. Transfer RNA

PROBLEM SET

molecules carrying this mutation can read the codon 5′-UGA-3′. Nucleotide analysis of normal and mutant molecules of this tRNA indicate that both forms have the anticodon 3′-ACC-5′.

a. Is the gene producing this tRNA a suppressor gene, that is, can it suppress the initial base substitution? Explain.

b. What codon would normally complement and be read by the anticodon of this tRNA?

c. How do you explain the fact that the anticodon of this tRNA is able to read the 5′-UGA-3′ codon? (Clue: Does the anticodon *alone* determine the mRNA codon that is read by a tRNA molecule?)

d. Is it possible, from the information given here, to pinpoint the site in the tRNA altered by the mutation?

e. The mRNA carrying the internal nonsense codon can be translated in the presence of the altered tRNA into a polypeptide. Does this polypeptide differ from that synthesized from the mRNA before either of these mutations occurred? Explain.

28-14. A strain of *E. coli* carries a substitution mutation causing the terminator codon 5′-UAG-3′ to occur at an interior position in a particular type of mRNA. This strain also carries a suppressor mutation that alters the anticodon of a type of tRNA molecule so that it reads 3′-AUC-5′. This type of tRNA reads this terminator codon and inserts the amino acid tyrosine.

a. What effect would the presence of both of these mutations together have on the polypeptide translated from the mRNA carrying the mutant terminator codon?

b. Do you expect that there might be other genes whose translation would be altered by the presence of the tRNA produced by this suppressor gene? If so, describe the circumstances under which translation would be altered and explain how it would be altered.

28-15. Assays carried out on the strain of bacteria described in problem 28-14 show that in addition to the mutant tyrosine tRNA carrying the anticodon 3′-AUC-5′, there is the wild-type tyrosine tRNA which carries the normal anticodon 3′-AUG-5′.

a. Can the same gene function in producing both of these types of tyrosine tRNA? Explain.

b. If most of the various kinds of mRNA molecules produced in the cell ending with the terminator codon 5′-UAG-3′ are found to be translated properly, what would you conclude about the relative amount of mutant tyrosine tRNA in the cell?

c. What would be the effect on the translation of the mRNA carrying the internal terminator codon (produced by the initial mutation described in problem 28-14) if the tyrosine tRNA carrying the mutant anticodon is present in low concentrations within the cell?

d. What would you predict about the general health and vitality of a bacterial cell with a relatively low concentration of the mutant tyrosine tRNA and a relatively high concentration of the wild-type tyrosine tRNA?

28-16. Mating a wild-type strain of an organism with a strain carrying both an original mutation and a second mutation that suppresses the original mutation will produce both wild-type progeny and some recombinants possessing the original mutation and others possessing the suppressor mutation.

PROBLEM SET

a. Upon what will the frequency of recombinant progeny depend?
b. Under what circumstance could the recombinant type that carries the suppressor mutation exhibit the wild-type phenotype?

28-17. The pyrimidine 5-bromouracil (5-BU), a base analog of thymine, carries bromine at the 5 position instead of the methyl (-CH_3) group found in thymine. 5-BU undergoes tautomeric shifts between its more common keto form (when it pairs with adenine) and its less common enol form (when it pairs with guanine).

a. A particular site within a duplex DNA molecule carries an AT base pair next to a CG base pair. This duplex replicates in the presence of the rare enol form of 5-BU. With which template base or bases will this 5-BU pair when it gets inserted into the newly synthesized DNA?
b. After incorporation, the enol form of 5-BU in the newly synthesized DNA shifts to its common keto form and the strand carrying it serves as a template for the synthesis of a new counterpart strand during the next round of DNA replication. Following a *third* round of replication, what would replace, in one of the product molecules, the CG base pair of the original duplex?
c. Assume that instead of the enol form, it is the keto form of 5-BU that is available for incorporation when the original duplex replicates. With which template base will it pair?
d. After incorporation, the keto form of 5-BU shifts to its rare enol form, and the strand carrying it serves as a template for the synthesis of a new counterpart strand during the next round of DNA replication. Following a *third* round of replication what would replace, in one of the product molecules, the AT base pair of the original duplex?
e. What can you conclude about the direction or directions in which 5-BU induces transition mutations?

28-18. The acridine dyes interact with DNA to give rise to mutations that generally involve the loss or addition of single bases. A particular wild-type gene produces an mRNA sequence that reads 5′AUGUUUUUUCCCAAACCCGGGUGA3′. A mutation of the gene producing this mRNA, induced using an acridine dye, results in a nonfunctional polypeptide product that consists of the following amino acid sequence: met-phe-phe-pro-lys-thr-arg-val. A subsequent mutation in the same gene, also induced using an acridine dye, restores the function of the polypeptide product, which now consists of the following amino acid sequence: met-phe-phe-pro-lys-thr-gly.

a. Describe, being as specific as you can, the most likely nature of the initial mutation.
b. Describe, being as specific as you can, the most likely nature of the second mutation.

28-19. As described in problem 28-18, acridine dyes interact with DNA to give rise to mutations that generally involve the loss or addition of single bases. A particular wild-type gene produces an mRNA sequence that reads 5′AUGCCCAAAUUUAAAGGGAAACCCUAA3′. An acridine dye–induced mutation in the gene producing this mRNA results in a nonfunctional polypeptide consisting of four amino acids. A second acridine dye–induced mutation occurs in a gene transcribing a certain type of tRNA. This second mutation restores the polypeptide to its original amino acid sequence.

PROBLEM SET

a. Describe, being as specific as you can, the most likely nature of the initial mutation.

b. Describe, being as specific as you can, the most likely nature of the second mutation.

28-20. Describe a pyrimidine dimer. What effect does the presence of a pyrimidine dimer have on the structure of a DNA duplex?

28-21. Describe the key molecular events believed to be involved in excision repair (dimer removal) in *E. coli*. Identify the major enzymes that catalyze these events.

28-22. In general terms, describe the daughter duplexes formed by the replication of a parental duplex that carries a pyrimidine dimer in one of its strands.

28-23. The postreplication gap found in a daughter duplex following the replication of a dimer-carrying parental duplex may be filled in by postreplication repair. Describe the key molecular events in one model of postreplication repair that may occur in *E. coli*.

28-24. Gene conversion (discussed in Chapter 19) occurs only when there is heterozygosity for two different alleles of gene. In addition, gene conversion is always directional: the allele converted is always converted to the other allele carried by the heterozygote. In light of these facts, could the process of gene conversion be explain by mutation?

CHAPTER

29

Gene Regulation

INTRODUCTION

Many genes can be **regulated,** that is, they can be turned on and off at various times for varying intervals in response to changes in a cell's internal and external environment. In addition to regulated genes, there are genes that are permanently switched on and in continual use throughout the cell's lifetime. These **nonregulated** genes guide the formation of enzymes and other polypeptides continually required by the cell. Enzymes that are always produced by a cell regardless of environmental conditions are designated as **constitutive.**

Gene regulation has been extensively studied in bacteria. The environmental conditions experienced by microorganisms can change abruptly. In response, many microorganisms have evolved the capacity to promptly modify their metabolic activities to utilize the nutritional resources that are available to them at any given time. This flexibility depends on (1) the possession of the appropriate genes and (2) the capacity to regulate these genes in response to the presence or absence of particular metabolites in the cellular environment. We will take a look at examples showing how this regulation occurs in two different situations. This chapter also considers transposable genetic elements—segments of double-stranded DNA with the ability to replicate and leave one copy at the original site and move the other copy to a different site within the genome. These elements may play a role in regulating gene action in both prokaryotes and eukaryotes.

AN INDUCIBLE SYSTEM: THE *lac* OPERON

Lactose-Utilization Enzymes and Induction

The bacterium *E. coli* requires an external supply of sugar as a source of energy and carbon. It has a preference for glucose but can utilize a number of other sugars in the absence of glucose. For example, if a cell utilizing glucose is transferred to a glucose-free environment containing an alternative sugar such as lactose, the genes coding for the enzymes necessary to utilize lactose are quickly turned on and the enzymes are synthesized.

Let us take a closer look at this process. Before lactose ($C_{12}H_{22}O_{11}$)—a sugar known as a disaccharide and also as a beta-galactoside—can be utilized by a cell, three events, each mediated by an enzyme, must occur. First, the lactose molecules must be moved inside the cell, a process catalyzed by the enzyme **beta-galactoside permease** (also known as **M protein).** Once inside the cell, some molecules of lactose must be converted into **allolactose,** an isomer of lactose that plays the key role in controlling the lactose-utilization genes, while most lactose molecules are broken into two monosaccharide units, glucose and galactose (see Figure 29-1). The galactose is subsequently converted into glucose and the glucose is used as a cellular source of energy and carbon. The reactions converting lactose into allolactose and into its component monosaccharides are both catalyzed by the enzyme **beta-galactosidase.** In addition to the two enzymes already mentioned, a third lactose-utilization enzyme, **beta-galactoside transacetylase,** whose role has yet to be determined, is also synthesized.

In order for these events to occur, the enzymes catalyzing them must be available within the cell in appreciable concentrations. Does it seem reasonable that bacteria that are metabolizing glucose in a lactose-free environment would expend energy and raw materials to synthesize appreciable quantities of the enzymes for lactose utilization? ________[1] Why or why not? ________________________[2] If no lactose is available, just a few molecules of these enzymes are synthesized by the bacteria, indicating that the genes coding for these enzymes have an extremely low level of expression. In bacteria growing on a glucose medium, do you think the genes coding for the lactose-utilization enzymes are turned on or off? ________[3] If cells growing on a glucose-containing medium are transferred to a medium containing lactose and no glucose, what do you think will happen to the genes coding for the lactose-utilization enzymes? ________[4] What substance, absent in the glucose medium and present in the lactose medium, do you think stimulates these genes to turn on? ________[5] What is the meaning of the term **induce?** ________________________[6] What is induced in the situation we have just considered? ________________________[7] What serves as the **inducing agent** or **inducer,** that is, what directly causes the induction to occur? ________[8]

FIGURE 29-1

Reactions catalyzed by the enzyme beta-galactosidase. Lactose brought into the cell by beta-galactoside permease is either converted to glucose and galactose *(top)* or to allolactose *(bottom)*, the true inducer for the lactose operon of *E. coli.*

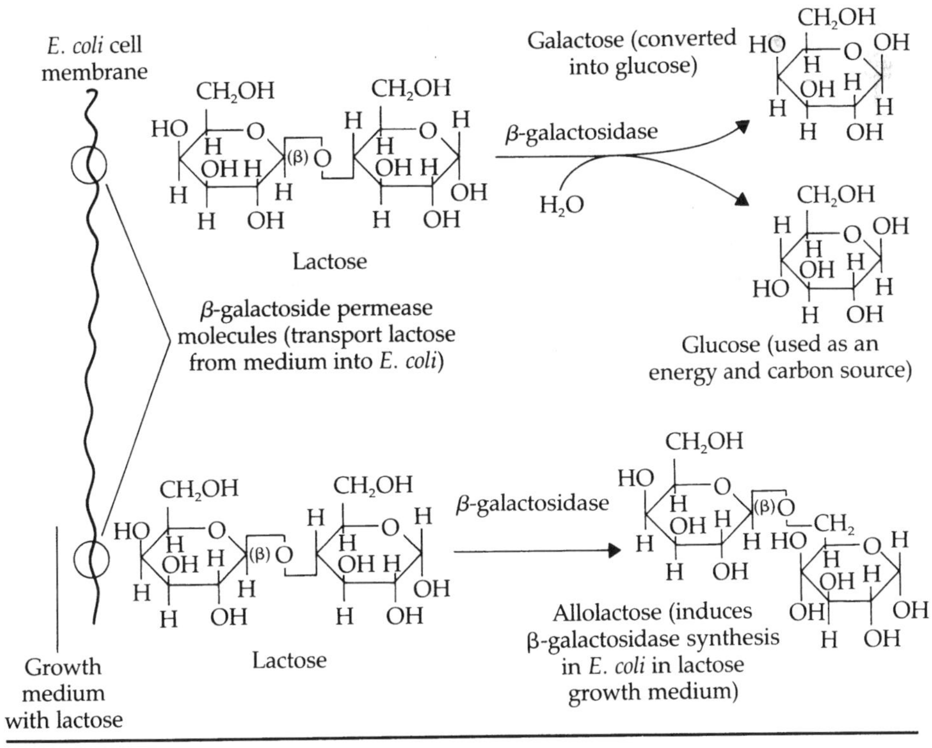

E. coli cell membrane

Lactose

β-galactosidase

H_2O

Galactose (converted into glucose)

Glucose (used as an energy and carbon source)

β-galactoside permease molecules (transport lactose from medium into *E. coli*)

Growth medium with lactose

Lactose

β-galactosidase

Allolactose (induces β-galactosidase synthesis in *E. coli* in lactose growth medium)

Peter J. Russell, *Genetics*, First Edition. Scott, Foresman/Little, Brown, 1986.

[1] No. [2] They have no use for these enzymes. [3] They are virtually off. [4] They will be turned on. [5] Lactose. [6] To cause the formation of something. [7] Lactose-utilization enzymes. [8] Allolactose.

Mapping the Structural Genes

At this point you have an overview of the induction of the lactose-utilization enzymes. Now we will look at some of the details of this induction, including how the genes for lactose-utilization enzymes are switched on and off. Most of this information was discovered by F. Jacob and J. Monod, who received a Nobel Prize for their work. As you will see, mutations in the various genetic elements involved played a vital role in the elucidation of the details of how induction operates. Jacob and Monod treated wild-type *E. coli* with mutation-inducing agents to produce an array of mutant strains. These strains were identified by measuring their production of each lactose-utilization enzyme when each of the mutants was supplied with lactose. Mutations mapping to a gene designated as *z* were found to cause the galactosidase enzyme to be either nonfunctional or not produced at all. Similar mutations mapped to genes *a* and *y* affected the permease and transacetylase enzymes, respectively. The research of Jacob and Monod indicated that each enzyme is coded for by its own cistron or polypeptide-coding structural gene and all three cistrons are located next to each other in a cluster on the *E. coli* chromosome in the sequence $z - y - a$.

As additional mutant strains were studied, it was noted that certain mutations mapping to the *z* or to the *y* loci influenced not only the expression of the gene carrying the mutation but the expression of other structural genes as well. Specifically, mutations in *z* also influenced the expression of *y* and *a*, while mutations in *y* influenced the expression of *a*. Mutations in *a*, however, only influenced the expression of *a*. These findings were interpreted to mean that all three genes were transcribed together to form a single polycistronic mRNA molecule carrying instructions for the synthesis of all three enzymes. A nonsense mutation (which changes a coding codon into a terminator codon) occurring in *z*, the gene translated first, could affect all three polypeptides: the enzyme produced from *z* could be nonfunctional and those from *y* (translated second) and *a* (translated third) not formed at all. Similarly, a nonsense mutation in *y* could interfere with production of the enzymes from *y* and *a*, while the enzyme from *z* would be normal. A similar mutation in *a* would affect only the enzyme formed by *a*; enzymes from *z* and *y* would be normal. Mutations of this sort—which prevent or reduce the synthesis of polypeptides coded for by wild-type alleles beyond, or downstream from, the allele carrying the mutation—are known as **polar mutations.**

Other mutations affecting the regulation of enzyme production did not map to these three structural genes but to genetic regions concerned with the control of the structural genes. For example, some mapped to a region designated as a regulatory gene (*i*) and others mapped to what is referred to as the operator (*o*).

The Promoter: The On-Off Switch

Since the three structural genes are always transcribed together on the same mRNA strand, how many promoter regions would you expect to find associated with these three genes? (Remember that the promoter is a section of DNA where RNA polymerase attaches prior to the initiation of transcription.) ________[9] The on-off control over these genes operates at the level of transcription. When the genes are turned off, there is exceedingly little transcription of them. When these genes are turned on, they are transcribed and the enzymes are synthesized. Consider the relationship between transcription and the interaction of the RNA polymerase and the DNA promoter segment. If RNA polymerase is blocked from attaching to the promoter, can transcription occur? ________[10] If RNA polymerase has access to the promoter, can transcription occur? ________[11] In light of this, what might you reasonably propose for a possible mechanism for regulating transcription? ________________________
________________________[12]

The Regulator Gene, Repressor Protein, and Operator

Whether the RNA polymerase can attach to the promoter depends on a protein designated as the **repressor** that is coded for by a **regulator** or **repressor gene.** This gene is continually transcribed and translated independently of the structural genes for the lactose-utilization enzymes, so that supplies of the repressor protein are always available in the cell. The repressor protein has an active site, known as the **operator site,** that allows it to bind to a short DNA segment adjacent to the promoter. When this binding occurs, access of the RNA polymerase to the promoter is blocked. What effect would this have on transcription? ________[13] Based on what you already know about the functioning of this system, do you think the repressor protein is likely to bind to the operator site in the absence of lactose? ________[14] Would RNA polymerase attach to

[9] One. [10] No. [11] Yes. [12] Allowing or blocking access of the RNA polymerase to the promoter region: if it has access, transcription occurs and if it is blocked, there is no transcription. [13] It would be prevented. [14] Yes.

the promoter under these conditions? _________ [15] Would transcription occur? _________ [16] Would translation of the lactose-utilization enzymes take place? _________ [17] Would the lactose-utilization enzymes be formed? _________ [18]

Now consider what would happen to this system when there is an abundance of lactose. Based on what you already know about this system, do you think that the repressor protein is likely to bind to the operator site? _________ [19] Would RNA polymerase attach to the promoter? _________ [20] Would transcription occur? _________ [21] Would translation of the lactose-utilization enzymes occur? _________ [22] Would lactose utilization enzymes be formed? _________ [23]

To summarize, the operator can be thought of as an on-off switch controlling the three structural genes. Their regulation depends on whether the repressor protein is bound or not bound to the operator site. At this point, we need to focus on what determines whether the repressor protein binds to the operator. One thing that should be apparent is that its ability to bind or not bind is related to the presence of a particular substance in the cell. What is that substance? _________ [24] In the absence of allolactose, the repressor protein binds to the operator; in its presence the repressor does not bind to the operator but instead complexes with the allolactose. Speculate on how the complexing of the repressor protein and allolactose might alter the repressor protein so that the repressor would be incapable of binding with the operator.

____ [25]

What do you think is the minimal number of active sites possessed by the repressor protein molecule? _________ [26]

Assume that *E. coli* bacteria are living in a medium containing a limited amount of lactose and they have synthesized enzymes for lactose utilization. What is going to happen to the concentration of lactose? _________ [27] As the lactose concentration changes, what will eventually happen to the number of molecules of repressor protein complexed with allolactose? _________ [28] If repressor molecules are not complexed with allolactose, what will they be able to do? _________ [29] What effect do you think this will have on the *lac* operon? _________ [30]

The Operon and Its Operation

The type of DNA unit we have been considering here, consisting of a cluster of structural genes and the adjacent controlling elements including the promoter and the operator, is known as an **operon.** Many different inducible operons have been studied in *E. coli* and all show similarities to the *lac* operon. Each operon has its own unique regulator gene and repressor protein. Note, however, that the regulator gene, which is often not contiguous with the operon, is not considered to be part of the operon since it is under separate transcriptional control. For the most part, inducible operons such as the *lac* operon generally code for enzymes participating in catabolic (degradation) processes carried out on substrates that may not be routinely present in the bacterial food medium. The major features of the working of the *lac* operon are presented in Figure 29-2.

To summarize the working of the *lac* operon, the operator controls the access of RNA polymerase to the promoter and can be considered as a switch. The repressor protein, which has the potential to bind with the operator, determines whether the switch is on or off. Whether the repressor protein blocks the operator is determined by the presence or absence of allolactose in the cell. When lactose is absent or present in very low concentrations, the switch is off since the repressor protein is bound to the operator. The switch gets turned on whenever lactose is available to be converted to allolactose which, in turn, complexes with the repressor protein. Before going further, take a moment to describe, in your own words and on a separate piece of paper, the sequence of *lac* operon events that occurs in an *E. coli* cell transferred from an environment free of lactose to one that contains lactose.

The Role of Partial Diploids in the Study of the Operon

Using matings between donor and recipient bacteria (see discussion of conjugation and sexduction in Chapter 22), Jacob and Monod produced partially diploid *E. coli* cells known as merozygotes. For example, the mating of an $F'(i^+)$ donor strain with an $F^-(i^-)$ recipient gives the recipient a genotype of i^+/i^-, making it diploid for this locus. When an $F'(i^+o^+z^+y^+a^+)$

[15] No. It is blocked by the repressor. [16] No, since access of the RNA polymerase to the promoter is blocked. [17] No, since there is no mRNA. [18] No. [19] No. [20] Yes. [21] Yes. [22] Yes. [23] Yes. [24] Allolactose. [25] The binding could alter the shape of the repressor protein so that it no longer binds with the operator. [26] At least two, one for recognition of the operator and another for recognition of the inducer, allolactose. [27] As more and more lactose molecules are utilized, their concentration will decline. [28] It will decline. [29] Combine with the operator. [30] Turn it off.

FIGURE 29-2

An inducible system: the *lac* operon. (a) The regulator gene produces the repressor protein. (b) The repressor protein by itself is active, that is, it can bind to the operator and block access of the RNA polymerase to the operator. (c) When the inducer molecule (•) is present, it complexes with the repressor molecule and changes the repressor molecule's shape. (d) The repressor's altered shape prevents it from binding with the operator. RNA polymerase can now bind to the promoter and transcription and translation of the structural genes occur to produce the enzymes.

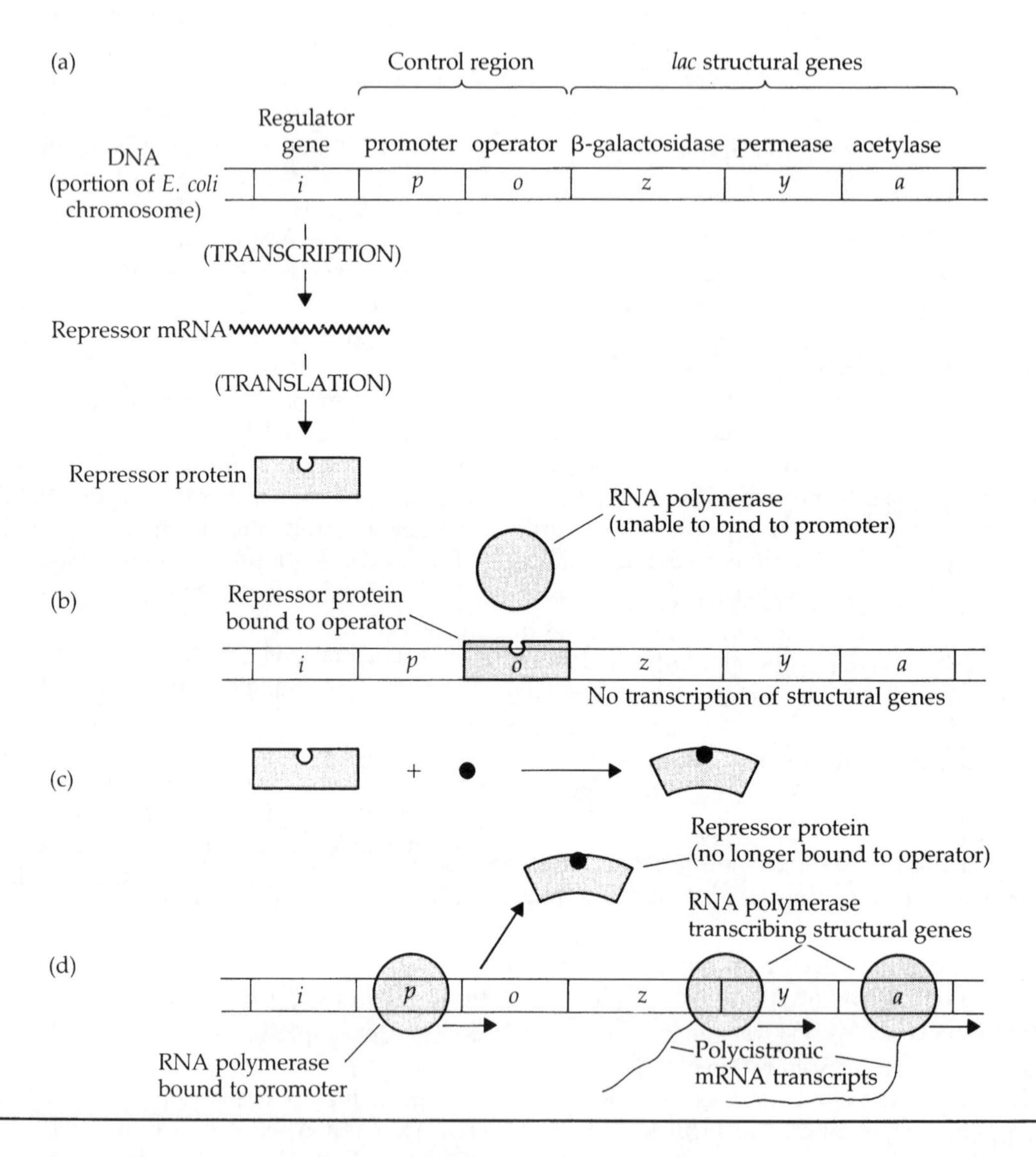

strain mates with an F^- recipient with genotype $i^+o^+z^-y^-a^+$, the recipient will have genotype $i^+o^+z^+y^+a^+/i^+o^+z^-y^-a^+$. Merozygotes were used to great advantage in the study of the *lac* operon. By separately culturing the various types of merozygotes on media with and without lactose, the function of the operon was evaluated by assaying for the lactose-utilization enzymes. As this Chapter's problem set illustrates, this technique was used to assess dominance relationships between alleles of same gene, to determine that the control exerted by regulator (*i*) gene involves a diffusible substance that it produced, and to gather other information that was basic to developing a model of the *lac* operon.

Catabolite (Glucose) Repression of the *lac* Operon

Earlier we mentioned that *E. coli* has a preference for glucose over other possible carbon sources. In other words, if glucose and another sugar such as lactose are present in the medium, the bacteria will metabolize glucose rather than the lactose. (This preference has an energetic basis: the glucose can be supplied directly to the cell's metabolic pathway while the lactose needs energy for processing before it can be used.) As we have just explored in some detail, when lactose is available the *lac* operon will be induced or turned on. If glucose is *also* present in the cell's environment, it can override this induction. We will take a look at the key events involved in this **glucose repression** (also known as **catabolite repression** or the **glucose effect**) of the *lac* operon.

A major role in glucose repression is played by a protein known as the **catabolite activating protein,** or **CAP,** and a section of the promoter adjacent to the RNA polymerase binding site designated as the **CAP site** (see Figure 29-3). RNA polymerase can bind to its section of the promoter and initiate transcription of the structural genes of the *lac* operon only when the CAP is bound to its portion of the promoter. This binding occurs only when the CAP is complexed with a cyclic form of adenosine monophosphate, cAMP. The availability of cAMP in a cell, as you may have guessed, depends on the concentration of glucose. Answering the following questions will show you the way these details fit together. However, before you do this, there are two additional things you need to know: (1) cAMP is produced in the cell from ATP and the reaction is catalyzed by an enzyme called **adenyl cyclase,** and (2) the activity of adenyl cyclase is *inversely* related to the concentration of glucose in the cell.

As you answer these questions, assume that, up to this point, the bacteria we are considering have been living on a lactose medium free of glucose and that the *lac* operon is switched on. When an abundant supply of glucose is added to the medium, what will happen to the activity of the adenyl cyclase enzyme? ________[31]

FIGURE 29-3

CAP site and RNA polymerase binding site of promoter.

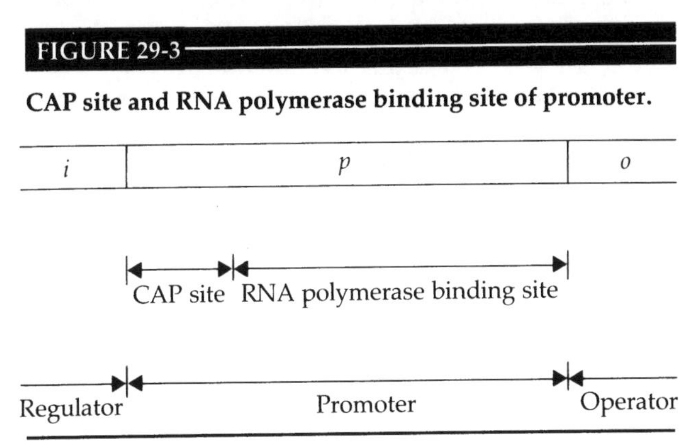

to the concentration of the cAMP? ________[32] to the concentration of the cAMP-CAP complex? ________[33] With the cAMP-CAP complex in low concentration, will it remain bound to the promoter's CAP section? ________[34] Will the RNA polymerase continue to be able to bind with the promoter? ________[35] Will transcription of the structural genes of the *lac* operon continue to occur? ________[36] With no lactose-utilization enzymes available, the bacteria will stop using lactose and will use glucose as their carbon and energy source.

Now assume that the bacteria under consideration have used up all the glucose in the medium although lactose is still available. In the absence of glucose, what will happen to the activity of the adenyl cyclase enzyme? ________[37] to the concentration of the cAMP? ________[38] to the concentration of the cAMP-CAP complex? ________[39] With the cAMP-CAP complex available in high concentration, will it bind with the promoter's CAP site? ________[40] Will the RNA polymerase be able to bind with the promoter? ________[41] Will transcription of the structural genes of the *lac* operon take place? ________[42] With lactose-utilization enzymes available, the bacteria will resume using lactose as their carbon and energy source.

Negative and Positive Control

Now let us compare the type of control exhibited by the repressor protein in regulating the *lac* operon and by the CAP in regulating the *lac* operon once it is induced. Which of these proteins acts by allowing transcription to occur? ________[43] Which of these proteins works by blocking transcription? ________[44] The control exerted by the repressor protein is referred to as **negative control** since it turns the system off. In contrast, the control shown by the CAP turns the system on and is an illustration of **positive control.**

A REPRESSIBLE SYSTEM: THE *trp* OPERON

Tryptophan-Biosynthesis Enzymes and Repression

Up to this point we have focused on the type of operon that becomes turned on or induced when a metabolite

[31] It will be reduced. [32] It will be reduced. [33] It will be reduced. [34] No. Only CAP complexed with cAMP can bind to the promoter's CAP site, and when the concentration of the complex is low, it will not bind to the CAP site.
[35] No. Without the CAP site filled, the RNA polymerase cannot bind to the promoter. [36] No. [37] It will be high. [38] It will increase.
[39] It will increase. [40] Yes. Only CAP complexed with cAMP can bind to the promoter's CAP site, and when the concentration of this complex is high, it binds to the CAP site. [41] Yes, since the CAP site is filled. [42] Yes. [43] CAP, when complexed with cAMP.
[44] Repressor protein.

not routinely present in the cell (and which needs to be broken down, or catabolized) becomes available to the cell. Genes that code for the enzymes necessary to utilize this catabolite are normally turned off and get switched on in its presence. Now we consider a different type of system in which genes that are normally expressed become switched off or **repressed** under certain environmental conditions. Generally, such genes code for enzymes that are routinely needed by the cell to make the particular end product (such as an amino acid) of a biosynthetic pathway. If that end product becomes available to the cell, the enzymes are no longer required and the genes coding for them are turned off. As an example of a repressible enzyme system we will take a look at the *E. coli trp* operon that codes for the enzymes necessary to synthesize the amino acid tryptophan. Many details of how this operon functions were discovered by C. Yanofsky and his coworkers through genetic analysis of bacterial strains carrying mutant forms of the operon's genetic elements.

Like all living things, *E. coli* must have all twenty biologically important amino acids available in order to synthesize the proteins necessary to carry out its life activities. Whenever these amino acids are present in the external medium, the bacteria simply absorb them. However, if one or more amino acids are not externally available, wild-type *E. coli* have the capacity to synthesize them from other substances in their food medium. The synthesis of each type of amino acid may require several stepwise biochemical reactions. Tryptophan synthesis, for example, requires a series of five reactions, each catalyzed by a different enzyme. Each of these enzymes must be present in a cell if tryptophan is to be synthesized and, of course, each enzyme is coded for by a particular structural gene.

If a cell were in an environment where there is little or no tryptophan, would it need the enzymes for tryptophan synthesis? ________[45] Would you expect that the genes coding for the five tryptophan-synthesis enzymes would be turned on or off? ________[46] If cells were in an environment with sufficient tryptophan to meet their requirements, would they need the tryptophan-synthesis enzymes? ________[47] With sufficient tryptophan available, would the genes coding for the five tryptophan-synthesis enzymes be turned on or off? ________[48] What is the substance whose presence or absence determines whether these genes are turned on or off? ________[49] What is being repressed in the situation we have just considered? ____[50]

Components of the Operon

Now let us take a look at some of the details of the repression of the genes coding for the tryptophan-synthesis enzymes. The five structural genes, each coding for one of the five enzymes, are positioned next to each other on the bacterial chromosome. All five genes are coordinately transcribed as a single large polycistronic mRNA molecule that carries instructions for the translation of all five enzymes. The transcription of these genes is controlled by adjacent regulatory sequences that consist of a promoter and an operator, with the operator sequence located within the promoter region. This assemblage of the operator, promoter, and the five structural genes makes up the *trp* operon.

The Regulator Gene, Repressor Protein, Operator, and Co-repressor

As is the case with inducible operons, whether RNA polymerase attaches to the promoter of the *trp* operon depends on a specific type of repressor protein that is coded for by a regulator gene, *trpR*. This gene, which has been mapped at a site some distance from the operon, is transcribed and translated independently of the operon's structural genes. When the repressor protein binds to the operator gene, access of the RNA polymerase to the promoter is blocked and transcription cannot occur.

Based on what you already know about repressible systems and their control, you should be able to answer the following questions. Assume the bacteria that we are considering live on a tryptophan-free medium. Earlier you noted that the tryptophan-synthesis enzymes are produced in the absence of tryptophan. What does this imply about RNA polymerase–promoter interaction; that is, is the RNA polymerase able to attach to the promoter in the absence of tryptophan? ________[51] What does this imply about the interaction of the operator and repressor protein in the absence of tryptophan; that is, is the repressor bound to the operator or not? ________[52] Is the operator switched on or off in the absence of tryptophan? ________[53]

[45] Yes. [46] On. [47] No. [48] Off. [49] Tryptophan. [50] The production of mRNA from the genes coding for the tryptophan-synthesis enzymes. [51] Yes. Since the enzymes for tryptophan synthesis are produced, the RNA polymerase must be able to attach to the promoter. [52] No. If it were bound to the operator, there would be no enzyme production. [53] On.

Now consider what would happen to these same bacteria when transferred to a medium containing tryptophan. Do the bacteria now need the tryptophan-synthesizing enzymes? ________[54] Will the bacteria continue to produce these enzymes now that tryptophan is available? ________[55] Do you think that the RNA polymerase is able to attach to the promoter in the presence of tryptophan? ________[56] What do you think is blocking the RNA polymerase from the promoter? ________[57] Is the operator switched on or off in the presence of tryptophan? ________[58]

Focus for a moment on the repressor protein. As you have just noted, the operator is switched off in the presence of tryptophan, implying that the repressor protein is bound to the operator. In the absence of tryptophan the operator is switched on, implying that the repressor protein is unable to bind to it. Describe a type of molecular interaction between the repressor protein and tryptophan that would permit the repressor protein to bind with the operator in the presence of tryptophan and prevent its binding with the operator in the absence of tryptophan. (Clue: The repressor protein has one active site that recognizes the operator sequence and another that recognizes tryptophan.)

____[59]

Since a repressible system is repressed only in the presence of the end product metabolite such as tryptophan, and since the metabolite itself is essential to the gene's repression, it is referred to as the **co-repressor.**

The Operon and Its Operation

The key features of the working of the *trp* operon are presented in Figure 29-4. To summarize the workings of the *trp* operon, the operator controls the access of RNA polymerase to the promoter and can be considered as a switch. The repressor protein, which has the potential to bind with the operator, determines whether the switch is on or off. Whether the repressor protein blocks the operator is determined by the presence or absence of tryptophan in the cell environment. The switch gets turned off whenever tryptophan is available in sufficient concentration to complex with the repressor protein; this changes the shape of the repressor protein, enabling it to bind with the operator and block transcription. The switch gets turned on whenever tryptophan is present in insufficient amounts to bind with the repressor protein; by itself, the repressor protein is unable to bind with the operator, thereby allowing transcription to occur.

Before going further, take a moment to describe, in your own words and on a separate piece of paper, the sequence of tryptophan-operon events that occurs in an *E. coli* cell transferred from an environment free of tryptophan to one containing tryptophan.

COMPARISON OF INDUCIBLE AND REPRESSIBLE SYSTEMS

At this point, we should compare the similarities and differences of inducible and repressible systems. First the similarities. Each type of system has the following features.

1. Enzyme-coding structural genes that can be regulated, that is, transcribed or not transcribed, depending on environmental conditions.
2. Controlling elements that are adjacent to the structural genes and consist of a promoter site (to which RNA polymerase can attach) and an operator site (to which the repressor protein can bind).
3. A regulator gene that synthesizes a repressor protein that is normally available in the cell at all times. The repressor protein can bind to the operator and complex with either the inducing agent in an inducible system or the end product metabolite (co-repressor) in a repressible system.
4. An operator serving as the on-off switch which controls the transcription of the structural genes. The operator gets blocked (that is, turned off) in either system when the repressor protein binds to it. When the operator is blocked, RNA polymerase cannot move beyond the promoter site and transcription cannot occur.

Now for the differences. Basically there is just one and it relates to the way in which the repressor protein functions. With an inducible system, the repressor protein, by itself, binds to the operator and keeps the system switched off. With a repressible system, the repressor protein, by itself, lacks the configuration necessary to bind with the operator; only when it is complexed with the co-repressor does its shape allow it to bind with the operator and block transcription.

[54] No, since tryptophan is absorbed from the medium. [55] No. [56] No. [57] The repressor protein, which is now bound to the operator. [58] Off. [59] The repressor protein complexes with tryptophan, and the shape of this complex allows it to bind with the operator, thereby turning the operon off. The repressor protein alone, that is, uncomplexed with tryptophan, lacks the correct shape for binding with the operator.

FIGURE 29-4

A repressible system: the *trp* operon. (a) The regulator gene guides the production of the repressor protein. (b) The repressor protein by itself is inactive, that is, it cannot bind with the operator. Consequently, RNA polymerase can join with the promoter and transcribe the structural genes. (c) When the co-repressor molecule (•) is present, it complexes with and alters the shape of the repressor molecule. (d) This altered shape allows the repressor to bind with the operator which, in turn, blocks the RNA polymerase from complexing with the promoter and prevents transcription.

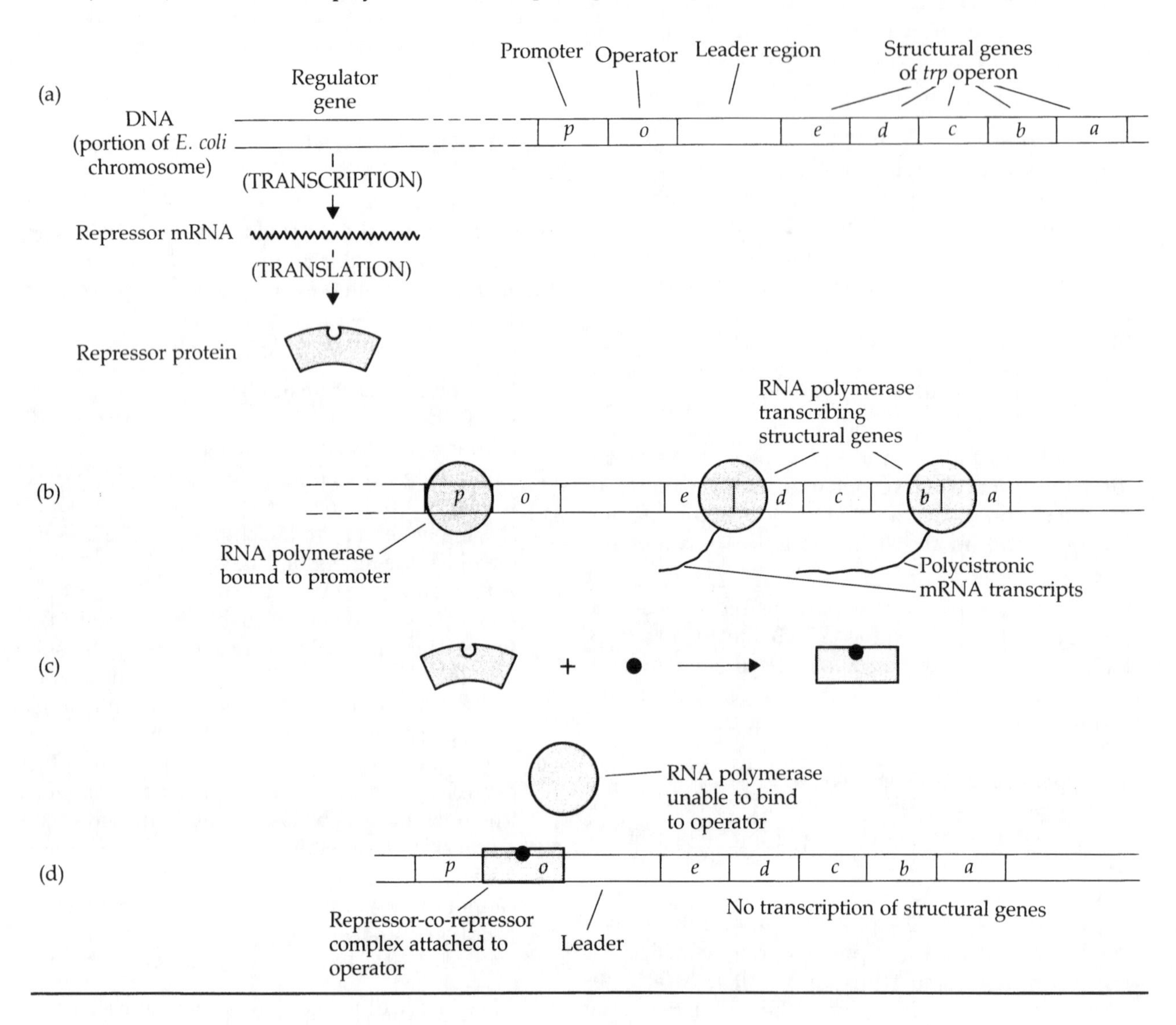

Under the conditions most routinely encountered by a cell, is an inducible system normally on or off? ________[60] What keeps the system off? ______________________________________[61] Under the conditions most routinely encountered by a cell, is a repressible system normally on or off? ________[62] What keeps the system on? ______________________________________[63] Is the repressor protein for a repressible system normally active or inactive? ________[64] How does the repressor protein become active, that is, capable of binding with the operator and blocking transcription? ______________________________________[65]

Both inducible and repressible systems may be regulated by environmental conditions. In the two systems, chemical substances in the environment interact

[60] Off. [61] The absence of the inducing agent; this absence allows the repressor protein to bind to the operator and switch if off.
[62] On. [63] The absence of the end product metabolite; this absence prevents the repressor protein from binding with the operator.
[64] Inactive. By itself it is incapable of binding to the operator. [65] By complexing with the co-repressor.

with the repressor molecule differently. An inducible system, normally off, is turned on by the presence of the inducing agent in the environment. A repressible system, normally on, is turned off by the presence of the end product metabolite (the co-repressor) in the environment.

ATTENUATION CONTROL OF REPRESSIBLE OPERONS

In addition to the system we have just been discussing, there is an additional control mechanism known as **attenuation** that influences the activities of the *trp* operon. Like the regulation exerted by the repressor protein, attenuation is keyed to the concentration of tryptophan in the cell and influences the rate of transcription of the operon. Attenuation can operate whenever the tryptophan operon is turned on, and it results in the concentration of the biosynthetic enzymes being more precisely attuned to the amount of amino acid present than could result from the repression/derepression control mechanism alone. Many of the details of attenuation were first worked out by C. Yanofsky and coworkers.

The Controlling Elements for Attenuation

There are two controlling elements for attenuation and both are found within the leader region of the *trp* operon. This region consists of a total of 162 DNA nucleotide pairs and is located between the promoter-operator region and the first of the five structural genes in the operon. Whenever the *trp* operon is switched on, RNA polymerase can attach to the promoter and begin transcription. The synthesis of the mRNA transcript begins with the first nucleotide in the leader region and normally continues until the terminator sequence is encountered at the far end of the last of the five structural genes. One of the control elements, the **attenuator,** functions in a way that is similar to that of the operator: it serves as an additional on-off switch for transcription. The other control element, the **leader protein region,** is a section of DNA coding for a molecule calling the **leader polypeptide.** These two control elements are shown in Figure 29-5.

The Attenuator

The attenuator is 28 nucleotide pairs long and runs from leader nucleotide pairs 114 through 141 (see Figure 29-5). The distinctive sequence of nucleotides making up the attenuator is very similar to the sequence signaling transcription termination at the end of many bacterial operons. It includes two sections rich in cytosine and guanine that complement each other, as well as a seven-nucleotide-long poly-U section. The attenuator serves as a switch controlling the activity of RNA polymerase. It can either (1) allow the RNA polymerase to proceed, as it normally would when the operator is derepressed, to the end of the fifth structural gene and complete transcription, or (2) block the RNA polymerase from proceeding much beyond the 140th nucleotide of the leader, thereby prematurely terminating transcription when the mRNA is about 140 nucleotides long.

Whether the attenuator allows or blocks transcription depends on the abundance of tryptophan in the cell. Intuitively, what effect do you think a relatively low tryptophan concentration would have on attenuation control of the production of mRNA transcripts from the operon? ______________________________ [66] With a relatively low tryptophan concentration, what effect would attenuation exert on

FIGURE 29-5

Two controlling elements (leader protein region and attenuator) in the leader region of the *trp* operon. Numbers designate the position, in nucleotide pairs, along the leader region (region is 162 nucleotide pairs long).

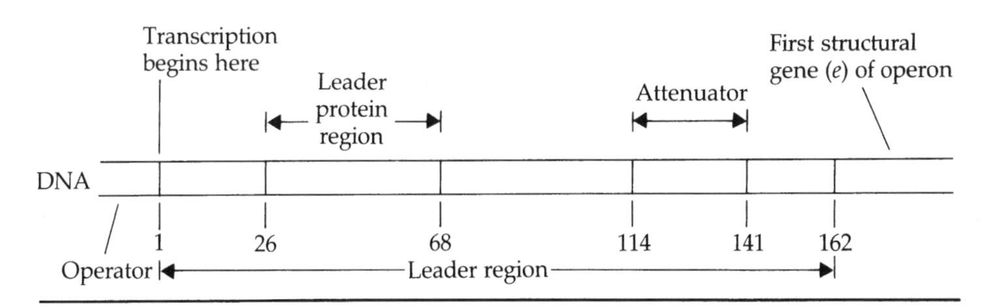

[66] More completed transcripts would be produced.

the production of the biosynthetic enzymes coded for by the operon? ________________________

____________[67] If the tryptophan concentration increases, what effect do you think it would have on the attenuation control of the production of mRNA transcripts from this operon? ________________

____________________[68] With a higher tryptophan concentration, what effect would attenuation exert on the production of the biosynthetic enzymes coded for by the operon? ____________________

____________________[69]

Attenuator Control of Transcription

At this point we will consider how the attenuator controls the transcription of the *trp* operon. (As we do this, remember that the attenuator makes up only a part of the leader region.) A key role in this control is played by the secondary structure assumed by the mRNA transcript of the leader region. This secondary structure arises because this mRNA transcript contains sections of bases that are complementary to each other and can thus form hydrogen bonds with each other. First we need to take a look at the secondary structure assumed by the leader region mRNA transcript *as it is being synthesized.* Within the leader transcript, nucleotides 74 through 85 complement nucleotides 108 through 119. Once enough mRNA synthesis has occurred so that both of these sections have been formed, hydrogen bonding can take place between them to produce the intramolecular configuration shown in Figure 29-6a.

In addition, the mRNA transcript of the leader region also has the potential to take on a different secondary configuration. If you examine Figure 29-6a, you will note that the sections from nucleotide 114 through 121 and from 126 through 134, provided that you ignore nucleotide 127 for the moment, complement each other (nucleotide 114 complements 134, nucleotide 115 complements 133, and so on.) Both of these complementary sections are transcribed from the attenuator. If these two sections are able to form hydrogen bonds with each other, the mRNA transcript of the leader takes on the secondary structure shown in Figure 29-6b.

We mentioned that the attenuator's sequence of nucleotides (nucleotides 114 through 141) is very similar to the sequence that signals transcription termination at the end of many bacterial operons. Since the second mRNA configuration that we considered develops in mRNA transcribed from the attenuator, it should come as no surprise that this molecular configuration is very similar to the configuration that develops at the ends of mRNA transcripts of bacterial operons. Put another way, this configuration serves as a transcription termination signal. In this mRNA, however, the signal arises in an interior portion of the molecule rather than at its end. Such termination signals are known as **hairpin terminators,** so named because their loop configuration resembles a hairpin (Figure 29-6b). When such a configuration arises behind (that is, upstream from) a "working" molecule of RNA polymerase, it causes a change in the configuration of the polymerase molecule which, in turn, prevents it from participating further in transcription.

Now take a comparative look at the two mRNA configurations shown in Figures 29-6a and b. Specifically, compare the nucleotides involved in the intramolecular hydrogen bonding of each configuration and note that some of the same nucleotides are involved in the hydrogen bonding in both configurations. What are those nucleotides? ____________

______________________________[70] Since the participation of nucleotides 114 through 119 in the intramolecular hydrogen bonding is essential to each configuration, can both configurations exist simultaneously in the same mRNA strand? __________[71] If the first configuration we considered is assumed by the molecule during its synthesis, can the second (or terminator) configuration develop? __________[72] Explain.

____[73]

At this point, you know how the RNA polymerase can be blocked from completing transcription of the leader region of the *trp* operon. If the mRNA molecule assumes the hairpin terminator configuration, RNA polymerase ceases its work and transcription comes to a premature halt. However, if the mRNA transcript retains its original, nontermination configuration, transcription proceeds to completion.

The Leader Polypeptide Region

Now we turn our attention to another question. What causes the hairpin terminator to come into existence? One thing that we already know is that the hairpin termination configuration arises when tryptophan is

[67] More would be produced. [68] Fewer complete transcripts would be produced. [69] Fewer molecules would be produced.
[70] Nucleotides 82 through 85 are paired with 108 through 111 in both configurations, and 114 through 119 are paired with 74 through 79 in the first and with 129 through 134 in the second. [71] No. One or the other configuration can exist, but not both. [72] No.
[73] If nucleotides 114 through 119 are hydrogen bonded to nucleotides 74 through 79 to produce the first configuration, they cannot form hydrogen bonds with nucleotides 129 through 134 to form the second configuration.

FIGURE 29-6

Alternative secondary structures assumed by mRNA transcript of the leader region.

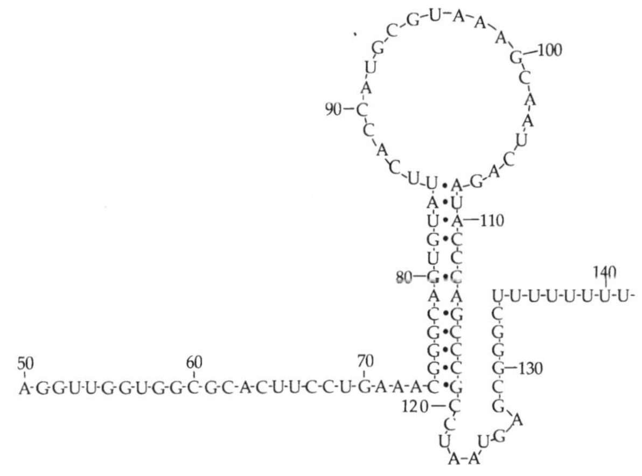

(a) Initial secondary structure

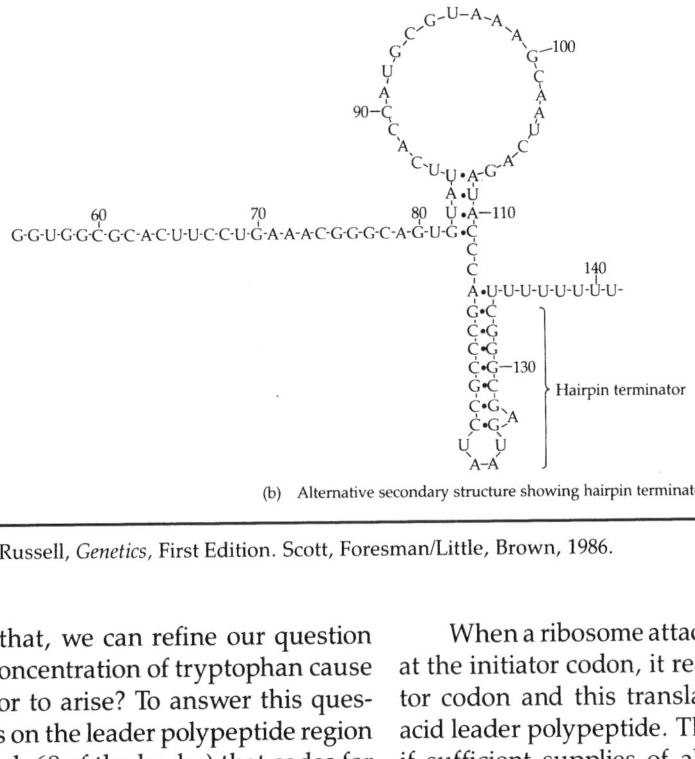

(b) Alternative secondary structure showing hairpin terminator

Peter J. Russell, *Genetics*, First Edition. Scott, Foresman/Little, Brown, 1986.

plentiful. In light of that, we can refine our question to: how does a high concentration of tryptophan cause the hairpin terminator to arise? To answer this question, we need to focus on the leader polypeptide region (nucleotides 27 through 68 of the leader) that codes for a short polypeptide known as the leader polypeptide. The portion of the mRNA transcript complementing the leader polypeptide region consists of an initiator codon (AUG) followed by 13 other codons, and a terminator signal (UGA), as shown in Figure 29-7. Note that two of these codons, which are side by side, code for tryptophan.

When a ribosome attaches to the mRNA transcript at the initiator codon, it reads through to the terminator codon and this translation produces a 14-amino-acid leader polypeptide. This synthesis can occur only if sufficient supplies of all of the component amino acids (or more precisely, their amino acid–tRNA complexes) are available and that, of course, includes tryptophan. What effect will a plentiful supply of tryptophan have on the availability of trp-tRNA^{trp} in the cell? ____________________

________[74] If tryptophan is plentiful, will the ribosome

[74] The trp-tRNA^{trp} should be abundant.

FIGURE 29-7

Portion of the mRNA strand that includes the transcript of the leader polypeptide region, and the polypeptide translated from nucleotides 27 through 71. Note that the 10th and 11th codons code for tryptophan.

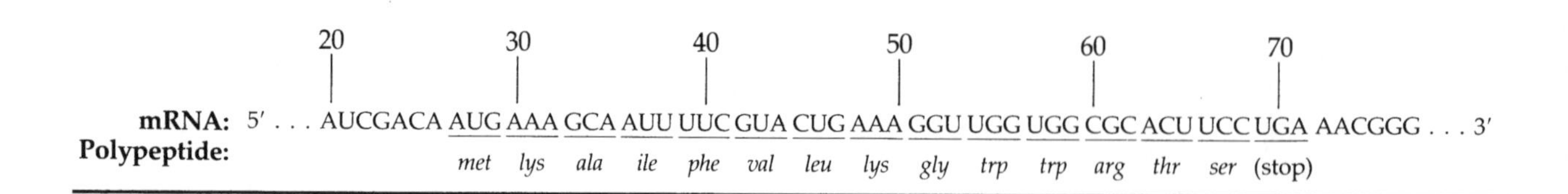

guiding the formation of the leader polypeptide be able to move beyond the two tryptophan codons and complete its reading of the mRNA coding for the leader polypeptide? ________[75] What effect will a scarcity of tryptophan have on the availability of trp-tRNA^{trp} in the cell? ________________________

________[76] If tryptophan is scarce, will the ribosome be able to move beyond the two tryptophan codons and complete its reading of this mRNA? ________[77] What effect do you think this will have on the synthesis of the leader polypeptide? ________________________

________________________[78]

At this point, let us summarize the two key points we have established about the mRNA strand transcribed from the leader region of the *trp* operon. First, there is a relationship between the secondary configuration assumed by the mRNA transcript and the ability of the RNA polymerase to continue the further synthesis of the transcript. Second, there is a relationship between the availability of tryptophan and the ability of a ribosome to complete its translation to the end (terminator signal) of the section of the mRNA strand coding for the leader protein. Now we need to tie these two things together: we must look at how the secondary configuration of the mRNA transcript and the RNA polymerase activity relate to the availability of tryptophan and ribosomal reading of the mRNA transcript.

Linkage of Transcription and Translation

The key to relating these two things lies in the fact that, in prokaryotic organisms, the processes of transcription and translation are linked: a strand of mRNA can be translated as it is being transcribed (see Chapter 26). In the case of the mRNA transcript of the leader region, a ribosome can attach to the initiator codon at the start of the leader polypeptide section while at the same time, a short distance ahead, RNA polymerase is adding nucleotides to create the very same mRNA strand.

Because these two processes are linked, events that occur in one can influence events in the other. Specifically, as the ribosome synthesizing the leader polypeptide moves down the strand of mRNA, it can alter the secondary configuration of the strand of mRNA. Here is how that occurs. If the ribosome can read all 14 of the codons that translate into the leader polypeptide, it will have moved far enough along the mRNA strand to cause nucleotides 114 through 119 to be released from the intramolecular hydrogen bonding of the initial configuration. Once these nucleotides are free, they base pair with nucleotides 126 through 134 to create the hairpin terminator configuration. What effect will this have on the transcription of the rest of the operon? ________________________

________[79] What effect will this have on the availability of the tryptophan-synthesis enzymes within the cell? ________________________

____[80]

Alternatively, if the ribosome gets stalled because of a shortage of tryptophan, it will be unable to read past the first nine codons of the leader polypeptide. As a consequence, the ribosome will not have moved far enough along the mRNA strand to release nucleotides 114 through 119 from the intramolecular hydrogen bonding of the initial mRNA configuration. Since these nucleotides are not available to base pair with nucleotides 126 through 134, the hairpin terminator

[75] Yes. The ribosome will be able to insert tryptophan at the two tryptophan codons and complete the translation.
[76] The trp-tRNA^{trp} will be reduced. [77] No. The ribosome will stall at the *trp* codons. [78] They will be prematurely terminated.
[79] It will prevent the transcription. [80] Their synthesis will be blocked and their concentration reduced.

configuration cannot be created. What effect will this have on the transcription of the rest of the operon? ______ 81

What effect will this have on the availability of the tryptophan-synthesis enzymes within the cell? ______ 82

A Summary of Attenuation

Complete Table 29-1 to summarize the process of attenuation.

Attenuation provides a mechanism for fine tuning the derepression of a repressible operon. It allows the production of biosynthetic enzymes to be more closely keyed to the cellular concentration of a particular amino acid. More than a seventy-fold adjustment in the concentration of the biosynthetic enzyme can occur through repression and derepression, with the system basically being on or off. Attenuation can produce a change of about ten-fold and adjusts enzyme concentrations between the extremes produced by repression and derepression. At this time, attenuation is known to be involved in the operons controlling the production of the enzymes involved in the synthesis of at least six other amino acids in addition to tryptophan.

TABLE 29-1

Summary of attenuation.

	Tryptophan Availability	
	Low	**High**
trp-tRNA^{trp} availability:	______ 83	______ 90
Ribosome reading of leader-protein codons (completed or stalled):	______ 84	______ 91
mRNA secondary structure (hairpin termination configuration or nontermination configuration):	______ 85	______ 92
RNA polymerase activity (transcription continued or terminated):	______ 86	______ 93
Operon structural genes (transcribed or not):	______ 87	______ 94
Tryptophan-synthesis enzymes (made or not):	______ 88	______ 95
Tryptophan (synthesized or not):	______ 89	______ 96

TRANSPOSABLE GENETIC ELEMENTS

Transposable genetic elements are a class of double-stranded DNA segments with the ability to replicate and leave one copy at the original site and move or transpose the other copy to a different site within the genome. These elements come in a range of sizes and are found in viruses, prokaryotes, and eukaryotes. Their existence was first deduced by B. McClintock in the 1940s as a consequence of her work with corn, and for this she received a Nobel prize in 1983. Transposable genetic elements have been identified through the abnormalities they produce in the expression of genes at or adjacent to their insertion sites. Each end of a transposable genetic element typically terminates with a short sequence of nucleotide pairs (approximately 20 to 40) carried in opposite orientations as inverted-repeat sequences. These terminal repeats appear to play a key role in the transposition process.

Prokaryotic Transposable Genetic Elements

The simplest prokaryotic transposable genetic elements, known as **insertion sequences,** or **IS elements,** consist of less than 2000 nucleotide pairs and are routinely found in bacterial chromosomes and plasmids. Each IS carries the genetic information necessary to guide its replication and transposition. A schematic representation of an insertion sequence is shown in Figure 29-8.

Transponsons, or **Tn elements,** are more complex transposable genetic elements. They tend to be much longer than 2000 nucleotide pairs and, in addition to

FIGURE 29-8

Insertion sequence.

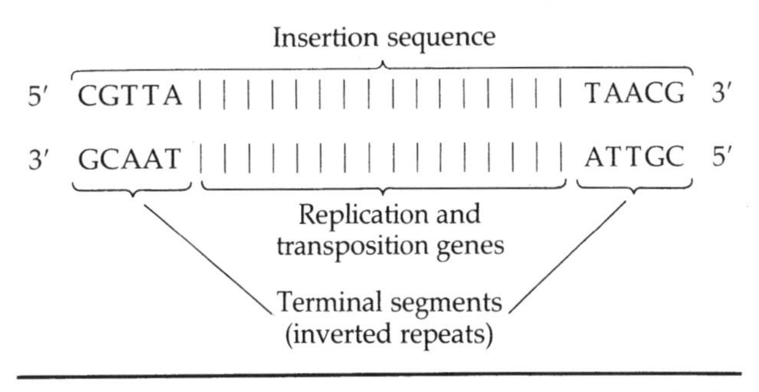

Peter J. Russell, *Genetics,* First Edition. Scott, Foresman/Little, Brown, 1986.

[81] It will allow transcription to be completed. [82] It will allow more synthesis and increase the concentration of enzymes. [83] Low. [84] Stalled. [85] Nontermination configuration. [86] Continued. [87] Transcribed. [88] Made. [89] Synthesized. [90] High. [91] Completed. [92] Termination configuration. [93] Terminated. [94] Not transcribed. [95] Not made. [96] Not synthesized.

genes governing their replication and transposition, carry genes for traits such as drug resistance. Transponsons are commonly found in plasmids and can be transposed to other plasmids as well as bacterial and viral chromosomes. A schematic diagram of a transposon is shown in Figure 29-9.

Both IS and Tn elements carry a gene that codes for a **transponase** enzyme that mediates transposition, along with the genetic elements that control the transcription of the transponase gene. Some transposable elements can insert into a large number of sites within the genome while others have a preference for a limited number of specific sites. Transposable elements are not inserted randomly, but rather into **target sites** within the receiving plastid or chromosome that are recognized by the transponase enzyme. Nucleotide composition of the target sequence for a particular transposable element can vary from one insertion site to another, but the number of nucleotide pairs comprising the sequence is constant.

The frequency of transposition varies with the particular transposable element, but generally falls in the range of 10^{-5} to 10^{-7} per generation. These elements can excise spontaneously but do so with a rate which is about 10^3 times less than the transposition frequency. Spontaneous excision does not include replication prior to removal.

Mechanism of Transposition A number of models of how transposition occurs have been developed although many details have yet to be worked out. Each model must take into account three points.

1. Transposition results in two copies of the transposable element, with one remaining at the original site and the other transposing to the target site.
2. Integration of the transposable element at the target site generates repeated sequences of target DNA that flank the transposable element following integration. These direct repeats are of constant length for each element, most commonly between five and nine nucleotide pairs.
3. Transposition causes the temporary formation of a **cointegrate,** a figure-eight configuration produced by joining the circular genetic unit that donates the transposable element with that receiving the replicate. When transposition is complete, the cointegrate carries two copies of the transposable element. The two joined circular units subsequently separate or **resolve** and assume separate existences, with both the donor and recipient units carrying a copy of the transposable element. This resolution is catalyzed by the enzyme **resolvase** which is coded for by a transposable-element gene.

FIGURE 29-9

Transposon.

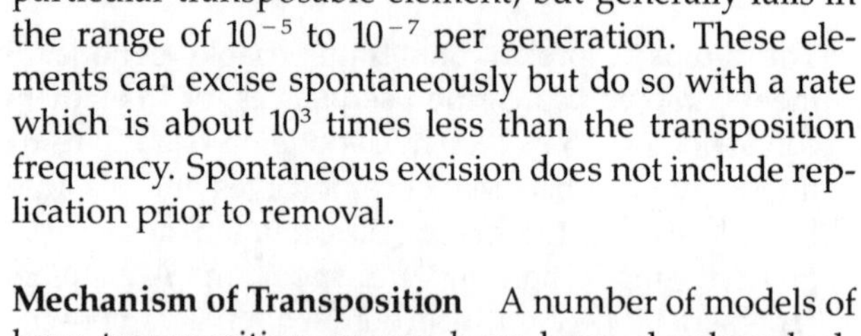

Transposon

5′ TAACT |||||||||||||||||||||||||||||||| AGTTA 3′

3′ ATTGA |||||||||||||||||||||||||||||||| TCAAT 5′

Replication and transposition genes

Antibiotic resistance gene

Terminal segments (inverted repeats)

Peter J. Russell, *Genetics,* First Edition. Scott, Foresman/Little, Brown, 1986.

The following description is one model of how transposition occurs. The donor unit with its transposable element and the recipient unit with its target sequence are shown in Figure 29-10. Both units are double-stranded, circular DNA molecules. The sugar phosphate backbones of both strands of each unit are nicked at the sites indicated by arrows. The nicks in each strand produce staggered cuts in both duplex molecules, as shown in Figure 29-11. The nick in each strand generates both a 3′ and a 5′ end. Note that the 3′ end of each donor-unit strand consists of transposable element nucleotides, and that the 5′ end of each target-unit strand consists of target-sequence nucleotides. The donor and the recipient units join together as shown in Figure 29-12. Note that the 3′ end of each donor-unit strand joins with a 5′ end of a recipient-unit strand. In addition, the 5′ end of each donor-unit strand joins with the 3′ end of each recipient-unit strand. This joining generates two replication forks. With the nucleotide sections of the combined transposable element–target sequence serving as templates, semiconservative replication at each replication fork produces two transposable elements, as shown in the cointegrate in Figure 29-13. The two joined circular units of the cointegrate separate or resolve by a process that involves crossing over between the two transposable elements as shown in Figure 29-14. Note that each end of the transposable element in the target unit is now flanked by direct repeats of the target sequence. Upon separation, both units assume separate existences, with each carrying a copy of the transposable element.

FIGURE 29-10

Donor unit with transposable element and recipient unit with target sequence.

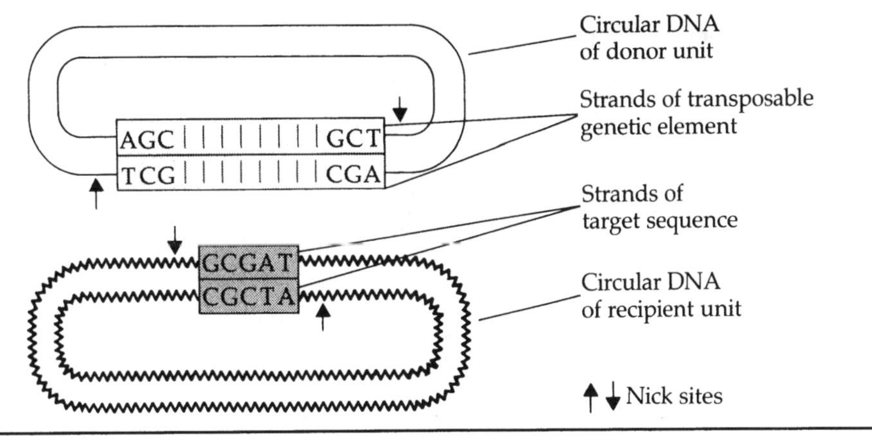

Peter J. Russell, *Genetics*, First Edition. Scott, Foresman/Little, Brown, 1986.

Consequences of Transposable-Element Insertion The incorporation of a transposable element into a prokaryotic gene may alter that gene's expression by acting like a switch and turning the gene on or off, or it may change the gene's rate of transcription. In other cases, the transposable element may alter the expression of neighboring genes, give rise to DNA rearrangements, or produce mutations in adjacent DNA sequences with the exact effect depending on the particular transposable element. For example, the *E. coli* IS1 element, when inserted into the operator-promoter region of the inducible *gal* operon (a series of adjacent genes concerned with galactose metabolism), reduces or turns off expression of all the structural genes "downstream" from the insertion site, regardless of the element's insertion orientation. In other cases the orientation of the transposable element determines the effect. For example, when the IS2 element is inserted into the operator-promoter region of the *gal* operon with one orientation, it turns on the transcription of the structural genes and keeps them on, resulting in constitutive enzyme production. When inserted into this same region with the opposite orientation, expression of the structural genes is reduced during induction. When a transposable element excises from a genetic unit, it may carry some of the adjacent DNA from the unit along with it, resulting in a deletion. Other transposition events can cause inversions and duplications.

Transposons and R Plasmids Transposons play an important role in the formation of R, or resistance,

FIGURE 29-11

Nicks in each strand produce staggered cuts in both the donor and recipient circular duplexes.

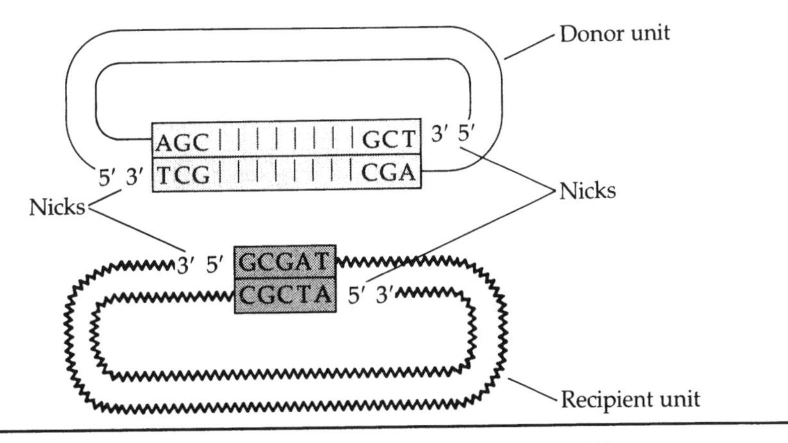

Peter J. Russell, *Genetics*, First Edition. Scott, Foresman/Little, Brown, 1986.

FIGURE 29-12

The donor and recipient units join together. This produces two replication forks.

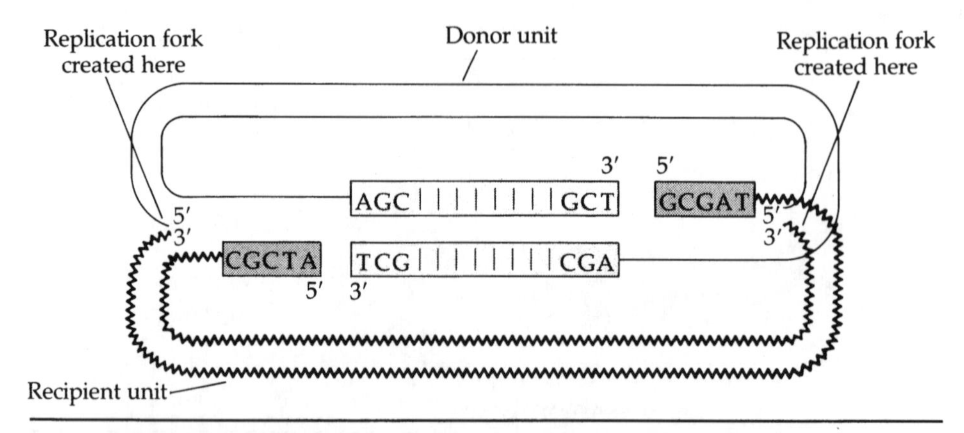

Peter J. Russell, *Genetics*, First Edition. Scott, Foresman/Little, Brown, 1986.

plasmids that are commonly found in bacteria. These plasmids carry one or more **r-determinant genes,** each of which confers resistance to a specific antibiotic. Each r-determinant gene is carried within a transposon and this facilitates its transfer to other plasmids or to the bacterial chromosome. R plasmids are readily transferred during conjugation to cells lacking them. Such transfers are of considerable medical importance since they can quickly convert a population of antibiotic-sensitive bacteria to antibiotic-resistant forms.

Transposons and the F Factor As discussed in Chapter 22, the F factor is a plasmid that controls *E. coli* mating type and can exist autonomously or integrated into the chromosome. The F factor carries four different IS units that are also carried at different sites and, in some cases, in different orientations on the chromosome of different *E. coli* strains. Homologous pairing between an IS in the F factor and another copy of the same IS unit in the chromosome allows the F factor to integrate into the chromosome at different sites and with different orientations to produce *Hfr* cells.

Eukaryotic Transposable Genetic Elements

Transposable elements have been found in corn, yeast, *Drosophila*, humans, and in eukaryotic viruses, and they show structural and functional similarities to the prokaryotic transposable elements. They can insert into many different sites within the host genome (with the production of short, flanking duplicates of target-sequence bases) and produce mutations when inserted into structural genes or controlling elements.

FIGURE 29-13

DNA replication at each replication fork produces two transposable elements in the cointegrate.

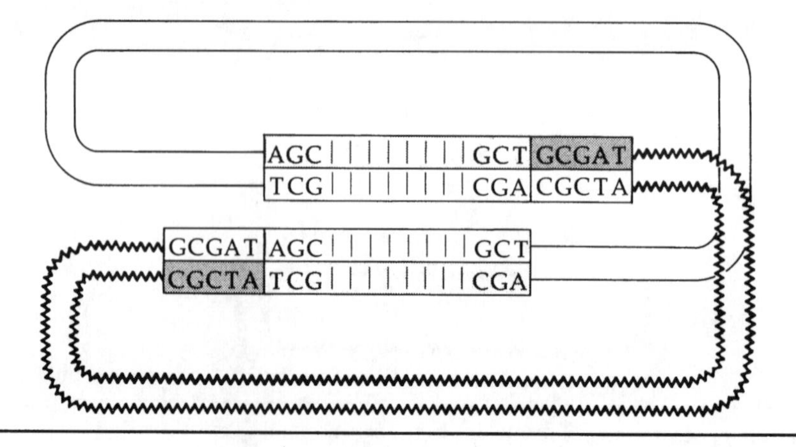

FIGURE 29-14

Crossing over between the two transposable elements separates the two units of the cointegrate.

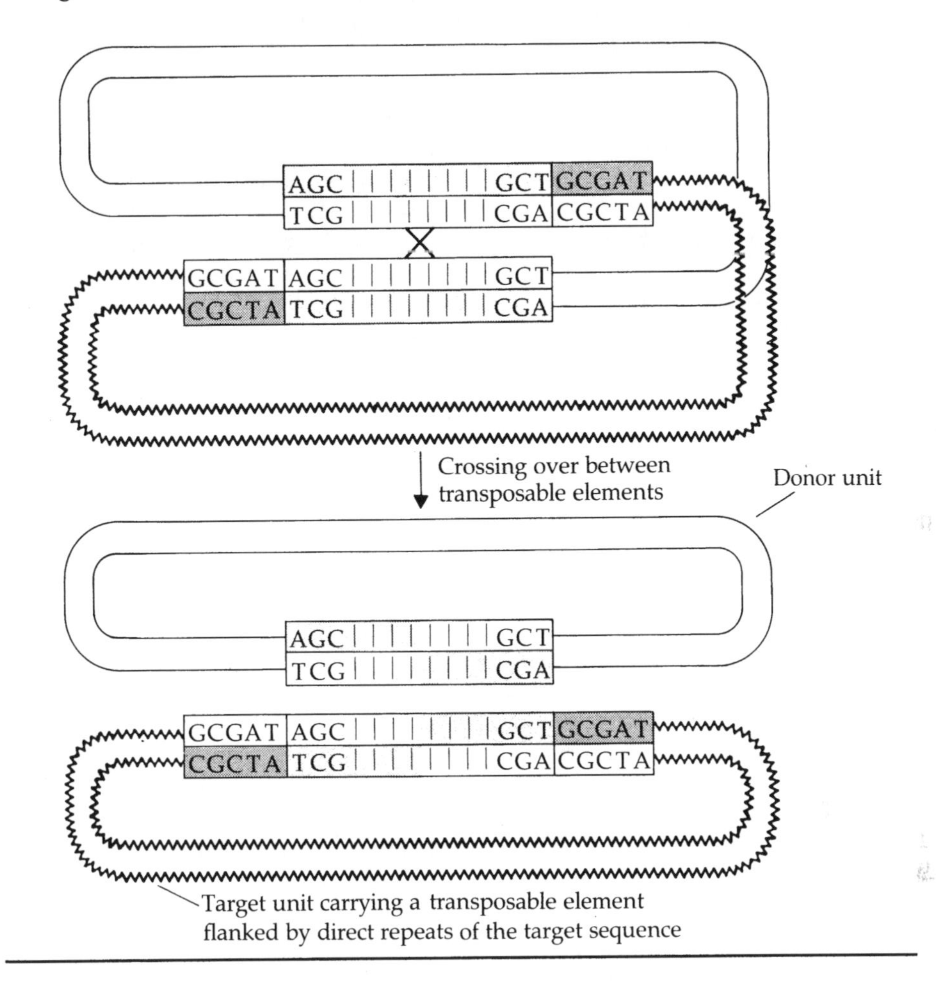

Some examples of these transposable elements will be briefly described.

Corn (Zea mays): Ds (dissociator) and *Ac* (activator) controlling elements are among the transposable elements in corn. The *Ac* units are about 4600 nucleotide pairs long, capable of autonomous transposition, and include a gene that codes for a transposase-type enzyme. *Ds* units are very similar to *Ac* units (for example, their terminal inverted repeats are the same as those of the *Ac* units) but they lack a segment of about 200 nucleotide pairs that apparently codes for transposase. The transposition of *Ds* occurs only when *Ac* is present on the same or different chromosomes to supply the necessary enzyme or enzymes, with the frequency of *Ds* transpositions increasing as the number of *Ac* units in the genome increases. Depending on its insertion site, a *Ds* element can produce chromosome breakage or inhibit gene expression. In the case of inhibition, excision of the *Ds* and *Ac* elements restores the gene to its original expression. In addition to the *Ds* and *Ac* units, the corn genome has an array of other kinds of transposable elements.

Yeasts (Saccharomyces cerevisiae): DNA segments known as *Ty* elements can transpose throughout the yeast genome. There are at least three types of *Ty* elements, and they are approximately 6000 nucleotide pairs long with direct-repeat termini. *Ty* elements occur in multiple copies in the yeast genome, with the most common, *Ty1*, present in about 35 copies. The insertion of a *Ty* element into a gene disrupts the gene's function.

Fruit fly (Drosophila melanogaster): Numerous types of repetitive DNA sequences are widely dispersed in the fruit-fly genome and can function like transposable elements. These sequences, which make up about 5% of the genome, include the mRNA-coding *copia* elements. More than 50 copies of *copia* elements can occur in a single cell at sites that vary from one fly to another. *Copia* elements are similar to yeast *Ty* units and cause mutations when inserted into genes.

Humans (Homo sapiens): Among the highly repetitive sequences of human DNA is a frequently encountered type containing a series of base pairs cleaved by the *Alu*1 restriction endonuclease enzyme. Most members of this *Alu* family are flanked by repeats and may be transposable elements. The human haploid genome carries more than 300,000 copies of *Alu* sequences that are dispersed over all the chromosomes. *Alu* sequences are also found in primate and rodent genomes.

Significance of Transposable Genetic Elements

At this point, the significance of transposable genetic elements is unclear. Two contrasting viewpoints prevail. The first of these is that transposable genetic elements have no functional role. The ability of these parasitic or "selfish" DNA units to insert at a wide range of sites within the genome assures their continued presence, despite the harmful consequences arising from the mutations they cause. The contrasting viewpoint is that transposable elements play a functional role by regulating gene action and facilitating the adaptedness of a population by promoting mutations at certain loci. Despite the present uncertainty over their significance, there is no question that transposable genetic elements have played an important role in the evolution of the genome of some bacteria as demonstrated by the rapid acquisition of antibiotic resistance by many pathogenic forms.

SUMMARY

Genes guiding the formation of constitutive polypeptides (those continually required by the cell) are permanently switched on. Many other genes can be regulated or switched on or off in response to various internal and external conditions. Inducible operons generally code for enzymes participating in catabolic (degradation) processes carried out on substrates not routinely present in the cell's environment. Inducible operons are normally turned off and become switched on or induced when the catabolite or inducer becomes available. Repressible operons generally code for enzymes routinely needed by the cell to make the particular end product of a biosynthetic pathway. Repressible operons are normally expressed and become switched off or repressed if the end product or corepressor becomes available. With both systems, an operator controls RNA polymerase access to the promoter and can be considered as a switch; a repressor protein, with the potential to bind with the operator, determines whether the switch is on or off. A summary comparing these two types of operons is included in the chapter.

If glucose is present and the inducible *E. coli lac* operon is switched on because lactose is available, catabolite repression will override the induction, allowing the cell to utilize glucose. Repressible operons have an additional control mechanism, attenuation, that fine tunes the derepression so that enzyme production is more closely related to the amount of corepressor available to the cell.

This chapter also considers transposable genetic elements—segments of double-stranded DNA with the ability to replicate and leave one copy at the original site and move the other copy to a different site within the genome. These elements may play a role in regulating gene action in both prokaryotes and eukaryotes.

PROBLEM SET

29-1. The *lac* operon in *E. coli* includes three structural genes, each of which codes for a different polypeptide. What prediction could you make regarding the relative amounts of these three polypeptides in the cell at any given time?

29-2. The inducer molecule for the *lac* operon in *E. coli* is a derivative of the lactose molecule known as allolactose. The conversion of lactose into allolactose is catalyzed by the enzyme beta-galactosidase, a product of one of the structural genes of the operon. In light of this, is it possible for the *lac* operon, while in the noninduced state, to be absolutely turned off? Explain.

29-3. State the function of each of the following DNA regions in the operation of the *E. coli lac* operon: promoter region, regulator gene, and operator. Which of these DNA regions codes for a polypeptide product?

29-4. Explain, in your own words, how the repressor protein and the inducer interact with each other and with the operator site to switch the *lac* operon on and off.

29-5. Five structural genes, designated as *d, e, f, g,* and *h*, are known to make up a particular inducible bacterial operon. Each guides the formation of an enzyme necessary to catabolize a particular metabolite. Assume that nothing is known about the sequence of these five genes nor about the manner in which they are transcribed. Mutations mapping to *d* disrupt the production of polypeptides from genes *d, e,* and *h*. Mutations mapping to *g* disrupt the production of polypeptides from genes *d, e, f, g,* and *h*. Mutations mapping to *h* disrupt the production of polypeptides from genes *e* and *h*. Mutations mapping to *f* disrupt the production of polypeptides from genes *f, d, e,* and *h*, while mutations mapping to *e* disrupt the production of polypeptides only from gene *e*. Based on this information, what conclusion, if any, can be drawn about the nature of these mutations (missense or nonsense), the manner of transcription of these loci, and their sequence within the operon?

29-6. The three structural genes making up the *lac* operon (the beta-galactosidase, permease, and transacetylase genes) are designated by the symbols *z, y,* and *a*, respectively, and the wild-type alleles of these genes are symbolized as z^+, y^+, and a^+. The operator, regulator, and promoter are designated by the letters *o, i,* and *p*, respectively, and their wild-type alleles are symbolized as o^+, i^+, and p^+. Strains of bacteria carrying mutations of the o^+, i^+, and p^+ alleles are known. Some of these mutations and their effects are as follows: o^c keeps the operator switched on; i^- keeps the operator switched on; i^s keeps the operator switched off; p^- prevents production of enzymes.

a. Speculate on a way in which each of these mutations might produce its effect.

b. For each bacterial genotype in the following list, indicate whether or not the structural genes of the operon will be expressed (that is, whether or not the enzymes will be produced) when lactose is present and when lactose is absent.
(1) $i^+p^+o^+z^+y^+a^+$ (2) $i^-p^+o^+z^+y^+a^+$ (3) $i^sp^+o^+z^+y^+a^+$
(4) $i^+p^+o^cz^+y^+a^+$ (5) $i^+p^-o^+z^+y^+a^+$

c. Assume that strains of bacteria are available with each of the five genotypes listed in 29-6b. Cells of each type are initially grown on a

PROBLEM SET

lactose-free medium and then transferred to a medium with lactose as the sole carbon source. Samples of cells of each genotype are taken before the exposure to lactose and 30 minutes after the exposure to lactose, and the concentration of their lactose-utilization enzymes is determined. For each genotype, predict the relative concentrations of the lactose-utilization enzymes present in the cells before and after the exposure to lactose.

29-7. Take yourself back to the time when the details of the *lac* operon regulation were being worked out. The existence of the regulator, operator, promoter, and the three structural genes had been documented by identifying strains of bacteria carrying mutations within each of these genetic elements that interfered with normal functioning. Valuable information about the operon and its regulation was obtained by studying strains of bacteria that were partial diploids (merozygotes). These strains were prepared by mating F' cells which served as genetic donors with F^- cells which acted as genetic recipients.

a. The mating of $F', i^+o^+z^+y^+a^+ \times F^-, i^+o^+z^-y^-a^-$ produced merozygotes of genotype $F', i^+o^+z^+y^+a^+/i^+o^+z^-y^-a^-$. These merozygotes behaved like wild-type cells, that is, they could be induced for lactose utilization. What does this outcome allow you to conclude about the dominance relationship of wild and mutant alleles of the structural genes?

b. The mating of $F', i^+o^+z^+y^+a^+ \times F^-, i^-o^+z^+y^+a^+$ produced merozygotes of genotype $F', i^+o^+z^+y^+a^+/i^-o^+z^+y^+a^+$. These merozygotes behaved like wild-type cells, that is, they could be induced for lactose utilization. What does this outcome allow you to conclude about the dominance relationship of the wild (i^+) and mutant (i^-) alleles of the regulator gene?

29-8. Again, take yourself back to the time when the details of *lac* operon regulation were being worked out. The manner in which the two regulatory elements of the operon (that is, the operator, o, and the regulator gene, i) carry out their activities was a subject of major interest to researchers. Two main possibilities were considered. The regulation occurred because the DNA of a regulatory element either (1) acts directly at a DNA site on the structural genes' DNA or (2) acts indirectly, producing a polypeptide product that, in turn, interacts with the structural genes' DNA.

a. Assume that the regulatory element's DNA acts directly at a DNA site on the structural genes' DNA to produce the regulation. Would it be necessary, in a merozygote, for the regulatory element to be carried on the same chromosome as the functional structural genes (that is, in the *cis* arrangement) in order to bring about their expression? Explain.

b. Assume that the regulatory element acts indirectly by producing a polypeptide that, in turn, interacts with the structural genes' DNA to produce the regulation. Would it be necessary, in a merozygote, for the regulatory element to be carried on the same chromosome as the structural genes (that is, in the *cis* arrangement) in order to bring about their expression? Explain.

c. Mating $F', i^+o^+z^+y^+a^+ \times F^-, i^-o^+z^-y^-a^-$ produces merozygotes of the genotype $F', i^+o^+z^+y^+a^+/i^-o^+z^-y^-a^-$ and mating $F', i^+o^+z^-y^-a^- \times F^-, i^-o^+z^+y^+a^+$ produces merozygotes of the genotype $F', i^+o^+z^-y^-a^-/i^-o^+z^+y^+a^+$. Determine whether each of these merozy-

PROBLEM SET

gotes carries the i^+ allele in the *cis* or *trans* arrangement relative to the wild-type alleles of the structural genes.

d. Studies indicate that *both* of the merozygotes described in 29-8c are inducible for the three structural genes. What does this information lead you to conclude about the nature of the control exerted by the i gene?

e. Mating $F', i^+o^+z^+y^+a^+ \times F^-, i^+o^cz^-y^-a^-$ produces merozygotes of the genotype $F', i^+o^+z^+y^+a^+/i^+o^cz^-y^-a^-$ and mating $F', i^+o^+z^-y^-a^- \times F^-, i^+o^cz^+y^+a^+$ produces merozygotes of the genotype $F', i^+o^+z^-y^-a^-/i^+o^cz^+y^+a^+$. Determine whether each of these merozygotes carries the o^+ allele in the *cis* or *trans* arrangement relative to the wild-type alleles of the structural genes.

f. Studies indicate that only the merozygotes described in 29-8e that show the *cis* arrangement are inducible for the three structural genes. What does this information lead you to conclude about the nature of the control exerted by the o gene?

g. Do these two regulatory elements, that is, the operator and the regulator, exert their control through the same basic type of mechanism? Explain.

29-9. Three linked loci carried by a bacterial species are known to play a role in a particular inducible operon where s designates the operator, t, the structural gene producing the enzyme catalyzing the catabolism of the inducer, and u, the repressor gene. Each locus is known to have one wild-type ($^+$) and one mutant ($^-$) allele. Each mutant allele disrupts the normal activities produced by its wild-type counterpart. The following are the genotypes for several haploid and merozygote strains. Indicate, as either high or low, the concentration of the structural-gene enzyme that would be expected for each of these genotypes in the presence and in the absence of the operon's inducer.

a. $s^+t^+u^+$
b. $s^-t^+u^+$
c. $s^+t^-u^+$
d. $s^+t^+u^-$
e. $s^+t^+u^+/s^-t^-u^-$
f. $s^-t^+u^+/s^+t^-u^-$
g. $s^+t^-u^+/s^-t^+u^-$
h. $s^+t^+u^-/s^-t^-u^+$

29-10. What is the value to an *E. coli* cell of the *lac* operon control system involving the catabolite activator protein (CAP)?

29-11. Assume that lactose is present in the environment of an *E. coli* cell and that the cell's *lac* operon is switched on.

a. Identify the proteins interacting with various portions of the DNA of the operon when no glucose is available to the cell. Specify the DNA regions with which each of these proteins interacts.

b. Identify the proteins interacting with various portions of the DNA of the *lac* operon when glucose is available to the cell. Specify the DNA regions with which each of these proteins interacts.

29-12. State the function that each of the following DNA regions plays in the operation of the repressible *trp* operon in *E. coli*: promoter region, regulator (i) gene, and operator (o) gene. Which of these DNA regions codes for a polypeptide product? Do these DNA regions play different roles in a repressible operon than they do in an inducible operon?

29-13. Explain, in your own words, how the repressor protein and the co-repressor interact with each other and with the operator site to switch a repressible operon on and off.

PROBLEM SET

29-14. Some of the preliminary studies on how attenuation operates compared the frequency with which various portions of the *trp* operon were transcribed. Approximately 15% of the time that the leader sequence (*trpL*) of the *trp* operon was transcribed, all five of the structural genes of the operon were also transcribed, with the transcription of both the first structural gene (*trpE*) and the fifth structural gene (*trpA*) occurring with essentially the same frequency. What does this information indicate about the manner in which attenuation operates?

29-15. Despite extensive work, researchers have been unable to identify a repressor protein for the operon controlling the production of enzymes for the histidine-biosynthesis pathway. Nonetheless, this operon is regulated, with the production of biosynthetic enzymes keyed to the cellular concentration of histidine. If, in fact, there is no repressor protein, speculate on how the regulation of this operon might take place.

29-16. Assume that an operon has both attenuation and repressor control. Can attenuation operate while RNA polymerase is blocked by repressor control? Explain.

29-17. A mutant strain of *E. coli* has been identified that lacks the ability to synthesize the repressor protein for the *trp* operon.

a. Based on what you know about the role of repressor proteins, what would you predict about the effect of this mutation when the strain carrying it is grown in the absence of tryptophan and when the strain is grown in the presence of tryptophan?

b. Laboratory studies on this strain show that there is an approximately ten-fold reduction in the production of tryptophan-biosynthesis enzymes when tryptophan is supplied to cells that have been growing in a tryptophan-free medium. Why does this occur? Briefly explain.

29-18. What role is played by the leader polypeptide in attenuation?

29-19. Two mutually exclusive secondary structures of the mRNA transcript of the leader region provide the basis for attenuation control. One configuration allows transcription to proceed, the other blocks it. Why are these structures mutually exclusive?

29-20. Attenuation is known to be involved in the control of several operons coding for the enzymes necessary for amino acid biosynthesis. Isolation and sequencing the complete leader polypeptide for each of these operons indicates that the polypeptide always contains the amino acid whose synthesis is controlled by the operon.

a. Is the inclusion of DNA triplets that code for that amino acid in the leader polypeptide gene critical to attenuation control?

b. Assume that the DNA triplets within the leader polypeptide which code for the operon's amino acid mutated so that they coded for a different amino acid. What effect would this have on attenuation? Explain.

29-21. Identify the similarities and differences between insertion sequences and transposons.

29-22. Assume that the duplex DNA of an insertion sequence is isolated and denatured. Renaturation is then allowed to occur. What configuration or configurations would you then expect to observe under an electron microscope. (Note: Individual DNA strands as well as duplexes can be distinguished under the electron microscope.)

PROBLEM SET

29-23. A mutation is detected in *E. coli* and you wish to determine whether it is due to the transposition of an IS unit or to a point mutation affecting a single base. Identify two features that would indicate that the change detected is due to the insertion of an IS unit rather than to a point mutation.

29-24. The *gal* operon of *E. coli* consists of three adjacent structural genes whose products are involved in galactose metabolism. The temperate λ phage inserts into the *E. coli* chromosome next to the *gal* operon and when λ excises from the chromosome, it may carry the *gal* operon along with it. Assume that λ phages are available that carry a normal, unmutated *gal* operon and others are available that carry a *gal* operon with an insertion sequence. DNA is extracted from both types of phages and denatured. The single-stranded DNA molecules from both types of phages are then incubated together and allowed to hybridize. What configuration or configurations would you expect to see when examining this hybridized DNA under an electron microscope? Explain the origin of each type of duplex molecule observed.

29-25. A transposon with an ampicillin resistance gene (Ap^r) is carried by an *E. coli* R plasmid and may be transferred to an R plasmid with genes for sulfonamide resistance (Su^r) and streptomycin resistance (Sm^r). This is done by isolating DNA from the Ap^r plasmid and using it to transform an ampicillin sensitive (Ap^s) *E. coli* strain carrying the Su^r and Sm^r R plasmid. Following this procedure, Ap^r transformants were selected and found to have the following genotypes: $Ap^rSu^rSm^r$, $Ap^rSu^rSm^s$, $Ap^rSu^sSm^r$, and $Ap^rSu^sSm^s$.

a. Describe how to select for the Ap^r transformants.

b. Speculate on how each transformant genotype arose.

CHAPTER

30

Recombinant DNA and DNA Sequencing

INTRODUCTION

Genetic engineering represents one of the major recent advances in genetics with far-reaching implications for the future. The technology is now available making it possible to insert a fragment of "foreign" DNA into a viral chromosome or into an extrachromosomal DNA molecule of bacterial DNA known as a plasmid. The hybrid molecule that results, known as a **recombinant DNA molecule,** can be introduced into a bacterial host where it replicates and gets transmitted to the next generation of bacteria. Once a large number of bacterial progeny have been produced, many copies of the recombinant DNA molecules can be extracted from these host cells and used for laboratory study. Alternatively, the recombinant DNA molecules can be left in the host cells so that the foreign genetic material can use the metabolic machinery of the bacteria to manufacture its protein product.

RESTRICTION ENDONUCLEASES

Essential to this wondrous feat is a group of enzymes known as endonucleases which break phosphodiester bonds within each of the sugar-phosphate backbones of double-stranded DNA molecules. One class of endonucleases breaks phosphodiester bonds randomly while another class, known as the **restriction endonucleases,** produces breaks at specific points within a DNA molecule. Each kind of restriction endonuclease recognizes a specific nucleotide sequence, often four to six nucleotide pairs long, within DNA molecules that are known as the **restriction,** or **recognition, site.** The breaks produced by the enzyme occur at specific points within these sequences and they arise wherever the restriction site occurs within a DNA molecule. A wide array of restriction endonuclease enzymes has been isolated from a variety of microorganisms. Note that although a restriction enzyme is obtained from a particular species, it will produce cleavages in DNA from *any* source, provided, of course, that the DNA carries the restriction site.

A study of the nucleotides making up restriction sites indicates that they often consist of palindromes, or inverted-repeat sequences, where the sequence of nucleotides making up one strand within the restriction site is inverted and repeated in the complementary strand. For example, in the following recognition site, one strand consists of the sequence AGATCT and its complement consists of this same sequence inverted. Notice that both strands of the recognition site read the same when read in the same direction, say, 5′ to 3′.

5′ . . . AGATCT . . . 3′
3′ . . . TCTAGA . . . 5′

This particular recognition site is recognized by the restriction enzyme designated as *Bgl*II obtained from the bacterium *Bacillus globigii*. A restriction endonuclease's name comes from the bacterium from which the enzyme is isolated: the designation consists of the first letter of the genus (*B*), followed by the first two letters of the species (*gl*), then the designation of strain if there is one (*Bacillus globigii* lacks this), and a chronologically assigned Roman numeral (II) to designate the specific restriction endonuclease isolated from this organism.

The cleavages produced by a restriction endonuclease usually occur at the same position on each strand within the recognition site. For our example, the cleavage in each strand occurs between the A and the G at its 5′ end as indicated by the arrows in the following sketch. Make a sketch in the space to the right of this sequence to show how these two strands would appear following these cleavages.

↓
5′ . . . AGATCT . . . 3′
3′ . . . TCTAGA . . . 5′
↑

Your sketch should show a staggered break that produces a "tail" or an unpaired single-stranded extension, four nucleotides long, on each segment.

5′ . . . A GATCT . . . 3′
3′ . . . TCTAG A . . . 5′

The staggered break produced by the restriction endonuclease results in single-stranded 5′ ends that complement each other. Verify this on the previous diagram.

JOINING DNA FRAGMENTS

Complementary, single-stranded ends of this type are said to be "sticky" since, under the appropriate *in vitro* conditions, such ends can be made to join. The joining is a two-step process that first involves the establishment of hydrogen bonds between the bases of the complementary ends and then the formation of phosphodiester bonds that close the nicks or breaks in the sugar-phosphate backbone of each strand. The enzyme DNA ligase catalyzes the formation or ligation of these covalent bonds between adjacent nucleotides.

Let us take a look at another restriction endonuclease, *Eco*RI (from *E. coli* strain RY13). This enzyme recognizes the following nucleotide sequence and cleaves the strands at the sites designated by the arrows.

↓
5′ . . . GAATTC . . . 3′
3′ . . . CTTAAG . . . 5′
↑

Does the restriction site for *Eco*RI involve inverted-repeat sequences? ________[1] Do the cleavages produced by this enzyme result in single-stranded tails on each of the segments? ________[2] Are the single-stranded ends sticky, that is, do they complement each other? ________[3]

Assume that you have two segments of double-stranded DNA, one from *E. coli* and the other from a chromosome of the fruit fly *D. melanogaster.* These are shown as follows, where each piece of DNA possesses a single restriction site for *Eco*RI and where N represents any nucleotide.

E. coli DNA:
5′ . . . NNNNNNNNNGAATTCNNNNNNNNNNN . . . 3′
3′ . . . NNNNNNNNNCTTAAGNNNNNNNNNNN . . . 5′

D. melanogaster DNA:
5′ . . . NNNNNNNNNGAATTCNNNNNNNNNNN . . . 3′
3′ . . . NNNNNNNNNCTTAAGNNNNNNNNNNN . . . 5′

Each type of DNA is treated with *Eco*RI. Following the *Eco*RI treatment, are the newly created single-stranded ends of the fruit fly DNA fragments complementary to the newly created single-stranded ends of the *E. coli* DNA? ________[4] Provided that DNA ligase is present in the appropriate *in vitro* setup, do you think that the fragments of fruit fly DNA can be joined to the fragments of bacterial DNA? ________[5] If such recombinant DNA molecules are subsequently treated with *Eco*RI, what do you think would happen? ________________[6]

The points we have just covered regarding restriction endonucleases give you some insight as to how these enzymes may be used to manipulate DNA molecules.

[1] Yes. [2] Yes. The cleavage produces a staggered break. [3] Yes. [4] Yes. [5] Yes. [6] Cleavage at the restriction site would release the fruit fly DNA from the bacterial DNA.

RECOMBINANT DNA CLONING PROCEDURES

Now we will consider the steps involved in producing a recombinant DNA molecule and getting a bacterial cell to **clone** or make many copies of it. The process of preparing and cloning recombinant DNA molecules can be divided into six basic steps.

1. Preparing the DNA fragments to be studied.
2. Incorporating this DNA into bacterial or viral DNA to make a recombinant DNA molecule.
3. Introducing the recombinant DNA molecule into the host cells.
4. Isolating host cells carrying the recombinant DNA molecule.
5. Replicating (cloning) of the recombinant DNA by the host cells.
6. Using the recombinant DNA molecules formed.

We will consider each of these steps.

Preparing DNA Fragments to Be Studied.

The DNA to be studied may be synthesized in the laboratory or isolated from a cell. Larger pieces are broken up by treatment with restriction endonucleases, and the resulting smaller pieces are separated from each other and collected for subsequent use. Figure 30-1 shows a fragment of DNA carrying two recognition sites for the restriction endonuclease *Bgl*II. Exposure to this enzyme produces staggered cleavages within each of these recognition sites to form duplex DNA fragments with sticky ends.

Making Recombinant DNA Molecules

The next step is to incorporate the DNA fragment to be cloned, referred to as the **insert,** into a segment of bacterial or viral DNA, known as the **vector.** Most often the vector is a bacterial plasmid, a circular duplex DNA molecule existing in the cell as an extrachromosomal molecule and replicating independently of the bacterial chromosome. As will be described shortly, plasmids can be constructed in the laboratory in a way that allows them to facilitate the screening of bacterial populations to identify cells that carry recombinant DNA molecules. Figure 30-2 shows a circular plasmid carrying a single recognition site for the same restriction endonuclease (*Bgl*II) used to prepare the DNA insert in Figure 30-1. How would you treat this vector DNA in order to convert it to linear form and simultaneously produce sticky ends complementing those of the insert DNA? ________[7] Following this treatment, the vector is incubated with the insert under conditions that cause the complementary single-stranded ends of the two types of DNA to join through hydrogen bonding. What would you then use to establish covalent phosphodiester bonds between the terminal nucleotides of the insert and the vector DNA? ________[8] What is the general term for the type of molecule produced here? ____________________

______[9] What configuration does this molecule possess? ________[10]

Introducing Recombinant DNA into Bacterial Hosts

Bacteria have the capacity to take up DNA molecules from their environment. This phenomenon, known as

FIGURE 30-1

Cleavage of DNA to form a sticky-ended fragment.

```
            ↓                          ↓
5′ . . . NNNNNAGATCTNNNNNNNNNNNNNNNNNAGATCTNNNNNN . . . 3′
3′ . . . NNNNNTCTAGANNNNNNNNNNNNNNNNNTCTAGANNNNNN . . . 5′
                ↑                          ↑

5′ . . . NNNNNA       GATCTNNNNNNNNNNNNNNNNNA        GATCTNNNNNN . . . 3′
3′ . . . NNNNNTCTAG       ANNNNNNNNNNNNNNNNNTCTAG        ANNNNNN . . . 5′
                      DNA fragment to be cloned
```

[7] Treat it with *Bgl*II. [8] DNA ligase. [9] Recombinant DNA molecule. [10] Circular.

FIGURE 30-2

Circular plasmid vector.

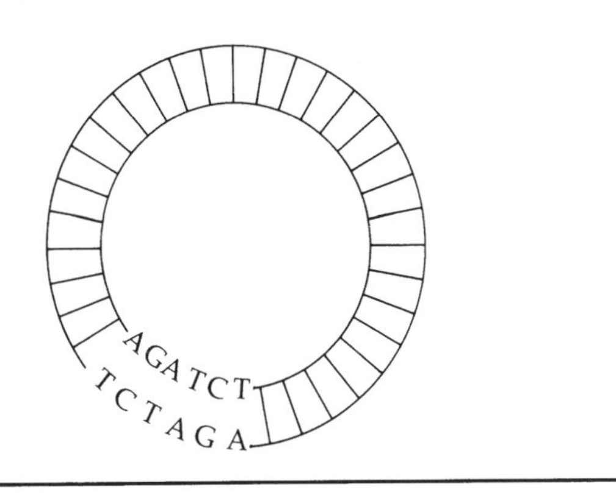

transformation (see Chapter 21), includes the uptake of plasmids that carry foreign DNA inserted through genetic engineering. *E. coli* is the bacterial host commonly used in these studies. Through an increase in the level of calcium ions in the environment, the cell wall of *E. coli* can be made more permeable to extraneous DNA, thereby increasing the cell's uptake rate for recombinant DNA plasmids. Even with heightened calcium levels, however, a very limited number of bacteria will take up a plasmid, and consequently, it is important to have some way to be able to identify those bacteria that have been transformed. Furthermore, some of the plasmids taken up by the bacteria may be "standard" plasmids that have failed to accept the foreign DNA insert. So, it is necessary to have a screening system to distinguish between three classes of bacteria—those with no plasmid, that is, untransformed bacteria; those with a standard plasmid; and those with a recombinant DNA plasmid.

Isolating Bacterial Hosts Carrying Recombinant DNA

The basis for this screening is provided by certain genes carried by the plasmid. Most commonly these confer antibiotic resistance. For example, assume that the type of plasmid used in our example was constructed (using DNA sections from other plasmids, restriction endonuclease, and DNA ligase) so that it contains genes for tetracycline and for ampicillin resistance, symbolized as *tet*ʳ and *amp*ʳ, respectively. Furthermore, this plasmid was designed so that it carries its single restriction site for *Bgl*II within one of the genes for antibiotic resistance, say, *tet*ʳ. Within plasmids that have accepted the insert, where do you think the insert would be located? ____________________ [11] What effect do you think that the incorporation of the foreign DNA at this site would have on the expression of the *tet*ʳ gene? ____________________ [12]

The bacteria to be screened are plated on agar containing ampicillin and allowed to grow into colonies. Which type or types of bacteria (untransformed, transformed with the standard plasmid, or transformed with the recombinant DNA plasmid) will grow on this medium? ____________________ [13]

Bacteria from each colony growing on the ampicillin medium are transferred by replica plating (see Chapter 20) to a medium containing tetracycline. Which type or types of bacteria will grow on this medium? ____________________ [14] How is it possible at this point to identify bacteria carrying the recombinant DNA plasmid? ____________________ [15]

Which type or types of bacteria would you select for cloning? ____________________ [16]

Replicating, or Cloning, Recombinant DNA in Bacterial Hosts

Once identified, samples of the bacteria carrying recombinant DNA plasmids are cultured to produce a large population of cells. Since a plasmid replicates independently of the main bacterial chromosome, each cell can possess a number of copies of the plasmid. Consequently, a large number of cells contain an even larger number of plasmids.

Using Recombinant DNA

When the bacteria are available in sufficient quantity, one of two things generally happens, depending on the researcher's interest in the foreign DNA. The goal, for example, may be to study the size and the nucleotide sequence of the foreign DNA. This requires the isolation of the foreign DNA segment and is accom-

[11] Between two sections of the *tet*ʳ gene. [12] It would block its expression. [13] Growth requires ampicillin resistance and that indicates the presence of the plasmid. Thus, the bacteria that grow are transformants of both types: those carrying the standard plasmid and others carrying the recombinant DNA plasmid. [14] Those with the functional *tet*ʳ allele; only the recombinant bacteria carrying the standard plasmid have this allele. [15] Remember that failure to grow on the tetracycline indicates a *tet*ʳ gene inactivated by the insertion of the foreign DNA. Compare the replica plate with the master plate; colonies growing on the ampicillin medium that do not grow on the tetracycline medium consist of bacteria carrying the recombinant plasmid. [16] Those carrying the recombinant DNA plasmid. They can be recovered from the master plate.

plished by extracting the plasmids from the bacterial cells and treating them with the same restriction endonuclease used in the initial preparation of the insert and the vector DNA. What will be the effect of this treatment on the isolated recombinant DNA plasmids?

______________________________ 17

Following the separation of the foreign and the plasmid DNA, many copies of the foreign DNA segment will be available for laboratory analysis.

Another possible goal is the study or the exploitation of the expression of the foreign DNA. Under certain conditions, a transformed bacterial cell can be made to translate the foreign DNA into the polypeptide product that it codes for, just as if it were bacterial genetic material. (Note that the foreign DNA must be located adjacent to the appropriate regulatory signals in the plasmid, including a transcriptional promoter and terminator sites, if the bacterial cell is to be able to make the foreign-gene product. If the foreign DNA is eukaryotic, additional genetic manipulation is generally required.) With a large number of bacteria carrying the foreign DNA, it may be possible to synthesize large quantities of the polypeptide product. Through genetic engineering, bacteria have been made to produce such medically useful substances as human insulin and the antiviral agent interferon.

Viruses as Vectors

Up to this point we have focused on plasmids as vectors. Certain viruses can also serve as vectors for transferring foreign DNA into bacterial or eukaryotic cells. For example, with certain forms of the λ bacteriophage of *E. coli*, it is possible, using the appropriate restriction endonuclease, to remove a central or so-called "stuffer" section from the linear viral chromosome. Treatment of foreign DNA with the same restriction endonuclease yields fragments that can join with the two end sections of the viral chromosome and become the new central portion of the chromosome. Under the appropriate conditions, this recombinant chromosome can be packaged inside of a viral coat to form a viral particle that, in turn, can infect a host cell. Although a significant portion of the viral chromosome is no longer present (it has been replaced with foreign DNA), the viral genetic material that is present enables the virus to complete its reproductive cycle and lyse the host cell. Obviously, with each round of viral reproduction, the number of viral particles carrying recombinant DNA molecules increases. Viral vectors have the advantage of being able to accommodate sections of foreign DNA considerably larger than those that can be inserted into plasmids: plasmid inserts range from a few base pairs to several kilobases in length while viral inserts can be between a few kilobases and (in certain viruses) 20 to 22 kilobases long. An additional advantage of using viral vectors is their much higher rate of transmission of the recombinant DNA molecule to host cells. As was noted earlier, plasmid transformation occurs in a very limited number of host cells while viral infection generally occurs at something close to 100%.

Joining Fragments That Lack Complementary Sticky Ends

The restriction endonucleases considered so far produce staggered cuts within their recognition sites, and the single-stranded sticky tails that result allow the fragments to join with other DNA fragments produced with the same restriction endonuclease. Another type of restriction endonuclease produces clean breaks, or blunt ends, within its recognition site. An example of this type of enzyme is *Hae*III (derived from the bacterium *Haemophilus aegyptius*) that cleaves within its recognition site as follows.

$$\begin{array}{c} \downarrow \\ 5'\ldots GGCC\ldots 3' \\ 3'\ldots CCGG\ldots 5' \\ \uparrow \end{array} \longrightarrow \begin{array}{cc} 5'\ldots GG & CC\ldots 3' \\ 3'\ldots CC & GG\ldots 5' \end{array}$$

Similar blunt-ended fragments can also be formed by the mechanical process of sheering that breaks double-stranded DNA molecules into shorter fragments. It is sometimes desirable to be able to join blunt-ended fragments together into recombinant DNA molecules. In some instances, this may be achieved by using a high concentration of the DNA ligase derived from the *E. coli* bacteriophage T4. In other cases, nucleotides are added onto the blunt-ended fragments to produce complementary single-stranded tails.

We will consider two ways of adding nucleotides to blunt ends. One way uses the enzyme *terminal transferase* to add single-stranded tails composed of a single type of nucleotide onto the 3′ end of DNA strands. For example, adding a strand of polyguanine to one double-stranded DNA fragment and a strand of polycytosine to another provides the fragments with complementary tails that can join the fragments together as shown in Figure 30-3.

Another method uses a high concentration of T4 ligase to join the DNA fragment under study to a synthetic, double-stranded DNA segment known as a **linker** molecule, that includes the recognition site for a particular restriction endonuclease. Subsequent treatment with the appropriate restriction endonu-

[17] It will release the foreign DNA from the plasmid.

FIGURE 30-3

The joining of two blunt-ended fragments to which complementary tails were added.

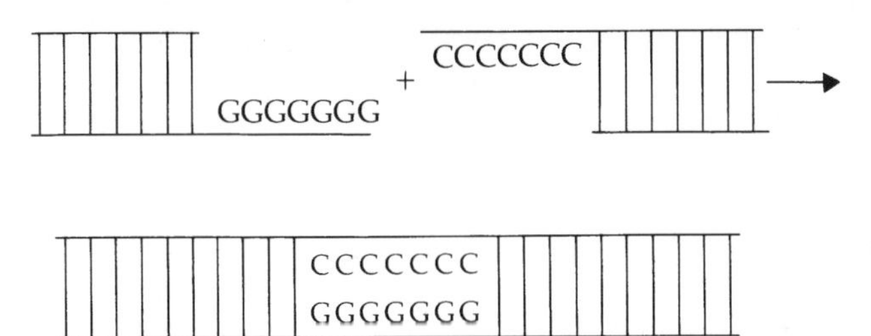

clease produces a staggered break within the recognition site of the linker, and the DNA segments that result have single-stranded tails. If a plasmid carrying the same recognition site is treated with the same restriction endonuclease, complementary sticky ends will be formed on the plasmid DNA, making it possible to incorporate the foreign DNA into the plasmid.

Electrophoresis and Autoradiography

Before going further, we need to take a brief look at two techniques that are widely used in recombinant DNA studies—**electrophoresis,** used to separate nucleic acid fragments according to their size, and **autoradiography,** used to localize radioactively labeled nucleic acid fragments.

Electrophoresis is an important tool for separating molecules with different physical characteristics. It takes advantage of the fact that charged molecules migrate when placed in an electrical field—positively charged molecules move toward the negative pole (cathode) and negatively charged molecules move toward the positive pole (anode). The molecules to be separated are suspended in a buffer (which stabilizes the pH) and then some of this suspension is applied to a porous medium (such as paper, starch, or gel) that is already saturated with the buffer. When an electrical field is applied across the medium, the molecules migrate through the medium at a rate determined by their physical characteristics (weight, size, and shape). By adjusting the pH of the buffer, the type of medium, the strength of the electrical field, and the length of time the setup is in operation, precise separations can be achieved.

Relatively short nucleic acid fragments can be readily separated using a medium of polyacrylamide gel. A thin layer of gel is prepared and the mixture of nucleic acid fragments to be separated is placed ("loaded") onto the gel at a site designated as the origin. With the application of an electrical field across the gel, the negatively charged DNA molecules begin migrating with their rate of movement inversely related to their molecular size. The smaller a fragment, the faster it moves and thus the farther it will travel from the origin in a given time period. Once separation has occurred, the location of the nucleic acid sites or bands (which, by themselves, are invisible) needs to be determined. This may be done by staining with a stain specific for nucleic acids or by exposing the gel to ultraviolet light (recall that nucleic acids absorb ultraviolet radiation). If the nucleic acid molecules have been labeled in advance with a radioactive tracer, the bands may be located through the detection of the radioactivity they give off using a method such as autoradiography (see the following paragraph). Resolution with polyacrylamide gels is excellent: nucleic acid fragments that differ in length by just a single nucleotide can be separated from each other and show up as separate bands. For example, a mixture of nucleic acid fragments two, three, five, and seven nucleotides long would separate into the four bands shown in the following diagram.

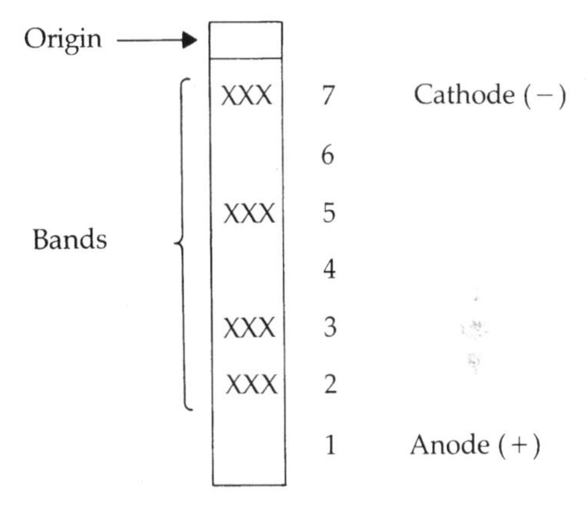

Autoradiography (introduced in Chapter 19) is a technique that uses photographic film to locate molecules labeled with a radioactive isotope. Autoradiography is often used following the electrophoresis of ^{32}P-labeled nucleic acid fragments to locate the sites to which the fragments migrate in the gel. The procedure is carried out by placing a sheet of film over the gel after electrophoresis is completed. During an interval of exposure, the radioactivity emitted by the isotope alters the photographic emulsion. Subsequent development of the film reveals darkened areas at the sites hit by the radiation that correspond to the location of the bands of labeled nucleic acid molecules in the gel.

Restriction Maps

As pointed out earlier, most restriction enzymes produce breaks in duplex DNA molecules at recognition sites that consist of a distinctive sequence of nucleotides. Consequently, a particular type of DNA molecule, exposed to the same type of restriction enzyme, consistently breaks up into fragments of the same number and length. Treatment of the same type of molecule with a different restriction enzyme yields a different characteristic set of fragments. The fragments formed by a particular restriction enzyme can be separated and their relative lengths estimated by electrophoresis. A map showing the linear arrangement of the recognition sites for a particular restriction enzyme found on a particular duplex DNA segment is a **restriction map.**

Isolating a Specific Gene for Cloning

Techniques are available for identifying and isolating a particular gene from a prokaryotic or eukaryotic genome. Once isolated, the gene may be cloned and used for further study. We will consider the key events involved in one approach to this task. As you read this discussion, keep in mind that many techniques are available for use in recombinant DNA studies, that other approaches could be used, and that many details have been omitted.

Our procedure begins by extracting the RNA from cells in which the gene we wish to clone is known to be active. This extract consists of an assortment of many kinds of RNA. If we are dealing with eukaryotic RNA, most of the molecules will carry a poly-A tail at their 3′ ends. Short chains of poly-T deoxyribonucleotides can be mixed with this RNA extract and, under the appropriate conditions, they will hybridize with the RNAs′ poly-A tails to create a short, double-stranded segment at the end of the otherwise single-stranded RNA molecules. The poly-T chains can then serve as primers for the enzyme **reverse transcriptase** which uses the RNA strands as a template for the synthesis of complementary DNA strands to form RNA-DNA hybrid molecules. Complementary DNA that is formed by reverse transcriptase on an RNA template is known as **copy DNA** or **cDNA.** The 3′ end of each cDNA strand is folded back on itself in a hairpin loop.

In the next step, the RNA strands of the hybrid molecules are destroyed and DNA polymerase is used to catalyze the formation of a new DNA strand to complement each cDNA strand, a process primed by the 3′ end of the hairpin loop of the cDNA strand. Since this procedure starts with an assortment of many kinds of RNA extracted from a cell, many kinds of double-stranded cDNA result. Next, using techniques described earlier in this chapter, the cDNA is inserted into plasmids that are incubated with a suitable *E. coli* host strain. Experimental conditions are designed so that each transformed cell receives a single cDNA molecule. *E. coli* cells transformed by cDNA recombinant plasmids are then cultured to clone the cDNA. This results in many different clones, each carrying a different cDNA molecule.

The next step is to figure out which clone carries the gene we seek. A sample of cells is obtained from each clone population and the cDNA-containing plasmids are removed and denatured. Some of the original RNA extract that we began with (this is the RNA mixture of many different types extracted from cells in which the gene is known to be active) is added to this single-stranded cDNA from each clone. RNA molecules in the sample that complement the cDNA hybridized with it, while RNA molecules that do not complement the cDNA remain single-stranded. Next, this mixture of cDNA and RNA is added to an *in vitro* setup containing radioactively labeled amino acids and all the other ingredients necessary for translation. The single-stranded RNA in the setup is available to guide polypeptide formation while any RNA hybridized with a cDNA strand is unavailable. Since the setup includes many different RNA molecules but only one type of cDNA molecule (that which was derived from the clone), many different polypeptides will be synthesized. After an appropriate interval, the setup is tested for the presence of the polypeptide product of the gene we wish to isolate. This may be done by adding antibodies specific for the polypeptide; if present, the polypeptide will react with the antibodies and precipitate out of solution.

Assume that this procedure is carried out on DNA derived from a particular clone and a precipitate is formed. Is the polypeptide product of the gene we seek present or absent? ________[18] If the polypeptide product is formed, what must have been available in the translation setup to guide its synthesis? ________ ________________________________[19] Was the RNA guiding this translation hybridized or unhybridized? ________[20] Explain. ________________ ________________________[21] Is the gene we seek

[18] Since there is a precipitate, the polypeptide must be present. [19] The corresponding RNA transcript. [20] Unhybridized.
[21] If the RNA had been hybridized, it could not have guided translation.

carried by this clone? ________[22] Explain. ________
____[23]

Now assume that this procedure is carried out on DNA derived from a different clone and a precipitate is not formed. Is the polypeptide product of the gene we seek present or absent? ________[24] Was the RNA coding for this polypeptide available to guide its translation? ________[25] Explain. ________
________________________[26] Is the gene we seek present in the DNA from this clone? ________[27] Explain. ________
______[28]

Why do the numerous other kinds of polypeptides formed in each translation setup not precipitate out?
____[29]
This procedure for identifying the gene will work only if the RNA mixture added to each setup contains RNA that complements the sense strand of the gene we seek. What assurance is there that this RNA is present in the mixture? ________
____________[30] The RNA mixture supplied to the translating setup is unpurified and could be contaminated with the polypeptide product of the gene we seek. How could we test to be sure that any of the polypeptide detected is newly synthesized? ________
____[31]

To summarize, we have just examined the key steps involved in a procedure designed to identify the DNA sequence of a particular gene. This involved extracting RNA from a cell where the gene is known to be active and using reverse transcriptase to prepare double-stranded cDNA that is then cloned. The clone that carries the gene we seek is identified by extracting the cDNA from each of the many clones and combining it with some of the original RNA extract in individual translation setups. Following translation, each setup is assayed for the polypeptide product of the gene we seek using antibodies specific for the polypeptide. The absence of the polypeptide indicates the presence of the gene that codes for the polypeptide; the RNA that would normally guide its formation has been inactivated by hybridizing with the DNA of the gene.

Identifying a Specific Gene on DNA Fragments: Southern Blotting

A number of procedures may be used to locate specific prokaryotic or eukaryotic genes on DNA fragments without cloning. One that is frequently used is known as **Southern blotting** (named after its developer, E. Southern). The DNA to be screened is broken down, using restriction enzymes, and the fragments are separated by gel electrophoresis. The gel is then treated with alkali to denature the DNA and is placed over a sheet of nitrocellulose filter. The DNA fragments are transferred by capillary action to the nitrocellulose where they become bound to the filter, producing a replica of the gel with the bands of DNA fragments located in the same relative positions. Next, the filter is exposed to a **probe:** labeled molecules of single-stranded RNA or cDNA that complement the gene we seek. Any DNA on the filter that complements the probe hybridizes with it. The filter is then washed to remove any unhybridized probe molecules and autoradiographed to identify the radioactive sites that indicate where DNA hybridized with the probe. If desired, double-stranded fragments of this same DNA may then be isolated from the corresponding site on a gel untreated with alkali; the section of the gel containing the DNA is cut out and its DNA is extracted.

Locating Specific Transcripts on RNA Fragments: Northern Blotting

A similar technique, known as **northern blotting,** is used to identify specific RNA molecules. This might be used, for example, to determine whether the transcript of a cloned gene under study is present in the RNA mixture extracted from a particular cell. The extracted RNA is separated using gel electrophoresis, transferred to nitrocellulose or another type of filter, and supplied with a labeled single-stranded DNA probe that complements the RNA transcript we wish to identify. Any of the probe that remains following washing has hybridized with the RNA transcript and is detected by autoradiography.

[22] No. [23] Had it been present, it would have hybridized with the appropriate RNA transcript and prevented it from guiding translation, blocking formation of the polypeptide we seek. [24] Since there is no precipitate, the polypeptide must be absent. [25] No. [26] It was hybridized and thus could not guide translation. [27] Yes. [28] It has hybridized with the RNA and thus is prevented from guiding translation. [29] The antibodies are specific for the polypeptide product of the gene we seek and only react with that polypeptide. [30] The RNA is the same RNA used initially to form the cDNA incorporated into the plasmids. [31] Since the amino acids supplied in the setup are labeled, the polypeptide product can be tested for the presence of the label to verify that it is newly synthesized.

DNA SEQUENCING

Determining the sequence of nucleotides making up a gene, once a tedious and time-consuming business, is the ultimate analysis of gene structure. Through cloning, it is relatively easy to produce many copies of a DNA molecule, and this has paved the way for the development of expedient approaches to sequencing DNA molecules. We will consider two techniques commonly used to sequence DNA segments of up to approximately 200 nucleotides: the Maxam-Gilbert chemical-cleavage method and the Sanger chain-termination method. Both methods are named after the researchers who developed them.

Maxam-Gilbert Chemical-Cleavage Method

For purposes of discussion, assume that the segment we are working with is single-stranded and seven nucleotides long with the sequence 5′ pGAACGTC 3′, where p designates the terminal 5′ phosphate. (Normally, of course, the nucleotides making up the strand would be unknown). The key steps in this method are as follows.

1. The first step is to produce, through cloning, a large number of copies of the DNA fragment. Since this sequencing procedure requires single-stranded molecules, the double-stranded fragments isolated after cloning are dissociated and electrophoretically separated into two fractions. The basis for the separation relates to a difference in the number of purines present in the two types of strands; one strand will have more and will therefore be heavier. The sequencing procedure can be carried out with either the heavier or lighter strand; once one strand has been sequenced, the makeup of its complement can readily be designated.

2. A batch of identical strands is labeled with a radioactive tracer (^{32}P). This is accomplished by enzymatically removing the 5′ terminal phosphate from the strands using bacterial alkaline phosphatase.

$$5'\ \text{pGAACGTC}\ 3' \xrightarrow{\text{Alkaline phosphatase}} 5'\ \text{GAACGTC}\ 3' + \text{p}$$

Then the phosphate is replaced with ^{32}P by supplying ^{32}P-labeled ATP in the presence of the enzyme polynucleotide kinase which transfers the ^{32}P to the 5′ end of the DNA molecule. This is shown in the following diagram where p* represents the labeled phosphate.

$$5'\ \text{GAACGTC}\ 3' + \text{p*} \xrightarrow{\text{Polynucleotide kinase}} 5'\ \text{p*GAACGTC}\ 3'$$

3. The batch of ^{32}P-labeled DNA strands is subdivided into four samples. Assume, for our discussion, that each sample contains, say, 1000 copies of our DNA sequence.

4. Each sample is then subjected to a different chemical procedure that "targets" one or two of the four kinds of bases present in the DNA segment. A number of procedures can be used, each designed to cause the release of the target nucleotide or nucleotides. One procedure, which we will presently inspect in depth, involves two steps: the first step alters the "target" base so that its nucleotide becomes unstable, and the second step removes the unstable nucleotide, producing DNA fragments. To illustrate this, say that the chemical treatment altered the internal C nucleotide in our sequence 5′ p*GAACGTC 3′. Subsequent treatment would divide the segment into 5′ p*GAA 3′ and 5′ GTC 3′ and release the altered C nucleotide.

$$5'\ \text{p*GAACGTC}\ 3' \longrightarrow 5'\ \text{p*GAA}\ 3' + 5'\ \text{GTC}\ 3' + \text{C}$$

Concentrations of the reagents involved in each chemical procedure are adjusted so that, on average, one base is altered (and therefore one cleavage produced) within each DNA strand present in the sample. Each of the potentially susceptible bases within a strand has an equal likelihood of being targeted by the chemical procedure, but usually only one of the target bases gets altered. We will take a look at how four chemical treatments that target A, G, C, and T and C, respectively, are used to sequence our DNA segment.

Chemical treatment with A as the target base: Our segment, 5′ p*GAACGTC 3′, contains two A nucleotides. On average, following chemical treatment, how many of these A nucleotides would be altered in each segment in our sample? ________[32] On average, how many of the 1000 segments in our sample would be expected to experience an alteration of the A nucleotide closest to their 5′ end? ________[33] What fragments are produced by the removal of this particular A nucleotide? ________[34] How many of these types of product fragments carry the ^{32}P label? ________[35] How many of the 1000 segments would be expected to experience an alteration of the A nucleotide farthest from their 5′ end? ________[36] What fragments are produced by the removal of this particular A nucleo-

[32] One. [33] 500. [34] Two fragments: 5′ p*G 3′ and 5′ ACGTC 3′. [35] One: 5′ p*G 3′. [36] 500.

tide? ________[37] How many of these types of fragments carry a ^{32}P label? ________[38]

As you have just demonstrated, the chemical treatment that targets the A nucleotide in our DNA gives rise to four kinds of fragments: 5′ p*G 3′, 5′ ACGTC 3′, 5′ p*GA 3′, and 5 CGTC 3′. Mark with an "X" on Figure 30-4 the sites to which these fragments would migrate when subjected to electrophoresis.[39] Following electrophoretic separation, how many of these four types of fragments could be detected by autoradiography? ________[40] Explain. ________________[41] Consider the two kinds of fragments that carry the label. Prior to chemical treatment, what nucleotide was covalently bonded to the nucleotide that now terminates the 3′ end of each fragment? ________[42] Mark with an "X" on Figure 30-5 the sites at which these labeled fragments would be found following electrophoresis.[43]

Chemical treatment with G as the target base: Our segment, 5′ p*GAACGTC 3′, contains two G nucleotides. Identify the fragment or fragments formed when the G closest to the 5′ end is removed. ________________[44] Identify the fragment or fragments formed when the G farthest from the 5′ end is removed. ________________[45] How many of these fragments could be electrophoretically separated? ________[46] How many of these fragments could be detected by postelectrophoresis autoradiography? ________[47] Consider the DNA fragments that carry the label. Prior to chemical treatment, what nucleotide was covalently bonded to the nucleotide that now terminates the 3′ end of each fragment? ________[48] Mark with an "X" on Figure 30-6 the site at which the labeled fragment would be found following electrophoresis.[49]

FIGURE 30-5

Sites shown in Figure 30-4 that are occupied by labeled DNA fragments.

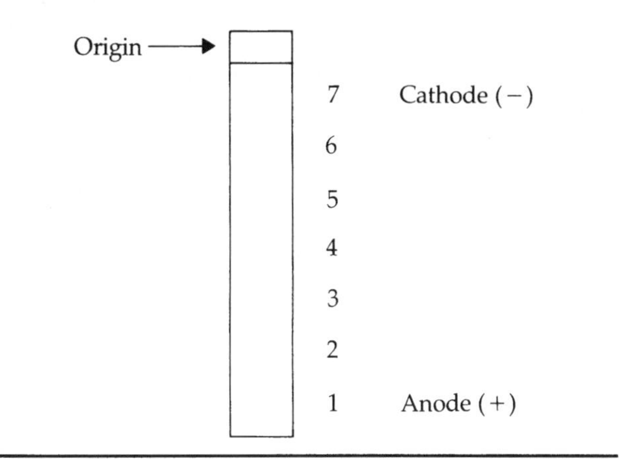

FIGURE 30-4

Sites occupied by the four fragments produced by the chemical treatment that targets the A bases in the DNA segment 5′ p*GAACGTC 3′.

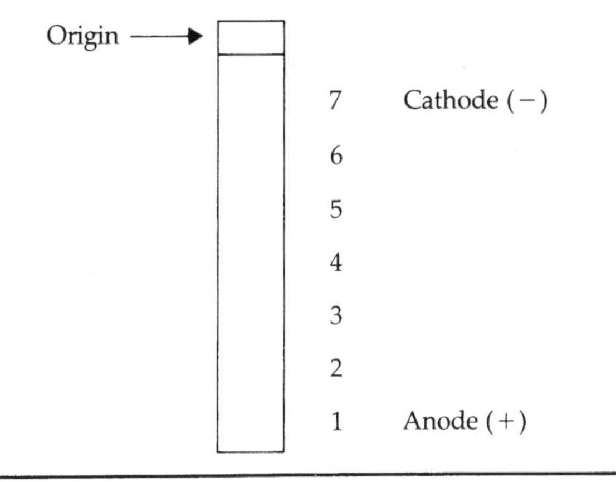

FIGURE 30-6

Sites occupied by labeled fragments produced by the chemical treatment that targets the G bases in the DNA segment 5′ p*GAACGTC 3′.

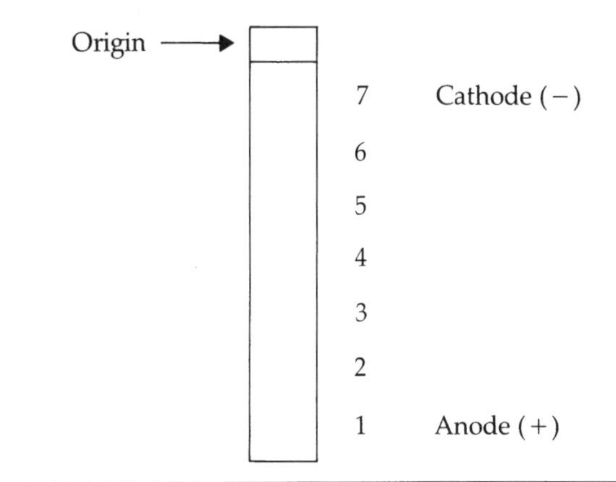

[37] 5′ p*GA 3′ and 5′ CGTC 3′. [38] One: 5′ p*GA 3′. [39] Each fragment migrates a distance that is directly related to the number of nucleotides it contains; fragments containing one, two, four, or five nucleotides migrate to those gel sites. [40] Two. [41] Only those carrying the ^{32}P label can be detected. [42] An A nucleotide. [43] Since the labeled fragments consist of one or two nucleotides, they will be found at sites 1 and 2, respectively. [44] 5′ AACGTC 3′. [45] 5′ p*GAAC 3′ and 5′ TC 3′. [46] All three. [47] One: only the fragment carrying the ^{32}P label, 5′ p*GAAC 3′. [48] G. [49] Since it consists of four nucleotides, it will be found at site 4.

Chemical treatment with C as the target base: Our segment, 5′ p*GAACGTC 3′, contains two C nucleotides. Identify the fragment or fragments formed when the C closest to the 5′ end is removed. ______________________________________[50] Identify the fragment or fragments formed when the C farthest from the 5′ end is removed. ______________________________[51] How many of these fragments could be detected by postelectrophoresis autoradiography? __________[52] Consider the fragments that carry the label. Prior to chemical treatment, what nucleotide was covalently bonded to the nucleotide that now terminates the 3′ end of each fragment? _____[53] Mark with an "X" on Figure 30-7 each site at which the labeled fragments will be found following electrophoresis.[54]

Chemical treatment with C and T as the target bases: The final chemical procedure targets both C and T bases. In addition to the two C nucleotides just considered, our segment 5′ p*GAACGTC 3′ contains one T nucleotide. Identify the fragments produced by this treatment that are detectable through postelectrophoresis autoradiography. (Remember that although this procedure targets both C and T, on average only a *single* nucleotide in any given strand is acted upon. Also remember that you have already identified the fragments arising when C is targeted; those same fragments would be produced here. ____________

FIGURE 30-7

Sites occupied by labeled fragments produced by the chemical treatment that targets the C bases in the DNA segment 5′ p*GAACGTC 3′.

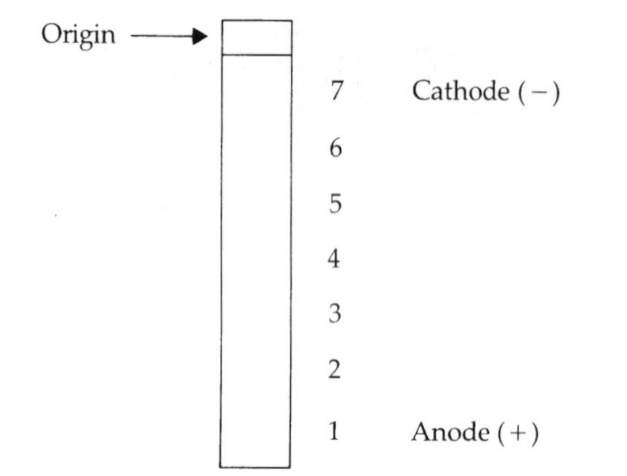

FIGURE 30-8

Sites occupied by labeled fragments produced by the chemical treatment that targets the C and T bases in the DNA segment 5′ p*GAACGTC 3′.

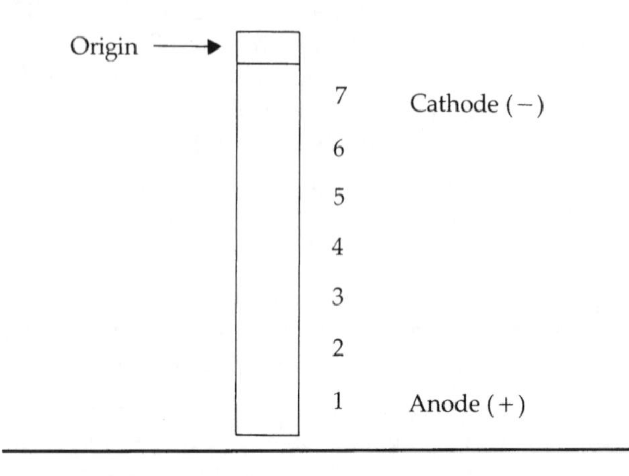

______________________________[55] Mark with an "X" on Figure 30-8 each site at which the labeled fragments would be found following electrophoresis.[56]

Up to this point we have separately considered the fragments that are produced by each chemical treatment and the banding pattern of the labeled fragments detected through postelectrophoresis autoradiography. Now we can consider how the migration patterns of the labeled fragments, taken collectively, give the sequence of our DNA segment. Usually, fragment separation for each of the four chemical procedures we have considered is carried out simultaneously. The fragments produced by each chemical procedure are "loaded" onto the same sheet of gel at separate, adjacent origin points and the fragments migrate downward from each origin in four separate parallel lanes. Following electrophoresis, the gel is autoradiographed. Figure 30-9 combines the four autoradiographic outcomes we considered separately.

What would any molecules appearing at site 7 in Figure 30-9 be? ______________________________[57] The key question for interpreting the autoradiograph is: *Prior* to the chemical treatment, what nucleotide was bonded to the nucleotide that now terminates the 3′ end of each labeled fragment? ______________________________[58]

This bit of information allows us to read the sequence of the DNA fragment directly from the autoradiograph. Beginning at the bottom, consider the single nucleotide fragment. This nucleotide is the first nucle-

[50] 5′ p*GAA 3′ and 5′ GTC 3′. [51] 5′ p*GAACGT 3′. [52] Two: 5′ p*GAA 3′ and 5′ p*GAACGT 3′. [53] C. [54] Sites 3, 5, and 6. [55] 5′ p*GAA 3′ and 5′ p*GAACGT 3′ when C is attacked, and 5′ p*GAACG 3′ when T is attacked. [56] Sites 3, 5, and 6. [57] Intact segments of DNA that escaped cleavage. [58] In each case, the nucleotide was the one removed by the specific chemical treatment.

FIGURE 30-9

Combined autoradiograph for the four chemical treatments considered for the Maxam-Gilbert method. With the exception of the 5′ terminal nucleotide, the sequence of the DNA segment under study can be read directly from the autoradiograph.

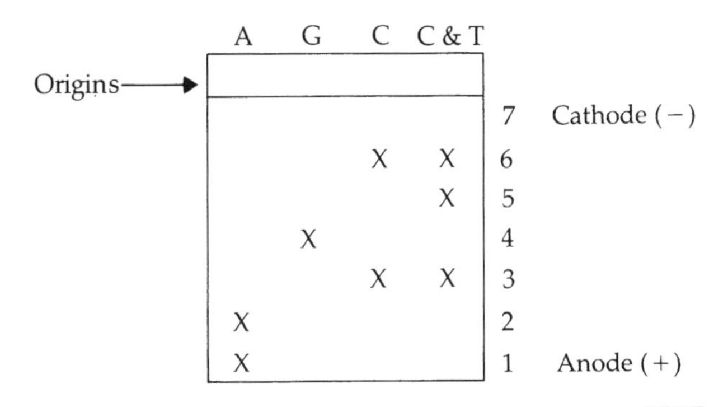

otide (that is, the nucleotide at the 5′ end) in our DNA segment. The nucleotide next to this 5′ terminal nucleotide in the original segment was removed by a specific chemical treatment. Which nucleotide was removed by this chemical treatment? ________[59] What is the second nucleotide in the segment? ________[60] Next we consider the two-nucleotide fragment. These two nucleotides making up this fragment are the first and second nucleotides (at the 5′ end) in our DNA segment. What nucleotide was targeted by the chemical treatment that removed the third nucleotide? ____[61]

What is the third nucleotide in the segment? ____[62] Which chemical procedure cleaved off the fourth nucleotide? ________________________________

________[63] What is the fourth nucleotide? ____[64] What nucleotides occupy the fifth, sixth, and seventh sites in our DNA segment? ________[65] Is there a limitation with this particular procedure? (Clue: Have we identified all the nucleotides in our DNA segment?)

________________________________[66]

As you have just demonstrated, with the exception of the 5′ terminal nucleotide, the sequence of the DNA segment under study can be read directly from the autoradiograph. Reading upward from the bottom takes us from the shortest fragment to the longest and gives the nucleotide sequence from the 5′ to the 3′ end of the DNA segment.

As noted earlier, the Maxam-Gilbert technique can be used to sequence segments of up to 200 or more nucleotides. Longer segments are sequenced by breaking them into shorter segments using an appropriate restriction endonuclease (illustrated by segments 1, 2, and 3 in Figure 30-10). Once these shorter segments have been sequenced, they, of course, need to be positioned relative to each other. This can be accomplished by cleaving other copies of the original long segment with a different restriction endonuclease. Segments produced by the second enzyme may overlap the DNA pieces produced by the first enzyme (for example, segments A and B in Figure 30-10). Relating the sequences of the overlapping segments to those of the first set of segments will establish the relative positions of the first set of segments.

FIGURE 30-10

The shorter segments (1, 2, and 3) produced by treating a long DNA molecule with a restriction endonuclease can be positioned relative to each other by treating other copies of the long molecule with a different restriction endonuclease. Some segments (A and B) produced by the second enzyme overlap the segments produced by the first enzyme. Relating the sequences of the overlapping segments to those of the first set of segments will establish the relative positions of the first set of segments.

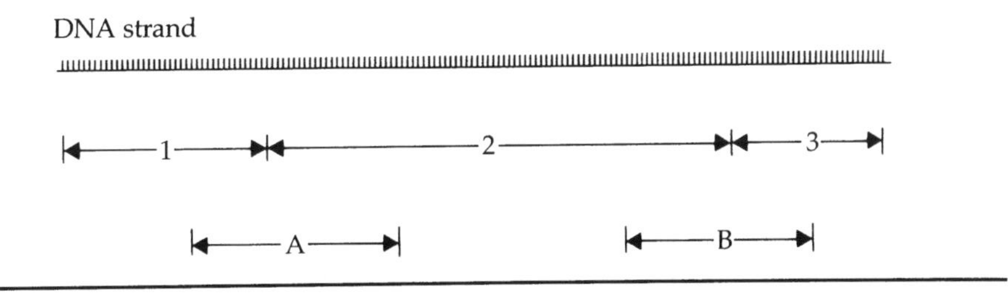

[59] A. [60] A. [61] A. [62] A. [63] It was cleaved by two treatments: the one specific for C *and* the one for T and C. [64] Any fragment appearing in the T and C column could have been cleaved at either a T or a C site. However, since this four-nucleotide fragment also appears in the C treatment column, we know that the fourth nucleotide is C. [65] G, T, and C, respectively. [66] It cannot identify the first nucleotide in the segment; other approaches must be used to identify this nucleotide.

Sanger Chain-Termination Method

Before looking at the second method for sequencing DNA, the Sanger chain-termination method, we need to review the way in which DNA polymerase works and consider two different forms of nucleotides.

DNA Polymerase Recall that this enzyme catalyzes the one-at-a-time addition of deoxyribonucleotides to a growing strand of DNA, and it requires a free hydroxyl (-OH) group at the 3′ end of a nucleic acid strand in order to form the phosphodiester bond that attaches each new nucleotide. As a consequence, the DNA polymerization must always begin with a previously formed primer molecule of nucleic acid with a free 3′ hydroxyl group. Furthermore—and this is the essential point to keep in mind as we consider this sequencing method—the addition of nucleotides can continue only as long as the growing DNA strand has a free hydroxyl group at its 3′ end.

Different Forms of Deoxyribonucleotides A cell's supply of DNA nucleotides available for incorporation into a growing DNA strand is in the form of 2′-deoxyribonucleoside triphosphates, an example of which is shown in Figure 30-11. Note that this type of nucleotide carries a free hydroxyl group attached to the 3′ carbon of its pentose sugar. As long as nucleotides of this type are incorporated into a growing DNA strand, do you think that the DNA polymerase would have the potential to continue to add nucleotides to the strand? __________[67]

DNA nucleotides can also occur in the form of 2′3′-dideoxyribonucleoside triphosphates as shown in Figure 30-12. Compare this structural diagram with that of the 2′-deoxyribonucleoside triphosphate in Figure 30-11. How do the two forms differ? ______________________________[68] What effect would the incorporation of a 2′3′-dideoxyribonucleoside triphosphate nucleotide have on DNA polymerase activity? ______________________________[69] What effect would this have on further elongation of the DNA chain? ______________________________[70] The method of DNA sequencing that we are about to consider makes use of the chain-terminating effects of these 2′3′-dideoxyribonucleoside triphosphate molecules.

FIGURE 30-11

The 2′-deoxyribonucleoside triphosphate form of an adenine nucleotide.

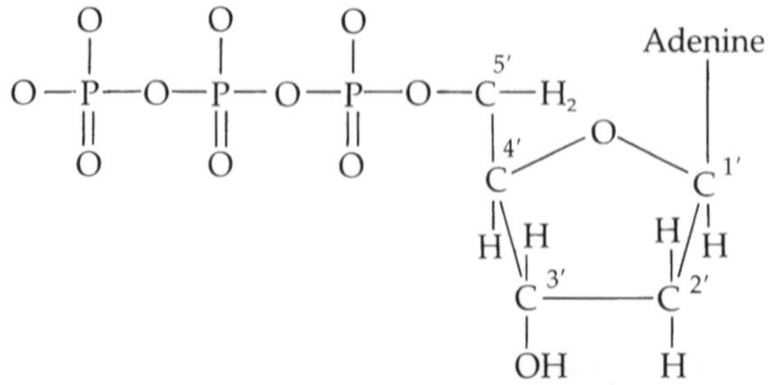

FIGURE 30-12

The 2′3′-dideoxyribonucleoside triphosphate form of an adenine nucleotide.

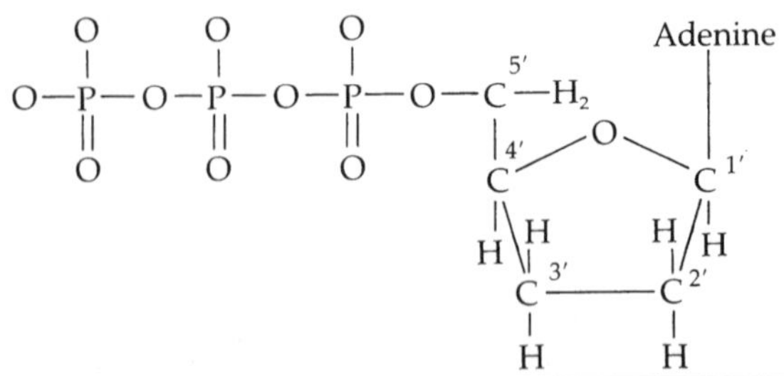

Procedure Followed in the Sanger Chain-Termination Method This approach to sequencing a DNA molecule involves setting up four *in vitro* DNA-synthesizing systems, each containing the following ingredients. (1) Identical DNA template and primer strands. The template strand is the DNA strand that is to be sequenced and it may be 200 or more nucleotides long. The primer is short and complements a portion of the template strand. (2) *E. coli* DNA polymerase. (3) The four kinds of nucleotides (ATP, TTP, CTP, and GTP) in their normal 2′-deoxyribonucleoside triphosphate form. All the molecules of at least one kind of nucleotide are labeled with ^{32}P so that the DNA fragments synthesized in these *in vitro* systems can be located on an autoradiograph. In addition, one kind of 2′-deoxyribonucleoside triphosphate is also available in its 2′3′-dideoxyribonucleoside triphosphate, or terminator, form. The nucleotide that is present in both normal and terminator forms is different in each of the four DNA-synthesizing setups.

Once the four setups are prepared, DNA synthesis begins in each. Let us focus on one of these setups—say, the one with adenine nucleotides present in both normal and terminator forms. However, before

[67] Yes, since the most recently incorporated nucleotide carries a free 3′ hydroxyl. [68] The 2′3′-dideoxyribonucleoside triphosphate nucleotide carries a hydrogen rather than a hydroxyl group on the 3′ pentose carbon. [69] In the absence of a free 3′ hydroxyl, the DNA polymerase could not function. [70] It would cease.

we consider how this system operates with the two forms of adenine nucleotide, let us consider how it would work with just normal adenine nucleotides and then with just terminator adenine nucleotides. If all the adenine nucleotides available in the setup were normal forms, what effect would this have on the synthesis of DNA strands? ______________________ ______________________[71] With just normal adenine nucleotides, would the newly synthesized DNA strands be of the same or of different lengths? _____[72] If all the adenine nucleotides available were of the terminator form, what effect would this have on the synthesis of DNA strands? ______________________ ______________________[73] With just terminator adenine nucleotides, would the newly synthesized DNA strands be of the same or of different lengths? _____[74] What would determine the length of these strands? ______________________[75]

Now back to the actual setup with its mixture of normal and terminator adenine nucleotides. What could potentially happen to the growing DNA strand each time a T nucleotide is encountered on the template? ______________________ _______[76] If a normal A nucleotide were incorporated, how would DNA synthesis be affected? ______________________ ______________________[77] If a terminator A nucleotide were incorporated, how would DNA synthesis be affected? ______________________ ______________________[78] Do each of the A nucleotide sites within the growing strand have an equal chance of being occupied by a terminator nucleotide? _____[79] What determines which particular A nucleotide site receives a terminator? _________[80] What determines the probability of incorporating the terminator form of adenine in the growing strand of DNA at any particular A nucleotide site? ______________________ ______________________[81]

Within the DNA-synthesizing setup, the relative concentrations of the terminator and the normal A nucleotides are adjusted so that there is, on average, one terminator inserted at some point into each of the DNA strands generated by the system. Would you predict that the large number of DNA strands synthesized in the setup would have the same or different lengths? _________[82] What determines the lengths of these fragments? ______________________ ______________________[83] Among the large number of fragments produced, would you expect to find some that terminated at each possible site? _________[84] What would the 3′ nucleotide be for each of the fragments produced? ______________________ ____________[85]

Assume that a complete reading of the template used in our DNA-synthesizing systems produces a DNA strand with the sequence 5′ CGAATTCGCGAC 3′. Identify the prematurely shortened DNA segments that would be produced in the system containing a mixture of normal and terminator A nucleotides? ______________________[86]

So far, we have considered what would happen in the setup with the A terminator and identified the DNA fragments it would form. Each of the other three DNA-synthesizing setups would contain the same template and a different type of terminator nucleotide. Identify the DNA fragments that would be synthesized in setups containing the following terminator nucleotides.

T terminator: ______________________[87]

C terminator: ______________________[88]

G terminator: ______________________[89]

After the completion of DNA synthesis, the DNA fragments produced in each of the four setups are treated with an appropriate restriction endonuclease to remove the primers and then are denatured to separate the template from the newly synthesized strands. The newly synthesized DNA fragments are then separated electrophoretically—with the fragments from each setup running in parallel lanes on the same gel.

[71] Each template would guide the formation of a complete complementary DNA strand. [72] The same length. [73] Synthesis of each strand would terminate when the first T nucleotide was encountered in the template. [74] The fragments would have the same, short length. [75] The location of the first A nucleotide in the template. [76] Either a normal or terminator adenine could be incorporated. [77] Synthesis of the strand would continue—at least until it reached the next T on the template. [78] It would cease. [79] Yes. [80] Random chance. [81] The concentration of the terminator nucleotide relative to the normal form. [82] Different lengths. [83] Location of the A sites: each is a potential site for termination. [84] Yes. [85] The terminator form of the A nucleotide. [86] With A nucleotides at the third, fourth, and eleventh positions, fragments would be formed that are 3, 4, and 11 nucleotides long and would read 5′ CGA 3′, 5′ CGAA 3′, and 5′ CGAATTCGCGA 3′. [87] 5′ CGAAT 3′, 5′ CGAATT 3′. [88] 5′ C 3′, 5′ CGAATTC 3′, 5′ CGAATTCGC 3′, 5′ CGAATTCGCGAC 3′. [89] 5′ CG 3′, 5′ CGAATTCG 3′, 5′ CGAATTCGCG 3′.

Using an "X," indicate, on Figure 30-13, the sites to which each of the fragments would migrate.[90]

Following electrophoresis, the gel is subjected to autoradiography to show the position of each type of fragment. (Remember that all of at least one of the nucleotides supplied to each setup is radioactively labeled so that the newly synthesized fragments are also labeled.) The autoradiograph is interpreted just like those we considered earlier in conjunction with the Maxam-Gilbert method, with the nucleotide sequence read directly from the autoradiograph. Reading upward from the bottom takes us from the shortest fragment to the longest and gives the nucleotide sequence from the 5′ to the 3′ end of the DNA segment. Remember, however, that the molecule that is being sequenced with this method is the template molecule, and the fragments showing up on the autoradiograph are parts of the strand that complement the template. Consequently, the sequence read from the autoradiograph needs to be converted into its complement to give the template sequence.

As noted earlier, the Sanger technique is used to sequence DNA segments with up to 200 or more nucleotides. As with the Maxam-Gilbert method, segments of greater length can be broken into shorter sections for sequencing. These adjacent sections can then be related by studying segments that overlap them.

FIGURE 30-13

Sites to which each of the DNA fragments produced in the four DNA-synthesizing setups would migrate during electrophoresis. Reading the autoradiograph upward from the bottom gives the nucleotide sequence from the 5′ to the 3′ end of the DNA segment.

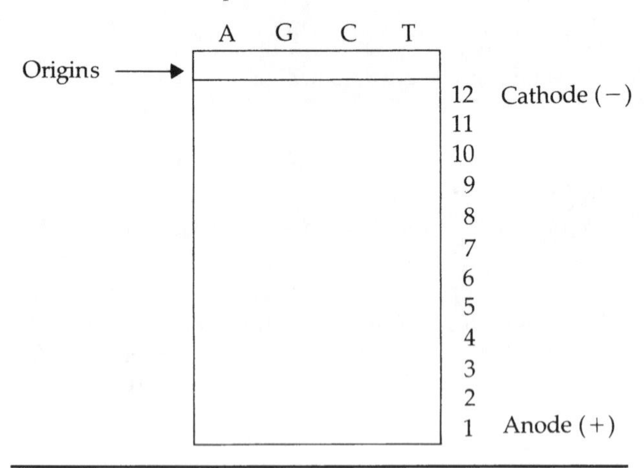

SUMMARY

Through the use of restriction endonucleases, a fragment of "foreign" DNA can be inserted into a viral chromosome or a bacterial plasmid vector, and the recombinant DNA molecule that results can be introduced into a bacterial host where it replicates and is transmitted to the next generation of bacteria. After several rounds of reproduction, many copies of the recombinant DNA molecule can be extracted from the bacterial host cells and used for laboratory study or left in the host cells so that the foreign genetic material can use the metabolic machinery of the bacteria to manufacture its protein product. Electrophoresis is used to separate nucleic acid fragments according to their size, and autoradiography is used to localize radioactively labeled nucleic acid fragments.

A procedure designed to identify the DNA sequence of a gene is reviewed. It involves extracting RNA from a cell where the gene is active and using reverse transcriptase to prepare double-stranded cDNA that is then cloned. Identifying the clone that carries the gene involves extracting the cDNA from each of the many clones formed and combining it with some of the original RNA extract in individual translation setups. Following translation, each setup is assayed for the polypeptide product of the gene we seek using antibodies specific for the polypeptide. The absence of the polypeptide indicates the presence of the gene that codes for the polypeptide (because the RNA that normally guides the formation of the polypeptide has been inactivated by hybridizing with the DNA of the gene).

Specific prokaryotic or eukaryotic genes can be located on DNA fragments by Southern blotting and specific transcripts can be located on RNA molecules by northern blotting. In each case, the nucleic acid to be screened is separated using gel electrophoresis, transferred to nitrocellulose or another type of filter, and supplied with a labeled single-stranded nucleic

[90] A total of twelve kinds of fragments, varying in length from one to twelve nucleotides, are formed in the four DNA-synthesizing setups. The A terminator produces fragments that are 3, 4, and 11 nucleotides long that migrate down the A lane to sites 3, 4, and 11, respectively. The T terminator produces fragments that are 5 and 6 nucleotides long that migrate down the T lane to sites 5 and 6, respectively. The C terminator produces fragments that are 1, 7, 9, and 12 nucleotides long that migrate down the C lane to sites 1, 7, 9, and 12, respectively. The G terminator produces fragments that are 2, 8, and 10 nucleotides long that migrate down the G lane to sites 2, 8, and 10.

acid probe that complements the nucleic acid molecule we wish to identify. Any of the probe that remains following washing has hybridized with the nucleic acid we seek and is detected by autoradiography.

DNA nucleotide sequencing is the ultimate analysis of gene structure. The Maxam-Gilbert chemical-cleavage and the Sanger chain-termination method are commonly used to sequence DNA segments of up to approximately 200 nucleotides. Both procedures allow DNA sequences to be read directly from autoradiographs.

PROBLEM SET

30-1. Bacterial plasmids and viral chromosomes serve as vectors in recombinant DNA studies. Identify three major functions served by the vector DNA.

30-2. You are asked to prepare a restriction map for a linear DNA segment. Copies of the segment are divided into three batches: one is treated with the restriction endonuclease *Eco*RI, a second with *Sal*I, and the third with a mixture of both enzymes. The patterns produced by electrophoretic separation of the DNA fragments after each treatment indicate the following: *Eco*RI yields fragments 1.5 kilobases, 1.2 kilobases, and 0.7 kilobases long; *Sal*I yields fragments 2.2 kilobases and 1.2 kilobases long; and *Eco*RI + *Sal*I yield fragments 1.2 kilobases, 1.0 kilobases, 0.7 kilobases, and 0.5 kilobases long.

a. How many recognition sites are present in the DNA segment for each of these restriction endonucleases? Explain.

b. How many kilobases are present in the DNA segment?

c. Can the *Sal*I recognition site be positioned accurately relative to the ends of the DNA segment?

d. Locate the *Sal*I recognition site relative to one of the fragments produced by the *Eco*RI treatment.

e. Draw a linear restriction map showing the relative locations of *Eco*RI and the *Sal*I recognition sites and the distances separating them.

30-3. A bacterial plasmid carries at least one recognition site for each of three different restriction endonucleases: *Eco*RI, *Hpa*I, and *Hind*III. Treatment of the plasmid with *Eco*RI yields a linear molecule 4.1 kilobases long. Three batches of this linear molecule are prepared. One is treated with *Hpa*I, a second with *Hind*III, and the third with a combination of *Hpa*I and *Hind*III. The DNA fragments produced in each of these treatments are electrophoretically separated and characterized as follows: *Hpa*I yields fragments 2.1 kilobases and 2.0 kilobases long; *Hind*III yields fragments 1.1 kilobases and 3.0 kilobases long; and *Hpa*I + *Hind*III yield fragments 1.0 kilobases, 1.1 kilobases, and 2.0 kilobases long.

a. How many recognition sites are carried by the plasmid for each of these restriction endonucleases?

b. How many kilobases are present in the plasmid?

c. Prepare a linear map showing the relative positions of the recognition sites for the enzymes used in this study and the distances separating them.

30-4. Assume that a bacterial plasmid vector carries a gene that confers resistance to the antibiotic ampicillin, amp^r. By using an appropriate restriction endonuclease, a segment of foreign DNA is inserted within this amp^r gene. The vector is then supplied to host bacterial cells which, prior to transformation, are sensitive to the ampicillin. A sufficient interval is allowed for transformation.

a. Is it possible to identify the bacterial cells that have acquired the recombinant DNA molecule by culturing all the bacteria on a medium containing ampicillin? Explain.

b. What general change would you make in the design of this experiment in order to readily distinguish between the transformed and the untransformed bacteria?

30-5. Viral chromosomes may be used as vectors for recombinant DNA molecules. Treatment of the chromosome of the *E. coli* bacteriophage λ with the appropriate restriction endonuclease cleaves the central portion (about

PROBLEM SET

35%) of the phage chromosome and produces sticky ends on the two outer, or tail, sections of the chromosome. (Note that the genetic material that allows the virus to infect the host and to complete its reproductive cycle is carried in these outer sections.) The tail sections are then isolated from the central section. Treating a foreign DNA molecule with the same restriction endonuclease yields segments with ends complementing those of the tail sections. These foreign DNA fragments are incubated with the tail sections to produce recombinant DNA molecules. Subsequent incorporation of these DNA molecules into intact phages depends on the size of the molecules. Studies indicate that DNA molecules that exceed 105% of the length of the viral chromosome or that are shorter than 78% of its length cannot be packaged into viral particles.

a. After carrying out this procedure to incorporate recombinant DNA molecules into viral particles, is it likely that most of the infective viruses produced will contain recombinant DNA?

b. The λ phage chromosome contains about 47,000 base pairs. What is the approximate size range of the foreign DNA that could be inserted into a viral vector?

c. What are two major advantages of using viruses as vectors (as opposed to using bacterial plasmids)?

30-6. *ColE1* plasmids, commonly used as a vector in recombinant DNA studies, carry genes that control the production of colicins. Colicins are secreted by *E. coli* cells and kill other bacteria lacking the ability to synthesize colicins. Why would these plasmids be particularly useful in recombinant DNA studies?

30-7. Assume that you are working in a laboratory where studies on a recombinant DNA preparation and cloning are in progress. Prokaryotic DNA that codes for the production of the industrially important chemical ethyl alcohol has been isolated. You are given the assignment of cloning this gene with the goal of producing appreciable quantities of ethyl alcohol. List the key steps that you would follow in order to get the bacterium *E. coli* to manufacture this alcohol.

30-8. Assume that you are working in a recombinant DNA laboratory and are given the task of constructing a DNA "library" for a particular starfish. This involves cloning the entire genome into phage particles. Outline the key steps that you would follow in carrying out this procedure.

30-9. Assume that you are working in a recombinant DNA laboratory. The DNA from a human chromosome that is known to carry the gene for insulin production has been broken into relatively short fragments by treatment with restriction endonucleases. You are assigned the task of identifying from among all the DNA fragments the one carrying this insulin gene so that the fragment can eventually be sequenced. Available to you is a supply of radioactively labeled mRNA transcripts of this gene. The laboratory is set up to carry out standard procedures including gel electrophoresis and autoradiography. Describe a procedure that would allow you to identify the DNA fragment carrying the insulin gene.

30-10. Why are the strands of DNA that are subjected to the Maxam-Gilbert chemical-cleavage method of sequencing DNA labeled with ^{32}P?

30-11. The Maxam-Gilbert chemical-cleavage method of sequencing can be used with DNA segments up to about 200 nucleotides long. Assume that the procedure is carried out on a sample of a DNA segment 200 nucleotides long.

PROBLEM SET

a. How many different kinds of bands would show up on the autoradiograph?
b. How many different kinds of DNA fragments would have produced these autoradiograph bands?
c. What would all of the fragments within a particular band have in common?
d. If you were to compare the labeled fragments from two different bands, what differences would you expect to find?
e. How many nucleotides would be present in the fragments making up the band farthest from the origin?

30-12. The DNA polymerase supplied to the DNA-synthesizing setups used in the Sanger chain-termination procedure for sequencing DNA is enzymatically modified using trypsin so that it no longer possesses its 5′-to-3′ exonuclease activity. This DNA polymerase, however, still retains its 3′-to-5′ exonuclease activity and its 5′-to-3′ polymerase capability.
a. Why do you think the polymerase used in this procedure is treated in this way?
b. Is this treatment essential to the success of the Sanger method of sequencing DNA?

30-13. Compare similarities and differences in the DNA fragments generated by the Maxam-Gilbert chemical-cleavage and the Sanger chain-termination methods of sequencing DNA.

30-14. The Maxam-Gilbert method for sequencing DNA cleaves each strand being sequenced into two fragments. However, only one of the two fragments shows up on an autoradiograph. Explain why this is so.

30-15. Both the Maxam-Gilbert and the Sanger methods of sequencing DNA give rise to DNA fragments of every possible length up to the length of the strand being sequenced. Briefly explain why this is so and why this is essential to sequencing the DNA molecule under study.

30-16. Assume that the Sanger chain-termination procedure is used to sequence a strand of DNA and the autoradiograph that results is as follows.
a. How many different kinds of fragments were produced from the chain-termination procedures?
b. What is the sequence of nucleotides in the DNA synthesized in this procedure?
c. What is the sequence of nucleotides in the DNA that was being sequenced by this procedure?
d. One of the four kinds of nucleotides supplied to the DNA-synthesizing setups involved in this procedure carry a radioactive label. What is that nucleotide?

PROBLEM SET

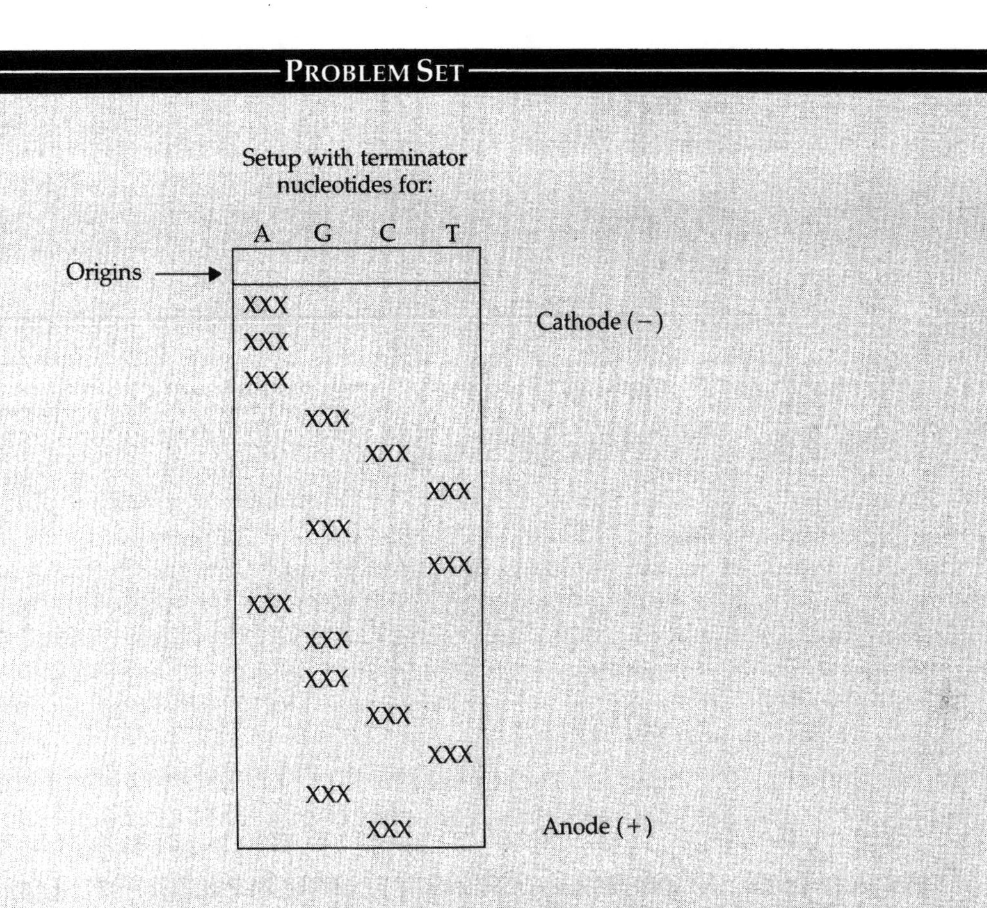

30-17. The Maxam-Gilbert chemical-cleavage procedure is used to sequence a strand of DNA, and the autoradiograph that results is as follows. What is the sequence of nucleotides in the DNA?

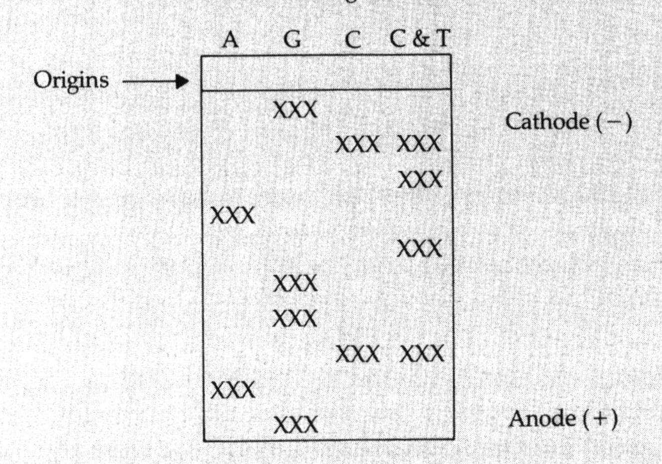

CHAPTER

31

Population Genetics I: The Hardy-Weinberg Equilibrium

INTRODUCTION

Most of this book has focused on the genetic makeup of individuals and the mechanisms governing parent-to-progeny transmission of genes. The basic principles of genetics can be related to evolutionary biology by applying them to populations of sexually reproducing, interbreeding organisms. Such populations, rather than individuals, are the basic units of evolution. Whereas individuals live for relatively short intervals, populations can exist for much longer periods. And whereas, apart from mutations that may occur, the genetic makeup of an individual is fixed from the time of conception until death, that of a population can change with time. This change results in the evolution of the population.

THE GENE POOL

The alleles carried by the individuals in a population make up the population's **gene pool.** With each round of interbreeding, this genetic material is recombined and passed to the next generation. A gene pool can be characterized by identifying the alleles that it contains and by determining their frequencies. Evolution is often defined as a shift in gene frequencies from one generation to the next. Analysis of such changes is an important part of population genetics since it permits the detection of evolution. If gene frequencies remain unchanged with the passage of generations, then the gene pool is stable and no evolution has occurred.

Evolutionary Forces

There are a number of **evolutionary forces** or mechanisms that can cause shifts in gene frequency. These forces are summarized here and covered in greater detail in Chapter 32. Each is illustrated using an autosomal locus with a dominant allele, *B*, and its recessive counterpart, *b*.

Mutation: If some *B* alleles in the gene pool mutate to *b* alleles, the frequency of *B* will be reduced and the frequency of *b* will be increased. In contrast, if *b* alleles mutate to *B*, the frequency of *b* will be reduced and the frequency of *B* will be increased.

Natural selection: If individuals expressing the dominant allele, those with genotypes *BB* and *Bb*, are less fit than those expressing the recessive allele, those with genotype *bb*, proportionately fewer of the less-fit individuals will reproduce and proportionately fewer of the *B* alleles will be passed on to the next generation. This will reduce the relative frequency of *B* alleles in the next generation's gene pool. Alternately, if selection works against the homozygous recessive individ-

uals, the relative frequency of *b* alleles will decline in the next generation's gene pool.
Migration or gene flow: If individuals migrating into a population have genotype *BB*, the frequency of the *B* allele in that population's gene pool will be increased. If individuals of this type leave a population, then the relative frequency of the *B* allele would decrease in that population's gene pool.
Genetic drift or sampling error: If only a very limited number of individuals of a large population give rise to the next generation and these individuals do not carry a representative sample of the population's gene pool, then gene frequencies in the next generation will be shifted accordingly.

In this chapter we will be focusing on populations that, for the most part, are free of these evolutionary forces and thus not undergoing evolution.

Characterizing a Gene Pool: Determining Allelic Frequencies

Population geneticists are often interested in describing the gene pool of a population. This can be easily done if we know the genes and their alleles that are present and the frequencies in which they occur. To illustrate this, we will consider an autosomal locus with two alleles that show complete dominance, where allele *A* is dominant to its recessive counterpart, *a*. Assume we have a population of 100 individuals: 20 are homozygous dominants (*AA*), 20 are heterozygous (*Aa*), and 60 are homozygous recessives (*aa*).

Genotype	Number of individuals	Percentage	Genotypic frequency
AA	20	20%	0.2
Aa	20	20%	0.2
aa	60	60%	0.6
Totals:	100	100%	1.0

From these data, the allelic frequencies for *A* and *a* can be calculated. First, it is necessary to determine the total number of *A* and *a* alleles in the gene pool. Since each individual carries two alleles for this gene, the gene pool will contain a total of 100 × 2 = 200 alleles. Next, the number of each type of allele is determined by totaling the number found in individuals with each of the three genotypes: the 20 *AA* individuals, each with two *A* alleles, carry a total of 20 × 2 = 40 *A* alleles; the 20 *Aa* individuals, each with one *A* allele and one *a* allele, carry a total of 20 × 1 = 20 *A* alleles and 20 × 1 = 20 *a* alleles; and the 60 *aa* individuals, each with two *a* alleles, possess a total of 60 × 2 = 120 *a* alleles.

From this information, the number of *A* and *a* alleles in the gene-pool can be determined: the number of *A* alleles is 40 (from the *AA* genotype) plus 20 (from the *Aa* genotype) for a total of 60; the number of *a* alleles is 20 (from the *Aa* genotype) plus 120 (from the *aa* genotype) for a total of 140.

From these totals, the relative allelic frequencies can be determined. The 60 *A* alleles are out of a total of 200, giving us a frequency of *A*, often written f(*A*), of 60/200 = 0.3, or 30%. The 140 *a* alleles out of 200 give a frequency of *a*, of f(*a*), of 140/200 = 0.7, or 70%. Note that since this gene has only two alleles, their relative frequencies total up to 0.3 + 0.7 = 1 and their percentages add up to 30% + 70% = 100%, as would be expected. Before going on, make sure you understand the distinction between gene, or allelic, frequencies and genotypic frequencies. In this example, the allelic frequencies are 0.3 and 0.7, for alleles *A* and *a*, respectively, and the genotypic frequencies are f(*AA*) = 20/100 = 0.2, f(*Aa*) = 20/100 = 0.2, and f(*aa*) = 60/100 = 0.6.

As a second example on determining allelic frequency, we will consider an autosomal locus with two alleles that show complete dominance, where allele *B* is dominant to its recessive counterpart, *b*. Assume that the population of 200 individuals consists of 20 homozygous dominants (*BB*), 40 heterozygotes (*Bb*), and 140 homozygous recessives (*bb*). What is the total number of alleles at this locus carried in the population's gene pool? ________[1] Now the number of *B* alleles can be determined. The *B* allele is present in genotypes *BB* and *Bb*. What is the number of *B* alleles carried by all the individuals with genotype *BB*? ____[2] How many *B* alleles are carried by all the individuals with genotype *Bb*? ________[3] What is the total number of *B* alleles in the gene pool? ________[4] What is their frequency? ______[5] their percentage? ______[6]

Next the number of *b* alleles is determined. The *b* allele is present in genotypes *Bb* and *bb*. What is the number of *b* alleles carried by all the *Bb* individuals? ________[7] How many *b* alleles are carried by all the *bb* individuals? ________[8] What is the total number of *b* alleles in the gene pool? ________[9] What is their frequency? ________[10] their percentage? ________[11]

Since there are only two types of alleles of the gene in the gene pool, their frequencies should add up

[1] Each of the 200 individuals carries two alleles at this locus; thus there are 400 alleles. [2] 20 *BB* individuals × 2 *B* alleles = 40. [3] 40 *Bb* individuals × 1 *B* allele = 40. [4] 80. [5] f(*B*) = 80/400 = 0.2. [6] 0.2 × 100 = 20%. [7] 40 *Bb* individuals × 1 *b* allele = 40. [8] 140 *bb* individuals × 2 *b* alleles = 280. [9] 320. [10] 320/400 = 0.8. [11] 0.8 × 100 = 80%.

to 1 as indeed they do: 0.8 + 0.2 = 1. (After calculating allelic frequencies for a population, it is a good idea to verify that they add up to 1. If you have made an arithmetic error, this check will often bring it to your attention.) Note that the gene-pool frequency for a particular allele can range from 0, if it is absent from the gene pool, to 1.0, or 100%, if it is the only form of the gene present in the gene pool. An allele that occurs in a gene pool with a frequency of 1.0 is said to be **fixed.**

An Alternative Way of Determining Allelic Frequencies

Another more direct approach to determining allelic frequencies is to begin by determining the frequency for each genotype. Consider once again the population of 200 consisting of 20 *BB*, 40 *Bb*, and 140 *bb* individuals. Determine f(*BB*). ______[12] Determine f(*Bb*). ______[13] Determine f(*bb*). ______[14] Allelic frequencies can be determined from these genotypic frequencies as follows.

Determining f(*B*): The *B* allele occurs in genotypes *BB* and *Bb*. Since f(*BB*) = 0.1, the alleles carried by *BB* individuals occur in the gene pool with a frequency of 0.1. Because all their alleles are *B* alleles, the frequency of *B* alleles contributed by *BB* individuals is 0.1. Since f(*Bb*) = 0.2, the alleles carried by *Bb* individuals occur in the gene pool with a frequency of 0.2. Because half of the alleles carried by the *Bb* individuals are *B*, the frequency of *B* alleles contributed by these individuals is (1/2)0.2 = 0.1. The frequency of *B* in the gene pool is 0.1 (from *BB*) + 0.1 (from *Bb*) = 0.2. The generalized expression for the frequency of *B* is f(*B*) = f(*BB*) + (1/2)f(*Bb*).

Determining f(*b*): The *b* allele occurs in genotypes *Bb* and *bb*. Since f(*Bb*) = 0.2 and half the alleles carried by the *Bb* individuals are *b*, the frequency of *b* contributed by *Bb* individuals is (1/2)0.2 = 0.1. Since f(*bb*) = 0.7 and all of the alleles carried by *bb* individuals are *b*, the frequency of *b* contributed by *bb* individuals is 0.7. The frequency of *b* in the gene pool is 0.1 (from *Bb*) + 0.7 (from *bb*) = 0.8. The generalized expression for the frequency of *b* is f(*b*) = (1/2)f(*Bb*) + f(*bb*).

Forming the Next Generation

The frequencies of alleles within a gene pool become important when we consider how these alleles contribute to forming the population's next generation. To keep our discussion as simple as possible, we will assume an idealized population that is free of evolutionary forces and that experiences random mating. Under these conditions, the relative frequency of an allele in a population's gene pool indicates the frequency with which it occurs in eggs and in sperm produced by that generation and the probability with which it participates in forming the next generation. For example, if f(*B*) = 0.2, then the probability of its occurring in a gamete, either egg or sperm, is 0.2. The probability of two such gametes joining together to form a homozygous dominant individual is given by the product law of probability and equals the product of their separate probabilities, that is, $0.2 \times 0.2 = (0.2)^2 = 0.04$.

Genotypic Frequencies Of the gametes produced by members of the population, some of the eggs and some of the sperm will carry the dominant allele, *B*, and the rest will carry the recessive allele, *b*. These gametes and their frequencies can be combined in a Punnett square to identify all possible genotypes produced by random union of the population's gametes and to determine the frequencies with which these genotypes occur.

	B(0.2)	*b*(0.8)
B(0.2)	*BB* 0.2 × 0.2 = 0.04	*Bb* 0.2 × 0.8 = 0.16
b(0.8)	*Bb* 0.8 × 0.2 = 0.16	*bb* 0.8 × 0.8 = 0.64

As the square indicates, the expected genotypic frequencies after a generation of random mating are as follows: f(*BB*) = 0.04, f(*Bb*) = 0.16 + 0.16 = 0.32, and f(*bb*) = 0.64. Note that these three genotypic frequencies add up to 1: 0.04 + 0.32 + 0.64 = 1.0.

Assume this new generation, like the parental generation, consists of 200 individuals. By multiplying the genotypic frequencies by 200, we can determine the expected numbers of individuals showing each genotype.

f(*BB*) = 0.04 × 200 = 8 *BB* individuals
f(*Bb*) = 0.32 × 200 = 64 *Bb* individuals
f(*bb*) = 0.64 × 200 = 128 *bb* individuals
Total: 200

Allelic Frequencies At this point we can consider the effect that the formation of this new generation has had on the frequencies of the two alleles. Begin by calculating the frequency of the *B* allele in the individ-

[12] 20/200 = 0.1. [13] 40/200 = 0.2. [14] 140/200 = 0.7.

uals of this new generation. ________[15] Now calculate f(*b*). ________[16]

How do these allelic frequencies compare with those allelic frequencies we started out with in the gene pool before random mating occurred and the new generation was formed? ________[17] In other words, the gene frequencies have not changed in going from generation 1 to generation 2. Complete the following Punnett square to show the random mating of members of this newly formed generation to produce a third generation.[18]

	B(0.2)	*b*(0.8)
B(0.2)		
b(0.8)		

A key point emerges here: in the absence of forces that would change gene frequencies (that is, in the absence of mutation, natural selection, migration, and genetic drift) and with random mating, gene frequencies remain constant and the ratio of genotypes remains unchanged, generation after generation. This principle, independently recognized shortly after the turn of the century by G. H. Hardy and W. Weinberg, is known as the **Hardy-Weinberg law.** The term **Hardy-Weinberg equilibrium** is often used to refer to the constancy of genotypic frequencies that arises under conditions of the law following a generation of random mating.

Generalized Expressions of the Hardy-Weinberg Law

The concept of the Hardy-Weinberg law can be expressed in general terms for genes that have two alleles, with one dominant to the other, by using the letters p and q to symbolize the frequency of the dominant and recessive alleles, respectively. These frequencies add up to 1, and this can be expressed as $p + q = 1$. Combining these symbols in a Punnett square gives the general expressions for the genotypic frequencies of the progeny produced through random mating of the parental population.

	p	q
p	$p \times p = p^2$	$p \times q = pq$
q	$p \times q = pq$	$q \times q = q^2$

Thus, following a round of random mating, the frequency of genotypes in the next generation can be generalized as follows: f(homozygous dominants) $= p^2$, f(heterozygotes) $= pq + pq = 2pq$, and f(homozygous recessives) $= q^2$. Since these genotypes represent all those produced, their frequencies will add up to 1, and this can be expressed as $p^2 + 2pq + q^2 = 1$.

To summarize, for an infinitely large, randomly mating population free of evolutionary forces, the Hardy-Weinberg Law states that (1) gene frequencies will remain constant, that is, in equilibrium, from one generation to the next, and (2) following one round of random mating, genotypic frequencies will remain constant from one generation to the next. For genes with two alleles, with one completely dominant to the other, these frequencies are given by the expression $p^2 + 2pq + q^2$, where p and q represent the frequency of the dominant and recessive alleles, respectively, and p^2, $2pq$, and q^2 represent the frequencies of the homozygous dominants, heterozygotes, and homozygous recessives, respectively.

The equilibrium of genotypic frequencies implies indefinite gene pool stability; that is, a population existing under Hardy-Weinberg conditions is not evolving. However, in the natural world, Hardy-Weinberg populations do not exist. Natural populations experience change: gene frequencies shift and populations evolve. Tracing changes in allelic frequencies from one generation to the next permits the detection of evolutionary change as it occurs and may provide important information about the mechanisms of evolution. The nonevolving Hardy-Weinberg population provides a standard or an index against which evolving populations can be compared.

APPLICATIONS OF THE HARDY-WEINBERG LAW

The remainder of this chapter looks at some of the ways in which population geneticists use the Hardy-

[15] The eight homozygous dominants, *BB*, have a total of 16 *B* alleles (8 individuals × 2 alleles/individual) and the heterozygotes, *Bb*, have 64 (64 individuals × 1 allele/individual) for a total of 16 + 64 = 80. This is 80 out of a total of 400 alleles (two alleles carried by each of the 200 individuals in the new generation), giving us f(*B*) = 80/400 = 0.2. [16] The 128 homozygous recessives, *bb*, have 128 × 2 = 256 *b* alleles, and the 64 heterozygotes have 64 *b* alleles, for a total of 256 + 64 = 320. This is 320 alleles out of 400, so f(*b*) = 320/400 = 0.80. [17] They are the same. [18] The genotypes occur in the same ratio of frequencies seen in generation 2, that is, 0.04 : 0.32 : 0.64, and the allelic frequencies remain unchanged: f(*B*) = 0.2 and f(*b*) = 0.8.

Weinberg law. First, we will consider its use in connection with autosomal loci having two alleles. From the two generalized expressions of the Hardy-Weinberg law, that is $p + q = 1$ and $p^2 + 2pq + q^2 = 1$, important deductions can be made about many populations. For example, if relative gene frequencies are known for a particular gene pool, the genotypic frequencies expected under Hardy-Weinberg conditions can be readily determined.

Predicting Future Genotypic Frequencies from Current Allelic Frequencies

Sometimes relative allelic frequencies are given to you, as in the following example.

PROBLEM: Sampling individuals from a fruit fly population indicates that a dominant allele and its recessive counterpart are present in the gene pool in frequencies of 0.1 and 0.9, respectively. Following a generation of random mating, what are the relative genotypic frequencies expected if the population exists under Hardy-Weinberg conditions? If 1000 individuals make up this new generation, how many would be expected to exhibit each genotype?

SOLUTION:

$$p = \text{f(dominant allele)} = 0.1$$
$$q = \text{f(recessive allele)} = 0.9$$

The relative genotypic frequencies are calculated from the general expressions. Determine the frequency of each genotype.

f(homozygous dominants) $= p^2 =$ ________[19]

f(heterozygotes) $= 2pq =$ ________[20]

f(homozygous recessives) $= q^2 =$ ________[21]

Verify your arithmetic by making sure that your genotypic frequencies add up to 1. ________[22] Now calculate the number of individuals expected to show each genotype in a generation of 1000 individuals.

Homozygous dominants = ________[23]

Heterozygotes = ________[24]

Homozygous recessives = ________[25]

As a check on your multiplication, verify that your numbers for the three genotypes add up to the total number of individuals in the population. ________[26]

Predicting Future Genotypic Frequencies from Current Genotypic Frequencies

In other problems you may be given basic information about the number of individuals in a population showing each genotype. From this, the relative frequencies of alleles in the gene pool can be determined and used to predict the number of individuals showing each genotype in a generation produced under Hardy-Weinberg conditions.

PROBLEM: The allele for gray fur color (*A*) in field mice is dominant to the recessive allele producing black fur (*a*). A population of 500 contains 250 homozygous dominants, 150 heterozygotes, and 100 homozygous recessives. Determine the frequency of dominant and recessive alleles in this population's gene pool and the relative frequencies of genotypes expected in the next generation. Assume Hardy-Weinberg conditions.

SOLUTION: First the number of dominant and recessive alleles in the gene pool and the relative frequency of each must be determined. Calculate the total number of *A* alleles in the gene pool and f(*A*). ________[27] Now calculate the total number of *a* alleles in the gene pool and f(*a*). ________[28] As a check on the arithmetic, verify that two allelic frequencies add up to 1. Next we calculate the genotypic frequencies expected in the next generation following random mating by setting $p = f(A) = 0.65$ and $q = f(a) = 0.35$, and substituting these values into the expression $p^2 + 2pq + q^2 = 1$. Calculate these frequencies.

$f(AA) = p^2 =$ ________[29]

$f(Aa) = 2pq =$ ________[30]

$f(aa) = q^2 =$ ________[31]

[19] $(0.1)^2 = 0.01$. [20] $2(0.1)(0.9) = 0.18$. [21] $(0.9)^2 = 0.81$. [22] $0.01 + 0.18 + 0.81 = 1.00$. [23] This is determined by multiplying the genotype's relative frequency by the total number of individuals in the population: $0.01 \times 1000 = 10$. [24] $0.18 \times 1000 = 180$. [25] $0.81 \times 1000 = 810$. [26] $10 + 180 + 810 = 1000$. [27] Two *A* alleles are found in each of the 250 homozygous dominants, giving us 500 *A* alleles, and each of the 150 heterozygotes carries one *A* allele, for $150 \times 1 = 150$ alleles. Thus, the number of *A* alleles in the gene pool is $500 + 150 = 650$. This is out of a total of 1000 alleles (500 individuals, each with two alleles: $500 \times 2 = 1000$), and $f(A) = 650/1000 = 0.65$. [28] Two *a* alleles are found in each of the 100 homozygous recessives, giving us 200 alleles. Combining these with the single *a* allele carried by each of the 150 heterozygotes gives a total of 350 recessive alleles out of a total of 1000 alleles in the gene pool. The $f(a) = 350/1000 = 0.35$. [29] $f(AA) = p^2 = (0.65)^2 = 0.4225$. [30] $f(Aa) = 2pq = 2 \times 0.65 \times 0.35 = 0.4550$. [31] $f(aa) = q^2 = (0.35)^2 = 0.1225$.

Verify your arithmetic by making sure that your genotypic frequencies add up to 1. ▬

Predicting Allelic Frequencies from the Frequency of Homozygous Recessives

When the alleles at a locus show complete dominance, it is impossible to identify the genotypes of homozygous dominants and heterozygotes since both express the same phenotype; only individuals showing the recessive trait can be accurately genotyped. Often nothing is known about a population other than its size and the number of individuals showing the recessive trait. Despite this limited information, the relative allelic frequencies and the frequencies of homozygous dominants and heterozygotes can be determined.

PROBLEM: The allele, *T*, for the ability to taste the chemical PTC (phenylthiocarbamide) is dominant to that for nontasting, *t*. Assume that in an isolated human population of 2000 individuals, 1820 are tasters and 180 are nontasters and that these individuals mate at random with regard to this trait. Determine the relative allelic frequencies for the population's gene pool and the genotypic frequencies for the homozygous dominant and heterozygous individuals within the gene pool.

SOLUTION: Of the population of 2000, the homozygous recessives, *tt*, occur with a frequency of 180/2000 = 90/1000 = 0.09. In a randomly mating population, the frequency of homozygous recessives is given by the expression q^2. Thus, $q^2 = 0.09$, and by taking the square root of each side of the equation, the value of q (the frequency of the recessive allele) can be determined. What is the value of q? ________[32] Once the value of q is known, it can be substituted in the equation $p + q = 1$, which is then solved for p (the frequency of the dominant allele). Determine the value of p. ________[33] Now that we have the relative frequencies for the alleles, the frequencies for the homozygous dominants (*TT*) and heterozygotes (*Tt*) can be determined. Calculate these frequencies. ________[34] At this point, verify that the three genotypic frequencies add up to 1.

Determining Frequencies of Codominant Alleles

The procedure for determining frequencies of codominant alleles is basically the same as that used for alleles that show dominance-recessiveness interactions. This can be illustrated by using a problem involving the human MN blood-antigen system. Two antigens, designated as M and N and produced by alleles *M* and *N*, respectively, may be found on the surfaces of red blood cells. Any particular individual will have one or both of these antigens. Individuals with two *N* alleles will produce antigen N only and will have blood type N; those with two *M* alleles will make antigen M only and have blood type M. Codominance occurs in individuals carrying both the *M* and *N* alleles: both antigens are produced resulting in type MN blood.

PROBLEM: The MN system blood type has been determined for all the individuals in a population of 800 and the genotypic frequencies are as follows: *MM*: 342/800 = 0.428; *MN*: 357/800 = 0.446; and *NN*: 101/800 = 0.126. Calculate the genotypic frequencies expected in this generation if the population is under Hardy-Weinberg conditions.

SOLUTION: This problem can be solved in three steps. Begin by determining the frequencies for alleles *M* and *N*. ________[35] Verify that these two allelic frequencies add up to 1. Next, determine the genotypic frequencies. With p = f(*M*) = 0.65 and q = f(*N*) = 0.35, the genotypic frequencies expected if the population is under Hardy-Weinberg conditions are given by the expression $p^2 + 2pq + q^2$. Calculate the genotypic frequencies.

f(*MM*) = p^2 = ________[36]

f(*MN*) = 2pq = ________[37]

f(*NN*) = q^2 = ________[38]

Finally, calculate the number of individuals in a population of 800 showing each genotype.

MM individuals = ________[39]

MN individuals = ________[40]

NN individuals = ________[41]

Verify your calculations by making sure that these numbers add up to 800. Because of the close agreement between the numbers of individuals expected to show the genotypes under Hardy-Weinberg conditions and those actually observed in the population, we can conclude that the population is under Hardy-Weinberg conditions. ▬

[32] $\sqrt{0.09} = 0.3$. [33] Substituting 0.3 for q in the equation $p = 1 - q$ gives $p = 1 - 0.3 = 0.7$.
[34] f(*TT*) = $p^2 = (0.7)^2 = 0.49$ and f(*Tt*) = 2pq = 2(0.7)(0.3) = 2(0.21) = 0.42. [35] f(*M*) = f(MM) + 1/2f(MN) = 0.428 + 1/2(0.446) = 0.428 + 0.223 = 0.65, and f(*N*) = f(NN) + 1/2f(MN) = 0.126 + 1/2(0.446) = 0.126 + 0.223 = 0.35. [36] $(0.65)^2 = 0.423$.
[37] 2(0.65)(0.35) = 0.455. [38] $(0.35)^2 = 0.123$. [39] (0.423)(800) = 338. [40] (0.455)(800) = 364. [41] (0.123)(800) = 98.

Determining Frequencies for a Multiple-Allelic Series

The problems we have considered so far have involved genes with two alleles. If a gene has more than two alleles, as do those controlling multiple-allelic traits, the Hardy-Weinberg law still applies, but additional symbols are needed to designate the additional alleles: r is generally used to represent the frequency of a third allele, s to represent that of a fourth allele, and so on. Additional terms must also be added to the binomial. For example, with three alleles, the genotypic frequencies arising from all possible combinations of the alleles under Hardy-Weinberg conditions are given by expanding the trinomial $(p + q + r)^2$. One way of carrying out this expansion is to place the three allelic frequencies into a Punnett square; use of the product law will produce an expression for the expected frequency of each possible genotype.

	p	q	r
p	p^2	pq	pr
q	pq	q^2	qr
r	pr	qr	r^2

Summing these expressions gives us the expanded trinomial: $p^2 + 2pq + 2pr + q^2 + 2qr + r^2$, where the terms p^2, q^2, and r^2 represent the frequencies of the homozygotes and the terms 2pq, 2pr, and 2qr represent the frequencies of the heterozygotes.

As an example, consider a multiple-allelic trait that we have looked at before (Chapter 8), the ABO blood system in humans. You will recall that this trait involves a multiple-allelic series with three major alleles, I^A, I^B, and i, which give rise to four different phenotypes, the A, B, AB, and O blood types. Using the symbols $p = f(I^A)$, $q = f(I^B)$, and $r = f(i)$ and expanding the trinomial $(p + q + r)^2$ to $p^2 + 2pq + 2pr + q^2 + 2qr + r^2$ gives us the distribution of genotypic frequencies. For each mathematical expression that follows, identify the genotype whose frequency is given by the expression and the phenotype produced by that genotype.

Expression	Genotype	Phenotype
p^2	________[42]	________[43]
2pq	________[44]	________[45]
2pr	________[46]	________[47]
q^2	________[48]	________[49]
2qr	________[50]	________[51]
r^2	________[52]	________[53]

PROBLEM: A survey of ABO blood types in a human population of 1000 indicates that 385 people have type A blood, 130 have type B, 46 have type AB, and 439 have type O. Determine the frequencies of the three alleles governing this trait. Is this population in Hardy-Weinberg equilibrium?

SOLUTION: One approach to calculating the allelic frequencies is to begin with the only blood type that is produced by a single homozygous genotype. Which blood type is this? ________[54] As you noted, $f(ii) = r^2$ can be set equal to the frequency of type O individuals, 439/1000: $f(ii) = r^2 = 439/1000 = 0.439$. Solving this equation for r gives the frequency of the i allele. What is the value of r? ________[55]

The next step is to determine the frequency of a second allele. Note that the $p^2 + 2pr + r^2$ portion of the expanded trinomial gives the frequencies of all the individuals with types A and O blood (genotypes I^AI^A, I^Ai, and ii), and furthermore, it is an expansion of the binomial $(p + r)^2$. Consequently, this binomial can be set equal to the combined frequencies of type A and type O individuals:

$$\begin{aligned} p^2 + 2pq + r^2 = (p + r)^2 &= f(\text{type A}) + f(\text{type O}) \\ &= 385/1000 + 439/1000 \\ &= 0.385 + 0.439 \\ &= 0.824. \end{aligned}$$

From this equation, derive the expression for p. ________[56] Since we know the value of r is 0.663, the expression can be solved to give the value of p. What is the value of p? ________[57]

[42] I^AI^A. [43] Type A. [44] I^AI^B. [45] Type AB. [46] I^Ai. [47] Type A. [48] I^BI^B. [49] Type B. [50] I^Bi. [51] Type B. [52] ii. [53] Type O. [54] Type O with genotype ii. [55] $\sqrt{0.439} = 0.663$. [56] Begin by taking the square root of both sides: $p + r = \sqrt{0.824}$. Since the $\sqrt{0.824} = 0.908$, then $p + r = 0.908$. Subtracting r from both sides of the equation gives $p = 0.908 - r$. [57] $0.908 - 0.663 = 0.245$.

At this point the frequencies for two of the three alleles, i and I^A, have been determined. The easiest way to get the frequency for the third allele, I^B, represented by q, is to place the values for r and p in the expression $p + q + r = 1$ and solve for q. Determine the value of q. ________[58]

Note that once the value of r has been determined, an alternative approach to identifying the other allelic frequencies would be to consider the $r^2 + 2qr + q^2$ portion of the expanded trinomial which gives the frequencies of all the individuals with blood types B and O. These three terms represent the expansion of the binomial $(q + r)^2$ which can be set equal to the combined frequencies of types B and O; since the value of r is known, the expression can be solved for q. Once r and q are known, p can readily be determined using the expression $p + q + r = 1$.

With the three gene frequencies identified, the genotypic frequencies expected under conditions of the Hardy-Weinberg equilibrium can be calculated.

$f(I^AI^A) = p^2 =$ ________[59]

$f(I^Ai) = 2pr =$ ________[60]

$f(ii) = r^2 =$ ________[61]

$f(I^AI^B) = 2pq =$ ________[62]

$f(I^Bi) = 2qr =$ ________[63]

$f(I^BI^B) = q^2 =$ ________[64]

Calculate the phenotypic frequencies expected under conditions of the Hardy-Weinberg equilibrium and the equivalent number of individuals per 1000 progeny.

f(type A) = ________[65]

f(type O) = ________[66]

f(type AB) = ________[67]

f(type B) = ________[68]

Because of the close agreement between the blood-type frequencies expected under Hardy-Weinberg conditions and those observed in the population, we can conclude that the population is under such an equilibrium. ▬

Determining Gene Frequencies for X-Linked Alleles

Some problems may involve genes that are carried on X chromosomes, that is, that are sex-linked. If such a gene has two alleles, for example, R and r, with R showing complete dominance to r, five possible genotypes can occur. Females could be X^RX^R, X^RX^r, or X^rX^r and males could be X^RY or X^rY. Because the males are hemizygous (they carry and express just one allele for any sex-linked trait), their allelic frequencies can readily be determined. For example, if 10% of the males in the population express the recessive trait, then 10% of the males possess the recessive allele and the remaining 90% have the dominant allele. Thus, among the males,

$$f(X^RY) = f(R) = p = 0.90$$
$$f(X^rY) = f(r) = q = \underline{0.10}$$
$$\text{Total} = 1.00.$$

Under normal circumstances, the frequencies of these alleles will be the same in the females of the population. By treating the females as a distinct population, their genotypic frequencies can be determined as follows.

$$f(X^RX^R) = p^2 = (0.9)(0.9) = 0.81$$
$$f(X^RX^r) = 2pq = 2(0.9)(0.1) = 0.18$$
$$f(X^rX^r) = q^2 = (0.1)(0.1) = \underline{0.01}$$
$$\text{Total} = 1.00$$

Note that the allelic frequencies are the same for each sex; the distinction between males and females arises because the males are hemizygous.

Testing for Hardy-Weinberg Equilibrium Using the Chi-Square Test

Problems dealing with population genetics sometimes ask you to determine whether a particular population is in a Hardy-Weinberg equilibrium. This is just another way of asking whether genotypes within the population occur in the frequencies predicted by the Hardy-Weinberg equilibrium. Close agreement between the genotypic frequencies observed in the population and those expected under a Hardy-Weinberg

[58] Substituting the values of r and p gives the equation $0.245 + q + 0.663 = 1$ which can be changed to $q = 1 - 0.245 - 0.663$, which gives $q = 0.092$. [59] $(0.245)^2 = 0.060$. [60] $2(0.245)(0.663) = 0.325$. [61] $(0.663)^2 = 0.440$. [62] $2(0.245)(0.092) = 0.045$. [63] $2(0.092)(0.663) = 0.122$. [64] $(0.092)^2 = 0.008$. [65] $0.060 + 0.325 = 0.385$, or 385/1000. [66] 0.440, or 440/1000. [67] 0.045, or 45/1000. [68] $0.122 + 0.008 = 0.130$, or 130/1000.

equilibrium indicate that the population is in such an equilibrium, while a lack of similarity tells us that the population is not in equilibrium. A statistical comparison of the observed and expected genotypes can be carried out using the chi-square test which is discussed in Chapter 6.

When testing genetic ratios with the chi-square test, degrees of freedom were defined as 1 less than the number of classes. In such cases, we arrived at the expected number for each class by determining how many of the total number of individuals in our population sample would fall into each class based on a particular theoretical ratio (for example, 3:1).

When dealing with gene-frequencies, we encounter a different situation, where the expected numbers are not based on a theoretical ratio but upon the actual observed numbers from our population sample. Specifically, we estimate gene frequencies from the observed number of individuals showing each genotype and use those gene frequencies to calculate how many of the total number of individuals in the sample would be expected to show each genotype. In the absence of a theoretical basis for determining the expected numbers, our definition of degrees of freedom has to be modified. Generally speaking, when gene frequencies are involved, one degree of freedom is lost for each allelic frequency that is estimated from the data. Put another way, the degrees of freedom equal the number of genotypic classes minus the number of alleles involved in producing those genotypic classes.

For example, in the problem considered earlier in this chapter involving the MN blood groups, there were three genotypes (*MM*, *MN*, and *NN*) represented in the population. From the observed number of individuals with each genotype, the allelic frequencies were estimated for *N* and *M*, and those frequencies were then used to determine the number of individuals expected to have each genotype. Had a chi-square test been carried out to assess the agreement between the observed and expected numbers, the degrees of freedom would be obtained by subtracting the number of gene frequencies estimated from the observed data (2) from the number of genotypic classes (3) to give one degree of freedom.

SUMMARY

The genetic material carried by the members of an interbreeding population that can be transmitted to the next generation makes up the population's gene pool. A gene pool can be characterized by identifying the genes it contains and determining the frequencies in which they occur. Evolution can be defined as a change in gene frequency that occurs over time.

The Hardy-Weinberg law applies to an infinitely large, randomly mating population of diploid individuals that is free of evolutionary forces (that is, free of mutation, genetic drift, natural selection, and migration). The law states that in the absence of these forces the population is nonevolving: gene frequencies remain unchanged from one generation to the next, and after one generation of random mating, genotypic frequencies are established at values that remain constant generation after generation.

The genotypic equilibrium frequencies for an autosomal locus with two alleles, where p and q represent the allelic frequencies, are given by an expansion of the binomial $(p + q)^2$, where p^2, $2pq$, and q^2 represent the frequencies of the homozygous dominants, heterozygotes, and homozygous recessives, respectively.

The Hardy-Weinberg law can be used to estimate allelic and genotypic frequencies whenever populations are assumed to be under equilibrium conditions. The law can readily be applied to loci involving multiple alleles as well as those that are sex-linked. Observed genotypic frequencies for a population can be compared with values expected under the Hardy-Weinberg law by using the chi-square test. When testing genotypic frequencies using this test, degrees of freedom are defined as the number of phenotypic classes less the number of alleles producing those classes.

PROBLEM SET

31-1. Examination of samples from a very large population of field mice indicates that a dominant allele for large ears, *E*, occurs in the population's gene pool with a frequency of 0.6 and its recessive counterpart for small ears, *e*, has a frequency of 0.4.

a. In the absence of evolutionary forces and after a generation of random mating, what frequencies would be predicted for genotypes *EE*, *Ee*, and *ee* under the Hardy-Weinberg equilibrium?

b. If this new generation consists of 10,000 individuals, what is the actual number of individuals exhibiting each genotype?

31-2. A population is known to be in Hardy-Weinberg equilibrium for an autosomal locus with two alleles, one of which is dominant to the other. Two-thirds of the population expresses the dominant trait.

a. What is the frequency of the recessive allele in the gene pool?

b. What percentage of the individuals showing the dominant trait are homozygous for the dominant allele?

31-3. The frequency of a recessive allele, *a*, in the gene pool of a population over a period of 250 generations is plotted in the following figure.

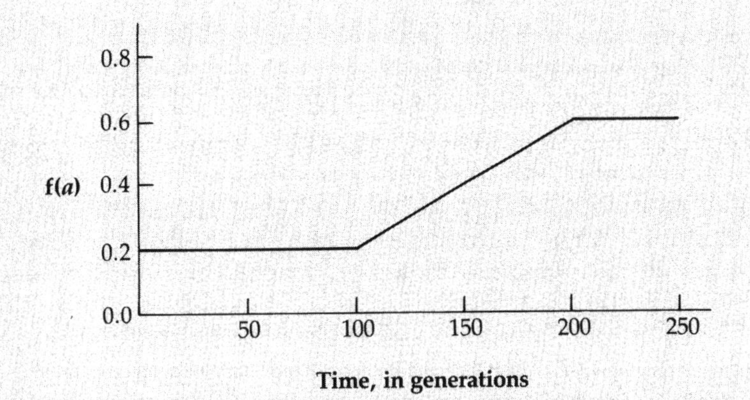

a. When during the time span shown is the population evolving with regard to the locus being considered? Explain.

b. When is the population most likely to have existed under Hardy-Weinberg conditions for the locus being considered? Explain.

c. Determine the genotypic frequencies for this population at generations 75 and 225. (Assume random mating.)

31-4. A dominant allele occurs with a frequency of 0.75 in the gene pool of a sexually reproducing, randomly mating population that exists under Hardy-Weinberg conditions. What is the expected frequency of heterozygous individuals in the next generation?

31-5. A population of 5000 has been under Hardy-Weinberg conditions for a number of generations and is known to contain a total of 450 individuals that are homozygous dominant at a particular locus.

a. What is the frequency of the recessive allele in the gene pool?

b. How many individuals in the population are homozygous recessives?

c. How many individuals are heterozygous?

PROBLEM SET

31-6. Two populations have the following genotypic frequencies.

	Population 1	Population 2
f(*BB*)	0.20	0.00
f(*Bb*)	0.00	0.40
f(*bb*)	0.80	0.60

a. What are the allelic frequencies for each of these populations?
b. Under Hardy-Weinberg conditions and with random mating, what genotypic frequencies would be expected in the next generation in each population?

31-7. A natural population of frogs consists of 96% brown and 4% green animals. Laboratory matings indicate that this skin color trait is determined by an autosomally based pair of alleles, with the allele for green skin, g, recessive to its brown counterpart, g^+.
a. Determine the frequencies of the dominant and recessive alleles in this population's gene pool.
b. Estimate the frequency of frogs in the population that are heterozygous for this trait.
c. What assumptions must be made to calculate the frequency of heterozygous individuals?

31-8. A wild population of grasshoppers exhibits two phenotypes with regard to leg length, long and short. The alleles involved are carried on a pair of autosomes with the allele for long legs, l, recessive to the allele for short legs, l^+. 49% of the population exhibits long legs.
a. Determine the frequencies of l and l^+ in the population's gene pool.
b. Assuming the population remains free of evolutionary forces and that mating is random, what genotypic frequencies would be expected in the next generation?

31-9. A pond contains a total of 3600 fish of the same species. One member of this population exhibits an abnormally shaped dorsal fin that reduces its swimming ability. The trait is known to be due to an autosomal recessive allele. How many fish in this population could be expected to carry this allele in the heterozygous condition if a Hardy-Weinberg equilibrium exists for the alleles involved?

31-10. Statistics gathered from a human population indicate that 80 males out of 1000 have a particular type of color blindness. The gene for this trait is sex-linked and the allele for color blindness, c, is recessive to the allele for normal color vision, C.
a. What are the frequencies for these two alleles in the population's gene pool?
b. What frequency of the women in the population are heterozygous carriers?
c. What frequency of the women have normal vision?
d. What frequency of the women are color-blind?

31-11. Coat color in shorthorn cattle is controlled by a gene with two alleles: C^R and C^W. The genotypes C^RC^R and C^WC^W result in red and white coats, respectively. With the C^RC^W genotype, codominant interaction produces a color known as roan, a mixture of red and white. In a herd of 420 adult cattle, 188 are red, 90 are white, and 142 are roan.

PROBLEM SET

a. Calculate the frequencies of the two alleles in this population's gene pool.

b. Assume that the population mates randomly to produce a generation of progeny. If the conditions of the Hardy-Weinberg law are met, what genotypic frequencies would be expected in this new generation?

31-12. Assume that a locus with three alleles determines coat color in rabbits. Expression of c^+ produces normal color, expression of c^h results in the Himalayan pattern (black feet and ears with white body) and expression of c produces the albino condition. The c^+ allele is dominant to c^h and c, and c^h is dominant to c. Use p, q, and r to represent the frequencies of c^+, c^h, and c, respectively. A randomly breeding population of rabbits includes each of the three phenotypes and exists under Hardy-Weinberg conditions. Describe, in terms of p, q, and r, the genotypic frequencies expected in the next generation.

31-13. Refer to the information given in problem 31-12. A population of 355 rabbits consists of 255 with normal coloration, 90 Himalayans, and 10 albinos. Determine the frequencies of c^+, c^h, and c in this population's gene pool.

31-14. Refer to the information given in problems 31-12 and 31-13. Assume the population of 355 rabbits referred to in 31-13 mates randomly under Hardy-Weinberg conditions to produce a generation of 1850 rabbits.

a. How many rabbits in this new generation would be expected to exhibit each genotype?

b. How many rabbits would be expected to exhibit each phenotype?

31-15. In fruit flies, the allele for vestigial wing, *vg*, is recessive to the allele for normal wings, vg^+. A laboratory population was founded with 1000 adult flies, 80% of which were homozygous for vestigial wings and 20% homozygous for wild-type wings. This population has been maintained for 200 generations in a screened population cage. (Note: The flies used to found the population are considered generation 0.) An examination of the 200th generation shows that 1% of the flies have vestigial wings.

a. What were the frequencies of vg^+ and *vg* in the gene pool of the flies used to found this population?

b. What are the frequencies of vg^+ and *vg* in the gene pool of the 200th generation? (Assume random mating.)

31-16. Refer to the information given in problem 31-15.

a. Assume that the first generation produced in the lab population (these would be the progeny of the flies used to found the population) was formed under Hardy-Weinberg conditions. What were the genotypic frequencies in this first generation?

b. Assume that generation 201 develops under Hardy-Weinberg conditions. What genotypic frequencies will be expected in this generation?

31-17. Refer to the information given in problem 31-15.

a. Has the population evolved during its 200-generation history? Explain.

b. Assume that the size of the population remained constant at 1000 individuals throughout its 200-generation history. What is the average number of vestigial alleles lost from the gene pool in each generation?

31-18. Phenylketonuria is a human metabolic disorder which is caused by an autosomally based recessive allele. Approximately 1 out of 40,000 individuals is affected with this disease. Assume that the population of the United States is in Hardy-Weinberg equilibrium for this locus.

PROBLEM SET

a. What is the frequency of the phenylketonuria allele in the U.S. gene pool?
b. What is the frequency of individuals expected to be heterozygous for this allele?

31-19. A particular type of red-green color blindness in humans is due to a sex-linked allele that is recessive to the allele for normal color vision. About 6% of the males of a particular population are affected with this disorder. Assume that the population is under Hardy-Weinberg conditions for this locus.
a. What is the frequency of females in this population who are expected to be color-blind?
b. What is the frequency of females who carry but do not express this allele?

31-20. Three alleles, designated as P^a, P^b, and P^c, occur at a single locus that controls the production of the human enzyme acid phosphatase. Each allele is codominant to the other two alleles. Six different combinations of these alleles are possible (P^aP^a, P^bP^b, P^cP^c, P^aP^b, P^bP^c, and P^aP^c), and each produces a distinctly different phenotype. In 1964 three researchers, L. Lai, S. Nevo, and A. G. Steinberg (*Science* 145: 1187–88) reported the number of individuals with each of these phenotypes in a population of 369 Brazilians: P^aP^a, 15; P^bP^b, 220; P^cP^c, 0; P^aP^b, 111; P^bP^c, 19; and P^aP^c, 4.
a. What is the frequency of each allele in the gene pool of this population?
b. Is the population in Hardy-Weinberg equilibrium with regard to this locus?

31-21. Testing an isolated population of 10,000 humans for ABO blood type indicates the following breakdown: type A, 2790; type B, 2460; type AB, 650; and type O, 4100. Assume this population is in Hardy-Weinberg equilibrium with regard to this locus. Determine the frequency of the three alleles, I^A, I^B, and i, responsible for these phenotypes. (Use p, q, and r, respectively, to represent the frequencies of these alleles.)

CHAPTER

32

Population Genetics II: Evolutionary Forces

INTRODUCTION

Evolution can be defined as a change in the genetic makeup of a population over time. Evolutionary changes are produced by the forces of mutation, migration, selection, and genetic drift, working individually or in combination. This chapter will look at how these evolutionary forces bring about changes in gene frequencies.

MUTATION

Forward Mutations

A mutation may change a gene to a different allelic form. Consider a locus with two alleles, say *D* and *d*, both of which occur in the gene pool of a population. From one generation to the next, some *D* alleles mutate to *d* ($D \rightarrow d$); this reduces the number of *D* alleles in the gene pool while increasing the number of *d* alleles. Assume this mutation occurs at a rate, symbolized by u, of 1 in 10,000, or 0.0001 gametes per generation. The actual decrease in the number of *D* alleles depends not only on this mutation rate but on the frequency of *D* alleles in the gene pool (that is, upon the number of alleles that are potential candidates for mutation). If the initial gene pool f(*D*), represented by p, is 0.7 and the mutation rate (u) is 1 in 10,000 gametes or 0.0001, the reduction in the number of *D* alleles in one generation is given by multiplying the allelic frequency (p) by the mutation rate (u). Calculate the reduction in f(*D*). ________[1] What is the new f(*D*)? ________[2]

Back-Mutations

f(*D*) is also influenced by back-, or reverse, mutations which convert *d* alleles into *D* ($d \rightarrow D$). If v represents the back-mutation rate and q represents f(*d*), the increase in f(*D*) is given by the expression vq. Initially, with f(*d*) = q = 0.3 and with a back-mutation rate of, say, 4 in 100,000 gametes (= 0.00004), what is the increase in f(*D*) in one generation? ________[3] If only back-mutations are considered, what f(*D*) will result? ________[4]

Forward and Back-Mutations Considered Together

Both forward and reverse mutations occur during each generation. The back-mutation ($d \rightarrow D$) increases the

[1] $0.7 \times 0.0001 = 0.00007$. [2] $0.7 - 0.00007 = 0.69993$.

[3] $vq = (0.00004)(0.3) = 0.000012$.

[4] The initial f(*D*) of 0.7 would be increased by 0.000012 to 0.700012.

number of *D* alleles, while the forward mutation ($D \rightarrow d$) reduces that number. What really matters, of course, is the net change in f(*D*), symbolized as Δp: Δp is the increase in f(*D*) from the back-mutation minus the decrease in f(*D*) from the forward mutation. Calculate Δp for the gene pool we have been considering. ________[5] Since the increase in f(*D*) is vq, and the decrease in f(*D*) is up, the generalized expression for Δp is

$$\Delta p = vq - up.$$

Generalized Expressions for Equilibrium Values of p and q

Eventually the two opposing mutation processes will reach an equilibrium where the number of *D* alleles removed by the forward mutation equals the number of *D* alleles added by the back-mutation. At that point, what is Δp? ________[6] Substituting 0 for Δp in the equation $\Delta p = vq - up$ gives $0 = vq - up$ which rearranges to $up = vq$. Since $p + q = 1$, q can be set equal to $1 - p$. Substituting this expression for q into the equation $up = vq$, gives $up = v(1 - p)$. Algebraic manipulations of this equation give

$$up = v - vp$$
$$up + vp = v$$
$$p(u + v) = v$$
$$p = \frac{v}{u + v}.$$

This equation gives the value of p when $\Delta p = 0$: this equilibrium frequency is customarily designated as $\hat{p}$ read "p hat").

Is $\hat{p}$ determined by the initial allelic frequencies? ________[7] What determines $\hat{p}$? ________________________[8] What is $\hat{p}$ for the example that we have been considering where $v = 4/100{,}000$ ($= 0.00004$) and $u = 1/10{,}000$ ($= 0.0001$)? ________[9] The frequencies of p and q at equilibrium add up to 1; if one allelic frequency is known, the other can readily be determined. For the gene pool we have been considering, what is the equilibrium f(*d*)? ____[10]

A generalized equation for q, expressed in terms of the two mutation rates, can readily be derived from the expression $\Delta q = up - vq$. Carry out this derivation in the space provided.[11]

Use this generalized expression to verify the numerical value of q (that is, 0.714) obtained previously. ______[12]

Consider this example. $f(A) = 0.9$ in the gene pool of a population and forward mutations occur at a rate of 1 per 20,000 gametes per generation. What effect on f(*A*) does this type of mutation have during a single generation? ________________________[13] Back-mutations at this locus occur at a rate of 1 per 100,000 gametes. What is the consequence for f(*A*) of the back-mutation in a single generation? ________________________[14] Considering both forward and back-mutations, what is f(*A*) following one generation of mutation? ________________________[15] With these mutation rates, what are f(*A*) and f(*a*) at the point where the loss of *A* alleles due to the forward mutation equals the gain of *A* alleles due to reverse mutation? ________[16]

The mutation rates that are known to occur in natural populations are so low that it would take an enormous number of generations for mutation working by itself to significantly alter allelic frequencies. Given enough generations, however, mutation alone could produce meaningful changes in allelic frequencies.

NATURAL SELECTION

Natural selection results in different genotypes reproducing with different degrees of success and, consequently, making different relative contributions to the next generation. To illustrate how natural selection changes gene frequencies, consider two alleles, *A* and its recessive counterpart *a*, both of which occur in the gene pool of a population. Prior to selection, genotypes *AA*, *Aa*, and *aa* occur in frequencies given by the expressions p^2, 2pq, and q^2, respectively. If both alleles occur in equal frequencies (that is, $f(A) = p =$

[5] $\Delta p = 0.000012 - 0.00007 = -0.000058$. [6] 0. [7] No. [8] Just the two mutation rates.
[9] $p = v/(u + v) = 0.00004/(0.0001 + 0.00004) = 0.00004/0.00014 = 0.286$. [10] $q = 1 - p = 1 - 0.286 = 0.714$. [11] The generalized equation is $q = u/(u + v)$. [12] With $u = 0.0001$ and $v = 0.00004$, $q = 0.0001/(0.0001 + 0.00004) = 0.0001/0.00014 = 0.714$. [13] It reduces f(*A*) by (0.9)(1/20,000) = 0.000045. [14] Back-mutations convert *a* to *A* and increase f(*A*) by (0.9)(1/100,000) = 0.000009. [15] The net change is $\Delta f(A) = 0.000009 - 0.000045 = -0.000036$ which reduces the initial f(*A*) to 0.899964. [16] F(*A*) at equilibrium $= \hat{p} = v/(u + v) = 0.00001/(0.00005 + 0.00001) = 0.00001/0.00006 = 0.1667$. F(*a*) $= 1.0 - 0.1667 = 0.8333$.

$f(a) = q = 0.5$), determine the preselection genotypic frequencies.

f(*AA*) = __________[17]

f(*Aa*) = __________[18]

f(*aa*) = __________[19]

What is the total of these genotypic frequencies? _____[20] Assume that selection does not affect the *AA* and *Aa* individuals but eliminates, say, 4 out of every 100 of the homozygous recessive (*aa*) individuals before they reproduce. Calculate the frequencies for the three genotypes following selection.

f(*AA*) = __________[21]

f(*Aa*) = __________[22]

f(*aa*) = __________[23]

What is the total of these genotypic frequencies? _____[24]

Once selection has occurred, the relative allelic frequencies, p and q, are changed. The new relative frequency for p, designated as p_1, is given by the following expression.

$$p_1 = \frac{p^2 + (1/2)(2pq)}{\text{Postselection total frequency}}$$

Use this expression to calculate the postselection f(*A*). __________[25] The new relative frequency for q, designated as q_1, is given by the following expression.

$$q_1 = \frac{(1/2)(2pq) + \text{Postselection value for } q^2}{\text{Postselection total frequency}}$$

Using this formula, calculate the postselection f(*a*). __________[26] What is the change in f(*p*) (that is, what is Δp) produced by selection? __________[27] To summarize, in this example we have determined the genotypic and allelic frequencies and the change in gene frequency following a generation of selection. Our next step will be to set up some generalized expressions for determining these values. Before we can do this however, we need to consider the concepts of fitness value and selection coefficients.

Fitness (Adaptive Value) and Selection Coefficients

The **fitness** of a genotype can be defined in terms of its capability for transmitting genes to the next generation. Fitness, also known as **Darwinian fitness** or **adaptive value** and symbolized by the letter W, can range from 1 for a genotype transmitting the greatest number of genes to the next generation (that is, producing the greatest number of progeny) to 0 for a genotype that transmits no genes (as would be the case, for example, with a lethal genotype). The reduced reproductive output for a genotype is expressed as its **selection coefficient,** symbolized by s. The relationship between s and W is given by the expression $W = 1 - s$. What is the selection coefficient for a genotype with the highest reproductive output? __________[28] for a genotype with no reproductive output? __________[29]

Postselection Genotypic Frequencies

The example of selection just considered shows a frequently encountered situation where two alleles show a dominance-recessiveness interaction and selection works against the homozygous recessive phenotype. The fitness value for the homozygous dominant and heterozygous genotypes is 1 and that for the recessive phenotype is $1 - s$. The frequency of each genotype in generation 1 (the first generation following a round of selection) is given by multiplying the generation-0 frequency by its fitness value. In generalized terms, give the generation 1 values for the following.

f(*AA*) = __________[30]

f(*Aa*) = __________[31]

f(*aa*) = __________[32]

In the equation $f(aa) = q^2 - sq^2$, what does the quantity ($-sq^2$) represent? ____________________[33] In generalized terms, what is the postselection total frequency for the three genotypes? __________[34]

[17] $p^2 = (0.5)^2 = 0.25$. [18] $2pq = 2(0.5)(0.5) = 0.50$. [19] $q^2 = (0.5)^2 = 0.25$. [20] 1. [21] f(*AA*) is unchanged. [22] f(*Aa*) is unchanged. [23] f(*aa*) is reduced by 4%, or by $0.25 \times 0.04 = 0.01$, giving a postselection f(*aa*) of $0.25 - 0.1 = 0.24$. [24] $0.25 + 0.50 + 0.24 = 0.99$. [25] $[0.25 + (1/2)(0.5)]/0.99 = (0.25 + 0.25)/0.99 = 0.50/0.99 = 0.505$. [26] $q_1 = [(1/2)(0.5) + 0.24]/0.99 = 0.495$. [27] Δp = Postselection f(p) − Preselection f(p) = $0.505 - 0.500 = 0.005$. [28] When $W = 1$, $s = 0$. [29] When $W = 0$, $s = 1$. [30] $p^2(1) = p^2$. [31] $2pq(1) = 2pq$. [32] $q^2(1 - s) = q^2 - sq^2$. [33] The f(*aa*) individuals lost through selection. [34] Since the sum of the preselection genotypic frequencies is $p^2 + 2pq + q^2 = 1$, the postselection total of the three frequencies, that is, $p^2 + 2pq + (q^2 - sq^2)$, is $1 - sq^2$.

Generalized Expressions for Postselection p

The preselection f(*A*) is as follows.

$$f(A) = p_0 = \frac{p^2 + (1/2)(2pq)}{1} = \frac{p^2 + pq}{1}$$

The postselection, or generation-1, value for f(*A*), designated as p_1 is obtained by dividing the sum of the postselection frequency of *A* (that is, $p^2 + (1/2)(2pq)$ by the postselection total frequency for all the genotypes (that is, $1 - sq^2$).

$$p_1 = \frac{p^2 + (1/2)(2pq)}{1 - sq^2} = \frac{p^2 + pq}{1 - sq^2}$$

Derivation of the Generalized Expression for Change in p

The change in p between generations 0 and 1, designated as Δp, is given by the expression $\Delta p = p_1 - p_0$. Inserting the expression obtained for p_1 changes this equation to

$$\Delta p = \frac{p^2 + pq}{1 - sq^2} - p.$$

Multiplying the p to the right of the minus sign by $(1 - sq^2)/(1 - sq^2)$ gives

$$\Delta p = \frac{p^2 + pq}{1 - sq^2} - \frac{p(1 - sq^2)}{1 - sq^2}$$
$$= \frac{p^2 + pq - p(1 - sq^2)}{1 - sq^2}$$
$$= \frac{p^2 + pq - [p - sq^2p]}{1 - sq^2}$$
$$= \frac{p^2 + pq - p + sq^2p}{1 - sq^2}.$$

Since $q = 1 - p$, the expression $1 - p$ can be substituted for the q in the quantity pq in the numerator, which gives us

$$\Delta p = \frac{p^2 + p(1 - p) - p + sq^2p}{1 - sq^2}$$
$$= \frac{p^2 + p - p^2 - p + sq^2p}{1 - sq^2}$$
$$= \frac{sq^2p}{1 - sq^2}.$$

Use this expression to verify the value of Δp (0.00505) obtained in the example considered above.

A close look at the generalized expression for Δp will highlight some important facts about selection. The impact of selection on allelic frequencies will, of course, be reflected in the magnitude, or absolute value, of Δp: the larger this value, the more rapidly selection operates to shift gene frequencies. Keep in mind that a round of selection against the homozygous recessives reduces both the frequency of the homozygous recessives and the sum of the frequencies of the three genotypes by the quantity sq^2. If that quantity is very small, the denominator in the expression $\Delta p = sq^2p/(1 - sq^2)$ is very close to 1 and the expression simplifies to $\Delta p = spq^2$.

Refer to the simplified expression for Δp and answer the following questions. If the selection coefficient, s, is small, what would happen to the rate of selection? ____________________[35] What would happen to the rate of selection if p was small? ____________________[36] What would happen to the rate of selection if q was small? ____________________[37] At what levels of p and q would selection pressure be most effective? __________[38] What would happen to Δp if p or q was equal to 0? __________[39] What would happen to Δp if $s = 0$? __________[40] Can selection operate if only one allele is present in the gene pool, that is, if either p or q = 0? __________[41] Can selection operate on a population that has only one allele in its gene pool if the environment showed extreme variability? __________[42] Can selection operate if each genotype in a population has the same fitness? __________[43]

Heterozygote Protection of Harmful Alleles

In the situation we have been considering, the recessive allele confers a selective disadvantage when expressed in homozygous recessive individuals, and selection eliminates some recessive alleles from the gene pool. However, the effectiveness of this selection is limited by the fact that it can act only against alleles carried by homozygous recessive individuals. The recessive alleles carried by the heterozygous individuals are "protected" from the effects of selection (remember that selection operates on the phenotype rather than on the genotype).

[35] It would result in a small Δp, indicating a low rate of selection. [36] It would result in a small Δp. [37] It would result in a small Δp. [38] At intermediate values of p and q. [39] Δp would equal 0. [40] Δp would equal 0. [41] No. [42] No. [43] No.

TABLE 32-1

Demonstration of the effectiveness of selection in eliminating a harmful recessive allele from a gene pool as the allele's frequency declines.

q	p	f(heterozygotes) = 2pq	f(homozygous recessives) = q^2	Ratio: $2pq/q^2$	Abundance of heterozygotes relative to homozygous recessives
0.5	0.5	2(0.5)(0.5) = 0.5	$(0.5)^2 = 0.25$	0.50/0.25	Two times
0.1	________[44]	________[45]	________[46]	________[47]	________[48]
0.05	________[49]	________[50]	________[51]	________[52]	________[53]
0.01	________[54]	________[55]	________[56]	________[57]	________[58]

Completing Table 32-1 will allow you to demonstrate the effectiveness of selection in eliminating a harmful recessive allele from a gene pool as its frequency declines. For each of the various frequencies of q listed, determine the values for p and the frequencies expected under Hardy-Weinberg conditions for the heterozygous individuals (2pq) and the homozygous recessive individuals (q^2). Then for each value of q, set up the ratio f(heterozygotes)/f(homozygous recessives). Use this ratio to determine the number of heterozygous individuals relative to the number of homozygous recessive individuals. For example, when q = 0.5, the ratio of f(heterozygotes) to f(homozygous recessives) is 0.50/0.25 which indicates that the individuals carrying the allele in the protected condition are twice as abundant as those carrying the allele exposed to selection.

Refer to Table 32-1 as you answer the following questions. As the frequency of the recessive allele decreases, what happens to the number of alleles carried by the heterozygotes relative to the number of alleles carried by the homozygotes? ________________[59] How effective would selection be at bringing about the complete removal of a harmful recessive allele from a gene pool? ________________[60] Identify a negative consequence to a population arising from the failure of selection to eliminate a rare recessive allele from its gene pool. ________________[61] Identify a positive consequence of selection failing to eliminate such an allele. ________________[62]

Selection against Dominant Alleles

Sometimes it is the dominant rather than the recessive allele which confers a selective disadvantage. In which genotype or genotypes will a harmful dominant allele be exposed to the effects of selection? ________________[63] What would you expect about the efficiency of selection against a harmful dominant allele? ________[64] Upon what will the rate of elimination of the allele depend? ________[65] If the selection was absolute, that is, if each individual expressing it made no contribution to the next generation, how many generations would be required to eliminate the allele from the gene pool? ________[66] With this extreme level of selection, how could the occurrence of an individual expressing the harmful dominant allele in a subsequent generation be explained? ________[67]

[44] p = 0.9. [45] 2pq = 0.18. [46] $q^2 = 0.01$. [47] 0.18/0.01. [48] 18 times. [49] p = 0.95. [50] 2pq = 0.095. [51] $q^2 = 0.0025$. [52] 0.095/0.0025. [53] 38 times. [54] p = 0.99. [55] 2pq = 0.0198. [56] $q^2 = 0.0001$. [57] 0.0198/0.0001. [58] 198 times.
[59] Proportionately more and more of the recessive alleles are carried by the heterozygous individuals and thus are protected from selection. [60] As the proportion of recessive alleles carried by the heterozygous individuals increases, natural selection would become increasingly ineffective at further reducing the frequency of that allele. [61] Some ill-adapted homozygous recessive individuals may be produced in each generation, thereby reducing the average fitness of the generation. [62] The continued presence of the allele contributes to the genetic variability of the population. [63] Homozygous dominants and heterozygotes. [64] With the allele exposed to selection in every individual possessing it, selection would be very efficient. [65] The selection coefficient.
[66] With s = 1, the allele would be eliminated in one generation. [67] The allele arose through mutation.

MIGRATION

Role in Changing Gene Frequencies

As organisms migrate from one population to another they carry their genes with them. The gene pool of the population supplying the migrants loses the genes the migrants carry, and the gene pool of the population receiving the migrants gains these genes. In both gene pools gene frequencies are altered. Migrants can also introduce new genetic variability to a population if their alleles were previously absent from the population they join.

The following illustration shows how migration can alter gene frequencies. Envision two populations of rodents, one on an island in the midst of a large river and the other on the river shore. f(allele *D*) is 0.1 in the island population and 0.3 in the shore population. Enough migrants from the shore population reach the island during the summer (when the river is low) so that these new arrivals come to make up 5% of the island population (with the premigration residents making up the remaining 95%). In this newly constituted or aggregate population, 95% of the rodents carry the *D* allele at the frequency of 0.1 while 5% carry it at the frequency of 0.3. f(*D*) in the aggregate population, symbolized as $f(D_a)$, is obtained by combining the frequency of *D* in the new residents with the frequency of *D* in the old residents, with these frequencies weighted in each case by their proportion in the aggregate population. Calculate $f(D_a)$. ________[68] Thus, as a result of migration, the frequency of the allele *D* in the island population has increased from 0.1 to 0.11.

Generalized Expression for Postmigration Allelic Frequencies

This procedure for calculating the postmigration allelic frequency can be put into generalized form, using p_c and p_r to represent f(*D*) in the populations contributing the migrants and receiving the migrants, respectively, and m to represent the proportion of the aggregate population made up of migrant individuals (this proportion is referred to as the **coefficient of replacement**). The portion of the population made up of original individuals can be symbolized as $1 - m$. Using these symbols, set up an expression for the frequency of the allele in the aggregate population, p_a.
________[69]

Returning to our rodent example, on another island in the same river, f(*D*) = 0.20. Immigrants to the island have f(*D*) = 0.1 and come to make up 15% of the island population. Determine the postmigration f(*D*) for the island population. ________[70]

Generalized Expression for Δp

In our example, the change in the value of p, that is, Δp, produced in the island population as a result of the migration is $0.185 - 0.2 = -0.015$. What, in general terms, is the expression for Δp? ________[71] Substitute the expression for p_a derived earlier into this equation and simplify. ________[72]

If the rodent migration continues, the frequency of the *D* allele in the aggregate population will continue to fall until it eventually equals the frequency of the allele in the shore population supplying the migrants. At that equilibrium point, that is, when $p_c = p_r$, Δp will equal 0. Another circumstance under which $\Delta p = 0$ would be whenever there is no migration, that is, when $m = 0$.

GENETIC DRIFT

Chance Changes in Gene Frequencies

One of the Hardy-Weinberg conditions is that the population be infinitely large. Natural populations, of course, are finite in size and many populations, at least during certain times, are small. In small populations and in other situations where a small group of parents gives rise to the next generation, gene frequencies can change from one generation to the next solely on the basis of chance. The smaller the group of parents contributing to the next generation, the greater the likelihood of these random fluctuations, which are referred to as **genetic drift.** Genetic drift can cause gene frequencies to change dramatically during the course of a few generations or even during a single generation. The eventual outcome of these random fluctuations is the loss of alleles from a gene pool which reduces the level of variability available to the population.

Sampling and Sampling Error

The inverse relationship between the occurrence of genetic drift and the size of the parental group producing a new generation can be understood by looking at

[68] $f(D_a) = (0.05)(0.3) + (0.95)(0.1) = 0.015 + 0.095 = 0.11$. [69] $p_a = m(p_c) + (1 - m)(p_r)$. Through algebraic manipulation, this expression can be simplified: $p_a = mp_c + p_r - mp_r = p_r + m(p_c - p_r)$. [70] $p_c = 0.1$, $p_r = 0.2$, and $m = 0.15$; $p_a = 0.15(0.10) + (1 - 0.15)(0.2) = 0.015 + 0.17 = 0.185$. [71] $\Delta p = p_a - p_r$. [72] Substitution gives $\Delta p = p_r + m(p_c - p_r) - p_r$, which simplifies to $\Delta p = m(p_c - p_r)$.

the process of sampling involved in coin flipping. A coin could be flipped an infinitely large number of times and, with an equally likely chance of getting a head (p = 1/2) or a tail (q = 1/2), the number of heads would equal the number of tails. In considering successively smaller samples of say 10,000, 1000, 100, and 10 flips, probability would also lead us to expect that in each case, half the flips would come up heads and the remainder as tails. Yet the actual outcomes do not always match up with those that are expected; deviations from expected are routinely encountered and generally attributed to chance. For example, few people would be surprised if a sample of 10 flips resulted in 7 heads and 3 tails. With increasing sample size, the magnitude of the deviation from expected diminishes and the observed values come closer to the expected. Intuitively, we might be very suspicious of a sample of 10,000 flips that had a ratio of heads to tails identical to that seen in the sample of ten flips, that is, 7000 heads to 3000 tails.

The suspicion could be supported from a statistical standpoint by considering the standard deviation, a statistic used to assess the variability or deviation from a mean or an expected value (see discussion in Chapter 17). By definition, about 2/3 of the time, the frequency with which heads, or tails, occurs in a large sample lies in the range of values determined by subtracting and adding the standard deviation to the expected frequency of heads. Standard deviation in this situation is given by the expression

$$\sqrt{\frac{p \times q}{N}}$$

where p is the probability of a head, q is the probability of a tail, and N is the number of flips. For our sample, p = q = 0.5 and N = 10,000. Calculate the standard deviation for this sample. __________[73] A standard deviation of 0.005 tells us that the frequency with which heads occur in a sample of 10,000 would be expected to fall between 0.5 − 0.005 = 0.495 and 0.5 + 0.005 = 0.505 approximately 2/3 of the time. In our sample, f(heads) with 10,000 flips is 0.7. Does this value fall within the expected range? __________[74]

Small Samples from a Gene Pool

Now we need to relate the sampling involved in coin flipping to sampling from a gene pool. Consider a gene pool that contains two alleles of a gene, *A* and *a*, which occur in equal frequencies. The gene pool could be represented by using a bucket of marbles. Assume that half the marbles are yellow and represent *A* alleles, while the other half are green and represent *a* alleles. The alleles in gametes that form the next generation could be represented by taking a random sample of the marbles in the pail. The size of the sample could, of course, vary from a few marbles to a great many, depending on the number of individuals that serve as

TABLE 32-2

The five outcomes that are possible with a sample of four marbles and the generalized expression for the probability of the occurrence of each.

	Outcome				
	1	2	3	4	5
Yellows:	4	3	2	1	0
Greens:	0	1	2	3	4
Proportion of yellows:	1	0.75	0.50	0.25	0
Probability (generalized expression):	a^4	$4a^3b$	$6a^2b^2$	$4ab^3$	b^4
Probability (numerical value):	__________[75]	__________[76]	__________[77]	__________[78]	__________[79]

[73] $\sqrt{\frac{0.5 \times 0.5}{10,000}} = \sqrt{\frac{0.25}{10,000}} = \sqrt{0.000025} = 0.005$. [74] No. It is well outside this range. [75] $(1/2)^4 = 1/16$. [76] $4(1/2)^3(1/2) = 4/16$. [77] $6(1/2)^2(1/2)^2 = 6/16$. [78] $4(1/2)(1/2)^3 = 4/16$. [79] $(1/2)^4 = 1/16$.

parents for the next generation. If the sample is large, there is a strong chance that the frequencies of yellow and green marbles (that is, *A* and *a*, respectively) drawn in the sample will be very close to the frequencies found in the bucket (that is, the gene pool). If, on the other hand, the sample is small as would be the case if, say, just a few parents contribute to the next generation, there is a good chance that its frequencies could differ considerably from those in the bucket.

To illustrate the effect that a small sample can have on gene frequencies, let us take a look at the outcomes that are possible when a sample of, say, four marbles (admittedly, an extremely small sample) is taken from our bucket. The sample would be one of five possible types: (1) four yellows; (2) three yellows, one green; (3) two yellows, two greens; (4) one yellow, three greens; or (5) four greens. The probability of the occurrence of each of these outcomes is given by expanding the binomial $(a + b)^4$, where a and b represent the frequency of yellow and green marbles, respectively. The expansion gives the expression $a^4 + 4a^3b + 6a^2b^2 + 4ab^3 + b^4$. The five outcomes and the generalized expression for the probability of the occurrence of each is given in Table 32-2. Calculate the numerical value for the expected probability of each outcome.

Refer to Table 32-2 as you answer the following questions. With what probability is a sample containing the same proportion of yellow and green marbles as is found in the bucket (or the same proportion of *A* and *a* alleles found in the gene pool) expected to occur?

__________[80]

Four of the five possible outcomes involve a proportion different from that found in the bucket. Two of these are samples containing just one of the two types of alleles. What is the probability of getting either one or the other of these mutually exclusive outcomes? __________[81] What would be the consequence for the gene pool of the next generation if the parents of that generation carried just one type of allele?

____[82]

The two remaining outcomes are samples with three marbles (or alleles) of one type and one marble (or allele) of the other type. What is the probability of getting one or the other of these two mutually exclusive outcomes? __________[83] With either of these samples what would happen to allelic frequencies in the next generation's gene pool? __________________

__________________________[84]

Considering all five outcomes, what is the expected probability that a sample will contain something other than the proportion of yellow and green marbles in the bucket (or something other than the same proportion of *A* and *a* alleles found in the gene pool)? __________[85]

To continue our example, assume that drift shifts the gene frequency in going from one generation to the next. If a small number of parents now give rise to a third generation, the relative frequencies of alleles might well shift again—maybe in the same direction as the original change or maybe in the opposite direction, perhaps moving frequencies back to or beyond their original values. Over a number of successive generations, each arising from a small group of parents, gene frequencies can drift considerably from their original values. Sooner or later through the continued occurrence of drift, a particular allele will either become lost or fixed.

The Occurrence of Drift

Genetic drift can operate in any circumstance where a limited number of parents contribute to a subsequent generation. This situation is, of course, encountered generation after generation in a population that remains small. It also occurs when a large population goes through a bottleneck, as might happen, for example, with an insect species where a population grows steadily larger throughout the warmer months of the year only to be drastically reduced with the onset of colder weather. If the "sample" of individuals that survives the winter and gives rise to the spring generation has gene frequencies that are representative of those of the previous year's population, then gene frequencies could remain free of the effects of drift. If, on the other hand, the sample producing the next generation was not representative, drift could occur.

Another opportunity for drift arises when a small group of adults leave one population and establish a new one, provided that the gene frequencies within this founding group differ from those of the population that supplied the founders. Gene frequency changes that arise because of a small founding population are said to be due to the **founder effect.** The smaller the reproducing group at the time of the bottleneck or at the founding of a new population, the greater the opportunity for random shifts in gene frequency.

[80] Less than half the time, with a probability of 6/16. [81] It is given by the sum of their separate probabilities: 1/16 + 1/16 = 2/16 = 1/8. [82] Homozygosity in the next generation with only one of the alleles represented in the gene pool. [83] The probability of getting a sample with the frequency of yellow, or *A*, at 0.75 and green, or *a*, at 0.25 is 4/16, or 1/4; the probability of a sample having yellow, or *A*, at 0.25 and green, or *a*, at 0.75 is also 4/16, or 1/4. The probability of getting one or the other of these outcomes is the sum of their separate probabilities: 1/4 + 1/4 = 1/2. [84] It would shift from 0.5 and 0.5 to 0.75 and 0.25 (or to 0.25 and 0.75). [85] More than half the time: 1/16 + 4/16 + 4/16 + 1/16 = 10/16.

SUMMARY

This chapter looks at each of the evolutionary forces. Natural selection is the major factor in producing evolutionary change—change that results in the adaptedness of a population. Genetic drift shifts allelic frequencies randomly, as a consequence of sampling error. Both of these forces reduce the genetic variability present in a gene pool. The forces of mutation (as the ultimate source of variation) and migration replenish the genetic variability within a gene pool, while also producing shifts in gene frequencies. Our approach, in separately considering each evolutionary force, is a simplistic one. Shifts in gene frequencies occurring in natural populations result from the collective interactions of these four evolutionary forces.

PROBLEM SET

32-1. a. Forward mutations occurring during a single generation are responsible for shifting the frequency of an allele from 0.8 to 0.799889. What is the forward mutation rate?

b. When both forward and back-mutations at this locus are considered, the gene frequency shift during a single generation is from 0.8 to 0.799942. What is the back-mutation rate?

32-2. The frequency of the *B* allele in a gene pool is 0.6. This allele mutates to its recessive form, *b*, with a frequency of 5×10^{-5} per generation.

a. What is the reduction in the frequency of *B* produced by one generation of *B*-to-*b* mutations?

b. Assume that the mutation of *B* to *b* is partially counterbalanced by the back-mutation of *b* alleles to *B* at a rate of 2×10^{-5} per generation. What is the net change in the frequency of *B* produced by both forward and back-mutations in the course of a generation?

32-3. A locus has two alleles, *S* and *s*, which occur in a gene pool with frequencies of 0.55 and 0.45, respectively. With a forward mutation rate of 1×10^{-4} and a back-mutation rate of 6.67×10^{-5}, what are the frequencies of *S* and of *s* when the two mutation processes reach equilibrium?

32-4. a. A genotype has a selection coefficient of 1. What does this indicate?

b. A genotype has an adaptive value of 1. What does this indicate?

c. What is the adaptive value for the genotype referred to in 32-4a?

d. What is the selection coefficient for the genotype referred to in 32-4b?

32-5. The allele *G* and its recessive counterpart, *g*, occur in a gene pool with frequencies of 0.7 and 0.3, respectively. Selection works against the homozygous recessive individuals, giving them an adaptive value (W) of 0.8. Both the homozygous dominant and heterozygous individuals have adaptive values of 1.

a. Determine the expected frequencies for the three genotypes before selection occurs.

b. Determine the relative frequencies of the three genotypes after selection occurs.

c. What are the frequencies of *G* and *g* after a generation's worth of selection?

32-6. A locus has two alleles (*T* and its recessive counterpart *t*), and f(*T*) and f(*t*) in the gene pool of a population are 0.65 and 0.35, respectively. The recessive homozygotes have an adaptive value of 0.7 while that for each of the other two genotypes is 1.0. The individuals in this population (consider it the first generation) interbreed to produce two successive generations.

a. What is the net change in f(*T*) in going from the first to second generation?

b. What is f(*T*) in generation 3?

32-7. Assume that a certain population is free of all evolutionary forces except for natural selection. A particular recessive allele is known to confer a selective disadvantage when expressed in recessive homozygotes. In going from one generation to the next, the frequency of the dominant counterpart to this allele shifts from 0.2 to 0.3.

a. What is the selection coefficient for the homozygous recessive genotype?

b. What is f(dominant allele) in generation 3?

32-8. a. Cystic fibrosis, a disorder arising from the expression of an autosomal recessive allele, occurs with a frequency of approximately 1 in 1600

PROBLEM SET

humans. Determine the frequency with which carriers, that is, heterozygous individuals, would be expected to occur in the human population and the ratio of carriers to homozygous recessive individuals.

b. Phenylketonuria, or PKU, a metabolic disorder produced by the expression of an autosomal recessive allele, occur in human populations with a frequency of about 1 in 40,000 individuals. Determine the frequency with which carriers, that is, heterozygous individuals, would be expected and determine the ratio of carriers to homozygous recessive individuals.

c. What happens to the ratio of heterozygous carriers to homozygous recessives as the frequency of a recessive harmful allele in a population's gene pool declines? As the proportion of heterozygous carriers increases, what happens to the effectiveness of selection against the recessive allele?

32-9. Migrating frogs come to make up 10% of an aggregate frog population. The frequency of a particular allele for skin pigmentation was 0.25 in the population prior to the migration and 0.6 among the migrants. Determine this frequency of the allele in the population following the migration.

32-10. Problem 32-9 involves a coefficient of replacement of 10%. Identify the coefficient of replacement that would be required to double the allelic frequency, that is, to change the allelic frequency from 0.25 to 0.5 in the frog population receiving the migrants.

32-11. Water snakes migrating from the mainland to an island make up 15% of the island's aggregate water-snake population. An allele causing banded skin pigmentation occurs in the mainland population with a frequency of 0.6. Following the migration, the frequency of this allele in the gene pool of the island population is 0.7. What was the frequency of this allele in the island population before migration occurred?

32-12. The allele for brown shell color occurs with a frequency of 0.10 in a population of land snails. During the rainy season, migrants from a neighboring population come to make up 16% of the population and increase the frequency of this shell-color allele to 0.13. What is the frequency of this allele in the population supplying the migrants? (Assume that the migrants are a representative sample of the snails in this population.)

32-13. The change in frequency of a recessive allele, *a*, in the gene pool of a population over a period of eight generations is plotted in the following figure.

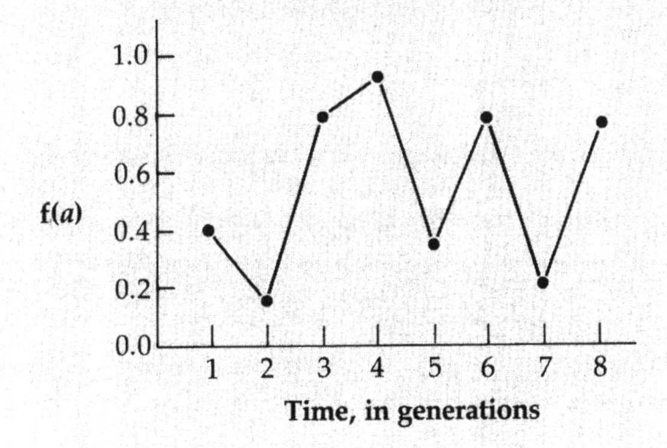

PROBLEM SET

a. Which evolutionary force is most likely responsible for the evolutionary change shown in the graph? Explain.
b. What prediction could be made about the number of parents giving rise to each of the generations shown here?
c. Additional data indicate that the number of adults in each generation never fell below 100,000. In light of this, how can the changes in gene frequency be explained?

32-14. Refer to the figure in problem 32-13. Under what circumstances could natural selection have been responsible for the allelic frequency changes seen over the eight generations?

32-15. A freshwater lake supports a large population of painted turtles. An allele that determines the number of stripes on the head occurs in the gene pool of this population with a frequency of 0.63. Several years before, the frequency of this allele in the population was 0.59. Ten years ago, small ponds were created on a number of farms surrounding the lake. A recent survey of these farm ponds indicates that many now support painted turtle populations and it is believed that these turtles or their ancestors came from the lake population. The present frequencies of the head-stripe allele in five of these farm pond populations are as follows: 0.82, 0.55, 0.72, 0.18, and 0.23. Assume that the same evolutionary force, that is, either mutation *or* drift *or* selection *or* migration, has been responsible for producing the changes seen in all five of these farm pond populations. Consider each of the four evolutionary forces and speculate on whether or not it could have produced the allelic frequencies found in the farm pond populations. Which evolutionary force seems to be the most likely cause of what is seen here?

32-16. A laboratory study is carried out on 48 small populations of the fruit fly *Drosophila melanogaster*, each of which is founded by mating two individuals heterozygous for a particular trait (say, $Dd \times Dd$).

a. Identify the genotypes that would be expected among the offspring of this mating and the frequency with which each genotype would be expected.

After the progeny from these $Dd \times Dd$ matings have developed, a male and a female are randomly selected from each of the 48 populations to serve as the parents of each population's second generation.

b. What is the probability that both of the flies chosen as parents of the second generation would have the *DD* genotype? the *dd* genotype?
c. How many of the 48 populations would be expected to lose either the *d* or *D* allele from their gene pools in producing the second generation?
d. How many of the 48 populations would be expected to retain the *D* and *d* allele in the frequencies with which they occurred in the first generation?
e. How many of the 48 populations would be expected to retain both *D* and *d* alleles in the second generation, but at a frequency different from those in which *D* and *d* occurred in the initial generation?
f. If the procedure used in producing the second generation is repeated for many generations, what will eventually happen in the gene pools of the populations?

APPENDIX A

Comprehensive Problem Set for Chapters 1–13

This set of comprehensive problems is designed to test your comprehension of many of the basic concepts of Mendelian genetics presented in Chapters 1 through 13. These problems consider the simultaneous inheritance of two or more traits that show different methods of inheritance. For example, a problem may involve an autosomal trait and a sex-linked trait or two sex-linked traits and an autosomal trait. Additional complexities can be introduced if the genes at one or more of the loci under consideration comprise a multiple-allelic series, exhibit codominance, involve lethality, or show gene interaction of one type or another. Problems involving three or more traits are best approached by considering different combinations of two of those traits at a time. These problems should not be attempted until you have mastered the material in Chapters 1 through 13 and successfully worked the problem sets at the ends of those chapters.

A-1. In corn, *Zea mays*, mutant genes for anther ear (*an*) and fine stripe (*f*) are carried at linked loci separated by approximately 18 map units. These mutant genes are recessive to their counterpart alleles, an^+ and f^+, which control the wild-type expressions of these traits. Identify the phenotypes of the progeny and their expected frequencies if both parents are heterozygous for each of these traits and

a. both carry the genes in repulsion.

b. both carry the genes in coupling.

c. one carries the genes in coupling and the other in repulsion.

d. The gene for colorless aleurone (*c*) is recessive to its dominant allele for colored aleurone (*C*) and is found on a chromosome different from that carrying the *an* and *f* loci. If both parents are heterozygous for each of these three loci and if one carries the linked genes in coupling and the other carries them in repulsion, what is the probability of producing a progeny plant that is

i. wild eared, wild striped, colored.

ii. anther eared, fine striped, colorless.

iii. anther eared, wild striped, colored.

A-2. In the fruit fly, the recessive genes for yellow body (*y*) and white eyes (*w*) are carried at separate loci on the X chromosome. In 1911 Thomas Hunt Morgan, a noted *Drosophila* geneticist, crossed pure-breeding females with yellow bodies and white eyes with wild-type males. Members of the F_1 were then crossed to produce an F_2 generation of 2205 flies. Of these, 29 expressed recombinant phenotypes.

a. Identify the phenotype and sex of each category of F_1 progeny.

b. Identify the phenotype and sex for each category of F_2 progeny and indicate whether each category is a parental or recombinant type and estimate the number of flies it contains.

c. What is the map distance separating the *y* and *w* loci?

d. In a similar pair of crosses studying white eyes and another sex-linked locus, miniature wings, caused by the expression of the recessive allele *m*, Morgan found that the F_2 generation with white-eyes, miniature-winged females and wild-type male grandparents consisted of 2441 flies, 900 of whom showed the recombinant phenotypes. Why is the frequency of recombinants in this F_2 generation significantly greater than the frequency of recombinants in the F_2 generation described earlier?

e. At an autosomal locus, the allele for dumpy wings, *dp*, is recessive to its wild-type counterpart for long wings, dp^+. Assume that the original pure-breeding female parents with yellow bodies and white eyes were homozygous for the *dp* allele and that the original wild-type male parents were homozygous for the dp^+ allele. What is the probability that

i. an F_1 male with yellow body and white eyes will have normal wings?

ii. an F_2 female with wild-type body and white eyes will have dumpy wings?

iii. an F_2 male with yellow body and wild-type eyes will have dumpy wings?

A-3. Green-weakness, or deutan, color blindness causes about 75% of all human color blindness and is characterized by difficulty in recognizing green color and distinguishing it from red. The allele responsible for this condition is recessive to that for normal vision and is sex-linked. Medical records have been examined for a number of families in which a female carrying the allele for green weakness but not expressing it (and who is therefore heterozygous) married a male with normal vision. The total number of offspring produced from these marriages was 160 of which 92 were female and 68 were males. All of the female progeny and 38 of the males had normal vision while the 30 remaining male progeny had deutan color blindness.

a. What is the theoretical phenotypic ratio expected among the progeny of this type of mating?

b. With a total of 160 progeny, what is the expected number of children in each of the three progeny classes?

c. What is the χ^2 value for this set of data?

d. How many degrees of freedom are involved?

e. Using the chi-square table (Table 6-5, p. 68), determine the probability value for the χ^2 value.

f. Can this deviation reasonably be attributed to chance? (Assume a level of significance of 0.05 in evaluating this.)

g. Are these data in support of the hypothesis that underlies the cross?

h. The metabolic disorder alkaptonuria is caused by the expression of a recessive allele, *a*, while expression of its dominant counterpart, *A*, results in normal metabolism. Brachydactyly, a disorder that results in shortened digits on the hands and feet, is due to the expression of a dominant mutant allele, *B*; its recessive counterpart, *b*, results in digits of normal length. The loci for alkaptonuria and brachydactyly are carried on different human autosomes. Assume that a female with normal vision and who carries the allele for deutan color-blindness is heterozygous at both the alkaptonuria and the brachydactyly loci. This female married a male with normal vision, normal metabolism, who carries the allele for alkaptonuria, and who has digits of normal length. Assume that the brachydactyly is 100% penetrant. Determine the probability of producing a

i. normal-vision male with normal metabolism and brachydactyly.

ii. color blind male with alkaptonuria and normal digits.

iii. normal-vision female who is a carrier of the color-blindness allele with normal metabolism and normal digits.

A-4. Assume that a new species of beetle has been recently discovered in a Central American rain forest. Genetic studies indicate that a dominant mutant allele, *W*, drastically reduces the wing size of the wild-type beetle (*ww*) to produce a wingless condition. Alleles *S* and *s* occur at an independently assorting locus and have no effect on wing size except when the dominant allele *S* occurs in the presence of *W*. Then a single copy of *S* suppresses the wingless phenotype and restores normal wing size. When either allele *W* or *S* occur in the homozygous condition, death of the beetle results.

a. Identify the expected phenotypic ratio among the progeny of a cross between two beetles with normal wings, each of which carries a suppressed *W* allele.

b. Identify the expected phenotypic ratio among the progeny of a cross between F_1 wingless beetles and F_1 normal-winged beetles from the most common type of F_1 progeny.

A-5. Assume that a perennial garden plant's flower color, which may be either yellow or white, is controlled by two genes, and flower size, which may be either large or small, is controlled by a third gene. Each of these genes is known to have two alleles. A cross between pure-breeding plants with large yellow flowers and other pure-breeding plants with small white flowers produces an F_1 with large yellow flowers. Two additional crosses are then carried out: F_1 plants are allowed to self-pollinate to produce the F_2, and other F_1 plants are crossed with pure-breeding plants with small white flowers. The outcomes of these two crosses are summarized as follows.

	Progeny from	
	$F_1 \times F_1 = F_2$	$F_1 \times$ Small white flowers
Large yellow:	153	57
Small yellow:	55	49
Large white:	128	153
Small white:	42	159
Totals:	378	418

Identify the method of inheritance for each locus. If linkage is involved, determine the map distance separating the loci.

A-6. In *Drosophila*, the loci for curled wings and ebony body are carried on the same autosome and are separated by a distance of about 20 map units; the mutant alleles at these loci, *cu* and *e*, respectively, are recessive to their wild-type counterparts. The locus for garnet eyes is carried on the *X* chromosome, and the mutant allele at this locus, *g*, is recessive to the wild-type allele for normal eye color. A curled-winged, ebony-bodied female that is homozygous for normal eye color is crossed with a male that is homozygous for wild-type wings and body color, and has garnet eyes.

a. Identify the progeny from this mating.

b. A female from this group of progeny is crossed with a curled-winged, ebony-bodied male with wild-type eyes. Identify the progeny expected from this cross and their relative frequencies.

A-7. The presence or absence of feathers on the lower part of the legs (shanks) of the Black Langsham breed of chickens is controlled by two independently assorting autosomal loci which we will designate as *f* and *g*. A cross between pure-breeding feathered and pure-breeding nonfeathered chickens produces an F_1 that is 100% feathered. Crossing members of the F_1 gives rise to 496 progeny, of which 34 lacked feathers and the remainder were feathered. The breeder carrying out these crosses concluded that the feathered condition is produced by the presence of at least one dominant allele at both of the loci. Do the data support the breeder's conclusion? Explain. Propose an alternative hypothesis if appropriate.

A-8. a. Verify your answer to problem A-7 before proceeding. The presence or absence of white bars on the black feathers of Black Langsham chickens is controlled by a sex-linked locus: expression of its dominant allele, *B*, causes bar production, and expression of its recessive counterpart, *b*, results in the absence of bars. A female that is nonbarred and heterozygous for both shank-feather loci is crossed with a male that is heterozygous for the barred condition and lacks shank feathers. Identify the phenotypes of the progeny expected from this mating and their frequencies.

b. Identify the phenotypes of the progeny of a cross between a nonfeathered, nonbarred female and a male that is heterozygous for the bar locus and the *f* locus, and homozygous recessive at the *g* locus.

A-9. Refer to the information regarding the sex-linked bar locus in problem A-8. An autosomal locus affects leg length in chickens and has a dominant allele, *C*, and a recessive allele, *c*. Heterozygosity at this locus produces a chicken with shortened legs that is designated as a creeper. Homozygosity of the dominant allele is lethal, while homozygosity of the recessive allele results in legs of normal length. For this problem, assume that both parents are homozygous dominant at the *f* and *g* loci.

a. Identify the genotypes and phenotypes of the progeny and their expected frequencies from a cross between a barred female with normal legs and a nonbarred, creeper male.

b. Mate creeper males and creeper females from the generation produced from the cross described in A-9a and identify the genotypes and phenotypes of the progeny and their expected frequencies.

A-10. Refer to the information given in problems A-8 and A-9. A series of matings between two chickens gives rise to 152 progeny. Of these, 46 are nonbarred, creeper females, 26 are nonbarred, normal females, 26 are barred, normal males, and 54 are barred, creeper males. Identify the genotypes of the parents for these two traits.

A-11. In humans, the length of the index finger relative to the length of the ring finger is believed to be controlled by an autosomal locus. The expression of the allele for short index fingers (S^1) is sex-influenced and acts as recessive in females and dominant in males; its counterpart allele is designated as S^2. A dominant allele, *B*, at a locus on another autosome is expressed in males but not in females and results in premature balding in males; expression of its recessive counterpart, *b*, results in nonbalding.

a. Identify the phenotypes of the male and female progeny of the mating of two individuals, each heterozygous for both loci.

b. Identify the phenotypes of the male and female progeny of the mating of a female heterozygous for both loci and a nonbald, long-fingered male.

A-12. Two major classes of pigments, brown ommochromes and brightly colored pterins, combine to produce the eye color of wild-type *Drosophila*. The synthesis of each type of pigment is controlled by a separate set of genes. Following their synthesis, the pigments molecules attach to protein granules and are deposited in the cells of the eyes. In the absence of ommochromes, the eyes are colored by the pterins and thus are brightly colored. This occurs, for example, with the expression of the recessive allele, *v*, at the sex-linked locus for vermilion eyes or with the expression of the recessive allele, *st*, at the autosomal locus for scarlet eyes (found on chromosome 3). In the absence of pterins, the eyes are colored by the brown ommochromes as seen, for example, with the expression of the recessive allele, *bw*, at the autosomal locus for brown eye color (found on chromosome 2). If both ommochromes and pterins are absent, as would be the case, for example, with double mutants such as scarlet brown (*st/st bw/bw*) or vermilion brown (X^vX^v *bw/bw* or X^vY *bw/bw*), the eyes appear white.

a. Determine the phenotypes expected among the F_1 and F_2 progeny of the mating of a scarlet-eyed female that is homozygous for the wild-type allele at the *bw* locus and a male that is white eyed because he expresses the recessive alleles at both of these loci.

b. Determine the phenotypes expected among the F_1 and F_2 progeny of the mating of a vermilion-eyed female that is homozygous for the wild-type allele at the *bw* locus and a male that is white eyed because he expresses the recessive alleles at both of these loci.

A-13. Refer to the information given in problem A-12.

a. Determine the expected phenotypes among the F_1 and F_2 progeny of a cross between a female that is white eyed because of the expression of recessive alleles at both the *v* and *bw* loci and a wild-type male derived from pure-breeding stock.

b. Determine the expected phenotypes among the F_1 and F_2 progeny of the reciprocal of the cross described in A-13a.

A-14. In *Drosophila*, the mutant genes for crossveinless (*cv*) wings and singed (*sn*) bristles are recessive to their wild-type counterparts. The loci for both of these genes are carried on the X chromosome and separated by a distance of about 6 map units. The gene for star eyes (*S*) is dominant to its wild-type counterpart and is carried on an autosome. Identify the phenotypes and their frequencies expected among the F_1 and F_2 progeny from a cross between a wild-type female derived from pure-breeding stock and a male with crossveinless wings, singed bristles, and that is homozygous for star eyes.

A-15. Assume that the inheritance of three loci, shrunken endosperm, dwarf plant height, and waxy endosperm, are under study in corn, *Zea mays*. The mutant alleles at these respective loci, *sh*, *d*, and *wx*, are recessive to their wild-type counterparts, sh^+, d^+, and wx^+, that produce full endosperm, tall plants, and starchy endosperm. The outcome of the cross between a plant heterozygous for each of these loci and one that is homozygous recessive at each of these loci produces a total of 7400 progeny and is summarized as follows.

Class	Phenotype	Frequency
1.	Full, tall, waxy	1430
2.	Shrunken, short, starchy	1487
3.	Shrunken, tall, waxy	397
4.	Shrunken, short, waxy	410
5.	Full, short, waxy	1400
6.	Full, tall, starchy	416
7.	Shrunken, tall, starchy	1455
8.	Full, short, starchy	405

Identify the method of inheritance for each locus. If linkage is involved, determine the map distance or distances separating the loci.

A-16. In mice, the loci for bent tail and tabby fur are carried on the sex, or X, chromosome and separated by a distance of 12 map units. The mutant alleles at these loci, *Bn* and *Ta*, respectively, are dominant to their recessive wild-type counterparts. The loci controlling behavioral patterns known as waltzer and jittery are carried on an autosome and separated by a distance of 18 map units. The mutant alleles at these loci, *v* and *ji*, respectively, are recessive to their wild-type counterparts. Females that are heterozygous at both the bent and tabby loci with the genes carried in coupling and that exhibit the waltzer and jittery traits are crossed with males that are wild-type with regard to all four traits and that are heterozygous at the waltzer and jittery loci, carrying the recessive genes in repulsion. Assume a 1:1 sex ratio among the progeny. Determine the probability of producing

a. a male offspring that is bent, tabby, waltzer, and jittery.
b. a male that shows wild-type traits at all four loci.
c. a female that is bent, wild-type at the tabby locus, waltzer, and wild-type at the jittery locus.
d. a litter of four, equally divided by sex, the males of which have bent tails and are wild-type for the other three traits, and the females of which have wild-type tails, tabby fur, are wild-type for the waltzer trait, and have jittery behavior.
e. a mouse with wild-type traits at all four loci as the fifth member of a litter consisting of four males with wild-type traits at all four loci.

A-17. a. The direction in which hairs grow on guinea pigs is determined by the interaction of two independently assorting gene pairs. At the *R* locus, the mutant allele, *R*, for rough coat, is dominant to the wild-type allele, *r*, for smooth coat. The alleles at the other locus, designated, say, as *M* and *M'*, show incomplete dominance, and modify the expression of the rough phenotype in the following way: *MM* modifies rough to smooth, *MM'* modifies rough to partly rough, and *M'M'* makes no modification in the rough phenotype. The genotype at the *M* locus does not modify the expression of the smooth phenotype. Several matings between individuals heterozygous for the two gene pairs (that is, *RrMM'* × *RrMM'*) produce a total of 112 progeny. Determine the progeny phenotypes and the number expected in each category.

b. Alleles C, c^k, c^d, and c^a are found at a coat-color locus in guinea pigs which assorts independently of the *R* and *M* loci described in A-17a. Each of these alleles exhibits complete dominance to the alleles to the right of it in the sequence listed. Expression of these four alleles produces black, sepia, cream, and albino colors, respectively. Identify the expected progeny phenotypes and their relative frequencies for a mating between a partly rough, cream coated guinea pig that is heterozygous at each of the three loci and one with genotype $rrM'M'c^ac^a$.

A-18. Three genetically determined fruit fly traits are studied in a series of crosses. Pure-breeding females with gray bodies, straight bristles, and straight wings are crossed with pure-breeding males with sable bodies, forked bristles, and bent wings. The female and male F_1 progeny have gray bodies, straight bristles, and straight wings. In subsequent separate crosses, F_1 females are crossed with pure-breeding males with sable bodies, forked bristles, and bent wings and F_1 males are crossed with pure-breeding females with sable bodies, forked bristles, and bent wings. The phenotypes of the first 400 progeny from each of these crosses is summarized in Table A-18. Identify the method of inheritance for each locus. If linkage is involved, determine the map distance or distances separating the loci.

A-19. Three genetically determined fruit fly traits are studied in a series of crosses. Pure-breeding females with gray bodies, complete wings, and red eyes are crossed with pure-breeding males with black bodies, cut wings, and purple eyes. The female and male F_1 progeny have gray bodies, complete wings, and red eyes. In subsequent separate crosses, F_1 females are crossed with pure-breeding males with black bodies, cut wings, and purple eyes, and F_1 males are crossed with pure-breeding females with black bodies, cut wings, and purple eyes. The phenotypes of the first 500 progeny from each of these crosses is summarized in Table A-19. Identify the method of inheritance for each locus. If linkage is involved, determine the map distance separating the loci.

A-20. In *Drosophila*, the mutant genes for curly wings (*Cy*), plum eye color (*Pm*), hairless (*H*), and stuble (*Sb*) are dominant to their respective wild-type counterparts. The *Cy* and *Pm* loci are linked as are the *H* and *Sb* loci. Each pair of linked loci is carried on a different pair of autosomes. Homozygosity of the mutant alleles at any of these four loci is lethal. The inversion of a section of the autosomes carrying each pair of these linked loci prevents crossing over between the loci.

TABLE A-18

	Phenotype			Cross A (involving F_1 females)		Cross B (involving F_1 males)	
Class	Body	Bristle	Wing	Female	Male	Female	Male
1.	Gray	Straight	Straight	48	39	93	0
2.	Gray	Forked	Bent	8	4	0	0
3.	Gray	Straight	Bent	42	40	105	0
4.	Sable	Straight	Straight	5	6	0	0
5.	Sable	Forked	Bent	44	50	0	104
6.	Gray	Forked	Straight	7	9	0	0
7.	Sable	Straight	Bent	9	6	0	0
8.	Sable	Forked	Straight	46	37	0	98

TABLE A-19

	Phenotype			Cross A (involving F_1 females)		Cross B (involving F_1 males)	
Class	Body	Wing	Eyes	Female	Male	Female	Male
1.	Gray	Complete	Red	57	60	134	0
2.	Black	Complete	Purple	57	52	128	0
3.	Black	Complete	Red	4	5	0	0
4.	Gray	Cut	Purple	3	4	0	0
5.	Black	Cut	Red	4	3	0	0
6.	Gray	Cut	Red	59	63	0	118
7.	Gray	Complete	Purple	2	5	0	0
8.	Black	Cut	Purple	66	56	0	120

a. A cross is carried out between flies that are heterozygous at all four loci; the genes on each pair of chromosomes under consideration are carried in repulsion. Identify the progeny genotypes and phenotypes along with their expected frequencies.

b. The cross described in 20a is repeated using parents that carry the genes on each pair of chromosomes under consideration in coupling. Identify the progeny genotypes and phenotypes along with their expected frequencies.

APPENDIX B

Solutions to Problems

Chapter 1: Mitosis and Meiosis

1-1. a. Chromatids are the two daughter strands of a duplicated chromosome that are held together at a centromere region. During both mitosis and meiosis, the two sister chromatids of each duplicated chromosome separate from each other; each is then designated as a chromosome. Both chromatids and chromosomes are composed of chromatin. A chromatid carries the same amount of DNA as its parent chromosome.

b. Homologous chromosomes are generally alike in terms of size, shape, and centromere placement. They usually carry the same sequence of genes and pair with each other during meiosis. Nonhomologous chromosomes do not carry the same linear sequences of genes and do not pair during meiosis.

1-2. Barring abnormalities that may occur, mitosis ensures that the genetic material carried by every cell in the body (except the sex cells) will be identical. In addition, when accompanied by cell division it results in the growth of the organism.

1-3. a. Four pairs of homologous chromosomes.

b. The greatest condensation, thus the most distinct appearance, occurs during late prophase, metaphase, and early anaphase.

c. When the chromosomes become distinct, each is short, thick, and longitudinally double, consisting of two sister chromatids joined at their centromeres.

d. Sixteen chromatids (eight chromosomes of two chromatids each).

e. Sixteen.

f. i. Each chromosome of the homologous pair carries the *B* allele. Since each homolog consists of two chromatids, there are four copies (2×2) of *B*.

ii. Following mitosis, each nucleus will have two copies of *B*.

g. i. The single chromosome carrying *A* consists of two genetically identical chromatids, so two copies of *A* are present.

ii. Following mitosis, each daughter nucleus will have one copy of *A*.

1-4. a. Four pairs of homologous chromosomes.

b. Throughout prophase I, each chromosome is longitudinally double, consisting of two sister chromatids joined at their centromeres. The chromatids start out elongate and thin and, by the end of the stage, condense to become thick and short. Early in prophase I, each duplicated chromosome synapses with its homolog to form a bivalent. By the end of this stage, each chromosome within a bivalent has begun to separate from its homolog except at the chiasmata, sites where crossing over has occurred.

c. Sixteen chromatids (eight chromosomes of two chromatids each).

d. Sixteen.

e. i. Each chromosome of the homologous pair carries the *B* allele. Since each homolog consists of two chromatids, there is a total of four copies (2×2) of *B*.

ii. After meiosis, each nucleus will have one copy of *B*.

f. i. The single chromosome carrying *A* consists of two genetically identical chromatids, so there are two copies of *A* present.

ii. Following meiosis, two of the four nuclei will have one copy of *A*, and the other two will have no copy of *A* (both carrying, instead, a copy of *a*).

1-5. Chief difference: In metaphase I of meiosis, each of the double-stranded chromosomes pairs up with its homolog while, in mitosis, there is no homologous pairing. Another difference: During the meiosis-I metaphase, each replicated chromosome of each bivalent is attached to a single spindle fiber from one or the other pole of the spindle. During metaphase of mitosis, each replicated chromosome is attached to two spindle fibers, one from each pole.

1-6. a. Each bivalent consists of four chromatids which make up a tetrad: 12 bivalents = 12 tetrads.

b. Each bivalent represents a synapsed *pair* of homologous chromosomes. 12 bivalents = 12 pairs of chromosomes. Since meiosis occurs in a diploid nucleus, the diploid ($2n$) number is 24.

c. See the answer to (b). The haploid (n) number is half the diploid number, or 12 (= 24/2).

1-7. a. During mitosis, the centromeres separate at the start of anaphase.

b. During meiosis, the centromeres separate at the start of anaphase II. (Note: During anaphase I, the two replicated chromosomes making up each bivalent separate from each other; this requires no centromere separation.)

1-8. a. The nuclei produced at the end of meiosis I each contain 23 replicated chromosomes, one member of each homologous pair present in the original diploid cell.

b. Twenty-three chromosomes is the haploid number for the species.

c. Two.

d. 46 ($= 23 \times 2$)

e. No. The chromosomes that exist at the end of mitosis would be single threaded, each consisting of a single chromatid. Since there is no crossing over during mitosis, the chromosomes would be free of genetic recombination. In addition, each chromosome within the nuclei would have a homolog.

1-9. **a.** Yes.
b. Yes.
c. Yes. Mitotic division maintains the chromosome number. Nuclei undergoing meiosis II have the same number of chromosomes as found in the haploid nuclei.
d. Yes. By definition, sister chromatids are identical. During meiosis, many of the chromatids present in a diploid cell participate in recombination during meiosis I and end up with new combinations of genetic material. As a consequence, the two chromatids of a replicated chromosome may no longer be alike. However, during mitosis there is no crossing over, and the chromatids participating in that process remain true sister chromatids.

1-10. The two phenomena: (1) crossing over which generates combinations of maternal and paternal genes within chromosomes, and (2) the assortment of chromosomes which generates combinations of maternal and paternal chromosomes.

1-11. **a.** Crossing over produces new combinations of genes within the chromatids because one member of each homologous pair can be traced back to a different parent. Crossing over exchanges a segment of maternal origin for one of paternal origin, and vice versa, forming recombinant strands with combinations of maternal and paternal genes.
b. Sister chromatids carry identical genetic material. Crossing over between sister chromatids would produce exactly the same combinations of genetic material that existed before crossing over.

1-12. Assortment produces new combinations of maternal and paternal *chromosomes* while crossing over produces new combinations of maternal and paternal *genes* within individual chromatids.

1-13. **a.** True.
b. False. The number of chromatids present in late prophase I is twice the number present at any one pole at the end of meiosis I.
c. False. The number of centromeres is reduced by one-half in the late prophase II nucleus.
d. False. The number of centromeres is reduced by one-half by the second meiotic division.
e. True.

1-14. **a.** Four different kinds of chromosomal combinations could occur: 1,2; 1*,2; 1,2*; and 1*,2*.
b. The probability of producing gametes that carry only chromosomes of paternal origin (1, 2) is 1/4.
c. The probability of producing gametes that carry only chromosomes of maternal origin (1*,2*) is 1/4.
d. The probability of producing gametes with a combination of maternal and paternal chromosomes (1*,2 and 1,2*) is 1/4 + 1/4 = 1/2.
e. We would expect each of the four types of gametes to be produced with the same frequency: 5000/4 = 1250.
f. Four different zygotic combinations would occur: 1*2 (egg) + 1,2 (sperm) = 1*,1,2,2 (zygote); 1*,2 + 1*,2 = 1*,1*,2,2; 1*,2 + 1,2* = 1*,1,2*,2; 1*,2 + 1*,2* = 1*,1*,2*,2.

1-15. **a.** Eight combinations are possible, depending on the alignment assumed by the dyads during metaphase of meiosis II. Four combinations are produced with one alignment and four arise with the alternative alignment as shown in Figure 1-15a.
b. Yes. An alternative tetrad arrangement, shown in Figure 1-15b, is equally likely during metaphase I, and eight additional combinations are possible, depending on the alignment of dyads that occurs during metaphase of meiosis II. Four combinations will

FIGURE 1-15a

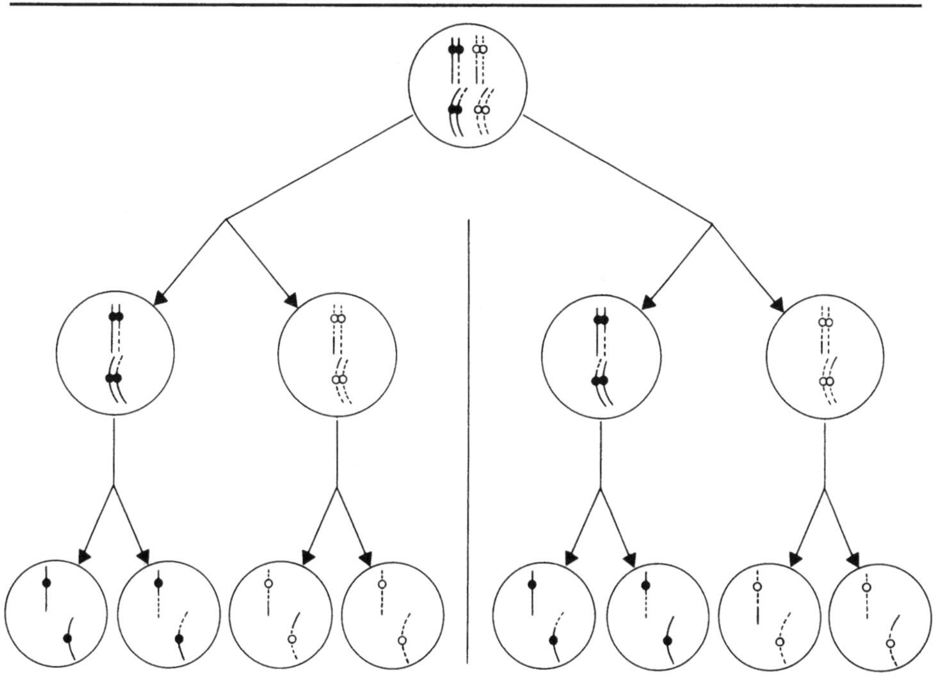

FIGURE 1-15b

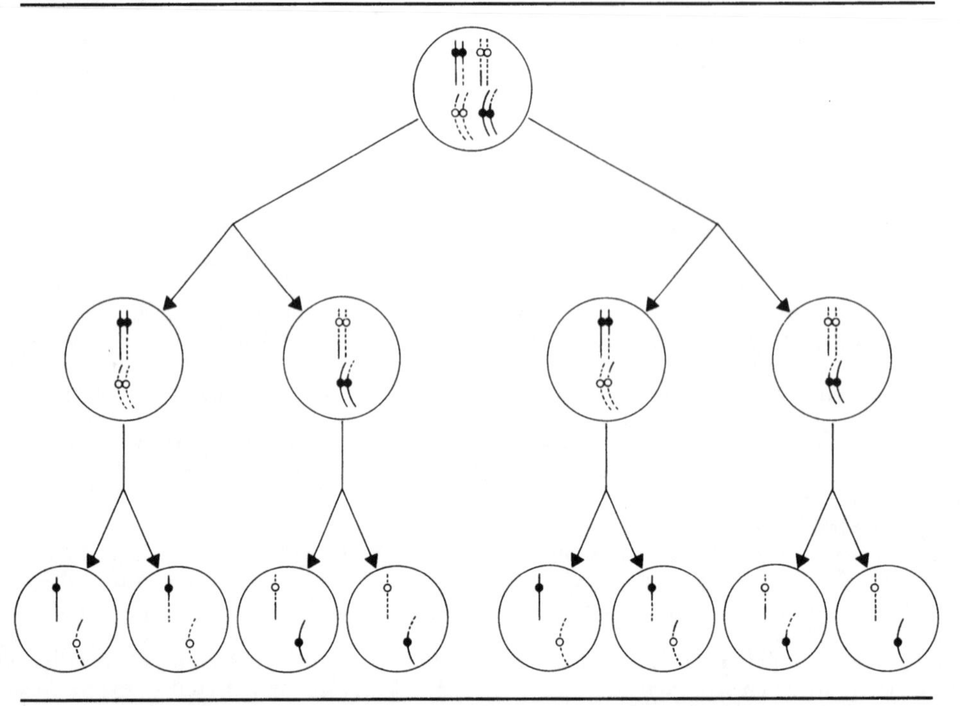

be produced with one dyad alignment and four others will be formed with the alternative alignment.

1-16. a. The haploid stage gives rise to haploid gametes by mitosis.

b. The diploid stage gives rise to haploid spores by meiosis.

Chapter 2: Mendel's Laws

2-1. a. A gene is a unit of inheritance found at a specific site on a chromosome. An allele is one of two or more alternative or contrasting forms of a gene.

b. Genotype refers to the genetic makeup of an organism. Phenotype refers to the observable expression or physical appearance produced by the genetic makeup of an organism.

2-2. a. Homozygous describes an allelic pair where both alleles of a gene are alike. Heterozygous refers to an allelic pair where the two alleles of a gene are different.

b. Dominant describes either characters or alleles whose expression is the same regardless of whether they are in the homozygous or heterozygous state. Recessive describes characters or alleles which are masked by dominant alleles and, therefore, expressed only in the homozygous condition.

2-3. a. No. An organism showing the dominant character may be either homozygous for the dominant allele or heterozygous.

b. The genotype for an organism showing the recessive character will always by homozygous for the recessive allele.

2-4. a. When the alleles are dominant.

b. When the alleles are recessive.

2-5. Pure-breeding plants express one or more characters in the same way generation after generation. Such plants are likely to be homozygous for these characters, a feature that Mendel realized was essential if one wishes to study the inheritance pattern of the alleles responsible for the characters.

2-6. Homozygous genotypes: both parental types (since they come from pure-breeding lines), some of the F_2 plants with purple petals, and all of the F_2 plants with white petals. Heterozygous genotypes: all of the F_1 plants and some of the F_2 plants with purple petals.

2-7. a. One of the two factors (alleles) in the F_1 plants is recessive and is masked by the dominant allele.

b. Each F_1 plant must have carried a factor from each parent since both factors express themselves in the F_2, with 75% of the progeny exhibiting one of the factors and 25% exhibiting the other.

c. The recessive factor exhibited by one parent is not expressed in the F_1, but reappears unchanged in 25% of the F_2 generation.

2-8. a. 100% will carry a single yellow factor.

b. Two types: 50% will carry the yellow factor, 50% the green factor.

2-9. a. All the gametes will carry *G*.

b. Half the gametes will carry *G*, and the other half *g*.

c. All the gametes will carry *a*.

d. Half the gametes will carry *A*, and the other half *a*.

2-10. He postulated two different kinds of factors. The two factors carried by the F_1 segregate when gametes are formed, with one factor going to each gamete. Random union of the two types of gamete gives the 3:1 F_2 ratio. Expression of the recessive trait occurs only when two recessive factors are combined.

2-11. P_1 organisms are pure breeding and therefore homozygous. The yellow parent's gametes all carry a single yellow allele and the green parent's gametes carry the green allele. F_1 plants are heterozygous, and half their gametes have the yellow allele and half have the green allele. Some of the F_2 yellow plants are heterozygous, and half the gametes from these plants will have the yellow allele and half will have the green allele. The remainder of the F_2 yellow plants are homozygous, and all their gametes will carry the yellow allele. All of the F_2 green plants are homozygous, and their gametes carry the green allele.

2-12. **a.** The height would be intermediate to that of the two parents.
b. (1) Phenotypically, the F_1 was not intermediate; all were like one of the parents. (2) In selfing the F_1, the character expression which was absent from the F_1 reappeared in 25% of the F_2.

2-13. The law of independent assortment came from studying the inheritance of two characters simultaneously. Crossing pure-breeding parents that differed in two pairs of alleles resulted in a phenotypic ratio of 9:3:3:1 in the F_2. This is the ratio expected if the characteristics are inherited independently of each other.

2-14. The two gene pairs segregate independently, so that each gamete contains one allele from *each* gene pair.
a. All gametes will be *aG*.
b. Two types of gametes in equal frequencies: *AG* and *aG*.
c. Four types of gametes in equal frequencies: *AG*, *aG*, *Ag*, and *ag*.
d. Two types of gametes in equal frequencies: *Ag* and *ag*.

Chapter 3: Crosses Involving Single-Gene Inheritance; Basic Probability

3-1. **a.** The genotype of the heterozygous yellow plant is o^+o; the genotype of the orange plant is *oo*.
b. The o^+o parent forms two kinds of gametes: half carry o^+ and half carry *o*. All gametes produced by the *oo* parent carry *o*.
c. The cross can be represented as $o^+o \times oo$. Progeny genotypes are shown in the following Punnett square.

♀ \ ♂	o^+	o
o	o^+o	oo

Progeny: 50% heterozygous with yellow fruit and 50% homozygous recessive with orange fruit.

3-2. **a.** Based on its phenotype, the red pigeon could be either homozygous, b^+b^+, or heterozygous, b^+b; however, since it received an allele from a brown (*bb*) parent, its genotype must be b^+b. The genotype of the brown pigeon is *bb*.
b. The *bb* pigeon produces one kind of gamete which carries *b*. The b^+b pigeon forms two kinds of gametes: half carry b^+ and half carry *b*.
c. The cross can be represented as $b^+b \times bb$. Progeny genotypes are shown in the following Punnett square.

♀ \ ♂	b^+	b
b	b^+b	bb

Half the progeny, those with genotype *bb*, would have brown feathers.
d. The probability of producing a red pigeon is 1/2. Since the production of each pigeon is an independent event, the probability of producing five red pigeons is found by multiplying together the probability for the occurrence of each event: $p = 1/2 \times 1/2 \times 1/2 \times 1/2 \times 1/2 = 1/32$.

3-3. **a.** The *P* allele for polydactyly is dominant to the *p* allele for normal fingers and toes. The genotype for the normal woman must be *pp*. Two genotypes are possible for the affected man: *PP* or *Pp*. Since this man's mother was not affected, her genotype must have been *pp*, and since she contributed one of her alleles to her son, his genotype must be *Pp*.
b. The cross can be represented as $pp \times Pp$. The woman would produce one kind of gamete carrying *p*. The man would produce two kinds of gametes, half carrying *P* and the other half carrying *p*. Progeny genotypes are shown in the following Punnett square.

♀ \ ♂	P	p
p	Pp	pp

Half the progeny would be heterozygous and affected with polydactyly and the other half would be homozygous recessive and normal. The chance of polydactyly in each offspring is 50%.

3-4. **a.** With their normal phenotypes, each parent could be either *CC* or *Cc*. Their child with cystic fibrosis must have genotype *cc*. Since each parent contributed an allele to that child, each parent must carry *c* and, therefore, each must have the *Cc* genotype.
b. The genotype of the afflicted child is obviously *cc*. Each of the three normal children could be either *CC* or *Cc* since both of these genotypes would be expected from this $Cc \times Cc$ mating.

3-5. **a.** Based on their phenotype, each parent could be either homozygous dominant, *WxWx*, or heterozygous, *Wxwx*. The starchy × starchy cross could be any of the following combinations of these genotypes: (1) $WxWx \times WxWx$, (2) $WxWx \times Wxwx$, or (3) $Wxwx \times Wxwx$. Cross 1 would give rise to all *WxWx* progeny with the starchy phenotype. Cross 2 would produce *WxWx* and *Wxwx* progeny in a 1:1 ratio, all of which would have the starchy phenotype. Cross 3 would produce *WxWx*, *Wxwx*, and *wxwx* progeny which would exhibit starchy and waxy phenotypes in a 3:1 ratio. Since the actual progeny are 100% starchy, cross 3 is ruled out. Either cross 1 or cross 2, however, would yield this outcome.
b. See the answer to problem 3-5a. The 506 starchy : 162 waxy ratio is very close to the 3:1 ratio expected if the cross were $Wxwx \times Wxwx$. This would give

starchy progeny genotypes *WxWx* and *Wxwx* and waxy progeny genotype *wxwx*.

c. Based on phenotype, the starchy parent could be either *WxWx* or *Wxwx*; the waxy parent must be *wxwx*. If the starchy parent were *WxWx*, the cross would be *WxWx* × *wxwx*, and all the progeny would be *Wxwx* and have the starchy phenotype. Alternatively, if the starchy parent were *Wxwx*, the cross would be *Wxwx* × *wxwx*, and two types of progeny, *Wxwx* and *wxwx*, with starchy and waxy phenotypes, respectively, would be formed in a 1:1 ratio. Since all the progeny are starchy, the starchy parent must be homozygous dominant.

d. See the answer to problem 3-5c. The 425 starchy : 409 waxy F_1 is close to the 1:1 ratio expected if the starchy parent had the heterozygous genotype, *Wxwx*. Thus, we can conclude that the starchy plant is heterozygous, which gives genotypes of *Wxwx* and *wxwx* for the starchy and waxy F_1 progeny, respectively.

3-6. a. Each albino parent must have the genotype of *aa*. The mating *aa* × *aa* produces *aa* progeny.

b. The albino woman must have genotype *aa*. The albino children have genotype *aa*, and since one of their *a* alleles must have come from each parent, the phenotypically normal father must carry the recessive allele and have genotype *Aa*. The mating, represented as *aa* × *Aa*, would produce *Aa* and *aa* progeny in a 1:1 ratio. The normal children must be heterozygous, with genotype *Aa*.

c. Since albino children (*aa*) are produced, each of the phenotypically normal parents must carry the recessive allele, giving them genotype *Aa*. The mating *Aa* × *Aa* would produce *AA*, *Aa*, and *aa* progeny in a 1:2:1 ratio, respectively. Thus, the normal progeny would have genotypes *AA* and *Aa*.

3-7. a. The probability of an individual with a normal phenotype carrying the allele for albinism is 3 out of 100, or 0.03. The probability of both members of a couple who are phenotypically normal carrying the allele is given by the product law: 0.03 × 0.03 = 0.0009.

b. The probability of the wife carrying the allele is 0.03.

3-8. a. The p^+ allele for normal enzyme production is dominant to the p allele for PKU. The heterozygous × heterozygous mating can be represented as p^+p × p^+p. Progeny genotypes are shown in the following Punnett square.

♀ \ ♂	p^+	p
p^+	p^+p^+	p^+p
p	p^+p	pp

Progeny genotypes: 25% homozygous dominant, 50% heterozygous, and 25% homozygous recessive (afflicted).

b. The homozygous dominant × heterozygous mating can be represented as p^+p^+ × p^+p. Progeny genotypes are shown in the following Punnett square.

♀ \ ♂	p^+	p
p^+	p^+p^+	p^+p

The progeny are 50% homozygous dominant and 50% heterozygous.

3-9. a. See the answer to problem 3-8a. If both parents are heterozygous, the probability of their producing a child affected with PKU is 1/4. Since each birth is an independent event, the probability of their producing three affected children is given by multiplying together the probability that each child will be affected: 1/4 × 1/4 × 1/4 = 1/64.

b. See the answer to problem 3-8b. If one parent is homozygous dominant and the other is heterozygous, the probability of producing one child with PKU is 0 since the progeny would be phenotypically normal. The probability of producing three afflicted children is 0.

3-10. a. A pink plant has a genotype of C^RC^W and the cross can be represented as C^RC^W × C^RC^W. F_1 progeny genotypes are shown in the following Punnett square.

	C^R	C^W
C^R	C^RC^R	C^RC^W
C^W	C^RC^W	C^WC^W

The expected F_1 plants would be red (C^RC^R), pink (C^RC^W), and white (C^WC^W) in a 1:2:1 ratio.

b. A white flowered plant has a genotype of C^WC^W and the cross can be represented as C^WC^W × C^WC^W. All of the progeny would be C^WC^W and have white flowers.

c. A pink-flowered plant has a genotype of C^RC^W and a white-flowered plant has a genotype of C^WC^W. The cross can be represented as C^RC^W × C^WC^W and progeny genotypes are shown in the following Punnett square.

	C^W
C^R	C^RC^W
C^W	C^WC^W

The expected F_1 plants would be pink (C^RC^W) and white (C^WC^W) in a 1:1 ratio.

3-11. The relatives are incorrect. Since each pregnancy is an independent event, the outcome of each of the three earlier pregnancies has no bearing on the outcome of the fourth. The probability of producing a male child in the fourth pregnancy is just what it is in any pregnancy: 1/2.

3-12. We are asked to determine the chance that a child will have normal bones. The husband is afflicted and we can infer that he is homozygous for the recessive allele and has genotype *dd*. The wife's phenotype is normal and we can infer that she has either genotype d^+d^+ or d^+d. Her afflicted father must have been homozygous for the recessive allele, *dd*, and since he must have contributed a d allele to his daughter, the daughter must be heterozygous, d^+d. The cross can be represented as $d^+d \times dd$ and progeny genotypes are shown in the following Punnett square.

♀ \ ♂	d
d^+	d^+d
d	dd

Half the progeny will be heterozygous, d^+d, and have normal bones while 50% will be homozygous recessive, dd, and have the disorder. The chance of a child escaping the disease is 1/2.

3-13. a. Since each birth is an independent event, the product law will give the probability of all five children being male: $1/2 \times 1/2 \times 1/2 \times 1/2 \times 1/2 = (1/2)^5 = 1/32$.

b. The probability of all five being female is determined using the product law: $1/2 \times 1/2 \times 1/2 \times 1/2 \times 1/2 = (1/2)^5 = 1/32$.

c. There are two mutually exclusive ways that the five babies can be of the same sex: either all can be male or all can be female. To get the probability of these two events combined we use the law of the sum: $1/32 + 1/32 = 2/32 = 1/16$.

3-14. a. In the absence of information to the contrary and because of the rarity of this disorder, we can assume that the man's father is normal (H^+H^+). If the husband's mother is heterozygous (HH^+), their mating can be represented as $HH^+ \times H^+H^+$, and the expectation is that 1/2 of their children would be afflicted, as shown in the following Punnett square.

♀ \ ♂	H^+
H	HH^+
H^+	H^+H^+

The probability of the man being afflicted is 1/2.

b. The man is afflicted (HH^+). His wife is normal (H^+H^+). The expected outcome of their mating, $H^+H^+ \times HH^+$, is shown in the following Punnett square.

♀ \ ♂	H	H^+
H^+	HH^+	H^+H^+

The probability of producing a normal child is 1/2; the probability of producing an afflicted child is 1/2.

i. The probability of the first child being afflicted is 1/2.

ii. The probability of one child have the harmful allele is 1/2. Since each birth is an independent event, the probability of three affected children is given by the product law: $1/2 \times 1/2 \times 1/2 = 1/8$.

iii. The probability of one child not having the harmful allele is 1/2. The probability of three children not having this allele is given by the product law: $1/2 \times 1/2 \times 1/2 = 1/8$.

3-15. a. The yellow parent plant must have the genotype yy. The red parent plant could be either YY or Yy. If the red plant is YY, the cross is $YY \times yy$, and all the F_1 plants would be Yy and have red tomatoes. If the red plant is Yy, the cross is $Yy \times yy$, and the F_1 would consist of Yy and yy plants with red and yellow tomatoes, respectively, in a 1:1 ratio. The ratio of 42 red : 37 yellow is very close to the 1:1 ratio expected if the red plant is heterozygous. Thus, we can conclude that the red plant is heterozygous.

b. See the answer to problem 3-15a. The crossing of red and yellow F_1 plants can be represented as $Yy \times yy$, and the progeny would be Yy and yy, with phenotypes of red and yellow, respectively, in a 1:1 ratio.

3-16. a. All of the progeny from the mating $ii \times ii$ would have the ii genotype and would exhibit interrupted veins.

b. The fact that 10% of the progeny carrying the ii genotype failed to express the genotype and instead exhibited the wild-type phenotype is most likely due to incomplete or reduced penetrance of the i allele. (Given the low frequency with which most mutations occur, and the relatively high frequency, 10%, with which the wild-type phenotype occurs in the F_1, it is not likely that mutation is responsible for the occurrence of the wild phenotype.)

c. Mating two F_1 flies can be represented as $ii \times ii$ and all the F_2 progeny are ii. With the i allele having 90% penetrance, 90% of the progeny would be expected to express the genotype, that is, 90% would have an interrupted vein, and 10% would show the wild-type phenotype.

3-17. a. Most of the individuals in this sample fall in the intermediate levels of severity.

b. Yes. The fact that the disorder exhibits different degrees of severity in different individuals tells us that the gene has variable expressivity.

c. No. A gene with variable penetrance is expressed in some individuals and not expressed in others. All the individuals in this sample are known to carry the gene for the disorder and are known to express it; in other words, the gene is penetrant in all the individuals in the sample. Nothing here indicates that there are humans carrying the dominant allele and not expressing it.

d. In general, the variation in expression might be due to environmental factors or to genetic differences at loci other than the one being considered.

e. Nonpenetrant individuals would fall at 0.

f. Individuals lacking the dominant allele would fall at 0.

3-18. a. The mating of two heterozygous black pigs would be expected to produce three black pigs for every one brown pig. Thus the probability of producing a single black pig from this mating is 3/4, and the probability of producing a single brown pig is 1/4. Each offspring arises from an independent fertilization, and thus the probability of getting three black and one brown offspring in the same litter is given by the product of their separate probabilities: $3/4 \times 3/4 \times 3/4 \times 1/4 = 27/256$.

b. See the answer to problem 3-18a. The probability of getting one black and three brown offspring together is given by the product of their separate probabilities: 3/4 × 1/4 × 1/4 × 1/4 = 3/256.

3-19. The information presented is consistent with the results that would be expected if the allele for the hairless condition is both dominant and lethal in the homozygous state. When paired with a normal, recessive allele in the heterozygous condition, this allele results in hairlessness. When it is absent altogether, as is the case when a dog is homozygous for the normal allele, hair is produced.

3-20. a. The mating of two creepers can be represented as *Cc* × *Cc* and genotypes *CC*, *Cc*, and *cc* would be formed. *CC* progeny would die during development and the living progeny would consist of *Cc* (creeper) and *cc* (normal) chickens in an expected ratio of 2:1.

b. Crossing the creeper (*Cc*) with a wild type (*cc*) in a testcross will verify that the creeper is heterozygous, since the testcross progeny will consist of creeper (*Cc*) and wild types (*cc*) in a 1:1 ratio.

c. The allele exerts its lethal effect only when present in a double dose, that is, when it is homozygous. Thus, it is acting like a recessive allele, which affects an organism's phenotype only in the homozygous state.

3-21. The yellow-coated mouse must be heterozygous (genotype *Yy*) while the dark-coated mouse has genotype *yy*. The mating, *Yy* × *yy*, would be expected to produce *Yy* and *yy* progeny in a 1:1 ratio.

Chapter 4: Crosses Involving Two Independently Assorting Traits (Dihybrid Crosses)

4-1. a. The gametes will carry one allele from each locus and show all possible combinations of the alleles at each locus: 1/4 *GH*, 1/4 *Gh*, 1/4 *gH*, and 1/4 *gh*.

b. All *gH*.

c. 1/2 *gH*, 1/2 *gh*.

d. 1/2 *GH*, 1/2 *Gh*.

e. All *gh*.

4-2. a. Homozygous green, homozygous full: *YYCC*; yellow, constricted: *yycc*.

b. Gametes of the *YYCC* plant will carry *YC*. Gametes of the *yycc* plant will carry *yc*.

c. F_1 progeny genotype: *YyCc*. Phenotype: green, full.

4-3. a. Each *YyCc* plant produces four kinds of gametes in equal numbers: *YC*, *Yc*, *yC*, *yc*.

b. To identify the F_2 phenotypes, the random union of all types of gametes can be shown in a Punnett square.

	YC	*Yc*	*yC*	*yc*
YC	*YYCC* (1)	*YYCc* (2)	*YyCC* (3)	*YyCc* (4)
Yc	*YYCc* (5)	*YYcc* (6)	*YyCc* (7)	*Yycc* (8)
yC	*YyCC* (9)	*YyCc* (10)	*yyCC* (11)	*yyCc* (12)
yc	*YyCc* (13)	*Yycc* (14)	*yyCc* (15)	*yycc* (16)

The phenotypes fall into four categories: (1) boxes 1, 2, 3, 4, 5, 7, 9, 10, and 13: green, full; 9 of 16 boxes; (2) boxes 6, 8, and 14: green, constricted; 3 of 16 boxes; (3) boxes 11, 12, and 15: yellow, full; 3 of 16 boxes; (4) box 16: yellow, constricted; 1 of 16 boxes. The phenotypic ratio is 9:3:3:1.

c. Expected frequencies: 450 plants (9/16 of 800) green, full; 150 (3/16 of 800) green, constricted; 150 (3/16 of 800) yellow, full; 50 (1/16 of 800) yellow, constricted.

4-4. a. The F_1 plant has genotype *YyCc*. The yellow, constricted plant has genotype *yycc*. The cross is *YyCc* × *yycc*. The *YyCc* parent would produce equal numbers of four kinds of gametes, *YC*, *Yc*, *yC*, and *yc*, each with an expected probability of 1/4. The *yycc* parent would produce one kind of gamete, *yc*, with a probability of 1. Gamete union is shown in the following branch diagram.

yycc gametes	*YyCc* gametes		Progeny genotypes	Expected probability
yc (1)	*YC* (1/4)	=	*YyCc*	(1)(1/4) = 1/4
	Yc (1/4)	=	*Yycc*	(1)(1/4) = 1/4
	yC (1/4)	=	*yyCc*	(1)(1/4) = 1/4
	yc (1/4)	=	*yycc*	(1)(1/4) = 1/4

Four genotypes occur in a 1:1:1:1 ratio.

b. Each of the four genotypes produces a different phenotype: *YyCc* = green, full; *Yycc* = green, constricted; *yyCc* = yellow, full; and *yycc* = yellow, constricted. Each would be expected to make up 1/4 of the progeny. If the total number of progeny is 1000, then 1000 × 1/4 = 250 would be expected to exhibit each phenotype.

4-5. a. Since all progeny are spotted, black, we know that each carries at least one dominant allele for each trait. From the phenotype, we cannot determine whether the other allele for each trait is dominant or recessive. We thus represent the progeny genotype as *S_B_* . Next consider the parents. The solid, black parent must be homozygous recessive at the *s* locus, and thus must have contributed an *s* allele to each of the progeny. This tells us that the progeny genotype must be *SsB_*. The spotted, brown parent must be homozygous recessive at the *b* locus and thus must have contributed a *b* allele to each of the progeny. This tells us that the progeny genotype must be *SsBb*.

b. *SsBb* × *SsBb*. The F_2 progeny would show four phenotypes: spotted, black; solid, black; spotted, brown; and solid, brown in a 9:3:3:1 ratio.

4-6. a. The four types of progeny show all possible phenotypic combinations for both traits. Listing them in order of decreasing numbers of dominant phenotypes, gives us spiny, purple: 256; spiny, white: 85; and smooth, purple: 93; and smooth, white: 27. Simplifying the ratio, by dividing each frequency by the lowest value in the series (27) gives 256/27 = 9.5, 85/27 = 3.1, 93/27 = 3.4, and 27/27 = 1, giving a ratio of 9.5:3.1:3.4:1 which is very close to the 9:3:3:1 ratio expected when two dihybrid individuals are crossed. The progeny support the conclusion that each parent plant has genotype *SsWw*.

b. The four types of progeny show all possible phenotypic combinations for both traits. Simplifying the ratio, by dividing each frequency by the lowest value in the series (69), gives 75/69 = 1.1, 82/69 = 1.2, 69/69 = 1, and 77/69 = 1.1, giving a ratio of 1.1:1.2:1:1.1 which is very close to the 1:1:1:1 ratio expected when a dihybrid is crossed with a plant homozygous recessive for both loci. The progeny support the conclusion that the parents' genotypes were either *SsWw* and *ssww*, or *Ssww* and *ssWw*; there is not enough evidence to choose one set over the other.

4-7. a. Consider each trait separately. The parental mating *Ss* × *ss* would give two progeny genotypes, *Ss* and *ss*, each with an expected probability of 1/2. The mating *Ww* × *Ww* gives genotypes *WW*, *Ww*, and *ww*, with expected probabilities of 1/4, 1/2, and 1/4, respectively. The probability of genotypic combinations is given by multiplying separate probabilities: p(*ssww*) = p(*ss*) × p(*ww*) = 1/2 × 1/4 = 1/8.

b. See the answer to problem 4-7a. p(*SsWw*) = p(*Ss*) × p(*Ww*) = 1/2 × 1/2 = 1/4.

c. The parental mating *Ss* × *ss* would give two progeny phenotypes, spiny (genotype *Ss*) and smooth (genotype *ss*), each with an expected probability of 1/2. The mating *Ww* × *Ww* gives phenotypes of purple (genotypes *WW* and *Ww*) and white (genotype *ww*), with expected probabilities of 3/4 and 1/4, respectively. The probability of phenotypic combinations is given by multiplying the separate probabilities: p(smooth, white) = p(smooth) × p(white) = 1/2 × 1/4 = 1/8.

d. See the answer to problem 4-7c. p(smooth, purple) = p(smooth) × p(purple) = 1/2 × 3/4 = 3/8.

4-8. a. The tall, hairy plant could have any of the following genotypes: *HHDD*, *HhDD*, *HHDd*, or *HhDd*.

b. The genotype for this plant could be determined by carrying out a testcross, that is, by crossing it with a plant homozygous recessive for the two loci under consideration (*hhdd*).

c. Two phenotypic classes in the testcross progeny indicate that the hairy, tall plant produced two kinds of gametes. This would occur if the parental plant had been homozygous dominant at one locus and heterozygous at the other, that is, if it is *HhDD* or *HHDd*.

d. Half the testcross progeny were hairy, tall and half were hairy, dwarf. In order for the height locus to exhibit two phenotypes in the testcross progeny, it must be heterozygous in the hairy, tall parental plant. This plant must have genotype *HHDd*.

4-9. The testcross is *HhDD* × *hhdd*. The *HhDD* parent would produce two kinds of gametes, *HD* and *hD*, each with a probability of 1/2. The double-recessive parent would produce one kind of gamete, *hd*, with a probability of 1. Combinations of these gametes would give rise to the following: *HD* + *hd* = *HhDd* with a probability of 1/2 × 1 = 1/2; *hD* + *hd* = *hhDd* with a probability of 1/2 × 1 = 1/2.

4-10. a. Since there were four types of testcross progeny, there must have been four types of gametes formed by the double-dominant parent.

b. The four phenotypes are hairy, tall; hairy, short; nonhairy, tall; and nonhairy, short.

c. The double-dominant parent must have been heterozygous at both loci, with genotype *HhDd*.

4-11. a. Male: *NnPp*; female: *NnPp*; mating: *NnPp* × *NnPp*. Gametes: Each parent will produce four kinds of gametes, each with a frequency of 1/4: *NP*, *Np*, *nP*, *np*. Progeny phenotypes: Normal vision, normal enzyme: 9/16; normal vision, phenylketonuria: 3/16; nearsightedness, normal enzyme: 3/16; nearsightedness, phenylketonuria: 1/16.

b. Since the traits are inherited independently, they may be considered separately. The parental mating for the PKU trait is *Pp* × *Pp*. This mating would be expected to produce normal children (genotypes *PP* and *Pp*) with a frequency of 3/4 and afflicted children (genotype *pp*) with a frequency of 1/4. The chance that the first (or any) child will have PKU is 1/4.

c. Considering this trait by itself, the parental mating is *Nn* × *Nn*. This would be expected to produce normal-vision children (genotypes *NN* and *Nn*) with a frequency of 3/4 and nearsighted children (genotype *nn*) with a frequency of 1/4.

d. The chance that the first (or any) child will have PKU and be nearsighted is given by the product of their separate probabilities: 1/4 × 1/4 = 1/16.

4-12. a. The husband has the normal phenotype for each trait and thus must carry at least one dominant allele at each locus: *N*_*P*__ . Since his father was nearsighted, the father's genotype at this locus must have been *nn*, and since the father could have only contributed an *n* allele to his son, the husband's genotype at this locus must be heterozygous: *Nn*. Since the husband's mother had phenylketonuria, her genotype at this locus must have been *pp* and since the mother could have contributed only a *p* allele to her son, the husband's genotype at this locus must be heterozygous: *Pp*. The husband's genotype is *NnPp*.

b. The mating is *nnPp* × *NnPp*. We can look at the inheritance of each trait separately and determine the probability of the phenotypic combinations using the product law. With cross *nn* × *Nn*, normal vision (genotype *Nn*) and nearsightedness (genotype *nn*) each occur with an expected frequency of 1/2. With the cross *Pp* × *Pp*, normal enzyme (genotypes *PP* and *Pp*) and PKU (genotype *pp*) occur with expected frequencies of 3/4 and 1/4, respectively. The probability of recessive expressions for both traits is 1/2 × 1/4 = 1/8.

c. See the answer to problem 4-12b. The probability of producing an individual with a dominant expression for both traits is 1/2 × 3/4 = 3/8.

d. See the answer to problem 4-12b. The probability of producing an individual with normal vision and PKU is 1/2 × 1/4 = 1/8.

e. See the answer to problem 4-12b. The probabilities for nearsightedness, phenylketonuria, and a girl among the progeny are 1/2, 1/4, and 1/2, respectively. Combining these probabilities using the product law gives a probability of 1/2 × 1/4 × 1/2 = 1/16.

4-13. a. One way to begin is to identify, to the extent possible, the genotypes of the progeny. The tall, smooth plants must carry at least one *S* and one *W* allele:

$S\_W\_$. The tall, wrinkled plants must carry at least one S and two w alleles: S_ww. From these incomplete genotypes, we can establish three requirements for the parental genotypes: (1) at least one parent must carry S, (2) at least one parent must carry W, and (3) both parents must carry w. All the proposed pairs of parents meet these requirements except for set 1, which fails to meet requirement 2.

b. The progeny from set 3 would be expected to fall into four genotypic classes of approximately equal size: 1/4 tall, wrinkled; 1/4 tall, smooth; 1/4 short, smooth; and 1/4 short, wrinkled. The progeny produced from the unknown parents are 1/2 tall, smooth and 1/2 tall, wrinkled. We can not say absolutely that set 3 did not produce these progeny, but the chance that they would produce nothing but tall, smooth and tall, wrinkled and none of the expected short, smooth and short, wrinkled is exceedingly small, especially in light of the large number of progeny. So, it is unlikely that set 3 are the parents.

c. Each of the two remaining sets, sets 2 and 4, would be expected to produce the same types and frequencies of progeny: 1/2 tall, smooth, 1/2 tall, wrinkled. Since the actual progeny are in agreement with these expectations, either set could be the parents.

4-14. One way to begin is to identify, to the extent possible, the genotypes of the progeny. The red, long flies must carry at least one se^+ allele and one vg^+ giving them a genotype of $se^+/\_vg^+/\_$. The red, vestigial must carry at least one se^+ and two vg alleles: $se^+/\_vg/vg$. The sepia, long must carry two se alleles and at least one vg^+ allele: $se/se vg^+/\_$. The sepia, vestigial must carry two se alleles and two vg alleles: $se/se vg/vg$. Based on these genotypes, we can establish four requirements for the parental genotypes: (1) at least one parent must carry se^+, (2) at least one parent must carry vg^+, (3) both parents must carry se, and (4) both parents must carry vg. The requirement which states that each parent must carry se, rules out sets 1, 2, and 3. Sets 4 and 5 meet all of the requirements, so we need to look at the progeny phenotypes and phenotypic ratio that each would produce. Since we are interested in just the phenotypic outcome, we can consider each trait separately, determine the probability with which each phenotype of that trait occurs, and then identify phenotypic combinations and determine their probabilities using the product law.

First consider set 4: $se^+/se vg^+/vg \times se^+/se vg/vg$. At the se locus, cross $se^+/se \times se^+/se$ would produce phenotypes of red (genotypes se^+/se^+ and se^+/se) and sepia (genotype se/se) with frequencies of 3/4 and 1/4, respectively. At the vg locus, cross $vg^+/vg \times vg/vg$ would produce phenotypes of long (genotype vg^+/vg) and vestigial (genotype vg/vg) in frequencies of 1/2 each. Combining these probabilities using the product law gives the following.

p(red, long)
 = p(red) × p(long) = 3/4 × 1/2 = 3/8
p(red, vestigial)
 = p(red) × p(vestigial) = 3/4 × 1/2 = 3/8
p(sepia, long)
 = p(sepia) × p(long) = 1/4 × 1/2 = 1/8
p(sepia, vestigial)
 = p(sepia) × p(vestigial) = 1/4 × 1/2 = 1/8

These progeny phenotypes and frequencies match those of the actual progeny and thus set 4 could be the parents.

Now we need to determine the progeny for set 5, $se^+/se vg/vg \times se/se vg^+/vg$. Cross $se^+/se \times se/se$ would produce phenotypes of red (genotype se^+/se) and sepia (genotype se/se), each with a frequency of 1/2. Cross $vg/vg \times vg^+/vg$ would produce long (genotype vg^+/vg) and vestigial (genotype vg/vg), each with a frequency of 1/2. Combining these probabilities using the product law gives the following.

p(red, long)
 = p(red) × p(long) = 1/2 × 1/2 = 1/4
p(red, vestigial)
 = p(red) × p(vestigial) = 1/2 × 1/2 = 1/4
p(sepia, long)
 = p(sepia) × p(long) = 1/2 × 1/2 = 1/4
p(sepia, vestigial)
 = p(sepia) × p(vestigial) = 1/2 × 1/2 = 1/4

Set-5 parents would produce the same classes of progeny, but their expected frequencies differ from those of the actual progeny. Thus set 4 is the most likely possibility.

4-15. a. The man has a genotype of $aamm$. One of his a alleles came from each of his parents. Thus his afflicted father must have been heterozygous.

b. The woman's genotype is $aaMm$. The woman's father was mm and he must have contributed the m allele to his daughter. Since her m came from her father, her M allele must have come from her mother. Thus the mother suffered from migraines.

c. Consider the inheritance of each allele separately. The probability of producing normal individuals from the mating $aa \times aa$ is 1. The probability of producing normal individuals from the mating $Mm \times mm$ is 1/2. The probability of producing an individual normal for both traits is 1 × 1/2 = 1/2.

d. The probability of producing an afflicted individual from the mating $aa \times aa$ is 0. Thus the probability of producing an individual afflicted with both traits is 0.

e. The probability of producing normal individuals from the mating $aa \times aa$ is 1. The probability of producing individuals with migraines from the mating $Mm \times mm$ is 1/2. The probability of producing an individual with normal vision and migraines is 1 × 1/2 = 1/2.

4-16. a. The black chicken with feathered legs has a genotype of F^BF^BFF and the white chicken with featherless legs has a genotype of F^WF^Wff. The progeny are all slate blue with feathered legs: F^BF^WFf.

b. The mating of two F_1 chickens is $F^BF^WFf \times F^BF^WFf$. Consider the inheritance of each locus separately. The mating $F^BF^W \times F^BF^W$ produces black, blue, and white chickens with frequencies of 1/4, 1/2, and 1/4, respectively. The mating $Ff \times Ff$ produces feathered legs and featherless legs with frequencies of 3/4 and 1/4, respectively. The branch diagram in Table 4-16b gives all possible phenotypic combinations and uses the product law to determine their frequencies.

TABLE 4-16b

Black (1/4)	/	Feathered (3/4)	=	Black, feathered	=	1/4 × 3/4 = 3/16
	\	Featherless (1/4)	=	Black, featherless	=	1/4 × 1/4 = 1/16
Blue (1/2)	/	Feathered (3/4)	=	Blue, feathered	=	1/2 × 3/4 = 3/8 = 6/16
	\	Featherless (1/4)	=	Blue, featherless	=	1/2 × 1/4 = 1/8 = 2/16
White (1/4)	/	Feathered (3/4)	=	White, feathered	=	1/4 × 3/4 = 3/16
	\	Featherless (1/4)	=	White, featherless	=	1/4 × 1/4 = 1/16

4-17. a. Consider the inheritance at each locus separately. The mating $F^BF^W \times F^BF^W$ produces black, blue, and white chickens with frequencies of 1/4, 1/2, and 1/4, respectively. The mating $Ff \times ff$ produces feathered legs and featherless legs with frequencies of 1/2 each. The probability of producing a blue chicken with featherless legs is 1/2 × 1/2 = 1/4.

b. See the answer to problem 4-17a. The probability of producing a white chicken with feathered legs is 1/4 × 1/2 = 1/8.

4-18. a. The mating is $LLRR' \times L'L'RR$. The probability of getting the oval progeny phenotype (genotype LL') from the mating $LL \times L'L'$ is 1. The probability of getting the purple phenotype (genotype (RR') from the mating $RR' \times RR$ is 1/2. The probability of getting the oval, purple phenotype is 1 × 1/2 = 1/2.

b. The mating is $LLR'R' \times LL'RR'$. The probability of getting the oval phenotype (genotype LL') from the mating $LL \times LL'$ is 1/2. The probability of getting the purple phenotype (genotype RR') from the mating $R'R' \times RR'$ is 1/2. The probability of getting oval, purple is 1/2 × 1/2 = 1/4.

Chapter 5: Crosses Involving Three or More Independently Assorting Traits

5-1. a. The number of types of gametes produced by a multihybrid parent is given by the expression 2^n, where n is the number of heterozygous loci. Here n = 3, and there are $2^3 = 8$ kinds of gametes. Each type would be produced with equal frequency.

b. Since the three traits assort independently, we can consider them separately and use the product law to combine probabilities. Each cross involves two heterozygous individuals and the probability of producing progeny with the dominant and recessive phenotypes is 3/4 and 1/4, respectively. The probability of progeny expected to be phenotypically dominant at all three loci is given by multiplying together the individual probabilities for dominance at each locus. p(dominance at all three loci) = 3/4 × 3/4 × 3/4 = 27/64.

5-2. a. Since p(homozygous recessiveness for any one trait) = 1/4, then p(homozygous recessiveness for all three traits) = 1/4 × 1/4 × 1/4 = 1/64.

b. Since p(heterozygosity for any one trait) = 1/2, then p(heterozygosity for all three traits) = 1/2 × 1/2 × 1/2 = 1/8.

c. Probability of getting $YYWwcc$: p(YY) from cross $Yy \times Yy$ = 1/4; p(Ww) from cross $Ww \times Ww$ = 1/2; and p(cc) from cross $Cc \times Cc$ = 1/4. p($YYWwcc$) = 1/4 × 1/2 × 1/4 = 1/32.

d. Probability of getting $yyWWcc$: p(yy) from cross $Yy \times Yy$ = 1/4; p(WW) from cross $Ww \times Ww$ = 1/4; and p(cc) from cross $Cc \times Cc$ = 1/4. p($yyWWcc$) = 1/4 × 1/4 × 1/4 = 1/64.

5-3. a. Each parent is heterozygous at the four loci under consideration. The generalized expression, 2^n, where n is the number of heterozygous loci, gives the number of different types of gametes produced by each parent. With n = 4, there are $2^4 = 16$ different gametic types. Each type would be produced with an equal frequency (1/16).

b. The probability of producing an individual homozygous dominant for any one of the four traits is 1/4, that is, p(YY) = p(WW) = p(CC) = p(DD) = 1/4. The probability of producing an individual homozygous dominant for all four traits is given by multiplying the probabilities for homozygous dominance at each locus: p($YYWWCCDD$) = 1/4 × 1/4 × 1/4 × 1/4 = 1/256.

5-4. a. The probability of a plant showing the dominant phenotype for any one of these traits is 3/4 and the probability of showing dominance at all four loci is 3/4 × 3/4 × 3/4 × 3/4 = 81/256.

b. The probability of producing an individual with the genotype YY from mating $Yy \times Yy$ = 1/4; the probability of ww or cc progeny from matings $Ww \times Ww$ and $Cc \times Cc$, respectively, is 1/4; and the probability of Dd from the mating $Dd \times Dd$ is 1/2. p($YYwwccDd$) = p(YY) × p(ww) × p(cc) × p(Dd) = 1/4 × 1/4 × 1/4 × 1/2 = 1/128.

c. The probability of a recessive genotype at any of the loci is 1/4 and the probability of its occurring at all four loci is 1/4 × 1/4 × 1/4 × 1/4 = 1/256.

5-5. a. Since each parent is either homozygous recessive or homozygous dominant, each would produce one type of gamete.

b. All the F_1 progeny would be genotypically and phenotypically identical, exhibiting heterozygosity at each of the seven loci and expressing the dominant trait at each locus.

5-6. a. The number of genotypes arising from crossing two individuals, both heterozygous at the same loci, is given by the expression 3^n, where n is the number of mutual heterozygous loci. In this case, n = 7 and there are $3^7 = 2187$ different genotypes.

b. The number of different phenotypes is given by the expression 2^n, where n is the number of heterozygous loci. There are $2^7 = 128$ different phenotypes.

5-7. a. Since the *B* and *C* loci are homozygous, they contribute no variability to the gamete, that is, every gamete carries *B* and *C*. Variability comes only from the heterozygous *A* locus and the two types of gametes are *ABC* and *aBC*. This number can be confirmed by using the generalized expression for the number of gametic types, 2^n, where n is the number of heterozygous loci. In this case n = 1 and there are $2^1 = 2$ kinds of gametes. Each gamete has an expected frequency of 1/2.

b. Two of three loci in the genotype *AabbCc* are heterozygous. Using the expression 2^n, where n = 2, gives $2^2 = 4$ different types of gametes: *AbC*, *Abc*, *abC*, and *abc*. Each gamete has an expected frequency of 1/4.

c. The genotype *AaBBCcDd* has three heterozygous loci. Using the expression 2^n, where n = 3, gives $2^3 = 8$ different types of gametes: *ABCD*, *aBCD*, *ABcD*, *aBcD*, *ABCd*, *aBCd*, *ABcd*, and *aBcd*. Each gamete is formed with an expected frequency of 1/8.

5-8. a. The male parent's genotype is *ttAaRhrh*. The female parent's genotype is *TtAArhrh*. Since the traits are inherited independently, we can consider them separately and use the product law to combine probabilities. The probability of producing a tongue roller from mating *tt* × *Tt* is 1/2; the probability of normal enzyme production from *Aa* × *AA* is 1; and the probability of Rh-negative blood from *Rhrh* × *rhrh* is 1/2. Multiplying these three probability values gives p(tongue rolling, normal enzyme, Rh-negative) = 1/2 × 1 × 1/2 = 1/4.

b. p(nonroller, normal enzyme, Rh positive) = 1/2 × 1 × 1/2 = 1/4.

5-9. a. Each pair of alleles involved in this mating can be considered separately: p(*Aa*) from *Aa* × *aa* = 1/2, p(*Bb*) from *BB* × *bb* = 1, p(*cc*) from *Cc* × *cc* = 1/2, p(*dd*) from *dd* × *dd* = 1, p(*Ee*) from *ee* × *Ee* = 1/2, and p(*Ff*) from *Ff* × *Ff* = 1/2. p(*AaBbccddEeFf*) = 1/2 × 1 × 1/2 × 1 × 1/2 × 1/2 = 1/16.

b. Considering the inheritance of each pair of alleles separately, p(dominance) from *Aa* × *aa* = 1/2, p(dominance) from *BB* × *bb* = 1, p(dominance) from *Cc* × *cc* = 1/2, p(recessiveness) from *dd* × *dd* = 1, p(recessiveness) from *ee* × *Ee* = 1/2, and p(recessiveness) from *Ff* × *Ff* = 1/4. Combining these gives p(dominance for A, B, C, and recessiveness for D, E, F) = 1/2 × 1 × 1/2 × 1 × 1/2 × 1/4 = 1/32.

c. p(recessiveness) from *Aa* × *aa* = 1/2, p(dominance) from *BB* × *bb*) = 1, p(recessiveness) from *Cc* × *cc* = 1/2, p(recessiveness) from *dd* × *dd* = 1, p(recessiveness) from *ee* × *Ee* = 1/2, and p(dominance) from *Ff* × *Ff* = 3/4. Combining these gives p(dominance for *B* and *F*, and recessiveness for *A, C, D, E*) = 1/2 × 1 × 1/2 × 1 × 1/2 × 3/4 = 3/32.

5-10. a. The number of different types of gametes with regard to the heterozygous loci is given by the expression 2^n, where n equals the number of heterozygous gene pairs. Since n = 11, there are $2^{11} = 2048$ different types of gametes.

b. Since the four additional loci are homozygous, they contribute no additional variability to the gametes. For example, if these additional loci were represented as *AAbbccDD*, every gamete produced by the individual would carry *AbcD*. Thus with 11 heterozygous loci and four homozygous loci, the number of different kinds of gametes with regard to these 15 loci is still 2^{11}.

5-11. a. The expression 2^n, where n equals the number of loci at which the parents are mutually heterozygous, gives the number of phenotypes among the progeny. Solving the expression $2^n = 32$ for n gives five heterozygous loci.

b. The proportion of homozygous recessives among the progeny is given by the expression $1/(4^n)$, where n = 5: $1/(4^5) = 1/1024$. With 3072 progeny, the number of homozygous recessives is 3027 × 1/1024 = 3072/1024 = 3.

5-12. a. The expression 3^n, where n equals the number of loci at which the parents are mutually heterozygous, gives the number of genotypes among the progeny. Solving the expression $3^n = 27$ for n gives three heterozygous loci.

b. The expression 2^n, where n equals the number of loci at which the parents are mutually heterozygous, gives the number of phenotypes among the progeny. There are $2^n = 2^3 = 8$ phenotypes.

Chapter 6: The Chi-Square Test

6-1. a. The cross between two heterozygous plants would be expected to yield normal and albino progeny in a 3:1 ratio. Since the total number of plants is 126 + 66 = 192 and the ratio has four parts to it, each part will have 192/4 = 48 plants. The expected number of normal plants is 3 × 48 = 144 and the number of albinos is 1 × 48 = 48.

b.

Class	Obs.	Exp.	Dev.	Dev.^{2}	Dev.^{2}/Exp.
Normal	126	144	+18	324	324/144 = 2.25
Albino	66	48	−18	324	324/48 = 6.75
					χ^2 = 9.00

6-2. a. Decreasing the discrepancy between the observed and expected numbers will reduce the χ^2 value.

b. Decreasing the size of the sample while holding the deviations constant will increase the size of the χ^2 value.

6-3. a. One less than the two classes involved gives 2 − 1 = 1 degree of freedom. The χ^2 value of 3.020, opposite one degree of freedom in Table 6-5, falls between 2.71 and 3.84, with a probability value between 0.10 and 0.05. Since the p-value is greater than 0.05, the deviation is probably due to chance. The deviation is nonsignificant and these results are considered to support the hypothesis.

b. Degrees of freedom equal 4 classes − 1 = 3. Opposite three degrees of freedom, the χ^2 value of 10.36 lies between 7.82 and 11.35, with a probability value between 0.05 and 0.01. Since this probability is less than 0.05, the probability is not high enough to attribute the results to chance. The deviation is significant and these data do not support the hypothesis.

c. Degrees of freedom: 3 − 1 = 2. Opposite two degrees of freedom, the χ^2 value of 1.555 lies between 1.39 and 2.41, with a probability value between 0.50 and 0.30. Since this probability is greater than 0.05, it is large enough to attribute the results to chance. The deviation is nonsignificant and these data support the hypothesis.

6-4. The deviation for problem 6-3b, significant at the 0.05 level, becomes nonsignificant at the 0.01 level. Deviations for 6-3a and 6-3c remain nonsignificant.

6-5. a. The hypothesis of incomplete dominance predicts three phenotypic classes in the F_2; red-, pink-, and white-flowered progeny in a theoretical ratio of 1:2:1, respectively.

b. The theoretical 1:2:1 ratio has 1 + 2 + 1 = 4 parts, and with 200 progeny, each part of the ratio is expected to have 200/4 = 50 individuals. The expected numbers of progeny are 50 red, 100 pink, and 50 white.

c.

Class	Obs.	Exp.	Dev.	Dev.^{2}	Dev.^{2}/Exp.
Red	42	50	−8	64	64/50 = 1.28
Pink	110	100	+10	100	100/100 = 1.00
White	48	50	−2	4	4/50 = 0.08
					χ^2 = 2.36

d. F_2 progeny fall into three classes; 3 − 1 = 2 degrees of freedom.

e. χ^2 value 2.36 falls between p = 0.5 and p = 0.3.

f. The p-value is greater than 0.05 and thus high enough to attribute the deviation to chance.

g. Since the discrepancy between observed and expected numbers is most likely due to chance, the deviation is considered nonsignificant and the data is considered to support the hypothesis underlying the theoretical F_2 ratio of 1:2:1.

6-6. a. The theoretical ratio of male to female is 1:1 and, with a total of 40 births, 20 males and 20 females would be expected.

b.

Class	Obs.	Exp.	Dev.	Dev.^{2}	Dev.^{2}/Exp.
Male	13	20	−7	49	49/20 = 2.45
Female	27	20	+7	49	49/20 = 2.45
					χ^2 = 4.90

c. With one degree of freedom, χ^2 = 4.90 lies between 3.84 and 6.64 with a probability value between 0.05 and 0.01. Since the p-value is less than 0.05, the odds are too low to attribute the deviation to chance and it is considered significant.

d. These results do not support the hypothesis underlying the theoretical 1:1 ratio.

e. Your belief should not be altered by this outcome. A sample of 40 births is small. A larger sample should be examined before deciding to set aside the hypothesis.

6-7. a. The expected phenotypic ratio for the progeny of mating two dihybrids together is 9:3:3:1. This ratio has 9 + 3 + 3 + 1 = 16 parts and, with 640 F_2 progeny, each part is expected to have 640/16 = 40 flies. The expected numbers of progeny are as follows: red, long: 9 × 40, = 360; sepia, long: 3 × 40 = 120; red, short: 3 × 40 = 120; and sepia, short: 1 × 40 = 40.

b.

Class	Obs.	Exp.	Dev.	Dev.^{2}	Dev.^{2}/Exp.
Red, long	344	360	−16	256	256/360 = 0.711
Red, short	134	120	14	196	196/120 = 1.633
Sepia, long	128	120	8	64	64/120 = 0.533
Sepia, short	34	40	−6	36	36/40 = 0.900
					χ^2 = 3.777

c. A χ^2 value of 3.777 with three degrees of freedom falls between 3.67 and 4.64 with a probability value between 0.30 and 0.20.

d. Since the p-value is greater than 0.05, the probability is high enough to consider the deviation as nonsignificant and to attribute it to chance.

e. Since the deviation is not considered statistically significant, these results support the hypothesis that gives the theoretical ratio of 9:3:3:1.

6-8. a. The errors include (1) using percent values rather than the actual raw numbers, (2) testing against a ratio derived from the data rather than from theoretical expectations based on the hypothesis, (3) basing the evaluation of the deviation on the magnitude of the χ^2 value rather than on its probability value, (4) making absolute statements about the role of chance in producing the outcome ("definitely due to chance") and about the hypothesis ("must be correct") rather than using statements like "the deviation can reasonably be attributed to chance" and "the data support the hypothesis," and (5) drawing a conclusion on the basis of a single mating.

b. The actual numbers of yellow- and brown-shelled snails in the F_2.

6-9. a.

Class	Obs.	Exp.	Dev.	Dev.^{2}	Dev.^{2}/Exp.
Brown	23	25	−2	4	4/25 = 0.16
Yellow	27	25	+2	4	4/25 = 0.16
					χ^2 = 0.32

With one degree of freedom, the p-value for χ^2 = 0.32 lies between 0.50 and 0.70. The probability is large enough to reasonably attribute the discrepancy to chance and the data are considered to support the hypothesis of a 1:1 ratio among the F_2 progeny.

b.

Class	Obs.	Exp.	Dev.	Dev.^{2}	Dev.^{2}/Exp.
Brown	23	37.5	−14.5	210.25	210.25/37.5 = 5.61
Yellow	27	12.5	+14.5	210.25	210.25/12.5 = 16.82
					χ^2 = 22.43

With one degree of freedom, the p-value for $\chi^2 = 22.43$ is less than 0.001. The odds are too small to reasonably attribute the discrepancy to chance and the data are not considered to support the hypothesis of a 3:1 ratio among the F_2 progeny.

c. After gaining a clear understanding of how to use the χ^2 test, including how to identify the ratio expected under the hypothesis being considered, the researcher should repeat the mating, using a larger group of parents to produce a larger F_2.

6-10. a. Since one of the classes of progeny has between 5 and 10 individuals, it is necessary to use the Yates correction factor to adjust for the small sample size. This correction is made by reducing the absolute value of the discrepancy for *each* class by 1/2. The chi-square calculation is as follows.

Class	Obs.	Exp.	Dev.	Adj. dev.	(Adj. dev.)^2	(Adj. dev.)^2/ Exp.
Brown	35	33	+2	1.5	2.25	2.25/33 = 0.068
Yellow	9	11	−2	1.5	2.25	2.25/11 = 0.205
						χ^2 = 0.273

b. With one degree of freedom, the p-value for $\chi^2 = 0.273$ lies between p-values of 0.7 and 0.5. The probability is high enough to reasonably attribute the discrepancy to chance. The data support the hypothesis.

c. Because of the small sample size, the researcher should once again repeat the mating in the hopes of producing a larger F_2 generation.

Chapter 7: More on Probability: Unordered Events and Binomial and Multinomial Distributions

7-1. Events are independent if the occurrence (or nonoccurrence) of one fails to influence the occurrence (or nonoccurrence) of the others. Events are mutually exclusive if the occurrence of one precludes the occurrence of the others.

7-2. The probability of two or more independent events occurring together is given by the product of their separate probabilities. The probability of two or more mutually exclusive events occurring together is given by the sum of their separate probabilities.

7-3. A sequence refers to the occurrence of two or more events, with those events occurring in a particular and specific order. A combination refers to the occurrence of two or more events, without regard to the order in which the events occur. In considering a combination, every possible sequence which could produce the combination needs to be taken into account.

7-4. The probability of the occurrence of a sequence is given by the product of the probabilities of each of the events in the sequence. The probability of the occurrence of a combination is given by the sum of the probabilities of each and every sequence which could give rise to the combination.

7-5. a. The probability of producing a brown guinea pig (genotype *bb*) from the cross *Bb* × *Bb* is 1/4. The probability of producing three brown guinea pigs is 1/4 × 1/4 × 1/4 = 1/64.

b. The probability of producing a black guinea pig from this cross is 3/4. The probability of producing a brown guinea pig is 1/4. The probability of producing a litter of brown, black, black in that order is 1/4 × 3/4 × 3/4 = 9/64.

c. Two other sequences are black, brown, black and black, black, brown. Each would occur with a probability of 9/64.

d. Three different sequences could give rise to a litter of one brown and two black. Since each sequence is a mutually exclusive event, the probabilities for the sequences are added: 9/64 + 9/64 + 9/64 = 27/64. Alternatively,

$$p(\text{1 brown, 2 black}) = \frac{3!}{1!2!}(3/4)^2(1/4)^1$$
$$= 3(9/16)(1/4) = 27/64.$$

e. This question is identical to problem 7-5d, phrased in a different way.

7-6. a. Each litter is independent of the other litters. The probability of these three successive litters is given by the product of their separate probabilities. The probability of each litter is 9/64. The probability of the three litters is 9/64 × 9/64 × 9/64 = 729/262,144 = 0.0028.

b. The probability of producing one such litter is 27/64. The probability of producing three of them in succession is 27/64 × 27/64 × 27/64 = 19,683/262,144 = 0.075.

7-7. a. The generalized expression for the probability of a sequence of black, brown, black is p × q × p or p^2q.

b. The generalized expression for the probability of the combination of two black and one brown is $3p^2q$.

c. The expression q^3 represents the probability of producing a litter of three brown guinea pigs, and $3pq^2$ represents the probability of producing a litter with the combination of one black and two brown guinea pigs.

7-8. a. Since no order is specified, we are dealing with a combination. The probability of this specific combination can be determined by using the expression

$$p(\text{specific combination}) = \frac{n!}{s!t!}p^sq^t$$

where n is the total number of guinea pigs in the combination (5), s is the number of black guinea pigs in the combination (4), t is the number of brown guinea pigs in the combination (1), p is the probability of producing a black guinea pig (3/4), and q is the probability of producing a brown guinea pig (1/4).

$$p(\text{4 black, 1 brown})$$
$$= \frac{5!}{4!1!}(3/4)^4(1/4)^1$$
$$= \frac{5 \times 4 \times 3 \times 2 \times 1}{(4 \times 3 \times 2 \times 1)(1)}(3/4 \times 3/4 \times 3/4 \times 3/4)(1/4)$$
$$= \frac{5}{1}(81/256)(1/4) = 5(81/1024)$$
$$= 405/1024 = 0.40$$

b. Since an order is specified, we are dealing with a sequence. The probability of the occurrence of a sequence is obtained by using the product law. The probability of black is 3/4; the probability of brown is 1/4. The probability of the black, black, brown, black, black sequence is $3/4 \times 3/4 \times 1/4 \times 3/4 \times 3/4 = 81/1024 = 0.079$.

7-9. a. Since no order is specified, and since we are dealing with two mutually exclusive events (boy versus girl), we can determine the probability of this specific combination by using the expression

$$p(\text{specific combination}) = \frac{n!}{s!t!}p^s q^t$$

where n is the total number of children in the combination (5), s is the number of boys in the combination (3), t is the number of girls in the combination (2), and p is the probability of producing a boy (1/2) and q is the probability of producing a girl (1/2).

$$\begin{aligned}
&p(3\text{ boys}, 2\text{ girls}) \\
&= \frac{5!}{3!2!}(1/2)^3(1/2)^2 \\
&= \frac{5 \times 4 \times 3 \times 2 \times 1}{(3 \times 2 \times 1)(2 \times 1)}(1/2 \times 1/2 \times 1/2)(1/2 \times 1/2) \\
&= \frac{5 \times 4}{2 \times 1}(1/8)(1/4) \\
&= \frac{20}{2}(1/32) = 10/32 = 5/16 = 0.31
\end{aligned}$$

b. There are two combinations which meet the qualifications of "at least four boys": either the combination of four boys and one girl or the combination of five boys. The probability of the four-boy, one-girl combination is given by the expression

$$p(\text{specific combination}) = \frac{n!}{s!t!}p^s q^t$$

where $n = 5$, $s = 4$, $t = 1$, and $p = q = 1/2$.

$$\begin{aligned}
&p(4\text{ boys}, 1\text{ girl}) \\
&= \frac{5!}{4!1!}(1/2)^4(1/2)^1 \\
&= \frac{5 \times 4 \times 3 \times 2 \times 1}{(4 \times 3 \times 2 \times 1)(1)}(1/2 \times 1/2 \times 1/2 \times 1/2)(1/2) \\
&= \frac{5}{1}(1/32) = 5/32 = 0.16.
\end{aligned}$$

The probability of the five-boy combination can be obtained by using either the equation we have just used

$$p(5\text{ boys}) = \frac{5!}{5!0!}(1/2)^5(1/2)^0$$

(remember that $0! = 1$, and that anything raised to the zeroth power is 1), or by using the product law (since this combination can arise from only one sequence: boy, boy, boy, boy, boy): $P = (1/2)^5$. Solving either expression gives the same answer: $1/32 = 0.031$. Since the requirement of "at least four boys" can be met by either the four-boy, one-girl combination or by the five-boy combination, the probability of one or the other occurring is given by adding their separate probabilities: $5/32 + 1/32 = 6/32 = 0.19$.

7-10. a. The probability of producing a male or female child is 1/2. The probability of two heterozygous parents ($Tt \times Tt$) producing a taster child is 3/4 and producing a nontaster child is 1/4. Since sex and tasting ability are independent events, we combine their probabilities using the product law: p(male taster) $= 1/2 \times 3/4 = 3/8$; p(male nontaster) $= 1/2 \times 1/4 = 1/8$; p(female taster) $= 1/2 \times 3/4 = 3/8$; p(female nontaster) $= 1/2 \times 1/4 = 1/8$.

b. Since no order is specified, we are concerned with all possible ways of getting this combination of three mutually exclusive events: one male taster, two male nontasters, and two female tasters. The formula to be used is

$$p(\text{specific combination}) = \frac{n!}{s!t!u!}p^s q^t r^u$$

where $n = 5$, p = probability(male taster) $= 3/8$, q = probability(male nontaster) $= 1/8$, r = probability(female taster) $= 3/8$, $s = 1$, $t = 2$, and $u = 2$. Substituting these values into the equation gives

$$\begin{aligned}
&p(1\text{ male taster}, 2\text{ male nontasters}, 2\text{ female tasters}) \\
&= \frac{5!}{1!2!2!}(3/8)^1(1/8)^2(3/8)^2 \\
&= \frac{5 \times 4 \times 3 \times 2 \times 1}{(1)(2 \times 1)(2 \times 1)}(3/8)(1/8 \times 1/8)(3/8 \times 3/8) \\
&= \frac{60}{2}(27/32{,}768) \\
&= 30(27/32{,}768) \\
&= 810/32{,}768 = 0.025.
\end{aligned}$$

c. Since no order is specified, we are concerned with all possible ways of producing this specific combination. We use the formula

$$p(\text{specific combination}) = \frac{n!}{s!t!u!\ldots}p^s q^t r^u\ldots$$

Substituting the appropriate values, we get

$$\begin{aligned}
&p(1\text{ male taster}, 1\text{ male nontaster}, 1\text{ female taster } 1\text{ female nontaster}) \\
&= \frac{4!}{1!1!1!1!}(3/8)^1(1/8)^1(3/8)^1(1/8)^1 \\
&= \frac{4 \times 3 \times 2 \times 1}{(1)(1)(1)(1)}(3/8)(1/8)(3/8)(1/8) \\
&= 24(9/4096) = 216/4096 = 0.053.
\end{aligned}$$

7-11. a. The probability of the first (or any other) child being galactosemic (genotype *gg*) from the mating $g^+g \times g^+g$ is 1/4.

b. The probability of the first (or any other) child being normal for galactose utilization (that is, having either genotype g^+g^+ or g^+g) from the mating $g^+g \times g^+g$ is 3/4.

c. The probability of any child being a girl is 1/2. The probability of any child being normal for galactose utilization is 3/4. The probability of these two independent events occurring together in the third (or in any child) is given by the product of the separate probabilities: $1/2 \times 3/4 = 3/8$.

d. Since we are interested in all the possible ways of producing mutually exclusive events, we use the formula

$$p(\text{specific combination}) = \frac{n!}{s!t!}p^s q^t.$$

The values for this combination are: n = 5, s = 4, t = 1, p = 3/8, and q = 1/8. Substitution gives us

$$\begin{aligned} &p(\text{4 normal males, 1 galactosemic female}) \\ &= \frac{5!}{4!1!}(3/8)^4(1/8)^1 \\ &= \frac{5 \times 4 \times 3 \times 2 \times 1}{(4 \times 3 \times 2 \times 1)(1)}(3/8 \times 3/8 \times 3/8 \times 3/8)(1/8) \\ &= \frac{5}{1}(81/4096)(1/8) = 5(81/32{,}768) \\ &= 405/32{,}768 = 0.012. \end{aligned}$$

7-12. a. Substituting in the expression

$$p(\text{specific combination}) = \frac{n!}{s!t!}p^s q^t$$

where n = 4, p = 1/2, q = 1/2, s = 2, and t = 2, gives

$$\begin{aligned} &p(\text{2 boys, 2 girls}) \\ &= \frac{4!}{2!2!}(1/2)^2(1/2)^2 \\ &= \frac{4 \times 3 \times 2 \times 1}{(2 \times 1)(2 \times 1)}(1/4)(1/4) \\ &= \frac{12}{2}(1/4)(1/4) = 6(1/16) = 6/16 = 0.375. \end{aligned}$$

b. Since the probability of the occurrence of the expected outcome is 3/8, then the likelihood of some outcome other than the results expected occurring is $1 - 0.375 = 0.625$.

7-13. a. Since tasting ability and albinism are inherited independently, we can consider their inheritance separately and combine probabilities using the product law. In the cross $Tt \times Tt$, the probability of producing a taster child is 3/4 and that of a nontaster child is 1/4. In the cross $a^+a \times aa$, the probability of producing a normally pigmented child is 1/2 and that of an albino child is 1/2. Combining these probabilities gives p(taster, normal pigmentation) = $3/4 \times 1/2 = 3/8$ and p(nontaster, albino) = $1/4 \times 1/2 = 1/8$. Since no order is specified, we are concerned with all possible ways of getting this combination of two mutually exclusive events. The formula to be used is

$$p(\text{specific combination}) = \frac{n!}{s!t!}p^s q^t$$

where n = 6, p = probability(taster, normal pigmentation) = 3/8, q = probability(nontaster, albino) = 1/8, s = 3 and t = 3. Substituting these values into the equation gives

$$\begin{aligned} &p(\text{3 normal tasters, 3 albino tasters}) \\ &= \frac{6!}{3!3!}(3/8)^3(1/8)^3 \\ &= \frac{6 \times 5 \times 4 \times 3 \times 2 \times 1}{(3 \times 2 \times 1)(3 \times 2 \times 1)}(3/8 \times 3/8 \times 3/8)(1/8 \times 1/8 \times 1/8) \\ &= \frac{120}{6}(27/262{,}144) \\ &= 20(27/262{,}144) = 540/262{,}144 = 0.0021. \end{aligned}$$

b. p(taster, albino) = $3/4 \times 1/2 = 3/8$. Since no order is specified, we are concerned with all possible ways of getting this combination of three mutually exclusive events. The formula to be used is

$$p(\text{specificcombination}) = \frac{n!}{s!t!u!}p^s q^t r^u$$

where n = 6, p = probability(taster, normal) = 3/8, q = probability(nontaster, albino) = 1/8, r = probability(taster, albino) = 3/8, s = 3, t = 2, and u = 1. These values are substituted into the equation.

$$\begin{aligned} &p(\text{3 normal tasters, 2 albino nontasters, 1 albino taster}) \\ &= \frac{6!}{3!2!1!}(3/8)^3(1/8)^2(3/8)^1 \\ &= \frac{6 \times 5 \times 4 \times 3 \times 2 \times 1}{(3 \times 2 \times 1)(2 \times 1)(1)}(3/8 \times 3/8 \times 3/8)(1/8 \times 1/8)(3/8) \\ &= \frac{120}{2}(81/262{,}144) \\ &= 60(81/262{,}144) = 4860/262{,}144 = 0.019 \end{aligned}$$

c. Since an order is specified, we are dealing with a sequence and use the product law to determine the probability of its occurrence. p(3 tasters with normal pigmentation, 2 nontaster albinos, 1 taster albino) = $(3/8 \times 3/8 \times 3/8 \times 1/8 \times 1/8 \times 3/8) = 81/262{,}144 = 0.00031$.

7-14. In the cross $Tt \times tt$, the probability of producing a taster child is 1/2 and that of a nontaster child is 1/2. In the cross $aa \times a^+a$, the probability of producing a normally pigmented child is 1/2 and that of an albino child is 1/2. For any birth, the probability of a male or female child is 1/2. Combining these probabilities gives us p(nontaster, albino, male) = $1/2 \times 1/2 \times 1/2 = 1/8$, p(nontaster, normal pigmentation, female) = $1/2 \times 1/2 \times 1/2 = 1/8$. Since no order is specified, we are concerned with all possible ways of getting this combination of two mutually exclusive events. The formula to be used is

$$p(\text{specific combination}) = \frac{n!}{s!t!}p^s q^t$$

where n = 5, p = probability(nontaster, albino, male) = 1/8, q = probability(nontaster, normal pigmentation, female) = 1/8, s = 3, and t = 2. Substituting these values into the equation gives

p(3 albino nontaster sons,
2 normal nontaster daughters)

$$= \frac{5!}{3!2!}(1/8)^3(1/8)^2$$
$$= \frac{5 \times 4 \times 3 \times 2 \times 1}{(3 \times 2 \times 1)(2 \times 1)}(1/8 \times 1/8 \times 1/8)(1/8 \times 1/8)$$
$$= \frac{20}{2}(1/32{,}768) = 10(1/32{,}768) = 0.00031.$$

Chapter 8: Multiple-Allelic Series

8-1. a. The silver-gray rabbit is $c^{ch}c^{ch}$ and the pure-breeding dark-gray rabbit is c^+c^+. All of the progeny would be c^+c^{ch} and would show the wild-type dark-gray phenotype.

b. The albino rabbit is cc and the pure-breeding Himalayan rabbit is c^hc^h. Their progeny would be c^hc and have the Himalayan phenotype.

c. All possible progeny genotypes and phenotypes from the mating $c^+c^{ch} \times c^hc$ are shown in the branch diagram that follows.

Gametes from c^+c^{ch}	Gametes from c^hc		Progeny genotypes	Progeny phenotypes
c^+ (1/2)	c^h (1/2)	=	c^+c^h (1/4)	Dark-gray
	c (1/2)	=	c^+c (1/4)	Dark-gray
c^{ch} (1/2)	c^h (1/2)	=	$c^{ch}c^h$ (1/4)	Light-gray
	c (1/2)	=	$c^{ch}c$ (1/4)	Light-gray

Four genotypes are expected in a 1:1:1:1 ratio. Two phenotypes are expected in a ratio of 1 dark-gray to 1 light-gray.

8-2. a. The rabbit with light-gray fur could have a genotype of either $c^{ch}c^h$ or $c^{ch}c$. The Himalayan rabbit could have a genotype of either c^hc^h or c^hc. Albinos (genotype cc) could be produced only if each parent possessed the c allele, that is, if the light-gray rabbit was $c^{ch}c$ and the Himalayan was c^hc.

b. All possible progeny genotypes and phenotypes from the mating $c^{ch}c \times c^hc$ are shown in the branch diagram that follows.

Gametes from $c^{ch}c$	Gametes from c^hc		Progeny genotypes	Progeny phenotypes
c^{ch} (1/2)	c^h (1/2)	=	$c^{ch}c^h$ (1/4)	Light-gray
	c (1/2)	=	$c^{ch}c$ (1/4)	Light-gray
c (1/2)	c^h (1/2)	=	c^hc (1/4)	Himalayan
	c (1/2)	=	cc (1/4)	Albino

In addition to the albino (cc) offspring, light-gray ($c^{ch}c^h$ and $c^{ch}c$) and Himalayan (c^hc) progeny would be expected.

8-3. a. Based on her phenotype, the woman's genotype could be either I^BI^B or I^Bi. Since her father has a type O phenotype, his genotype must be ii. Since the woman received an i allele from her father, we know her genotype is I^Bi. Since her husband has type O blood, his genotype must be ii.

b. Half the woman's eggs carry I^B and the other half carry i. All the sperm produced by her husband carry i.

c. All possible progeny genotypes and phenotypes from the mating are shown in the branch diagram that follows.

Gametes from male	Gametes from female		Progeny genotypes	Progeny phenotypes
i (1)	I^B (1/2)	=	I^Bi (1/2)	Type B
	i (1/2)	=	ii (1/2)	Type O

Expected progeny: half type B, half type O.

8-4. a. The woman is type A and has two possible genotypes: I^AI^A and I^Ai. The man is type B and has two possible genotypes: I^BI^B and I^Bi.

b. The first child is type O and therefore has a genotype of ii. Since each parent has contributed an allele to the child, each parent must carry the i allele. Therefore, the mother's genotype is I^Ai and the father's genotype is I^Bi.

c. All possible phenotypes from the mating $I^Ai \times I^Bi$ are shown in the following branch diagram.

Gametes from male	Gametes from female		Progeny genotypes	Progeny phenotypes
I^B (1/2)	I^A (1/2)	=	I^AI^B (1/4)	Type AB
	i (1/2)	=	I^Bi (1/4)	Type B
i (1/2)	I^A (1/2)	=	I^Ai (1/4)	Type A
	i (1/2)	=	ii (1/4)	Type O

Expected progeny phenotypes: in addition to type O, types AB, B, and A are possible, all in a 1:1:1:1 ratio.

8-5. The man has type AB blood and his genotype must therefore be I^AI^B. The child has type A blood and two genotypes are possible: I^AI^A and I^Ai. Each parent contributed an allele to the child. The mother has B blood and two genotypes are possible: I^BI^B and I^Bi. Since the type A child carries no I^B allele, the mother must have contributed allele i. This would make the child's genotype I^Ai and mean that the child's I^A allele must have come from the father. Since the man in question carries this allele, it is therefore possible for him to be the father. As an expert, however, your testimony should also point out that *any* man carrying the I^A allele could have fathered the child. It is impossible to prove that this particular man is responsible.

8-6. The woman has type AB blood and must therefore have genotype I^AI^B. Her husband has type B blood and, based solely on this, could have genotype I^BI^B or I^Bi. The husband's father has type O blood with a genotype of *ii*. Since the husband received one of his alleles from his father, his genotype must be I^Bi. Genotypes and phenotypes possible from the mating $I^AI^B \times I^Bi$ are shown in the following branch diagram.

Gametes from female	Gametes from male		Progeny genotypes	Progeny phenotypes
I^A (1/2)	I^B (1/2)	=	I^AI^B (1/4)	Type AB
	i (1/2)	=	I^Ai (1/4)	Type A
I^B (1/2)	I^B (1/2)	=	I^BI^B (1/4)	Type B
	i (1/2)	=	I^Bi (1/4)	Type B

Half of the progeny would have type B blood (half of these would be homozygous and the other half heterozygous).

8-7. a. Since neither allele carried by the pollen, that is S_5 or S_6, is possessed by the S_2S_3 plant, normal pollen grain development and fertilization could occur.

b. Half the pollen from the S_2S_3 plant carries S_2 and the other half carries S_3. No S_3 pollen could develop on the S_1S_3 plant, but all of the S_2 pollen could; thus we would expect that 50% of the pollen could develop.

c. A homozygous plant with genotype, say, S_2S_2, could arise only if pollen carrying the S_2 allele developed on the stigma of a plant carrying S_2. Because of the incompatibility, no such development could occur.

d. The cross $S_3S_4 \times S_1S_2$ would involve no incompatibility and works out as follows.

Pollen gametes	Egg gametes		Progeny genotypes
S_3 (1/2)	S_1 (1/2)	=	S_1S_3 (1/4)
	S_2 (1/2)	=	S_2S_3 (1/4)
S_4 (1/2)	S_1 (1/2)	=	S_1S_4 (1/4)
	S_2 (1/2)	=	S_2S_4 (1/4)

Four genotypes would be expected in a 1:1:1:1 ratio.

e. The cross $S_3S_4 \times S_2S_3$ involves S_3 incompatibility. Only S_4 pollen could develop on the stigmas of the S_2S_3 plant, giving rise to equal numbers of S_2S_4 and S_3S_4 progeny as follows.

Pollen gametes	Egg gametes		Progeny genotypes
S_4 (1)	S_2 (1/2)	=	S_2S_4 (1/2)
	S_3 (1/2)	=	S_3S_4 (1/2)

8-8. The B, MN child could have either genotype I^BI^BMN or I^BiMN. Her AB, M mother would have a genotype of I^AI^BMM. Because these traits are inherited independently, they can be considered separately, beginning with the ABO type. Since the child has no I^A, the mother must have contributed I^B to her. The other allele possessed by the daughter (either I^B or i) must have come from her father. All of the men whose blood types are listed could supply either I^B or i, and thus any one is a possible father. With regard to the MN blood type, since the mother has only allele M to contribute, the daughter's N allele must have come from the father. All of the men whose blood types are listed could supply allele N. Therefore the father could have any of the blood types listed.

8-9. The type A, Rh-negative child could have the following genotypes: I^AI^Arhrh or I^Airhrh. Since the two traits are inherited independently, they can be considered separately, beginning with the ABO type. The parents in set 1 would produce type A (I^Ai) and type O (ii) progeny, and those in set 2 would produce type A (I^AI^A, I^Ai) progeny. The parents in set 3 would produce type AB (I^AI^B) and type B (I^Bi) progeny, and those in set 4 would produce type B (I^BI^B)(I^Bi) and type O (ii) progeny. Thus, only sets 1 and 2 could produce a type A child and need be considered further. Set 1 would produce Rh-positive ($Rhrh$) and Rh-negative ($rhrh$) progeny, and set 2 would produce Rh-positive ($Rhrh$) progeny. Only the parents in set 1 could produce a type A, Rh-negative child.

8-10. Since each trait is inherited independently, we can consider each separately, beginning with the ABO type. The child is type B and could have a genotype of either I^BI^B or I^Bi. The father is type O and must therefore have genotype ii. Since the father could only supply an i allele, the child must be I^Bi and her other allele, I^B, must have come from the mother. Mothers 1 (type A), 4 (type A), and 5 (type O) could not have contributed the I^B allele, since they carry alleles I^A and/or i; thus we can eliminate them from further consideration and focus on choices 2 and 3. Next consider the MN blood type. The child is type MN and thus has the MN genotype. The father is type N and must have genotype NN. Since the child's N allele must have come from the father, her M allele must have come from her mother and it could have been supplied by mother 2 (type M) or mother 3 (type MN). Now consider the Rh trait. The child is Rh-positive and has a genotype of either $RhRh$ or $Rhrh$. The father is Rh-negative and must have a genotype of $rhrh$. Since the father could only supply the daughter with the rh allele, the genotype of the daughter must be $Rhrh$. This leaves the daughter's Rh allele to be supplied by her mother. Mother 2 (Rh-negative with genotype $rhrh$) could not supply the Rh allele; mother 3 (Rh-positive with genotype $Rh\_$), however, could supply it. Thus, of the five possibilities, only mother 3 could supply the alleles necessary to be the mother of this child.

8-11. a. Since each individual has a pair of chromosomes with this locus, the smallest number of allelic types that could be carried by an individual is one, and that would occur in homozygous individuals.

b. The largest number of different allelic types that could be carried by an individual is two. These individuals would be heterozygous at this locus.

c. The total number of different genotypes is given by the expression $n(n + 1)/2$, where n is the number of alleles in the series. In this case n = 6, and the number of different genotypes is 6(6 + 1)/2 = 6(7)/2 = 42/2 = 21.

d. The number of homozygous genotypes is equal to n, or 6.

e. The number of heterozygous genotypes is given by the expression n(n − 1)/2. Since n = 6, the number of heterozygous genetic combinations is 6(6 − 1)/2 = 6(5)/2 = 30/2 = 15.

8-12. a. The expected genotypes and phenotypes that result from the mating $s^cs \times Ss$ are shown as follows.

Gametes from cow	Gametes from bull	Progeny genotypes	Progeny phenotypes
s^c (1/2)	S (1/2)	= Ss^c (1/4)	Banded
	s (1/2)	= s^cs (1/4)	Solid
s (1/2)	S (1/2)	= Ss (1/4)	Banded
	s (1/2)	= ss (1/4)	Irregular spotted

Genotypes: Ss^c, s^cs, Ss, and ss in a 1:1:1:1 ratio. Phenotypes: banded, solid, and irregular spotted in a 2:1:1 ratio.

b. The expected genotypes and phenotypes that result from the mating $s^hs^c \times Ss^c$ are shown as follows.

Gametes from cow	Gametes from bull	Progeny genotypes	Progeny phenotypes
S (1/2)	s^h (1/2)	= Ss^h (1/4)	Banded
	s^c (1/2)	= Ss^c (1/4)	Banded
s^c (1/2)	s^h (1/2)	= s^hs^c (1/4)	Regular spotted
	s^c (1/2)	= s^cs^c (1/4)	Solid

Genotypes: Ss^h, s^hs^c, Ss^c, and s^cs^c in a 1:1:1:1 ratio. Phenotypes: banded, regular spotted, and solid in a 2:1:1 ratio.

8-13. a. The bull could be s^hs^c or s^hs.

b. A testcross, between the bull and a cow with either genotype ss or s^cs^c, would provide some evidence about the bull's genotype. Since a cow rarely produces more than a single calf, it would be best to mate the bull with at least several ss or s^cs^c cows. There is evidence that the bull is homozygous if all the calves are phenotypically like him. There is evidence that the bull is heterozygous if one or more calves are phenotypically different from him.

8-14. Each cow would have the genotype s^cs^c and each produces a calf with regular spotting. If the bull is heterozygous, the chance of producing a calf with regular spotting in any one of these matings is 1/2. The probability that the bull will be heterozygous and father five calves with regular spotting in five independent births is 1/2 × 1/2 × 1/2 × 1/2 × 1/2 = 1/32. The probability of the bull being homozygous is 1 − 1/32 = 31/32 = 0.9688, or 0.9688 × 100 = 96.88%.

Chapter 9: Modified Dihybrid Ratios: Interaction of Products of Nonallelic Genes

9-1. The genotype of each parent is *BbIi*. The two loci show dominant epistasis, that is, the presence of a dominant allele at the *I* locus blocks the expression of whatever alleles are present at the *B* locus. Since these loci assort independently, we can consider their inheritance separately and use the product law to combine probabilities. In the cross $Bb \times Bb$, the expected probability of producing progeny with a dominant genotype (either *BB* or *Bb*) is 3/4 and the probability of producing progeny with the recessive genotype is 1/4. Similarly, in the cross $Ii \times Ii$, the expected probability of producing progeny with a dominant genotype is 3/4 and the probability of producing progeny with the recessive genotype is 1/4. (See Table 9-1.) Results: 12/16 of the progeny have at least one dominant allele at the *I* locus and thus will have white fur; 4/16 have recessive alleles at the *I* locus and are thus pigmented: of these, 3/16 have a dominant genotype at the *B* locus and are black and 1/16 have the recessive genotype at the *B* locus and are

TABLE 9-1

Dominant genotype at *I* locus (3/4)	Dominant genotype at *B* locus (3/4)	=	Dominance at both loci 3/4 × 3/4 = 9/16
	Recessive genotype at *B* locus (1/4)	=	Dominance at *I*, recessiveness at *B* 3/4 × 1/4 = 3/16
Recessive genotype at *I* locus (1/4)	Dominant genotype at *B* locus (3/4)	=	Recessiveness at *I*, dominance at *B* 1/4 × 3/4 = 3/16
	Recessive genotype at *B* locus (1/4)	=	Recessiveness at both loci 1/4 × 1/4 = 1/16

brown. This phenotypic ratio of 12:3:1 is customarily seen when dominant epistasis at one locus operates on another.

9-2. A complementary interaction of dominant genes at two loci could be inferred from the F_2 9:7 phenotypic ratio. For this interaction, at least one dominant gene is necessary at both loci for pigment production. A dominant gene at one locus without its "complement" at the other locus will not allow pigment production.

9-3. The ratio of 219:15 is very close to a ratio of 15:1 which would be expected with duplicate interaction, that is, if the dominant gene at either locus is all that is required to produce the feathered-leg phenotype. Thus, only individuals which are homozygous recessive for both of the loci involved will have featherless legs.

9-4. The progeny of individuals heterozygous for both loci would be expected in a 9:6:1 ratio if (1) the dominant genes at each locus complement each other to product the red phenotype, (2) the dominant gene at either one of the two loci without a dominant "complement" at the other locus produces the sandy phenotype, and (3) the absence of dominant genes from both loci produces the white phenotype. In the mating *AaBb* × *AaBb*, nine of 16 progeny will be expected to have at least one dominant allele at the *A* locus *as well as* at least one dominant allele at the *B* locus. A dominant gene at the *A* locus would complement a dominant gene at the *B* locus to produce a red phenotype. Six of the 16 progeny would have at least one dominant gene at the *A* locus *or* at the *B* locus, which would result in the sandy color. One of the 16 progeny would be expected to show the absence of a dominant gene from both the *A* and the *B* loci and would have a white coat.

9-5. a. Progeny from *PPrr* × *ppRR* would all be *PpRr*. Interaction between the dominant allele at each locus produces a walnut comb.

b. Progeny from *PpRr* × *PpRr*: 9/16 with at least one dominant allele at both loci would have walnut combs, 3/16 with at least one dominant allele at the *P* locus and recessive alleles at the *R* locus would have pea combs, 3/16 with at least one dominant allele at the *R* locus and recessive alleles at the *P* locus would have rose combs, and 1/16 with recessive alleles at both loci would have single combs.

9-6. a. A dominant genotype at the inhibitory locus would result in white feathers, regardless of the genotype at the color locus. White feathers would also result whenever the recessive genotype occurs at the color locus. The three genotypic combinations that follow give a white phenotype.

Color locus	Inhibitory locus
Dominant: colored	Dominant: inhibition
Recessive: white	Dominant: inhibition
Recessive: white	Recessive: noninhibition

b. Colored feathers would result only with a dominant genotype at the color locus and a recessive genotype at the inhibitory locus.

Color locus	Inhibitory locus
Dominant: colored	Recessive: noninhibition

c. A cross between two individuals heterozygous at both loci would result in a phenotypic ratio of 13 white to three colored among their offspring.

9-7. First, we need to determine the ratio for the three fruit shapes that occur in the F_2. The smallest phenotypic class contains 19 plants, and dividing the number of plants in each of the other two classes by 19 will give us the parts of the ratio: 106/19 = 5.6 (sphere plants) and 170/19 = 8.9 (oval plants). Thus, our ratio is 8.9:5.6:1 which is very close to a 9:6:1, the ratio that results when complete dominance operates at each locus and interaction between dominant genotypes at both loci give rise to one phenotype, a dominant genotype at either locus produces a second phenotype, and the absence of a dominant genotype from both loci gives rise to a third phenotype. These outcomes are summarized in the following table.

One locus	Other locus	Phenotype	Fraction of F_2
Dominant	Dominant	Oval	9/16
Dominant	Recessive	Sphere	6/16
Recessive	Dominant	Sphere	
Recessive	Recessive	Long	1/16

At each locus, the allele for sphere shape is dominant to the allele for long shape.

9-8. a. The F_1 progeny have genotype *AaBb*, where the heterozygosity at the *B* locus modifies the expression of the *Aa* genotype to produce a partly rough coat.

b. i. Since the two loci are inherited independently, the inheritance of each locus can be considered separately and the product law can be used to combine probabilities. In the cross *Aa* × *Aa*, the genotypes *AA*, *Aa*, and *aa* would be expected to occur with frequencies of 1/4, 1/2, and 1/4, respectively. In the cross *Bb* × *Bb*, the genotypes *BB*, *Bb*, and *bb* would be expected to occur with frequencies of 1/4, 1/2, and 1/4, respectively. Smooth-coated animals would result under two general genotypic conditions: (1) an *AA* or *Aa* genotype coupled with a *BB* genotype, and (2) an *aa* genotype coupled with any genotype at the *B* locus. The probability of producing *AABB* is 1/4 × 1/4 = 1/16, *AaBB* is 1/2 × 1/4 = 1/8, *aaBB* is 1/4 × 1/4 = 1/16, *aaBb* is 1/4 × 1/2 = 1/8, and *aabb* is 1/4 × 1/4 = 1/16. Adding up the probability of all possible ways of producing a smooth-coated animal gives 1/16 + 1/8 + 1/16 + 1/8 + 1/16 = 7/16.

ii. Rough-coated animals would result from a dominant genotype at the *A* locus coupled with a *bb* genotype. Expected probabilities for rough-coated animals are as follows: p(*bbAA*) = 1/4 × 1/4 = 1/16 and p(*bbAa*) = 1/4 × 1/2 = 1/8. Combining these two probabilities gives an expected frequency of 1/16 + 1/8 = 3/16.

iii. Partly rough-coated animals would result from a dominant genotype at the *A* locus coupled with a *Bb* genotype. Expected probabilities of partly rough-coated animals are as follows: p(*BbAA*) = 1/2 × 1/4 = 1/8 and p(*BbAa*) =

$1/2 \times 1/2 = 1/4$. Combining these two probabilities gives an expected frequency of $1/8 + 1/4 = 3/8$, or 6/16.

c. The F_2 would be expected to contain smooth-, partly rough-, and rough-coated animals in a 7:6:3 ratio.

9-9. The F_2 generation consists of three phenotypes: 144 tan-spotted, 47 black-spotted, and 63 unspotted animals. This is close to the ratio of 9:3:4 which arises when complete dominance operates at each locus and the recessive genotype at one of the two loci blocks the expression of whatever genotype is present at the other locus. With that type of interaction, the genotype *S_T_* produces tan spots, *S_tt* produces black spots and genotypes *ssT_* and *sstt* result in the absence of spots.

9-10. **a.** Three genotypes are possible: *GGcc*, *ggcc*, and *ggCC*. Crossed with green plants of genotype *GGCC*, they would form F_1 plants of genotypes *GGCc*, *GgCc*, and *GgCC*, respectively.

b. Crossing two members of the F_1, *GgCc* × *GgCc*, gives an expected F_2 of green- and white-seeded plants in a 9-to-7 ratio. Among the F_2 plants, those with a dominant genotype at both loci, that is, *G_C_*, would have green seeds, those with a recessive genotype at either or both loci, that is, *ggC_*, *G_cc*, and *ggcc*, would have white seeds.

9-11. The cross *Ggcc* × *ggCc* would produce progeny with genotypes *GgCc*, *ggCc*, *Ggcc*, and *ggcc*, with respective phenotypes of green, white, white, and white. Since each genotype would be expected in equal frequency, the probability of producing a white plant is $1/4 + 1/4 + 1/4 = 3/4$.

9-12. The cross *Ggcc* × *ggcc* would produce progeny with genotypes *Ggcc* and *ggcc* and both would have white phenotypes. The probability of producing a white plant is 1.

9-13. **a.** *aaBb*: dark-sooty (*Bb* specifies sooty, and the interaction with *aa* darkens the color to dark-sooty).

b. *aabb*: black.

c. *Aabb*: black (*bb* masks *A* locus and is expressed).

d. *AaBB*: red (*BB* masks *A* locus and is expressed).

e. *AaBb*: sooty (*Bb* masks *A* locus and is expressed).

f. *aaBB*: sooty (*BB* specifies red, and the interaction with *aa* darkens this color to sooty).

9-14. **a.** The inheritance of each locus can be considered independently. In the cross *Aa* × *Aa*, the genotypes *AA*, *Aa*, and *aa* would be expected to occur with frequencies of 1/4, 1/2, and 1/4, respectively. Similarly, the cross *Bb* × *Bb* would produce genotypes *BB*, *Bb*, and *bb* with frequencies of 1/4, 1/2, and 1/4, respectively. The black phenotype results from genotypes *aabb* and the combination of *bb* with a dominant genotype at the *A* locus, that is, *AAbb* and *Aabb*. The product law gives the following probabilities of these genotypes: p(*aabb*) $= 1/4 \times 1/4 = 1/16$; p(*AAbb*) $= 1/4 \times 1/4 = 1/16$; p(*Aabb*) $= 1/2 \times 1/4 = 1/8$. Adding these probabilities gives us $1/16 + 1/16 + 1/8 = 4/16$.

b. The dark-sooty phenotype arises from the combination *aaBb*. The probability of this combination is $1/4 \times 1/2 = 1/8$.

c. The red phenotype arises from the combination of *BB* with a dominant allele at the *A* locus, that is, genotypes *AABB* and *AaBB*. The probability of *AABB* $= 1/4 \times 1/4 = 1/16$; p(*AaBB*) $= 1/4 \times 1/2 = 1/8$; therefore, p(red) $= 1/16 + 1/8 = 3/16$.

d. The only remaining phenotype is sooty. With 4/16 of the progeny black, 1/8 dark-sooty, and 3/16 red, the remaining progeny (7/16) would be expected to be sooty. This gives sooty, black, red, and dark-sooty individuals in a ratio of 7:4:3:2.

9-15. **a.** *AaBb* × *aabb* produces four genotypes: *AaBb*, *Aabb*, *aaBb*, and *aabb*, with phenotypes of sooty, black, dark-sooty, and black, respectively. Thus, the probability of producing black offspring is 1/2 and of producing red offspring is 0.

b. *aaBB* × *AAbb* produces one genotype, *AaBb*, which has a sooty phenotype. The probability of producing red or black offspring is 0.

9-16. A recessive genotype at the *G* locus, combined with *HH* or *Hh*, blocks the expression of the *H* genes, with the phenotype determined by *gg* (medium-gray). Otherwise, the genotype at the *H* locus blocks the expression of the *G* locus genotype and determines the phenotype.

9-17. **a.** Since the three loci are independently assorting, the inheritance of each can be considered separately and probabilities can be combined using the product law. The cross *cc* × *Cc* produces genotypes *cc* and *Cc* with expected frequencies of 1/2 each. The cross *BB* × *bb* produces genotypes *BB*, *Bb*, and *bb*, with frequencies of 1/4, 1/2, and 1/4, respectively. The cross *Aa* × *AA* produces genotypes *AA* and *Aa* with frequencies of 1/2 each. These genotypes and their frequencies are combined in the branch diagram in Table 9-17a. Expected phenotypic outcome: 1/2 albino, 3/8 black agouti, and 1/8 brown agouti.

b. Since the albino is derived from the F_2 of the cross worked through in answering problem 9-17a, its genotype must be one of the six listed in the answer to 9-17a. From their phenotypes alone, we know that the albino parent has alleles *cc* at the *C* locus and that the black agouti parent has at least one dominant allele at each of the three loci: *C_B_A_*. The albino progeny of this cross must have genotype *cc* and this tells us that each parent must carry at least one *c* allele. Similarly, the brown progeny (genotype *bb*) and the nonagouti progeny (genotype *aa*) tell us that each parent has at least one copy of both allele *b* and allele *a*. Thus, we know that the genotype of the black agouti parent is *CcBbAa* and that the genotype of the albino parent is *cc_b_a_*. Of the six possible albino genotypes listed in the answer to 9-17a, we can eliminate four, leaving genotypes *ccBbAa* and *ccbbAa* as possibilities.

c. The albino's genotype is either *ccBbAa* or *ccbbAa*. Testcrossing it with any mouse that is homozygous for the recessive allele at the *B* locus (for example, brown or brown agouti) should reveal its genotype. If all testcross progeny are brown, there is support for the *bb* genotype; if the progeny are black and brown, its genotype is *Bb*. Note that if all the progeny are brown, it would be wise to repeat this mating a second or even third time to provide a large sample of progeny upon which to base the identification of the albino's genotype as *bb*.

9-18. **a.** In the order listed in the problem, the phenotypes

TABLE 9-17a

C-locus genotypes	*B*-locus genotypes	*A*-locus genotypes	Expected Progeny	
			Genotypes	Phenotypes
cc (1/2)	*BB* (1/4)	*AA* (1/2) =	*ccBBAA* (1/16)	Albino
		Aa (1/2) =	*ccBBAa* (1/16)	Albino
	Bb (1/2)	*AA* (1/2) =	*ccBbAA* (1/8)	Albino
		Aa (1/2) =	*ccBbAa* (1/8)	Albino
	bb (1/4)	*AA* (1/2) =	*ccbbAA* (1/16)	Albino
		Aa (1/2) =	*ccbbAa* (1/16)	Albino
Cc (1/2)	*BB* (1/4)	*AA* (1/2) =	*CcBBAA* (1/16)	Black agouti
		Aa (1/2) =	*CcBBAa* (1/16)	Black agouti
	Bb (1/2)	*AA* (1/2) =	*CcBbAA* (1/8)	Black agouti
		Aa (1/2) =	*CcBbAa* (1/8)	Black agouti
	bb (1/4)	*AA* (1/2) =	*CcbbAA* (1/16)	Brown agouti
		Aa (1/2) =	*CcbbAa* (1/16)	Brown agouti

are brown, nonspotted, normal intensity; albino; and black agouti, spotted, normal intensity.

b. All progeny would carry genotype *cc* and thus would be albino.

c. Since the loci are independently assorting, the inheritance of each can be considered separately and probabilities can be combined using the product law. The cross *cc* × *CC* produces the pigmented phenotype with an expected frequency of 1. The cross *bb* × *Bb* produces black and brown with frequencies of 1/2 each. The cross *Aa* × *Aa* produces agouti and nonagouti with frequencies of 3/4 and 1/4, respectively. The cross *Ss* × *ss* produces nonspotted and spotted with frequencies of 1/2 each. The cross *DD* × *dd* produces milky with an expected frequency of 1. These phenotypes and their frequencies are combined in the branch diagram in Table 9-18c. Since each of the progeny will be pigmented (genotype *Cc*) and milky (genotype *Dd*), we need not include these loci in our branch diagram as long as we include their contributions when we list the progeny phenotypes.

Chapter 10: Sex Chromosomes, Sex-Chromosome Systems, and Sex Linkage

10-1. Key: *r* is the allele for red weakness; *R* is the allele for normal vision. Cross: $X^rX^r \times X^RY$.

Female gametes	Male gametes	Progeny genotypes
X^r	X^R =	X^RX^r
	Y =	X^rY

Progeny: 50% heterozygous (carrier) females with normal vision; 50% hemizygous males with red weakness.

10-2. a. Key: m^+ is the dominant allele for normal wings; *m* is the recessive allele for miniature wings. Cross: $X^mX^m \times X^{m^+}Y$.

Female gametes	Male gametes	Progeny genotypes
X^m	X^{m^+} =	$X^{m^+}X^m$
	Y =	X^mY

Progeny: 50% heterozygous (carrier) females with normal wings; 50% hemizygous males with miniature wings.

TABLE 9-18c

B-locus phenotypes	*A*-locus phenotypes	*C*-locus phenotypes		Expected progeny phenotypes and frequencies
Black (1/2)	Agouti (3/4)	Nonspotted (1/2)	=	Black, agouti, nonspotted, milky (3/16)
		Spotted (1/2)	=	Black, agouti, spotted, milky (3/16)
	Nonagouti (1/4)	Nonspotted (1/2)	=	Black, nonspotted, milky (1/16)
		Spotted (1/2)	=	Black, spotted, milky, (1/16)
Brown (1/2)	Agouti (3/4)	Nonspotted (1/2)	=	Brown, agouti, nonspotted, milky (3/16)
		Spotted (1/2)	=	Brown, agouti, spotted, milky (3/16)
	Nonagouti (1/4)	Nonspotted (1/2)	=	Brown, nonspotted, milky (1/16)
		Spotted (1/2)	=	Brown, spotted, milky, (1/16)

b. Cross: F_1 female × father, or $X^{m+}X^{m} \times X^{m+}Y$.

Male gametes	Female gametes		Progeny genotypes
X^{m+} (1/2)	X^{m+} (1/2)	=	$X^{m+}X^{m+}$ (1/4)
	X^{m} (1/2)	=	$X^{m+}X^{m}$ (1/4)
Y (1/2)	X^{m+} (1/2)	=	$X^{m+}Y$ (1/4)
	X^{m} (1/2)	=	$X^{m}Y$ (1/4)

Progeny: 1/2 normal-winged females, 1/4 normal-winged males, 1/4 miniature-winged males.

c. Cross: $X^{m+}X^{m} \times X^{m}Y$.

Female gametes	Male gametes		Progeny genotypes
X^{m+} (1/2)	X^{m} (1/2)	=	$X^{m+}X^{m}$ (1/4)
	Y (1/2)	=	$X^{m+}Y$ (1/4)
X^{m} (1/2)	X^{m} (1/2)	=	$X^{m}X^{m}$ (1/4)
	Y (1/2)	=	$X^{m}Y$ (1/4)

Progeny: 1/4 normal-winged females, 1/4 miniature-winged females, 1/4 normal-winged males, and 1/4 miniature-winged males.

10-3. a. Key: *B* is the allele for black coat; *B'* is the allele for yellow coat. Cross: $X^{B}X^{B'} \times X^{B'}Y$.

Female gametes	Male gametes		Progeny genotypes
X^{B} (1/2)	$X^{B'}$ (1/2)	=	$X^{B}X^{B'}$ (1/4)
	Y (1/2)	=	$X^{B}Y$ (1/4)
$X^{B'}$ (1/2)	$X^{B'}$ (1/2)	=	$X^{B'}X^{B'}$ (1/4)
	Y (1/2)	=	$X^{B'}Y$ (1/2)

Progeny: 1/4 tortoise-shell females, 1/4 hemizygous black males, 1/4 homozygous yellow females, 1/4 hemizygous yellow males.

b. Cross: $X^{B'}X^{B'} \times X^{B}Y$.

Female gametes	Male gametes		Progeny genotypes
$X^{B'}$ (1)	X^{B} (1/2)	=	$X^{B}X^{B'}$ (1/2)
	Y (1/2)	=	$X^{B'}Y$ (1/2)

Progeny: 1/2 tortoise-shell females, 1/2 hemizygous yellow males.

c. No. Both B and B' alleles must be present to produce the tortoise-shell phenotype. Since the alleles involved are carried on the X chromosome and the male has a single X chromosome, no males will exhibit the tortoise-shell phenotype.

10-4. Cross: $X^BX^{B'} \times X^?Y$. Progeny: Among the males, there are equal numbers of X^BY and $X^{B'}Y$; among the females, there are equal numbers of $X^BX^{B'}$ and X^BX^B. One possible approach to identify the color of the father is to place the parental gametes, to the extent that they can be identified, in a Punnett square. Then place the genotypes of the progeny in the square, taking care to position the female offspring across from the male gamete carrying the X chromosome and the male offspring across from the male gamete carrying the Y chromosome, as follows.

	X^B	$X^{B'}$
$X^?$	X^BX^B	$X^BX^{B'}$
Y	X^BY	$X^{B'}Y$

Since the black females must have a genotype of X^BX^B, and since each parent must have contributed an X^B, the tomcat must be X^BY and have a black coat. Union of the X^B gamete from the tomcat with the $X^{B'}$ gamete from the mother produces the tortoise-shell females.

10-5. Key: h is the recessive allele for hemophilia, H is the dominant allele for normal blood clotting. Father of woman: X^hY. Mother of woman: X^HX^H (she is assumed to be homozygous for the normal allele since there is no history of hemophilia in her family). Woman: Since she is receiving an X chromosome from each parent, she must be X^HX^h. Her husband is X^HY.

Cross: $X^HX^h \times X^HY$.
Gametes: X^H (1/2), X^h (1/2), X^H (1/2), Y (1/2).

a. To be hemophiliac, a daughter must have two X^h chromosomes and thus must have received an X^h chromosome from each parent. The X^HX^h mother could contribute an X^h chromosome but the X^HY father could not. Consequently the probability of a hemophiliac daughter is 0.

b. A hemophiliac son must have genotype X^hY and thus must receive the mother's X^h chromosome. The probability of an egg carrying this chromosome is 1/2, and this egg must unite with a Y-carrying sperm, also produced with a probability of 1/2. Multiplying these probabilities gives the probability of a hemophiliac son: $1/2 \times 1/2 = 1/4$.

10-6. Cross: $X^HX^H \times X^hY$.

Female gametes	Male gametes		Progeny genotypes
X^H (1)	X^h (1/2)	=	X^HX^h (1/2)
	Y (1/2)	=	X^HY (1/2)

Progeny: None of the progeny is expected to be hemophiliac.

10-7. a. Moths exhibit the ZW sex-chromosome system: the males are ZZ and the females, ZW. The mating of a light male and dark female can be symbolized as $Z^lZ^l \times Z^LW$.

Male gametes	Female gametes		Progeny genotypes
Z^l (1)	Z^L (1/2)	=	Z^LZ^l (1/2)
	W (1/2)	=	Z^lW (1/2)

Progeny: All males are Z^LZ^l and, because the allele for dark wings is dominant, will have dark wings. The females are all Z^lW and will have light wings.

b. The mating of an F_1 male with an F_1 female can be symbolized as $Z^LZ^l \times Z^lW$.

Female gametes	Male gametes		Progeny genotypes
Z^l (1/2)	Z^L (1/2)	=	Z^LZ^l (1/4)
	Z^l (1/2)	=	Z^lZ^l (1/4)
W (1/2)	Z^L (1/2)	=	Z^LW (1/4)
	Z^l (1/2)	=	Z^lW (1/4)

Progeny: Half the F_2 males (Z^LZ^l) will have dark wings and half (Z^lZ^l) will have light wings. Half the F_2 females (Z^LW) will have dark wings and half (Z^lW) will have light wings.

c. A comparison of phenotypes of the male and the female parents and progeny for each cross is as follows.

Cross producing F_1:

Parents: light male × dark female
Progeny: dark males and light females

Cross producing F_2:

Parents: dark male × light female
Progeny: light and dark males, light and dark females

The crisscross inheritance pattern is shown by the cross producing the F_1: the phenotype of the male parent appears in the female progeny while that of the female parent is exhibited by the male progeny.

10-8. a. With this locus carried on both the X and Y chromosomes, the mating is $X^{bb^+}Y^{bb^+} \times X^{bb}X^{bb}$.

Female gametes	Male gametes	Progeny genotypes
X^{bb} (1)	X^{bb^+} (1/2)	= $X^{bb^+}X^{bb}$ (1/2)
	Y^{bb^+} (1/2)	= $X^{bb}Y^{bb^+}$ (1/2)

Progeny: All will have normal bristles: half are heterozygous females ($X^{bb^+}X^{bb}$) and half are heterozygous males ($X^{bb}Y^{bb^+}$). The outcome is the same as would be expected if the locus were carried on a pair of autosomes.

b. The reciprocal of the cross in 10-8a is $X^{bb}Y^{bb} \times X^{bb^+}X^{bb^+}$.

Female gametes	Male gametes	Progeny genotypes
X^{bb^+} (1)	X^{bb} (1/2)	= $X^{bb^+}X^{bb}$ (1/2)
	Y^{bb} (1/2)	= $X^{bb^+}Y^{bb}$ (1/2)

Progeny: Same as the outcome in 10-8a. Identical outcomes for reciprocal crosses are expected for autosomal loci and for loci present on *both* the X and Y chromosomes. With X-linked loci, the outcomes of reciprocal crosses differ.

c. Cross: $X^{bb^+}X^{bb} \times X^{bb}Y^{bb^+}$.

Female gametes	Male gametes	Progeny genotypes	Progeny phenotypes
X^{bb^+} (1/2)	X^{bb} (1/2)	= $X^{bb^+}X^{bb}$ (1/4)	Normal female
	Y^{bb^+} (1/2)	= $X^{bb^+}Y^{bb^+}$ (1/4)	Normal male
X^{bb} (1/2)	X^{bb} (1/2)	= $X^{bb}X^{bb}$ (1/4)	Bobbed female
	Y^{bb^+} (1/2)	= $X^{bb}Y^{bb^+}$ (1/4)	Normal male

The outcome would differ if the locus were carried on a pair of autosomes: both the normal and bobbed groups would consist of equal numbers of males and females. With the locus carried on the sex chromosomes, the bobbed flies are all female and the males predominate, two to one, among the normal flies.

10-9. Key: T = dominant autosomal allele for normal color vision;
t = recessive autosomal allele for total color blindness;
G = dominant sex-linked allele for normal vision;
g = recessive sex-linked allele for green weakness color blindness.

Genotypes: Woman: TtX^gX^g; man: TtX^GY. Since the two traits are independently assorting, their inheritance can be considered separately and the product law can be used to combine probabilities. In the mating $Tt \times Tt$, the expected probabilities of normal vision and total color blindness among the progeny are 3/4 and 1/4, respectively. In the mating $X^gX^g \times X^GY$, the progeny fall into two groups: green color blind males and normal vision females, each with expected probability of 1/2. These probabilities are combined in the branch diagram in Table 10-9. Progeny: 3/8 will be green color blind males, 3/8 normal vision females, 1/8 totally color blind males (the alleles for green weakness will not separately manifest themselves because of the total color blindness), and 1/8 totally color blind males.

10-10. Begin by writing out, to the extent possible, the genotypes of the progeny. Since both traits are sex-linked, each X chromosome carries an allele for eye color *and* an allele for body color.

Progeny	Phenotypes	Progeny genotypes
Females:	1/2 red eyes, yellow body	$X^{w^+y}X^{?y}$
	1/2 white eyes, normal body	$X^{wy^+}X^{w?}$
Males:	1/2 red eyes, yellow body	$X^{w^+y}Y$
	1/2 white eyes, normal body	$X^{wy^+}Y$

Inspection of the genotypes of the progeny gives us some information about the genotypes of the parents. Begin by looking at the male progeny. The X chromosomes carried by both types of males must have come from the female parent. Thus, the X chromosomes carried by the female parent must have been X^{w^+y} and X^{wy^+}, giving her a genotype of

TABLE 10-9

Normal vision (3/4)	Male, green color blind (1/2)	= Male, green color blind (3/8)
	Female, normal vision (1/2)	= Female, normal vision (3/8)
Totally color blind (1/4)	Male, green color blind (1/2)	= Male, totally color blind (1/8)
	Female, normal vision (1/2)	= Female, totally color blind (1/8)

$X^{w^+y}X^{wy^+}$. Now inspect the female progeny. One of the X chromosomes carried by each female offspring was contributed by the female parent. Since the genotype of the female parent is already established, the X chromosomes contributed by this parent can be identified. The other X chromosome carried by each type of female offspring, that is, $X^{?y}$ and $X^{w?}$, must have come from the male parent. The male parent has but a single X chromosome to contribute, and thus the paternal X chromosomes carried by each daughter must be identical. From one type of female offspring, we know this X carries the *y* allele, and from the other type of female offspring, we know it carries the *w* allele. Thus, the genotype of the male parent is $X^{wy}Y$.

10-11. **a.** The daughter must be MmX^iX^i.

b. With regard to the cleft-iris trait, one of the X chromosomes carried by the daughter must have come from the father. Thus, the father must be X^iY. The migraine-free mother must have the genotype of *mm*. Since the daughter suffers from migraine headaches, the daughter must have inherited the dominant allele for migraine from her father. Thus, the father must have the *M* allele. Based on the information provided here, we cannot tell whether he is homozygous or heterozygous for this condition.

c. We are told that the mother is migraine-free. Thus, her genotype for this trait is *mm*. One of the X^i chromosomes carried by the daughter was contributed by the mother, and thus the mother carries the allele for cleft iris; we do not know whether the mother is homozygous or heterozygous for this allele.

10-12. **a.** Chickens exhibit the ZW sex-chromosome system; the females are heterogametic, that is, ZW, and the males are homogametic, that is, ZZ. Cross: $Z^gZ^g \times Z^GW$.

Male gametes	Female gametes		Progeny genotypes	Progeny phenotypes
Z^g	Z^G	=	Z^GZ^g	Males: silver
	W	=	Z^gW	Females: gold

b. Cross: $Z^gW \times Z^GZ^g$.

Female gametes	Male gametes		Progeny genotypes	Progeny phenotypes
Z^g	Z^G	=	Z^GZ^g	Males: 1/2 silver, 1/2 gold
	Z^g	=	Z^gZ^g	
W	Z^G	=	Z^GW	Females: 1/2 silver, 1/2 gold
	Z^g	=	Z^gW	

10-13. **a.** Cross: $Z^gW \times Z^GZ^G$.

Male gametes	Female gametes		Progeny genotypes	Progeny phenotypes
Z^g	Z^g	=	Z^GZ^g	Males: all silver-plumed
	W	=	Z^GW	Females: all silver-plumed

b. Cross: $Z^GW \times Z^GZ^g$.

Female gametes	Male gametes		Progeny genotypes	Progeny phenotypes
Z^G	Z^G	=	Z^GZ^G	Males: all silver, (1/2 homozygous, 1/2 heterozygous)
	Z^g	=	Z^GZ^g	
W	Z^G	=	Z^GW	Females: 1/2 silver, 1/2 gold
	Z^g	=	Z^gW	

10-14. Since affected individuals die during the teenage years and since both the mother and father have reached adulthood, we can assume that neither is affected. Since the son is affected, his X chromosome, which came from his mother, must carry the *d* allele: X^d. Thus, the carrier mother is not affected because her other X chromosome carries the dominant normal allele; her genotype is X^DX^d. Since the father is phenotypically normal, his single X chromosome must carry the normal-type allele; his genotype is X^DY. Thus, the mating is $X^DX^d \times X^DY$. Gamete union and the expected progeny are shown in the branch diagram that follows.

Male gametes	Female gametes		Progeny genotypes	Progeny phenotypes
X^D	X^D	=	X^DX^D	Females: all normal phenotype
	X^d	=	X^DX^d	
Y	X^D	=	X^DY	Males: 1/2 normal, 1/2 affected
	X^d	=	X^dY	

The chance that the other son will develop the disorder is 1/2. There is no chance that their daughters will be affected.

10-15. **a.** The original mating can be symbolized as $X^BY \times X^bX^b$. All daughters will be X^BX^b. The marriage of the daughter to a normal man can be symbolized as $X^BX^b \times X^bY$. The probability that one of their sons will have brown tooth enamel is 1/2.

b. The probability that a daughter from the $X^BX^b \times X^bY$ mating will have normal tooth enamel is 1/2.

10-16. Since the gene for antigen production is carried only on the Y chromosome, antigen production can occur only in males. The affected male will transmit this gene to all of his male progeny—all of whom will make the antigen—but to none of his female progeny.

10-17. a. Cross: $X^{bb^+}X^{bbl} \times X^{bbl}Y^{bb^+}$.

Female gametes	Male gametes	Progeny genotypes	Progeny phenotypes
X^{bb^+} (1/2)	X^{bbl} (1/2)	$= X^{bb^+}X^{bbl}$ (1/4)	Normal (heterozygous) females
	Y^{bb^+} (1/2)	$= X^{bb^+}Y^{bb^+}$ (1/4)	Normal (homozygous) males
X^{bbl} (1/2)	X^{bbl} (1/2)	$= X^{bbl}X^{bbl}$ (1/4)	Dead females
	Y^{bb^+} (1/2)	$= X^{bbl}Y^{bb^+}$ (1/4)	Normal (heterozygous) males

Outcome: All the living progeny would be normal. The sex ratio would be one female to two males.

b. Yes. The lethal allele would be transmitted to male progeny rather than female progeny and consequently half of the male zygotes would fail to develop. The phenotypes of the living progeny would be the same (all would have normal bristles), but the sex ratio would be changed to one male to two females.

Chapter 11: Human Pedigree Analysis

11-1. a. Since she has the normal phenotype, she lacks the dominant allele and must be homozygous recessive: *dd*.

b. His affected phenotype indicates he would have at least one copy of the dominant allele and therefore have either genotype *DD* or *Dd*. Since both II-2 and II-5 must be *dd*, each must have received a recessive allele from each parent and the father must be *Dd*.

c. The expected children from the mating *Dd* × *dd* would be *Dd* (affected) and *dd* (normal) in a 1:1 ratio.

d. Yes.

11-2. a. Based on her normal phenotype she must be either *DD* or *Dd*. Since all of her sons are affected and thus have genotype *dd*, she must have contributed a *d* allele to each of them. Consequently she must be heterozygous, *Dd*.

b. His normal phenotype indicates the absence of the dominant allele. He must have genotype *dd*.

c. *dd*

d. The progeny from the mating *Dd* × *dd* would be *Dd* (normal) and *dd* (affected) in a 1:1 ratio.

e. Yes.

11-3. If the allele is recessive, the pedigree requires that *both* mother and father possess the allele for the trait. If it is dominant, only the father requires a copy. Because of the allele's rarity, the chance that two unrelated individuals will possess it is very small. In light of this, it is more reasonable to support the hypothesis that the allele is dominant.

11-4. See the answer to problem 11-3. The chance that two *related* individuals carry the same allele is considerably higher than it would be for a single individual from the population at large to have the allele. Under these circumstances it is more reasonable to support the hypothesis that the allele is recessive.

11-5. Yes, but only if the trait is due to a sex-linked recessive allele *and* the mother is a carrier *and* she transmits her *d*-carrying X chromosome to her three sons *and* her *D*-carrying X chromosome to her two daughters. Note that this situation is unlikely unless the two parents are cousins.

11-6. Since the parents are normal, each could have a genotype of either *EE* or *Ee*. Since they have affected children who must have genotype *ee*, and since each parent must have contributed an *e* allele to these children, each parent must thus have genotype *Ee*. (Further support for *Ee* genotypes for III-7 and III-8 comes from the fact that each had a parent who was *ee*.) Their mating is *Ee* × *Ee* and expected progeny would have genotypes *EE*, *Ee*, and *ee* in a 1:2:1 ratio. The chance of producing an affected child, *ee*, in the next pregnancy is 1/4.

11-7. III-7 married his first cousin, III-8, and since each has the *Ee* genotype (see answer to the problem 11-6) they could produce *ee* children. Male III-2 married a woman from outside the family who, because of the rarity of the recessive allele, can be assumed to be homozygous for the dominant allele, *EE*, rather than heterozygous. Male III-2, like his brother had an affected parent and is heterozygous, *Ee*, but since his wife is *EE*, III-2's children will have genotypes *EE* or *Ee* and all will be normal.

11-8. Since III-10 is normal, he may have either genotype *EE* or *Ee*. His affected mother must have genotype *ee* and must have contributed an *e* allele to him. Thus, he must have genotype *Ee*. The woman he may marry is affected and thus must have genotype *ee*. The mating is *Ee* × *ee* and their expected children will have genotypes *Ee* and *ee* in a 1:1 ratio. Thus, the chance that any given child will be affected is 1/2.

11-9. a. Since the trait skips the first generation but appears in the second and third, we could tentatively conclude that the mutant allele is recessive. (Note: this explanation requires that individual II-1 be heterozygous for the recessive allele.)

b. Provided that the allele responsible for the trait is recessive, the woman's genotype could be either homozygous dominant or heterozygous. One of the woman's daughters, III-1, expresses the trait and thus must carry two alleles for it. One of these alleles must have come from her mother, who, therefore, must be heterozygous.

c. Based on phenotype alone, each of the original parents could be either homozygous dominant or heterozygous. Their child, II-2, expresses the trait and thus must have two alleles for it. Since one of these alleles was contributed by each parent, each must possess the allele for the trait and each is heterozygous.

d. I-2's X chromosome has the normal allele and I-1 must be a carrier. Note that the trait is not likely to be sex-linked since II-1 would also have to be a carrier in order to produce affected daughter, III-1; this is not a likely occurrence if the allele is rare.

11-10. a. Both I-1 and I-2 would carry at least one recessive allele since each must have contributed this allele to produce their affected progeny. In addition, at least one of them would have to carry the dominant allele in order to produce normal progeny. Thus, one parent must be heterozygous and the other could be either heterozygous or homozygous recessive.

b. The affected progeny must have at least one copy of the dominant allele and thus at least one of the parents must carry this allele. In addition, the normal progeny must have two recessive (normal) alleles. Since one of these alleles came from each parent, each parent must carry at least one recessive allele. Thus, one of the parents must be heterozygous and the other could be either heterozygous or homozygous recessive.

c. Yes. Since both II-2 and II-3 are affected, both would be homozygous recessive for the allele and all their progeny would be homozygous recessive.

d. We cannot say with certainty. Two genotypic combinations are possible for the mating of II-2 and II-3: (1) homozygous dominant × heterozygous results in all dominant progeny, and (2) heterozygous × heterozygous results in 3/4 dominant and 1/4 recessive progeny.

11-11. The allele appears to be recessive. The pedigree shows that (1) the trait skipped generation II (a dominant allele would not skip a generation), (2) most of the affected individuals are male (a dominant allele would affect approximately equal numbers of males and females), and (3) affected males come from carrier mothers (a dominant allele would show affected males coming from affected mothers).

11-12. a. Yes. The affected son, III-3, tells us that II-1 must have been a carrier.

b. Since she produced an affected son, IV-6, III-12 must have been a carrier. This conclusion is further supported by the presence of the affected daughter, who must have received one of her two alleles for this trait from her mother.

c. The mother is a carrier (see the answer to 11-12b) and the father is affected; the mating could be shown as X*X × X*Y, where the asterisk represents the recessive allele responsible for the trait. The expected progeny would be X*X*, X*X, X*Y, and XY in a 1:1:1:1 ratio. Half the expected progeny, X*X* and X*Y, would be affected. Thus, the chance of the fifth child being affected is 1/2.

d. Carrier females make up 1/4 of the expected progeny (see the answer to 11-12c). The chance that a sixth child will be a carrier female is 1/4.

11-13. a. The mating of IV-8 and a normal male could be shown as X*X* × XY, where the asterisk represents the recessive allele for the trait. All the sons would be affected and all the daughters would be carriers with the normal phenotype. The chance that a son will be affected is 1.

b. Although she has four normal children, one of whom is a son, the woman could still be a carrier of the allele. The odds of producing a normal son in a mating between a carrier woman and a normal male is 1/2.

11-14. It is unlikely, since the allele is very rare. For example, say that in the population at large, one in 1000 individuals is heterozygous. This means that the chance that individual I-1 or II-1 or II-10 is heterozygous is 1/1000. In order for this pedigree to be due to an autosomal recessive allele, all three of these individuals would have to be heterozygous for this allele. The probability of that occurring is given by the product law: 1/1000 × 1/1000 × 1/1000 = 1/1,000,000,000. If, in contrast, the allele were relatively common, such that, say, one in two individuals is heterozygous, the probability of getting this pedigree is 1/2 × 1/2 × 1/2 = 1/8, a distinct possibility.

11-15. Yes. Autosomal dominant inheritance is characterized by the occurrence of the trait in each generation, a nearly equal distribution of the trait among males and females, and the production of affected and normal offspring in a 1:1 ratio from affected × normal matings. The pedigree shows all of these characteristics. Expanding the pedigree, through the addition of another generation for example, might allow us to be more precise in identifying the type of allele responsible for the trait.

11-16. The pedigree provides support for the dominance of the allele. It shows that (1) the trait occurs in each of the first four generations, and its absence from generation V can be attributed to the fact that there are only two individuals in that generation (with a recessive allele the trait may skip generations); (2) there are normal × affected matings that produce both normal and affected in approximately the expected 1:1 ratio (with a recessive allele, this type of mating would normally give rise to all normal progeny); (3) there are affected × affected matings that produce some normal progeny (with a recessive allele, all progeny would be affected).

11-17. a. Based on her affected progeny, IV-2 could be homozygous dominant (*GG*) or heterozygous (*Gg*). Her children are normal and thus homozygous for the recessive allele (*gg*). Since the mother must have contributed a recessive allele to each child, the mother must be heterozygous: *Gg*.

b. IV-1 is normal and must have genotype *gg*. IV-2 must have genotype *Gg* (see the answer to 11-17a). Thus, the mating is *Gg* × *gg* and the expected progeny will be *Gg* (affected) and *gg* (normal) in a 1:1 ratio. The chance of producing an affected child is 1/2. The chance of producing a male is 1/2. The chance of a child being both affected *and* male is given by the product of the separate probabilities: 1/2 × 1/2 = 1/4.

c. She has the normal phenotype and thus does not possess the allele for the trait. If she marries a normal man, the chance of their having an affected child is 0.

11-18. a. IV-5's father, III-4, is affected and based on this phenotype could be homozygous dominant (*GG*) or heterozygous (*Gg*). The father's mother, II-4, was normal and thus homozygous for the normal recessive allele. She must have contributed a recessive allele to III-4 who therefore must be heterozygous. Because of the rarity of the allele for the trait, III-4's affected wife, III-3, is assumed to be heterozygous. Thus, the mating of IV-5's parents can be symbolized as *Gg* × *Gg* and their expected progeny will have genotypes *GG*, *Gg*, and *gg* in a 1:2:1 ratio. Genotypes *GG* and *Gg* are affected and one out of every three *affected* children will be homozygous for the dominant allele. Since IV-5 is affected, the chance that he is homozygous dominant is 1/3.

b. They could have normal children only if IV-5 is heterozygous. The chance of that is 2/3 (see the answer to 11-18a). If he is heterozygous, the mating can be written as *gg* × *Gg*. Expected progeny would be *gg* (normal) and *Gg* (affected) in a 1:1 ratio. The chance of a normal child is 1/2. The chance that IV-5 is heterozygous *and* will have a normal child is given by the product of these probabilities: 2/3 × 1/2 = 2/6 = 1/3.

11-19. It is unlikely, since it requires that II-5, II-7, and II-9 all be heterozygous. In addition, the pedigree shows that the only affected individuals are males. (With an autosomal allele, affected males and females would be found in approximately equal numbers). Furthermore, the pedigree shows that carrier female × normal male matings give rise to progeny which are in accord with the expected outcome of this type of mating: 1/2 males are affected and 1/2 males are normal, all the females are normal. (With an autosomal allele, all the generation-II individuals would be carriers; the mating of a normal individual, assumed to be homozygous because of the rarity of the recessive allele, and a carrier, that is, a heterozygote, would give all normal progeny).

11-20. a. The mother, II-10, is a carrier (since her father, I-2, was affected) and the father, II-9, is normal. The mating could be written as X*X × XY, where the asterisk represents the recessive allele. Expected progeny: 1/4 normal males, 1/4 affected males, and 1/2 normal females. The chance of producing a normal child, regardless of sex, is 1/4 + 1/2 = 3/4.

b. See the answer to 11-20a. Since all females from this mating will be normal, the chance is 1.

c. See the answer to 11-20a. Since half the males from this mating will be normal, the chance is 1/2.

Chapter 12: Linkage, Crossing Over, and the Two-Point Testcross

12-1. a. At the end of the first meiotic division, the result is the following.

C D
C d

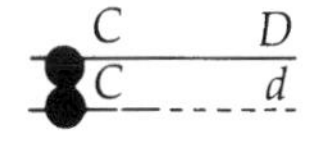

(Note that each double-stranded chromosome generally occurs in a different nucleus.) At the end of the second meiotic division, the result is the following.

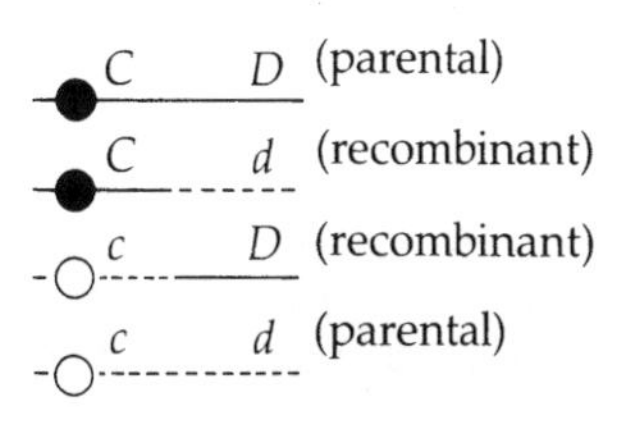

(Note that each single-stranded chromosome generally occurs in a different nucleus.)

b. *CD*, coupling; *Cd*, repulsion; *cD*, repulsion; and *cd*, coupling.

12-2. No. Recombination can be detected only when a crossover occurs *between* the loci being considered. Crossovers outside this region will produce recombinations but, since they do not involve our marker genes, they will not be detected in a *CcDd* × *ccdd* testcross.

12-3. With coupling, the parentals are *CD* and *cd*, and the recombinants are *Cd* and *cD*. With repulsion, the parentals are *Cd* and *cD*, and the recombinants are *CD* and *cd*.

12-4. a. Only two of the four tetrad strands are involved in each crossover event. With 18% of the tetrads showing crossing over, 9% of the product chromosomes will be recombinants and 9% will be nonrecombinants. Combining these nonrecombinants with the 82% nonrecombinants arising from the noncrossover tetrads gives a total of 9 + 82 = 91% nonrecombinants.

b. The percentage of tetrads showing crossing over is twice the percentage of the recombinant chromosomes.

c. The frequency of recombinants equals half the frequency of crossing over because only two of the four chromatids in each tetrad participate in a crossover event and become recombinants. The remaining strands retain the parental combinations.

12-5. a. The greater the distance separating two loci, the greater the frequency of crossing over between the two loci.

b. With the probability of crossing over equal to 1, half the resultant chromosomes would be recombinants and half would be nonrecombinants. Since there are two types of recombinants and two types of nonrecombinants, the gametic ratio would be 1:1:1:1.

12-6. a. All the chromosomes (or gametes) that result would be noncrossover types.

b. No. The second crossover in the *c*-to-*d* region cancels out the effects of the first; that is, the *Cd* and *cD* crossover types that arise from the first crossover are converted back to *CD* and *cd* by the second crossover. So, although two crossover events have occurred, they go undetected since no recombinant strands can be detected.

c. Since each crossover is an independent event, their separate probabilities are combined using the product law: 0.15 × 0.15 = 0.0225, or 2.25%.

12-7. a. Half the chromosomes (or gametes) would be crossover types and half would be noncrossover types.

b. A third crossover in the *c*-to-*d* region restores the strands to the arrangement seen following the first crossover. The presence of recombinant strands verifies that crossing over has occurred, but does not tell us the specific number of crossover events that took place. The best we can do is to state that an odd number of crossover events occurred.

c. Since each crossover is an independent event, their separate probabilities are combined using the product law: 0.15 × 0.15 × 0.15 = 0.003375, or 0.34%.

12-8. The P_1 mating crosses a gray-bodied, straight-winged fly, homozygous for each trait, with a black-bodied, curved-winged fly. This mating can be symbolized as: $b^+cu^+/b^+cu^+ \times bcu/bcu$. (Note that any crossing over that occurs between these loci will not be detectable since the alleles carried by each set of homologs are identical and the strands generated by crossing over would be identical to the precrossover strands. The F_1 progeny will all be b^+cu^+/bcu. The testcross mating of an F_1 fly with a homozygous recessive fly could be symbolized as: $b^+cu^+/bcu \times bcu/bcu$.

a. If there is no crossing over, the F_1 fly would produce two kinds of gametes: b^+cu^+ and bcu. The homozygous recessive fly would produce one kind of gamete: bcu. The following branch diagram shows the union of these gametes and the progeny genotypes and phenotypes.

Gametes		Genotypes	Phenotypes
bcu	b^+cu^+ =	b^+cu^+/bcu	Gray body, straight wings (wild-type)
	bcu =	bcu/bcu	Black body, curved wings

Two phenotypes would be produced: 50% gray-bodied with straight wings (wild-type) and 50% black-bodied with curved wings.

b. With some crossing over between these two loci, two additional types of gametes would be produced by the F_1 fly: b^+cu and bcu^+. The following branch diagram shows the union of gametes and the progeny formed.

Gametes		Genotypes	Phenotypes
bcu	b^+cu^+ =	b^+cu^+/bcu	Gray body, straight wings (wild-type)
	bcu =	bcu/bcu	Black body, curved wings
	b^+cu =	b^+cu/bcu	Gray body, curved wings
	bcu^+ =	bcu^+/bcu	Black body, straight wings

c. No. If crossing over occurred only some of the time, the phenotypes would not occur in equal frequencies because the four kinds of gametes produced by the F_1 parent would not be formed in equal numbers. The recombinant gametes, b^+cu and bcu^+, would be produced with lower frequency than the gametes showing the parental combinations and, consequently, the progeny produced from these gametes, b^+cu/bcu and bcu^+/bcu, would occur with a lower frequency.

12-9. a. The fact that two categories of progeny are relatively large and two are relatively small supports the hypothesis that these loci are linked. If there was no linkage, the four categories of progeny would be expected to occur with approximately equal frequencies. Support for this conclusion would be provided by carrying out a chi-square test. The deviations from the numbers expected under the hypothesis of independent assortment give a χ^2 value of 266.78. With three degrees of freedom, this value gives a p-value of less than 0.001. The deviation is considered highly significant, failing to support the hypothesis that the genes are independently assorting. (For details of the chi-square test, see Chapter 6.)

b. The recombinants are the two categories occurring with the lower frequencies: black color, normal wings (52 flies), and normal color, vestigial wings (48 flies). The total number of recombinants is 52 + 48 = 100 out of a total of 252 + 52 + 48 + 248 = 600 flies. The percentage of recombinants is 100/600 = 0.167, or 0.167 × 100 = 16.7%.

12-10. a. No. Since the number of plants in each of the four categories is nearly equal, there is support for the hypothesis that the two loci assort independently. Statistical support for this conclusion would be provided by carrying out a chi-square test. The deviations from the numbers expected under the hypothesis of independent assortment give a χ^2 value of 2.21. With three degrees of freedom, this χ^2 value gives a p-value between 0.50 and 0.70 and, since this is greater than 0.05, the deviation is considered not statistically significant and the data support the hypothesis of independent assorting loci. (For details of the chi-square test, see Chapter 6.) (There is a slight chance that the two loci could be linked but separated by more than 50 map units, leading to the 1:1:1:1 ratio.)

b. If linked, the percentage of recombination is 246/500 = 0.508, or 0.508 × 100 = 50.8%.

12-11. Recombinants make up 8% of the progeny. Since a crossover frequency of 1% equals one unit of map distance, the two loci are 8 map units apart.

12-12. a. The genotype of the female, homozygous for both red eye color and vestigial wings, is represented as cn^+vg/cn^+vg. The genotype of the male, homozygous for both cinnabar eyes and normal wings, is $cnvg^+/cnvg^+$. In producing the F_1, the parental flies each make one kind of gamete: the eggs carry cn^+vg and the sperm carry $cnvg^+$ and all the F_1 progeny will be $cn^+vg/cnvg^+$.

b. An F_1 female, $cn^+vg/cnvg^+$, would produce four kinds of gametes: two with parental combinations, cn^+vg and $cnvg^+$, and two with recombinants, cn^+vg^+ and $cnvg$. The proportion in which these gametes are produced can be predicted because we know the map distance separating the two loci. Since one map unit equals 1% crossover frequency, and two loci are separated by 67 − 57 = 10 units of map distance, the crossover frequency is 10%, and 10% of the gametes would be recombinants. With two kinds of recombinants, each would be expected to occur with a frequency of 5%. The remaining gametes (90%) would carry one of the two parental combinations, each type with an expected frequency of 45%.

c. The F_1 female $cn^+vg/cnvg^+$ is crossed with a male $cnvg/cnvg$. The following branch diagram shows the union of gametes from these parents.

Gametes			Progeny
	cn^+vg (45%)	=	$cn^+vg/cnvg$ (45%)
	$cnvg^+$ (45%)	=	$cnvg^+/cnvg$ (45%)
$cnvg$ (100%)			
	cn^+vg^+ (5%)	=	$cn^+vg^+/cnvg$ (5%)
	$cnvg$ (5%)	=	$cnvg/cnvg$ (5%)

12-13. In making a map, the sequence of the loci and the distance between them must be determined. Loci *d* and *f*, with a recombination frequency of 22%, are 22 map units apart. Loci *d* and *e*, with a recombination frequency of 6%, are six map units apart. And loci *e* and *f*, with a recombination frequency of 16%, are 16 map units apart. Begin the mapping with the two loci separated by the greatest distance, *d* and *f*, which are 22 units apart.

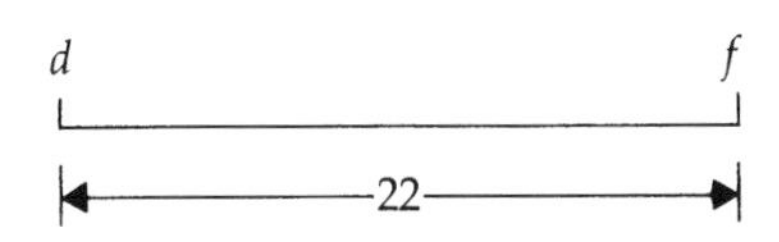

Two alternatives exist with regard to the placement of the *e* locus. It could be to the right of locus *d*

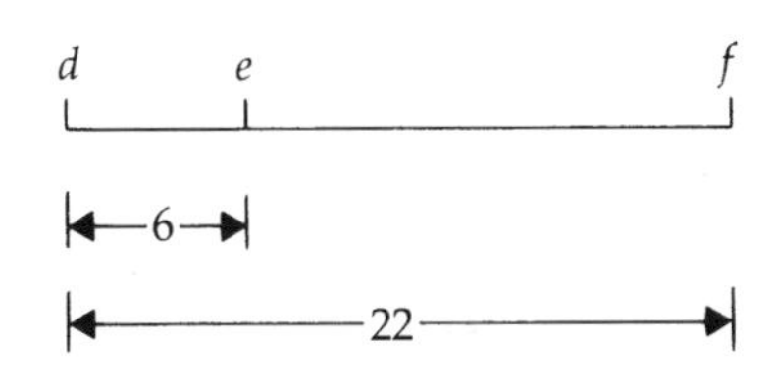

or to the left of locus *d*.

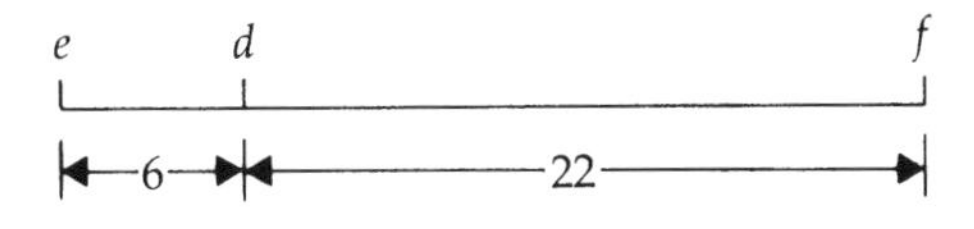

If locus *e* is to the left, the distance between *e* and *f* would be 28 units. This is not consistent with the data which tell us that loci *e* and *f* are 16 units apart. This *e*-to-*f* distance of 16 units, however, is consistent with *e* positioned to the right of *d*, so, the first alternative is correct.

12-14. The combination of black bodies and purple eyes could be generated by crossing over. The two loci are six map units apart and therefore the crossover frequency is 6%. Thus, 6% of the gametes are recombinants, equally divided into two categories, b^+p^+ and bp. Since the black-bodied, purple-eyed flies represent just one recombinant type, they would be expected to occur with a frequency of 3%.

12-15. a. Begin mapping with the two loci separated by the greatest distance: *t* and *u* are three map units apart. Since *s* and *t* are separated by two map units, *s* could be positioned either two map units to the right or to the left of *t*. Placing *s* two units to the right of *t* gives an *s*-to-*u* distance of one unit which is in accord with the 1% recombination frequency given for *s* and *u*.

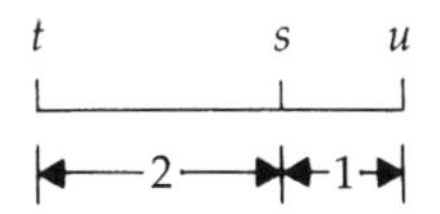

b. Loci *u* and *s* are 11 map units apart. Since *s* and *t* are separated by four units, *t* could be four map units either to the right or the left of *s*. Placing *t* four units to the left of *s* gives a *t*-to-*u* distance of seven units which matches the 7% recombination frequency given for *u* and *t*.

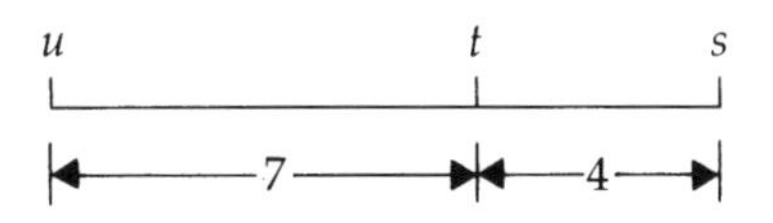

c. Loci *s* and *t* are 15 map units apart. Since *s* and *u* are separated by two map units, *u* could be two units to the right or to the left of *s*. Placing *u* two units to the right gives a *u*-to-*t* distance of 13 units which is in accord with the 13% recombination frequency given for *u* and *t*.

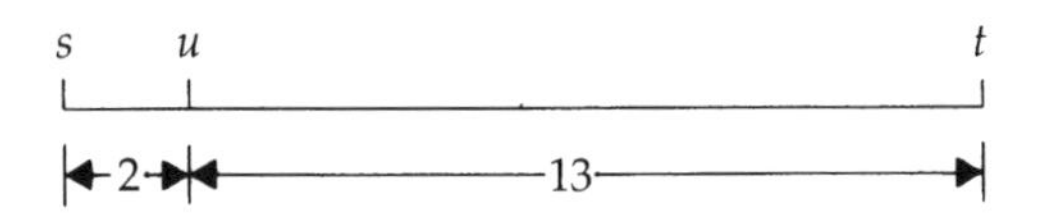

12-16. a. The linkage arrangement can be determined by examining the testcross progeny. The two most common classes of progeny, wild-type (114) and light eyes, vestigial wings (126), arise from the two most common types of gametes produced by the dihybrid parent, the parental allelic combinations. The alleles under consideration must thus show the coupling arrangement, with the two recessive alleles on one chromosome and the two dominant alleles on the other.

b. The two recombinant classes make up $20/260 = 0.077$, or $0.077 \times 100 = 7.7\%$ of the progeny. The recombination frequency is 7.7%.

c. The crossover frequency is two times the recombination frequency (since, with each crossover event, only two of the four product strands are recombinants): $2 \times 7.7\% = 15.4\%$.

12-17. With a crossover rate of 10.8%, the recombination frequency is 10.8%/2 = 5.4%. The expected number of recombinants is 5.4% × 298 = 16.09, or about 16. Since this is the *total* number of recombinants, eight are expected in each class. The remaining 298 − 16 = 282 progeny are nonrecombinants that fall into two classes, with 282/2 = 141 expected in each. Since the alleles carried by the dihybrid parent are in repulsion, m^+s and ms^+ are the parental allelic combinations, and the recombinant combinations are m^+s^+ and ms. Thus, the expected testcross outcome would be m^+s, 141; ms^+, 141; m^+s^+, 8; and ms, 8.

12-18. a. In the absence of crossing over, the male will produce two types of gametes in equal numbers, b^+p and bp^+.

b. The female, because of crossing over, will produce four types of gametes, two parental types, b^+p^+ and bp, with a frequency of 47% each, and two recombinant types, b^+p and bp^+, with a frequency of 3% each.

c. The progeny are summarized in the branch diagram in Table 12-18c. The 1st, 4th, 5th, and 7th genotypes, in the order listed in the branch diagram above, have at least one dominant allele at each loci and thus have normal body color and red eyes. This phenotype occurs with a frequency of 0.235 + 0.015 + 0.235 + 0.015 = 0.5. The 2nd and 3rd genotypes show at least one dominant allele at the b locus, are homozygous recessive at the p locus, and have normal body color and purple eyes. This phenotype occurs with a frequency of 0.235 + 0.015 = 0.25. The 6th and 8th genotypes are homozygous recessive at the b locus, have at least one dominant allele at the p locus, and exhibit black body color and red eyes. This phenotype occurs with a frequency of 0.235 + 0.015 = 0.25.

12-19. a. The two loci are 75 − 62 = 13 map units apart. The cross can be represented as $srcd^+/sr^+cd$ (female) × $srcd/srcd$ (male). The female will make four kinds of gametes: two parental types, $srcd^+$ and sr^+cd, each making up 43.5% of her gametes; and two recombinant types, sr^+cd^+ and $srcd$, each making up 6.5% of her gametes. The male will make a single type of gamete: $srcd$.

Male gamete	Female gametes		Progeny genotypes	Progeny phenotypes
$srcd$ (100%)	$srcd^+$ (43.5%)	=	$srcd^+/srcd$	Striped, wild-type eyes (43.5%)
	sr^+cd (43.5%)	=	$sr^+cd/srcd$	Wild-type body, cardinal eyes (43.5%)
	sr^+cd^+ (6.5%)	=	$sr^+cd^+/srcd$	Wild-type body and eyes (6.5%)
	$srcd$ (6.5%)	=	$srcd/srcd$	Striped, cardinal eyes (6.5%)

b. The cross can be represented as $srcd/srcd$ (female) × $srcd^+/sr^+cd$ (male). The female will make one kind of gametes: $srcd$ (100%). The male, because of the absence of crossing over, will make two types of gametes: $srcd^+$ (50%) and sr^+cd (50%).

Female gamete	Male gametes		Progeny genotypes	Progeny phenotypes
$srcd$ (100%)	$srcd^+$ (50%)	=	$srcd^+/srcd$	Striped, wild-type eyes (50%)
	sr^+cd (50%)	=	$sr^+cd/srcd$	Wild-type body, cardinal eyes (50%)

12-20. a. The cross can be represented as $srcd^+/sr^+cd$ (female) × $srcd^+/sr^+cd$ (male). The female will make four kinds of gametes: two parental types, $srcd^+$ and sr^+cd, each making up 43.5% of her gametes;

TABLE 12-18c

Male gametes	Female gametes		Progeny genotypes	Expected frequency	
b^+p (0.50)	b^+p^+ (0.47)	=	b^+p^+/b^+p	(0.50)(0.47) = 0.235	(1)
	bp (0.47)	=	bp/b^+p	(0.50)(0.47) = 0.235	(2)
	b^+p (0.03)	=	b^+p/b^+p	(0.50)(0.03) = 0.015	(3)
	bp^+ (0.03)	=	bp^+/b^+p	(0.50)(0.03) = 0.015	(4)
bp^+ (0.50)	b^+p^+ (0.47)	=	b^+p^+/bp^+	(0.50)(0.47) = 0.235	(5)
	bp (0.47)	=	bp/bp^+	(0.50)(0.47) = 0.235	(6)
	b^+p (0.03)	=	b^+p/bp^+	(0.50)(0.03) = 0.015	(7)
	bp^+ (0.03)	=	bp^+/bp^+	(0.50)(0.03) = 0.015	(8)

and two recombinant types, $srcd$ and sr^+cd^+, each making up 6.5% of her gametes. The male will make two types of gametes: $srcd^+$ (50%) and sr^+cd (50%).

Male gametes	Female gametes	Progeny genotypes	Progeny phenotypes
$srcd^+$ (50%)	$srcd^+$ (43.5%)	= $srcd^+/srcd^+$	Striped, wild-type eyes (21.8%)
	sr^+cd (43.5%)	= $sr^+cd/srcd^+$	Wild-type eyes and body (21.8%)
	$srcd$ (6.5%)	= $srcd/srcd^+$	Striped, wild-type eyes (3.3%)
	sr^+cd^+ (6.5%)	= $sr^+cd^+/srcd^+$	Wild-type eyes and body (3.3%)
sr^+cd (50%)	$srcd^+$ (43.5%)	= $srcd^+/sr^+cd$	Wild-type eyes and body (21.8%)
	sr^+cd (43.5%)	= sr^+cd/sr^+cd	Wild-type body, cardinal eyes (21.8%)
	$srcd$ (6.5%)	= $srcd/sr^+cd$	Wild-type body, cardinal eyes (3.3%)
	sr^+cd^+ (6.5%)	= sr^+cd^+/sr^+cd	Wild-type eyes and body (3.3%)

The progeny fall into three phenotypic classes: wild-type eyes and body with a frequency of 21.8% + 3.3% + 21.8% + 3.3% = 50.2%; striped body, normal eyes with a frequency of 21.8% + 3.3% = 25.1%; and wild-type body, cardinal eyes with a frequency of 21.8% + 3.3% = 25.1%. (Note that these percents do not add to 100 because of rounding off.) The phenotypic ratio is 2:1:1.

b. No. Because of the absence of crossing over in the male, the progeny will always be formed in a 2:1:1 phenotypic ratio regardless of the degree of linkage. No individuals homozygous recessive for both loci will be formed.

12-21. a. The cross can be represented as $CyPm^+/Cy^+Pm \times CyPm^+/Cy^+Pm$. Because there is no crossing over, both sexes will make two kinds of gametes, $CyPm^+$ and Cy^+Pm, in equal frequencies.

Male gametes	Female gametes	Progeny genotypes	Progeny phenotypes
$CyPm^+$ (50%)	$CyPm^+$ (50%)	= $CyPm^+/CyPm^+$	Lethal (25%)
	Cy^+Pm (50%)	= $CyPm^+/Cy^+Pm$	Curly, plum (25%)
Cy^+Pm (50%)	$CyPm^+$ (50%)	= $Cy^+Pm/CyPm^+$	Curly, plum (25%)
	Cy^+Pm (50%)	= Cy^+Pm/Cy^+Pm	Lethal (25%)

Half the progeny die and the remainder are heterozygous for both loci, carry their genes in repulsion, and have the curly, plum phenotype.

b. No. Such a cross can be represented as $Cy^+Pm^+/CyPm \times Cy^+Pm^+/CyPm$. Because there is no crossing over, both sexes will make two kinds of gametes, Cy^+Pm^+ and $CyPm$, in equal frequencies.

Male gametes	Female gametes	Progeny genotypes	Progeny phenotypes
Cy^+Pm^+ (50%)	Cy^+Pm^+ (50%)	= Cy^+Pm^+/Cy^+Pm^+	Lethal (25%)
	Cy^+Pm^+ (50%)	= $Cy^+Pm^+/CyPm$	Curly, plum (25%)
$CyPm$ (50%)	$CyPm$ (50%)	= $CyPm/Cy^+Pm^+$	Curly, plum (25%)
	$CyPm$ (50%)	= $CyPm/CyPm$	Wild-type wings and eyes (25%)

Twenty-five percent of the progeny die; 50% are heterozygous for both loci with the curly, plum phenotype, carrying their genes in repulsion; and 25% are wild-type. The living progeny occur in a ratio of 2:1, curly, plum to wild-type.

Chapter 13: The Three-Point Testcross and Chromosomal Mapping

13-1. a. The parental linkage arrangements are STU and stu. These are found in classes 1 and 2, respectively.

b. This chromosome was contributed by the triple–homozygous recessive parent.

c. Double-crossover progeny are produced from the two-point testcross, but they cannot be detected since the second crossover cancels the effects of the first and restores the s and u loci to their original, precrossover linkage arrangement. Considering a third locus (such as t) positioned between the other two makes possible the detection of many double crossovers.

d. The s-to-u map distance based on the two-point testcross data will be shorter because the double crossovers go undetected and thus the recombination frequency is underestimated.

e. Classes 3 and 4: crossovers between s and t.

f. Classes 5 and 6: crossovers between t and u.

g. Classes 7 and 8: one crossover between s and t and another between t and u.

13-2. a. The s-to-t distance is based on the total number of crossovers occurring in that region. In addition to the 18% from classes 3 and 4 (single crossovers between s and t), all of the individuals in classes 7 and 8 (double crossovers) are derived from gametes with a crossover in the s-to-t region, giving a total of 18% + 3% = 21%, or 21 map units. The t-to-u distance is based on 13% from classes 5 and 6 (single crossovers between t and u) plus 3% from classes 7 and 8 giving a total recombination frequency of 13% + 3% = 16%, or 16 map units.

b. The s-to-u distance can be obtained by adding the s-to-t and t-to-u distances: 21 + 16 = 37 map units.

c. With the two-point testcross, double-crossover progeny seen in the three-point results would be recorded as parental types. The *s*-to-*t* distance would be 18 map units and the *t*-to-*u* distance would be 13 units, giving an *s*-to-*u* distance of 18 + 13 = 31 units.

13-3. Outcome (a): (1) The parental types are always the two most common classes, and the double crossovers are always the two least common classes.

Parentals: $\frac{D\,e\,f}{d\,e\,f}$ (36%) $\frac{d\,E\,F}{d\,e\,f}$ (40%)

Double crossovers: $\frac{D\,E\,f}{d\,e\,f}$ (1%) $\frac{d\,e\,F}{d\,e\,f}$ (1%)

(2) The alleles which are interchanged when comparing the two double-crossover types with the two parental types belong to the middle locus. In this case, this is the *e* locus and the order of the loci is *d-e-f* (or *f-e-d*).
(3) The chromosomes carried by the trihybrid parent are *Def* and *dEF*, and since the dominant (or recessive) alleles are not grouped on the same chromosome, they are arranged in repulsion.
Outcome (b): (1)

Parentals: $\frac{D\,E\,F}{d\,e\,f}$ (37%) $\frac{d\,e\,f}{d\,e\,f}$ (35%)

Double crossovers: $\frac{D\,e\,f}{d\,e\,f}$ (1%) $\frac{d\,E\,F}{d\,e\,f}$ (1%)

(2) Comparing the two double-crossover types with the two parental types shows that alleles at the *d* locus are interchanged. *d* is the middle locus and the correct order of the loci is *e-d-f* (or *f-d-e*).
(3) The chromosomes carried by the triple-heterozygous parent are *DEF* and *def*, and since the dominant (or recessive) alleles are grouped on the same chromosome, they are arranged in coupling.
Outcome c: (1)

Parentals: $\frac{D\,e\,f}{d\,e\,f}$ (31%) $\frac{d\,E\,F}{d\,e\,f}$ (34%)

Double crossovers: $\frac{d\,E\,f}{d\,e\,f}$ (1%) $\frac{D\,e\,F}{d\,e\,f}$ (1%)

(2) The alleles which appear to be interchanged when comparing the double-crossover types with the parentals are found at the *f* locus. Thus, *f* is the middle locus, and the order of the loci is *e-f-d* (or *d-f-e*).
(3) The chromosomes carried by the triple-heterozygous parent are *Def* and *dEF*, and since the dominant (or recessive) alleles are not grouped on the same chromosome, the alleles are arranged in repulsion.

13-4. a. No. Each class of testcross progeny reflects a different type of gamete produced by the triple-heterozygous parent, and the number of progeny in each class reflects the relative frequency with which each type of gamete is formed. If the loci assort independently, the triple-heterozygote would be expected to produce $2^3 = 8$ kinds of gametes formed in approximately *equal* frequencies. Although this cross produces eight classes of progeny, they show very different frequencies, with the two most common classes having 17 times the number of progeny found in the two least frequent classes. Consequently, it is highly unlikely that the three loci are assorting independently.

b. Yes. Eight classes of testcross progeny occurring in *different* frequencies support the hypothesis that three loci are linked. The different frequencies for the classes indicate that the different kinds of gametes are produced in different frequencies, depending on whether they experience no, one, or two crossover events.

c. The chi-square test. Deviations from the numbers of progeny in each class expected under the hypothesis of independent assortment give a χ^2 value of 950.02. With seven degrees of freedom, the probability value is less than 0.001, indicating lack of support for the hypothesis of independent assortment. (Details of the chi-square test are found in Chapter 6.)

d. The four classes arising from gametes formed through single-crossover events each occur in about the same frequency. This indicates that the crossover frequencies are about the same in the two chromosome regions that separate the middle locus from the other two loci under study. Similar crossover frequencies imply similar distances separating the loci.

13-5. a. The classes with parental-type chromosomes are the two largest: wild, wild, wild (413) and thread, curled, striped (398).

b. Since double-crossover events are less frequent than single crossovers, the classes with double-crossover chromosomes are always the two smallest: wild, curled, wild (2) and thread, wild, striped (3).

c. The middle locus is identified by comparing the double-recombinant types with the parentals and determining the locus that has its alleles interchanged. In this case, it is the curled (*cu*) locus and the sequence is *th-cu-sr* (or *sr-cu-th*).

13-6. a. Single crossovers in the *cu*-to-*th* region produced two classes: thread, wild, wild (35) and wild, curled, striped (29). In addition to the 35 + 29 = 64 individuals in these classes, each of the five double-crossover individuals also experienced a crossover in the *th*-to-*cu* region, bringing the total to 64 + 5 = 69, which is 69/1000 = 0.069, or 0.069 × 100 = 6.9%, of all the progeny.

b. Single crossovers in the *cu*-to-*sr* region produced two phenotypic classes: thread, curled, wild (66) and wild, wild, striped (54). In addition to the 66 + 54 = 120 individuals in these classes, each of the five double-crossover individuals also experienced a crossover in this region, bringing the total to 120 + 5 = 125, which is 125/1000 = 0.125, or 0.125 × 100 = 12.5%, of all the progeny.

c. Since a 1% crossover frequency equals one unit of map distance, there are 6.9 map units between *th* and *cu* and 12.5 map units between *cu* and *sr*.

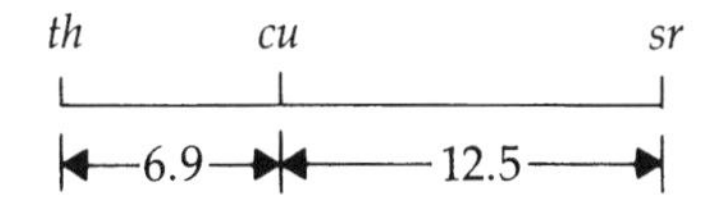

13-7. **a.** Parental-, or nonrecombinant-, type gametes are the two most common types of gametes produced by the triple-heterozygous parent. These gametes, in turn, give rise to the two most common classes of progeny: curved, wild, black (529) and wild, purple, wild (509).

b. The parental-type chromosomes produced by the triple heterozygote are *c+b* and *+pr+*. With two recessive alleles, *c* and *b*, on one chromosome and the third, *pr*, on the homolog, the alleles are carried in repulsion.

c. There are three alternative arrangements. (1) With all the recessive alleles on one chromosome and all the dominants on the other, the parental-type gametes would be *cprb* and + + + and the two most common progeny classes would be curved, purple, black and wild, wild, wild. (2) With *c+ +/+prb*, the parental-type gametes would be *c+ +* and *+prb* and the two most common progeny classes would be curved, wild, wild and wild, purple, black. (3) With *+ +b/cpr+*, the parental-type gametes would be *+ +b* and *cpr+* and the two most common progeny classes would be wild, wild, black and curved, purple, wild.

13-8. A comparison of the two double-crossover types (wild, wild, wild and curved, purple, black) with the two parental types (curved, wild, black and wild, purple, wild) indicates that the purple locus is in the middle. The curved, wild, wild (52) and wild, purple, black (60) classes have experienced a crossover between the purple and black loci, as have the nine progeny in the two classes of double-crossover types. The total number of progeny showing a crossover in the *pr*-to-*b* region is 52 + 60 + 9 = 121, which makes up 121/1467 = 0.082, or 0.082 × 100 = 8.2%, of all progeny. With a recombination frequency of 8.2%, these two loci are 8.2 map units apart. The wild, wild, black (148) and curved, purple, wild (160) classes have experienced a crossover between curved and purple loci, as have the nine progeny in the two classes of double-crossover types. The total number of progeny showing a crossover in the *c*-to-*pr* region is 148 + 160 + 9 = 317, which makes up 317/1467 = 0.216, or 0.216 × 100 = 21.6%, of all the progeny. With a recombination frequency of 21.6%, these two loci are 21.6 map units apart.

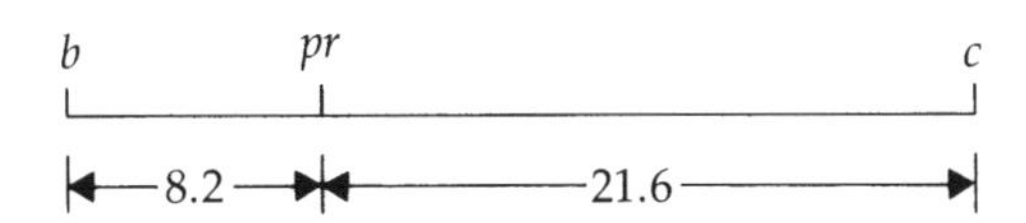

13-9. **a.** The two most common classes are the parental types $sh^+wx^+c^+$ (1210) and *shwxc* (1373) and the two least common classes are the double-crossover types sh^+wxc (4) and $shwx^+c^+$ (5). A comparison of the double-crossover classes with the parental classes shows that the alleles at the *sh* locus are interchanged, indicating that this locus is in the middle, and the actual sequence is *wxshc* (or *cshwx*).

b. Plants in the wx^+shc (354) and $wxsh^+c^+$ (318) classes have experienced a single crossover in the *wx*-to-*sh* region, as have each of the nine double-crossover plants, giving a total of 354 + 318 + 9 = 681, which is 681/3400 = 0.200, or 0.200 × 100 = 20.0%, of all progeny. Plants in the wx^+sh^+c (74) and $wxshc^+$ (62) classes have experienced a single crossover in the *sh*-to-*c* region, as have each of the nine double-crossover plants, giving a total of 74 + 62 + 9 = 145, which is 145/3400 = 0.043, or 0.043 × 100 = 4.3%, of all progeny.

c.

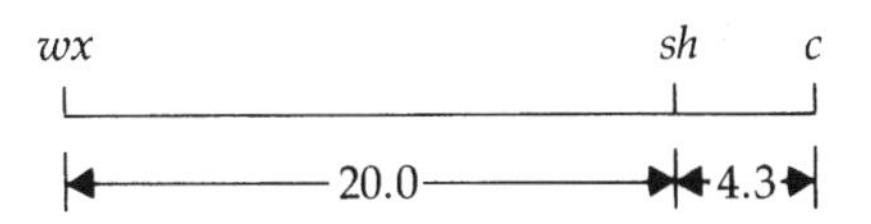

13-10. **a.** The product law can be used to determine the expected double-crossover frequency. Multiplying the frequency of a crossover in the *wx*-to-*sh* region (0.20) by the frequency of a crossover in the *sh*-to-*c* region (0.043) gives an expected double-crossover frequency of 0.20 × 0.043 = 0.0086. With 3400 progeny, about 29 (3400 × 0.0086) would be expected to carry the double crossovers.

b. $$\text{Coefficient of coincidence} = \frac{\text{Observed double crossovers}}{\text{Expected double crossovers}} = 9/29 = 0.31.$$

This indicates that 31% of the expected double crossovers actually occurred. Interference = 1 − Coefficient of coincidence = 1 − 0.31 = 0.69. This indicates that 69% of the expected double crossovers do not occur.

13-11. **a.** The two most common classes are parental types ylz^+cv^+ (463) and y^+lzcv (488), and the two least common classes are the double-crossover types ylz^+cv (8) and y^+lzcv^+ (7). A comparison of doubles with the parentals shows that the *cv* locus is in the middle, giving a sequence of *ycvlz* (or *lzcvy*). To reflect the correct sequence, the parentals should be rewritten as ycv^+lz^+ and y^+cvlz, and the doubles as $ycvlz^+$ and y^+cv^+lz.

b. All individuals in classes *ycvlz* (77) and $y^+cv^+lz^+$ (87) have experienced a single crossover in the *y*-to-*cv* region, as have the 15 double recombinants, for a total of 77 + 87 + 15 = 179, which is 179/1300 = 0.138, or 0.138 × 100 = 13.8%, of all progeny. All individuals in classes y^+cvlz^+ (89) and ycv^+lz (81) have experienced a single crossover in the *cv*-to-*lz* region, as have the 15 double recombinants, for a total of 89 + 81 + 15 = 185, which is 185/1300 = 0.142, or 0.142 × 100 = 14.2%, of all progeny.

c.

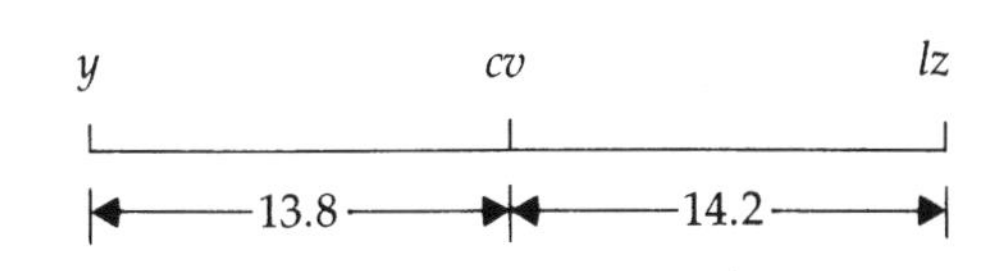

13-12. **a.** The expected frequency of double crossovers is obtained by multiplying the probabilities of crossovers in each region: 0.138 × 0.142 = 0.020. With 1300 progeny, 0.020 × 1300 = 26 double-crossover types would be expected.

b. $\frac{\text{Coefficient of coincidence}}{} = \frac{\text{Observed doubles}}{\text{Expected doubles}} = 15/26 = 0.58.$

Of the expected doubles, 58% actually occurred. Interference = 1 − Coefficient of coincidence = 1 − 0.58 = 0.42, meaning that 42% of the expected doubles do not occur.

13-13. **a.** Each map distance gives the probability of a crossover occurring in that region. The expected frequency of double crossovers is obtained by multiplying the probability of a crossover in each region: 0.112 × 0.158 = 0.0177.

b. With 1000 progeny, the expected number of double-crossover progeny is 0.0177 × 1000 = 18. However, because of interference, only 0.3 of these occur: (18)(0.3) = 5 (approximately).

c. The 11.2 units of map distance between the *a* and *b* loci indicate that the probability of a crossover occurring in that region is 0.112. With 1000 progeny, 0.112 × 1000 = 112 would experience crossovers in this region. Some of these crossovers, however, would have occurred in producing double-crossover individuals. Subtracting the number of double-crossover individuals from the total number of individuals with crossovers in this region gives 112 − 5 = 107 individuals who fall into the single-crossover class.

d. The 15.8 units of map distance between the *b* and *c* loci indicate that the probability of a crossover occurring in that region is 0.158. With 1000 progeny, 0.158 × 1000 = 158 would experience crossovers in this region. Subtracting the number of double-crossover progeny gives 158 − 5 = 153 individuals in this single-crossover class.

e. Of the 1000 progeny, five are double-crossover types, 107 + 153 = 260 are single-crossover types, and the remainder, 1000 − 265 = 735, are parental, or nonrecombinant, types.

13-14. A good approach to combining these four maps into one is to arrange the four segment maps so that the genes that are common to them are in alignment.

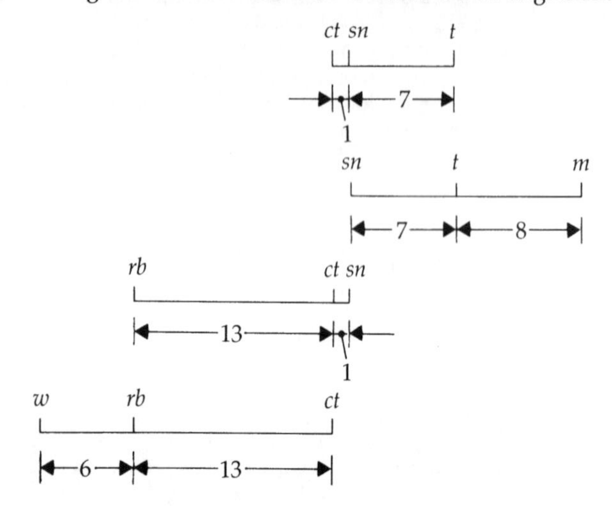

By setting the site of the *w* locus at 0, the four segment maps can be combined into one map showing all loci and the distances between them.

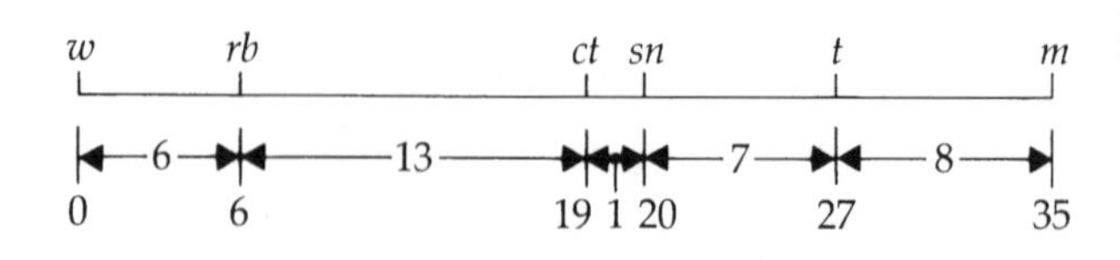

Chapter 14: Haploid Genetics: Tetrad Analysis I

14-1. In most organisms it is impossible to identify the haploid products arising from a single diploid cell that undergoes meiosis. However, the gametic products can be sampled through the mating of organisms in a testcross. From the analysis of the genotypes of the testcross progeny, deductions can be made about the segregation of alleles.

14-2. The ascospores making up each pair are genetically identical to each other because they arise mitotically from the same meiosis-II nucleus.

14-3. **a.** The alleles in the ascus on the left show a 4-4 pattern which is produced whenever alleles segregate during meiosis I. The ascus on the right shows a 2-2-2-2 pattern which results when alleles segregate during meiosis II.

b. Second-division, or M-2, segregation pattern.

14-4. As the distance increases, the frequency of crossing over increases and proportionately more of the asci would exhibit the 2-2-2-2, or M-2, segregation pattern and fewer would exhibit the 4-4, or M-1, pattern.

14-5. **a.** Ascus classes 1 and 2 show a 4-4 distribution of alleles, indicating segregation during the first meiotic division (M-1). Ascus classes 3, 4, 5, and 6 show patterns that indicate segregation during the second meiotic division (M-2).

b. Ascus classes 1 and 2 show M-1 segregation, indicating the absence of crossing over.

c. Ascus classes 3, 4, 5, and 6 show M-2 segregation, indicating that crossing over occurred during their formation.

14-6. **a.** Classes 1 and 2 are random variants of each other as are classes 3, 4, 5, and 6. Variants arise because of the random alignment of chromosomes during meiosis.

b. No. One would expect to see more noncrossover types (classes 1 and 2) because crossover events are relatively rare. The relationship between the total number of asci in classes 1 and 2 and the total number of asci in classes 3, 4, 5, and 6 depends on the number of crossovers that take place.

14-7. **a.** Yes. An increase in distance would increase the amount of crossing over and that would mean proportionately fewer asci in classes 1 and 2, and more asci in classes 3, 4, 5, and 6.

b. Only half of the ascospores in each ascus in classes 3, 4, 5, and 6 carry recombinants because only two of four chromatids have participated in crossing over.

c. One crossover in the *fl*-to-centromere region could produce either a class-3, -4, -5, or -6 ascus while a second crossover in the same region and involving the same two strands would result in a class-1 or -2 parental-type ascus. So, although two crossovers had occurred, no recombinants would be detected from them and the calculated map distance would be underestimated.

14-8. The 151 + 161 = 312 asci in classes 1 and 2 are parental types and have experienced no crossing over. The 8 + 6 + 7 + 7 = 28 asci in classes 3, 4, 5, and 6 have experienced crossing over and contain recombination products; they make up 28/340 = 0.0824, or 0.0824 × 100 = 8.24%, of all the asci formed. Since only half the ascospores in these asci are recombination products, the percentage of recombination is half of this percentage, or 8.24% × 1/2 = 4.12%, which equals 4.12 map units.

14-9. Of the 410 asci studied, 107, or 107/410 = 0.261 (26.1%) of all the asci, have the M-2 pattern and thus have experienced crossing over. Since only half the ascospores in these asci are recombinants, the percentage of recombination is half this percentage or 26.1% × 1/2 = 13.05%. The centromere-to-*suc* distance is 13.05 map units.

14-10. A map distance of 10.6 units indicates a recombination frequency of 10.6%. The percentage of asci carrying recombinant ascospores is twice this value or 10.6% × 2 = 21.2%, and the number of asci is 0.212 × 230 = 49 asci.

14-11. **a.** The tetrad type for each class is followed by the segregation pattern for the *bn* and *his-5* loci, respectively: (1) PD, M-1, M-1; (2) NPD, M-1, M-1; (3) T, M-2, M-1; (4) T, M-1, M-2; (5) PD, M-2, M-2; (6) NPD, M-2, M-2; (7) T, M-2, M-2.

b. A comparison of the numbers of parental ditypes (186) and nonparental ditypes (199) shows that they are very similar, indicating that the loci are on separate chromosomes.

14-12. **a.** Classes 3, 5, 6, and 7 show an M-2 segregation pattern for the *bn* locus.

b. Classes 3, 5, 6, and 7 contain a total of 49 + 1 + 1 + 1 = 52 asci which make up 52/620 = 0.084, or 0.084 × 100 = 8.4%, of the total number of asci. Since only half of the spores within these asci are recombinants, the percentage of recombination is 8.4% × 1/2 = 4.2%, and there are 4.2 map units between the centromere and *bn*.

c. Classes 4, 5, 6, and 7 show an M-2 segregation pattern for the *his-5* locus.

d. Classes 4, 5, 6, and 7 contain a total of 185 + 1 + 1 + 1 = 188 asci which make up 188/620 = 0.3032, or 0.3032 × 100 = 30.32%, of all the asci. The percentage of recombination is 30.32% × 1/2 = 15.16%, and there are 15.16 map units between the centromere and the *his-5* locus.

14-13. There is less than a 1% chance that the loci are linked. The nonsignificant difference indicates the loci are assorting independently; they are on different chromosomes.

Chapter 15: Haploid Genetics: Tetrad Analysis II

15-1. Since most of the crossovers involving the locus closest to the centromere (the *a* locus) also involve the *b* locus, the loci are on the same arm of the chromosome, separated by a distance of 12 − 10 = 2 units.

15-2. **a.** For each class, the tetrad type is listed first, followed by the segregation patterns for the *lys* and *sp* loci, respectively: (1) PD, M-1, M-1; (2) NPD, M-1, M-1; (3) T, M-1, M-2; (4) T, M-2, M-1; (5) PD, M-2, M-2; (6) NPD, M-2, M-2; (7) T, M-2, M-2.

b. The PD frequency greatly exceeds the NPD frequency, indicating that the loci are linked.

15-3. **a.** Classes 4, 5, 6, and 7 show the M-2 segregation pattern for the *lys* locus. The number of asci in these classes is 8 + 135 + 1 + 9 = 153. These make 153/850 = 0.18, or 0.18 × 100 = 18%, of all the asci. Since only half the ascospores in these asci have recombination involving the *lys* locus, the percentage is reduced to 18% × 1/2 = 9%. The *lys*-to-centromere distance is nine map units.

b. Classes 3, 5, 6, and 7 show M-2 segregation for the *sp* locus. The number of asci in these classes is 130 + 135 + 1 + 9 = 275 which make up 275/850 = 0.3235, or 0.3235 × 100 = 32.35%, of all the asci. The percentage of recombination is half this value, or 32.35% × 1/2 = 16.2%. The *sp* locus is 16.2 map units from the centromere.

c. The *lys* locus is closest to the centromere. Classes 4, 5, 6, and 7 show the M-2 segregation pattern for this locus and contain a total of 8 + 135 + 1 + 9 = 153 asci. Three of these same classes, 5, 6, and 7, containing a total of 135 + 1 + 9 = 145 asci, show the M-2 pattern for the *sp* locus. Thus, 145/153 = 0.948, or 0.948 × 100 = 94.8%, of the recombinations involving the *lys* locus also involve the *sp* locus. This indicates that the loci are on the same arm of the chromosome.

15-4. **a.** Subtraction of the *lys*-to-centromere distance (nine map units) from the *sp*-to-centromere distance (16.2 map units) gives us a distance of 7.2 map units between the two loci.

b. Classes 2 and 6 each experienced two crossovers in the *lys*-to-*sp* interval, and all the ascospores in these asci carry recombinants arising from these crossovers. Classes 3, 4, and 7 experienced single crossovers in this region, and half the ascospores in these asci carry recombinants. To determine the *lys*-to-*sp* distance, we count all the asci in classes 2 and 6 (2 + 1 = 3) and half the asci in classes 3, 4, and 7 (130 + 8 + 9 = 147, 147 × 1/2 = 73.5). The recombination frequency is 3 + 73.5 = 76.5, 76.5/850 = 0.09, and there are 0.09 × 100 = 9 map units between the two loci.

c. The estimate of *lys*-to-*sp* distance based on the number of crossovers is more accurate since it picks up crossovers that would otherwise go undetected.

d.

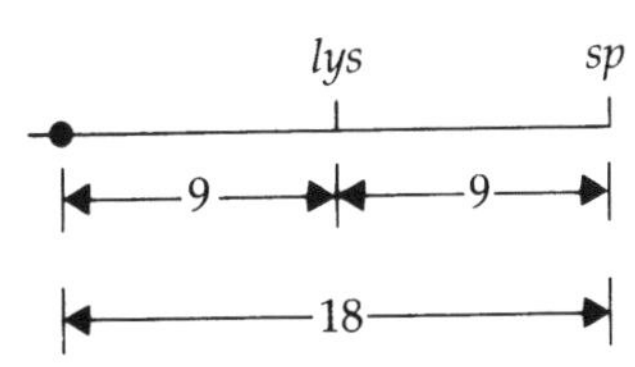

15-5. **a.** Begin by identifying the tetrad type for each class and the segregation pattern shown by each locus. A comparison of the number of PD (865) and the

NPD (5) asci shows a great excess of PD asci, indicating linkage.

b. Classes 4, 5, 6, and 7 show the M-2 segregation pattern for the *fi* locus and include 11 + 110 + 2 + 10 = 133 asci, which are 133/1100 = 0.121, or 0.121 × 100 = 12.1%, of all the asci; the percentage of the recombination is half this, or 12.1% × 1/2 = 6.05%, and the *fi*-to-centromere distance is six map units. Classes 3, 5, 6, and 7 show M-2 segregation for the *cys* locus. These classes include 209 + 110 + 2 + 10 = 331 asci, which are 331/1100 = 0.30, or 0.30 × 100 = 30% of all the asci; the percentage of the recombination is half this, or 30% × 1/2 = 15%, and 15 map units separate the *cys* locus from the centromere.

c. The *fi* locus is closest to the centromere. Classes 4, 5, 6, and 7 show M-2 segregation for this locus and contain a total of 11 + 110 + 2 + 10 = 133 asci. Three of these classes (5, 6, and 7), containing a total of 110 + 2 + 10 = 122 asci, also show the M-2 pattern for the *cys* locus. Thus, 122/133 = 0.917, or 0.917 × 100 = 91.7%, of the recombinations involving the *fi* locus also involve the *cys* locus. This high value indicates that the loci are on the same arm of the chromosome.

15-6. a. Classes 2 and 6 each experienced two crossovers in the *fi*-to-*cys* region, and all the ascospores in these asci carry recombinants arising from these crossovers. Classes 3, 4, and 7 experienced single crossovers in this region, and half the ascospores in these asci carry recombinants. To determine the distance between the two loci, we first count all the asci in classes 2 and 6 (3 + 2 = 5) and half the asci in classes 3, 4, and 7 (209 + 11 + 10 = 230, 230 × 1/2 = 115). The recombination frequency is 5 + 115 = 120, 120/1100 = 0.109, and the map distance between the two loci is 0.109 × 100 = 10.9 units.

b.

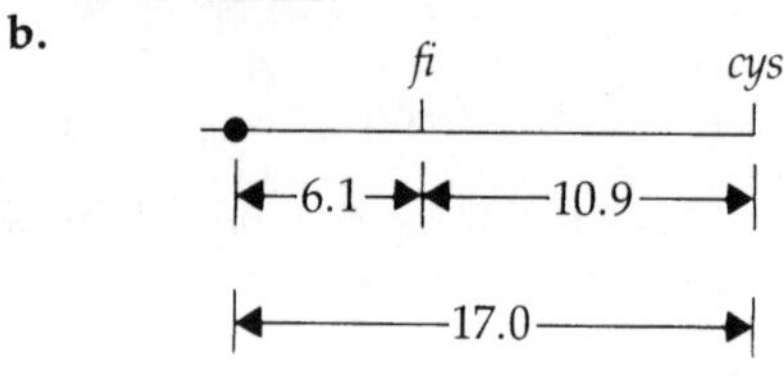

15-7. a. A comparison of the number of PD (411) and NPD (3) tetrads shows a large excess of PD tetrads, indicating linkage.

b. Classes 4, 5, 6, and 7 show M-2 segregation for the *cr* locus and include 63 + 5 + 1 + 5 = 74 asci, which are 74/620 = 0.119, or 0.119 × 100 = 11.94%, of all the asci; the recombination frequency is half this, or 11.94% × 1/2 = 5.97%, and 5.97 map units separate the *cr* locus and the centromere. Classes 3, 5, 6, and 7 show M-2 segregation for the *arg* locus. These classes include 138 + 5 + 1 + 5 = 149 asci, which are 149/620 = 0.24 or 24%, of all the asci; the percentage of recombination is half this, or 24% × 1/2 = 12%, and 12 map units separate the *arg* locus from the centromere.

c. The locus-to-locus distance is given by the following equation:

$$\text{Distance} = \frac{(1/2)\text{T} + \text{NPD}}{\text{Total asci}} \times 100.$$

Tetratypes are found in classes 3, 4, and 7 which contain a total of 138 + 63 + 5 = 206 progeny, and nonparental ditypes are found in classes 2 and 6 with a total of 2 + 1 = 3 progeny. Substituting these values into the expression gives

$$\frac{(1/2)206 + 3}{620} \times 100 = \frac{106}{620} \times 100$$
$$= 17.1 \text{ map units.}$$

d. The *cr* locus is closest to the centromere; classes 4, 5, 6, and 7 show M-2 segregation for this locus and contain a total of 63 + 5 + 1 + 5 = 74 asci. Three of these classes (5, 6, and 7), containing a total of 5 + 1 + 5 = 11 asci, also show the M-2 pattern for the *arg* locus. Thus, 11/74 = 0.149, or 0.149 × 100 = 14.9%, of the recombinations involving the *cr* locus also involve the *arg* locus. This relatively low value would be expected if the loci were on different arms of the chromosome.

e.

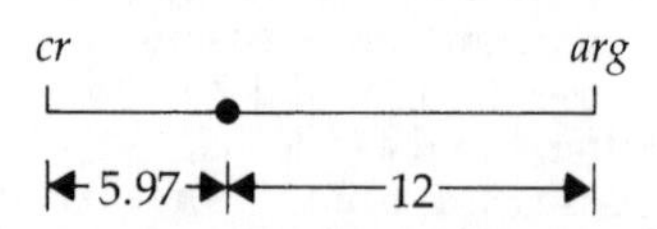

15-8. a. If crossing over occurs before chromosomal replication, when a homologous pair exists as two single-stranded chromosomes, the events of meiosis could be represented as shown in Figure 15-8a. PD (++ ++ ++ ++ *uv uv uv uv*) tetrads would be produced from diploid fusion nuclei that experienced no crossing over, and NPD (+*v* +*v* +*v* +*v u*+ *u*+ *u*+ *u*+) tetrads would be formed from diploid fusion nuclei that experienced crossing over.

FIGURE 15-8a

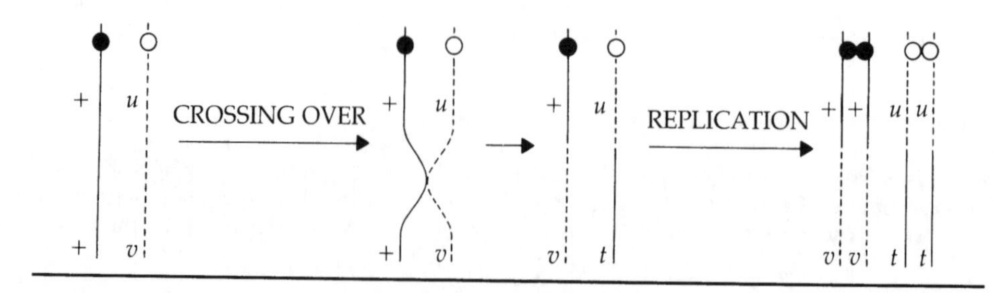

b. There would be more NPD tetrads and fewer PD tetrads.

c. No tetratype tetrads could be produced; crossing over at the two-strand stage will never yield more than two types of tetrads.

15-9. a. If crossing over takes place after replication, when a homologous pair consists of four chromatids, the events of meiosis could be represented as shown in Figure 15-9a. PD tetrads would be produced from diploid fusion nuclei that experienced no crossing over, and tetratype tetrads would be formed from diploid fusion nuclei that experienced crossing over.

b. With more crossing over, the frequency of tetratype tetrads would increase and the frequency of PD tetrads would decrease.

c. NPD tetrads asci would be produced if double-crossover exchanges took place.

d. It occurs after chromosomal replication.

15-10. The cross is $a+ \times +g$ and it produces a diploid fusion nucleus of genotype $+a+g$. First, we must identify for each class the tetrad type and segregation pattern shown by the two loci. For each class, the tetrad type is listed first, followed by the segregation pattern for the *a* locus and the *g* locus, respectively: (1) NPD, M-2, M-2; (2) T, M-2, M-1; (3) NPD, M-1, M-1; (4) T, M-2, M-2; (5) PD, M-2, M-2; (6) PD, M-1, M-1; and (7) T, M-1, M-2. Next, we determine whether the loci are linked. A comparison of the number of PD tetrads (classes 5 and 6: 316 + 3 = 319) and NPD tetrads (classes 1 and 3: 1 + 2 = 3) shows a large excess of PD tetrads, indicating linkage. Next, we determine the locus-to-centromere distances. Classes 1, 2, 4, and 5 show M-2 segregation for the *a* locus and include 37 + 3 + 1 + 3 = 44 asci, which are 44/437 = 0.101, or 0.101 × 100 = 10.1%, of all the asci; the recombination percentage is half this, or 10.1% × 1/2 = 5.05%, and 5.05 map units separate the *a* locus and the centromere. Classes 1, 4, 5, and 7 show M-2 segregation for the *g* locus. These classes include 1 + 3 + 3 + 75 = 82 asci, which are 82/437 = 0.188, or 0.188 × 100 = 18.76%, of all the asci; the percentage of recombination is half this, or (18.76% × 1/2 = 9.38%, and 9.38 map units separate the *g* locus from the centromere. Now we determine whether the loci are on the same or different chromosome arms. The *a* locus is closest to the centromere; classes showing M-2 segregation for this locus (1, 2, 4, and 5) contain a total of 44 asci. Three of these classes (1, 4, and 5), containing a total of 1 + 3 + 3 = 7 asci, also show the M-2 pattern for the *g* locus. Thus, 7/44 = 0.159, or 0.159 × 100 = 15.9%, of the recombinations involving the *a* locus also involve the *g* locus. This relatively low value would be expected if the loci were on different arms of the chromosome.

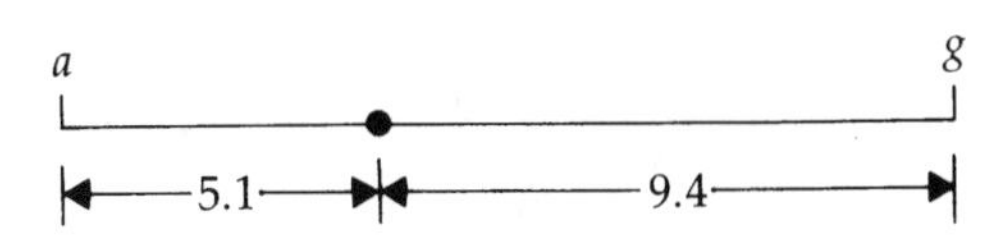

15-11. a. Classes 1, 2, and 3 are classified as PD, T, and NPD, tetratypes, respectively. The fact that the number of NPD tetrads (6) is very small relative to the number of PD tetrads (762) indicate linkage.

b. The map distance is given by the expression

$$\frac{1/2(\text{T}) + \text{NPD}}{\text{Total number of asci}} \times 100$$

$$= \frac{1/2(132) + 6}{762 + 132 + 6} \times 100$$

$$= \frac{72}{900} \times 100$$

$$= 0.08 \times 100 = 8 \text{ map units.}$$

15-12. a. Classes 1, 2, and 3 are classified as PD, NPD, and T tetratypes, respectively. The similarity in the number of PD (519) and NPD (498) tetrads indicates that these loci are independently assorting. Note that the statistical similarity of this ratio to a 1:1 ratio could be confirmed using the chi-square test.

b. The two loci are on separate chromosomes.

15-13. a. Equal frequencies of the four types of spores indicate that the two loci are independently assorting.

b. The loci are on separate chromosomes.

15-14. a. Types $pr+$ and $+ac$, with respective frequencies of 1602 and 1710, are parentals, and types *prac* and $++$, with respective frequencies of 130 and 158, are recombinants. The absence of a 1:1:1:1 ratio among these four types of spores indicates that the loci are linked.

b. The map distance is given by the expression

$$\frac{\text{Number of recombinant spores}}{\text{Total number of spores}} \times 100$$

$$= \frac{130 + 158}{1602 + 1710 + 130 + 158} \times 100$$

$$= \frac{288}{3600} \times 100 = 8 \text{ map units.}$$

FIGURE 15-9a

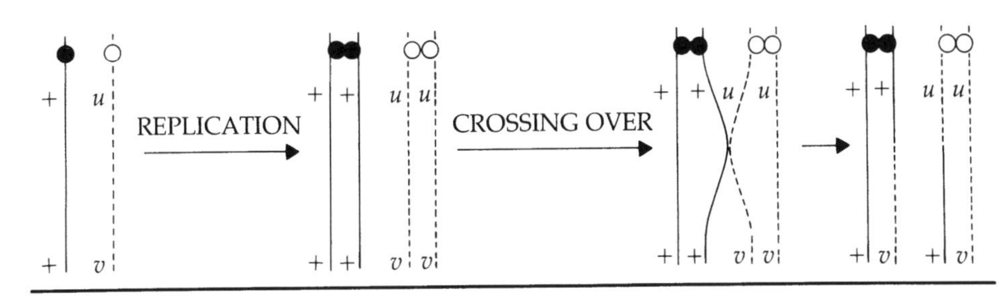

Chapter 16: Changes in Chromosome Structure and Number

16-1. **a.** Duplication or receiving a nonreciprocal translocation.
b. Deficiency or donating a nonreciprocal translocation.
c. Inversion or an intrachromosomal translocation.
d. Interchromosomal translocation.

16-2. **a.** $2n - 1 = 75$.
b. $2n + 1 = 77$.
c. 76.
d. Although material from 76 chromosomes would be present, 75 chromosomes would be counted.
e. 76.
f. $2n - 2 = 74$.
g. $4n = 152$.

16-3. The normal chromosome shows a loop while the chromosome carrying the deficiency does not.

Normal chromosome: ——————Ω——————
Deficient chromosome: ——————————————

16-4. No. Since the section no longer has a counterpart with which to cross over, there will be no disruption.

16-5. The similarity in the two polypeptides suggests that possibly a duplication of a gene coding for a hemoglobin polypeptide occurred in the evolutionary past; subsequent mutations in each gene produced the differences in the polypeptide products that exist today.

16-6. **a.** Standard Down syndrome: 47 chromosomes (three separate copies of chromosome 21); translocational Down syndrome: 46 chromosomes (three copies of chromosome 21 are present: two copies are separate and the third is translocated onto another chromosome).
b. One parent is normal; the other will carry the 15/21 translocation and have 45 chromosomes.
c. Half will have chromosome duplication or deficiencies, and will not develop.
d. Only half of the zygotes formed would survive. If a child develops, the probability is 1/3 that it will be affected with Down syndrome, 1/3 that it will be normal, and 1/3 that it will be a translocation carrier.

16-7. The two chromosome pairs will take on a cross-shaped configuration as follows.

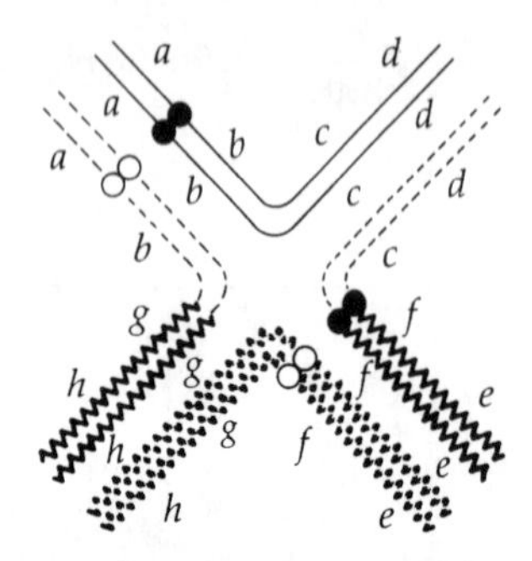

16-8. The primary oocyte that undergoes meiosis is XX. Primary nondisjunction will result in either an egg carrying two X chromosomes or carrying no sex chromosomes. Normal sperm will carry either X or Y. The following are possible combinations.

Egg		Sperm		Zygote
XX	+	X	=	XXX
XX	+	Y	=	XXY
O	+	X	=	XO
O	+	Y	=	YO

16-9. **a.** The loop indicates either a duplication or a deficiency. Since the recessive alleles are expressed, the loop most likely can be attributed to a deletion of the segment carrying these two loci. The "heterozygote" is not a true heterozygote for these two loci since it has only one copy of each locus. The lost section carried the dominant alleles for both loci.
b. The loop consists of the section of the normal chromosome which corresponds to the deficiency. The loop arose because the loci it carries have no counterparts with which to synapse.
c. Both loci are located within the lost section or, put another way, within the segment that forms the loop.
d. Since the loop is small, the two loci must be fairly close to each other.

16-10. Failure of meiosis in the formation of both the egg and the sperm: XX + XY = XXXY. Multiple fertilization of a normal egg (X) by two X sperm and one Y sperm.

16-11. Mitotic nondisjunction of the Y chromosome resulted in a somatic XO cell which, in turn, gave rise to other XO cells.

16-12. **a.** The F_1 plants were sterile because their chromosomes lacked homologs. Consequently synapsis could not occur and meiosis was disrupted.
b. Somatic doubling gave each chromosome a homolog and thus the 36-chromosome plants could form fertile progeny.
c. Such a mating gave rise to triploids which, because of meiotic problems, formed unbalanced gametes.

16-10. Sex in animals is usually determined by certain chromosomal mechanisms which are thrown off by polyploidy. Such mechanisms are not generally found in plants. Furthermore, many plants may reproduce asexually or through self-fertilization, thereby allowing a single polyploid individual to give rise to others. Sexual reproduction with cross-fertilization is the rule for most animals, and two individuals of the same polyploidy are required to give rise to polyploid descendants.

16-11. **a.** No. Crossing over does occur within the inversion but because the recombinant chromosomes have imbalances in centromeres and genes, the gametes containing them give rise to inviable zygotes. Since no progeny are recombinant for loci within the inversion, it appears that the inversion suppresses crossing over.
b. Yes. Since each chromosome would have a centromere and a standard complement of loci, the gametes would be genetically balanced.
c. No. Synapsis of the homologs occurs normally without the twisting of one of the chromosomes.

Crossing over within the inversion will give rise to recombinant chromatids with single centromeres and the normal number of loci.

16-15. **a.** Two possible causes: translocation or inversion.

b. Two suggestions: (1) Carry out matings to check for linkage. If new linkage arrangements are detected, then a translocation may be the cause. If the linkage groups are unchanged, then the cause may be an inversion. (2) Carry out a cytological examination of synapsed chromosomes. A cross-shaped figure supports translocation while a loop indicates an inversion.

16-16. Autopolyploidy arises through the addition of one or more copies of the same genome, that is, a genome from the same species, while allopolyploidy arises through the addition of one or more copies of a different genome, that is, a genome from another species.

16-17. With odd-numbered polyploids, the "extra" homologs are unable to pair during meiosis and unbalanced gametes (with chromosomal deficiencies or surpluses) are formed. With even-numbered polyploids, each chromosome has a homolog with which to pair and some balanced gametes are formed; their union gives rise to zygotes that will develop.

16-18. **a.** A pericentric inversion includes the centromere whereas a paracentric inversion does not.

b. Only crossing over within a paracentric inversion produces recombinant strands with an excess or deficiency of centromeres. Following crossing over within a pericentric inversion, all of the product chromatids have a single centromere.

c. Yes. Crossovers within both types of inversions cause the recombinant chromatids to be genetically unbalanced; only the nonrecombinant chromatids give rise to progeny.

16-19. Yes. If the inversion were paracentric, the cells in anaphase I would show a bridge, or dicentric, chromosome attached to both poles of the spindle apparatus and a fragment, or acentric, chromosome attached to neither pole. If the inversion were pericentric, there would be no bridge and no fragment.

16-20. With a reciprocal translocation, unbalanced gametes arise as a consequence of chromosomal disjunction while with an inversion they are the consequence of crossing over.

Chapter 17: Quantitative Inheritance, Statistics, and Heritability

17-1. **a.** The continuous distribution of seed lengths supports the idea that this is a quantitative trait. The discontinuous distribution of plant heights, that is, the two distinct phenotypic classes, would be expected for a nonquantitative trait.

b. The continuous F_2 distribution for seed length supports a polygenic basis. The discontinuous F_2 phenotypic distribution for plant height would be expected for a trait determined by a single pair of alleles, one dominant and one recessive, operating at a single locus. Note that this distribution could depict a 3:1 ratio.

17-2. Since the genotype is the same, environmental factors produce the variation within each phenotypic class. For example, plants might receive different amounts of sunlight, minerals, or water.

17-3.

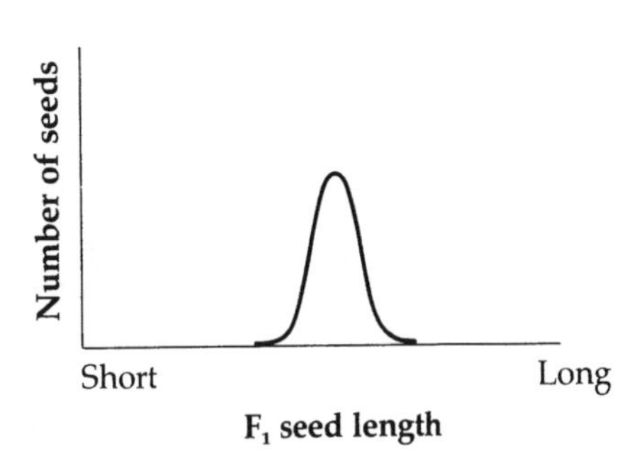

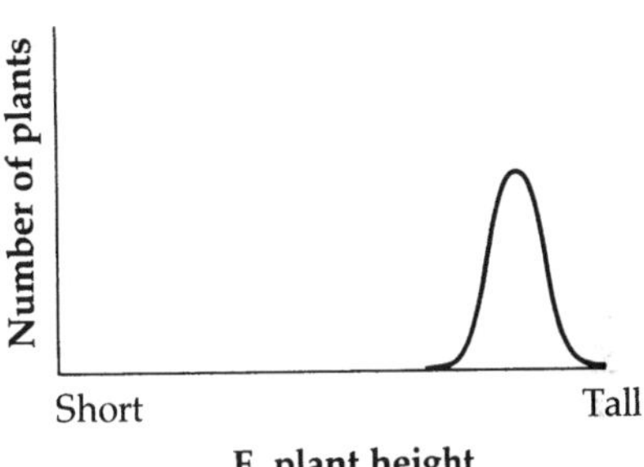

17-4. **a.** Five phenotypes: black (due to four pigment alleles), dark (three pigment alleles), intermediate (two pigment alleles), light (one pigment allele), and white (no pigment alleles).

b. Black: *AABB*; white: *aabb*; their children: *AaBb* (intermediate color).

c. The mating is *AaBb* × *AABB*. *AaBb* makes four kinds of gametes and *AABB* makes one kind. Random union of these gametes is shown as follows.

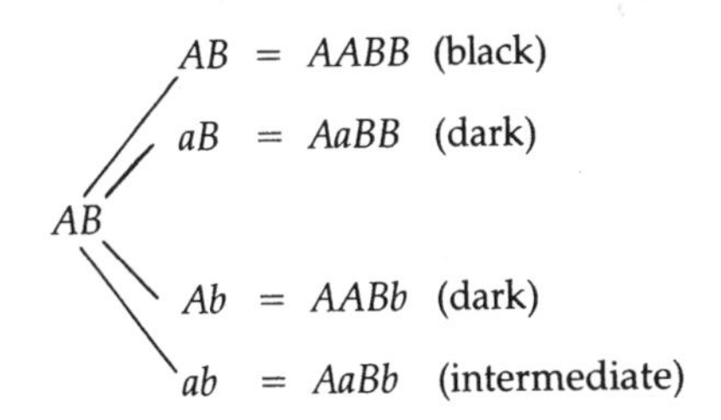

d. Yes, provided both parents have genotype *AaBb*. The mating *AaBb* × *AaBb* could produce *AABB* (black) children among others.

17-5. The number of allelic pairs can be determined since the values in the ratio correspond to the coefficients in an expansion of the binomial $(a + b)^{2n}$, where n is the number of gene pairs. The coefficients 1, 6, 15, 20, 15, 6, and 1 arise when the binomial is raised to the sixth power. Thus, $2n = 6$, and $n = 6/2 = 3$ gene pairs. Another approach involves the expression $1/(4^n)$, where n is the number of loci at which the F_1 parents are heterozygous for the trait. This expression gives the fractions of individuals in the F_2 showing the phenotypic extreme of one of the original parents. Since one out of 64 show one of the parental extremes, the expression $1/(4^n)$ can be set equal to 1/64 and $4^n = 64$. Since $4^3 = 64$, three pairs of alleles must be involved.

17-6. a. *aabbcc* × *aabbcc* gives all white progeny.

b. To get white progeny, both parents must carry at least one recessive allele at each locus. If both parents are heterozygous at all three loci, the probability of producing individuals that are homozygous recessive at all three loci (that is, that have the white phenotype) is 1/4 × 1/4 × 1/4 = 1/64. The parents are *AaBbCc* × *AaBbCc*.

c. To get white progeny, both parents must carry at least one recessive allele at each locus. If both parents are homozygous recessive at one locus and heterozygous at the other two, the probability of producing individuals that are homozygous recessive at all three loci is 1 × 1/4 × 1/4 = 1/16. The parents could be, for example, *AaBbcc* × *AaBbCc*.

17-7. a. The expression $1/(4^n)$, where n is the number of loci at which the F_1 parents are heterozygous for the trait, gives the fraction of individuals in the F_2 that show the phenotypic extreme of one of the original parents. Since two out of 496 show one of the parental extremes, the expression $1/(4^n)$ can be set equal to 2/496 = 1/248, and can be solved for n. Since $1/(4^n) = 1/248$, then $4^n = 248$. Since $4^4 = 256$, which is very close to 248, we can conclude that four pairs of alleles are involved in determining this trait.

b. The conclusion is based on a limited amount of data. As the number of loci involved in determining a quantitative trait increases, a proportionately larger sample is necessary to get a reliable estimate of the number of individuals showing an extreme parental phenotype. In addition, environmental factors could cause enough variation to prevent the clear distinction of the phenotypes.

c. Two F_2 individuals would be expected to show the other parental extreme.

17-8. Four loci are actively involved in determining this trait. Assuming that each has two alleles, one of which makes a contribution to length and the other makes no contribution, an individual with a 7-inch tail would have eight active alleles with each making a contribution to length, while an individual with a minimal tail length of 4 inches has no active alleles. The 3-inch difference between the extremes can be attributed to the additive effect of the eight contributing alleles. The contribution of each allele to tail length is (3 inches)/(8 alleles) = 3/8 inch per allele.

17-9. The fraction of F_2 individuals that is as extreme as either parent is given by the general expression $1/(4^n)$, where n is the number of allelic pairs involved in determining the trait. With four pairs, 1/256 of the F_2 would be expected to show the extreme of either parent, and with five pairs, 1/1024 would be expected. Since over 1000 progeny were surveyed and no individuals as extreme as either parent were found, it could be tentatively concluded that more than five pairs of alleles are involved in determining this trait. Examining a larger sample could provide additional information on this.

17-10. The expression for the number of F_2 phenotypic classes is 2n + 1, where n is the number of allelic pairs determining the trait. With six pairs of alleles, 2(6) + 1 = 13 phenotypic classes would be expected. The expression for the number of F_2 genotypes is 3^n, where n is the number of allelic pairs; six pairs of alleles would give $3^6 = 729$ expected genotypic classes.

17-11. a. Distributions 2 and 3 have the same variance.

b. Distributions 1, 2, and 3 have the same mean.

17-12. a. The frequency distribution gives the number of individuals showing each of the values in the distribution, as shown in Table 17-12a.

b. The mean is obtained by multiplying the value for each class (v) by its frequency (f), adding these (v)(f) values, and dividing by the total number of values:

$$\text{Mean} = \frac{\text{Sum of (v)(f) values}}{\text{Total number of values}} = \frac{486}{21}$$
$$= 23.1 \text{ centimeters.}$$

c. Variance is determined by summing the squared differences between each measurement (x) and the mean ($\bar{x}$) and dividing by the degrees of freedom (total number of individuals minus 1; that is, n − 1):

$$V = \frac{\Sigma (x - \bar{x})^2}{n - 1}.$$

This calculation is shown in Table 17-12c.

d. Standard deviation (SD) is the square root of the variance: $SD = \sqrt{5.33} = 2.31$ centimeters.

17-13. Standard deviation indicates the extent to which values making up a sample are distributed around the mean of that particular sample. The standard error indicates the relationship between the mean of a particular sample and the expected true mean of the entire population.

17-14. a. Standard deviation = SD = $\sqrt{\text{Variance}} = \sqrt{1.21} = 1.1$ centimeters.

b. 68% of the sample values will fall one standard deviation on either side of the mean: 5.27 ± 1.1 centimeters, or from 4.17 to 6.37 centimeters.

c. 99% of the sample values will fall three standard deviations on either side of the mean: 5.27 ± (3)(1.1) = 5.27 ± 3.3 centimeters, or from 1.97 to 8.57 centimeters.

d. SE = $SD/\sqrt{n} = 1.1/\sqrt{60} = 1.1/7.75 = 0.14$ centimeters.

17-15. a. SE = $SD/\sqrt{n}$, where SD is the standard deviation and n is the sample size. A larger sample would cause the standard deviation to be divided by a larger number which would make the standard error smaller.

TABLE 17-12a

Class value, v (length):	18	19	20	21	22	23	24	25	26	27	28
Frequency, f:	1	0	1	2	4	5	3	2	1	1	1
(v)(f) =	18	0	20	42	88	115	72	50	26	27	28

TABLE 17-12c

Class value	Frequency, (f)	$(x - \bar{x})$	$(x - \bar{x})^2$	$(f)(x - \bar{x})^2$
18	1	−5.1	26.01	26.01
19	0	0	0	0
20	1	−3.1	9.61	9.61
21	2	−2.1	4.41	8.82
22	4	−1.1	1.21	4.84
23	5	−0.1	0.01	0.05
24	3	0.9	0.81	2.43
25	2	1.9	3.61	7.22
26	1	2.9	8.41	8.41
27	1	3.9	15.21	15.21
28	1	4.9	24.01	24.01
			Total:	106.61
			Variance = 106.61/20 =	5.33 cm²

b. The 95.5% confidence interval covers two standard error values on either side of the mean: 37.6 ± (2)(0.7) = 37.6 ± 1.4 eggs per fly, or from 36.2 to 39.0 eggs per fly.

c. Correct 95.5% of the time.

d. Substituting into the equation SE = SD/$\sqrt{n}$ gives 0.7 = SD/$\sqrt{100}$ = SD/10. SD = 0.7 × 10 = 7 eggs per fly.

17-16. a. Monozygotic, or identical, twins arise from a single zygote and thus have the same genetic makeup. Dizygotic, or fraternal, twins arise from two separate zygotes produced when two simultaneously released eggs are fertilized; dizygotic twins show no more genetic similarity than would be expected for any two siblings.

b. Twins show concordance for a trait if both express it or if both fail to express it. Discordance for a trait occurs if one expresses it and the other does not.

17-17. a. Broad heritability measures the extent to which phenotypic variation is produced by genetic differences, including additive effects, dominance, and epistasis. Narrow heritability measures the phenotypic variation produced by additive genes only.

b. Narrow heritability measures the variance contributed by additive genes and is thus more useful to breeders since the traits they are concerned with are generally controlled by genes of this type.

17-18. a.

$$\text{Realized heritability} = \frac{\text{Gain because of selection}}{\text{Degree of selection}} = \frac{Y_o - Y}{Y_p - Y}$$

where Y is the average yield for the generation from which the parents are selected, Y_o is the average yield for the next generation, and Y_p is the average yield for the selected parents. The gain in selection is 28.0 − 25.7 = 2.3, and the degree of selection is 29.1 − 25.7 = 3.4. Realized heritability is 2.3/3.4 = 0.676.

b. Yes, since the realized heritability is high.

17-19. (See the equation for realized heritability in the answer to problem 17-18.) For this problem, heritability = 0.11, Y = 9.2, Y_p = 11.1, and the equation must be solved for Y_o, the average litter size in the next generation.

$$0.11 = \frac{Y_o - 9.2}{11.1 - 9.2} = \frac{Y_o - 9.2}{1.9}$$

$$(0.11)(1.9) = Y_o - 9.2$$

$$Y_o = (0.11)(1.9) + 9.2 = 0.21 + 9.2 = 9.41$$

17-20. a. 67% of the 45 twin pairs surveyed, or 45 × 0.67 = 30 pairs, showed concordance.

b. The trait has a significant genetic basis.

17-21. a. Since the variance is 9, the standard deviation is $\sqrt{9}$ = 3 tomatoes. The range of 24 to 30 tomatoes represents one standard deviation on either side of the mean and would be expected to include 68.3% of the plants.

b. The expected percentage of plants producing between 27 and 30 tomatoes (see answer to 17-21a) is 68.3%/2 = 34%, approximately. Half (50%) of all the plants in the distribution produce more tomatoes than the mean (27). Since the plants producing between 27 and 30 tomatoes make up 34% of all plants, we get the percentage of plants producing more than 30 tomatoes by subtracting 34% from 50% to get 16%.

c. Two standard deviations separate 33 from the mean. The expected percentage of plants falling in this range is 95.5%/2 = 47.75%.

d. Two standard deviations separate 21 from the mean. The expected percentage of plants falling in the 21-to-27 range is 95.5%/2 = 47.75%. Subtracting this from the half (50%) of the plants in our distribution that produce fewer tomatoes than the mean (27) gives 2.25% as the expected percentage of plants producing fewer than 21 tomatoes.

17-22. a. Very little, since this is a pure-breeding or genetically uniform line. The phenotypic variation is environmentally induced.

b. Variance = (Standard deviation)² = (2.39)² = 5.71 sq. days. Since the phenotypic variance is environmentally induced, all this variance may be considered as environmental variance. Environmental variance = 5.71 sq. days.

c. Heritability is the proportion of total phenotype variation due to genetic factors. Since the population is genetically uniform, heritability is 0.

d. The time to maturity would be unchanged from that of the parental generation: 77 days. Without genetic variability, selection will not change the maturation time.

17-23. a. The F_1 is a genetically uniform population and thus its phenotypic variation can be attributed to environmental variance: $V_e = 4.85$ sq. days. The F_2 is a genetically heterogeneous population and its phenotypic variation is due to a combination of genetic and environmental variances: $V_g + V_e = 9.67$ sq. days. Substituting the value of V_e into this equation gives: $V_g = 9.67 - 4.85 = 4.82$ sq. days.

b. Broad heritability: H_B, $= V_g/V_p = 4.82/9.67 = 0.498 = 0.5$.

Chapter 18: Nucleic Acid Structure

18-1. a. 3′ TTAGCGGGTAACGTCAAG 5′

b. 3′ GCTAACCGAAT 5′

18-2. a. AT base pairs are joined by two hydrogen bonds whereas GC pairs are joined by three. The additional bond at the GC pairs makes them more resistant to denaturation than the AT pairs. As a consequence, the higher the percentage of GC pairs in a DNA duplex, the greater the amount of heat required for denaturation. Each duplex has the same number of base pairs. Since molecule 2 has nine CG pairs, it will have the highest t_m; molecule 3, with five pairs, will have an intermediate t_m; and molecule 1, with three pairs, will have the lowest t_m.

b. The greater the percentage of AT pairs in a molecule, the smaller the amount of energy required for melting and the lower the t_m.

18-3. Although the DNA molecules include the same frequencies of bases and the same number of base pairs, they would also have to have their bases arranged in identical sequences in order for the molecules to be identical. We have no information indicating that the sequences are identical and thus cannot conclude that the molecules are identical.

18-4. a. The presence of thymine indicates that this is DNA. Thus, the nucleotides would contain deoxyribose sugar.

b. Adenine, guanine, and cytosine would be expected. However, we can be certain only of the presence of adenine.

c. Since the molecule is double-stranded, we can predict the frequencies of the other bases. Since T is always paired with A, the percentage of T equals the percentage of A, and A and T comprise 13% + 13% = 26% of the bases in the molecule. This leaves 100% − 26% = 74% of the bases to be C and G. Since C and G are always paired in a double-stranded DNA. they will occur with equal frequencies. Thus, the frequency of C or G equals half of their combined frequency, or 74% × 1/2 = 37%.

d. No. The bases in one section of a single strand have no chemical relationship to other bases in the same strand.

18-5. a. The presence of uracil indicates that this is RNA. Thus, the nucleotides would contain ribose sugar.

b. Adenine, guanine, and cytosine would be expected. However, we can be certain only of the presence of adenine.

c. If the molecule is double-stranded, we can predict the frequencies of the other bases. Since U is always paired with A, the percentage of U equals the percentage of A, and U and A comprise 23% + 23% = 46% of the bases in the molecule. This leaves 100% − 46% = 54% of the bases to be C and G. Since C and G are always paired in a double-stranded molecule, they will occur with equal frequencies. Thus, the frequency of C or G equals half of their combined frequency, or 54% × 1/2 = 27%.

d. No. The bases in one section of a single strand have no chemical relationship to other bases in the same strand.

18-6. a. Each duplex has 15 A bases and 15 T bases, giving an A-to-T ratio of 15/15 = 1.

b. Each duplex has 10 G bases and 10 C bases, giving a G-to-C ratio of 10/10 = 1.

c. The (A + T)-to-(G + C) ratio for each molecule = (15 + 15)/(10 + 10) = 30/20 = 1.5.

d. Yes. The identical (A + T)-to-(G + C) ratios for each molecule confirm the fact that they have the same nucleotide composition.

18-7. Virus 1: Since T is present, it may be assumed that the molecule is DNA. Since A = T and C = G, it is, most likely, double-stranded. Virus 2: Since T is present, the molecule is DNA. Lacking information about the frequencies of the other bases, we cannot tell the number of strands in the molecule. Virus 3: Since U is present it may be assumed that the molecule is RNA. Since A = U and C = G it is, most likely, double-stranded. Virus 4: Since T is present, the molecule is DNA. Since A does not equal T and C does not equal G, the molecule is single-stranded. Virus 5: Since U is present, the molecule is RNA. Lacking information about the frequencies of the other bases, we cannot tell the number of strands in the molecule. Virus 6: Since U is present, the molecule is RNA. Since A does not equal U and C does not equal G, the molecule is single-stranded. Virus 7: Since T is present, the molecule is DNA. Since A does not equal T and C does not equal G, the molecule is single-stranded. Note that the percentage of G can be determined by subtracting the sum of the percentages of the other three bases (23 + 18 + 28 = 69) from 100%: 100% − 69% = 31%.

18-8. a. Since there is one base in each nucleotide, the total number of bases gives us the total number of nucleotides. Since each of the 4200 base pairs contains two bases, there is a total of 4200 × 2 = 8400 bases for a total of 8400 nucleotides.

b. Since each of the 8400 nucleotides has one sugar, there are 8400 sugars.

c. With a complete turn of the double helix every 10 base pairs, there would be 4200/10 = 420 turns.

d. A complete turn requires 3.4 nanometers. With 420 complete turns, the molecule would be 420 × 3.4 = 1428 nanometers long.

18-9. Virus 3 is double-stranded and virus 4 is single-stranded. Since each has the same number of nucleotides, the molecules would be of different lengths with the length of each strand of the molecule in virus 3 half the length of the strand in virus 4.

18-10. a. Multiplying the grams of DNA by the number of nucleotides in a gram gives the total number of nucleotides in the DNA: $(3.2 \times 10^{-12})(2.0 \times 10^{21}) = 6.4 \times 10^{9}$ nucleotides.

b. Dividing the number of nucleotides by 2 gives the number of nucleotide pairs: $(6.4 \times 10^{9})/2 = 3.2 \times 10^{9}$ nucleotide pairs.

c. Dividing the number of nucleotide pairs by the number of pairs in 1 millimeter of duplex DNA gives the total length of the duplex DNA in the haploid cell: $(3.2 \times 10^{9})/(2.9 \times 10^{6}) = (1.103 \times 10^{3}$ millimeters $= 1.103$ meters.

18-11. a. Multiplying the grams of DNA by the number of nucleotides in a gram gives the total number of nucleotides in the DNA: $(4.7 \times 10^{-15})(2.0 \times 10^{21}) = 9.4 \times 10^{6}$ nucleotides.

b. Dividing the number of nucleotides by 2 gives the number of nucleotide pairs: $(9.4 \times 10^{6})/2 = 4.7 \times 10^{6}$ nucleotide pairs.

c. Dividing the number of nucleotide pairs by the number of pairs in 1 millimeter of duplex DNA gives the total duplex length: $(4.7 \times 10^{6})(2.9 \times 10^{6}) = 1.62$ millimeters.

Chapter 19: DNA Replication and Recombination

19-1. Two bands would occur, one containing the heavy (^{15}N) strands and the other, the light (^{14}N) strands. The light band would be wider, since there are three times as many ^{14}N strands as there are ^{15}N strands. This is predicted by semiconservative replication where one of the four strands is an ^{15}N parent strand, one of four is an ^{14}N parent strand, and two of the four are ^{14}N daughter strands.

19-2. a. The banding pattern would have been the same. Of the duplexes, 1/4 would have both strands labeled with ^{15}N and 3/4 would have both strands labeled with ^{14}N. When heat denatured, 1/4 of the strands would contain ^{15}N and 3/4 would contain ^{14}N.

b. Each strand of each duplex would contain some ^{15}N and some ^{14}N, with the ^{14}N predominating. One broad band would result, positioned between the sites expected for strands fully labeled with ^{15}N and those fully labeled with ^{14}N, but closer to the ^{14}N site.

19-3. a. See Table 19-3a. If you do not understand where these proportions come from, consider the following. All nucleotides in the duplexes prior to the first round of replication contain nonradioactive P. Both strands of each duplex would serve as templates and all the nucleotides incorporated into their newly synthesized complementary strands would contain ^{32}P. Thus, following the first round of replication, one strand of each new duplex would consist of ^{32}P nucleotides and the other strand, nonradioactive P nucleotides. The following diagram shows the production of these duplexes as well as those produced in the second and third round of replication, where | represents a strand with nonradioactive P nucleotides and |*, a strand with ^{32}P nucleotides.

Prereplication: ||

After replication 1: ||* |*|

After replication 2: ||* |*|* |*|* |*|

After replication 3: ||* |*|* |*|* |*|* |*|* |*|* |*|* |*|

After replication 2, both strands of half the duplexes would contain all ^{32}P nucleotides. The other half of the duplexes would have one strand of ^{32}P nucleotides and one strand of nonradioactive P nucleotides. After replication 3, both strands of 3/4 of the duplexes would consist of all ^{32}P nucleotides, and 1/4 of the duplexes would have one strand of ^{32}P nucleotides and one strand with nonradioactive P nucleotides.

b. In duplexes containing some ^{32}P, half the nucleotides would contain ^{32}P and those would comprise all the nucleotides in one strand. (The other strand would consist solely of normal P nucleotides.)

19-4. After the second replication on the ^{32}P medium and before transfer to the medium with nonradioactive P, the duplexes could be represented as follows, where | is a strand containing nonradioactive P nucleotides and |* is a strand containing ^{32}P nucleotides.

After replication 2: ||* |*|* |*|* |*|

After the next replication (on the nonradioactive P medium), the duplexes could be represented as follows.

After replication 3: || ||* |*| ||* |*| ||* |*| ||

Three-fourths of the duplexes would have one strand of ^{32}P nucleotides and one strand with nonradioactive P nucleotides. One-fourth of the duplexes would have both strands containing nonradioactive nucleotides.

19-5. The duplexes present prior to the first round of replication would contain nothing but nonradioactive P nucleotides. If the DNA replicated under the conservative scheme, one of the duplex molecules arising from the first round of replication would be the original, fully conserved duplex, and the other duplex would be newly synthesized. Thus, following the first round of replication, one duplex would contain only nonradioactive nucleotides and the other duplex only radioactive nucleotides. The following diagram shows these duplexes and those produced in the second and third rounds of conservative replication, where | represents a strand containing nonradioactive

TABLE 19-3a

After replication	Proportion of Duplexes with		
	Only ^{32}P nucleotides	Some ^{32}P nucleotides and some normal P nucleotides	Only normal P nucleotides
1	0	All	0
2	1/2	1/2	0
3	3/4	1/4	0

P nucleotides and |* represents a strand containing ^{32}P nucleotides.

Prereplication: | |
After replication 1: | | |*|*
After replication 2: | | |*|* |*|* |*|*
After replication 3: | | |*|* |*|* |*|* |*|* |*|* |*|* |*|*

After replication 2, 1/4 of the duplexes would contain only nonradioactive P nucleotides and 3/4 would contain only ^{32}P nucleotides. After replication 3, 1/8 would contain only nonradioactive P nucleotides and 7/8 would contain only ^{32}P nucleotides.

19-6. **a.** DNA polymerase I has a dual role: as an exonuclease, it removes the ribonucleotides making up the RNA primer segments; and as a polymerase, it fills, with deoxyribonucleotides, the gap created by the removal of each RNA primer. DNA polymerase III guides the synthesis of strands of DNA. RNA primase guides the synthesis of RNA primers. DNA ligase guides the formation of phosphodiester bonds between adjacent segments of DNA.

b. RNA primase, DNA polymerase III, DNA polymerase I (as an exonuclease), DNA polymerase I (as a polymerase), and DNA ligase.

c. Discontinuous DNA replication involves the formation of many short fragments of DNA (Okazaki fragments) which are then joined to form a long DNA strand. Since the synthesis of each of these fragments requires an RNA primer, there will be much use of all four of these enzymes during discontinuous DNA synthesis. Continuous replication requires a single RNA primer to initiate the synthesis of a very long strand of DNA, so the enzymes involved in primer synthesis and removal, gap filling, and the end-to-end joining of DNA fragments are called upon much less frequently for the continuous synthesis of DNA.

19-7. **a.** Single-stranded DNA-binding protein associates with the individual strands of the duplex after they have separated from each other and maintains the separation. DNA helicase is the enzyme catalyzing the breaking of the hydrogen bonds joining the two strands of the double helix. DNA gyrase is an enzyme that relieves the twist tension in the intact sections of the duplex molecule beyond the replication fork by creating nicks in strands, allowing those strands to rotate to relieve the supercoiling, and then closing the nicks.

b. Helicase, single-stranded DNA-binding protein, and DNA gyrase.

c. These three proteins do not participate directly in DNA replication. Rather, they pave the way for it by preparing the replication fork so that DNA replication, continuous and discontinuous, can occur. There is no difference in the frequency of use of these three proteins by continuous or discontinuous replication.

19-8. A template is an already existing nucleic acid strand that guides, in complementary fashion, the synthesis of its nucleotide image. A strand of DNA is the template for a DNA complementary strand. The template ensures that the newly synthesized DNA is accurately replicated and therefore meaningful. A primer is a short, newly synthesized nucleic acid segment that is hydrogen bonded to a complementary portion of the DNA strand serving as its template. It provides the essential, free 3′ hydroxyl end required by polymerase III for the initiation of DNA synthesis.

19-9. One strand must be synthesized continuously while the other strand is synthesized discontinuously. Synthesis along the 5′-to-3′ template strand is initiated by a single RNA primer and occurs continuously. Synthesis along the 3′-to-5′ template strand occurs discontinuously and generates short DNA pieces, each with its own RNA primer, that are joined to make a single strand following primer removal.

19-10. **a.** The replication fork moves to the left.

b. Continuous synthesis occurs along the lower, or 5′-to-3′, template strand.

c. Okazaki fragments will be assembled to make up the discontinuous strand, and the upper template strand guides their formation.

d. Each Okazaki fragment requires an RNA primer. Since the fragments are synthesized along the upper template strand, numerous RNA primers will be on the upper template strand.

e. The continuously assembled strand is the leading strand. It forms on the lower template strand.

19-11. **a.** With a single replication fork having a replication rate of 2600 nucleotide pairs per minute, $6 \times 10^7 = 60{,}000{,}000$ nucleotide pairs would require $60{,}000{,}000/2600 = 23{,}076.9$ minutes, or $23{,}076.9/60 = 384.6$ hours, or $384.6/24 = 16$ days. Since there are two replication forks at the single origin, with each fork "working" at the rate of 2600 nucleotide pairs per minute, the replication would require about $16/2 = 8$ days.

b. At the rate of 2600 nucleotide pairs per minute, each replication fork could, in four minutes, replicate $2600 \times 4 = 10{,}400$ nucleotide pairs; dividing this value into the total number of nucleotide pairs in the molecule gives us the number of replication forks: $60{,}000{,}000/10{,}400 = 5769$. Since each origin has two replication forks, the number of origins is $5769/2 = 2884.5$, or about 2885.

19-12. **a.** The DNA fragments produced during the 15-second exposure represent the beginning of the continuously assembled leading strands and the discontinuously assembled lagging strands. Eventually both types of strands will reach the length of the bacterial chromosome. As the interval of exposure to the tracer lengthens, more tracer-carrying short fragments (Okazaki fragments) are synthesized and more of these fragments become incorporated into the lagging strands. Consequently, increasing amounts of the tracer will be found in longer fragments, but some will also be found in the newly formed short fragments.

b. Yes. Since replication is discontinuous along one of the original DNA strands at each replication fork, the short fragments will be produced until the end of DNA replication.

c. Yes. If replication occurred in continuous fashion along both strands at each replication fork, a short fragment of newly synthesized DNA would exist along each template strand for only a short time at the beginning of synthesis. With new nucleotides continually being added, each of these strands would grow progressively longer and longer, and short fragments would not be found in a cell again during that round of DNA replication.

d. Of the four kinds of nucleosides that are deoxyribonucleotide precursors, three of them are also precursors of ribonucleotides and could also be incorporated into RNA. Thymidine is the only one that is incorporated into DNA only.

19-13. Three general types of DNA molecules would be found. There would be a large number of short Okazaki fragments that contain the labeled thymidine; many of these would not have been assembled into longer molecules because of the ligase deficiency. The second type would be of intermediate length and would contain the labeled thymidine; most of these molecules would be the newly synthesized leading strands. The third type of DNA molecule would be unlabeled and contain many thousands of nucleotides; this would be the original DNA present in the cell prior to this round of DNA replication.

19-14. The double helix of the chromosome makes a complete turn about its axis every 3.4 nanometers, or 0.0034 micrometers. With a length of 1300 micrometers, the chromosome has a total of 1300/0.0034 = 382,353 turns. Since there are two replication forks involved and each is preceded by a swivel point, half of these turns, or 382,353/2 = 191,176 turns, are untwisted through each swivel point in a 40-minute period. The rate of untwisting at each swivel is 191,176/40 = 4779 turns per minute.

19-15. a. Note that a correct drawing could show either the inner or the outer strand as the labeled strand.

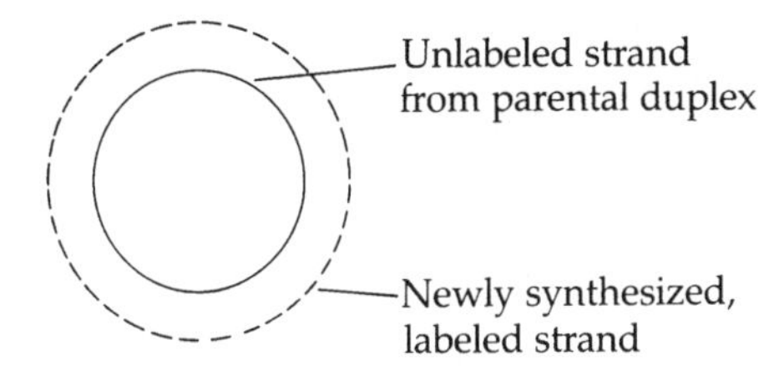

An autoradiograph of the chromosome consists of a single strand of dots produced by radioactivity from the labeled strand.

b. Theta-form chromosome:

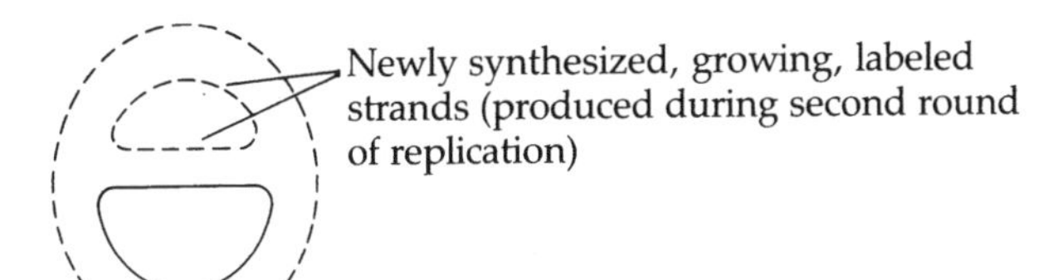

Autoradiograph of this chromosome:

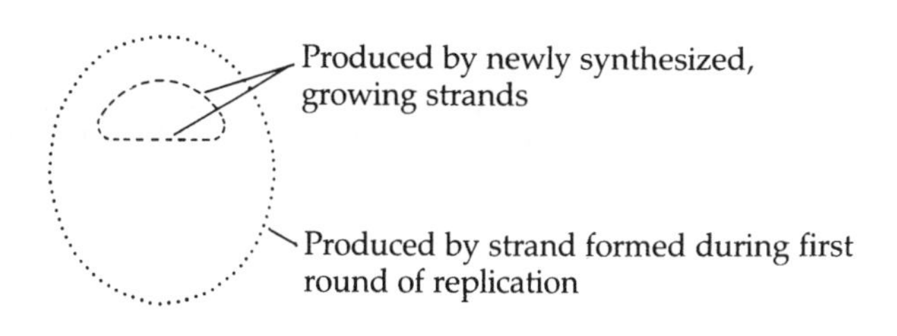

19-16. a. The chromosome rotates counterclockwise as more of the original positive strand peels off to form a longer tail.

b. Discontinuous synthesis occurs along the exposed, or tail, section of the original positive strand, beginning at its 5′ end.

c. Okazaki fragments will join to make up the discontinuous strand that is assembled on the original positive strand template.

d. No RNA primer is required. The 3′ hydroxyl end created by the initial nicking of the original positive strand serves as the site for the addition of deoxyribonucleotides that will make up the continuous strand.

19-17. The frequency of mismatch repair would be reduced, thereby reducing the frequency of gene conversion.

19-18. The duplex that does not experience mismatch repair carries information for both the s^+ and s alleles. When this duplex replicates semiconservatively prior to mitosis, one daughter duplex will carry both strands of allele s^+ and the other daughter duplex will carry both strands of allele s. In the ascus that is formed, alleles a^+ and a would be expected in the ascospores in a 4:4 ratio, while alleles s^+ and s would occur in either a 5:3 or a 3:5 ratio, depending on whether the duplex that underwent mismatch repair initially carried s^+ or s.

19-19. Branch migration causes the elongation of the heteroduplex region so that it includes both intermediate loci.

Chapter 20: The Basis of Prokaryotic Inheritance

20-1. A broth is a liquid food medium generally used for mass culturing, whereas agar is a semisolid, gel-like medium generally used for culturing individual colonies.

20-2. a. Both a colony and a lawn are aggregations of bacteria visible to the naked eye. A colony is a relatively small circular patch of bacteria which presumably develops from a single bacterium and covers a limited area of the agar. A lawn is a smooth layer of bacteria covering all of the surface of the agar in the petri dish. A lawn develops from the reproduction of numerous cells.

b. Overnight growth of the bacteria on a petri dish is generally sufficient time to allow for colony production. A lawn is produced by adding a larger number of cells (about 10^8, or two drops from an overnight broth culture) to liquid agar which is then poured on a plate, or alternatively, by spreading the bacteria over the surface of an agar plate. The overnight growth of these cells will form a continuous layer of cells.

20-3. The key to this is sufficient dilution of the sample used to inoculate the agar plate.

20-4. a. A minimal medium usually contains a sugar (or sometimes an amino acid) which serves as a carbon and energy source, and an assortment of salts (for some species of bacteria, a few other supplements are needed as well). A complete medium contains preformed quantities of all of the metabolites essential for bacteria growth and reproduction.

b. Prototrophic, or nutritionally competent, bacteria can reproduce on both the complete and minimal media.

c. Auxotrophic, or nutritionally deficient, bacteria can reproduce on the complete medium only.

20-5. Bacteria that can live on a minimal medium are able to synthesize from the constituents of the medium all the amino acids and vitamins necessary for their maintenance, growth, and reproduction.

20-6. Yes. Auxotrophic bacteria can generally be grown on a minimal medium supplemented with the particular substance or substances that they are unable to synthesize.

20-7. a. The auxotrophic cells most likely arose when a gene coding for a biosynthetic enzyme experienced a spontaneous mutation. The bacterium carrying this mutation subsequently reproduced.

b. The auxotrophs were able to survive because the medium is complete and thus supplies the metabolite they are unable to synthesize.

c. The auxotrophs would lack the activity of at least one of the enzymes made by the prototroph.

20-8. In principle, a sample of bacteria from the culture tube could be used to inoculate a petri dish containing a complete medium. Overnight culturing of this plate would allow both the prototrophic and auxotrophic bacteria to grow into colonies. This plate could then serve as the master plate for replica plating onto a petri dish containing a minimal medium. Overnight culturing of this replica plate would allow just the prototrophs to grow into colonies. Any master plate colony not appearing on the replica plate would consist of the auxotrophic bacteria. This approach to isolation would be difficult, however, because of the very limited number of auxotrophic bacteria that would be present (mutants would occur with a rate of about 1 in 10^6 bacteria). A better approach would be to enrich for mutants by inoculating a penicillin-containing minimal medium with a sample from the culture tube. Here the prototrophs could divide and would be killed off, while the autotrophs would survive. Removal of the penicillin and transferral of the cells to a complete medium would allow the auxotrophs to form colonies.

20-9. a. A sample of bacteria from the culture tube could be used to inoculate a petri dish containing a complete medium. Overnight culturing of this plate would allow both the prototrophic and auxotrophic bacteria to grow into colonies. This plate could then serve as the master plate for replica plating onto a series of petri dishes containing a minimal medium, each supplemented with a different amino acid. The plate with the minimal medium and amino acid that developed the same distribution and number of colonies as the complete-medium master plate would contain the amino acid that the auxotroph was unable to synthesize. All the other minimal-medium-plus-amino-acid plates would allow just the growth of the prototrophic bacteria. The penicillin-enrichment procedure could be used to facilitate this process.

b. The auxotrophic bacteria could be cultured by growing them on a complete medium or a minimal medium supplemented with the amino acid they were unable to synthesize.

20-10. a. Four of the master-plate colonies, 2, 6, 9, and 11, failed to grow on the minimal medium, most likely because they were incapable of synthesizing all their essential molecules from the constituents of the minimal medium.

b. The single colony that failed to grow on the minimal medium but grew on the minimal medium plus proline (colony 9) lacked the ability to synthesize proline.

c. The two colonies that failed to grow on the minimal medium but grew on the minimal medium plus valine (colonies 2 and 11) lacked the ability to synthesize valine.

d. The ten bacterial types that grew on the minimal medium were obviously able to synthesize aspartic acid. Since the four types that did not grow on the minimal medium also failed to grow on the minimal medium plus aspartic acid, it is reasonable to conclude that the bacteria from all 14 of the master-plate colonies were able to synthesize this amino acid.

20-11. a. We could tentatively conclude that these bacteria were unable to synthesize an essential nutrient and that the nutrient was not aspartic acid, proline, or valine.

b. Since the bacteria grew only when both amino acids were supplied, we can conclude that the bacteria lacked the ability to synthesize both valine and proline.

c. If these two amino acids are synthesized separately, this could mean that the bacteria carried two separate mutations, one affecting the biosynthetic pathway for each amino acid. If the synthesis of the two amino acids is related, such that one amino acid is a precursor for the synthesis of the second, a single mutation that blocks the synthesis of the precursor amino acid might be responsible.

20-12. a. Strain 3 is the wild-type. (The wild strain can synthesize both enzymes and thus would be expected to grow on both types of media.) Strain 2 is deficient for enzyme A. (The strain deficient for enzyme A would not be able to produce the intermediate substance from the minimal medium. Since it has enzyme B, it can, if supplied with the intermediate, synthesize the amino acid. Thus, it will not grow on the minimal medium but can grow on the supplemented minimal medium.) Strain 1 is deficient for enzyme B. (The strain deficient for enzyme B will not be able to convert the intermediate substance into the amino acid and thus will not be able to grow on either medium.)

b. A strain deficient in both enzymes would be unable to grow on either medium. This strain cannot be unambiguously distinguished since the same results were found with the strain deficient in enzyme B.

20-13. The key steps are as follows. (1) Spread an appropriately diluted sample from the mixed culture on minimal agar medium supplemented with phenylalanine. This will allow both the wild-type Phe$^+$ and the mutant Phe$^-$ cells to grow. The count of colonies following overnight incubation will give the total number of Phe$^+$ and Phe$^-$ bacteria in the inoculating sample. (2) Replica plate the colonies in this dish onto a minimal medium where only the wild-type Phe$^+$ cells will grow. The number of colonies following overnight incubation gives the number of Phe$^+$ cells in the inoculating sample. (3) The difference between the number of colonies growing on the minimal medium plus phenylalanine and the number of colonies growing on the minimal medium equals the number of Phe$^-$ cells in the inoculating sample. Dividing this number by the total number of bacteria in the sample gives the relative proportion of Phe$^-$ cells in the sample.

20-14. a. The wild-type cells will be able to grow on this medium since they can make their own tyrosine. The penicillin, however, interferes with cell-wall synthesis and thus kills these actively growing cells. The Tyr$^-$ cells will not be able to grow on this tyrosine-deficient medium and will not be killed by the penicillin. Thus, this medium selects for the Tyr$^-$ cells.

b. Washing away the penicillin and transferring the survivors to a penicillin-free medium containing tyrosine will allow them to reproduce.

Chapter 21: Transformation and Mapping the Bacterial Chromosome

21-1. Transformation is a two-step process which involves (1) the uptake of the DNA fragment by the bacterial cell followed by (2) the integration of the fragment into the bacterial chromosome. Integration of the fragment requires that it be homologous with a section of the recipient's chromosome. The eukaryotic DNA lacks this homology.

21-2. Since cotransformation of the two loci is a very rare event, it is likely that the cotransformants arise from a double-transformation event. Therefore, it seems likely that the two loci are separated by more than 20,000 base pairs.

21-3. Note that the insertion of each donor DNA fragment requires two crossovers.

a. One crossover in zone 1 and one in zone 2.

b. One crossover in zone 1 and one in zone 3.

c. One crossover in zone 2 and one in zone 3.

21-4. Yes. Double transformants can also arise through the uptake of two separate DNA fragments, each carrying one of the two loci involved.

21-5. a. Cotransformation frequency =

$$\frac{\text{Number of } e \text{ and } f \text{ cotransformants}}{\text{Total number transformed for } e}.$$

Cotransformant class e^+f^+ has 106 bacteria. Classes e^+f^- (158) and e^+f^+ (106) are both transformed for the e locus. The frequency of cotransformation of f with e is $106/(158 + 106) = 106/264 = 0.402$.

b. Cotransformation frequency =

$$\frac{\text{Number of } f \text{ and } g \text{ cotransformants}}{\text{Total number transformed for } f}.$$

Cotransformant class f^+g^+ has 177 bacteria. Classes f^+g^- (45) and f^+g^+ (177) are both transformed for the f locus. The frequency of cotransformation of g with f is $177/(45 + 177) = 177/222 = 0.797$.

c. Cotransformation frequency is inversely related to the distance between the loci under study, that is, the greater the cotransformation frequency, the closer the loci. The f and g loci are much closer than are the e and f loci, with the distance between f and g (transformation frequency 0.797) about half that of the distance separating the e and f loci (transformation frequency 0.402).

21-6. a. As the DNA concentration is reduced to half and then to one-quarter of its initial level, the relative number of single transformants declines to $56/100 = 0.56$, $0.56 \times 100 = 56\%$, and to $30/100 = 0.30$, $0.30 \times 100 = 30\%$, respectively, of its initial level. The relative number of double transformants declines to $6/10 = 0.6$, $0.6 \times 100 = 60\%$, and to $2/10 = 0.20$, $0.20 \times 100 = 20\%$, respectively, of its initial level. A comparison of the decline in single transformants with the decline in double transformants indicates that they are declining at about the same rate. This implies that the double transformants are produced through the uptake of single pieces of DNA.

b. Since the double transformants arise primarily through the uptake of single pieces of DNA, we could conclude that the two loci under study are closely linked.

21-7. a. As the DNA concentration is reduced to half and then to one-quarter of its initial level, the relative number of single transformants declines to $38/81 = 0.469$, $0.469 \times 100 = 47\%$, and to $19/81 = 0.234$, $0.234 \times 100 = 23\%$, respectively, of its initial level. The relative number of double transformants declines to $8/75 = 0.107$, $0.107 \times 100 = 10.7\%$, and to $0.8/75 = 0.0107$, $0.0107 \times 100 = 1.07\%$, respectively, of its initial level. A comparison of the decline in single transformants with the decline in double transformants indicates that the double transformants decline at a much faster rate. This implies that the double transformants are produced through the uptake of two pieces of DNA.

b. Since the double transformants arise through the uptake of two pieces of DNA, the two loci under study are relatively distantly linked.

21-8. a. Since the vast majority of bacteria in this mixture are untransformed (genotype c^-d^-), any approach used to identify the transformants should block the growth of these cells. One possible approach is as follows. A diluted sample from the mixture is plated onto a minimal medium supplemented with amino acid C which would allow both c^-d^+ and c^+d^+ to grow. A second sample is plated onto a minimal medium supplemented with amino acid D which would allow the growth of c^+d^- and c^+d^+ cells. Following overnight incubation, the colonies growing on each of these two types of media are separately replica plated onto the opposite type of medium; that is, minimal medium + amino acid C colonies are replica plated onto minimal medium + amino acid D, and vice versa. Following overnight incubation, the arrangement of colonies on each replica plate is compared with that found on its master plate. Any colony found to be growing on one of the master plates and on its replica is a double transformant with genotype c^+d^+. Colonies found on the master plate with minimal medium + amino acid C but absent from its replica plate (minimal medium + amino acid D) are c^-d^+ transformants. Colonies found on the master plate with minimal medium + amino acid D but absent from its replica plate (minimal medium + amino acid C) are c^+d^- transformants.

b. The low frequency of double transformants suggests two transformation events, with one piece of DNA carrying c^+ and the other carrying d^+. (Note that there are no donor strands carrying both c^+ and d^+.)

21-9. a. Since the frequency of the double-transformant class is similar to the frequency of the single-transformant classes, it is likely that the double-transformant class arose from the uptake of a single piece of DNA. Note that the donor strands carry both c^+ and d^+.

b. There are many more double transformants produced in the second experiment. The frequency in the second experiment (0.418) is about 54 times the frequency in the first experiment (0.0077).

c. The high-frequency in the second experiment suggests close linkage—close enough to have both loci readily transmitted on a single piece of transforming DNA.

d. Yes. If the double transformants in the second experiment have arisen through the simultaneous uptake of two pieces of DNA, their frequency would be much lower.

21-10. a. With distantly linked loci, the most common transformants would be single transformants. These three classes of transformants ($s^+t^-u^-$, $s^-t^+u^-$, and $s^-t^-u^+$) would be expected to occur in roughly equal frequencies.

b. In order for a transformant to acquire wild-type alleles at two of the three loci under consideration, two transformation events, each involving a separate fragment of transforming DNA, would have to occur. The probability of this is given by the product law and thus the frequency of this type of transformant will be considerably less than the frequency of single-locus transformants (which arise through single transformation events).

21-11. a. Double transformants, $s^-t^+u^+$, and single transformants, $s^+t^-u^-$, would be expected to be the most common since each can arise through uptake of a single piece of transforming DNA.

b. Each class would be expected to occur in roughly equal frequencies.

21-12. Cotransformation frequency =

$$\frac{\text{Number of } o \text{ and } n \text{ cotransformants}}{\text{Total number transformed for } n}.$$

Cotransformant class n^+o^+ has 283 bacteria. Classes n^+o^- (200) and n^+o^+ (283) are transformed for the n locus. The frequency of cotransformation of o with n is 283/(200 + 283) = 283/483 = 0.586.

21-13. a. Transformant classes 1, 5, and 7, are single recombinants. Classes 2, 4, and 6 are double recombinants. Class 3 is a triple recombinant.

b. The triple-transformant class (3) has the highest frequency, indicating that the three loci were transmitted together on a single piece of DNA. This cotransmission could occur only if the loci were very closely linked.

c. If the loci were not tightly linked, the triple transformant would arise through the simultaneous uptake of separate pieces of DNA. This occurs with a frequency that is extremely low relative to the frequency of the uptake of a single piece. Consequently, if the loci were not tightly linked, the triple-transformant class would occur with a very low frequency.

d. Class 2, with the lowest frequency, arose through a quadruple crossover.

e. A comparison of the genotypes of the quadruple-crossover class (*met*$^+$ *ile*$^-$ *thr*$^+$) and the recipient parent (*met*$^-$ *ile*$^-$ *thr*$^-$) shows that the *ile* locus carries the same allele (*ile*$^-$) in both, indicating that *ile* is the middle locus.

21-14. a. The cotransformation frequencies can be determined by inspecting the loci two at a time in each of the seven classes. Begin by identifying the classes that are transformed for *met* and then identify the *met* classes that are also cotransformed for *ile*. Dividing the number of cotransformants by the total number of *met* transformants gives the cotransformation frequency. Classes 2, 3, 6, and 7 with a total of 65 + 8012 + 1002 + 2106 = 11,185 bacteria are transformed for *met*, and classes 3 and 6 with a total of 8012 + 1002 = 9014 bacteria are cotransformed for *ile*. The cotransformation frequency is 9014/11,185 = 0.806, or 0.806 × 100 = 80.6%. Regarding the cotransformation of *thr* with *met*, classes 2, 3, 6, and 7 with a total of 65 + 8012 + 1002 + 2106 = 11,185 bacteria are transformed for *met*, and classes 2 and 3 with a total of 65 + 8012 = 8077 bacteria are cotransformed for *thr*. The cotransformation frequency is 8077/11,185 = 0.722, or 0.722 × 100 = 72.2%. Regarding the cotransformation of *ile* with *thr*, classes 2, 3, 4, and 5 with a total of 65 + 8012 + 2800 + 815 = 11,692 bacteria are transformed for the *thr* locus, and classes 3 and 4 with a total of 8012 + 2800 = 10,812 bacteria are cotransformed for *ile*. The cotransformation frequency is 10,812/11,692 = 0.925, or 0.925 × 100 = 92.5%.

b. The cotransformation map for these three loci is as follows. Remember that the greater the cotransformation frequency, the closer the linkage.

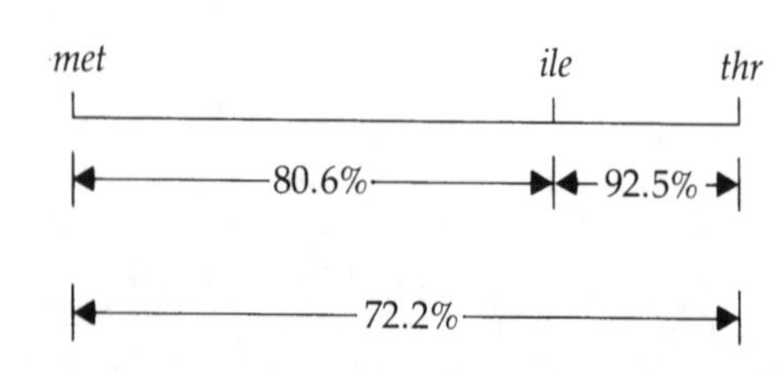

21-15. The triple-transformant class occurs with a very low frequency relative to the double-transformant classes and with a low frequency relative to the single-transformant classes. This implies that the triple transformant arose not through the uptake of a single piece of DNA, but rather through the simultaneous occurrence of two or, possibly, three transformation events. This leads us to conclude that none of the loci are tightly linked.

21-16. **a.** Transformant classes 3, 6, and 7 are single transformants, classes 1, 2, and 5 are double transformants, and class 4 is a triple transformant.

b. The class with the lowest frequency, class 5, with a frequency of 11, arose through a quadruple crossover.

c. A comparison of the genotypes of the recipient parent, *aro⁻his⁻try⁻*, and the quadruple-crossover transformant, *aro⁻his⁺try⁺*, indicates that the *aro* locus carries the same allele (*aro⁻*) in both and is thus the middle locus.

Chapter 22: Conjugation and Mapping the Bacterial Chromosome

22-1. Following an interval of conjugation, the key task is the identification of recombinant cells. The antibiotic or phage sensitivity of the Hfr cells makes it possible to selectively eliminate the Hfr cells from the mixture of Hfr, F^-, and F^- recombinant cells. Since both the F^- and the F^- recombinants are resistant, they will survive.

22-2. At the start of the conjugation study, the F^- recipient strain carries mutant alleles at the loci that are being studied. At the end of the conjugation interval, the F^- recombinants will have acquired donor alleles at the loci experiencing recombination. Media designed to select bacteria with donor alleles at these loci will allow the recombinant cells to grow while blocking growth of the nonrecombinants.

22-3. Antibiotic or phage sensitivity makes it possible to selectively eliminate the Hfr cells from the mixture of Hfr, F^-, and F^- recombinants. If the gene conferring this sensitivity has been transferred to the F^- recipients, they too will be eliminated. If the locus is very distant to the origin, it will be transferred to the recipients only after a very long conjugation interval.

22-4. **a.** The transfer of an F factor from an F^+ donor to an F^- recipient converts the F^- cell to F^+.

b. The insertion of the autonomous F factor carried by an F^+ into the chromosome of the F^+ cell converts the cell to Hfr.

c. An $F^- \times$ Hfr mating which results in the complete transfer of all of the F factor and the chromosome to the F^- cell converts the F^- cell to Hfr status. Alternately, an F^- cell could receive an F factor from an F^+ cell, with the F factor subsequently integrating into the recipient's chromosome.

d. Excision of the F factor from the Hfr chromosome converts the Hfr cell to F^+.

e. Excision of the F factor from the Hfr chromosome with the F factor removing a small amount of the chromosome's genetic material in the process converts the Hfr cell to F′.

22-5. Conversion of F^- cells to Hfr requires complete transfer of the F factor. The leading portion of the F factor is transferred at the start of conjugation but the remainder is transferred only after the entire bacterial chromosome has passed to the recipient; conjugating cells seldom remain in contact long enough for this to happen.

22-6. The F factor carried by an Hfr cell is integrated into the bacterial chromosome and once the leading portion of the F factor is transferred, the bacterial chromosome follows along. The F factor of an F^+ cell, in contrast, is separate from the bacterial chromosome and is transferred independently of it.

22-7. **a.** Nonrecombinant F^- cells lack the ability to synthesize arginine and histidine and to utilize galactose as a carbon source. A medium deficient in arginine will select for cells recombinant for *arg⁺*. A medium with galactose as the only carbon source will select for cells recombinant for *gal⁺*. A medium deficient in histidine will select for cells recombinant for *his⁺*.

b. The order is F factor-*gal*-*his*-*arg*. Colonies arising from *gal⁺* cells first appear in samples taken at the 15-minute interval. This would indicate that this locus is transferred between 10 and 15 minutes of conjugation. (Since the number of colonies jumps from 120 to 150 in the 20-minute sample, transfer to some cells presumably occurs after 15 minutes of conjugation; the transfer time could be estimated as about 15 minutes). Colonies arising from *his⁺* cells appear in the 45-minute sample with the number increasing in the 50-minute sample, indicating that the *his* locus is transferred between 40 and 50 minutes. We could estimate the transfer time at 45 minutes. No colonies are found on the medium selecting for *arg⁺*, indicating that this locus is transmitted after the 55 minutes covered by these data. The distance between the F factor and the *gal* locus is about 15 map minutes, and between the *gal* locus and the *his* locus the distance is about 30 map minutes. The *his*-to-*arg* distance cannot be determined from these data, but it must be more than 55 map minutes from the F factor.

c. This Hfr strain has its F factor inserted into the chromosome at a site 15 minutes past the F-factor insertion site for the first Hfr strain, in the direction away from the *gal* locus, so that the interval between *gal* and the F factor for this strain is 30 minutes. The F factor is integrated with the same orientation as in the first Hfr strain.

22-8. **a.** The r^+ gene was transferred during the first 10 minutes.

b. Transfer of the u^+ gene began between 5 and 10 minutes after the start of conjugation and was completed by 20 minutes.

c. Two factors: (1) not all of the recipient cells begin conjugation at the same time and (2) once transferred, recombination is a chance event the probability of which will increase with time.

d. Some of the conjugating cells separate before the u^+ gene is transferred to them.

22-9. **a.** No. In both cases the recombination levels have plateaued, indicating that there has been sufficient time for the gene to have entered the recipient cells. The lower levels of recombination reflect the fact that the greater the distance of a locus from the F-factor insertion site, the lower its chances of being incorporated into the recipient's chromosome.

b. The longer it takes for a locus to be transferred after the start of conjugation, the more distant the locus is from the origin. Locus *u* is closest to the origin, *s* is farthest from it, and *t* is intermediate to these two loci.

c. Extrapolating each curve back to 0% allows us to estimate the time when each gene was first transferred: u^+ at about 8 minutes, t^+ around 18 minutes, and s^+ around 28 minutes. These estimates indicate that u^+ and t^+ are separated by approximately $18 - 8 = 10$ map minutes and that t^+ and s^+ are separated by approximately $28 - 18 = 10$ map minutes.

22-10. a. Any cell growing on this medium must synthesize all five of the amino acids from the minimal medium. Only prototrophs (genotype $trp^+his^+gly^+met^+arg^+str^r$) can do this.

b.

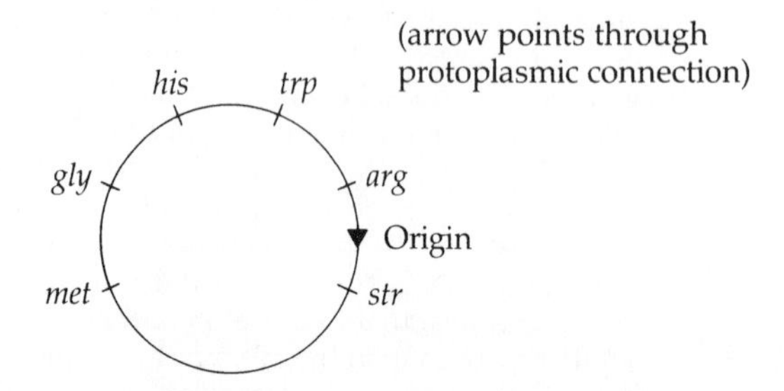

c. Cells that grow on this medium must have the genotype $trp^+his^+gly^+met^+arg^+$. Any F^- cells that ceased conjugation after the transfer of the gly^+ and before the transfer of the met^- (that is, after 40 minutes) would possess this genotype and would be able to grow. For F^- cells that continue to conjugate past 50 minutes, the met^- gene of the Hfr donor will have been transferred to them converting the genotype of many to $trp^+his^+gly^+met^-arg^+$. These cells would be incapable of growing on a minimal media because they would be unable to synthesize methionine.

22-11. a. The first gene transferred is always located adjacent to the leading portion of the F factor. Since each Hfr strain transfers a different gene first, each must have its F factor inserted at a different site.

b. and **c.** (Note that the spacing between loci is arbitrary since no information is given about the distances between loci.)

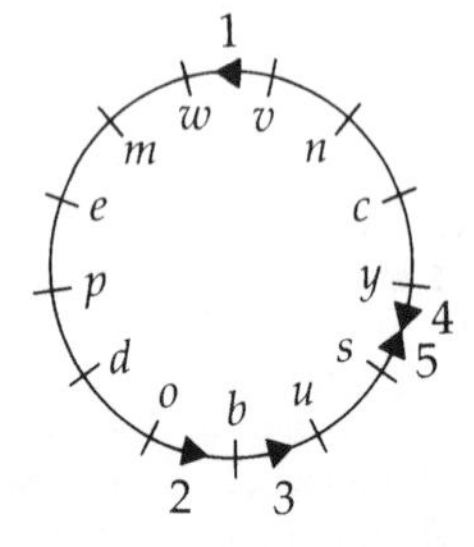

d. Hfr strains 1, 2, 3, and 5 transfer in the same direction. Hfr strain 4 transfers in the opposite direction. The different directions of transfer are due to the different orientations of the inserted F factor.

22-12. a. In order to produce Hfr recombinants, it is necessary that the Hfr parent transfer a complete F factor and its entire chromosome. Of the seven loci studied in the cross involving Hfr strain 4, locus *e* is the last transferred. Among the recombinants, cells receiving the *e* gene from the Hfr cells would have the greatest likelihood of being Hfr.

b. No. Some of the recombinant cells would have experienced a spontaneous disruption of conjugation after locus *e* had been transferred and before the rest of the chromosome and the balance of the F factor had been transferred; these cells would not be Hfr.

22-13. a. None. The streptomycin would block the growth of the streptomycin-sensitive Hfr cells.

b. No. Nonrecombinant F^- cells would be l^- and thus lack the capacity to make amino acid l. The medium lacks this amino acid and thus no growth would occur.

c. The recombination percentage of each locus depends upon its chromosomal position. Loci that are closer to the leading portion of the F factor will be transferred with a higher frequency than those that are further away. The sequence in which the loci are transferred is *l-g-d-n-a-t-m*. The *l*-to-*g* distance: 13 map units; *g*-to-*d* distance: 33 units; *d*-to-*n* distance: 5 units; *n*-to-*a* distance: 12 units; *a*-to-*t* distance: 14 units; *t*-to-*m* distance: 20 units.

22-14. a. The *s* locus is known to be transferred first and the fact that the $s^+t^-u^-$ genotype is the most common recombinant verifies this. That the second most common genotype is $s^+t^-u^+$ indicates that the *u* locus is transferred next. Only 10 colonies show the $s^+t^+u^+$ genotype, indicating that *t* is the last transferred locus. The order is *s-u-t*.

b. The *u* locus is in the middle. The only way that the two loci on either side of it could be included without the middle *u* locus being present is through the occurrence of two separate recombinations, each involving two crossovers. One would recombine the segment carrying the *s* locus and the other, the segment carrying the *t* locus. Based on these data, one could conclude that the occurrence of such double crossovers is a relatively rare event.

22-15. a. The following sketches show that the F^-, $pro^+met^+pur^+$ recombinants could arise in both matings with equal frequencies from two crossovers. A crossover is indicated by an "X."

Hfr: pro^+ met^+ pur^-
X X
F^-: pro^- met^- pur^+

Hfr: pro^- met^- pur^+
X X
F^-: pro^+ met^+ pur^-

b. The following sketches show that the F^-, $pro^+met^+pur^+$ recombinants would arise in one mating from two crossovers and in the other mating from four crossovers. Therefore, the frequencies will be very different depending on the configuration of Hfr and F^- in the mating.

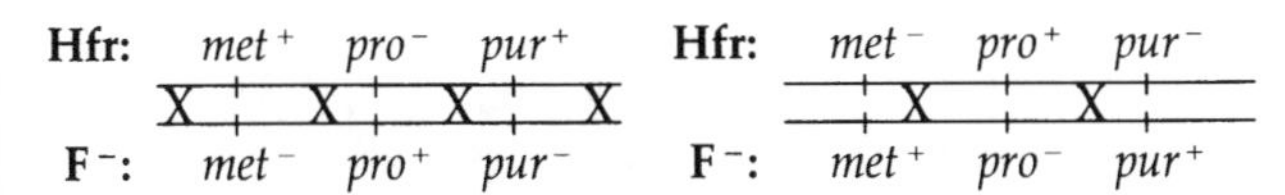

c. With the *pro-met-pur* sequence, wild-type recombinants would arise in both crosses through double crossovers and, consequently, the recombinants would be expected in approximately equal frequencies. With the *met-pro-pur* sequence, recombinants carrying wild-type alleles at all three loci would arise in one cross through a quadruple crossover and in the reciprocal cross through a double crossover. Since quadruple crossovers occur with a much lower frequency than double crossovers, we would expect a big difference in the number of wild-type recombinants produced in the two crosses. Since the outcomes of the reciprocal crosses show this type of recombinant in approximately equal frequencies, the sequence is *pro-met-pur*.

Chapter 23 Transduction and Mapping the Bacterial Chromosome

23-1. a. Classes arg^+met^- and arg^-met^+ are single transductants. Class arg^+met^+ is a double transductant.

b. There are a total of 99 + 1289 = 1388 transductants for arg^+ and of these, 1289 are cotransductants for met^+. The cotransduction frequency is 1289/1388 = 0.929, or 0.929 × 100 = 92.9%.

23-2. The failure to detect cotransduction between two loci indicates that the distance separating the loci is greater than the length of the DNA fragment included in a transducing phage. By itself, that knowledge does not contribute much, but in conjunction with additional information about other loci that are cotransduced with each of these loci, it might be useful in preparing a map.

23-3. a. The high frequency of cotransduction of *s* with *u* and of *t* with *u* indicates that the distances between *t* and *s* and between *t* and *u* are relatively short. The very low level of cotransduction of *s* and *t* tells us they are relatively far apart. From this information we can conclude that the order of the loci is *s-u-t*.

b. The distance between *v* and *s* is relatively short while the distances between *v* and *u* and between *v* and *t* are relatively great. The only position for *v* that is consistent with this information is on the side of *s* away from the *u* and *t* loci, producing the order *v-s-u-t*.

23-4. a. The results of the first study tell us that the *u* and *w* loci are relatively close to each other and that *v* is relatively distant from *u*. Two maps are consistent with these results: *u w* ——— *v* or *v* ——— *u w*.

b. The second study makes it possible to identify the correct map from the two possibilities. This study tells us that *v* and *u* are closer together than are *v* and *w*; the map consistent with this is *v* ——— *u w*.

c. The third study indicates that a DNA fragment bearing both the *v* and *u* loci never carries the *w* locus. We could speculate that the distance between the *v* and *u* loci might be close to the length of the DNA fragment that is incorporated into the P1 phage. Mapping additional loci to this region would allow a better estimate of the size of the piece of DNA responsible for the transduction.

23-5. a. The relative similarity of the frequency with which each of the four loci is cotransduced with the *a* locus indicates that these four loci are relatively close to each other on the bacterial chromosome and about the same distance from the *a* locus.

b. The greater the frequency of cotransduction between two loci, the shorter the distance between them. The data indicate that the *e* and *d* loci are closest to the *a* locus and that the *b* locus is most distant. Several sequences are consistent with this data, and without further information, the loci cannot be ordered.

c. The difference in cotransduction frequencies for *a* with *e* and for *a* with *d* is so small that it is impossible to reliably position the *e* and *d* loci.

d. A three-factor transduction study involving the *a*, *e*, and *d* loci could be carried out. The outcome would allow us to correctly position both the *d* and the *e* loci relative to the *a* locus.

23-6. a. The quadruple crossover is as follows. Each crossover is indicated by an "X."

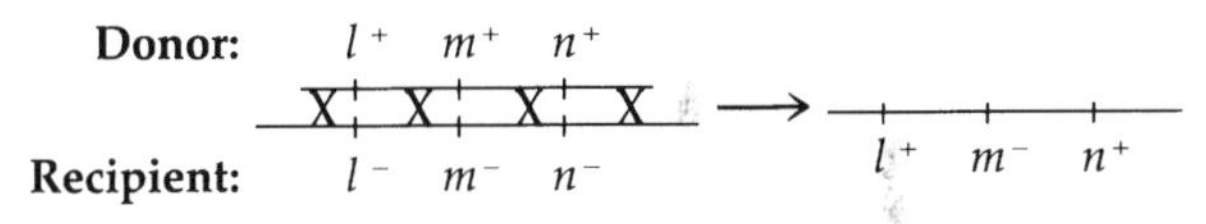

b. The transductant class having the lowest frequency is always the quadruple crossover. It arises as follows.

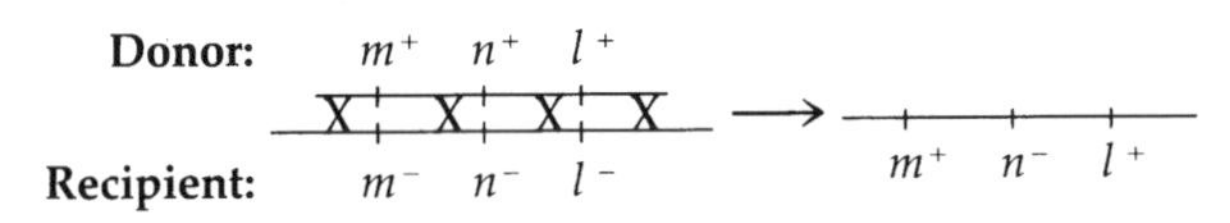

c. The locus that carries the same allele in the recipient parent and in the quadruple-crossover type is the middle locus. A comparison of the recipient genotype ($l^-m^-n^-$) and the quadruple-crossover type ($l^-m^+n^+$) shows that the *l* locus carries the same allele and is therefore the middle locus.

23-7. a. Yes, the middle locus can be identified. A comparison of the genotypes of the recipient parent $a^-d^-e^-$, and the transductant class with the lowest frequency, $a^+d^+e^-$), indicates that the locus carrying the same allele in both is the *e* locus. This tells us that the *e* locus is in the middle position. Thus, the correct order is *a-e-d*.

b. The *a* and *d* loci are cotransduced in classes 1 and 2 which have a combined frequency of 161 + 6 = 167. The total number of transductants is 1107. The frequency of cotransduction is 167/1107 = 0.151.

c. The a and e loci are cotransduced in classes 1 and 3 which have a combined frequency of 161 + 120 = 281. The total number of transductants is 1107. The frequency of cotransduction is 281/1107 = 0.254.

23-8. a. With the order i-j-k, the donor and recipient DNA would appear as follows.

Donor: i^- j^+ k^-
1 2 3 4
Recipient: i^+ j^- k^+

The four regions of the DNA are numbered in this sketch and in those that follow. The transductant classes would arise from crossovers as follows: $i^-j^+k^+$, two crossovers (regions 1 and 3); $i^-j^+k^-$, two crossovers (regions 1 and 4); $i^-j^-k^+$, two crossovers (regions 1 and 2); $i^-j^-k^-$, four crossovers (regions 1, 2, 3, and 4).

b. With the order i-k-j, the donor and recipient DNA would appear as follows.

Donor: i^- k^- j^+
1 2 3 4
Recipient: i^+ k^+ j^-

The transductant classes would arise from crossovers as follows: $i^-j^+k^+$, rewritten as $i^-k^+j^+$, four crossovers (regions 1, 2, 3, and 4); $i^-j^+k^-$, rewritten as $i^-k^-j^+$, two crossovers (regions 1 and 4); $i^-j^-k^+$, rewritten as $i^-k^+j^-$, two crossovers (regions 1 and 2); and $i^-j^-k^-$, rewritten as $i^-k^-j^-$, two crossovers (regions 1 and 3).

c. With the order j-i-k, the donor and recipient DNA would appear as follows.

Donor: j^+ i^- k^-
1 2 3 4
Recipient: j^- i^+ k^+

The transductant classes would arise from crossovers as follows: $i^-j^+k^+$, rewritten as $j^+i^-k^+$, two crossovers (regions 1 and 3); $i^-j^+k^-$, rewritten as $j^+i^-k^-$, two crossovers (regions 1 and 4); $i^-j^-k^+$, rewritten as $j^-i^-k^+$, two crossovers (regions 2 and 3); and $i^-j^-k^-$, rewritten as $j^-i^-k^-$, two crossovers (regions 2 and 4).

d. The crossover events are either double or quadruple. Double-crossover events involve breaks in two regions with the subsequent exchange of one segment of DNA (which may carry one, two, or all three of the loci under consideration). Quadruple-crossover events involve breaks in four regions, with the subsequent exchange of two segments of DNA, each bearing one of the three loci under consideration. The quadruple crossover would be expected to occur with a much lower frequency, given by the product of the probabilities of the two separate double crossovers.

23-9. a. If the order is i-j-k the category of progeny $i^-j^-k^-$ could only have arisen through a quadruple crossover. The quadruple-crossover class always has the lowest frequency, but the data show that genotype $i^-j^-k^-$ has the highest frequency. Therefore, i-j-k is not the correct order.

b. With the j-i-k order, double crossovers can produce each of the four classes of transductants. Although this order cannot be eliminated as readily as the i-j-k order, the fact that double crossovers produce class sizes ranging from 24 to 1810 (with one of these frequencies about 75 times greater than the other) make it an unlikely choice. The most likely order is i-k-j. With this order, the least frequent group of transductants ($i^+k^+j^+$, with 24 cells) arises from a quadruple crossover.

23-10. If the order were j-i-k, the $i^-j^-k^+$ combination with 847 bacteria (rewritten as $j^-i^-k^+$) would arise only through a quadruple crossover. The other genotype for which data are given, $i^+j^-k^-$, with 50 bacteria (rewritten as $j^-i^+k^-$), could arise from a double crossover. With the frequency of the quadruple-crossover class being about 17 times greater than that of a double crossover class, it seems very unlikely that the j-i-k order is the correct one. On the other hand, if the order was i-k-j, the $i^-j^-k^+$ combination (rewritten as $i^-k^+j^-$) would arise from a double crossover and the $i^+j^-k^-$ combination (rewritten as $i^+k^-j^-$) would arise only through a quadruple crossover. With the frequency of the quadruple-crossover class being about 1/17 of the double-crossover class, it seems highly likely that i-k-j is the correct order.

23-11. a. Yes, it covers all possibilities for cotransduction of the other two loci with the g locus.

b. No. The data are incomplete since they include some of the situations in which h and i are cotransduced, but not all of them. For example, they do not include h^+i^+ transductants that are untransduced at the g locus.

c. Selecting initially for h would provide a complete set of data for cotransductance of h with i, and of h with g.

23-12. a. The wild-type transductants arise from the cross $a^+b^+c^-$ × $a^-b^-c^+$ through a double crossover as follows. Each crossover is indicated by an "X."

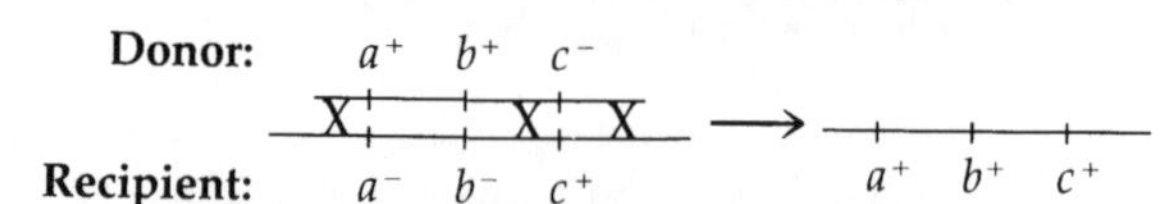

b. The wild-type transductants arise from the cross $a^-b^-c^+$ × $a^+b^+c^-$ through a double crossover as follows.

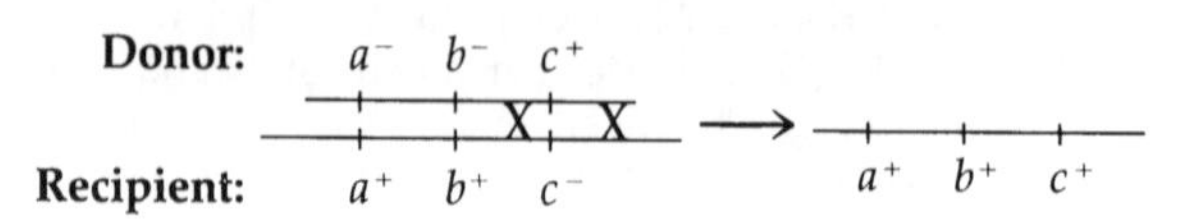

c. Since both crosses produce the wild-type transductant through a double crossover, we would expect that the frequency of wild-type transductants produced in each cross would be roughly the same.

d. The wild-type transductants arise from the cross $a^+c^-b^+$ × $a^-c^+b^-$ through a quadruple crossover as follows.

Donor: a^+ c^- b^+ → a^+ c^+ b^+

Recipient: a^- c^+ b^-

e. The wild-type transductants arise from the cross $a^-c^+b^-$ × $a^+c^-b^+$ through a double crossover as follows.

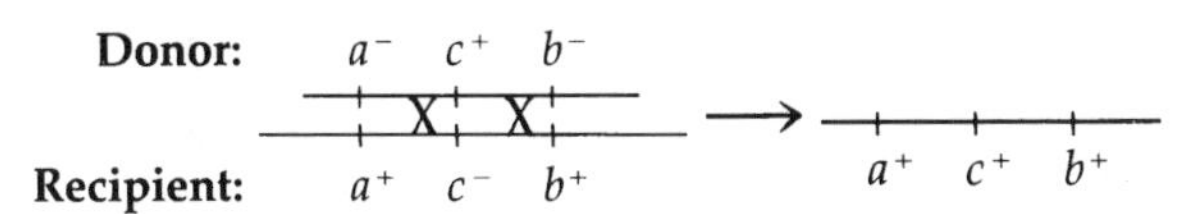

f. Since a quadruple crossover (cross 1) is a rare event compared with the frequency of a double crossover (cross 2), cross 1 would be expected to produce considerably fewer wild-type transductants than cross 2.

23-13. The marked difference in the number of wild-type transductants produced in the two crosses supports the idea that *c* is the middle locus and that the order is *a-c-b*.

Chapter 24 Viral (Bacteriophage) Genetics

24-1. A plaque is a clearing produced in an opaque layer (lawn) of bacteria by the activity of bacteriophages. As the phages reproduce, they bring about the destruction of bacterial cells. Following the release of the phages, the phages infect additional bacteria in the same region, and this progressive destruction creates and enlarges the clearing.

24-2. A high concentration of both of the phage types that are being crossed guarantees that most of the host cells are infected with both types of viruses.

24-3. A "concentration appropriate for plating" is one which is sufficiently dilute so that each plaque is assumed to be produced by the descendants of a single virus and so that the viruses added to each plate produce a number of plaques which can be easily and accurately counted (usually between 30 and 300 plaques). For example, if the bacteria were added initially to the plate in a concentration of 10^8 per milliliter and each infected bacterium on lysis produced 100 viruses, there could be as many as 10^{10} virus particles per milliliter. To see the discrete plaque produced by a single virus, the 10^{10} per milliliter should be diluted to about 10^2 per plate to produce approximately 100 individual plaques on a plate.

24-4. a. In addition to parental-type phages, two recombinant classes with the parental genotypes, a^-b^- and a^+b^+, would be present.

b. In addition to parental-type phages, two recombinant classes with genotypes a^-b^+ and a^+b^- would be present.

c. In addition to parental-type phages, two recombinant classes with parental-type genotypes, a^-b^- and a^+b^+, would be present.

24-5. a. The total number of recombinant viruses (a^-b^+ and a^+b^-) is 38 + 43 = 81. This is out of a total of 1086 + 38 + 43 + 1190 = 2357 viruses. The percentage of recombinants is 81/2357 = 0.0344, or 0.0344 × 100 = 3.4%.

b. 3.4 map units separate the *a* and *b* loci.

24-6. No. The two types of recombinant progeny viruses released from a single bacterium would not necessarily be present in equal frequencies. This is due to the fact that the replicated viral chromosomes have multiple opportunities to experience recombination during the infective cycle. The data in problem 24-5 show approximately equal frequencies of the recombinant classes. These equal frequencies were found among the enormous number of viral progeny produced from the infection of a very large population of bacterial cells and arose through random pairing, crossing over, and recombination.

24-7. One unit of map distance represents a recombination frequency of 1%. With the two loci 12.8 map units apart, we would expect 12.8% of the 1600 plaques to arise from recombinant viruses: 12.8% × 1600 = 0.128 × 1600 = 204.8, or approximately 205 plaques. Since the phages arise from a very large population of bacteria, the two categories of recombinants would be expected to occur in equal numbers.

24-8. The production of two types of plaques is just what would be expected if a single locus were involved. If two loci were involved, at least some genetic recombinants would be expected. To illustrate this, we could designate the two loci as *s* and *t*; the initial cross could be represented as s^-t^- (mutant parent) × s^+t^+ (wild-type parent) and we would expect that some s^-t^+ and s^+t^- recombinants would be formed. These recombinants might have a plaque type which differed from the wild and mutant types, although no such plaques were reported. Alternatively, if a mutant allele at either locus could produce the mutant (tiny) phenotype, both types of recombinants (along with genotype s^-t^-) would show this mutant phenotype. The data suggests this may be the case since there is an excess of tiny plaque types. A careful statistical analysis and further experiments would be needed to clarify this point.

24-9. a. Different. The recombinants from the s^+c^+ × s^-c^- cross are s^-c^+ and s^+c^-, while those from the s^-c^+ × s^+c^- cross are s^+c^+ and s^-c^-.

b. Similar. The frequency with which recombinants arise depends on the frequency of crossing over between the two loci which, in turn, depends on the distance separating the two loci. In this case, that distance is the same, regardless of whether the mutant alleles are carried in coupling or repulsion. Thus, we would expect the recombination frequencies calculated from each cross to be the same.

24-10. a. There are a total of 16 + 12 = 28 recombinants. Expressing this total as a proportion of the total number of plaques of all types (2188) gives us the recombination frequency of 28/2188 = 0.0128 = 0.013. There are 0.013 × 100 = 1.3 map units separating these two loci.

b. There are a total of 103 + 119 = 222 recombinants. Expressing this total as a proportion of the total number of plaques of all types (2412) gives us the recombination frequency of 222/2412 = 0.092. There are 0.092 × 100 = 9.2 map units separating these two loci.

c. Two maps are consistent with the data from these two crosses:

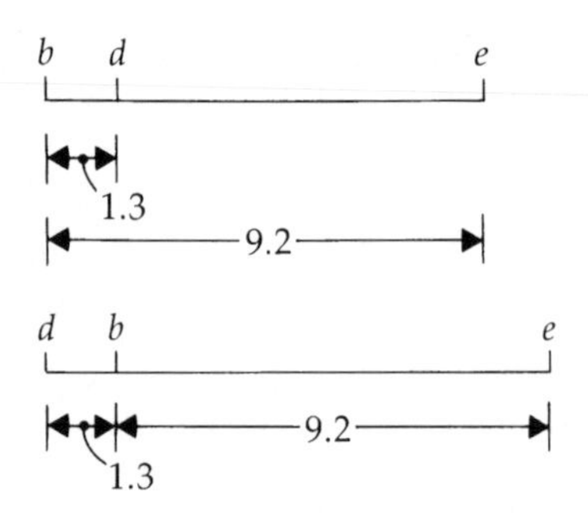

24-11. The best way to set up this map is to start with the greatest distance, 9.8 units between loci *b* and *e*. In answering problem 24-10, you noted that the *d* locus could be positioned either to the right or to the left of the *b* locus. The *d*-to-*e* distance of 7.9 units tells us that locus *d* is to the right of locus *b*.

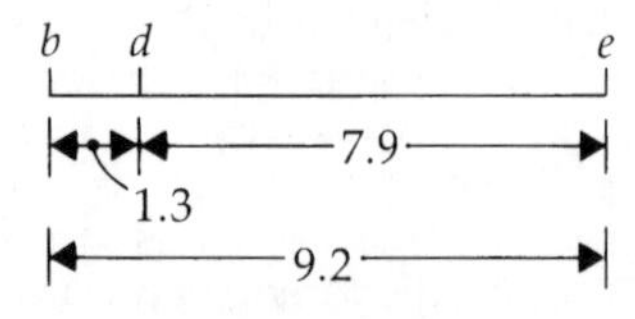

The *b*-to-*a* distance tells us that locus *a* is either 4.0 units to the right or to the left of *b*, but since the *a*-to-*d* distance is 2.9 units, we know that the *a* locus must be positioned to the right of *b*.

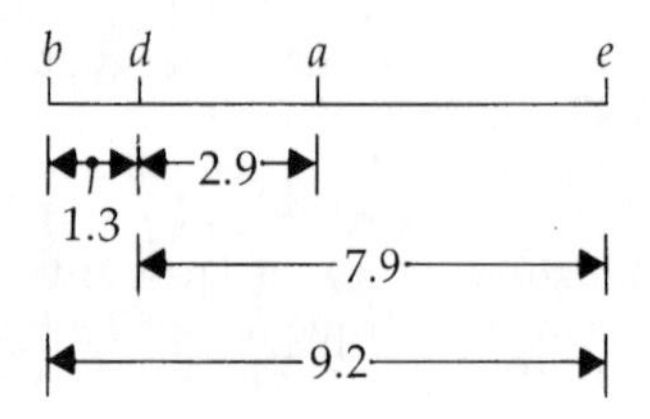

With a *c*-to-*a* distance of 4.8, *c* could be located either to the right or left of *a*, but since the *c*-to-*e* distance is 0.4 units, we know that *c* must be located between *a* and *e*.

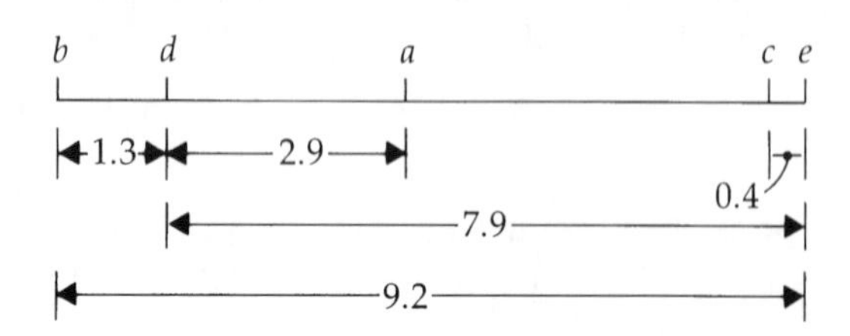

24-12. The two most frequent classes of progeny (1 and 2) represent the parental types. The two least frequent classes (7 and 8) are the products of double crossovers. Double-crossover events appear to transpose the middle locus. A comparison of the parentals with the double-crossover products indicates that the *t* locus appears to have changed position and thus is in the middle. Single crossovers in the chromosomal region between the *s* and *t* loci have produced classes 3 and 4 which contain a total of 360 + 321 = 681 progeny. In addition, each of the double-crossover progeny (classes 7 and 8) has experienced a crossover in this same region. The total number of double-crossover progeny is 81 + 72 = 153. Thus, 681 + 153 = 834 progeny out of a total of 7724, or 834/7724 = 0.108, 0.108 × 100 = 10.8%, have experienced a crossover in the *s*-to-*t* region. The map distance between these two loci is 10.8 units. Crossovers in the *t*-to-*u* region have been experienced by progeny in single-crossover classes 5 and 6 as well as in the double crossover classes 7 and 8. The total number of progeny with crossovers in the *t*-to-*u* region is 683 + 729 + 81 + 72 = 1565, which is 1565/7724 = 0.2026, or 20.3%, of the total number of progeny. Thus, the map distance between *t* and *u* is 20.3 units.

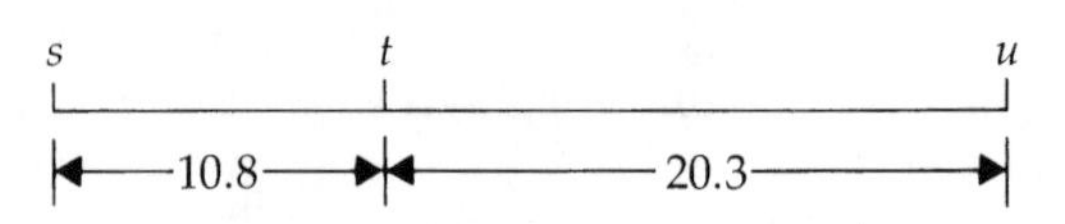

Chapter 25 Complementation Testing and Fine-Structure Mapping of the Gene

25-1. **a.** The coinfection of *E. coli* B cells provides an opportunity for intragenic recombinants to form.

b. The plaques would be produced by wild-type phages, the result of recombination between the parental phages within the region between the two mutation sites, as only wild-type phages can grow in the K12(λ) strain.

c. Mutant viruses could mutate back, or revert, to the wild type. Revertants could be detected by separately supplying samples of each *rIIB* mutant to *E. coli* B cells, collecting the lysate, and supplying it to K12(λ) cells. Any wild-type viruses that were detected would be revertants. Wild-type revertants would be formed at a much lower frequency than wild-type recombinants. To avoid ambiguity, the minimum frequency of wild-type recombinants in experiments such as these is usually set at 0.01%.)

d. Wild-type recombinants would arise only if the two mutations were at different sites within the gene and if crossing over occurred between these two sites. So the two mutations are separated by a distance sufficient to allow some crossing over between them.

e. Both recombinant types (wild-type and double-mutant) and both parental types could multiply on *E. coli* B.

f. There are four wild-type recombinant plaques. Doubling that number gives an estimate of the total number of recombinants (that is, the double mutants combined with the wild types) out of a total of 1110 phages added to each plate. Recombination frequency = 8/1110 = 0.0072.

g. Map distance = 0.0072 × 100 = 0.72 map units (percent recombination).

25-2. **a.** Progeny phages are produced, that is, complementation occurs, only if the mutations carried by two mated strains are found in different genes. Mutation *1* and mutation *2* do not complement each other (no progeny are formed), indicating that they are in the same gene or cistron. Mutation *3* complements both mutations *1* and *2*, indicating that it belongs to a gene different from that containing mutations *1* and *2*. Mutation *4* complements *1* and *2*; its failure to complement *3* leads to the conclusion that it belongs to the gene containing mutation *3*. Mutations *5* and *6* complement only mutations *3* and *4*, indicating that *5* and *6* are not in the gene containing mutations *3* and *4*. The failure of mutations *5* and *6* to complement *1* and *2* indicates that they are in the gene containing mutations *1* and *2*. The fact that mutation *5* fails to complement *6* further confirms the fact that *5* and *6* belong to the same cistron. The six mutations are in two genes.

b. See the answer to 25-2a. One gene contains mutations *1, 2, 5,* and *6* and the other contains mutations *3* and *4*.

25-3. The absence of progeny phages indicates that the mutant phages did not complement one another. Therefore this group of *rII* mutants all belong to the same gene.

25-4. **a.** Three genes. The reasoning is as follows: *B* complements *A*; therefore, mutations *B* and *A* are in different genes. *C* complements both *B* and *A*; therefore, *C, B,* and *A* are each in different genes. *D* fails to complement *A*, and can therefore be placed in *A*'s cistron. (*D*'s complementation of *B* and *C* further confirms its presence in a cistron different from that for *B* and *C*.) *E*'s failure to complement *B* places it in the *B* cistron (further confirmed by its complementation of *A, C,* and *D*). *F*'s failure to complement *A* and *D* places it in the *A* cistron. Thus, there are three genes.

b. See the answer to 25-4a. One gene contains *A, D,* and *F*; another, *B* and *E*; and the third, *C*.

25-5. The absence of plaques on the K12(λ) indicates that no wild-type recombinants were formed during the coinfection. This implies that the mutations carried by the two mutants occur at the same site within the DNA sequence of the *rIIB* gene. These mutations might be the same point mutation, an overlapping point mutation and a deletion, or two overlapping deletion mutations.

25-6. **a.** Two types of recombinants would be expected from the mating: wild-type and double-mutant. An estimate of the total recombinant frequency is obtained by doubling the frequency of wild-type recombinants: $0.0002 \times 2 = 0.0004$. The percentage of recombinants is $0.0004 \times 100 = 0.04\%$ and the two mutation sites are separated by 0.04 map units.

b. The genome contains 1500 map units and 200,000 base pairs. One map unit contains 200,000/1500 = 133.3 base pairs.

c. With one map unit containing 133.3 base pairs, 0.04 map units contains $0.04 \times 133.3 = 5.33$ base pairs. The two sites are separated by about five base pairs.

25-7. **a.** Since strain B is the permissive host, both mutants could reproduce. As their chromosomes were replicated, recombination could occur. The phages released following recombination would be wild-type (+ +) and double-mutant (*st*). Since not all the viral chromosomes would participate to yield recombinants in this chromosome region, some parental types would be released as well.

b. With strain K12(λ), neither mutant could reproduce without complementation. The phages released following complementation would have the genotypes of the parental phages (*s*+ and +*t*). Some recombinant phages might also be formed.

c. No. The occurrence of complementation indicates that the two mutations are in different genes.

25-8. **a.** The very low rate of phage production strongly indicates the absence of complementation. Thus, the mutant genes carried by each virus belong to the same locus.

b. In the absence of complementation, the progeny would be wild-type viruses.

c. The wild-type viruses could arise in two possible ways: (1) an *rII* mutant could experience a reversion mutation and revert back to the wild-type genotype (this would occur with a very low frequency and account for only a minute fraction of the yield) or, (2) if the DNA alteration carried by each mutant occurred at a different site within the gene and if crossing over occurred between the two sites, some wild-type recombinants could be formed. You could test for the presence of wild-type phages by plating the progeny on more K12(λ) at a low density so that only one phage infects a cell. Note that if + + recombinants were formed, they could complement the infecting mutant phages, and some of the progeny released could be mutant types as well.

25-9. In both cross 2 and cross 5, the deletions carried by the parental phages do not overlap, and consequently crossing over at any point between the deletion sites would produce some wild-type recombinants. Crosses 1, 3, 4, and 6 involve deletions that partially or completely overlap, and thus these crosses would not give rise to wild-type recombinants.

25-10. Zone 2 lies outside deletions *D-3, D-4, D-5,* and *D-7*; crossovers occurring in the region between the point mutation site in zone 2 and these deletions could give rise to wild-type recombinants. Zone 2 falls within deletions *D-1, D-2,* and *D-6*, and crosses involving these deletions would not produce wild-type recombinants.

25-11. The production of wild-type recombinants with *D-1, D-2, D-3, D-5,* and *D-6* deletions tells us that the point mutation lies outside these deletions, that is, outside of zones 1–5 and zones 7–10. This leaves zone 6 as the only possible site for the unknown point mutation. The absence of wild-type recombinants with *D-4* and *D-7*, the only two deletions that incorporate zone *6*, is consistent with the zone *6* location.

25-12. **a.** Generally, the greater the frequency with which two sites recombine, the greater the distance between them. The distance from the newly mapped *6-1* point mutation to mutation *6-2* is relatively short; the distance to mutation *6-4* is relatively far; and the distance to mutation *6-3* is intermediate.

b. More information is needed in order to definitely determine the sequence.

25-13. The absence of recombinants indicates that all of the point mutations are within the *B* gene.

25-14. Point mutation *F*: The absence of recombinants with *D-14* indicates that the mutation lies within the *11–15* region. The presence of recombinants in crosses with *D-12* (which overlaps with zone *11*) and *D-15* (which spans zones *13, 14*, and *15*) tells us that the mutation cannot lie within these regions. The only zone that is in *D-14* and outside of *D-12* and *D-15* is zone 12.
Point mutation *G*: The absence of recombinants with *D-11* tells us that the mutation lies between zones *1* and *5*. The absence of recombinants with *D-12* is compatible with *G* falling within zones *1–5*. The presence of recombinants with *D-13* indicates that *G* is not within *D-13*. Thus, *G* is not in the zones of *D-11* that overlap with *D-13*; that is, *G* is not in zones *3, 4*, or *5*. This leaves zones *1* or *2* as the site of *G*.
Point mutation *H*: The absence of recombinants with *D-12* tells us that *H* must fall within zones *1–11*. The presence of recombinants with *D-11, D-13*, and *D-14* tells us that *H* cannot be in the zones of these mutations that overlap with *D-12*; that is, *H* is not in zones *1–5*, nor in zones *3–9*, nor in zone *11*. This places *H* in zone *10*.
Point mutations *I* and *J*: Identical outcomes of the crosses with the five deletion mutations tell us that these two point mutations will be found in the same region or regions of gene *B*. The absence of recombinants with *D-12* tells us that the two mutations fall within zones *1* and *11*; and the absence of recombinants with *D-13* narrows that to zones *3–9*. The presence of recombinants in crosses with *D-11* (which eliminates zones *1–5* as possible sites) places these mutations in zones *6–9*. (Note that *I* and *J* are not necessarily at the same site within the *6–9* zone.)

25-15. Two possibilities exist. (1) If some reference deletions are known that have been mapped within the *6–9* zone, setting up separate crosses between viruses carrying these mutations and *I* mutant viruses and *J* mutant viruses could enable us to distinguish between the site of *I* and the site of *J*. (2) If other point mutations have already been mapped to sites within the *6–9* zone, then the relative frequency of recombinants arising from matings (in B strain bacteria) between viruses carrying these known point mutations and viruses carrying the unknown point mutations (that is, *I* and *J*) would allow a precise identification of the site for each.

Chapter 26 Gene Expression: Transcription and Translation

26-1. a. A transcriptional unit is a section of a DNA molecule carrying information that is incorporated into a single molecule of RNA.
b. The sense strand and the antisense strand complement each other and together make up the section of a double-stranded DNA molecule which serves as a unit of transcription.
c. During transcription, the sense strand serves as a template for the formation of a complementary strand of RNA. During DNA replication, it serves as a template for the formation of a new double helix.
d. During transcription, the antisense strand serves no function. During DNA replication, it serves as a template for the formation of a new double helix.

26-2. a. In the upstream portion (that is, towards the 3′ end) of the DNA sense strand.
b. The promoter serves at the site to which the RNA polymerase binds and at which the separation and unwinding of the two DNA strands begins.
c. Not usually (an exception: promoters for RNA polymerase III).
d. The reduction in the rate of transcription after altering the sequence suggests that the sequence is important for the recognition and initiation of transcription. The alteration at one position in the sequence may markedly reduce the "affinity" of the site for the RNA polymerase, and consequently, significantly fewer RNA transcripts are formed.

26-3. a. The polypeptide would have ^{14}C cysteine at each position corresponding to a cysteine codon in the mRNA because the ^{14}C cys-tRNA^{cys} recognizes the cysteine codons.
b. ^{14}C alanine would be incorporated at each position corresponding to a cysteine codon in the mRNA because the ^{14}C ala-tRNA^{cys} recognizes the cysteine codons.
c. No alanine would be incorporated because there are no mRNA codons coding for alanine.
d. This experiment verifies that neither the amino acid nor the tRNA^{ala} part of the ala-tRNA^{ala} complex recognizes the cysteine codons.
e. The labeling of the amino acids makes it possible to readily determine whether they have been incorporated into the polypeptides produced in the *in vitro* setups.
f. Yes. If the amino acid portion of the amino acid–tRNA complex had interacted with the mRNA, no alanine would have been incorporated into the polypeptide produced in the setup supplied with the ^{14}C ala-tRNA^{cys}. Since the alanine incorporation occurred in that setup, we can conclude that the tRNA portion of the complex interacts with the mRNA. The experiment shows that the tRNA recognizes the codon, and whatever it carries will be incorporated into the polypeptide.

26-4. a. In microorganisms, the production of many kinds of polypeptides is keyed to environmental conditions that may change rapidly. Short-lived mRNA makes it possible for the organism to adjust to such changes: they can promptly cease production of polypeptides they no longer need.
b. Messenger RNA coding for polypeptides for which there is an ongoing need will have to be continually synthesized. This disadvantage is outweighed by the advantage cited in the answer 26-4a.
c. Since tRNA and rRNA molecules are constantly being used in the cell, one could predict that they would have much longer life spans, and that is in fact the case.

26-5. Disagree. Despite their small size, molecules of tRNA are highly specific. They must be able to bind with the appropriate aminoacyl-tRNA synthetase molecule in order to form amino acid–tRNA complexes. They must also be able to recognize the appropriate mRNA codon and occupy the A and P ribosomal binding sites.

26-6. a. The enzyme peptidyl transferase catalyzes the formation of peptide bonds. Blocking its activity will prevent the joining of amino acids thereby preventing the initiation of polypeptide synthesis (if the antibiotic is supplied early enough) or chain elongation (if the antibiotic is supplied after initiation of polypeptide synthesis).

b. Puromycin can occupy the A site and accept the growing amino acid chain from the tRNA molecule in the P site. When this happens, puromycin establishes a bond with the carboxyl end of the polypeptide strand. In order for the puromycin to be incorporated (like an amino acid) into an interior position within the polypeptide chain, it would need a terminal carboxyl group to establish a peptide bond with the amino acid that would follow it. Since it lacks a terminal carboxyl group, its addition to a polypeptide chain brings about premature termination by blocking further elongation.

c. Blockage of the A site on the ribosome will prevent the next amino acid–tRNA complex from entering the A site, thereby blocking chain elongation.

d. Blocking mRNA from attaching to the ribosomes interferes with the formation of the initiation complex and blocks the initiation of translation.

26-7. No. The initiating aminoacyl-tRNA complex is the exception since it binds at the P site. All the other aminoacyl-tRNA complexes that participate in polypeptide synthesis do so by first binding at the A site.

26-8. The difference in length between the primary transcripts of eukaryotic mRNA molecules and the mature mRNA molecules is due to the fact that the primary transcripts contain noncoding introns. Removal of the introns and joining together of the coding exons produces mature mRNA with a considerably shorter length. Prokaryotes, in contrast, lack introns and the primary transcript and the mature mRNA show little difference in length.

26-9. Among the differences: (1) many types of eukaryotic mRNA are long lived whereas most prokaryotic mRNA are short lived; (2) all eukaryotic mRNA is monocistronic (that is, it carries information for the synthesis of a single polypeptide) whereas many kinds of prokaryotic mRNA molecules are polycistronic (that is, they carry information for the synthesis of two or more polypeptides); (3) the different kinds of eukaryotic mRNA are synthesized by three kinds of RNA polymerase enzymes whereas all prokaryotic mRNA is formed by one type of RNA polymerase; (4) eukaryotic mRNA is capped with a 7-methylguanosine at its 5′ end and has had a poly-A tail added to its 3′ end while prokaryotic mRNA lacks both of these modifications.

26-10. a. In general, a mutation in an intron would have no effect on gene expression since introns are cut out of the mRNA transcript before translation.

b. An intron mutation interfering with removal of the intron could have a significant effect on the polypeptide synthesized from the strand, since the normally noncoding intron would now become coding. This could significantly alter the primary sequence of the polypeptide.

26-11. The sections removed from prokaryotic tRNA and rRNA primary transcripts are generally external (that is, at one or the other or at both ends) to the portions of the molecules that eventually become functional. The introns of eukaryotic primary transcripts are internal sections of the transcripts that interrupt the coding sequences, or exons, of a gene.

26-12. a. The primary mRNA transcript produced from this transcription unit consists of 5318 nucleotides.

b. During post-transcriptional modification, the addition of a cap to the 5′ end and a poly-A tail to the 3′ end increase the length of the mRNA.

c. The excision of several introns reduces the transcript length.

d. Since each amino acid is coded for by a triplet of mRNA nucleotides, the total number of nucleotides necessary to code for the polypeptide is $262 \times 3 = 786$.

e. In addition to the coding region, the mature mRNA transcript has a leader (including the cap) at its 5′ end, a three-nucleotide stop codon (to signal the termination of polypeptide synthesis), and a trailer (including a poly-A tail) at its 3′ end.

26-13. When it is necessary for *E. coli* to synthesize histidine, it is essential that all the enzymes needed for the process be available at the same time and in about the same quantities. The polycistronic arrangement efficiently coordinates the production of these metabolically related enzymes through the synthesis of a single type of mRNA molecule.

26-14. a. RNA polymerase adds nucleotides to the 3′ end of the molecule; thus, RNA strands grow in the 5′-to-3′ direction.

b. RNA strands grow only by adding nucleotides to their 3′ ends. Since each of the antiparallel strands of the duplex serves as the sense strand for different transcriptional units, the RNA molecules formed along one DNA strand grow in the direction opposite to that in which molecules are growing along the other DNA strand. Synthesis occurs in both directions.

26-15. a. Found in DNA of prokaryotes in rho-independent terminators; constitutes a signal causing RNA polymerase to cease functioning and to be released (along with RNA strand) from the DNA.

b. Found in DNA of prokaryotes (specifically in *E. coli*) in the promoter section; constitutes a recognition site for RNA polymerase.

c. Found in DNA of prokaryotes (specifically in *E. coli*) in the promoter section; constitutes a recognition site for RNA polymerase.

d. Found in mRNA of eukaryotes at the 5′ end of the leader section; orients the mRNA strand with regard to ribosome.

e. Found in DNA of eukaryotes in the promoter section; constitutes a recognition site for RNA polymerase II.

f. Found in mRNA of prokaryotes in the portion of the leader near the first translated codon; binds to the sequence at the 3′ end of 16S ribosomal RNA.

26-16. a. (1) An error made in the synthesis of a single polypeptide influences just that molecule; the consequence of this kind of error should be undetectable as many other normal copies of the polypeptide should be available. (2) A transcription error made in a single mRNA molecule influences all polypeptide molecules synthesized from it; since there are likely to be additional error-free copies of this mRNA available in the cell, normal polypeptide molecules should be plentiful and the consequence for the cell should be negligible. (3) An error made in the replication of the sense strand of an mRNA transcriptional unit means that all of the mRNA molecules transcribed from it will reflect the error and, in turn, that all molecules of the polypeptide coded for by the transcriptional unit will also reflect the error. Assuming that there is just a single copy of the transcriptional unit present in the cell, the consequences of every molecule of a particular (and presumably vital) polypeptide carrying an error could be severe.

b. Only the error made in the replication of the sense strand of an mRNA transcriptional unit is potentially heritable. To be heritable, this error would have to occur in a germ cell.

26-17. a. True.

b. False. Formylmethionine is modified by conversion to methionine or removed altogether before the polypeptide takes up its functional role in the cell.

c. False. In some instances the initiation code is GUG.

d. False. AUG codons in the initiation (first translated) position always code for the modified form of methionine whereas internal AUG codons always code for the standard, unmodified methionine.

e. True.

26-18. a. Transcription; temporarily complexes with the RNA polymerase core enzyme and is essential for the recognition of DNA promoter sites.

b. Translation; catalyzes the transfer of the amino acid (or growing polypeptide chain) carried by the tRNA molecule in the P site to the amino end of the amino acid carried by the tRNA molecule in the A site and catalyzes the formation of a peptide bond between them.

c. Translation; comprise three types of protein molecules that facilitate the formation of an initiation complex.

d. Transcription; temporarily complexes with the DNA and the core RNA polymerase enzyme and plays a role in the recognition of the termination signal. Note that the rho factor is essential for termination with rho-dependent units of transcription only.

e. Translation; interact with amino acid–tRNA complexes enabling them to pair with an internal codon of an mRNA strand.

f. Translation; interacts with the terminator codon once it has lined up with the A site thereby blocking the A site and causing the release of the polypeptide from the ribosome and the ribosome from the mRNA strand.

26-19. a. False. There is at least one type of aminoacyl-tRNA synthetase enzyme for each of the 20 amino acids.

b. False. An activated amino acid is one that has interacted with ATP, acquired a high-energy bond, and is ready to complex with an appropriate tRNA molecule.

c. False. A peptide bond joins the amino group of one amino acid to the carboxyl group of another amino acid.

d. True.

e. True.

26-20. The three events are (1) a new aminoacyl-tRNA complex becomes bound to the A site of the ribosome; (2) the peptide chain associated with the tRNA molecule in the P site becomes attached by a peptide bond to the amino group of the aminoacyl-tRNA complex carried in the A site; and (3) translocation moves the ribosome one codon along the mRNA strand, releasing the tRNA molecule from the P site and shifting the tRNA molecule carrying the growing peptide chain from the A site to the P site.

26-21. a. True.

b. False. At least 50 different types of tRNA molecules have been identified in both prokaryotes and in eukaryotes.

c. True.

d. False. A strand of mRNA with a number of ribosomes bound to it is known as a polyribosome.

e. True.

Chapter 27: The Genetic Code

27-1. A system of triplet code words adequately codes for the 20 amino acids and then some. Four and five letter code words could *also* have coded for the 20 amino acids. However, the most reasonable approach in proposing an explanation for a phenomenon is to begin with the simplest scheme that adequately explains things, and then if necessary (that is, if subsequent information warrants), move to a more elaborate explanation. The initial speculation was that the code words were triplets and the information subsequently gathered supported this.

27-2. The results given here are insufficient to identify the number of bases in a code word. The fact that no amino acids require all four kinds of bases for polypeptide incorporation works against the idea of a four-letter code word. Nothing given here allows us to distinguish between the other possibilities—for example, code words with one, two, three, five, or even six bases, or those with a variable number of bases.

27-3. The code is described as degenerate because most (18 out of the 20) of the amino acids are each coded for by more than a single codon. Put another way, there are 20 amino acids but more than twenty codons. If the code lacked this feature, the 20 amino acids would be coded for by exactly 20 codons.

27-4. Synonymous codons code for the same amino acids. They would not occur in a code that lacked degeneracy.

27-5. a. A single codon would be altered.

b. A single amino acid has the potential to be affected.

27-6. a. Because of the overlapping reading, the single-base substitution would occur in three codons.

b. Three amino acids have the potential to be affected.

27-7. a. Two amino acids would be coded for; AAA codes for lysine and CCC codes for proline.

b. Four amino acids would be coded for; AAA codes for lysine, AAC for asparagine, ACC for threonine, and CCC for proline.

c. No. With overlapping reading, two codons, AAC and ACC, would be read between AAA and CCC and thus two amino acids would be positioned between the one coded for by AAA and that coded for by CCC.

27-8. A nonoverlapping code places no restrictions on the amino acids that can be positioned next to any of the amino acids in a polypeptide. An overlapping code, in contrast, imposes restrictions. For an example of this, refer to your answer to problem 27-7c.

27-9. Advantage: overlapping reading of the code is efficient; it allows a significant increase in the amount of information carried without increasing the amount of the genetic material. Disadvantage: a mutation involving the substitution of a single base could, as you noted in your answer to problem 27-6, potentially alter three amino acids, and this could alter the polypeptide product in a significant way, possibly making it nonfunctional. Alternatively, the mutation could alter more than a single gene product.

27-10. **a.** Eight codons were present: UUU, UUG, UGU, GUU, GGU, GUG, UGG, and GGG.

b. Only UUU codes for phenylalanine.

c. The key points involved in determining the relative concentrations of U and G are as follows. The frequency with which phenylalanine occurs in the polypeptide depends on the frequency with which UUU codons occur in the RNA. This frequency is, in turn, a function of the concentration in which U was supplied to the RNA-synthetizing system. The probability of U occurring in a codon is given by its relative concentration. For example, if the concentration of U was 50%, then the likelihood of its occurring in a codon would be 1/2. Since the placement of any ribonucleotide in an RNA molecule is an independent event, the probability of three uracils occurring together in the same codon equals the relative concentration of uracil raised to the third power. (To continue our example, with a concentration of 1/2, the likelihood of three uracils occurring together would be $(1/2)^3$.) To figure out the actual concentration of U, we begin with what we know: the concentration of phenylalanine is 8/27. This tells us that the relative frequency of U occurring in a codon in the RNA was 8/27. We know that the UUU frequency equals the concentration of U raised to the third power: $[U]^3 = 8/27$, so $[U] = \sqrt[3]{8/27} = 2/3$. Since the relative concentrations of U and G must add up to 1, we know that the $[G] = 1/3$.

27-11. **a.** The 26 amino acids would require a minimum of 26 code words. With three kinds of bases, a code word containing three bases would give us $3^3 = 27$ different combinations which could adequately code for the 26 amino acids.

b. Two terminator codons plus 26 codons for the amino acids gives a total of 28. A triplet code would give us 27 codons, one short of the required number. A code of four bases would be necessary.

27-12. **a.** The 12 amino acids and a terminator signal would require a minimum of 13 code words. With two kinds of bases, a code word containing four bases would give us $2^4 = 16$ different combinations, adequate for the 13 code words necessary.

b. With a four-base codon, a single-stranded molecule containing 156 nucleotides would contain $156/4 = 39$ code words. If one of these is a terminator codon, then 38 could code for amino acids. Thus, the biggest polypeptide that could be made from this strand would be 38 amino acids long.

c. A double-stranded molecule containing 272 nucleotides would have $272/2 = 136$ nucleotides in each strand. With a four-base codon, there would be $136/4 = 34$ code words; if one is a terminator, then the maximum polypeptide size would be 33 amino acids.

27-13. **a.** The codon is 5′ UAC 3′.

b. Tyrosine.

c. The base at the 5′ end of the anticodon is G which can, of course, form hydrogen bonds with C; because of wobble, it can also bond with U. Thus, this anticodon can pair with either 5′ UAC 3′ or 5′ UAU 3′ codons, both of which code for tyrosine.

d. Wobble does not alter the amino acid associated with a particular tRNA molecule, nor does it alter the amino acid specified by a code word. Wobble may, depending on the base carried at the 5′ end of the anticodon, reduce the specificity governing anticodon-codon pairing such that a particular anticodon (and the tRNA molecule carrying it) may be able to match up with more than a single codon.

27-14. **a.** The mRNA strand contains eight codons, the last of which is a terminator codon; thus, the polypeptide will have seven amino acids.

b. Three kinds of amino acids (methionine, leucine, and tyrosine) will be present.

c. The amino acid sequence is *met-leu-tyr-leu-leu-tyr-leu*.

d. In the absence of wobble, the anticodon of any particular kind of tRNA molecule would be able to pair with just a single type of codon. Since there are seven codons that call for amino acids in this mRNA and since each is different, seven different kinds of tRNA would be needed.

e. With wobble, the anticodons of certain types of tRNA molecules can match up with more than just a single type of codon. Anticodons with U at the 5′ end can pair with A or G in the third position of a codon (provided, of course, that the other bases of the anticodon and codon complement each other). Thus, the leucine-carrying tRNA molecule bearing the anticodon GAU can pair with CUA and CUG codons (which occupy the second and seventh sites in the mRNA strand). Anticodons with G at their 5′ end can pair with U or C in the third position of a codon. Thus, the leucine-carrying tRNA molecule that bears the anticodon GAG can pair with the CUC and CUU codons in the fourth and fifth sites of the mRNA strand, and the tyrosine-carrying tRNA that bears the anticodon AUG can pair with the UAC and UAU. So, with wobble, three kinds of tRNA could translate six of the codons in the mRNA strand. Adding in the tRNA for methionine, the minimal number of tRNA molecules that could translate the strand is four.

27-15. Wobble is involved. Normally, when wobble occurs, it occurs between the base at the 3′ end of the codon and the base at the 5′ end of the anticodon (remember that the codon and the anticodon match up in antiparallel fashion). In this case, however, the wobble is taking place between the base at the 5′ end of the codon and the base at the 3′ end of the anticodon. Because of this wobble, the anticodon of the tRNA that carries (N-formyl) methionine can pair with AUG (usually found) and GUG (sometimes found) whenever either of these codons occurs in an initiator site.

27-16. a. The second mutation compensated for the initial mutation and restored the reading frame to its original pattern from the site of the second mutation onward.

b. No. The mRNA nucleotides between the sites of the initial mutation and the second mutation would be read with the shifted reading frame, and thus the amino acids coded for by codons read in this interval could most likely be different from those found in the corresponding section of the original polypeptide.

c. The greater the number of nucleotides between the sites of the first and second mutations, the greater the number of codons read with the altered reading frame, the larger the garbled section of the polypeptide, and the greater the chance of the polypeptide failing to function like the original.

27-17. a. A strain carrying either two additions or two deletions will have a reading frame that will be normal up to the site of the first mutation; the frame will be thrown off by a single nucleotide between the site of the first and the site of the second mutation and thrown off by two nucleotides from the site of the second mutation onward. A strain carrying either three deletions or three additions would have its reading frame altered between the sites of the first and second mutations and between the sites of the second and third, but would have its reading frame restored to its original pattern from the site of the third mutation onward.

b. No. Amino acids in the polypeptide would differ from the original in a section corresponding to the mRNA section in which triplets were read with the altered reading frame.

27-18. The fact that viral strains with one or two additions or deletions fail to exhibit the normal phenotype, but that strains with three additions or deletions have the normal phenotype, supports the idea of triplet code words.

Chapter 28 Mutations

28-1. a. The addition to the gene causes an additional nucleotide to be inserted into the mRNA transcript. When that mRNA is translated, its reading frame will be thrown off from the point of the insertion onward.

b. Since the DNA code words are triplets, the insertion of two additional nucleotides into the DNA would restore the reading frame.

c. No. The amino acid sequence would be translated normally up to the point of the first addition and beyond the point of the third addition. The section of the polypeptide coded for by the region between the first and the third additions would be garbled.

28-2. By definition, a mutation is a change in the genetic material. The alteration described here arises in RNA during its transcription; the DNA guiding the formation of this RNA is unchanged. Only changes in the DNA qualify as mutations.

28-3. a. Changing the third nucleotide in the triplet from T to A would restore the triplet to its original GTA sequence and would thus qualify as a true reversion.

b. The original triplet, GTA, codes for the amino acid histidine. The initial mutation changes the triplet to GTT which codes for glutamine. A second mutation changes the triplet GTT to one synonymous with the original triplet (that is, to one coding for histidine) and qualifies as an operational reversion. Changing GTT to GTG would do this.

c. The two polypeptides would be identical.

28-4. a. Two different polypeptides are required to form S hemoglobin and thus two different genes are involved.

b. The mutation occurred in the gene coding for the beta polypeptide and could have been a base substitution altering a single codon.

c. Two codons code for glutamic acid (GAA and GAG) and four codons code for valine (GUU, GUC, GUA, and GUG). The glutamic acid codon GAG and the valine codon GUG differ in a single base as do the glutamic acid codon GAA and the valine codon GUA. Within each of these pairs, the glutamic acid codon could be converted to the valine codon by changing the middle A to U. Such changes would occur in the mRNA if there had been a T-to-A alteration in the middle base of the glutamic acid triplet in the DNA of the gene.

28-5. HbA has glutamic acid at the position occupied by lysine in HbC. There are two glutamic acid codons (GAG and GAA) and two lysine codons (AAA and AAG). If the mRNA producing the beta chain of HbA carried the codon GAG, the complementary DNA triplet in the beta chain gene would be CTC. Substituting a T for the first C would convert the DNA triplet to TTC which would transcribe into the codon AAG which codes for lysine. In a similar way, GAA could be converted into AAA.

28-6. A frameshift mutation arising because of the addition or deletion of a nucleotide could have altered the reading frame, thereby causing the terminator codon to be overlooked and allowing ribosomes to translate beyond the normal termination point. Alternatively, a substitution causing a base change in a terminator codon could have permitted an amino acid to be incorporated in that position, resulting in translation until the next terminator codon is found.

28-7. This mutation would cause the insertion of an additional G into the mRNA transcript, producing a reading-frame shift. This would garble the translation of the polypeptide downstream of the GGG codon.

28-8. The tRNA with the CCCC anticodon would read the GGGG portion of the mRNA and (1) insert glycine at that point and (2) since the ribosome advances four nucleotides along the strand, restore the reading frame to its original pattern. The polypeptide synthesized would have the wild-type sequence. Additionally, if there are any other GGGG sections in the mRNA, this tRNA could insert glycine at those sections and throw off subsequent reading by one base from that point onward.

28-9. Since the suppressor gene caused the insertion of an additional C nucleotide in the tRNA anticodon, it arose through the addition of a G nucleotide in the anticodon region of the tRNA gene.

28-10. Leucine is coded for by a total of six codons: UUA, UUG, CUU, CUC, CUA, and CUG. One of these codons is repeated three times to make up the group of identical codons in the original mRNA. Tryptophan is coded for by a single codon: UGG. Since each mutation involved just a single base substitution, the codon we seek to identify must differ from UGG by a single nucleotide; of the six leucine codons, only UUG meets this requirement. A different single nucleotide change would convert UUG to the serine codon UCG. No change would be required in the third codon. Each of the original codons was UUG.

28-11. **a.** Lysine is coded for by two codons: AAA and AAG. Arginine is coded for by six codons: AGA, AGG, CGU, CGC, CGA, and CGG. Substitution of G for A in the second position of either of the two lysine codons will produce arginine codons. This codon alteration would arise from the substitution of a C for a T in the DNA strand guiding the mRNA transcription. Since a pyrimidine replaces a pyrimidine, the mutation is a transition.

b. Leucine is coded for by six codons: CUU, CUC, CUA, CUG, UUA, and UUG. Proline is coded for by CCU, CCC, CCA, and CCG. Substitution of C for U in the second position of any of the first four leucine codons will produce a proline codon. This codon alteration would arise from the substitution of a G for an A in the DNA strand guiding the mRNA transcription. Since a purine replaces a purine, the mutation is a transition.

c. Arginine is coded for by AGA, AGG, CGU, CGC, CGA, and CGG. Tryptophan is coded for by UGG. Substitution of U for A in the first position of the arginine codon AGG will produce the tryptophan codon. This codon alteration would arise from the substitution of a T for an A in the DNA strand guiding the transcription of the mRNA. Since a pyrimidine replaces a purine, the mutation is a transversion. A second mutation involving the substitution of U for C in the arginine codon CGG would also produce the tryptophan codon. This change could arise from the substitution of A for G in the DNA strand. Since a purine replaces a purine, the mutation is a transition.

28-12. **a.** A true back-mutation changes the nucleotide altered by the initial mutation back to the same nucleotide present prior to the initial mutation. A suppressor mutation causes a change in a nucleotide other than the one altered by the original mutation.

b. The progeny would differ, depending on the type of mutation. Crossing organisms that have experienced a true back-mutation with the wild-type will produce nothing but wild-type progeny. Crossing organisms that have experienced a suppressor mutation with the wild-type will produce wild-type progeny and two types of recombinants, some of which will possess and show the original mutant phenotype and others of which will possess the suppressor mutation.

28-13. Yes, this is a suppressor gene. Normally the codon 5′-UGA-3′ is a terminator codon and its presence within the mRNA strand will cause the premature termination of the polypeptide. However, the altered tRNA's anticodon complements this codon causing tryptophan to be incorporated into the polypeptide, thereby suppressing the mutation generating the terminator codon.

b. The anticodon 3′-ACC-5′ would normally complement and read the codon 5′-UGG-3′ which codes for tryptophan.

c. In this case at least, another portion of the tRNA molecule must also be influential; that is, it is not just the anticodon portion of the tRNA molecule that determines whether or not translation occurs. Note that wobble is not involved in this pairing; check the wobble pairing rules in Chapter 27, if necessary, to verify this.

d. The only thing that can be said with certainty is that the mutation has not altered the anticodon of the tRNA and therefore the site altered by the mutation must be elsewhere in the molecule.

e. The polypeptides synthesized before and after the two mutations would differ in a single amino acid. The premutation polypeptide would have glycine at the position corresponding to the 5′-GGA-3′ codon while the postmutation polypeptide would have tryptophan at that position. The rest of the molecules would be the same.

28-14. **a.** The polypeptide would be of normal length but would carry tyrosine at an interior site most likely occupied, prior to the mutation, by another amino acid. The effect of this amino acid substitution on the functioning of the polypeptide could vary greatly, depending on how different chemically the tyrosine is from the amino acid it replaces and where the substitution occurs relative to the active site of the polypeptide. The polypeptide might be capable of at least some of its original activity.

b. The tRNA produced by the suppressor gene would recognize *any* 5′-UAG-3′ terminator codon. Thus, any mRNA with this terminator codon could have it read by the mutant tyrosine tRNA and have tyrosine inserted at that point, with translation extending beyond the point where it would normally end. This could potentially alter the functioning of a large number of polypeptides in the cell.

28-15. **a.** No. There are two different gene products and therefore two different genes would be involved.

b. Normal translation of most of these mRNA molecules would imply that the relative concentration of mutant tyrosine tRNA was low.

c. Many of the translations of this mRNA would be terminated prematurely when the internal mutant terminator was encountered. In some instances however (and the frequency of this would depend on the relative number of mutant tyrosine tRNAs in the cell), the entire strand would be transcribed.

d. Bacteria of this type would be able to translate mRNA molecules ending with the UAG terminator most of the time and would be able to occasionally translate the mRNA strand carrying the internal mutant terminator, and consequently might be able to survive but perhaps only under the ideal conditions that can be supplied through laboratory culture.

28-16. a. The frequency of recombinants would depend upon whether the loci were carried on separate chromosomes or on the same chromosome. If the loci were linked, the recombinant frequency would be influenced by the distance separating the loci (closer loci would show less recombination than more distant loci).

b. One possible way that the recombinant type carrying the suppressor mutation could exhibit the wild-type phenotype is if the amino acid change it produced was remote from the active site of the polypeptide and therefore had no effect on the polypeptide's biological properties.

28-17. a. The enol form of 5-BU pairs with guanine.

b. Once the enol form has shifted to the keto form, it pairs with adenine when the strand carrying it serves as a template during the next DNA replication. Following a third round of replication, one of the product molecules would contain a TA pair at the site occupied by the CG pair in the original DNA molecule.

c. The keto form of 5-BU pairs with adenine.

d. Once the keto form has shifted to the enol form, it pairs with guanine when the strand carrying it serves as a template during the next DNA replication. Following a third round of replication, one of the product molecules would contain a GC pair at the site occupied by the AT pair in the original DNA molecule.

e. Depending on 5-BU's tautomeric form at the time it serves as a template base, it can induce transitions in both directions, that is, from AT to GC or from GC to AT.

28-18. a. A mutation resulting in the insertion of an additional A nucleotide after the AAA codon would shift the reading frame from that point onward, causing the newly inserted nucleotide and the remaining nucleotides to be read as codons ACC, CGG, and GUG (with the final A nucleotide left over). These three codons would translate into thr, arg, and val, respectively.

b. Following the second mutation, the functioning of the polypeptide is restored and all the amino acids in the polypeptide sequence are identical to those in the initial, premutation, version of the polypeptide except for the thr which replaces a pro. A mutation resulting in the loss of one of the C nucleotides from the CCC sequence preceding the GGGUGA 3′ section would cause the nucleotides from the lys codon onward to be read as codons ACC, GGG, and UGA which would translate into thr and gly, followed by a terminator signal.

28-19. a. A mutation causing the addition of a U nucleotide following the UUU codon (fourth codon in from the 5′ end) would cause a shift in the reading frame from that point onward. The newly inserted U would be read with the next two nucleotides, AA, to produce the terminator codon UAA.

b. One of the tRNA molecules involved in the formation of the polypeptide carries an anticodon that complements the codon UUU. If an additional T nucleotide were inserted into the portion of the DNA transcribing into the anticodon of this tRNA, the anticodon would consist of four nucleotides (AAAA). This anticodon could read the additional U inserted by the initial mutation along with the UUU triplet which precedes it in the mRNA sequence. This would restore the original reading frame and that, in turn, would restore the polypeptide product to its original amino acid sequence.

28-20. A pyrimidine dimer consists of two adjacent pyrimidines in the same DNA strand joined together by stable covalent linkages. The dimer causes a bulge in the duplex at the site of the dimer which may have lethal consequences unless repaired.

28-21. The key events are as follows. (1) An endonuclease attaches at the dimer site and breaks a phosphodiester bond in the strand carrying the dimer. (2) An exonuclease attaches at the 5′ end produced by the endonuclease and creates a gap by removing the dimer and some neighboring nucleotides. (3) DNA polymerase I fills this gap by adding nucleotides one at a time to the 3′ end created by the endonuclease, using the exposed complementary portion of the intact duplex strand as the template. (4) DNA ligase establishes a phosphodiester bond between the 3′ end of the new section and the adjacent 5′ end of the original section.

28-22. One daughter duplex is normal. The other is defective: its newly synthesized strand has a postreplication gap opposite the dimer of the template strand. The gap arises because DNA polymerase is unable to function when it encounters the distortion at the dimer site in the parental duplex and skips over the site before resuming replication.

28-23. The repair involves a recombination event between the newly synthesized gap-carrying strand of the defective daughter duplex and the strand of the normal daughter duplex derived from the parental duplex. An endonuclease nicks the normal daughter duplex strand and a section of this strand adjacent to the nick separates from its complement and is transferred to patch the gap found in the defective strand of the other duplex. The gap created in the strand contributing this fragment is now filled with DNA nucleotides by repair synthesis and the strand is restored to its original state.

28-24. If gene mutation were responsible, gene conversion would be expected to occur with equal frequency in both heterozygotes and homozygotes. In addition, the process would lack directionality; that is, the converted allele would be converted to a number of alternative forms rather than to the single alternative form carried by the heterozygote.

Chapter 29 Gene Regulation

29-1. The three structural genes of the *lac* operon are controlled by the same operator and are cotranscribed as a single polycistronic mRNA. For that reason, the relative amounts of the polypeptide products of these structural genes would be expected to be the same.

29-2. If the *lac* operon is to have the potential to be induced, the cell must always have the potential to convert lactose into allolactose. It can achieve that potential if a few molecules of beta-galactosidase are available in the cell at all times. For that reason the *lac* operon is never completely switched off; in the noninduced state a few molecules of beta-galactosidase can be synthesized. (Note that there must also be molecules of permease available to get the lactose into the cell.)

29-3. The promoter region is the site to which RNA polymerase must bind in order to initiate transcription of the structural genes of the operon. The regulator gene codes for the repressor protein. The operator region is the site to which the repressor protein binds. Of these three sites, only the regulator gene codes for a polypeptide product.

29-4. The repressor protein, by itself, can bind to the operator and thereby block the access of the RNA polymerase to the structural genes; this turns the operon off. When the inducer (allolactose) is available, it complexes with the repressor protein, altering the repressor protein's configuration so that it is unable to bind with the operator; this gives the RNA polymerase access to the structural genes and the operon is switched on.

29-5. If the genes are transcribed together as a polycistronic mRNA strand, a nonsense mutation (changing a sense codon to a terminator codon) in any one of the genes could disrupt not only its own expression, but the expression of *all* the genes that are distal or downstream to it. (In contrast, a missense mutation, resulting in an amino acid substitution, only affects the expression of the gene carrying the mutation.) Mutations in d disrupt e and h; therefore, e and h are downstream of d. Mutations in g disrupt d, e, f, and h; therefore d, e, f, and h are downstream of g. Since mutations in g disrupt f, but mutations in d do not, f must be upstream of d. Mutations in h disrupt e; therefore e is downstream of h. The evidence from these mutants supports the idea of a polycistronic transcription unit with the loci in the order g, f, d, h, e.

29-6. a. The mutation giving rise to the o^c allele may have altered the operator so that it is permanently unable to bind with the repressor protein. The mutation forming the i^- allele may have altered the repressor gene so that the repressor protein is unable to bind with the operator gene. The mutation producing i^s might have generated a repressor that binds to the operator, but that cannot recognize the inducer; this would result in the permanent attachment of the repressor to the operator, permanently blocking transcription. The mutation producing p^- may have altered the promoter so that it is no longer recognized by RNA polymerase; this would prevent RNA polymerase from initiating transcription, thereby blocking all enzyme production.

b. See Table 29-6b. Note: *Yes* in Table 29-6b indicates that the structural genes will be expressed and *No* indicates that they will not.

c. See Table 29-6c.

29-7. a. The wild-type alleles are dominant to their mutant counterparts.

b. The wild-type allele, i^+, is dominant to its mutant counterpart (i^-).

29-8. a. Yes. If direct interaction provides the basis for the regulation, then both the regulatory element and the structural genes would need to be next to each other on the same chromosome. The regulatory element cannot act on the DNA of another chromosome because this would require it to physically move to another chromosome.

b. If regulation depended on the interaction of the regulatory element's polypeptide product with the structural genes' DNA, then the regulatory element and the structural genes would not need to be on the same chromosome. A polypeptide product of the regulatory element could diffuse through the cytoplasm to the structural genes on any chromosome.

c. The first listed merozygote shows the *cis* arrangement, with the i^+ and $z^+y^+a^+$ on the same chro-

TABLE 29-6b

Genotype	Structural Gene Expression: Lactose present	Structural Gene Expression: Lactose absent	Comments
(1) $i^+p^+o^+z^+y^+a^+$	Yes	No	Wild-type operon
(2) $i^-p^+o^+z^+y^+a^+$	Yes	Yes	No repressor, therefore synthesis is constitutive
(3) $i^sp^+o^+z^+y^+a^+$	No	No	Repressor not inactivated by inducer and thus always bound to operator
(4) $i^+p^+o^cz^+y^+a^+$	Yes	Yes	Operator altered and repressor cannot bind to it
(5) $i^+p^-o^+z^+y^+a^+$	No	No	Promoter altered which prevents RNA polymerase from transcribing operon

TABLE 29-6c

Genotype	Lactose Utilization: Before	Lactose Utilization: After	Comments
(1) $i^+p^+o^+z^+y^+a^+$	Very low	High	Wild-type, lactose inducible
(2) $i^-p^+o^+z^+y^+a^+$	High	High	Constitutive expression
(3) $i^sp^+o^+z^+y^+a^+$	Very low	Very low	Uninducible
(4) $i^+p^+o^cz^+y^+a^+$	High	High	Constitutive expression
(5) $i^+p^-o^+z^+y^+a^+$	Very low	Very low	Transcription initiation blocked

mosome; the second listed merozygote shows the *trans* arrangement with the i^+ and $z^+y^+a^+$ on different chromosomes.

d. Since both merozygotes are inducible, the control mechanism must involve the interaction of a polypeptide product with the DNA of the structural genes. (This is the only mechanism that could explain the induction with i^+ carried in *trans*.)

e. The first listed merozygote shows the *cis* arrangement, with the o^+ and $z^+y^+a^+$ on the same chromosome; the second listed merozygote shows the *trans* arrangement with the o^+ and $z^+y^+a^+$ on different chromosomes.

f. This indicates that the regulatory effects of the operator must involve direct interaction with that of the structural genes. No protein is produced by the operator.

g. No. The DNA of the operator controls by interacting directly with the DNA of the structural genes, while the DNA of the regulator controls by interacting indirectly through its polypeptide product.

29-9. See Table 29-9.

29-10. The control system involving CAP ensures that whenever both glucose and lactose are available, glucose will be preferentially utilized. This is energetically efficient from the standpoint of the cell: glucose can be processed just as it is while lactose requires energy to be broken down into glucose and galactose before utilization.

29-11. a. With no glucose, cyclic AMP levels are high as are those of the CAP–cyclic AMP complex which binds to the CAP site within the promoter. RNA polymerase can attach to the promoter and transcribe the structural genes of the operon.

b. With glucose, cyclic AMP levels are low as are those of the CAP–cyclic AMP complex. RNA polymerase is attached to the promoter but no transcription occurs because CAP is not bound to the promoter's CAP site.

29-12. The promoter region is the site to which RNA polymerase must bind in order to initiate transcription of the structural genes of the operon. The regulator gene codes for the *trp* repressor protein. The operator is the site to which the *trp* repressor protein, when complexed with tryptophan, binds. Of these three sites, only the regulator gene segment codes for a polypeptide product. These three DNA regions play the same roles in a repressible operon that they play in an inducible operon.

29-13. The repressor protein, by itself, cannot bind to the operator. Under these conditions the operator is on, RNA polymerase binds to the promoter region, and the structural genes are transcribed. When the corepressor is available, it complexes with the repressor protein, altering its configuration so that the repressor protein can now bind with the operator; this blocks the access of the RNA polymerase to the structural genes making transcription impossible and thereby switching the operon off.

29-14. About 85% of the time that the leader sequence is transcribed, the structural genes are not. That would indicate that attenuation operates by blocking transcription after it has started but before it reaches the structural gene region of the operon.

29-15. This operon could be controlled solely by attenuation.

29-16. No. Attenuation control operates only when RNA polymerase can bind with the promoter site and initiate transcription. If access of RNA polymerase to the promoter is blocked by the repressor-corepressor complex, attenuational control cannot come into play.

29-17. a. In the absence of the repressor, RNA polymerase would always be able to bind to the promoter and transcription of the structural genes of the operon would occur. The cell would synthesize enzymes for the biosynthesis of tryptophan, regardless of the amount of tryptophan in the cell.

b. Attenuation control is operating: with an abundance of tryptophan, the leader protein can be

TABLE 29-9

Genotype	Concentration of Enzyme: With inducer	Concentration of Enzyme: Without inducer	Comments
a. $s^+t^+u^+$	High	Low	Wild-type, fully inducible
b. $s^-t^+u^+$	High	High	Defective, operator permanently on
c. $s^+t^-u^+$	None	None	Defective structural gene, enzyme absent or defective
d. $s^+t^+u^-$	High	High	Defective regulator gene produces defective repressor; system permanently on
e. $s^+t^+u^+/s^-t^-u^-$	High	Low	Wild phenotype, mutant alleles recessive to wild-type alleles
f. $s^-t^+u^+/s^+t^-u^-$	High	High	Defective operator gene on same chromosome as wild-type structural gene—*cis* arrangement; system permanently on
g. $s^+t^-u^+/s^-t^+u^-$	High	High	Defective operator gene on same chromosome as wild-type structural gene—*cis* arrangement; system permanently on
h. $s^+t^+u^-/s^-t^-u^+$	High	Low	Wild-type operator next to wild-type structural gene, wild-type repressor protein available: wild phenotype

synthesized (since ample amounts of trp-tRNA^{trp} are available). Completion of this synthesis causes the secondary configuration to be assumed by the mRNA which blocks the RNA polymerase from completing its transcription of the operon.

29-18. The leader polypeptide acts as a sensor of the level of a particular amino acid–tRNA complex in a cell and controls the amount of enzyme synthesis according to the level of that amino acid–tRNA complex.

29-19. Each configuration arises because of intramolecular hydrogen bonding between sections of bases that are complementary. The two configurations are mutually exclusive because some of the same bases are involved in the internal hydrogen bonding for each configuration. Those bases can participate in one configuration or the other, but not both.

29-20. **a.** Yes. Attenuation is responsive to the concentration of the amino acid in the cell. It takes place when the operon's products are not needed by the cell, that is, when the amino acid end product is plentiful. Transcription is terminated when an mRNA transcript of the leader region assumes the hairpin terminator configuration that prevents further activity by the RNA polymerase. This terminator configuration is assumed only if the ribosome can completely translate the leader polypeptide section of the mRNA. The only way for the translation of the leader polypeptide section to be responsive to the concentration of the amino acid in the cell is if it includes codons for that amino acid. When the amino acid is available, translation can be completed. In its absence, translation will be blocked.

b. The attenuation mechanism would respond to changes in the level of amino acids other than the one specified by the operon.

29-21. Both types of transposable elements terminate in inverted-repeat nucleotide sequences and both carry genes governing their replication and transposition. Transposons are distinguished from insertions sequences in that transposons are considerably longer and carry genes for traits unrelated to their replication and transposition, such as drug resistance.

29-22. Some strands that complemented each other would join to reform DNA duplexes. Other strands would experience intramolecular hydrogen bonding between their complementary end sections, with the section in between remaining unpaired. These single-stranded molecules would have a stem-and-loop, or lollipop, configuration.

29-23. A mutation caused by the insertion of an IS unit into a gene will significantly increase the gene's length by the number of nucleotide pairs making up the IS unit. A point mutation could change the gene's length by no more than one nucleotide pair. In addition, nucleotide sequencing of the gene carrying the IS would show the presence of a direct repeat of a sequence of host DNA on either end of the IS. Prior to insertion, only one copy of this sequence would exist.

29-24. Some duplexes would appear normal while others would show a loop in one strand. The normal duplexes would consist of two complementary strands of DNA from either nonmutant phages or from mutant phages. The duplexes with a loop would have one strand from a nonmutant phage and the other from a mutant phage, with the strand carrying the IS looping out at the IS site because the nonmutant strand would have no sequence to complement it.

29-25. **a.** Plate the bacteria on a medium containing ampicillin. Any cells forming colonies must carry the Ap^r gene.

b. The Ap^r transposon inserts into the recipient genome at various points. If it inserts outside of the transposon carrying Su^rSm^r, those genes would function normally to produce the $Ap^rSu^rSm^r$ transformants. If it inserts within this transposon, the effect varies depending on the exact insertion site: it may disrupt the functioning of one of the resistance genes (to produce the $Ap^rSu^rSm^s$ and $Ap^rSu^sSm^r$ transformants) or both resistance genes (to produce the $Ap^rSu^sSm^s$ transformants).

Chapter 30: Recombinant DNA and DNA Sequencing

30-1. Since these vectors can readily get into host cells, they provide a way of getting the foreign DNA under study into host cells. Since the vectors are replicating systems, they provide a way of making many copies of the foreign DNA that they carry (whenever the vector replicates, the foreign DNA replicates). Since certain genes (for example, those conferring resistance to selected antibiotics) can be incorporated into some vector DNA molecules, they provide a way of identifying the host cells that have picked up the recombinant DNA molecule.

30-2. **a.** Three DNA fragments produced with two *Eco*RI indicate two *Eco*RI recognition sites. Two fragments produced with *Sal*I indicate one *Sal*I recognition site.

b. The three fragments produced with *Eco*RI or the two fragments produced with *Sal*I have a total of 3.4 kilobases, indicating that the DNA segment contains 3.4 kilobases.

c. No. Two locations are possible with the *Sal*I recognition site located 2.2 kilobases from one end of the strand and 1.2 kilobases from the other end.

d. When the DNA is treated with both enzymes simultaneously, the 1.5-kilobase fragment produced with *Eco*RI alone is not present; two new fragments, 1.0 kilobases and 0.5 kilobases, are present. This tells us that the 1.5-kilobase fragment must have been cleaved into the 1.0-kilobase and 0.5-kilobase fragments by the *Sal*I enzyme. Thus, the *Sal*I recognition site is located within the 1.5-kilobase fragment.

e. The following is one approach to preparing the restriction map. The *Sal*I recognition site is located 2.2 kilobases from one end of the strand and 1.2 kilobases from the other end. In addition, the *Sal*I site lies within the 1.5-kilobase fragment produced by the *Eco*RI treatment. There are two possible ways of orienting the 1.5-kilobase fragment relative to the *Sal*I recognition site, as shown in Figure 30-2e(1). In addition to the 1.5-kilobase fragment, treatment with *Eco*RI produces two other fragments that are 1.2 kilobases and 0.7 kilobases in length. Since the 1.5-kilobase fragment came out of the interior of the DNA segment, these other two fragments must have come from the ends. The lengths of these two end fragments are compatible only with the first of the two possible orientations shown in the previous sketches.

FIGURE 30-2e(1)

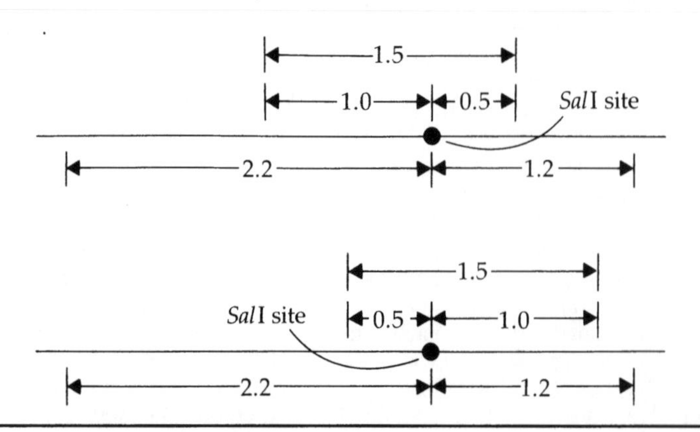

FIGURE 30-2e(2)

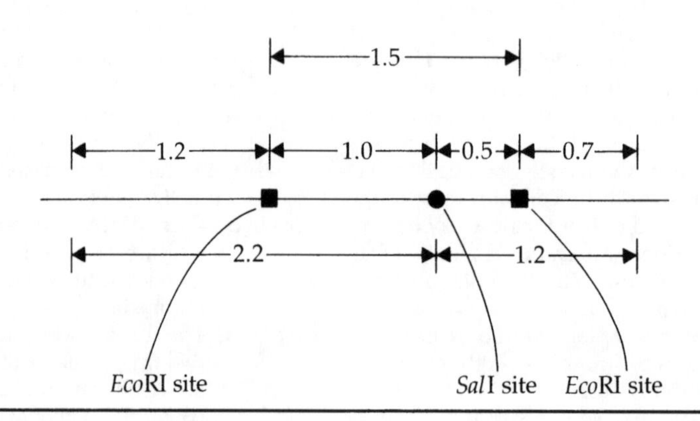

Thus, our restriction map is as shown in Figure 30-2e(2).

30-3. a. *Eco*RI converts the plasmid into a single linear molecule, indicating cleavage at just one recognition site. Treating separate batches of this molecule with *Hpa*I and with *Hind*III result, in each case, in the production of two fragments, indicating that each of these enzymes also has just one recognition site in the plasmid.

b. The linear molecule produced by the *Eco*RI treatment is 4.1 kilobases long and thus there must be 4.1 kilobases in the plasmid.

c. The 4.1-kilobase molecule we begin with is shown in the following sketch. Since it was linearized with *Eco*RI, the *Eco*RI site is placed at the ends of the full-length plasmid. (Note that there is only one *Eco*RI site—it has been split into two sections by the *Eco*RI enzyme.)

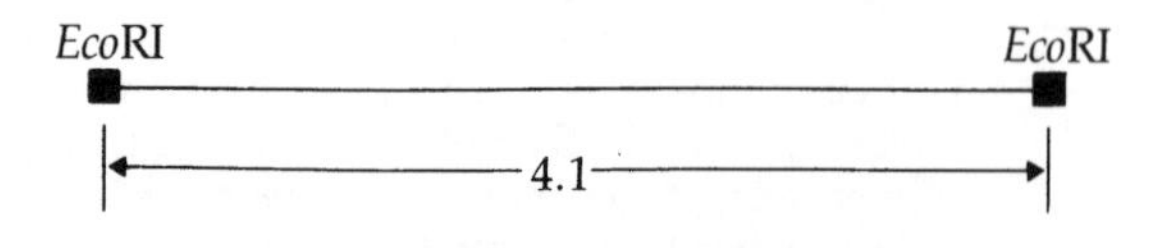

Treatment with *Hpa*I divides it into two fragments 2.1 kilobases and 2.0 kilobases long, respectively. Treatment with *Hind*III divides it into two fragments 1.1 kilobases and 3.0 kilobases long, respectively. The *Hpa*I and the *Hind*III sites can be positioned on the fragment in either of two possible ways; both are consistent with the fragment sizes produces by treatment with the individual enzymes.

a.

*Hpa*I
*Eco*RI — 2.1 — ▼ — 2.0 — *Eco*RI
1.1 — ▲ — 3.0
*Hind*III

b.

*Hpa*I
*Eco*RI — 2.1 — ▼ — 2.0 — *Eco*RI
3.0 — ▲ — 1.1
*Hind*III

The outcome of the joint treatment with *Hpa*I and *Hind*III allows us to identify the correct orientation. Notice that with both possibilities, the *Hpa*I recognition site falls within the 3.0-kilobase fragment produced from the *Hind*III treatment. With the two recognition sites positioned as they are in map (a), *Hpa*I would cut the 3.0-kilobase fragment into 2.0-kilobase and 1.0-kilobase pieces. With the two recognition sites positioned as they are in map (b), *HpaI* would cut the 3.0-kilobase fragment into 2.1-kilobase and 0.9-kilobase pieces. According to our data, treatment with both enzymes produced segments that are 1.0 kilobases, 1.1 kilobases, and 2.0 kilobases in length. This is consistent with map (a) but not with (b), so (a) is the correct map.

30-4. a. No. Insertion of the foreign DNA within the *amp*^{r} gene renders that gene inactive. Consequently, transformed bacteria carrying the recombinant DNA molecule are identical to the untransformed bacteria in their lack of ampicillin resistance. Thus, they would be killed.

b. A better design would be to use a vector with genes conferring resistance to two different antibiotics; one which will be unaltered and act as a selectable gene for the presence of the plasmid and the other into which the target DNA can be inserted and disruption of its function can be assayed. If the plasmid carried a gene for resistance to a second type of antibiotic to which the pretransformation bacteria are sensitive, it would be possible to distinguish between the transformed and the untransformed bacteria. The untransformed bacteria, lacking resistance to this second antibiotic, would be killed off by it while the transformed bacteria would be able to survive in its presence.

30-5. a. Yes. Three types of DNA fragments are available for potential incorporation into the viral heads: foreign DNA fragments, tail sections, and recombinant DNA molecules. Appropriately sized foreign DNA fragments could get incorporated but the virus would not be infective. Tail sections, either individually or joined into pairs, are too small to be incorporated. Recombinant DNA molecules that fall into the appropriate size range could be incorporated and the viruses would be infective.

b. The central section of the chromosome carries about 35% of the viral chromosome. Thus, the two tail sections must carry about 65% of the base pairs, or about $0.65 \times 47{,}000 = 30{,}550$ base pairs. The minimal amount of DNA required for packaging is equivalent to 78% of the viral chromosome or $47{,}000 \times 0.78 = 36{,}660$ base pairs, and the maximum is equivalent to 105% of the viral chromosome, or $47{,}000 \times 1.05 = 49{,}350$ base pairs. Subtracting the number of base pairs in the two tail sections (30,550) from these two limits gives us the size range for the amount of foreign DNA that can be incorporated into a viral vector: from about 6110 to 18,800 base pairs.

c. Viral vectors (1) transmit recombinant DNA molecules with a much higher frequency than do plasmids and (2) can accommodate sections of foreign DNA that are considerably larger than those that can be inserted into plasmids.

30-6. These plasmids make it possible to readily differentiate between transformed and untransformed bacteria. Untransformed bacteria, that is, those lacking the plasmid, will be killed by the colicins secreted by the bacteria that have acquired the plasmid.

30-7. The key steps of one approach are as follows. (1) Treat the DNA coding for the production of ethyl alcohol with an appropriate restriction endonuclease. (2) Treat a bacterial plasmid with the same restriction endonuclease to convert it into linear form. The bacterial plasmid that is used should contain genes that make it possible to screen for bacterial cells that subsequently acquire the plasmid. (3) Incubate the foreign DNA with the plasmid under conditions that promote the joining of the sticky ends of the two types of DNA to form a recombinant DNA molecule. Be sure to include DNA ligase. (4) Supply the recombinant DNA molecules to a suitable bacterial host under conditions that favor transformation. (5) Identify the bacteria that have picked up the recombinant DNA molecule and allow those bacteria to reproduce. Once there are a great many bacteria, ethyl alcohol should be produced in appreciable amounts. This approach assumes that the foreign DNA carries regulatory signals that can be recognized by the host bacteria; in the absence of these signals, expression of the foreign DNA is not likely.

30-8. (1) Isolate the DNA from starfish cells, and by using an appropriate restriction endonuclease, break it up into segments that fall into the size range of DNA fragments that can be incorporated into the viral vector. (2) Using the same restriction endonuclease, remove the central section from the vector chromosome. (3) Incubate the starfish DNA fragments with the tail sections of the viral chromosome. (4) Incorporate the recombinant DNA molecules that result into intact phage particles. (5) Infect *E. coli* cells with the phages. What results is a series of *E. coli* cells that carry the entire starfish genome. Maintaining cultures of these cells will preserve the "library" of starfish DNA.

30-9. (1) Electrophoretically separate the human DNA fragments. (2) Denature the DNA fragments in the gel. (3) Blot the single-stranded fragments that result onto a nitrocellulose filter. (4) Expose the filter to the radioactively labeled mRNA. Because of base complementarity, the mRNA will hybridize with the DNA sequence coding for the insulin. (5) Autoradiography of the filter will detect the label carried by the mRNA thereby indicating the location of the DNA fragment carrying the insulin gene. Because of the blot method of transfer, the location of this DNA fragment on the filter can readily be related to the banding pattern on the gel, and the particular band containing the insulin gene can be identified.

30-10. The ^{32}P labeling makes it possible to identify, through autoradiography, the 5′ DNA fragments released during the chemical cleavage of the DNA segments undergoing sequencing.

30-11. a. If we assume that some of the DNA segments escaped cleavage (and thus remained 200 nucleotides long), the autoradiograph would show 200 different bands.

b. Each band would be produced by fragments of different lengths; thus there would be 200 kinds of fragments.

c. All the fragments within a particular band would have the same number of nucleotides and thus the same cleavage point.

d. The fragments in two different bands would have a different number of nucleotides as reflected in their different positions on the gel. The two sets of fragments would have different cleavage points.

e. The DNA fragments of the band farthest from the origin would consist of a single nucleotide. This would be the first nucleotide in the sequence.

30-12. a. If the DNA polymerase retained its 5′-to-3′ exonuclease activity, it would be capable of cleaving nucleotides from the 5′ ends of the primer strand and the template strand.

b. The Sanger method of sequencing DNA relies on the length of the newly synthesized DNA fragments to provide the sequence of nucleotides in these fragments and that, in turn, gives the sequence of the template strand. The length of the newly formed DNA fragments is determined by a single factor: the sites at which terminator forms of the various nucleotides are inserted as the DNA strands are being synthesized. If the DNA polymerase retained its 5′-to-3′ exonuclease activity, its ability to cleave nucleotides from the 5′ end of the primer strand would introduce a second factor which would influence the lengths of the fragments. If this approach to sequencing is to work, all the fragments that are produced must have primer strands of the same length, that is, each fragment must have its 5′ end at the same nucleotide site.

30-13. All the DNA fragments generated in the sequencing of a particular DNA segment by the Maxam-Gilbert procedure have the same 5′ end in common; the same can be said for the fragments generated by the Sanger method. Each procedure produces DNA fragments with all possible lengths, that is, fragments with 3′ ends at each possible nucleotide site. Each procedure produces DNA fragments that are labeled: each Maxam-Gilbert fragment carries a single labeled nucleotide whereas the longer fragments produced with the Sanger method are more highly labeled, that is, it can have several to many labeled nucleotides. Sanger-method fragments have, at their 3′ ends, the nucleotide that is present in terminator form in the synthesizing setup from which they are formed. Labeled Maxam-Gilbert fragments have, at their 3′ ends, whatever nucleotide happened to be next to the nucleotide targeted by the chemical cleavage procedure forming the fragments.

30-14. With the Maxam-Gilbert method, the radioactive label is attached at the 5′ end of the copies of the DNA strand being sequenced. Only one of the two fragments produced by the chemical cleavage of each strand carries the label. Fragments with and without the label are applied to the gel and both migrate during electrophoresis, but only those carrying the label can be detected by autoradiography.

30-15. Both the Maxam-Gilbert and the Sanger methods are designed to produce either cleavages (with the Maxam-Gilbert method) or terminator nucleotide insertion (with the Sanger method) at every nucleotide position in the DNA strand being studied. The production of a nested set of DNA fragments of every possible length, which have a common beginning and a unique end, makes it possible to read the sequence of the nucleotides comprising the DNA directly from the autoradiograph.

30-16. a. A total of fifteen kinds of fragments varying in length from one to fifteen nucleotides are formed in this procedure.

b. The DNA sequence is read upward from the bottommost band (the shortest fragment, giving the nucleotide at the 5′ end) to the uppermost band (the longest fragment, giving the nucleotide at the 3′ end). The DNA synthesized in this procedure has the sequence 5′ CGTCGGATGTCGAAA 3′.

c. The DNA being sequenced in this procedure is the template DNA for the DNA-synthesizing system which complements the synthesized strand. Thus, the template sequence must be 3′ GCAGCCTACAGCTTT 5′.

d. The shortest DNA fragment produced here contains a single nucleotide which is C. Since that DNA fragment shows up on the autoradiograph, we know that the C nucleotide must carry the label. Thus it seems reasonable to conclude that all the C nucleotides supplied to the DNA synthesizing systems carry the label.

30-17. The sequence of nucleotides in the DNA is determined by reading the autoradiograph from the bottom to the top which gives the nucleotides in the 5′-to-3′ direction. The sequence is 5′ ?GACGGTATCG 3′. This sequencing technique fails to identify the first nucleotide in the sequenced DNA segment.

Chapter 31 Population Genetics I: The Hardy-Weinberg Equilibrium

31-1. a. Allelic frequencies: $f(E) = p = 0.6$ and $f(e) = q = 0.4$. After random mating, the genotypic frequencies expected would be $f(EE) = p^2 = (0.6)^2 = 0.36$, $f(Ee) = 2pq = (2)(0.6)(0.4) = 0.48$, and $f(ee) = = q^2 = (0.4)^2 = 0.16$. (As an arithmetic check, note that these frequencies add up to 1.)

b. The number of mice showing each genotype is determined by multiplying the genotypic frequency by the total population size (10,000): *EE* individuals = (0.36)(10,000) = 3600; *Ee* individuals = (0.48)(10,000) = 4800; *ee* individuals = (0.16)(10,000) = 1600.

31-2. a. Since 2/3 show dominant trait, $1 - 2/3 = 1/3$, or 0.333, show the recessive trait. f(recessive allele) = q and f(recessive genotype) $= f(q)^2 = 0.333$. Taking the square root of both sides gives $f(q) = \sqrt{0.333} = 0.577$.

b. Substituting the value of q into the expression $p + q = 1$ and solving for p gives $p + 0.577 = 1$ and $p = 1 - 0.577 = 0.423$. The frequency of homozygous dominants is $p^2 = (0.423)^2 = 0.179$, and these make up $0.179/0.667 = 0.268$, or $0.268 \times 100 = 26.8\%$, of the dominant-trait individuals.

31-3. a. The population is evolving when $f(a)$ is changing, and that occurs between generations 100 and 200.

b. The population is under Hardy-Weinberg conditions for two intervals when $f(a)$ is constant: from generation 0 to 100 and from 200 to 250.

c. At generation 75, $f(a) = q = 0.2$. Substituting this value into the expression $p + q = 1$ and solving for p gives $p + 0.2 = 1$ and $p = 1 - 0.2 = 0.8$. f(homozygous dominants) $= p^2 = (0.8)^2 = 0.64$, f(heterozygotes) $= 2pq = 2(0.8)(0.2) = 0.32$, and f(recessives) $= q^2 = (0.2)^2 = 0.04$. At generation 225, $f(a) = q = 0.6$. Substituting into the expression $p + q = 1$ and solving for p gives $p + 0.6 = 1$ and $p = 1 - 0.6 = 0.4$ f(homozygous dominants) $= p^2 = (0.4)^2 = 0.16$, f(heterozygotes) $= 2pq = 2(0.4)(0.6) = 0.48$, and f(recessives) $= q^2 = (0.6)^2 = 0.36$.

31-4. f(dominant allele) $= p = 0.75$. Since $p + q = 1$, $0.75 + q = 1$, and $q = 1 - 0.75 = 0.25$. f(heterozygotes in next generation) $= 2pq = 2(0.75)(0.25) = 0.375$.

31-5. **a.** Under Hardy-Weinberg conditions, f(homozygous dominants) $= 450/5000 = 0.09 = p^2$. Taking the square root of both sides of the equation $p^2 = 0.09$ gives $p = \sqrt{0.09} = 0.3$. f(recessive allele) $= q$ is given by subtracting p from 1: $q = 1 - p = 1 - 0.3 = 0.7$.
b. f(homozygous recessives) $= q^2 = (0.7)^2 = 0.49$. The total number of individuals is 5000, therefore $5000(0.49) = 2450$ are homozygous recessive.
c. f(heterozygotes) $= 2pq = 2(0.3)(0.7) = 0.42$. The number of heterozygotes is $5000(0.42) = 2100$.

31-6. **a.** In population 1, there are no heterozygotes. All the *B* alleles are carried by *BB* individuals and $f(B) = 0.2$; all the *b* alleles are carried by *bb* individuals and $f(b) = 0.8$. In population 2, $f(b) = 0.60 + (1/2)(0.40) = 0.60 + 0.20 = 0.8$ and $f(B) = 0.2$. Both populations have the same allelic frequencies.
b. $f(B) = 0.2$ and $f(b) = 0.8$ in both populations and the expected genotypic frequencies would be the same for both populations. The expected $f(BB)$ is $(0.2)^2 = 0.04$; $f(Bb) = 2pq = 2(0.2)(0.8) = 0.32$; $f(bb) = (0.8)^2 = 0.64$.

31-7. **a.** f(homozygous recessives) $= q^2$ and 4% of the frogs have this genetic makeup. Thus $f(gg) = q^2 = 4\% = 0.04$. Solve for q by taking the square root of each side of the equation $q^2 = 0.04$. This gives $q = \sqrt{0.04} = 0.2$. Since $p + q = 1$, $p = 1 - q$. Since $q = 0.2$, the value of p can be determined by subtraction: $p = 1 - 0.2 = 0.8$. Thus, $f(g^+) = p = 0.8$ and $f(g) = q = 0.2$.
b. The frequency of heterozygotes can be determined by substituting values for p and q into expression 2pq. $f(g^+g) = 2pq = (2)(0.2)(0.8) = 0.32$.
c. The basic assumption is that the population is in Hardy-Weinberg equilibrium. Secondary assumptions are that the population is free of evolutionary forces and is mating randomly.

31-8. **a.** $f(ll) = 0.49 = q^2$. If $q^2 = 0.49$, then $q = \sqrt{0.49} = 0.7$. Since $p + q = 1$, and $q = 0.7$, we can solve the equation $p + 0.7 = q$ for p: $p = 1 - 0.7 = 0.3$. Thus, gene frequencies are as follows: $f(l) = q = 0.7$ and $f(l^+) = p = 0.3$.
b. Genotypic frequencies predicted after a generation of random mating are as follows: $f(l^+l^+) = p^2 = (0.3)^2 = 0.09$, $f(l^+l) = 2pq = (2)(0.3)(0.7) = 0.42$, and $f(ll) = q^2 = (0.7)^2 = 0.49$.

31-9. One out of 3600 individuals shows this recessive trait. Thus, the frequency of the homozygous recessive individuals, q^2, $= 1/3600$. The frequency of the recessive allele, $q = \sqrt{1/3600} = 1/60 = 0.017$. Since $p + q = 1$, $p = 1 - q$ and substituting in the value for q gives $p = 1 - 0.017 = 0.983$. The proportion of heterozygous carriers is 2pq or $(2)(0.983)(0.017) = 0.0334 = 3.34\%$. The actual number of heterozygous carriers is given by multiplying their frequency (0.0334) times the population size (3600): $3600 \times 0.0334 = 120$ individuals.

31-10. **a.** 8% of the males in the population are color-blind and 92% have normal color vision, with genotypes X^cY and X^CY, respectively. Since each male carries a single allele for the trait, the frequencies of the alleles carried by the males can readily be identified: $f(c) = q = 8\% = 0.08$ and $f(C) = p = 92\% = 0.92$. Since these allelic frequencies apply not just to the males but to females as well, these are the allelic frequencies for the entire population's gene pool.
b. f(heterozygous female carriers) $= 2pq$. Since $p = 0.92$ and $q = 0.08$, then $f(X^CX^c) = (2)(0.92)(0.08) = 0.147$.
c. The total frequency of women with normal vision will be the sum of the frequency of heterozygous carriers (X^CX^c) and the frequency of women who are homozygous for the normal allele (X^CX^C). The frequency of the carriers is 0.147 (see the answer to 31-10b). f(homozygous dominants) $= p^2$, and since $p = 0.92$, then $f(X^CX^C) = (0.92)^2 = 0.846$. The combined frequency for these two genotypes is $0.147 + 0.846 = 0.993$.
d. The frequency of color-blind women (X^cX^c) is given by the expression q^2. Since $q = 0.08$, $q^2 = (0.08)^2 = 0.0064$.

31-11. **a.** Since each of the 420 cattle carries two alleles for this trait, the population has a total of 840 alleles. Each of the 188 red cattle carries two C^R alleles and each of the 142 roan carries one C^R allele, for a total of: $2(188) + 142 = 518$. $f(C^R) = p = 518/840 = 0.617$. Each of the 90 white cattle carries two C^W alleles and each of the 142 roan carries one C^W, for a total of $2(90) + 142 = 322$. $f(C^W) = q = 322/840 = 0.383$.
b. The expected genotypic frequencies are $f(C^RC^R) = p^2 = (0.617)^2 = 0.381$, $f(C^RC^W) = 2pq = 2(0.617)(0.383) = 0.473$, and $f(C^WC^W) = q^2 = (0.383)^2 = 0.147$.

31-12. The frequencies of expected genotypes are given by expanding the trinomial $(p + q + r)^2$ to $p^2 + 2pq + 2pr + q^2 + 2qr + r^2$. The frequencies of the homozygous genotypes, c^+c^+, c^hc^h, and cc, are given by p^2, q^2, and r^2, respectively. The frequencies of the heterozygous genotypes, c^+c^h, c^+c, and c^hc, are given by 2pq, 2pr, and 2qr, respectively.

31-13. Begin with the phenotype that arises from a single genotype, that is, the albino phenotype with genotype *cc*. $f(cc) = r^2 = 10/355 = 0.028$ and $r = \sqrt{0.028} = 0.167$. The frequency of individuals with the Himalayan (c^hc^h and c^hc) genotypes (see answer to problem 31-12) is $q^2 + 2qr$ and the frequency of albinos (cc) is r^2; combining these gives f(Himalayans) + f(albinos) $= q^2 + 2qr + r^2$. The expression $q^2 + 2qr + r^2$ represents an expansion of the binomial $(q + r)^2$ and thus f(Himalayans) + f(albinos) $= (q + r)^2$. Substituting in the actual frequencies for the two genotypes gives

$$90/355 + 10/355 = (q + r)^2$$
$$0.254 + 0.028 = (q + r)^2$$
$$0.282 = (q + r)^2$$
$$\sqrt{0.282} = q + r$$
$$0.531 = q + r$$
$$0.531 - r = q.$$

Substituting the value for r = 0.167,

$$0.531 - 0.167 = q$$
$$0.363 = q.$$

Substituting the values for r and q in the expression $p + q + r = 1$ gives

$$p + 0.363 + 0.167 = 1$$
$$p + 0.53 = 1$$
$$p = 1 - 0.53 = 0.469.$$

Thus, $f(c^+) = p = 0.469$, $f(c^h) = q = 0.363$, and $f(c) = r = 0.167$.

31-14. a. In the generation of 1850 white rabbits, the expected number showing each genotype is as follows: $f(c^+c^+) = p^2 = (0.469)^2 = 0.220$, and $0.220 \times 1850 = 407$, $f(c^+c^h) = 2pq = 2(0.469)(0.363) = 0.340$, and $0.340 \times 1850 = 629$, $f(c^+c) = 2pr + 2(0.469)(0.168) = 0.158$, and $0.158 \times 1850 = 292$, $f(c^hc^h) = q^2 = (0.363)^2 = 0.132$, and $0.132 \times 1850 = 244$, $f(c^hc) = 2qr = 2(0.363)(0.168) = 0.122$, and $0.122 \times 1850 = 226$, and $f(cc) = r^2 = (0.168)^2 = 0.0282$, and $0.0282 \times 1850 = 52$.

b. The expected number showing each phenotype is as follows: f(normal) = $f(c^+c^+) + f(c^+c^h) + f(c^+c) = 0.220 + 0.340 + 0.158 = 0.178$, and $0.178 \times 1850 = 1328$, f(Himalayan) = $f(c^hc^h) + f(c^hc) = 0.132 + 0.122 = 0.254$, and $0.254 \times 1850 = 470$, and f(albino) = $f(cc) = 0.0282$, and $0.0282 \times 1850 = 52$.

31-15. a. Initially, all the *vg* alleles are carried by the 800 homozygous recessive founders; each carries two *vg* alleles for a total of 1600 out of 2000 alleles in the gene pool. $f(vg) = 1600/2000 = 0.8$. Rearranging the equation $p + q = 1$ gives $p = 1 - q$. Substituting 0.8 for q gives $p = 1 - 0.8 = 0.2$. Thus, the founding allelic frequencies are: $f(vg) = 0.8$ and $f(vg^+) = 0.2$.

b. $f(vgvg) = q^2 = 1\% = 0.01$ and $q = \sqrt{0.01} = 0.1$. Since $p + q = 1$, $p = 1 - q$; substituting 0.1 for q gives $p = 1 - 0.1 = 0.9$. Thus, the current allelic frequencies are $f(vg) = 0.1$ and $f(vg^+) = 0.9$.

31-16. a. If the first generation developed under Hardy-Weinberg conditions, $f(vg^+vg^+) = p^2 = (0.2)^2 = 0.04$; $f(vg^+vg) = 2pq = 2(0.2)(0.8) = 0.32$; $f(vgvg) = q^2 = (0.8)^2 = 0.64$.

b. If generation 201 develops under Hardy-Weinberg conditions, $f(vg^+vg^+) = p^2 = (0.9)^2 = 0.81$; $f(vg^+vg) = 2pq = 2(0.9)(0.1) = 0.18$; $f(vgvg) = q^2 = (0.1)^2 = 0.01$.

31-17. a. Evolution may be defined as a change in gene frequency. During the 200 generations the $f(vg)$ shifted from 0.8 to 0.1 and thus the population has evolved.

b. Initially, there were 1600 *vg* alleles in the gene pool. At generation 200 $f(vg) = 0.1$, or 10%. With 1000 individuals, each with two alleles, the gene pool has a total of 2000 alleles; 10% of that is 200. Net loss over 200 generations is $1600 - 200 = 1400$ alleles. Over a period of 200 generations, the loss per generation is $1400/200 = 7$ alleles.

31-18. a. f(homozygous recessives) = $q^2 = 1/40{,}000$. f(recessive allele) = $q = \sqrt{1/40{,}000} = 1/200 = 0.005$.

b. Since $p + q = 1$, and $q = 0.005$, $p + 0.005 = 1$ and $p = 1 - 0.005 = 0.995$. The f(heterozygotes) $= 2pq = 2(0.005)(0.995) = 0.00995$, or about 1 in 100.

31-19. a. Because the males are hemizygous, the frequency of the recessive allele in the population equals the frequency of males expressing the trait: f(recessive allele) = $q = 0.06$. Color-blind females carry the recessive allele on both of their X chromosomes and f(color-blind females) = $q^2 = (0.06)^2 = 0.0036$, or 0.36%.

b. First, the value of p must be determined: $p = 1 - q = 1 - 0.06 = 0.94$. f(female carriers) = $2pq = (2)(0.94)(0.06) = 0.1128$, or about 113 females per 1000 individuals.

31-20. a. Since the population consists of 369 individuals, each carrying two alleles for this trait, there are a total of $369 \times 2 = 738$ alleles in the gene pool. $f(P^a) = (2)f(P^aP^a) + f(P^aP^b) + f(P^aP^c) = 2(15) + 111 + 4 = 145$, and $145/738 = 0.196$. $f(P^b) = (2)f(P^bP^b) + f(P^aP^b) + f(P^bP^c) = 2(220) + 111 + 19 = 570$, and $570/738 = 0.772$. $f(P^c) = (2)f(P^cP^c) + f(P^bP^c) + f(P^aP^c) = 2(0) + 19 + 4 = 23$, and $23/738 = 0.0312$.

b. If the population is in Hardy-Weinberg equilibrium, the genotypic frequencies would be given by the Hardy-Weinberg expressions. $f(P^aP^a) = (f(P^a))^2 = (0.196)^2 = 0.0384$, $f(P^bP^b) = (f(P^b))^2 = (0.772)^2 = 0.596$, $f(P^cP^c) = (f(P^c))^2 = (0.0312)^2 = 0.00097$, $f(P^aP^b) = (2)f(P^a)f(P^b) = 2(0.196)(0.772) = 0.303$, $f(P^aP^c) = (2)f(P^a)f(P^c) = 2(0.196)(0.0312) = 0.0122$, and $f(P^bP^c) = (2)f(P^b)f(P^c) = 2(0.772)(0.0312) = 0.048$. Expressed in terms of the numbers of individuals in a population of 369, the expected frequencies are $P^aP^a = 0.0384 \times 369 = 14.2$, $P^bP^b = 0.596 \times 369 = 220$, $P^cP^c = 0.00097 \times 369 = 0.4$, $P^aP^b = 0.303 \times 369 = 111.81$, $P^aP^c = 0.0122 \times 369 = 4.50$, and $P^bP^c = 0.048 \times 369 = 17.7$ These expected numbers are very close to the numbers actually found in the population, indicating that the population is in Hardy-Weinberg equilibrium. This agreement could be confirmed using the chi-square test.

31-21. The only phenotype arising from a single homozygous genotype is type O, produced by genotype *ii*. With $r = f(i)$, type O occurs with an expected frequency of r^2. $r^2 = 4100/10{,}000 = 0.410$ and $r = \sqrt{0.410} = 0.640$. The combined numbers of A and O genotypes $(2790 + 4100 = 6890)$ are given by the expression $p^2 + 2pr + r^2$, an expansion of the binomial $(p + r)^2$. Taking the square root of both sides of the equation $(p + r)^2 = 0.6890$, gives $p + r = \sqrt{0.6890} = 0.830$. Substituting the value for r gives the expression $p + 0.640 = 0.830$. Solving for p, we get $p = 0.830 - 0.640 = 0.190$. Since $p + q + r = 1$, the values for r and p can be substituted in the equation and it can be solved for q: $0.190 + q + 0.640 = 1$. $q = 1 - 0.640 - 0.190 = 0.170$.

Chapter 32 Population Genetics II: Evolutionary Forces

32-1. a. The relationship between the initial allelic frequency (0.8), the change in allelic frequency (Δp) produced by forward mutation, and the postmutation allelic frequency (0.799889) is $0.8 - \Delta p =$

0.799889. Solving for Δp gives $\Delta p = 0.8 - 0.799889 = 0.000111$. The relationship between Δp, the initial allelic frequency (p), and the mutation rate (u) is given by $\Delta p = up$; substituting the two known quantities gives $0.000111 = u(0.8)$ which can be solved for u. $u = 0.000111/0.8 = 0.000139$.

b. The net change is $0.8 + (-0.799942) = -0.000058$. Forward mutation produces a loss of 0.000111 and back-mutation a gain of $(-0.000058 + 0.000111 = +0.000053$. The relationship between Δp due to back-mutations, the initial allelic frequency (q), and the back-mutation rate (v) is $\Delta p = vq$; substituting the two known quantities gives $0.000053 = v(0.2)$ which can be solved for v. $v = 0.000053/0.2 = 0.000265$.

32-2. a. The initial $f(B) = p = 0.6$ and the mutation rate (u) is $5 \times 10^{-5} = 0.00005$. The reduction in the number of *B* alleles is given by multiplying the allelic frequency (p) by the mutation rate (u): $pu = (0.6)(0.00005) = 0.00003$. In one generation $f(B)$ is reduced to $0.6 - 0.00003 = 0.59997$.

b. The net change (Δp) in $f(B)$ in one generation equals the increase in $f(B)$ from back-mutations less the decrease in $f(B)$ from forward mutations. The increase from back-mutations is given by multiplying the back-mutation rate $v = 2 \times 10^{-5} = 0.00002$ by the frequency of $b = q = 0.4$: $vq = (0.00002)(0.4) = 0.000008$. The net change is $\Delta p = vq - up = 0.000008 - 0.000030 = -0.000022$. The effects of both forward and back-mutations in one generation would change $f(B)$ from 0.6 to $0.6 - 0.000022 = 0.599978$.

32-3. The equilibrium frequency for p is given by the following expression: $p = v/(u + v)$, where u and v are the forward and back-mutation rates, respectively. Substituting the values for v and u gives $p = 6.67 \times 10^{-5}/(1 \times 10^{-4} + 6.67 \times 10^{-5}) = 0.40012$. Since $p + q = 1$, $q = 1 - p$. Substituting the value of p gives $q = 1 - 0.40012 = 0.59988$.

32-4. a. A selection coefficient of 1 indicates maximum level of selection against the genotype; that is, the genotype has no reproductive output.

b. An adaptive value of 1 indicates maximum transmission of genetic material to the next generation.

c. When the selection coefficient is 1, the adaptive value is 0.

d. When the adaptive value is 1, the selection coefficient is 0.

32-5. a. Prior to selection, the expected genotypic frequencies are given by the expression $p^2 + 2pq + q^2$ Since $p = 0.7$ and $q = 0.3$, $f(GG) = p^2 = (0.7)^2 = 0.49$· $f(Gg) = 2pq = 2(0.7)(0.3) = 0.42$; and $f(gg) = q^2 = (0.3)^2 = 0.09$

b. Since there is selection only against the homozygous recessive, the postselection values for $f(GG)$ and $f(Gg)$ will remain at 0.49 and 0.42, respectively. Since *gg* has an adaptive value (W) of 0.8, the selective coefficient for the genotype is $1 - W$, or 0.2. Selection reduces $f(gg)$ by a factor of 0.2, or 20% (reduction of $0.09 \times 0.2 = 0.018$), giving a postselection $f(gg) = 0.09 - 0.018 = 0.072$. Since the population size is now $1 - q^2$ or $1 - 0.018 = 0.982$, the relative postselection frequencies are as follows: $f(GG) = 0.49/0.982 = 0.499$ $f(Gg) = 0.42/0.982 = 0.428$, and $f(gg) = 0.72/0.982 = 0.073$.

c. The postselection $f(G)$ is given by the expression $(p^2 + pq)/(1 - sq^2)$. Substituting the values for p ($= 0.7$), q ($= 0.3$), and s ($= 0.2$) gives

$$f(G) = \frac{(0.7)^2 + (0.7)(0.3)}{1 - (0.2)(0.3)^2} = \frac{0.49 + 0.21}{1 - 0.018} = \frac{0.7}{0.982} = 0.713.$$

For the new relative frequencies, $p + q = 1$, and the postselection $f(g)$ ($= q$) can be obtained by subtracting the postselection $f(G)$ ($= p = 0.713$) from 1: $q = 1 - p = 1 - 0.713 = 0.287$.

32-6. a. Adaptive value $W = 0.7$ and $s = 1 - 0.7 = 0.3$ for genotype *tt*. The net change in $f(T)$ (that is, Δp) is $sq^2p/(1 - sq^2) = (0.3)(0.35)^2(0.65)/[1 - (0.3)(0.35)^2] = (0.3)(0.1225)(0.65)/[1 - (0.3)(0.1225)] = 0.02389/0.9633 = 0.0248$.

b. $f(T)$ in generation 2 is $0.65 + 0.0248 = 0.675$. The generation producing generation 3 has $f(T) = 0.675$ and $f(t) = 1 - f(T) = 0.325$. Selection coefficient(s) for the *tt* genotype is still 0.3. The net change in $f(T)$ between generations 2 and 3 is $\Delta p = sq^2p/(1 - sq^2) = (0.3)(0.325)^2(0.675)/[1 - (0.3)(0.325)^2] = (0.3)(0.1056)(0.675)/[1 - (0.3)(0.1056)] = 0.02138/0.9683 = 0.02209$ $f(T)$ in generation 3 is $0.675 + 0.02209 = 0.697$.

32-7 a. In the initial population, f(dominant allele) $= p = 0.2$, f(recessive allele) $= q = 0.8$, and $\Delta p = 0.1$. These values can be substituted in the equation $\Delta p = sq^2p/(1 - sq^2)$ which can then be solved for s, the selection coefficient.

$$\begin{aligned} 0.1 &= s(0.8)^2(0.2)/[1 - s(0.8)^2] \\ &= s(0.64)(0.2)/[1 - s(0.64)] \\ 0.1(1 - 0.64s) &= 0.128s \\ 0.1 - 0.064s &= 0.128s \\ 0.1 &= 0.128s + 0.064s = 0.192s \\ 0.1/(0.192) &= s = 0.5208. \end{aligned}$$

b. In generation 2, f(dominant allele) $= p = 0.3$, f(recessive allele) $= q = 0.7$ and $s = 0.5208$, and we wish to determine Δp for generation 3. These values can be placed in the equation $\Delta p = sq^2p/(1 - sq^2)$ which can then be solved for Δp. $\Delta p = sq^2p/(1 - sq^2) = (0.5208)(0.7)^2(0.3)/[1 - (0.5208)(0.7)^2] = (0.5208)(0.49)(0.3)/[1 - (0.5208)(0.49)] = 0.07656/[1 - 0.255] = 0.0766/0.745 = 0.103$. $f(T)$ in generation 3 is $0.3 + 0.103 = 0.403$.

32-8. a. Cystic fibrosis occurs with a frequency of 1/1600 $= 0.000625$ and is due to a homozygous recessive genotype. Consequently, the frequency can be set equal to q^2·

$$\begin{aligned} q^2 &= 0.000625 \\ q &= \sqrt{0.000625} = 0.025. \end{aligned}$$

Before we can determine the frequency of carriers, we must determine the value of p by using the expression $p + q = 1$: $p = 1 - q = 1 - 0.025 = 0.975$. The frequency of carriers is $2pq = 2(0.975)(0.025) = 0.04875$, which is close to

1 in 20 individuals. The ratio of carriers to homozygous recessives is approximately 80:1 (0.04875:0.000625 = 78:1).

b. PKU frequency = 1/40,000 = 0.000025 = q^2· q = $\sqrt{0.000025}$ = 0.005; p = 1 − q = 1 − 0.005 = 0.995. f(carriers) = 2pq = 2(0.995)(0.005) = 0.00995, which is close to 1 in 100. The ratio of carriers to homozygous recessives is approximately 400:1 (0.00995:0.000025 = 398:1).

c. As the allelic frequency declines, the ratio gets larger; proportionately more and more of the recessive alleles are carried by heterozygous individuals and are thus protected from the effects of selection. Consequently, selection is less effective.

32-9. a. The postmigration value of p in the aggregate population is $p_a = p_r + m(p - p_r)$, where p and p_r are the allelic frequencies of the population contributing the migrants and the population receiving the migrants, respectively, and m is the proportion of the aggregate population that consists of migrants. Substituting p = 0.6, p_r = 0.25, and m = 0.10 into the expression for p_a gives p_a = 0.25 + 0.1(0.6 − 0.25) = 0.25 + 0.1(0.35) = 0.25 + 0.035 = 0.285.

32-10. The question asks us to determine the value of the coefficient of replacement (m) necessary to change the allelic frequencies in the recipient population from 0.25 to 0.5. Substituting the values into the equation $p_a = p_r + m(p - p_r)$ gives 0.5 = 0.25 + m(0.6 − 0.25), which can be solved for m:

$$0.5 - 0.25 = m(0.35)$$
$$m = \frac{0.5 - 0.25}{0.35} = \frac{0.25}{0.35} = 0.714.$$

The frequency of the allele would double if migrants make up 71.4% of the population.

32-11. The coefficient of replacement (m) is 0.15, the post-migration allelic frequency in the aggregate island population, p_a, is 0.7, the allelic frequency in the mainland population, p_c, is 0.6, and we need to determine the allelic frequency for the island population prior to migration, p_r. These values can be substituted into the expression $p_a = m(p_c) + (1 - m)(p_r)$ which is solved for p_r.

$$0.7 = (0.15)(0.6) + (1 - 0.15)(p_r)$$
$$0.7 = 0.09 + 0.85(p_r)$$
$$0.61 = 0.85(p_r)$$
$$p_r = 0.61/0.85 = 0.718$$

32-12. The coefficient of replacement (m) is 0.16, the post-migration allelic frequency in the receiving population, p_a, is 0.13, the allelic frequency for the receiving population prior to migration, p_r is 0.1, and we need to determine the allelic frequency in the contributing population, p_c. These values can be substituted into the expression $p_a = p_r + m(p - p_r)$ which is solved for p_c.

$$0.13 = 0.1 + 0.16(p - 0.1)$$
$$0.13 = 0.1 + 0.16p - 0.016$$
$$0.13 - 0.1 + 0.016 = 0.16(p_c)$$
$$0.046 = 0.16(p_c)$$
$$p = 0.046/0.16 = 0.2875$$

32-13. a. Since f(*a*) shifts around so much in the course of the eight generations, it is most reasonable to assume that genetic drift is responsible for the changes.

b. If drift is responsible, it is likely that a small number of parents give rise to each generation.

c. Although the number of individuals making up each generation was large, drift could operate if, for some reason, the number of individuals participating in reproduction was small.

32-14. Drastic changes in selective, or environmental, conditions would be required in order for natural selection to produce the allelic frequency changes seen here.

32-15. Mutation could be responsible only if it occurred at exceedingly high rates; rates in natural populations are so low that they could not produce changes of this magnitude in the time available. Drift could be the cause if each population had been founded by a very limited number of lake turtles; sampling error could have established the pond-population variation. Selection could be responsible if the environmental conditions of the ponds differ from one pond to the next, but this seems unlikely given the close proximity of the ponds to each other. Migration could be occurring with some regularity since the ponds are close to the lake; however, regular migration into each pond population would tend to make allelic frequencies in the various pond populations similar to each other and to those found in the lake population. Since the data shows major differences in allelic frequency, migration could be ruled out. Based on the information available, drift seems the most likely cause.

32-16. a. Genotypes *DD Dd*, and *dd* would be expected in a ratio of 1.2:1.

b. The probability that a fly chosen from this new generation would be *DD* is 1/4 and the probability that two such flies would be selected as parents of the next generation is given by the product law· 1/4 × 1/4 = 1/16. The probability of selecting two *dd* parents is also 1/4 × 1/4 = 1/16.

c. If both parents are *DD* the *d* allele will be lost from the next generation. This would be expected in (1/16)(48) = 3 populations. If both parents are *dd*, the *D* allele will be lost from the next generation. This would be expected in 1/16 × 48 = 3 populations. Thus, a total of six populations would be expected to lose one allele and fix the other.

d. *D* and *d* would be retained in the same frequencies if the randomly selected parents were both *Dd*. The probability of that occurring is 1/2 × 1/2 = 1/4. Out of 48 populations, we would expect that 1/4 × 48 = 12 populations would be this way. The allelic frequencies would also be the same if the parents were *DD* × *dd* or *dd* × *DD* The probability of this is 2 × 1/4 × 1/4 = 2/16 = 1/8. Out of 48 populations, the expectation is that 1/8 × 48 = 6 populations would be this way. The total number of populations retaining the same frequencies is 12 + 6 = 18.

e. Two possibilities exist: f(*D*) = 0.75, f(*d*) = 0.25 and f(*D*) = 0.25, f(*d*) = 0.75. The first would arise from matings *DD* × *Dd* and *Dd* × *DD*· the probability here is 2 × 1/4 × 1/2 = 2/8 = 1/4. The second would arise from matings $Dd \times dd$ and $dd \times Dd$; the probability here is $2 \times 1/2 \times 1/4 = 2/8 = 1/4$. $1/4 + 1/4 = 2/4 = 1/2$ of the 48 populations, or 24, would show these gene frequencies.

f. Sooner or later, each population will possess just one or the other allele. About half would be expected to possess just the *D* allele while the rest would have only the *d* allele.

Appendix A: Comprehensive Problem Set for Chapters 1–13

A-1. a. The map distance of 18 units indicates a crossover frequency between the two loci of 18%. The cross can be represented as $an^+f/anf^+ \times an^+f/anf^+$. Each parent will make four kinds of gametes: two parental types, an^+f and anf^+, each making up 41% of the gametes, and two recombinant types, an^+f^+ and *anf*, each making up 9% of the gametes. The random union of these gametes is shown in the Punnett square in Figure A-1a. Four progeny phenotypes result: wild eared, wild striped: 0.5081, or 50.81%; wild eared, fine striped: 0.2419, or 24.19%; anther eared, wild striped: 0.2419, or 24.19%; anther eared, fine striped: 0.0081, or 0.81%.

b. The cross can be represented as $an^+f^+/anf \times an^+f^+/anf$. Each parent will make four kinds of gametes: two parental types, an^+f^+ and *anf*, each making up 41% of the gametes, and two recombinant types, an^+f and anf^+, each making up 9% of the gametes. The random union of these gametes is shown in the Punnett square in Figure A-1b. Four progeny phenotypes result: wild eared, wild striped: 0.6681, or 66.81%; anther eared, fine striped: 0.1681, or 16.81%; wild eared, fine striped: 0.0819, or 8.19%; anther eared, wild striped: 0.0819, or 8.19%.

c. The cross can be represented as $an^+f^+/anf \times an^+f/anf^+$. Each parent will make four kinds of gametes. The coupling parent makes two parental types, an^+f^+ and *anf*, each making up 41% of the gametes, and two recombinant types, an^+f and anf^+, each making up 9% of the gametes. The repulsion parent makes two parental types, an^+f and anf^+, each making up 41% of the gametes, and two recombinant types, an^+f^+ and *anf*, each making up 9% of the gametes. The random union of these gametes is shown in the Punnett square in Figure A-1c. Four progeny phenotypes result: wild eared, wild striped: 0.5369, or 53.69%; wild eared, fine striped: 0.2131, or 21.31%; anther eared, wild striped: 0.2131, or 21.31%; anther eared, fine striped: 0.0369, or 3.69%.

d. Since the locus for aleurone color is inherited independently of the two linked loci, its inheritance can be considered separately. The cross can be represented as $Cc \times Cc$, with the probability of producing colored (*CC* and *Cc*) and colorless (*cc*) progeny equal to 0.75 and 0.25, respectively. The probabilities for combinations of the two linked traits are obtained from the answer to A-1c.

i. p(wild eared, wild striped, colored) = $0.5369 \times 0.75 = 0.403$

ii. p(anther eared, fine striped, colorless) = $0.0369 \times 0.25 = 0.009$

iii. p(anther eared, wild striped, colored) = $0.2131 \times 0.75 = 0160$

A-2. a. The cross can be represented as $X^{yw}X^{yw} \times X^{++}Y$. The female will produce one type of gamete, X^{yw}, and the male will produce two types of gametes, X^{++} and Y. Two types of F_1 progeny result: half will be $X^{++}X^{yw}$ (females with wild-type phenotypes that are carriers of the recessive allele for each trait), and half will be $X^{yw}Y$ (males with yellow bodies, and white eyes).

b. The mating of members of the F_1 to produce the F_2 can be represented as $X^{++}X^{yw} \times X^{yw}Y$. The female will produce four kinds of gametes: two parental types, X^{++} and X^{yw}, and two recombinant types, X^{+w} and X^{y+}. The male makes two kinds of gametes: X^{yw} and Y. Union of these gametes is shown

FIGURE A-1a

	an^+f (0.41)	anf^+ (0.41)	an^+f^+ (0.09)	anf (0.09)
an^+f (0.41)	an^+f/an^+f (0.1681) Wild eared, fine striped	anf^+/an^+f (0.1681) Wild eared, wild striped	an^+f^+/an^+f (0.0369) Wild eared, wild striped	anf/an^+f (0.0369) Wild eared, fine striped
anf^+ (0.41)	an^+f/anf^+ (0.1681) Wild eared, wild striped	anf^+/anf^+ (0.1681) Anther eared, wild striped	an^+f^+/anf^+ (0.0369) Wild eared, wild striped	anf/anf^+ (0.0369) Anther eared, wild striped
an^+f^+ (0.09)	an^+f/an^+f^+ (0.0369) Wild eared, wild striped	anf^+/an^+f^+ (0.0369) Wild eared, wild striped	an^+f^+/an^+f^+ (0.0081) Wild eared, wild striped	anf/an^+f^+ (0.0081) Wild eared, wild striped
anf (0.09)	an^+f/anf (0.0369) Wild eared, fine striped	anf^+/anf (0.0369) Anther eared, wild striped	an^+f^+/anf (0.0081) Wild eared, wild striped	anf/anf (0.0081) Anther eared, fine striped

FIGURE A-1b

	an^+f^+ (0.41)	anf (0.41)	an^+f (0.09)	anf^+ (0.09)
an^+f^+ (0.41)	an^+f^+/an^+f^+ (0.1681) Wild eared, wild striped	anf/an^+f^+ (0.1681) Wild eared, wild striped	an^+f/an^+f^+ (0.0369) Wild eared, wild striped	anf^+/an^+f^+ (0.0369) Wild eared, wild striped
anf (0.41)	an^+f^+/anf (0.1681) Wild eared, wild striped	anf/anf (0.1681) Anther eared, fine striped	an^+f/anf (0.0369) Wild eared, fine striped	anf^+/anf (0.0369) Anther eared, wild striped
an^+f (0.09)	an^+f^+/an^+f (0.0369) Wild eared, wild striped	anf/an^+f (0.0369) Wild eared, fine striped	an^+f/an^+f (0.0081) Wild eared, fine striped	anf^+/an^+f (0.0081) Wild eared, wild striped
anf^+ (0.09)	an^+f^+/anf^+ (0.0369) Wild eared, wild striped	anf/anf^+ (0.0369) Anther eared, wild striped	an^+f/anf^+ (0.0081) Wild eared, wild striped	anf^+/anf^+ (0.0081) Anther eared, wild striped

in the branch diagram in Table A-2b. Four of the eight categories of F_2 progeny are recombinants and they include a total of 29 flies, giving an average of 29/4 = 7.25, or about 7, flies per category. The other four categories are parental types and they contain the remaining 2205 − 29 = 2176 flies, giving an average of 2176/4 = 544 flies per category.

c. The percentage of recombinants among the F_2 progeny is 29/2205 = 0.0132, or 0.0132 × 100 = 1.32%, indicating a distance of 1.32 map units between the two loci.

d. The percentage of recombinants among the F_2 progeny in this cross is 900/2441 = 0.3687, or 0.3687 × 100 = 36.87%, indicating a distance of 36.87 map units between the two loci. Since these loci are considerably further apart than the *y* and *w* loci, there is a much greater opportunity for crossing over to occur between them, generating a much higher number of recombinants.

e. The *dp* locus is inherited independently of the sex-linked loci and thus can be considered separately. The mating producing the F_1 can be represented as $dpdp \times dp^+dp^+$, and all the F_1 progeny will be dp^+dp and have wild-type wings. The mating producing the F_2 is $dp^+dp \times dp^+dp$, and 3/4 of the F_2 progeny (genotypes dp^+dp^+ and dp^+dp) will have

FIGURE A-1c

		Gametes of coupling parent			
		an^+f^+ (0.41)	anf (0.41)	an^+f (0.09)	anf^+ (0.09)
Gametes of repulsion parent	an^+f (0.41)	an^+f^+/an^+f (0.1681) Wild eared, wild striped	anf/an^+f (0.1681) Wild eared, fine striped	an^+f/an^+f (0.0369) Wild eared, fine striped	anf^+/an^+f (0.0369) Wild eared, wild striped
	anf^+ (0.41)	an^+f^+/anf^+ (0.1681) Wild eared, wild striped	anf/anf^+ (0.1681) Anther eared, wild striped	an^+f/anf^+ (0.0369) Wild eared, wild striped	anf^+/anf^+ (0.0369) Anther eared, wild striped
	an^+f^+ (0.09)	an^+f^+/an^+f^+ (0.0369) Wild eared, wild striped	anf/an^+f^+ (0.0369) Wild eared, wild striped	an^+f/an^+f^+ (0.0081) Wild eared, wild striped	anf^+/an^+f^+ (0.0081) Wild eared, wild striped
	anf (0.09)	an^+f^+/anf (0.0369) Wild eared, wild striped	anf/anf (0.0369) Anther eared, fine striped	an^+f/anf (0.0081) Wild eared, fine striped	anf^+/anf (0.0081) Anther eared, wild striped

TABLE A-2b

Male gametes	Female gametes		Progeny: Genotypes	Progeny: Phenotypes
X^{yw}	X^{++}	=	$X^{++}X^{yw}$	Female, wild-type (parental type)
	X^{yw}	=	$X^{yw}X^{yw}$	Female, yellow body, white eyes (parental type)
	X^{+w}	=	$X^{+w}X^{yw}$	Female, wild body, white eyes (recombinant type)
	X^{y+}	=	$X^{y+}X^{yw}$	Female, yellow body, wild eyes (recombinant type)
Y	X^{++}	=	$X^{++}Y$	Male, wild-type (parental type)
	X^{yw}	=	$X^{yw}Y$	Male, yellow body, white eyes (parental type)
	X^{+w}	=	$X^{+w}Y$	Male, wild body, white eyes (recombinant type)
	X^{y+}	=	$X^{y+}Y$	Male, yellow body, wild eyes (recombinant type)

wild-type wings and 1/4 (genotype *dpdp*) will have dumpy wings.

i. The probability that an F_1 male with yellow body and white eyes (or any other F_1 fly) will have normal wings is 1.

ii. The probability that an F_2 female with wild-type body and white eyes (or any other F_2 fly) will have dumpy wings is 1/4.

iii. The probability that an F_2 male with yellow body and wild-type eyes (or any other F_2 fly) will have dumpy wings is 1/4.

A-3. a. The mating described can be represented as XX* × XY where the asterisk represents the recessive allele for color blindness. Phenotypically, the expected progeny fall into three categories: normal-vision females, normal-vision males, and color blind males in a 2:1:1 ratio, respectively. (Although all the females would be phenotypically normal, half would be expected to be homozygous for the normal allele and half would carry the allele for color blindness in the heterozygous condition.)

b. The 2:1:1 ratio has four parts, and with a total of 160 progeny, each part of the ratio is expected to represent 40 children (160/4 = 40). The expected number of progeny in the phenotypic classes is: normal female, 2 × 40 = 80; normal male, 1 × 40 = 40; and color blind male, 1 × 40 = 40.

c.

Class	Obs.	Exp.	Dev. (Obs. − Exp.)	Dev.²	Dev.²/Exp.
Normal female	92	80	12	144	144/80 = 1.80
Normal male	38	40	−2	4	4/40 = 0.10
Color blind male	30	40	−10	100	100/40 = 2.50
					$\chi^2 = 4.40$

d. Degrees of freedom = 3 classes − 1 = 2.

e. With two degrees of freedom, the χ^2 value of 4.40 falls between 3.22 and 4.61 with a probability value between 0.20 and 0.10.

f. Yes. Since the probability value is greater than 0.05, the odds are great enough to attribute the deviation to chance and to label the deviation as not statistically significant.

g. Yes. These data support the hypothesis that underlies the 2:1:1 theoretical ratio.

h. Since the loci for the two autosomally based traits are inherited independently of the sex-linked trait and independently of each other, the inheritance of each trait can be considered separately. With regard to the alkaptonuria locus, the cross can be represented as *Aa* × *Aa* and the probability of producing normal (*AA* and *Aa*) and alkaptonuric (*aa*) children is 3/4 and 1/4, respectively. With regard to the brachydactyly locus, the cross can be represented as *Bb* × *bb* and affected (*Bb*) and normal (*bb*) children are each produced with a probability of 1/2.

i. p(normal-vision male, normal metabolism, brachydactyly) = 1/4 × 3/4 × 1/2 = 3/32.

ii. p(color blind male, alkaptonuria, normal digits) = 1/4 × 1/4 × 1/2 = 1/32.

iii. p(normal-vision female who is a carrier of the color blindness allele, normal metabolism, normal digits) = 1/4 × 3/4 × 1/2 = 3/32.

A-4. a. The cross can be represented as *WwSs* × *WwSs*. Since the loci assort independently, we can consider the inheritance of each gene separately and then combine probabilities using the product law. The cross *Ww* × *Ww* produces genotypes *WW*, *Ww*, and *ww* in expected frequencies of 1/4, 1/2, 1/4, respectively. Similarly, the cross *Ss* × *Ss* produces genotypes *SS*, *Ss*, and *ss* in expected frequencies of 1/4, 1/2, and 1/4, respectively.

WW (1/4)	*SS* (1/4)	= *WWSS* (1/16)	Lethal
	Ss (1/2)	= *WWSs* (1/8)	Lethal
	ss (1/4)	= *WWss* (1/16)	Lethal
Ww (1/2)	*SS* (1/4)	= *WwSS* (1/8)	Lethal
	Ss (1/2)	= *WwSs* (1/4)	Normal wings
	ss (1/4)	= *Wwss* (1/8)	Wingless
ww (1/4)	*SS* (1/4)	= *wwSS* (1/16)	Lethal
	Ss (1/2)	= *wwSs* (1/8)	Normal wings
	ss (1/4)	= *wwss* (1/16)	Normal wings

Beetles with normal wings comprise 1/4 + 1/8 + 1/16 = 7/16 of all progeny (living and dead), and the wingless beetles make up 2/16. Considering just the living progeny gives a ratio of 7:2, normal wings to wingless.

b. The F_1 wingless progeny have genotype *Wwss*, and the most common type of beetle with normal wings has genotype *WwSs*. The cross *Ww* × *Ww* produces genotypes *WW*, *Ww*, and *ww*, with expected frequencies of 1/4, 1/2, 1/4, respectively. Similarly, the cross *Ss* × *ss* produces genotypes *Ss* and *ss* in expected frequencies of 1/2 each. Since genotype *WW* is lethal, it is not considered in the branch diagram that follows.

Ww (1/2)	*Ss* (1/2)	= *WwSs* (1/4)	Normal wings
	ss (1/2)	= *Wwss* (1/4)	Wingless
ww (1/4)	*Ss* (1/2)	= *wwSs* (1/8)	Normal wings
	ss (1/2)	= *wwss* (1/8)	Normal wings

Beetles with normal wings comprise 1/4 + 1/8 + 1/8 = 2/4 of all progeny (living and dead), and the wingless beetles make up 1/4. Considering just the living progeny gives a ratio of 2:1, normal wings to wingless.

A-5. Start with the assumption of no linkage. In the absence of gene interaction, three independently assorting genes, each with two alleles and standard dominance-recessiveness allelic interaction, would be expected to give rise to 2^3 = 8 F_2 phenotypes. Only four are found in this F_2 generation, indicating that interaction is occurring between two or all three of the genes. Considering the F_2, there are 153 + 128 = 281 large-flowered plants and 55 + 42 = 97 small-flowered plants, and the ratio of 281:97 is very close to 3:1. This is the expected outcome for a locus where the alleles show dominance-recessiveness interaction and where the locus is operating by itself and assorting independently of the other loci.

With regard to color, there are 153 + 55 = 208 yellow-flowered plants and 128 + 42 = 170 white-flowered plants, and the 208:170 ratio is very close to 9:7 (this agreement could be confirmed using the chi-square test). This 9:7 ratio can be explained if yellow-pigment production requires at least one dominant allele at *both* loci (that is, A__B__) with white produced when there is no dominant allele at one or the other or both loci. This conclusion is strengthened by the testcross outcome. Here there are 57 + 153 = 210 large-flowered plants and 49 + 159 = 208 small-flowered plants, and this ratio of 210:208 is very close to the 1:1 ratio expected in a testcross for a locus operating by itself and assorting independently of the other loci. Considering color, there are 57 + 49 = 106 yellow-flowered plants and 153 + 159 = 312 white-flowered plants, and the 106:312 ratio is very close to the 1:3 ratio expected in a testcross if the two loci interact as we have predicted.

A-6. a. The female parent has genotype X^+X^+ *cue/cue* and the male parent has genotype X^gY cu^+e^+/cu^+e^+. The female will make one type of gamete (X^+*cue*) and the male will make two types of gametes ($X^gcu^+e^+$ and Ycu^+e^+). Union of these gametes would be expected to produce two types of progeny in equal numbers: X^+X^g cu^+e^+/cue (females with wild-type wings, bodies, and eyes) and X^+Y cu^+e^+/cue (males with wild-type wings, bodies, and eyes).

b. The female in this cross has genotype X^+X^g cu^+e^+/cue and the male has genotype X^+Y *cue/cue*. With the sex-linked locus assorting independently of the two linked autosomal loci and with crossing over occurring between the two autosomal loci, the female makes eight types of gametes, four that are parental types with respect to the *cu* and *e* loci and four that are recombinant types. Since the two loci are separated by a distance of about 20 map units, the crossover frequency between them is about 20%. The recombinant gametes are collectively expected to make up 20% of all gametes produced by the female, with each of the four types occurring with an expected frequency of 5%. The parental-type gametes are expected to make up 80% of the female's gametes, with each of the four types occurring with an expected frequency of 20%. The parental types are X^+*cue* (20%), $X^+cu^+e^+$ (20%), X^g*cue* (20%), $X^gcu^+e^+$ (20%). The recombinant types are X^+cue^+ (5%), X^+cu^+e (5%), X^gcue^+ (5%), and X^gcu^+e (5%). The male is expected to produce two types of gametes in equal numbers: X^+*cue* (50%) and *Ycue* (50%). The union of these gametes is shown in the branch diagram in Table A-6b.

A-7. No. The F_1 progeny are heterozygous for both of the loci involved. Of the F_2 produced by their interbreeding, 9/16 would be expected to have at least one dominant allele at both loci. If the breeder's conclusion is correct, the F_2 progeny would be expected to consist of feathered and nonfeathered individuals in a 9:7 ratio. The cross outcome of 453:34 feathered to nonfeathered is not in agreement with this prediction but is close to the 15:1 ratio expected if the feathered condition is due to a dominant allele at either or both of the two loci. (Agreement between the data and the expected 15:1 ratio could be confirmed using the chi-square test.)

A-8. a. The female has genotype Z^bWFfGg and the male has genotype Z^BZ^bffgg. Since the traits under consideration are controlled by loci that assort independently, we can consider each separately and

TABLE A-6b

Male gametes	Female gametes	Progeny genotypes	Progeny phenotypes
X^+cue (0.5)	X^+cue (0.2)	$= X^+X^+\ cue/cue$ (0.1)	Female, wild eyes, curled wings, ebony body
	$X^+cu^+e^+$ (0.2)	$= X^+X^+\ cu^+e^+/cue$ (0.1)	Female, wild eyes, wings and body
	X^gcue (0.2)	$= X^+X^g\ cue/cue$ (0.1)	Female, wild eyes, curled wings, ebony body
	$X^gcu^+e^+$ (0.2)	$= X^+X^g\ cu^+e^+/cue$ (0.1)	Female, wild eyes, wings, and body
	X^+cue^+ (0.05)	$= X^+X^+\ cue^+/cue$ (0.025)	Female, wild eyes, curled wings, and wild body
	X^+cu^+e (0.05)	$= X^+X^+\ cu^+e/cue$ (0.025)	Female, wild eyes, wild wings, and ebony body
	X^gcue^+ (0.05)	$= X^+X^g\ cue^+/cue$ (0.025)	Female, wild eyes, curled wings, and wild body
	X^gcu^+e (0.05)	$= X^+X^g\ cu^+e/cue$ (0.025)	Female, wild eyes, wild wings, and ebony body
$Ycue$ (0.5)	X^+cue (0.2)	$= X^+Y\ cue/cue$ (0.1)	Male, wild eyes, curled wings, and ebony body
	$X^+cu^+e^+$ (0.2)	$= X^+Y\ cu^+e^+/cue$ (0.1)	Male, wild eyes, wild wings, and wild body
	X^gcue (0.2)	$= X^gY\ cue/cue$ (0.1)	Male, garnet eyes, curled wings, and ebony body
	$X^gcu^+e^+$ (0.2)	$= X^gY\ cu^+e^+/cue$ (0.1)	Male, garnet eyes, wild wings, and wild body
	X^+cue^+ (0.05)	$= X^+Y\ cue^+/cue$ (0.025)	Male, wild eyes, curled wings, and wild body
	X^+cu^+e (0.05)	$= X^+Y\ cu^+e/cue$ (0.025)	Male, wild eyes, wild wings, and ebony body
	X^gcue^+ (0.05)	$= X^gY\ cue^+/cue$ (0.025)	Male, garnet eyes, curled wings, and wild body
	X^gcu^+e (0.05)	$= X^gY\ cu^+e/cue$ (0.025)	Male, garnet eyes, wild wings, and ebony body

then use the product law to combine probabilities. The cross $Z^BZ^b \times Z^bW$ is expected to produce barred males (Z^BZ^b), nonbarred males (Z^bZ^b), barred females (Z^BW), and nonbarred females (Z^bW), each with a frequency of 1/4. The cross *FfGg* × *ffgg* will be expected to produce feathered and nonfeathered progeny with frequencies of 3/4 and 1/4, respectively. The phenotypes expected among the progeny are given in the branch diagram in Table A-8a.

b. The female parent has genotype Z^bWffgg and the male parent has genotype Z^BZ^bFfgg. The cross $Z^bW \times Z^BZ^b$ will be expected to produce barred males (Z^BZ^b), nonbarred males (Z^bZ^b), barred females (Z^BW), and nonbarred females (Z^bW), each with a frequency of 1/4. The cross *Ffgg* × *ffgg* will be expected to produce feathered (*Ffgg*) and nonfeathered (*ffgg*) progeny, each with a frequency of 1/2. The phenotypes expected among the progeny are given in the branch diagram in Table A-8b.

A-9. a. Since both parents are *FFGG*, we need only consider the sex-linked *b* and the autosomal *c* loci. The cross can be represented as $Z^BWcc \times Z^bZ^bCc$. Since the two traits are inherited independently, we can consider the inheritance of each separately and combine probabilities using the product law. The cross $Z^BW \times Z^bZ^b$ produces two types of progeny, Z^bW and Z^BZ^b, each with a frequency of 1/2. The cross *cc* × *Cc* produces two types of progeny, *Cc* and *cc*, each with a frequency of 1/2. Results are shown in Table A-9a.

b. The cross can be represented as $Z^bWCc \times Z^BZ^bCc$. The cross $Z^bW \times Z^BZ^b$ produces four types of progeny: Z^BW, Z^bW, Z^BZ^b, and Z^bZ^b, each with a frequency of 1/4. The cross *Cc* × *Cc* produces three types of progeny: *CC*, *Cc*, and *cc*, with frequencies of 1/4, 1/2, and 1/4, respectively. Results are shown in Table A-9b. Considering only the surviving progeny (ignoring the lethal genotypes), we get barred, creeper female (1/6); barred, normal-legged female (1/12); nonbarred, creeper female (1/6); nonbarred, normal-legged female (1/12); barred, creeper male (1/6); barred, normal-legged male (1/12); nonbarred, creeper male (1/6); and nonbarred, normal-legged male (1/12).

A-10. The two traits under consideration can be considered separately. All the female progeny are nonbarred and must have genotype Z^bW. Each female's W chromosome must have come from the mother and the Z chromosome from the father. Since the Z chromosomes all carry the *b* allele, the father's genotype is Z^bZ^b. Each male carries a Z chromosome from each

TABLE A-8a

Bar locus	Feather loci	Progeny phenotypes and frequencies
Barred males (1/4)	Feathered (3/4)	= Barred, feathered males (3/16)
	Nonfeathered (1/4)	= Barred, nonfeathered males (1/16)
Nonbarred males (1/4)	Feathered (3/4)	= Nonbarred, feathered males (3/16)
	Nonfeathered (1/4)	= Nonbarred, nonfeathered males (1/16)
Barred females (1/4)	Feathered (3/4)	= Barred, feathered females (3/16)
	Nonfeathered (1/4)	= Barred, nonfeathered females (1/16)
Nonbarred females (1/4)	Feathered (3/4)	= Nonbarred, feathered females (3/16)
	Nonfeathered (1/4)	= Nonbarred, nonfeathered females (1/16)

parent. All male progeny are barred and, based on their phenotype, could be either Z^BZ^B or Z^BZ^b. If their father is Z^bZ^b, then each male must be heterozygous (Z^BZ^b), indicating that the Z chromosome contributed by the mother carries *B*. Thus, the mother has genotype Z^BW. With regard to the leg length trait, three parental matings are possible: (1) *cc* × *cc*, giving all normal progeny; (2) *cc* × *Cc*, producing normal and creeper progeny in a 1:1 ratio; and (3) *Cc* × *Cc*, giving lethal, creeper, and normal progeny in a 1:2:1 ratio. Since the progeny include both creeper and normal types, we can rule out possibility 1. Within both the female and male progeny, the ratios of creeper to normal (46:26 and 54:26, respectively) are in much closer agreement with the 2:1 ratio expected for the living progeny in possibility 3 than with the 1:1 ratio expected in possibility 2, indicating that each parent is heterozygous. (Agreement between the observed and the 2:1 expected ratios could be statistically confirmed using the chi-square test.) Thus, the parents have genotypes Z^BWCc and Z^bZ^bCc.

A-11. **a.** The mating can be represented as S^1S^2Bb (female) × S^1S^2Bb (male). Since these traits are inherited independently, each can be considered separately

TABLE A-8b

Bar locus	Feather loci	Progeny phenotypes and frequencies
Barred males (1/4)	Feathered (1/2)	= Barred, feathered males (1/8)
	Nonfeathered (1/2)	= Barred, nonfeathered males (1/8)
Nonbarred males (1/4)	Feathered (1/2)	= Nonbarred, feathered males (1/8)
	Nonfeathered (1/2)	= Nonbarred, nonfeathered males (1/8)
Barred females (1/4)	Feathered (1/2)	= Barred, feathered females (1/8)
	Nonfeathered (1/2)	= Barred, nonfeathered females (1/8)
Nonbarred females (1/4)	Feathered (1/2)	= Nonbarred, feathered females (1/8)
	Nonfeathered (1/2)	= Nonbarred, nonfeathered females (1/8)

TABLE A-9a

Bar genotype	Leg-length genotype		Progeny Genotypes	Phenotypes	Frequencies
Z^bW (1/2)	*cc* (1/2)	=	Z^bWcc	Nonbarred, normal-legged females	1/4
	Cc (1/2)	=	Z^bWCc	Nonbarred, creeper females	1/4
Z^BZ^b (1/2)	*cc* (1/2)	=	Z^BZ^bcc	Barred, normal-legged males	1/4
	Cc (1/2)	=	Z^BZ^bCc	Barred, creeper males	1/4

and probabilities can then be combined using the product law. The mating $S^1S^2 \times S^1S^2$ will produce genotypes S^1S^1, S^1S^2, and S^2S^2 with expected frequencies of 1/4, 1/2, and 1/4, respectively. The mating *Bb* × *Bb* will produce genotypes *BB*, *Bb*, and *bb* with frequencies of 1/4, 1/2, and 1/4, respectively. These genotypes are combined in the branch diagram in Table A-11a. Half the progeny with each genotype will be expected to be female and the other half, male. There would be two classes of female progeny (long, nonbald and short, nonbald) in a 3:1 ratio, and four classes of male progeny (short, bald; short, nonbald; long, bald; and long, nonbald) in a 9:3:3:1 ratio.

b. The mother's genotype is S^1S^2Bb and the father's is S^2S^2bb, and the mating can be represented as $S^1S^2Bb \times S^2S^2bb$. $S^1S^2 \times S^2S^2$ produces genotypes S^1S^2 and S^2S^2 each with an expected frequency of 1/2, and *Bb* × *bb* produces genotypes *Bb* and *bb* each with a frequency of 1/2. These genotypes are combined in the branch diagram in Table A-11b. Half the progeny with each genotype will be expected to be female and the other half, male. All female progeny would be long, nonbald and the males would fall into four classes (short, bald; short, nonbald; long, bald; and long, nonbald) in a 1:1:1:1 ratio.

A-12. a. The cross producing the F_1 can be represented as $st/st\ bw^+/bw^+ \times st/st\ bw/bw$, with the female and male each producing a single type of gamete ($stbw^+$ and $stbw$, respectively). All the F_1 progeny have genotype $st/st\ bw^+/bw$ and will have scarlet eyes. Each member of the F_1 will produce two types of gametes: $stbw^+$ and $stbw$. The union of these gametes to produce the F_2 is shown as follows.

TABLE A-9b

Bar genotype	Leg-length genotype		Progeny Genotypes	Phenotypes	Frequencies
	CC (1/4)	=	Z^BWCC	Barred, lethal females	1/16
Z^BW (1/4)	*Cc* (1/2)	=	Z^BWCc	Barred, creeper females	1/8
	cc (1/4)	=	Z^BWcc	Barred, normal-legged females	1/16
	CC (1/4)	=	Z^bWCC	Nonbarred, lethal females	1/16
Z^bW (1/4)	*Cc* (1/2)	=	Z^bWCc	Nonbarred, creeper females	1/8
	cc (1/4)	=	Z^bWcc	Nonbarred, normal-legged females	1/16
	CC (1/4)	=	Z^BZ^bCC	Barred, lethal males	1/16
Z^BZ^b (1/4)	*Cc* (1/2)	=	Z^BZ^bCc	Barred, creeper males	1/8
	cc (1/4)	=	Z^BZ^bcc	Barred, normal-legged males	1/16
	CC (1/4)	=	Z^bZ^bCC	Nonbarred, lethal males	1/16
Z^bZ^b (1/4)	*Cc* (1/2)	=	Z^bZ^bCc	Nonbarred, creeper males	1/8
	cc (1/4)	=	Z^bZ^bcc	Nonbarred, normal-legged males	1/16

TABLE A-11a

Finger-length locus	Balding locus		Progeny: Genotype	Female phenotype	Male phenotype
S^1S^1 (1/4)	*BB* (1/4)	=	S^1S^1BB (1/16)	Short, nonbald	Short, bald
	Bb (1/2)	=	S^1S^1Bb (1/8)	Short, nonbald	Short, bald
	bb (1/4)	=	S^1S^1bb (1/16)	Short, nonbald	Short, nonbald
S^1S^2 (1/2)	*BB* (1/4)	=	S^1S^2BB (1/8)	Long, nonbald	Short, bald
	Bb (1/2)	=	S^1S^2Bb (1/4)	Long, nonbald	Short, bald
	bb (1/4)	=	S^1S^2bb (1/8)	Long, nonbald	Short, nonbald
S^2S^2 (1/4)	*BB* (1/4)	=	S^2S^2BB (1/16)	Long, nonbald	Long, bald
	Bb (1/2)	=	S^2S^2Bb (1/8)	Long, nonbald	Long, bald
	bb (1/4)	=	S^2S^2bb (1/16)	Long, nonbald	Long, nonbald

			Genotype	Phenotype
$stbw^+$	$stbw^+$	=	$st/st\ bw^+/bw^+$	Scarlet
	$stbw$	=	$st/st\ bw^+/bw$	Scarlet
$stbw$	$stbw^+$	=	$st/st\ bw^+/bw$	Scarlet
	$stbw$	=	$st/st\ bw/bw$	White

The F_2 progeny will consist of scarlet- and white-eyed flies in a 3:1 ratio.

b. The cross producing the F_1 can be represented as $X^v/X^v\ bw^+/bw^+ \times X^vY\ bw/bw$ with the female producing a single type of gamete, X^vbw^+, and the male, two types, X^vbw and Ybw. All the F_1 females will have genotype $X^v/X^v\ bw^+/bw$ (vermilion eyes) and all the males will have genotype $X^v/Y\ bw^+/bw$ (vermilion eyes). The F_1 females will produce two types of gametes, X^vbw^+ and X^vbw, and the males, X^vbw^+ X^vbw, Ybw^+, and Ybw. The union of these gametes to produce the F_2 is shown as follows.

			Genotype	Phenotype
X^vbw^+	X^vbw^+	=	$X^vX^v\ bw^+/bw^+$	Vermilion female
	X^vbw	=	$X^vX^v\ bw^+bw$	Vermilion female
	Ybw^+	=	$X^vY\ bw^+/bw^+$	Vermilion male
	Ybw	=	$X^vY\ bw^+/bw$	Vermilion male
X^vbw	X^vbw^+	=	$X^vX^v\ bw^+/bw$	Vermilion female
	X^vbw	=	$X^vX^v\ bw/bw$	White female
	Ybw^+	=	$X^vY\ bw^+/bw$	Vermilion male
	Ybw	=	$X^vY\ bw/bw$	White male

TABLE A-11b

Finger-length locus	Balding locus		Progeny: Genotype	Female phenotype	Male phenotype
S^1S^2 (1/2)	*Bb* (1/2)	=	S^1S^2Bb (1/4)	Long, nonbald	Short, bald
	bb(1/2)	=	S^1S^2bb(1/4)	Long, nonbald	Short, nonbald
S^2S^2(1/2)	*Bb* (1/2)	=	S^2S^2Bb(1/4)	Long, nonbald	Long, bald
	bb(1/2)	=	S^2S^2bb (1/4)	Long, nonbald	Long, nonbald

Within both the male and female F_2 progeny, vermilion- and white-eyed flies will occur in a 3:1 ratio.

A-13. a. The cross producing the F_1 can be represented as $X^vX^v\ bw/bw \times X^+Y\ bw^+/bw^+$ with the female producing a single type of gamete, $X^v bw$, and the male, two types, X^+bw^+ and Ybw^+. All the F_1 females will have genotype $X^+X^v\ bw^+/bw$ (wild-type eyes) and all the males will have genotype $X^vY\ bw^+/bw$ (vermilion eyes). The F_1 females will produce four types of gametes (X^+bw^+, X^+bw, X^vbw^+ and X^vbw) as will the males (X^vbw^+, X^vbw, Ybw^+, and Ybw). The union of these gametes to produce the F_2 is shown as follows.

		Genotype	Phenotype
X^+bw^+	X^vbw^+	$= X^+X^v\ bw^+/bw^+$	Wild-type female
	X^vbw	$= X^+X^v\ bw^+/bw$	Wild-type female
	Ybw^+	$= X^+Y\ bw^+/bw^+$	Wild-type male
	Ybw	$= X^+Y\ bw^+/bw$	Wild-type male
X^+bw	X^vbw^+	$= X^+X^v\ bw^+/bw$	Wild-type female
	X^vbw	$= X^+X^v\ bw/bw$	Brown female
	Ybw^+	$= X^+Y\ bw^+/bw$	Wild-type male
	Ybw	$= X^+Y\ bw/bw$	Brown male
X^vbw^+	X^vbw^+	$= X^vX^v\ bw^+/bw^+$	Vermilion female
	X^vbw	$= X^vX^v\ bw^+/bw$	Vermilion female
	Ybw^+	$= X^vY\ bw^+/bw^+$	Vermilion male
	Ybw	$= X^vY\ bw^+/bw$	Vermilion male
X^vbw	X^vbw^+	$= X^vX^v\ bw^+/bw$	Vermilion female
	X^vbw	$= X^vX^v\ bw/bw$	White female
	Ybw^+	$= X^vY\ bw^+/bw$	Vermilion male
	Ybw	$= X^vY\ bw/bw$	White male

Within both the male and female F_2 progeny, wild-type, vermilion-eyed, brown-eyed, and white-eyed flies will occur in a 3:3:1:1 ratio.

b. The cross producing the F_1 can be represented as $X^+X^+\ bw^+/bw^+ \times X^vY\ bw/bw$ with the female producing a single type of gamete, X^+bw^+, and the male, two types, X^vbw and Ybw. All the F_1 females will have genotype $X^+X^v\ bw^+/bw$ (wild-type eyes), and all the males will have genotype $X^+Y\ bw^+/bw$ (wild-type eyes). The F_1 females will produce four types of gametes: X^+bw^+, X^+bw, X^vbw^+, and X^vbw; and the males will produce four types: X^+bw^+, X^+bw, Ybw^+, and Ybw. The union of these gametes to produce the F_2 is shown as follows.

		Genotype	Phenotype
X^+bw^+	X^+bw^+	$= X^+X^+\ bw^+/bw^+$	Wild-type female
	X^+bw	$= X^+X^+\ bw^+/bw$	Wild-type female
	Ybw^+	$= X^+Y\ bw^+/bw^+$	Wild-type male
	Ybw	$= X^+Y\ bw^+/bw$	Wild-type male
X^+bw	X^+bw^+	$= X^+X^+\ bw^+/bw$	Wild-type female
	X^+bw	$= X^+X^+\ bw/bw$	Brown female
	Ybw^+	$= X^+Y\ bw^+/bw$	Wild-type male
	Ybw	$= X^+Y\ bw/bw$	Brown male
X^vbw^+	X^+bw^+	$= X^+X^v\ bw^+/bw^+$	Wild-type female
	X^+bw	$= X^+X^v\ bw^+/bw$	Wild-type female
	Ybw^+	$= X^vY\ bw^+/bw^+$	Vermilion male
	Ybw	$= X^vY\ bw^+/bw$	Vermilion male
X^vbw	X^+bw^+	$= X^+X^v\ bw^+/bw$	Wild-type female
	X^+bw	$= X^+X^v\ bw/bw$	Brown female
	Ybw^+	$= X^vY\ bw^+/bw$	Vermilion male
	Ybw	$= X^vY\ bw/bw$	White male

The F_2 females will be wild-type and brown in a 3:1 ratio, and the males will be wild-type, vermilion, white, and brown in a 3:3:1:1 ratio.

A-14. The female's genotype is $X^{++}X^{++}\ s/s$ and the male's is $X^{cvsn}Y\ S/S$. The cross can be represented as $X^{++}X^{++}\ s/s \times X^{cvsn}Y\ S/S$. The female will make a single type of gamete, $X^{++}s$, and the male will make two types, $X^{cvsn}S$ and YS. Two types of F_1 progeny, $X^{++}X^{cvsn}\ S/s$ and $X^{++}Y\ S/s$, will be expected in equal numbers and both the males and females will have wild-type wings and bristles, and star eyes. The cross producing the F_2 can be represented as $X^{++}X^{cvsn}\ S/s \times X^{++}Y\ S/s$. Since the sex-linked traits are inherited independently of the autosomal trait, inheritance of the two types of traits can be considered separately and then probabilities can be combined using the product law to generate the F_2. With regard to the sex-linked traits, the female will produce two parental-type gametes, X^{++} and X^{cvsn}, and two recombinant-type gametes, X^{cv+} and X^{+sn}. With the *cv* and *sn* loci separated by 6 map units, the crossover frequency between them is 6%, indicating that each type of recombinant gamete will be produced with a frequency of 0.03. The parental-type gametes will make up 94%

of the gametes, giving each type a frequency of 0.47. The male will produce two types of gametes, X^{++} and Y, each with a frequency of 0.5. The union of these gametes is shown in the branch diagram in Table A-14(1).

Phenotypic outcome: Females with wild-type wings and bristles, (0.5); male, wild wings, wild bristles (0.235); male, crossveinless wings, singed bristles (0.235); male, crossveinless wings, wild bristles (0.015); and male, wild wings, singed bristles (0.015). With regard to the autosomal trait, both the female and the male will produce two types of gametes, *S* and *s*, in equal frequencies. Combining these gametes results in the following progeny: *SS* (0.25), *Ss* (0.50), and *ss* (0.25). The phenotypic outcome is star eyes (0.75) and wild-type eyes (0.25). These phenotypes are combined in the branch diagram in Table A-14(2).

Results: Half the progeny will be females that fall into two phenotypic classes: wild, wild, star (0.375) and wild, wild, wild (0.125). The other half of the progeny will be males that fall into eight phenotypic classes: wild, wild, star (0.176); crossveinless, singed, star (0.176); crossveinless, wild, star (0.01125); wild, singed, star (0.01125); wild, wild, wild (0.05875); crossveinless, singed, wild (0.05875); crossveinless, wild, wild (0.00375); and wild, singed, wild (0.00375).

A-15. The absence of a 1:1:1:1:1:1:1:1 ratio tells us that the three loci are not independently assorting relative to each other and that at least some linkage is involved. The linkage can be evaluated by comparing the loci, considering them two at a time.

Shrunken endosperm and height loci:
wild ear, tall: 1430 (class 1) + 416 (class 6) = 1846
full, short: 1400 (class 5) + 405 (class 8) = 1805
shrunken, short: 1487 (class 2) + 410 (class 4) = 1897
shrunken, tall: 397 (class 3) + 1455 (class 7) = 1852

Shrunken endosperm and starchy endosperm loci:
full, waxy: 1430 (class 1) + 1400 (class 5) = 2830
full, starchy: 416 (class 6) + 405 (class 8) = 821
shrunken, starchy: 1487 (class 2) + 1455 (class 7) = 2942
shrunken, waxy: 397 (class 3) + 410 (class 4) = 807

Height and starchy endosperm loci:
tall, waxy: 1430 (class 1) + 397 (class 3) = 1827
tall, starchy: 416 (class 6) + 1455 (class 7) = 1871
short, starchy: 1487 (class 2) + 405 (class 8) = 1892
short, waxy: 410 (class 4) + 1400 (class 5) = 1810

This pairwise comparison of loci indicates an approximate 1:1:1:1 ratio with the shrunken endosperm and height loci, indicating that these loci are independently assorting. The same approximate ratio is found with the height and starchy endosperm loci. However, the comparison of the shrunken endosperm and starchy endosperm loci shows a significant departure from a 1:1:1:1 ratio, indicating linkage of these two loci. (The statistical significance of this departure can be confirmed using the chi-square test.) The parental-type combinations are full, waxy (2830) and shrunken, starchy (2942), indicating that the mutant genes are carried in repulsion in the heterozygous parent that had genotype $sh^+wx/shwx^+$ at these loci. The recombinant types are full, starchy (821) and shrunken, waxy (807), and they make up (1628/7400 = 0.22, or 0.22 × 100 = 22%, of the progeny. This recombination frequency tells us that the two loci are separated by a distance of 22 map units. The dwarf locus is located on a different chromosome from that carrying the shrunken and starchy loci. The cross under consideration can be represented as $sh^+wx/shwx^+\ d^+/d \times shwx/shwx\ d/d$.

A-16. Since the sex chromosome is inherited independently of the autosome, we can consider each separately and then combine probabilities using the product law. With regard to the sex-linked loci, the female's genotype is $X^{BnTa}X^{bnta}$ and the male's is $X^{bnta}Y$. The female produces two parental-type gametes, X^{BnTa} and X^{bnta}, and two recombinant-type gametes, X^{Bnta} and X^{bnTa}. Since 12 map units separate these loci, recombinants will make up a total of 12% of the gametes and each type will occur with a frequency of 0.06. The remaining 88% of the gametes will be equally divided between the two parental types, with each having a

TABLE A-14(1)

Male gametes	Female gametes		Progeny: Genotype	Progeny: Phenotype
X^{++} (0.5)	X^{++} (0.47)	=	$X^{++}X^{++}$ (0.235)	Female, wild wings, wild bristles
	X^{cvsn} (0.47)	=	$X^{++}X^{cvsn}$ (0.235)	Female, wild wings, wild bristles
	X^{cv+} (0.03)	=	$X^{++}X^{cv+}$ (0.015)	Female, wild wings, wild bristles
	X^{+sn} (0.03)	=	$X^{++}X^{+sn}$ (0.015)	Female, wild wings, wild bristles
Y (0.5)	X^{++} (0.47)	=	$X^{++}Y$ (0.235)	Male, wild wings, wild bristles
	X^{cvsn} (0.47)	=	$X^{cvsn}Y$ (0.235)	Male, crossveinless wings, singed bristles
	X^{cv+} (0.03)	=	$X^{cv+}Y$ (0.015)	Male, crossveinless wings, wild bristles
	X^{+sn} (0.03)	=	$X^{+sn}Y$ (0.015)	Male, wild wings, singed bristles

TABLE A-14(2)

Autosomal locus	Sex-linked loci		Progeny phenotype
Star eyes (0.75)	Female, wild wings, wild bristles (0.5)	=	Female, wild, wild, star (0.375)
	Male, wild wings, wild bristles (0.235)	=	Male, wild, wild, star (0.176)
	Male, crossveinless wings, singed bristles (0.235)	=	Male, crossveinless, singed, star (0.176)
	Male, crossveinless wings, wild bristles (0.015)	=	Male, crossveinless, wild, star (0.01125)
	Male, wild wings, singed bristles (0.015)	=	Male, wild, singed, star (0.01125)
Wild eyes (0.25)	Female, wild wings, wild bristles (0.5)	=	Female, wild, wild, wild (0.125)
	Male, wild wings, wild bristles (0.235)	=	Male, wild, wild, wild (0.05875)
	Male, crossveinless wings, singed bristles (0.235)	=	Male, crossveinless, singed, wild (0.05875)
	Male, crossveinless wings, wild bristles (0.015)	=	Male, crossveinless, wild, wild (0.00375)
	Male, wild wings, singed bristles (0.015)	=	Male, wild, singed, wild (0.00375)

frequency of 0.44. The male will produce two types of gametes, X^{bnta} and Y, each with a frequency of 0.5. The random union of these gametes is shown in the branch diagram in Table A-16(1).

With regard to the autosomal loci, the female's genotype is *vji/vji* and the male's is *v*+/+*ji*. The female will make a single type of gamete, *vji*. The male produces two parental-type gametes, *v*+ and +*ji*, and two recombinant-type gametes, *vji* and + +. Since 18 map units separate these loci, recombinants will make up a total of 18% of the gametes and each type will occur with a frequency of 0.09. The remaining 82% of the gametes will be equally divided between the two parental types, with each having a frequency of 0.41. The random union of these gametes is shown in the branch diagram that follows.

Female gamete	Male gametes		Progeny Genotype	Progeny Phenotype
vji (1.0)	*v*+ (0.41)	=	*v*+/*vji* (0.41)	Waltzer, wild
	+*ji* (0.41)	=	+*ji*/*vji* (0.41)	Wild, jittery
	vji (0.09)	=	*vji*/*vji* (0.09)	Wild, wild
	+ + (0.09)	=	+ +/*vji* (0.09)	Waltzer, jittery

a. p(male, bent, tabby; waltzer, jittery) = (0.22)(0.09) = 0.0198.

TABLE A-16(1)

Male gametes	Female gametes		Progeny Genotype	Progeny Phenotype
X^{bnta} (0.5)	X^{BnTa} (0.44)	=	$X^{BnTa}X^{bnta}$ (0.22)	Female, bent, tabby
	X^{bnta} (0.44)	=	$X^{bnta}X^{bnta}$ (0.22)	Female, wild tail, wild fur
	X^{Bnta} (0.06)	=	$X^{Bnta}X^{bnta}$ (0.03)	Female, bent, wild fur
	X^{bnTa} (0.06)	=	$X^{bnTa}X^{bnta}$ (0.03)	Female, wild tail, tabby fur
Y (0.5)	X^{BnTa} (0.44)	=	$X^{BnTa}Y$ (0.22)	Male, bent, tabby
	X^{bnta} (0.44)	=	$X^{bnta}Y$ (0.22)	Male, wild tail, wild fur
	X^{Bnta} (0.06)	=	$X^{Bnta}Y$ (0.03)	Male, bent, wild fur
	X^{bnTa} (0.06)	=	$X^{bnTa}Y$ (0.03)	Male, wild tail, tabby fur

b. p(male, wild, wild; wild, wild) = (0.22)(0.09) = 0.0198.

c. p(female, bent, wild; waltzer, wild) = (0.03)(0.41) = 0.0123.

d. p(male, bent, wild; wild, wild) = (0.03)(0.09) = 0.0027; p(female, wild, tabby; wild, jittery) = (0.03)(0.41) = 0.0123. Since no birth order is specified, we need to consider all combinations of birth order which is given by $[4!/(2!2!)](0.0027)^2(0.0123)^2 = 6.62 \times 10^{-9}$.

e. The genotype of the fifth mouse is independent of the genotypes of the other four mice. Since no sex is specified, the fifth mouse could be either a male or a female. The probability of a wild-type female is 0.0198 and the probability of a wild-type male is 0.0198. Adding the probability of these two mutually exclusive events gives the probability that the fifth mouse will be wild-type: 0.0198 + 0.0198 = 0.0396. The four other mice in the litter are wild-type males, each occurring with a probability of (0.22)(0.09) = 0.0198. The probability of four such males in a row is $(0.0198)^4$. The probability of getting the five mice as specified is $(0.0198)^4(0.0396) = 6.09 \times 10^{-9}$.

A-17. a. Each parent will produce four types of gametes *RM*, *RM'*, *rM*, and *rM'*. The union of these gametes is shown as follows.

		Progeny	
		Genotype	Phenotype
RM	*RM*	= *RRMM*	Smooth
	RM'	= *RRMM'*	Partly rough
	rM	= *RrMM*	Smooth
	rM'	= *RrMM'*	Partly rough
RM'	*RM*	= *RRM'M*	Partly rough
	RM'	= *RRM'M'*	Rough
	rM	= *RrMM'*	Partly rough
	rM'	= *RrM'M'*	Rough
rM	*RM*	= *RrMM*	Smooth
	RM'	= *RrMM'*	Partly rough
	rM	= *rrMM*	Smooth
	rM'	= *rrMM'*	Smooth
rM'	*RM*	= *RrM'M*	Partly rough
	RM'	= *RrM'M'*	Rough
	rM	= *rrMM'*	Smooth
	rM'	= *rrM'M'*	Smooth

We would expect smooth, partly rough, and rough phenotypes in a 7:6:3 ratio. With a total of 112 progeny, we would expect 49 smooth, 42 partly rough, and 21 rough.

b. The partly rough, cream coated guinea pig that is heterozygous at each of the three loci will have genotype $RrMM'c^dc^a$ and will produce eight types of gametes: RMc^d, RMc^a, $RM'c^d$, $RM'c^a$, rMc^d, rMc^a, $rM'c^d$, and $rM'c^a$. The parent with genotype $rrM'M'c^ac^a$ will produce one type of gamete: $rM'c^a$. The union of these gametes is shown as follows.

		Progeny	
		Genotypes	Phenotypes
$rM'c^a$	RMc^d	= $RrMM'c^dc^a$	Partly rough, cream
	RMc^a	= $RrMM'c^ac^a$	Partly rough, albino
	$RM'c^d$	= $RrM'M'c^dc^a$	Rough, cream
	$RM'c^a$	= $RrM'M'c^ac^a$	Rough, albino
	rMc^d	= $rrMM'c^dc^a$	Smooth, cream
	rMc^a	= $rrMM'c^ac^a$	Smooth, albino
	$rM'c^d$	= $rrM'M'c^dc^a$	Smooth, cream
	$rM'c^a$	= $rrM'M'c^ac^a$	Smooth, albino

Six phenotypes will result (smooth, cream; smooth, albino; partly rough, cream; partly rough, albino; rough, cream; and rough, albino) in a 2:2:1:1:1:1 ratio.

A-18. Since the original parents are pure breeding and all the F_1 progeny have gray bodies, straight bristles, and straight wings, we may conclude that the genes for sable body (*s*), forked bristles (*f*), and bent wings (*bt*) are recessive to their respective counterparts for gray body (s^+), straight bristles (f^+), and straight wings (bt^+). The reciprocal crosses involving the F_1 progeny are testcrosses, and since each gives a different outcome, we may conclude that at least one of these genes is sex-linked. Straight wings and bent wings occur among the male and female progeny in approximately equal numbers in both crosses, as would be expected in the segregation of an autosomal locus. Such agreement is not seen however between gray bodies and sable bodies nor between straight bristles and forked bristles and either or both of these loci could be sex-linked. In testcross A, where the male parent is heterogametic and expresses recessive traits for the loci under consideration, combining the male and female progeny will allow us to identify the recombinant-type gametes formed by the female parent. The linkage can be evaluated by comparing the loci, considering them two at a time.

Body and bristle loci:

gray, straight: 48 + 39 (class 1) and 42 + 40 (class 3) = 169

gray, forked: 8 + 4 (class 2) and 7 + 9 (class 6) = 28

sable, straight: 5 + 6 (class 4) and 9 + 6 (class 7) = 26

sable, forked: 44 + 50 (class 5) and 46 + 37 (class 8) = 177

Bristle and wing loci:
straight, straight: 48 + 39 (class 1) and 5 + 6 (class 4) = 98
straight, bent: 42 + 40 (class 3) and 9 + 6 (class 7) = 97
forked, straight: 7 + 9 (class 6) and 46 + 37 (class 8) = 99
forked, bent: 8 + 4 (class 2) and 44 + 50 (class 5) = 106
Body and wing loci:
gray, straight: 48 + 39 (class 1) and 7 + 9 (class 6) = 103
gray, bent: 8 + 4 (class 2) and 42 + 40 (class 3) = 94
sable, straight: 5 + 6 (class 4) and 46 + 37 (class 8) = 94
sable, bent: 44 + 50 (class 5) and 9 + 6 (class 7) = 109
This pairwise comparison of loci indicates an approximate 1:1:1:1 ratio with the body-color and wing loci, indicating that these loci are independently assorting. The same approximate ratio is found with the bristle and wing loci, indicating that these loci are independently assorting. However, the comparison of the body-color and bristle loci shows a significant departure from a 1:1:1:1 ratio, indicating linkage of these two loci (note that the statistical significance of this departure can be confirmed using the chi-square test). Earlier we concluded that one of these two loci was sex-linked, and since we have just shown that the two of them are linked, both must be sex-linked. The parental-type combinations are gray, straight (169) and sable, forked (177), indicating that the recessive genes are carried in coupling in the heterozygous female parent who had genotype s^+f^+/sf at these loci. The recombinant types are gray, forked (28) and sable, straight (26) and they make up 54/400 = 0.135, or 0.135 × 100 = 13.5%, of the progeny. This recombination frequency tells us that the two loci are separated by a distance of 13.5 map units on the X chromosome. The wing locus is located on an autosome.

A-19. Keep in mind that there is no crossing over in *Drosophila* males. Since the original parents are pure breeding and all the F_1 progeny have gray bodies, complete wings, and red eyes, we may conclude that the genes for black body (b), cut wings (ct), and purple eyes (pr) are recessive to their respective counterparts for gray body (b^+), complete wings (ct^+), and red eyes (pr^+). The reciprocal crosses involving the F_1 progeny are testcrosses, and since each gives a different outcome, we may conclude that at least one of these genes is sex-linked. Gray bodies and black bodies occur among the male and female progeny in approximately equal numbers in both crosses, as would be expected in the segregation of an autosomal locus. The same could be said about red eyes and purple eyes. Such agreement is not seen however between complete wings and cut wings, and this locus could be sex-linked. In testcross A, where the male parent is heterogametic and expresses recessive traits for the loci under consideration, combining the male and female progeny will allow us to identify the recombinant-type gametes formed by the female parent. The linkage can be evaluated by comparing the loci, considering them two at a time.
Body and wing loci:
gray, complete: 57 + 60 (class 1) and 2 + 5 (class 7) = 124
gray, cut: 3 + 4 (class 4) and 59 + 63 (class 6) = 129
black, complete: 57 + 52 (class 2) and 4 + 5 (class 3) = 118
black, cut: 4 + 3 (class 5) and 66 + 56 (class 8) = 129
Wing and eye loci:
complete, red: 57 + 60 (class 1) and 4 + 5 (class 3) = 126
complete, purple: 57 + 52 (class 2) and 2 + 5 (class 7) = 116
cut, red: 4 + 3 (class 5) and 59 + 63 (class 6) = 129
cut, purple: 3 + 4 (class 4) and 66 + 56 (class 8) = 129
Body and eye loci:
gray, red: 57 + 60 (class 1) and 59 + 63 (class 6) = 239
gray, purple: 3 + 4 (class 4) and 2 + 5 (class 7) = 14
black, red: 4 + 5 (class 3) and 4 + 3 (class 5) = 16
black, purple: 57 + 52 (class 2) and 66 + 56 (class 8) = 231
This pairwise comparison of loci indicates an approximate 1:1:1:1 ratio with the body and wing loci, indicating that these loci are independently assorting. The same approximate ratio is found with the wing and eye loci, indicating that these loci are also independently assorting. However, the comparison of the body and eye loci shows in a significant departure from a 1:1:1:1 ratio, indicating linkage of these two loci (note that the statistical significance of this departure can be confirmed using the chi-square test). Earlier we concluded that these two loci were autosomal, and since we have just shown that the two of them are linked, both must be found on the same autosome. The parental-type combinations are gray, red (239) and black, purple (231), indicating that the recessive genes are carried in coupling in the heterozygous female parent who has genotype b^+pr^+/bpr at these loci. The recombinant types are gray, purple (14) and black, red (16), and they make up 30/500 = 0.06, or 0.06 × 100 = 6%, of the progeny. This recombination frequency tells us that the two loci are separated by a distance of 6 map units on the autosome. The wing locus is located on the X chromosome.

A-20. **a.** The cross can be represented as $CyPm^+/Cy^+Pm$ HSb^+/H^+Sb × $CyPm^+/Cy^+Pm$ HSb^+/H^+Sb. Because there is no crossing over, each parent makes equal frequencies of just four kinds of gametes which arise by independent assortment of the autosomes involved: $CyPm^+$ HSb^+, $CyPm^+$ H^+Sb, Cy^+Pm HSb^+, and Cy^+Pm H^+Sb. These gametes are shown combined in the Punnett square in Table A-20a. Three-fourths of the progeny die (boxes 1, 2, 3, 5, 6, 8, 9, 11, 12, 14, 15, and 16) and 1/4 (boxes 4, 7, 10, and 13) are genotypically identical to the parents with regard to the four traits considered.

b. The cross can be represented as $CyPm/Cy^+Pm^+$ HSb/H^+Sb^+ × $CyPm/Cy^+Pm^+$ HSb/H^+Sb^+. Because there is no crossing over, each parent makes equal frequencies of just four kinds of gametes which arise by independent assortment of the autosomes involved: $CyPm$ HSb, $CyPm$ H^+Sb^+, Cy^+Pm^+ HSb, and Cy^+Pm^+ H^+Sb^+. These gametes are shown combined in the Punnett square in Table A-20b. Of all the progeny, 7/16 (those in boxes 1, 2, 3, 5, 6, 9, and 11) possess lethal combinations; 4/16 (boxes 4, 7, 10, and 13) are curly, plum, hairless, stuble; 2/16 (boxes 8 and 14) are curly, plum;

TABLE A-20a

	$CyPm^+ HSb^+$	$CyPm^+ H^+Sb$	$Cy^+Pm Hsb^+$	$Cy^+Pm H^+Sb$
$CyPm^+ HSb^+$	$CyPm^+/CyPm^+$ HSb^+/HSb^+ (1)	$CyPm^+/CyPm^+$ H^+Sb/HSb^+ (2)	$Cy^+Pm/CyPm^+$ HSb^+/HSb^+ (3)	$Cy^+Pm/CyPm^+$ H^+Sb/HSb^+ (4)
$CyPm^+ H^+Sb$	$CyPm^+/CyPm^+$ HSb^+/H^+Sb (5)	$CyPm^+/CyPm^+$ H^+Sb/H^+Sb (6)	$Cy^+Pm/CyPm^+$ HSb^+/H^+Sb (7)	$Cy^+Pm/CyPm^+$ H^+Sb/H^+Sb (8)
$Cy^+Pm HSb^+$	$CyPm^+/Cy^+Pm$ HSb^+/HSb^+ (9)	$CyPm^+/Cy^+Pm$ H^+Sb/HSb^+ (10)	Cy^+Pm/Cy^+Pm HSb^+/HSb^+ (11)	Cy^+Pm/Cy^+Pm H^+Sb/HSb^+ (12)
$Cy^+Pm H^+Sb$	$CyPm^+/Cy^+Pm$ HSb^+/H^+Sb (13)	$CyPm^+/Cy^+Pm$ H^+Sb/H^+Sb (14)	Cy^+Pm/Cy^+Pm HSb^+/H^+Sb (15)	Cy^+Pm/Cy^+Pm H^+Sb/H^+Sb (16)

TABLE A-20b

	$CyPm HSb$	$CyPm H^+Sb^+$	$Cy^+Pm^+ Hsb$	$Cy^+Pm^+ H^+Sb^+$
$CyPm HSb$	$CyPm/CyPm$ HSb/HSb (1)	$CyPm/CyPm$ HSb/H^+Sb^+ (2)	$CyPm/Cy^+Pm^+$ HSb/HSb (3)	$CyPm/Cy^+Pm^+$ HSb/H^+Sb^+ (4)
$CyPm H^+Sb^+$	$CyPm/CyPm$ HSb/H^+Sb^+ (5)	$CyPm/CyPm$ H^+Sb^+/H^+Sb^+ (6)	$CyPm/Cy^+Pm^+$ HSb/H^+Sb^+ (7)	$CyPm/Cy^+Pm^+$ H^+Sb^+/H^+Sb^+ (8)
$Cy^+Pm^+ HSb$	$CyPm/Cy^+Pm^+$ HSb/HSb (9)	$CyPm/Cy^+Pm^+$ HSb/H^+Sb^+ (10)	Cy^+Pm^+/Cy^+Pm^+ HSb/HSb (11)	Cy^+Pm^+/Cy^+Pm^+ HSb/H^+Sb^+ (12)
$Cy^+Pm^+ H^+Sb^+$	$CyPm/Cy^+Pm^+$ HSb/H^+Sb^+ (13)	$CyPm/Cy^+Pm^+$ H^+Sb^+/H^+Sb^+ (14)	Cy^+Pm^+/Cy^+Pm^+ HSb/H^+Sb^+ (15)	Cy^+Pm^+/Cy^+Pm^+ H^+Sb^+/H^+Sb^+ (16)

2/16 (boxes 12 and 15) are hairless, stuble; and 1/16 (box 16) are wild-type. Considering only the surviving progeny (that is, ignoring the lethal genotypes) gives 4/9 curly, plum, hairless, stuble; 2/9 curly, plum; 2/9 hairless, stuble; and 1/9 wild-type.

Correlation Guide for Several Genetics Texts

Chapters from several leading genetics texts that correlate with the chapters in this book are listed in the following table.

Nickerson	Burns and Bottino, 6th edition	Gardner and Sunstad, 7th edition	Hartl, Freifelder, and Snyder	Klug and Cummings, 2nd edition	Rothwell, 4th edition	Russell, 2nd edition	Snyder, Freifelder, and Hartl	Strickberger, 3rd edition	Suzuki, Griffiths, Miller, and Lewontin, 4th edition	Weaver and Hedrick	Zubay
1	4	3	2	2	2, 3	1	2	2	3	4	2
2	2, 3	2	1	3	1	2	1	6, 7	2	2	1
3	2	2	1	3	1	2	1	6	2	3	1
4	3	2	1	3	1	2	1	7	2	3	1
5	3	2	1	3	1	2	1	7	2	3	1
6	5	2	2	3	6	5	2	8	5	2	1
7	5	2	2	—	6	23	2	8	3	2	Appen.
8	17	10	1	4	4	4	1	9	4	3	1
9	3	15	1	4	4	4	1	11	3	3	1
10	7, 8	4	2	5	5	3	2	12	1, 3	3	2
11	—	2, 4	1	3	6	3	1	6, 12	5	3	15
12	6	6	3	6	8	5	3	16	5	5	2, 14
13	6	6	3	6	8	5	3	17	6	5	14
14	9	6	3	6	9	6	3	18, 7	6	5	14
15	9	6	3	6	9	6	3	18, 7	8, 9	5	14
16	14, 15	12, 13	6	12	10	17	6	21, 22	23	4	9
17	18	15	10	7	7	23	16	14, 15	11	3	1
18	8	5	4	8, 9	11	8	4	4	11, 18	6	3
19	8	5	4	10	11	10	4, 8	5	10	7	4, 10
20	8	7	11	14	15	7	7	19	10	13	8
21	8	7	11	14	15	7	7	19	10	13	13
22	8	7	11	14	15	7	7	19	10	13	13
23	8	7	11	14	15	7	7	19	10	13	13
24	8	9	11	14	16	20	7	20	10	13	12
25	11	10	11	19	16	14	7	25	12	13	12
26	9	8	12	16, 18	12, 13	11, 12	9	26, 27	13	10, 9	5, 6
27	10	8	12	17	12	13	9	28	13	10	6
28	13	9	13	13	14	16	10	23, 24	7, 17	11	8, 9
29	12	11	14, 5	21	17	18, 20	11, 7	29	16, 19	12, 8	16, 21
30	20	10	15	20	18	15	12	30	15	16	11
31	19	16	8	25	21	24	11	31	24	18	20
32	19	16	9	25	21	24	15	32	24	18, 19	20

Index